# Lecture Notes in Control and Information Sciences

Edited by M. Thoma

Vol. 43: Stochastic Differential Systems
Proceedings of the 2nd Bad Honnef Conference of the SFB 72 of the DFG at the University of Bonn
June 28 – July 2, 1982
Edited by M. Kohlmann and N. Christopeit
XII, 377 pages. 1982.

Vol. 44: Analysis and Optimization of Systems
Proceedings of the Fifth International Conference on Analysis and Optimization of Systems
Versailles. December 14–17, 1982
Edited by A. Bensoussan and J. L. Lions
XV, 987 pages, 1982

Vol. 45: M. Arató
Linear Stochastic Systems with Constant Coefficients
A Statistical Approach
IX, 309 pages. 1982

Vol. 46: Time-Scale Modeling of Dynamic Networks with Applications to Power Systems
Edited by J. H. Chow
X, 218 pages. 1982

Vol. 47: P.A. Ioannou, P.V. Kokotovic
Adaptive Systems with Reduced Models
V, 162 pages. 1983

Vol. 48: Yaakov Yavin
Feedback Strategies for Partially Observable Stochastic Systems
VI, 233 pages, 1983

Vol. 49: Theory and Application of Random Fields
Proceedings of the IFIP-WG 7/1 Working Conference held under the joint auspices of the Indian Statistical Institute
Bangalore, India, January 1982
Edited by G. Kallianpur
VI. 290 pages. 1983

Vol. 50: M. Papageorgiou
Applications of Automatic Control Concepts to Traffic Flow Modeling and Control
IX, 186 pages. 1983

Vol. 51: Z. Nahorski, H.F. Ravn, R.V.V. Vidal
Optimization of Discrete Time Systems
The Upper Boundary Approach
V, 137 pages 1983

Vol. 52: A. L. Dontchev
Perturbations, Approximations and Sensitivity Analysis of Optimal Control Systems
IV, 158 pages. 1983

Vol. 53: Liu Chen Hui
General Decoupling Theory of Multivariable Process Control Systems
XI, 474 pages. 1983

Vol. 54: Control Theory for Distributed Parameter Systems and Applications
Edited by F. Kappel, K. Kunisch, W. Schappacher
VII, 245 pages. 1983.

Vol. 55: Ganti Prasada Rao
Piecewise Constant Orthogonal Functions and Their Application to Systems and Control
VII, 254 pages. 1983.

Vol. 56: Dines Chandra Saha, Ganti Prasada Rao
Identification of Continuous Dynamical Systems
The Poisson Moment Functional (PMF) Approach
IX, 158 pages. 1983.

Vol. 57: T. Söderström, P.G. Stoica
Instrumental Variable Methods for System Identification
VII, 243 pages. 1983.

Vol. 58: Mathematical Theory of Networks and Systems
Proceedings of the MTNS-83 International Symposium
Beer Sheva, Israel, June 20–24, 1983
Edited by P.A. Fuhrmann
X, 906 pages. 1984

Vol. 59: System Modelling and Optimization
Proceedings of the 11th IFIP Conference
Copenhagen, Denmark, July 25-29, 1983
Edited by P. Thoft-Christensen
IX, 892 pages. 1984

Vol. 60: Modelling and Performance Evaluation Methodology
Proceedings of the International Seminar
Paris, France, January 24–26, 1983
Edited by F. Bacelli and G. Fayolle
VII, 655 pages. 1984

Vol. 61: Filtering and Control of Random Processes
Proceedings of the E.N.S.T.-C.N.E.T. Colloquium
Paris, France, February 23–24, 1983
Edited by H. Korezlioglu, G. Mazziotto, and J. Szpirglas
V, 325 pages. 1984

For information about Vols. 1–42 please contact your bookseller or Springer-Verlag.

# Lecture Notes in Control and Information Sciences

Edited by M. Thoma and A. Wyner

IFIP

113

M. Iri, K. Yajima (Editors)

# System Modelling and Optimization

Proceedings of the 13th IFIP Conference
Tokyo, Japan, August 31 – September 4, 1987

Springer-Verlag London Ltd.

**Editors**

Masao Iri
Dept. of Mathematical Eng. & Instrumentation Physics
Faculty of Engineering
University of Tokyo
7-3-1 Hongo, Bunkyo-ku
Tokyo 113
Japan

Keiji Yajima
Institute of JUSE
4-30-3 Sendagaya, Shibuya-ku
Tokyo 151
Japan

ISBN 978-3-540-19238-1 ISBN 978-3-540-39164-7 (eBook)
DOI 10.1007/978-3-540-39164-7

Library of Congress Cataloging in Publication Data
System modelling and optimization : proceedings of the 13th IFIP Conference, Tokyo, Japan, August 31–September 4, 1987 / M. Iri, K. Yajima, editors.
(Lecture notes in control and information sciences ; 113)
Papers selected from the 13th IFIP Conference on System Modelling and Optimization, held Aug. 31–Sept. 4, 1987, Tokyo, Japan; conference organized for Technical Committee 7 of the IFIP, co -sponsored by Ino the Information Processing Society of Japan.

1. System analysis--Data processing--Congresses. 2. Digital computer simulation--Congresses. 3. Mathematical optimization--Congresses. I. Iri, Masao. II. Yajima, Kaiji
III. IFIP Conference on System Modelling and Optimization (13th : 1987 : Tokyo, Japan) IV. IFIP TC-7 (Organization) V. Jōhō Shori Gakkai (Japan) VI. Series.
QA402.S95823 1988 88-12351


Originally published by Springer-Verlag London Berlin Heidelberg in 2001



2161/3020-543210

PREFACE

These Proceedings cotain 73 contributed papers selected from among those presented at the 13th IFIP Conference on System Modelling and optimization, Tokyo, Japan, August 31-September 4, 1987, as well as 4 papers for the plenary sessions of the Conference.

The Conference was organized for Technical Committee 7 of the International Federation for Information Processing (IFIP) and was co-sponsored by the Information Processing Society of Japan (IPSJ), the International Federation of Automatic Control (IFAC), the International Federation of Operational Research Societies (IFORS), and the Association of Asian-Pacific Operational Research Societies (APORS).

The main work of the organization of the conference was done by the members of the Local Organizing Committee, namely,

| | |
|---|---|
| S. Hitotumatu | Kyoto University |
| M. Iri (Chairman) | University of Tokyo |
| T. Kawai | Keio University |
| H. Kobayashi | IBM, Japan |
| H. Morimura | Tokyo Institute of Technology |
| Y. Murotsu | University of Osaka Prefecture |
| K. Tone | Saitama University |
| J. Tsunekawa | Institute of JUSE |
| K. Yajima (Secretariat) | Institute of JUSE |
| A. Yamada | Chuo University |
| K. Yamashita | Toshiba Ltd |
| M. Yoshida | Chuo University |

following the instructions of the International Program Committee whose members are:

*A. V. Balakrishnan*, U.S.A.
*J. Dolezal*, C.S.S.R.
*Y. Evtushenko*, U.S.S.R.
*M. Florian*, Canada
*M. Iri*, Japan
*P. Kall*, Switzerland
*R. Kluge*, German Democratic Republic
*J. L. Lions*, France
*M. Lucertini* (Chairman), Italy

*K. Malanowshi*, Poland
*M. J. D. Powell*, United Kingdom
*A. Prekopa*, Hungary
*H. Scolnik*, Argentina
*J. Stoer*, Federal Republic of Germany
*P. Thoft-Christensen*, Denmark
*G. C. Vansteenkiste*, Belgium
*K. Wakayama*, Japan (IFORS/APORS)
*R. J. B. Wets*, U.S.A.
*K. Yajima*, Japan
*H. Zhou*, China

About 150 papers were submitted to the Conference, and were preliminarily screened by the Local Organizing Committee. Among them about 140 were accepted for presentation by the International Program Committee, and 103 papers were actually presented at the Conference. The Conference programme which was actually run is attached in the appendix.

The papers which the authors wanted to be included in the Proceedings were reviewed during the Conference by the chairpersons of the respective sessions and some anonymous referees who attended each session.

Based on their evaluations the Local Organizing Committee decided which of those papers to include in the Proceedings, taking account of the number of available pages.

The conference was financially supported by

Central Research Institute of Electric Power Industry
Forster, Ltd.
Fujutsu Ltd.
Hitachi, Ltd.
IBM, Japan Ltd.
Information and Mathematical Science Laboratory, Inc.
Institute of JUSE
Japan Business Automation Co., Ltd.
JUSE Press Ltd.
Mitsubishi Electric Corporation
Mitsubishi Oil Co., Ltd
Mitsubishi Research Institute, Inc.
NEC Corp.

Nippon Telegram and Telephone, Corp.
System Brains Co., Ltd.
The Tokyo Electric Power Co., Inc.
Tokyo Gas Ltd.
Toshiba Corporation
Union of Japanese Scientists and Engineers

Their support is gratefully acknowledged by the Conference organizer.

January 1988

Masao Iri
Keiji Yajima

## Contents

### Invited Lectures



### Control and Estimation

## Continuous Estimation



## Distributed Parameter Systems



## Combinatorial Optimization

## Computational Geometry



## Simulation



## Modelling and Methodology in Social and Economic Systems



## Optimization and Reliability of Structural Systems

**Energy Systems**



**Operational Research in Forestry**



**Biological Systems**



**Industrial Applications**

Computers

# TRUST REGIONS AND PROJECTED GRADIENTS

Jorge J. Moré †
Mathematics and Computer Science Division
Argonne National Laboratory
Argonne, Illinois

## 1 Introduction

The numerical solution of large scale linearly constrained problems by algorithms which use the gradient projection method is a promising research area. Algorithms based on the gradient projection method are able to drop and add many constraints at each iteration, and this ability gives them an important advantage in large scale problems. In contrast, active set strategies for linearly constrained problems restrict the change in the dimension of the working subspace by only dropping or adding one constraint at each iteration. This can be a serious disadvantage for large scale problems because it implies that if there are $k_1$ constraints active at the solution and $k_2$ constraints active at the starting point, then at least $|k_2 - k_1|$ iterations are required for convergence.

We consider algorithms which combine gradient projection and trust region methods and which are designed for the numerical solution of the general minimization problem

$$\min \{f(x) : x \in \Omega\}, \tag{1.1}$$

where $f : R^n \to R$ is a continuously differentiable mapping on the closed convex set $\Omega$. Of special interest is the case where $\Omega$ is defined by the bound constraints

$$\Omega = \{x \in R^n : l \leq x \leq u\} \tag{1.2}$$

for some vectors $l$ and $u$, or the case where $\Omega$ is defined by general linear constraints.

Algorithms which combine gradient projection and trust region methods have been proposed by Conn, Gould and Toint [1986a] for the numerical solution of bound constrained problems, and by Toint [1987] for the general problems (1.1). One of the reasons for interest in algorithms of this type is that previous experience with trust region methods for unconstrained problems leads to the conjecture that these algorithms will perform well on constrained problems; support for this conjecture is provided by the numerical results of Conn, Gould and Toint [1986b] for bound constrained problems.

The aim of this paper is to show how the ideas from the gradient projection method combine with trust region methods, and to give an indication of the powerful convergence results that are available for algorithms of this type. Section 2 introduces the notion of the projected gradient and discusses the relationship of the projected gradient with the gradient projection method. The results of this section are mainly based on the work of Calamai and Moré [1987] and Burke and

† Work supported in part by the Applied Mathematical Sciences subprogram of the Office of Energy Research of the U.S. Department of Energy under Contract W-31-109-Eng-38.

Moré [1987]. Other recent convergence results are discussed by Dunn [1987a]. Section 3 incorporates the gradient projection method into a trust region method by following the ideas proposed by Toint [1987]. The interested reader should compare this algorithm with other algorithms based on the gradient projection method, for example, those proposed by Bertsekas [1982], Gafni and Bertsekas [1984], Gawande and Dunn [1987], and Dunn [1987b]. Section 4 contains an analysis of some of the global convergence properties of this algorithm. In this section we only discuss those results that are closely related to the projected gradient. A more extensive development with rate of convergence results will be published elsewhere.

## 2 The Projected Gradient

A reason for introducing the projected gradient is to provide a function $\nu : R^n \to R$ such that if $\{x_k\}$ is a sequence in $\Omega$ which converges to some $x^* \in \Omega$, and if $\{\nu(x_k)\}$ converges to zero, then

$$(2.1) \qquad \langle \nabla f(x^*), x - x^* \rangle \geq 0, \qquad x \in \Omega.$$

This is the standard first order condition for a minimizer of $f$; if $x^* \in \Omega$ satisfies (2.1) then $x^*$ is a *stationary point* of problem (1.1). Other functions satisfy this requirement, but only the projected gradient satisfies Theorem 2.3 below.

Let $f : R^n \to R$ be differentiable on $\Omega$ and let $\nabla f$ be the gradient of $f$ with respect to an inner product $\langle \cdot, \cdot \rangle$. Recall that a direction $v$ is *feasible* at $x \in \Omega$ if $x + \tau v$ belongs to $\Omega$ for all $\tau > 0$ sufficiently small, and that the *tangent cone* $T(x)$ is the closure of the cone of all feasible directions. The *projected gradient* $\nabla_\Omega f$ of $f$ is defined by

$$\nabla_\Omega f(x) \equiv argmin\{\|v + \nabla f(x)\| : v \in T(x)\},$$

where the norm $\|\cdot\|$ is generated by the inner product $\langle \cdot, \cdot \rangle$. Since $T(x)$ is a nonempty closed convex set, $\nabla_\Omega f(x)$ is uniquely defined.

The following result of Calamai and Moré [1987] gives some of the basic properties of the projected gradient.

**Lemma 2.1** *Let $\nabla_\Omega f$ be the projected gradient of $f$.*

*(a) The point $x \in \Omega$ is a stationary point of problem (1.1) if and only if $\nabla_\Omega f(x) = 0$.*

*(b) If $f : R^n \to R$ is continuously differentiable on $\Omega$ then the mapping $\|\nabla_\Omega f(\cdot)\|$ is lower semicontinuous on $\Omega$.*

*(c) $min\{\langle \nabla f(x), v \rangle : v \in T(x), \|v\| \leq 1\} = -\|\nabla_\Omega f(x)\|$.*

The term projected gradient is not entirely appropriate because at an interior point $x$ of $\Omega$ the projected gradient reduces to $-\nabla f(x)$. Part $c$ of Lemma 2.1 shows that it might be more appropriate to call $\nabla_\Omega f(x)$ the *projected steepest descent direction.*

An important consequence of parts $a$ and $b$ of Lemma 2.1 is that if $\{x_k\}$ converges to $x^*$ and $\nabla_\Omega f(x_k)$ converges to zero, then $x^*$ is a stationary point for problem (1.1). Also note that part $c$ of Lemma 2.1 suggests that an iterative scheme for problem (1.1) should drive $\nabla_\Omega f(x_k)$ to zero.

However, this may not be possible because $\nabla_\Omega f$ is usually discontinuous at a stationary point. Thus, it is surprising that the gradient projection algorithm produces a sequence $\{x_k\}$ such that $\nabla_\Omega f(x_k)$ converges to zero.

Recall that given an inner product norm $\|\cdot\|$, the projection into $\Omega$ is the mapping $P : R^n \rightarrow \Omega$ defined by requiring that

$$\|P(x) - x\| \leq \|z - x\|, \qquad z \in \Omega.$$

Since $\|\cdot\|$ is an inner product norm, $P(x)$ is uniquely defined. The projection $P$ can also be defined in terms of the inner product by requiring that

$$\langle P(x) - x, P(x) - z\rangle \leq 0, \qquad z \in \Omega, \tag{2.2}$$

where $\langle\cdot,\cdot\rangle$ is the inner product associated with the norm $\|\cdot\|$. The *gradient projection* algorithm is defined by

$$x_{k+1} \equiv P(x_k - \alpha_k \nabla f(x_k)),$$

where the step $\alpha_k > 0$ satisfies two requirements. The first requirement is that

$$f(x_{k+1}) \leq f(x_k) + \mu\langle\nabla f(x_k), x_{k+1} - x_k\rangle, \tag{2.3}$$

for some constant $\mu \in (0,1)$. This is a *sufficient decrease* condition which guarantees, in particular, that $f(x_{k+1}) < f(x_k)$. We need another condition on $\alpha_k$ because the sufficient decrease condition allows all $\alpha_k \geq 0$ sufficiently small. The second requirement on $\alpha_k$ is that for positive constants $\gamma_1$ and $\gamma_2$

$$\alpha_k \geq \gamma_1 \quad \text{or} \quad \alpha_k \geq \gamma_2 \bar{\alpha}_k \tag{2.4}$$

where $\bar{\alpha}_k > 0$ satisfies

$$f(x_k + s_k(\bar{\alpha}_k)) \geq f(x_k) + \mu\langle\nabla f(x_k), s_k(\bar{\alpha}_k)\rangle$$

and the mapping $s_k : R \rightarrow R^n$ is defined by

$$s_k(\alpha) \equiv P(x_k - \alpha\nabla f(x_k)) - x_k. \tag{2.5}$$

Conditions (2.3) and (2.4) on $\alpha_k$ were proposed by Calamai and Moré [1987] with the aim of proving convergence results without specifying a particular method for the computation of $\alpha_k$. Note that these conditions can be satisfied in a finite number of steps. For example, for given constants $\beta \in (0,1)$ and $\gamma > 0$, we can follow Bertsekas [1976] and choose

$$\alpha_k = \gamma\beta^{m_k}$$

where $m_k$ is the smallest nonnegative integer such that (2.3) holds. In this case (2.4) holds with $\gamma_1 = \gamma$ and $\gamma_2 = \beta$.

Calamai and Moré [1987] established the following convergence result for the gradient projection algorithm under assumptions (2.3) and (2.4) on the step.

**Theorem 2.2** *Let $f : R^n \rightarrow R$ be continuously differentiable on $\Omega$, and let $\{x_k\}$ be the sequence generated by the gradient projection algorithm. Assume that $\{\alpha_k\}$ is bounded above. If $f$ is bounded below on $\Omega$ and $\nabla f$ is uniformly continuous on $\Omega$ then*

$$\lim_{k\rightarrow\infty} \nabla_\Omega f(x_k) = 0.$$

Theorem 2.2 is interesting because most convergence results for problem (1.1) only show that every limit point of $\{x_k\}$ is a stationary point. Note that this result is a consequence of parts *a* and *b* of Lemma 2.1.

The implications of Theorem 2.2 for linearly constrained problems are of interest. In a linearly constrained problem $\Omega$ is a *polyhedral* set and thus there is no loss of generality in assuming that $\Omega$ is defined by the set of linear constraints

$$\Omega = \{x \in R^n : \langle c_j, x\rangle \geq \delta_j,\ j = 1, ..., m\},$$

for some vectors $c_j \in R^n$ and scalars $\delta_j$. In the linearly constrained case, $x^* \in \Omega$ is a stationary point if and only if $x^*$ is a Kuhn-Tucker point. Thus,

$$\nabla f(x^*) = \sum_{j\in A(x^*)} \lambda_j^* c_j, \qquad \lambda_j^* \geq 0, \tag{2.6}$$

where the set of *active constraints* is defined by

$$A(x) \equiv \{j : \langle c_j, x\rangle = \delta_j\}.$$

Note that we have not made any linear independence assumptions on the active constraints; if the active constraints are linearly dependent then there is an infinite number of multiplier sets $\{\lambda_j^*\}$ which satisfy (2.6).

**Definition.** The stationary point $x^*$ is *nondegenerate* if there is a set of multipliers $\lambda_j^* > 0$ which satisfy the Kuhn-Tucker conditions (2.6).

The following result shows, in particular, that the gradient projection algorithm identifies the active constraints at nondegenerate stationary points.

**Theorem 2.3** *Let $f : R^n \to R$ be continuously differentiable on a polyhedral $\Omega$, and let $\{x_k\}$ be an arbitrary sequence in $\Omega$ which converges to $x^*$. If $\{\nabla_\Omega f(x_k)\}$ converges to zero, and $x^*$ is nondegenerate, then $A(x_k) = A(x^*)$ for all $k \geq 0$ sufficiently large.*

Calamai and Moré [1987] established Theorem 2.3 under the assumption that the active constraint normals were linearly independent. Burke and Moré [1987] were able to drop this assumption and to extend this result to certain non-polyhedral sets $\Omega$; details are provided by Burke and Moré [1987].

Theorems 2.2 and 2.3 indicate that we can use the gradient projection algorithm on linearly constrained problems with the aim of identifying the active constraints at a solution. The next section shows how the gradient projection algorithm can be incorporated into a trust region method.

## 3 Trust Region Methods

At each iteration of a trust region method there is an approximation $x_k \in \Omega$ to the solution, a bound $\Delta_k$, and a model $\psi_k : R^n \to R$ of the possible reduction $f(x_k + w) - f(x_k)$ for $\|w\| \leq \Delta_k$. We assume that $\psi_k$ is defined in a neighborhood of $x_k$, and that

$$\psi_k(0) = 0, \qquad \nabla\psi_k(0) = \nabla f(x_k).$$

There are several reasonable choices of the model. In many situations the model $\psi_k$ is a quadratic, and thus

$$\psi_k(w) = \langle \nabla f(x_k), w \rangle + \tfrac{1}{2}\langle w, B_k w \rangle, \tag{3.1}$$

for some symmetric matrix $B_k$. One of the nice features of Toint [1987] is that he allows a general $\psi_k$, for example, it is possible to choose $\psi_k(w) = f(x_k + w)$. It is also worthwhile noting that Toint [1987] only requires that there are positive constants $\kappa_1$ and $\kappa_2$ such that

$$\|\nabla \psi_k(0) - \nabla f(x_k)\| \le \min\{\kappa_1, \kappa_2 \Delta_k\}.$$

In order to simplify the presentation we do not allow this level of generality.

Given the model $\psi_k$ and the bound $\Delta_k$, a trust region method computes a step $s_k$ which is an approximate minimizer of the subproblem

$$\min\left\{\psi_k(w) : x_k + w \in \Omega,\ \|w\| \le \Delta_k\right\}. \tag{3.2}$$

We assume that $\psi_k(s_k) < \psi_k(0)$ and that $x_k + s_k \in \Omega$. Later on we shall impose more requirements on the step $s_k$.

The iterate $x_k$ and the bound $\Delta_k$ are updated according to the same rules as in trust region methods for unconstrained minimization. See, for example, Moré [1983]. These rules depend on the ratio

$$\rho_k = \frac{f(x_k + s_k) - f(x_k)}{\psi_k(s_k) - \psi_k(0)}$$

of the actual reduction in the function to the predicted reduction in the model. Since the step $s_k$ is chosen so that $\psi_k(s_k) < \psi_k(0)$, a step with $\rho_k > 0$ yields a reduction in the function. The iterate $x_k$ is updated as follows:

If $\rho_k > \eta_0$ then $x_{k+1} = x_k + s_k$.

If $\rho_k \le \eta_0$ then $x_{k+1} = x_k$.

In general $\eta_0$ is chosen fairly small with $\eta_0 = 0.0001$ a typical value. The updating rules for $\Delta_k$ depend upon constants $\eta_1$ and $\eta_2$ such that

$$0 < \eta_0 < \eta_1 < \eta_2 < 1.$$

If $\rho_k \le \eta_1$ then $\Delta_k$ is decreased, while if $\rho_k \ge \eta_2$ then $\Delta_k$ is not decreased. Typical values for $\eta_1$ and $\eta_2$ are $\eta_1 = 1/4$ and $\eta_2 = 3/4$. The updating rules which govern the rate at which $\Delta_k$ is either increased or decreased depend on constants $\sigma_1, \sigma_2$ and $\sigma_3$ such that

$$0 < \sigma_1 < \sigma_2 < 1 < \sigma_3.$$

The trust region bound $\Delta_k$ is updated as follows:

If $\rho_k \le \eta_1$ then $\Delta_{k+1} \in [\sigma_1 \min\{\|s_k\|, \Delta_k\}, \sigma_2 \Delta_k]$

If $\rho_k \in (\eta_1, \eta_2)$ then $\Delta_{k+1} \in [\sigma_2 \Delta_k, \Delta_k]$

If $\rho_k \ge \eta_2$ then $\Delta_{k+1} \in [\Delta_k, \sigma_3 \Delta_k]$

Typical values for these parameters are $\sigma_1 = 1/10$, $\sigma_2 = 1/2$, and $\sigma_3 = 4$.

We are now ready to specify the step. We incorporate the gradient projection ideas by following the suggestions of Toint [1987] and choosing a step $s_k$ that gives as much reduction in the model $\psi_k$ as one step of the gradient projection method applied to (3.2). In analogy with the terminology used with trust region methods for unconstrained minimization, the step produced by the gradient projection method applied to (3.2) is called a *Cauchy step*. In this algorithm the Cauchy step is of the form $x_k + s_k(\alpha_k)$ where the function $s_k(\cdot)$ is defined by (2.5) and $\alpha_k$ satisfies the following two requirements. The first requirement is that

$$\psi_k(s_k(\alpha_k)) \le \mu_0 \langle \nabla f(x_k), s_k(\alpha_k)\rangle \quad \text{and} \quad \|s_k(\alpha_k)\| \le \mu_2 \Delta_k, \tag{3.3}$$

while the second requirement is that

$$\alpha_k \ge \gamma_1 \quad \text{or} \quad \alpha_k \ge \gamma_2 \bar{\alpha}_k, \tag{3.4}$$

where $\bar{\alpha}_k > 0$ satisfies

$$\psi_k(s_k(\bar{\alpha}_k)) > \mu_0 \langle \nabla f(x_k), s_k(\bar{\alpha}_k)\rangle \quad \text{or} \quad \|s_k(\bar{\alpha}_k)\| \ge \mu_1 \Delta_k.$$

We assume that

$$0 < \mu_0 < 1, \quad 0 < \mu_1 < \mu_2.$$

The step $s_k$ is required to satisfy a requirement similar to (3.3). We assume that

$$\psi_k(s_k) \le \mu_0 \langle \nabla f(x_k), s_k(\alpha_k)\rangle, \qquad \|s_k\| \le \mu_2 \Delta_k, \qquad x_k + s_k \in \Omega. \tag{3.5}$$

In particular, this allows the choice of $s_k = s_k(\alpha_k)$.

These requirements on $s_k$ and on $\alpha_k$ differ from those required by Toint [1987]. Instead of (3.5), Toint assumes that the step $s_k$ satisfies

$$\psi_k(s_k) \le \mu_0 \psi_k(s_k(\alpha_k)), \qquad \|s_k\| \le \mu_2 \Delta_k, \qquad x_k + s_k \in \Omega.$$

In view of (3.3), it is clear that if this condition holds then (3.5) also holds. Another advantage of (3.5) is that it allows a step such that

$$\psi_k(s_k) \le \mu_0 \min\left\{\langle \nabla f(x_k), w\rangle : x_k + w \in \Omega, \ \|w\| \le \mu_2 \Delta_k\right\}.$$

If $\Omega$ is defined by linear constraints and the norm $\|\cdot\|$ is either the $l_1$ or $l_\infty$ norm, then a step that satisfies this condition can be obtained by solving a linear programming problem.

The requirements on $\alpha_k$ imposed by Toint [1987] are weaker than (3.3) and (3.4). Instead of allowing $\|s_k(\bar{\alpha}_k)\| \ge \mu_1 \Delta_k$, Toint allows

$$\alpha_k \ge \frac{\gamma_3 \Delta_k}{\|\nabla f(x_k)\|} \tag{3.6}$$

for some $\gamma_3 > 0$. This requirement is weaker because if $\alpha_k \ge \gamma_2 \bar{\alpha}_k$ and $\|s_k(\bar{\alpha}_k)\| \ge \mu_1 \Delta_k$ then (3.6) holds with $\gamma_3 = \gamma_2 \mu_1$. This remark holds because the projection operator is non-expansive. In this paper we do not allow (3.6) for two reasons. First of all, (3.3) and (3.4) can be satisfied quite easily and naturally; see the proof of Theorem 4.2. The other reason is that the analysis of the algorithm with (3.6) requires that the sequence of gradients $\{\nabla f(x_k)\}$ be bounded.

## 4 Convergence Analysis

The convergence analysis of the trust region method presented in Section 3 requires the following properties of the projection operator.

**Lemma 4.1** *If $P$ is the projection into $\Omega$ then the function $\phi_1$ defined by*

$$\phi_1(\alpha) = \|P(x + \alpha d) - x\|, \qquad \alpha > 0,$$

*is isotone (nondecreasing) for all $x \in R^n$ and $d \in R^n$, and the function $\phi_2$ defined by*

$$\phi_2(\alpha) = \frac{\|P(x + \alpha d) - x\|}{\alpha}, \qquad \alpha > 0,$$

*is antitone (nonincreasing) for all $x \in R^n$ and $d \in R^n$.*

The results in Lemma 4.1 are not standard in the literature on projection operators as discussed, for example, by Zarantonello [1971]. The isonicity of $\phi_1$ is established by Toint [1987], and the result on $\phi_2$ is due to Gafni and Bertsekas [1984]. Moré and Calamai [1987] provide an alternate proof of this last result. Also note that the basic inequality (2.2) implies that

$$-\langle \nabla f(x_k), s_k(\alpha)\rangle \geq \frac{\|s_k(\alpha)\|^2}{\alpha}, \qquad \alpha > 0, \tag{4.1}$$

where $s_k(\cdot)$ is defined by (2.5). As we shall see, this inequality is crucial to the convergence analysis. We also need the result that if $x_k \in \Omega$ then the derivative $s_k'(0)$ exists, and is given by

$$s_k'(0) = \lim_{\alpha \to 0} \frac{s_k(\alpha)}{\alpha} = \nabla_\Omega f(x_k). \tag{4.2}$$

This result is a consequence of Lemma 4.6 in Zarantonello [1971]. Also see Proposition 2 in McCormick and Tapia [1972].

**Theorem 4.2** *If $x_k$ is not a stationary point then there is a step $\alpha_k > 0$ which satisfies (3.3) and (3.4). This step can be determined with a finite number of evaluations of $s_k(\cdot)$.*

**Proof.** We first claim that (3.3) holds for all $\alpha_k > 0$ sufficiently small. This claim follows from (4.2) and part $c$ of Lemma 2.1 because they show that

$$\lim_{\alpha \to 0} \frac{\psi_k(s_k(\alpha)) - \psi_k(0)}{\alpha} = \langle \nabla f(x_k), s_k'(0)\rangle = -\|\nabla_\Omega f(x_k)\|^2 < 0.$$

Now choose constants $\beta \in (0, 1)$ and $\gamma > 0$, and set

$$\alpha_k = \gamma \beta^{m_k}$$

where $m_k$ is the smallest nonnegative integer such that (3.3) holds. The above argument shows that $m_k$ exists, and it is clear that (3.4) holds with $\gamma_1 = \gamma$ and $\gamma_2 = \beta$. ∎

The next step in the convergence analysis is to obtain an estimate on the predicted decrease by the gradient projection step. This estimate is expressed in terms of the function $\omega_k : R^n \to R$ defined by

$$\omega_k(s) = \frac{\psi_k(s) - \psi_k(0) - \langle \nabla \psi_k(0), s\rangle}{\|s\|^2}.$$

Note that $\omega_k(\cdot)$ is defined for all $s \neq 0$ in a neighborhood of $x_k$. The analysis of trust region methods in terms of $\omega_k(\cdot)$ is another nice feature of Toint [1987].

**Lemma 4.3** *If $x_k$ is not a stationary point and*

$$\psi_k(s_k(\alpha)) > \mu_0\langle\nabla f(x_k), s_k(\alpha)\rangle$$

*then $\omega_k(s_k(\alpha))$ is positive and*

$$\alpha \geq \frac{(1-\mu_0)}{\omega_k(s_k(\alpha))}.$$

**Proof.** Since $\psi_k(0) = 0$ and $\nabla\psi_k(0) = \nabla f(x_k)$, inequality (4.1) implies that

$$\omega_k(s_k(\alpha)) > (\mu_2 - 1)\frac{\langle\nabla f(x_k), s_k(\alpha)\rangle}{\|s_k(\alpha)\|^2} \geq \frac{1-\mu_0}{\alpha}.$$

This inequality implies that $\omega_k(s_k(\alpha))$ is positive and also yields the desired result. ∎

This result implies that $\alpha_k$ is bounded away from zero if $\alpha_k \geq \gamma_2\bar{\alpha}_k$ and $\omega_k(\cdot)$ is uniformly bounded. This happens, for example, if $\psi_k$ is the quadratic (3.1) and the matrices $\{B_k\}$ are uniformly bounded, or if $\psi_k$ is defined by $\psi_k(w) = f(x_k + w)$ and $\nabla f$ is Lipschitz continuous. In the following result we define $\beta_k$ by

$$\beta_k = 1 + \sup\{|\omega_k(s)| : 0 < \|s\| \leq \mu_2\Delta_k\},$$

and establish the promised estimate of the predicted decrease in terms of $\beta_k$. Note that in the following result we do not rule out the possibility that $\beta_k = \infty$.

**Theorem 4.4** *If $\alpha_k$ satisfies (3.3) and (3.4) then there is a constant $\mu_3 > 0$ such that*

$$-\langle\nabla f(x_k), s_k(\alpha_k)\rangle \geq \mu_3\left[\frac{\|s_k(\alpha_k)\|}{\alpha_k}\right]\min\left\{\Delta_k, \frac{1}{\beta_k}\left[\frac{\|s_k(\alpha_k)\|}{\alpha_k}\right]\right\}. \tag{4.3}$$

**Proof.** If $x_k$ is a stationary point then $s_k(\alpha_k) = 0$ for any $\alpha_k > 0$, and thus (4.3) is trivially satisfied. In the remainder of the proof we assume that $x_k$ is not a stationary point.

We first consider the case in which there is an $\bar{\alpha}_k$ such that $\alpha_k \geq \gamma_2\bar{\alpha}_k$. If $\bar{\alpha}_k$ satisfies the assumption of Lemma 4.3 and $\|s_k(\bar{\alpha}_k)\| \leq \mu_2\Delta_k$, then inequality (4.1) implies that

$$-\langle\nabla f(x_k), s_k(\alpha_k)\rangle \geq \left[\frac{\|s_k(\alpha_k)\|}{\alpha_k}\right]^2\alpha_k \geq \gamma_2\frac{(1-\mu_0)}{\beta_k}\left[\frac{\|s_k(\alpha_k)\|}{\alpha_k}\right]^2$$

so that (4.3) holds if $\mu_3 \leq \gamma_2(1-\mu_0)$. Assume now that $\|s_k(\bar{\alpha}_k)\| \geq \mu_1\Delta_k$. Without loss of generality we also assume that $\gamma_2 \leq 1$ because (3.4) holds with $\gamma_2$ replaced by $\min\{1, \gamma_2\}$. Since $\phi_2$ in Lemma 4.1 is antitone, we obtain that

$$\frac{\|s_k(\gamma_2\bar{\alpha}_k)\|}{\gamma_2\bar{\alpha}_k} \geq \frac{\|s_k(\bar{\alpha}_k)\|}{\bar{\alpha}_k},$$

and since $\phi_1$ in Lemma 4.1 is isotone, this implies that

$$\|s_k(\alpha_k)\| \geq \|s_k(\gamma_2\bar{\alpha}_k)\| \geq \gamma_2\|s_k(\bar{\alpha}_k)\| \geq \gamma_2\mu_1\Delta_k.$$

Inequality (4.1) now yields that

$$-\langle\nabla f(x_k), s_k(\alpha_k)\rangle \geq \left[\frac{\|s_k(\alpha_k)\|}{\alpha_k}\right]\|s_k(\alpha_k)\| \geq \gamma_2\mu_1\Delta_k\left[\frac{\|s_k(\alpha_k)\|}{\alpha_k}\right],$$

and thus (4.3) holds if $\mu_3 \le \gamma_2\mu_1$. The last case that needs to be considered is when $\alpha_k \ge \gamma_1$. In this case inequality (4.1) implies that

$$-\langle \nabla f(x_k), s_k(\alpha_k)\rangle \ge \left[\frac{\|s_k(\alpha_k)\|}{\alpha_k}\right]^2 \alpha_k,$$

and thus (4.3) holds if $\mu_3 \le \gamma_1$. ∎

In most situations Theorem 4.4 is adequate, but we are also interested in the case when $\alpha_k$ is arbitrarily large. This situation can be handled by improving inequality (4.3). We claim that if we define

$$\hat{\alpha}_k = \min\{\alpha_k, \gamma_3\}$$

for any $\gamma_3 > 0$, then

$$-\langle \nabla f(x_k), s_k(\alpha_k)\rangle \ge \mu_3 \left[\frac{\|s_k(\hat{\alpha}_k)\|}{\hat{\alpha}_k}\right] \min\left\{\Delta_k, \frac{1}{\beta_k}\left[\frac{\|s_k(\hat{\alpha}_k)\|}{\hat{\alpha}_k}\right]\right\}. \tag{4.4}$$

Since $\phi_2$ in Lemma 4.1 is antitone, (4.3) implies (4.4) when $\alpha_k \le \gamma_3$. We prove (4.4) when $\alpha_k \ge \gamma_3$ by first noting that inequality (2.2) implies that the mapping $-\langle \nabla f(x_k), s_k(\cdot)\rangle$ is isotone. Hence, if $\alpha_k \ge \gamma_3$ then (4.1) shows that

$$-\langle \nabla f(x_k), s_k(\alpha_k)\rangle \ge -\langle \nabla f(x_k), s_k(\gamma_3)\rangle \ge \frac{\|s_k(\gamma_3)\|^2}{\gamma_3},$$

and thus (4.4) holds if $\mu_3 \le \gamma_3$.

There are several differences between estimate (4.4) and the estimate of Toint [1987]. He proves that there is a constant $\mu_4 > 0$ such that

$$-\psi_k(s_k(\alpha_k)) \ge \mu_4 \|s_k(1)\|^2 \min\left\{\Delta_k, \frac{\|s_k(1)\|^2}{\beta_k}\right\}.$$

Note, in particular, that this estimate is weaker than (4.4) when $\alpha_k \in (0,1)$ because Lemma 4.1 implies that

$$\frac{\|s_k(\hat{\alpha}_k)\|}{\hat{\alpha}_k} \ge \|s_k(1)\|$$

for $\alpha_k \in (0,1)$. This difference is of importance to the following result.

**Theorem 4.5** *Assume that $f : R^n \to R$ is continuously differentiable and bounded below on $\Omega$. If $\{\beta_k\}$ is uniformly bounded above then*

$$\liminf_{k\to\infty} \frac{\|s_k(\hat{\alpha}_k)\|}{\hat{\alpha}_k} = 0.$$

**Proof.** Assume, on the contrary, that there is an $\epsilon > 0$ such that

$$\frac{\|s_k(\hat{\alpha}_k)\|}{\hat{\alpha}_k} \ge \epsilon$$

for all $k$ sufficiently large. We claim that this assumption implies that

$$\sum_{k=1}^{\infty} \Delta_k \tag{4.5}$$

is a convergent series. The proof of this claim proceeds along familiar lines; the key is to consider those iterations with $\rho_k > \eta_1$. If there is a finite number of iterations with $\rho_k > \eta_1$, then $\Delta_{k+1} \leq \sigma_2 \Delta_k$ for all $k$ sufficiently large and thus (4.5) is a convergent series. Assume now that there is an infinite sequence $\{k_i\}$ of iterations with $\rho_{k_i} > \eta_1$. Note that if $\rho_k > \eta_1$, then Theorem 4.4 and the choice of step (3.5) shows that

$$f(x_k) - f(x_{k+1}) \geq \eta_1 (-\psi_k(s_k)) \geq \eta_1 \mu_0 \mu_3 \epsilon \min\{\Delta_k, \frac{\epsilon}{\beta_{max}}\},$$

where $\beta_{max}$ is an upper bound on the sequence $\{\beta_k\}$. Thus, since $\{f(x_k)\}$ is bounded below,

$$\sum_{i=1}^{\infty} \Delta_{k_i}$$

is a convergent series, and since the updating rules of $\Delta_k$ imply that

$$\sum_{k=k_1}^{\infty} \Delta_k \leq \left(\frac{\sigma_3}{1-\sigma_2}\right) \sum_{i=1}^{\infty} \Delta_{k_i},$$

the series (4.5) is convergent.

We now show that convergence of (4.5) implies that $\{|\rho_k - 1|\}$ converges to zero. First note that the sequence $\{x_k\}$ converges because (4.5) is a convergent series and

$$\|x_{k+1} - x_k\| \leq \|s_k\| \leq \mu_2 \Delta_k.$$

Since $f$ is continuously differentiable, this implies that there is a sequence $\{\epsilon_k\}$ converging to zero such that

$$|f(x_k + s_k) - f(x_k) - \langle \nabla f(x_k), s_k \rangle| \leq \epsilon_k \|s_k\|.$$

This estimate and the definition of $\beta_k$ imply that

$$|f(x_k + s_k) - f(x_k) - \psi_k(s_k)| \leq \epsilon_k \|s_k\| + \beta_k \|s_k\|^2.$$

Another estimate is needed in order to show that $\{|\rho_k - 1|\}$ converges to zero. Note that $\{\Delta_k\}$ converges to zero, and thus Theorem 4.4 and the choice of step (3.5) imply that

$$-\psi_k(s_k) \geq \mu_0 \mu_3 \epsilon \Delta_k.$$

The last two estimates yield that $\{|\rho_k - 1|\}$ converges to zero. However, the updating rules for $\Delta_k$ show that $\Delta_k$ is not decreased if $\rho_k \geq \eta_2$. Thus $\{\Delta_k\}$ cannot converge to zero if $\{|\rho_k - 1|\}$ converges to zero. This contradicts the convergence of the series (4.5) and establishes the result. ∎

We have assumed that $\{\beta_k\}$ is bounded because this assumption holds for several important choices of the model $\psi_k$ and because this assumption simplifies the proof of Theorem 4.5. A weaker assumption is that if we define

$$\hat{\beta}_k = \max\{\beta_j : 0 \leq j \leq k\},$$

then the series

(4.6)
$$\sum_{k=1}^{\infty} \frac{1}{\hat{\beta}_k}$$

diverges. It is possible to use the proof techniques of Powell [1984] and Toint [1987] to establish Theorem 4.5 under this assumption, but this lengthens the proof.

If we compare Theorem 4.5 with the analogous result of Toint [1987] we find other differences. Toint assumes that $\Omega$ is bounded and that $\nabla f$ is Lipschitz continuous on $\Omega$. Also note that Toint proves that

$$\liminf_{k\to\infty} \|s_k(1)\| = 0. \tag{4.7}$$

This result follows from Theorem 4.5 because Lemma 4.1 implies that

$$\frac{\|s_k(\hat{\alpha}_k)\|}{\hat{\alpha}_k} \ge \max\left\{\frac{\|s_k(\alpha_k)\|}{\alpha_k}, \frac{\|s_k(\gamma_3)\|}{\gamma_3}\right\},$$

and thus we obtain that

$$\liminf_{k\to\infty} \|s_k(\gamma_3)\| = 0 \tag{4.8}$$

for any $\gamma_3 > 0$. In this connection it is of interest to note that Lemma 4.1 shows that (4.7) and (4.8) are equivalent.

We now prove that the projected gradient converges to zero at points near $x_k$. For this purpose we define the *Cauchy point* by

$$x_k^C = x_k + s_k(\hat{\alpha}_k).$$

Note that $x_k^C$ belongs to $\Omega$ and that $\|x_k^C - x_k\| \le \mu_2 \Delta_k$.

**Theorem 4.6** *Assume that $f : R^n \to R$ is bounded below on $\Omega$ and that $\{\beta_k\}$ is uniformly bounded above. If $\nabla f$ is uniformly continuous on $\Omega$ then*

$$\liminf_{k\to\infty} \|\nabla_\Omega f(x_k^C)\| = 0$$

**Proof.** The basic inequality (2.2) implies that

$$\hat{\alpha}_k \langle \nabla f(x_k), x_k^C - z_k \rangle \le -\langle x_k^C - x_k, x_k^C - z_k \rangle \le \|x_k^C - x_k\| \, \|x_k^C - z_k\|$$

for all $z_k \in \Omega$. Hence, if $v_k$ is a feasible descent direction at $x_k^C$ with $\|v_k\| \le 1$, then $x_k^C + \tau_k v_k$ belongs to $\Omega$ for some $\tau_k > 0$, and thus setting $z_k = x_k^C + \tau_k v_k$ yields

$$-\langle \nabla f(x_k), v_k \rangle \le \frac{\|x_k^C - x_k\|}{\hat{\alpha}_k}.$$

Hence,

$$-\langle \nabla f(x_k^C), v_k \rangle \le \|\nabla f(x_k^C) - \nabla f(x_k)\| + \frac{\|x_k^C - x_k\|}{\hat{\alpha}_k},$$

and by virtue of part c of Lemma 2.1 ,

$$\|\nabla_\Omega f(x_k^C)\| \le \|\nabla f(x_k^C) - \nabla f(x_k)\| + \frac{\|x_k^C - x_k\|}{\hat{\alpha}_k}.$$

The proof follows from this estimate and the uniform continuity of $\nabla f$ by noting that since $\{\hat{\alpha}_k\}$ is bounded above, Theorem 4.5 implies that $\{\|x_k^C - x_k\|\}$ converges to zero. ∎

Theorem 4.6 establishes a connection between trust region methods for the minimization problem (1.1) and the projected gradient. At first sight this result does not seem to be the appropriate result because Theorem 4.6 is expressed in terms of $x_k^C$ and not in terms of $x_k$. However, without further assumptions on the relationship between $x_k + s_k$ and $x_k^C$, this is the appropriate result.

Theorem 4.6 yields some useful results even without further assumptions on $s_k$. For example, if the sequence $\{x_k\}$ converges to some $x^* \in \Omega$ then we claim that Theorem 4.6 shows that $x^*$ is a stationary point. We establish this claim by noting that since $\{\hat{\alpha}_k\}$ is bounded, Theorem 4.5 shows that $\{x_k^C\}$ also converges to $x^*$. The claim now follows from Lemma 2.1.

There are many other interesting applications and consequences of the results of this paper, but due to space limitations, these shall be developed in another paper.

**Acknowledgement.** The analysis presented in this paper benefited from discussions with Jim Burke and Gerardo Toraldo.

## References

Bertsekas, D. P. [1976]. On the Goldstein-Levitin-Polyak gradient projection method, IEEE Trans. Automat. Control 21, 174-184.

Bertsekas, D. P. [1982]. Projected Newton methods for optimization problems with simple constraints, SIAM J. Control Optim. 20, 221-246.

Burke, J. V. and J. J. Moré [1986]. On the identification of active constraints, Argonne National Laboratory, Mathematics and Computer Science Division Report ANL/MCS-TM-82, Argonne, Illinois.

Calamai, P. H. and J. J. Moré [1987]. Projected gradient methods for linearly constrained problems, Mathematical Programming, 39, 93-116.

Conn, A. R., Gould, N. I. M. and Ph. L. Toint [1986a]. Global convergence of a class of trust region algorithms for optimization problems with simple bounds, University of Namur, Department of Mathematics Report 86/1, Namur, Belgium.

Conn, A. R., Gould, N. I. M. and Ph. L. Toint [1986b]. Testing a class of methods for solving minimization problems with simple bounds on the variables, University of Namur, Department of Mathematics Report 86/3, Namur, Belgium.

Dunn, J. C. [1987a]. On the convergence of projected gradient processes to singular critical points, J. Optim. Theory Appl. 55, 203-216.

Dunn, J. C. [1987b]. Projected Newton methods for nonlinearly constrained minimization problems, preprint.

Gafni, E. M. and D. P. Bertsekas [1984]. Two-metric projection methods for constrained optimization, SIAM J. Control Optim. 22, 936-964

Gawande, M. and J. C. Dunn [1987]. Variable metric gradient projection processes in convex feasible sets defined by nonlinear inequalities, preprint, to appear in J. Appl. Math. Optim.

McCormick, G. P. and R. A. Tapia [1972]. The gradient projection method under mild differentiability conditions, SIAM J. Control 10, 93-98.

Moré, J. J. [1983]. Recent developments in algorithms and software for trust region methods, in Mathematical Programming Bonn 1982 - The State of the Art, A. Bachem, M. Grötschel, B. Korte, eds., Springer-Verlag.

Powell M. J. D. [1984]. On the global convergence of trust region algorithms for unconstrained minimization, Math. Programming 29, 297-303.

Toint, Ph. L. [1987]. Global convergence of a class of trust region methods for nonconvex minimization in Hilbert space, University of Namur, Department of Mathematics Report 87/6, Namur, Belgium.

# HOW TO COPE WITH GREY PART OF MANAGEMENT
## —SYSTEM MODELLING AND OPTIMIZATION IN JAPANESE INDUSTRY—

**Hajime Karatsu**
**Research and Development Institute**
**Tokai University**
**2-28-4, Tomigaya, Shibuya-ku, Tokyo 151, Japan**

The only way to draw back from the specter of trade protectionism soaking developed nations is to revitalize the competitive power of each enterprise in private sectors, as President Reagan spoke of in the text for U.S. Congress last February.

We have never experienced a time when a new way for improving management activity is requested for purpose of getting high productivity and quality of management.

The success story of NUMMI, joint venture between General Motors and Toyota in Fremont, Calif., gives us an impressive suggestion of the possibility to change the factory innovatively in productivity and quality by introducing new management systems.

The Business Week magazine reported that although the poor quality of the UAW union members was blamed for the shoddy cars being produced in the U.S., the GM-Toyota joint venture shows us that the union is not at fault, but the management is.

Basically, management is a battle against thousands of different possible breakdowns. Therefore if you are not aware of all the aspects in management, you will fail easily. In other words, management is a battle against errors: mistake in planning schedules, incorrect designing, accidental mixture of materials other than the one originally arranged, machines not always working uniformly, etc. Moreover, it is quite possible that employees make misunderstandings. If these errors accumulate, that result will be a pile of unexpected problems.

We cannot predict where and how such errors will occur. However, everyone in the firm must work in cooperation looking for any potential problems and take care of them in order to pervent future trouble.

I would like to explain the way to cope with the grey part of the management in Japan through my experience.

### 1. Quality management

Early March, 1985 I was requested to attend a committee of the Department of Defence of the U.S.A. for the purpose of presenting some recommendations for the way to revitalize American semiconductor industries, which were depressed by the Japaese in quality and cost.

At the beginning of my presentation, I showed a table of the accelerated life test data of the standard type Integrated Circuit of 4000 series.

We pick up 100 samples from each manufacturer and test them under the atmosphere of 85℃ and 85% humidity, which is quite a hard condition for plastic packaged I.C.s.

The data shows a big difference in failure rate between each firm.

However, the manufacturing apparatus or equipment of each firm is almost the same. The silicon wafer as the material is procured from few and limited suppliers. The design of the mask pattern is also quite similar to each other. The ability of engineers and worker on the manufacturing shop floor may have no difference, either. Then, why has such a big difference been brought up?

You might easily understand the reason why such difference occurred. It all depends on the skill of management — how to organize and to operate men, machine, materials, which is like driving a car that becomes a comfortable and convenient vehicle driven by a well trained and skillful driver, or on the contrast become a most dangerous, terrible one, when the handle is gripped by a poor and careless driver.

## 2. Solving problems

Another point I mentioned was the difference in assumptions between U.S. and Japanese engineers. If I talk with U.S. engineers, their discussions tend to become "digital". They always think in terms of black or white and yes or no. However, productivity is not so simple and clear. It may sometimes be neither black nor white. This "gray" part, as it were, might be important. If you unnaturally divide these gray aspects into black or white, you will hardly be able to succeed in anything.

Basically, manufacturing is a battle against a million different possible breakdowns. Therefore, if you are not aware of all the aspects in production, you will fail easily. In other words, manufacturing is a battle against errors, such as mistakes in planning schedules, incorrect designing or an accidental mixture of materials other than the one originally arranged. Machines do not always work uniformly. Moreover, it is quite possible that factory workers will make mistakes. If these errors accumulate, the result will be a pile of defective goods.

We cannot predict where and how such errors will occur, however. Everyone in the factory must cooperate, looking for any potential problems, and take care of them in order to prevent future trouble. Basically, this is Japan's total quality control (TQC) system.

Using the following story, I then explained the difference in attitude between Japanese and U.S. workers toward product quality. When we purchased semiconductors from a U.S. -affiliated factory in Singapore, they received some defective goods. When we opened a package, we found a small defect that caused the chips' alumimum pattern to corrode.

After the factory manager came to Japan, we asked him to be more careful about his workers' clothing and masks. He replied that his men applied the passivation covers to protect the chips from dust contamination, so there must have been some other cause for the chips' defects. The factory manager had developed a tightly constructed theory, and he would not budge from it. His attitude could be paraphrased, "since we tried as hard as possible, why are these Japanese noisily complaining about trifles ?"

If such defects had occurred in a Japanese factory, I wonder what the response would have been. Since dust is intrinsically harmful to semiconductors, even if the passivation covers were coated on the chips, if a defect such as dust contamination was found, all the employees would be determined to try hard not to let it happen again and would be more careful the next time. Without this demanding attitude toward quality, which stresses re-evaluation and improvement even in quality control processes that are believed "perfect," it could not be said that Japanese employees take advantage of the grey parts of production.

I cited another, similar story to drive my point home. There are Japanese belt manufacturers who import specialty rubber belts from the United States and then sell them in Japan. However, claims for damages from these rubber belts snapping or breaking had been increasing. This was a problem. Complaints were made to the U.S. manufacturers and engineers quickly came to investigate the problem. Their conclusion was: "These belts conform to all our standards. They are not inferior goods. No doubt if the belts are breaking, the user must be handling them improperly." After absolving themselves of all blame, the engineers returned to the United States.

The engineers in Japan had different ideas. There are belts that break and belts that don't break. There should be a way of ensuring that none break. They searched tor a damaged belt and after studying it for only two weeks they understood the reason for its defectiveness. They had solved the problem.

This attitude of deciding that something is either correct or incrrect without considering other possibilities, is definitely not conducive to solving problems.

After making my point, I moved on to a different, yet related issue. In the United States, specialists are highly valued. This in itself is not bad at all; however, due to this attitude, sometimes problems may crop up. By having experts working solely in their specialized fields, they are likely to develop a compartmentalized, self-serving attitude that may inhibit admitting their own mistakes. In addition, such specialists may try to prove that they have not made any mistakes at all by citing any petty excuse they can. These explanations, even if they are truthful and logically persuasive, will not help to improve production efficiency and/or quality. Furthermore, the more sophisticated and persuasive the explanation is, the less chance there is for improvement.

I then emphasized the following point with a strong voice: Even if you can deceive people about a product through misleading statements, sooner or later the product will speak for itself.

### 3. Better the quality, lower the cost

In the process of rebuilding Japanese industries destroyed during the war, a large part of our efforts was concentrated on improving the quality of our products. Some of them achieved top-level quality in the world before we knew it. Quality products will always sell, because good quality best serves the consumer. Through our experience, we have learned another thing. That is the fact that, as quality control forms an integral part of the manufacturing process to improve product quality by reducing production of defective goods, the cost of production will decrease without exception.

"The better the quality, the lower the cost ?" Many people might think this is too good to be true. While this seems to be a contradiction in terms, it is a very natural consequence in the view of Q.C. experts. Walter A. Shewhatrrt, who first proposed "Statistical Quality Control," pointed out this fact in his first book "Economic Control of Manufactured Products" published in 1931.

As inferior products are eliminated through innovation in the manufacturing process, materials, labor and energy oterwise used can be saved while producing an equivalent amount of product value, which means lower costs. In addition, when a large volume of inferior goods is produced, a machine must be stopped frequently for adjustment or the material must be replaced often to produce satisfactory products. This reduces the operation rate. If inferior products can be eliminated, the machine, once started in the morning, can be run until closing time, making the total production larger. As the rate of inferior products is reduced, the cost is made lower and lower.

In general, QC method was developed as the technology to improve the quality of products in the manufacturing plant.

But, today, in Japan, QC is accepted not only as the method to improve the quality but also to level up all kinds of job site: manufacturing plans, savings banks, department stores, telecommunications, the government and even restaurants and so on. Multiplying the effect of each other realized high productivity of Japanese economical power.

I made a series of TV programs concerning QC about 20 years ago. That period was just the dawn of QC in Japan, and when the last program was over, a party was held at a restaurant near by the studio. I still remember the impressive words expressed by the producer of this program at the party.

He said, "Almost all of this program was no other than common sense. It is a very natural matter. But I am impressed at one point in QC. The procedure to solve the problem and to improve the quality is arranged very skilfully; that means QC theory is not the knowledge but the way to get better result in the shop-floor easily, even by each blue collar worker."

He was right and gave me quite a new view of Quality Control.

## 5 . TQC (Total Quality Control)

Nevertheless, it is not a very easy task for a factory to actually stop the manufacture of defective products and merely begin producing quality products. This is due to the fact that there are numerous causes for a defect. The first of such sources is miscalculation in planning. An innovated facility would serve no good, if a machine is not well designed; it would not operate effectively and the defective rate would rise. But, on the other hand, even if the design is good, if the designated machine type in the specifications is not supplied, defective goods will once again be manufactured as a consequence. Thus, it is also important to execute TQC even with the suppliers. Adequate care of manufacturing facilities to process the purchased machinery also influences the defective rate. The laborers' quality of work is also another factor which may give birth to defective products. Moreover, even if there is nothing wrong with the finished product, a user may mishandle it and result in problems. Thus, the method of after-sales-service is another important factor. The purpose of the design may be incorrect. There are also cases of trouble due to inadequate responses to questions by users over the phone.

In this way, in order to successfully supply genuinely satisfactory products to the users, the entire function of a company should systematically work to guarantee its products' quality. This is the only way to overcome, grey part in shop floors.

It is from this concept that the currently common knowledge of TQC or total quality control originated; that without it quality control in its real sense could not be accomplished in Japan. And this concept served as one of the driving forces for the success of the Japanese economy today.

In Japan, there exists a system which presents an award known as the "Deming Award" to companies which have successfully and effectively introduced quality control. Many of the world-renowned first class Japanese companies have been awarded the "Deming Award." Such companies consist of various industries such as electrical, automobiles, steel, constructing company, etc.

In the selection of candidates for this award, company presidents attend the judges' conference to explain the top policies of their respective quality control programs. The criteria consist of all aspects of a company, and a president is questioned on his factory, its deign, management and labor policies, accounting, etc. Thus, in Japan, the "Deming Award" is significant for a company's efficient TQC operation and the award improves the company's image. Likewise, the receiving of this award is utilized extensively in a company's publicity activities.

## 4 . QC circle

The QC circle is one of the conspicuous elements of TQC in Japan. It is a voluntary group engaged in discovering and drawing up solution plans for problems at a working level.

On the occasion of an international conference on quality control held in Tokyo in 1969, some representatives visited our factory. Following presentations by our four circles on improvement measures, one foreigner brought up a question to a female worker, one of the speakers, during the question and answer session. He asked her, "What you presented on the impromevent of the factory is the work of technicians and not of a common laboroer like you. Isn't this interfering with the boundaries of the division of labor ? What is your opinion on this ?" To this, one of the members of the circle replied, "I understand your question. Nevertheless, we are best versed on the work within the factory. In the course of our daily work, we discovered a problem in the process, discussed it among ourselves and came up with a countermeasure which lowered the defect rate by 1/3. Is there anything wrong with this ?"

This is the very attitude of the QC circle. Its members often stay late even after working hours to carry heated discussions on improving their work efficiency. They seem to be enjoying solving difficult problems. It is one of their pleasures, and may be termed as a "game for improvement." This situation is comparable to volleyball players of a company who do not demand extra pay for practicing after work hours.

They would study data on the quality of their daily work and are able to actually see the effects of their efforts. This may imply a return to the days of craftsmanship.

Scientific management for Taylor and Gilbraith was effective for the aims of mass production, but it deprived the laborer of the pleasure of "making" things. The QC circle has now rediscovered this pleasure.

Another important element to point out here is also statistical quality control which allows the QC circle to come up with improvement measure. In Japan, this is known as the "paraphernalia" for improvement. Although the science of statistics is very difficult, it was the Japanese statistical specialists who facilitated and summarized the scientific procedures for novices and educated QC circles.

A tool is indispensable. It was with the important tool of the telescope that enabled Galieo Galilei to discover the rings around the planet Saturn. And it was the statistical means which served as paraphernalia that now enables a member of the QC circle, be it a young girl with only a high school education, to come up with an improvement measure that even a specialized technician may have never thought of. This certainly was very exciting for these girls and this is "the" reason for an extensive infiltration of the QC circle among Japanese companies.

## 5. Technical analysis and statistical analysis

If you see female Japanese assembly line workers without any technical background whatsoever making suggestions even engineers haven't been able to think of, you will ask what makes it possible for those women to acquire their technical knowledge. Theanswer is statistical methods.

We have two methods of analyzing and eliminating trouble in the manufacturing shop. One is by technological analysis; the other is by statistical analysis. QC uses statistical methods to analyze and improve the quality of products.

In the color TV factory of our company, a female employee of the Quality Assurance Section found that the failure rate of TV tuners differs depending on the type of RV, even when the same type of TV tuner is installed in the sets.

She thought that there must be some reason for this difference in failure rates of TV tuners. Therefore she drew diagrams which showed the relation between the failure rate of tuner and the length of the shaft, the temperature of the set, the diameter of the tuner knob, size of the cabinet and so on. At last she discovered a correlation between the failure rate and the distance from tuner to speaker; in other words, the failure rate of the tuner is quite low when the tuner is attached far from the speaker. On the other hand, when the tuner is attached near the speaker the set doesn't work well.

Such a conclusion would be hard to draw through technical analysis alone. But by accumulating market data we can discover such a phenomenon. We call this the law of large numbers.

Therefore, you may understand that we have two ways to find out the cause of defectives, one is by the use of anlysis based on technology and the other is through statistics.

One of QC's specialties is the use of statistical methods to eliminate trouble in the shop.

In order to find causes for a defect, you don't always need sophisticated techniacal expertise. What you need to do is analyze data. And quality control circles have learned to use statical tools; this is what makes the circles so successful. Statical analysis can be used to solve problems not only in manufacturing but also in sales, accounting, personnel management and service.

Spurred on by this method, QC circles are in demand among various fields including manufacturing, construction, financing, rastaurants and department stores. The same thing may produce varying degrees of results. Companies introducing the method seem to get better results in their work.

## 6. Automation and QC

During the first oil shock, every Japanese company suffered from the double pinch of falling sales and rising wages. Each company had to help itself because there was no one else to help it. Companies made desperate efforts to improve their productivity, and one way to do this was, naturally, to make an automated and unmannid plant. In the course of their efforts, they discovered an important rule, and that rule has become generally accepted. That is, in order to have successful automatic operations, it is necessary to drastically reduce the rejection rate.

A high rate of substandard products indicates that we have not yet discovered a way to produce only excellent ones. If machines are automated without first reducing the rejection rate, these machines will efficiently produce a mountain of inferior goods. According to my experience, if the rejection rate is more than a few percent, mechanization will produce very poor results.

Some people might think that automation itself reduces the rejection rate. On the contrary, a reduction of the inferior rate prior to automation raises productivity, making it possible to eliminate many workers completely. And since automation eliminates the errors caused by worker mistakes, the inferior rate is further decreased after automation.

Automation has certainly helped Japanese industry raise productivity. Wages increased 2.2-fold during the ten year period from 1976 to 1986. Furthermore, Japan successfully held prices of industrial products to a minimum during the period.

Hearing about automation, some people instantly link it with unemployment. This, however, is too short-sighted. People are now able to live in comfort thanks to a rise in productivity due to automation. Nowadays, everyone can use commodities which even noblemen could not afford in the past. People today also enjoy more leisure time, producing new waves of artistic activity. Japanese industry has witnessed a more than 2.7-fold increase in productivity over the last ten years, while at the same time, the jobless rate has remained about two percent. The jobless rate is low because so many new types of occupations have been created. This also substantiates the theory that productivity growth creates new enployment. In a country where there is no growth in productivity, inflation inflicts great hardship upon the people while pushing pu the rate of unemployement. It is important therefore, to understand correctly the importance of productivity. This is why Japan has been able to raise the income level and absorb oil price hikes while at the same time holding the inflation rate low. Japan was more successful in doing this than any other country of the world during the two oil crises of the 1970's. What made that success possible was increased automation, and higher productivity. For that purpose, quality control serves as a most powerful weapon.

# OPTIMAL CONTROL DISCRETE-TIME SYSTEMS

Jaroslav Dolezal

Institute of Information Theory and Automation Czechoslovak Academy of Sciences, 182 08 Prague, Czechoslovakia

## Introduction

The aim is to present a brief survey concerning the optimization of disrete-time /multistage/ systems described by a nonlinear vector difference-like equation. As the number of stages is assumed to be given, the pertinent optimization problem is finite-dimensional and thus, at least in principle, the methods of mathematical programming theory can be applied. On the other hand, special structure of such mathematical programming problems can be often exploited to derive more meaningful optimality conditions either necessary or sufficient.

From the point of view of necessary optimality conditions which are maybe most interesting subject from practical considerations, one has to keep in mind that there is no full analogy with the continuous-time system. Namely, without additional assumptions regarding convexity of the problem in question the discrete maximum principle is neither necessary nor sufficient condition for optimality of general nonlinear systems. Therefore, the desired reduction, of overall optimization to optimization stage-by-stage is possible only for a considerably narrow class of problems. It is seemingly the main reason, why there are still attempts to give to the discrete maximum spinciple a more constructive form.

It will be shown, that in the discrete-time case many different problems can be treated in unified way. For example, one can assume problems with various control and state constraints, with isoperimetric contraints, with various delays, with parameters, etc. It is also formally possible to deals with abstract constraining sets described by their conical approximations. However, in most concrete problems these constraining sets are given by a system of equalities and/or inequalities, which can be handled directly. It is worth mentioning that also the so-called discrete inclusion formulation is possible in the case of discrete control systems. Situation is analogical as in the continuous-time case.

Recently much interest was devoted to the analysis of nondifferentiable optimization problems, namely to the Lipschitz continuous case. Using the notions of a generalized gradient and a generalized normal cone and using several results of convex analysis its possible to formulate necessary optimality conditions in a quite general setting. These conditions include in a straightforward way the classical differentiable case. One can expect further development in this direction.

Further it is reasoble to investigate also problems with several objectives. For a given hierarchy of objectives the corresponding necessary optimality conditions can be derived in particular cases. The frequently studied solution type is the so-called Pareto optimum. Also for this vector optimization problem it is not difficult to find necessary optimality conditions enabling to determine a Pareto solution set.

If, moreover, several decision-makers are admitted, it is possible to describe dynamic situations with conflicting interests usually denoted as multistage games. In the case of two decision-makers /players/ with opposite interests one has zero-sum multistage games, the solution of which is the well-known saddle-point, the corresponding necessary optimality conditions for this two-sided optimization problem can be derived. Finally, it is not very difficult to modify these results to general N-player nonzero-sum multistage games and to postulate necessary optimality conditions for the Nash equilibrium solution.

Practically in all the mentioned problems of discrete-time system optimization the resulting necessary optimality conditions has the form of a discrete two-point boundary-value problem. Its solution cannot be obtained, except of few special cases, in an analytic way. Very often it is therefore assumed that discrete-time system is linear and objectives quadratic as functions of state and control. The solution is then given in closed form using discrete Riccati equations.

## Discrete Optimal Control Problems

Problems of these type are extensively studied in a number of monographs and papers. For all let us recall /1-4/. As the respective optimization problem is a finite dimensional one, the methodology of mathematical programming is widely used. In /2/ an alternative geometrical approach is suggested for optimization of discrete-time systems. In /5/ the scheme originally developed in /1/ is generalized to include also cases with the so-called state-dependent region of admissible controls. This is the way how to deal with more general formulation of /4/ under milder assumptions of /1/.

Consider the formulation of discrete optimal control given in /5/ which admits general type of state constraints and state-dpendent control constraints given as general sets. As a special case these constraints are described by a system of equalities and inequalities, this being very often the case in practical applications. Discrete analogy of an integral functional is assumed, i.e. sum over the prescribed number of stages.

This formulation enables, after the appropriate transcription, to deal with a number of other problems which have quite different original setting. The analogy with both, the respective mathematical programming problems and the continuous-time optimal control problems is source of a number of interesting results. In this way it is possible to include terminal objective functional, problems with delays in control and/or state /6, 7/, problems with isoperimetric constraints /8/, periodic systems /9/, problems with parameters, minimax problems /10/, etc.

## Existence of Optimal Controls

Again, invoking the respective results of mathematical programming theory one can easily postulate basic sufficient assumptions /continuity of functions, compactness of constraining sets/ for the case of constant region of admissible controls /4/. However, for state-dependent control region more subtle approach using some basic facts of set-valued mapping theory is necessary /11, 12/. It is interesting to note that in discrete case no additional convexity assumptions, used in continuous-time cases are necessary, i.e. no relaxation procedure is needed.

## Necessary Optimality Conditions

General formulation of a discrete-time control problem requires two crutial steps to derive necessary optimality conditions. First, the general set constraints are approximated by certain convex sets, usually cones, the polars of which are used to postulate necessary optimality conditions. One proceeds in a quite analogical way as in mathematical programming.

Second step concernes set-valued mappings which define state-dependent regions of asmissible controls. Namely it is necessary assume that these mappings admit locally smooth selections /4, 5/. For the corresponding exact form of necessary optimality conditions the reader can consult /5/. Anyhow, see also the closely related contribution /13/. As it was possible to expect by an analogy with a continuous-time case, also in discrete-time case adjoint system of multipliers is to be solved. Necessary optimality conditions result in a discrete nonlinear two-point boundary-value problem which must be solved by numerical algorithms, in general. Only in the case of liner system and quadratic functional the analytic solution having form of a matrix discrete Riccati equation is obtained /5/.

It is not difficult to derive the corresponding necessary optimality conditions for above mentioned modifications of this basic problem. For example, for isoperimetric type constraints and problems with delays it is only necessary to augment the existing state vector in appropriate way to apply the results of /5/ and to evaluate them for the original problem. Minimax problems of discrete-time optimal control are treated in /10/.

## Discrete Maximum Principle

The maximum principle formulation of necessry optimality conditions for continuous -time problems /14/ inspired many attempts to obtaian its discrete analogy. However, especially some early results in this respect were wrong. It has shown that without aditional assumption regarding convexity the discrete maximum principle does not constitute part of necessary optimality conditions for discrete systems. Such assumption substitutes the convexifying effect of continuous time as recognized in /15/, where also weaker assumption of the so-called directional convexity was suggested and used later /1, 2, 5/. Survey of pertinent Soviet results is given in /16/. Recent attempt to augment the class of discrete system for which discrete maximum principle is valid is reported in /17/, as the idea of stage-by-stage optimization is attractive. The notion of a generalized Hamiltonian together with geometrical "upper boundary approach" avoided explicit directional coonvexity assumption.

The question of importance of discrete maximum principle as a necessary optimality condition seem to be appropriate at this place. In fact the practical impacts of maximum principle formulation are somewhat less evident than in the continuous case. For computation only mathematical programming formulation is mostly sufficient. Observe also to certain extent complementary assumptions for existence and maximum principle in discrete and continuous systems.

## Sufficient Optimality Conditions

Apparent assumptions to guarantee optimal solutions of discrete-time problems are linearity of system equations and convexity of constraining sets and an objective functional as implied by mathematical programming theory. Another "parameter-free" approach is followed in /13, 18/ for the so-called difference inclusion formulation of discrete control problems. In /18/ sufficient optimality conditions were obtained using convex analysis approach and multivalued mapping theory. Convex analysis was also mainly used to give sufficient optimality conditions for discrete-time optimization problem of Bolza type /19/.

## Nonsmooth Problems

A rapid development of mathematical programming theory in the last decade contributed highly to the understanding and development of the area of nonsmooth optimization. This includes convex problems and Lipschitz problems. Introduction of a generalized gradient /20/ was basic step in this respect. Existing necessary optimality conditions and other results of nonsmooth analysis can be used when studying nonsmooth discrete-time optimal control problems. Then role of a convex approximation plays the generalized tangent cone. Necessary optimality conditions are formulated in a general form in /21-23/ using general constraining sets. Also more concrete case of explicitly given constraints is analyzed there. The obtained conditions exhibit the same overall structure as in differentiable case. Further generalizations in this respect can be expected, however, the more general the respective formulation, the less constructive the corresponding necessary optimality conditions.

## Problems with Several Objectives

There are several ways to treat the problems with several objectives, the so-called multiobjective optimization. Most of available methods can be modified also for discrete-time optimal control problems. One way to go is to exploit some given structure of the multiobjective problem and investigate the respective hierarchical solution /24/. The popular Pareto solution was studied for discrete-time systems since /25/. Anyhow, observe the existing analogy of multiobjective and isoperimetric problems /8/. For problems with state-dependent regions of admissible controls the respective necessary optimality conditions are formulated in /26/.

Also in multiobjective case it was possible to obtain deeper results for linear problems with quadratic objectives. As for multiobjective mathematical programming problems there is a number of results dealing with nonsmooth cases /27-29/, one can derive the corresponding form of necessary optimality conditions for nonsmooth multiobjective discrete-time optimal control problems in a straightforward way.

## Multistage Games

The study of dynamic situations under conflicting interests resulted in a number of results for the so-called differential games. Their discrete-time analogy - multistage games - were investigated using geometric approach /2/ or results for discrete-time control systems /30/. For the case two players with opposite objectives one obtains necessary saddle-point conditions for two-player zero-sum multistage games. In general case of N-player nonzero-sum multistage games necessary conditions for the so-called Nash equilibrium can be postulated for this N-sided optimization problem. The respective existence theory is investigated in /31/.

Other frequently studied solution types are minimax /security/ solution and various kind of cooperations. If biased information sructure is assumed the so-called Stackelberg solution can be appropriate /32/. Again, for linear multistage games with quadratic objective functionals it is possible to obtain explicit form of strategies for each participating player. As for control problems /one-player game/, one can analyze nonsmooth multistage game, e.g. along the lines of /33/.

## Numerical Algorithms

Almost every described problem, after application of the corresponding optimality conditions, has the form of a discrete-time nonlinear two-point boundary-value problem. There is full analogy with continuous-time cases. Also here it is necessary to use iterative numerical techniques to obtain approximate solutions /34-37/. One can use either indirect methods based on necessary optimality conditions /quasilinearization type methods/ or direct approach based on decreasing the respective objective /gradient type methods/. In fact, most algorithms developed for the solution of mathematical programming problems and continuous-time control problems are directly applicable for discrete-time problems without principal troubles.

## Conceusions

The study of discrete-time problems is important either as sometimes the only accesible way to deal with approximated original continuous-time problems or to deal with intrinsically discrete-time situations. The presented survey is by far not complete, only selected deterministic problems are pointed out. More concrete areas and examples of application can be found in the mentioned references.

References


/1/ Canon M. D., Cullum C. D., Polak E.: Theory of Optimal Control and Mathematical Programming. McGraw-Hill, New York 1970.

/2/ Blaquiere A., Gerard F., Leitmann G.: Quantitative and Qualitative Games. Academic Press, New York 1969.

/3/ Propoi A. I.: Elements of Optimal Discrete Processes. Nauka, Moscow 1973. In Russian.

/4/ Boltjanskij V. G.: Optimal Control of Discrete Systems, Nauka, Moscow 1973. In Russian.

/5/ Dolezal J.: Necessary optimality conditions for discrete systems with state-dependent control region. Kybernetika 11 (1975), 423-450.

/6/ Dolezal J.: Necessary condition for discrete dynamical systems with delays and general constraints. Kybernetika 17 (1981), 425-432.

/7/ Dolezal J.: Optimal control of discrete systems involving delays. JOTA 39 (1983), 59-66.

/8/ Dolezal J.: Discrete optimal control problems with isoperimetric constraints and discrete multiobjective optimization. Preprints of the 2nd IFAC Symposium on Optimization Methods, Varna 1979, 7-13.

/9/ Dolezal J.: Optimal control of discrete systems: periodic systems and systems with delays. Proceedings of the 5th Symposium Computer in Chemical Engineering, High Tatras 1977, 831-834.

/10/ Dolezal J.: Existence of optimal solutions in general discrete systems. Kybernetika 11 (1975), 301-312.

/12/ Dolezal J.: Optimal control existence results for general discrete systems. Appl. Math. Optimiz. 3 (1976), 51-63.

/13/ Hautus M. L. J.: Necessary conditions for multiple constraint optimization problems. SIAM J. Control 11 (1973), 653-669.

/14/ Pontrjagin L. S., Bltjanskij V. G., Gamkrelidze R. V., Mishchenko E. F.: The Mathematical Theory of Optimal Processes. Interscience, New York 1962.

/15/ Holtzman J. M., Halkin H.: Directional convexity and the maximum principle for discrete systems. SIAM J. Control 4 (1966), 265-275.

/16/ Gabasov R., Kirillova F.: On discrete maximum principle. Math. Operationsforsch. Statist., Ser. Optimization 10 (1979), 543-553.

/17/ Nahorski Z., Ravn H. F., Vidal R. V. V.: The discrete-time maximum principle: a survey and some new results. Int. J. Control 40 (1984), 533-554.

/18/ Dolezal J.: On the sufficient optimality conditions for difference inclusions. Systems Science 5 (1979), 187-198.

/19/ Rockafellar R. T., Wets R. J.-B.: Deterministic and stochastic optimization problems of Bolza type in discrete time. Working Paper WP-81-69, IIASA, Laxenburg 1981.

/20/ Clarke F. H.: Generalized gradients and applications. Trans. Amer. Math. Soc. 205 (1975), 247-262.

/21/ Dolezal J.: Non-smooth and non-convex problems in discrete optimal control. Int. J. Systems Sci. 13 (1982), 969-978.

/22/ Dolezal J.: Necessary conditions for Lipschitz continous discrete control problems. In "Mathematical Control Theory", J. Zabczyk (Ed.), Banach Center Publications, Vol. 14, Polish Scientific Publishers, Warsaw 1985, 171-179.

/23/ Mordukhovich B. Sh.: On necessary conditions for an extremum in nonsmooth optimization. Soviet Math. Dokl. 32 (1985), 215-219.

/24/ Dolezal J.: Hierarchical solution concept for static and multistage decision problems with two objectives. Kybernetika 12 (1976), 363-385.

/25/ DaCunha N. O., Polak E.: Constrained minimization under vector-valued criteria in finite dimensional spaces. J. Math. Anal. Appl. 19 (1976), 103-124.

/26/ Dolezal J.: Necessary optimality conditions for discrete systems with state-dependent control region and vector-valued objective function. Systems Science 3 (1977), 171-184.

/27/ Goffin J. L., Haurie A.: Pareto optimality with nondifferentiable cost functions. In "Multiple Criteria Decision Making", H. Thiriez, S. Zionts (Eds.), Springer-Verlag, Berlin 1976, 232-246.

/28/ Kanniappan P.: Necessary conditions for optimality of nondifferentiable convex multiobjective programming. JOTA 40 (1983), 167-174.

/29/ Dolezal J.: Necessary conditions for Pareto optimality in nondifferentiable problems. Problem of Control Theory 14 (1985), 131-141.

/30/ Dolezal J.: Necessary optimality conditions for N-player nonzero-sum multistage games. Kybernetika 12 (1976), 268-295.

/31/ Kleindorfer P. R., Sertel M. R.: Equilibrium existence results for simple dynamic games. JOTA 14 (1974), 613-631.

/32/ Dolezal J.: Stackelberg solution concept for general multistage games. Kybernetika 14 (1978), 369-380.

/33/ Aubin J.-P.: Locally Lipschitz cooperative games. J. Math. Economics 8 (1981), 241-262.

/34/ Polak E.: Computational Methods in Optimization: Unified Approach. Academic Press, New York 1971.

/35/ Dolezal J.: On a certain type of discrete two-point boundary-value problem arising in discrete optimal control. Kybernetika 15 (1979), 215-221.

/36/ Ritch P. S.: Discrete optimal control with multiple constraints I: Constraint separation and transformation technique. Automatica 9 (1973), 415-429.

/37/ Dolezal J.: Gradient method for discrete-time optimal control. In "5th Symp. on Algorithms", Part Algorithms, High Tatras 1981, 249-270. In Czech.

# CONTROL PROBLEMS UNDER INSUFFICIENT INFORMATION

Yu. S. Osipov

Institute of Mathematics and Mechanics
Kovalevskoi Street 16
Sverdlovsk, USSR

Control under incomplete information is one of the most urgent questions of the modern control theory. This question arises in numerous technological problems and needs specific mathematical tools for its investigation. The present paper is devoted to an application of some mathematical tools of this kind to the inverse problems for dynamical systems. The results are obtained by A. V. Kryazhimskii and the autor and published partly in [1−6]. The present paper is divided into three parts. We begin with the existence results, then pass on to constructive solution methods and conclude with numerical examples.

Consider a control system

$$\dot{x}=f\ (t,\ x,\ v),\ x(t_0)\in X_0,$$

where $x\in R$, $t\in T=[t,\ \theta]$, $v\in R$, the set $X_0$, is a given compactum, the function $f$ is continuous and Lipschitz in $x$. The system is considered on a bounded time interval T. Control $v$ is a measurable function of time with values in a given compactum $V$. We suppose that control $v$ is not known to us. Thus $v$ might be a disturbance, or some unknown parameter of the system, or counteracting control not known beforehand. Below we keep on calling functions $v$ controls, though we have in mind all the above mentioned cases. It is assumed that the actual initial state $x_0=x\ (t_0)$ of the system lies in a given compactum $X_0$.

Any pair

$$(x,\ v),\quad x=x\ (t),\ v=v\ (t),\ x\ (t_0)\in X_0$$

will be called a process. Let $(x_*,\ v_*)$ be the process which actually takes place. Suppose that at every moment $t$ the vector

$$z\ (t)\ =Gx_*\ (t)$$

is measured, G being a given n×m matrix. In particrlar, $z$ may be formed by some of coordinates of $x$. Any process giving the same realization of $z$ will be called compatible with $z$. We denote by $P$ the set of all processes compatible with the measurement $z$. It is clear that we are not able to select the process wich actually has taken place among the collection $P$ of all the processes compatible with $z$. Thus any process belonging to $P$ may in fact be considered as the real one.

Suppose now that the measurements of $z$ are not precise. Namely, for the measurement results $\zeta(t)$ the following inequality is true

$$|\zeta(t)-z(t)| \leq h.$$

Let us put the question if it is possible to construct a stable approximation $R$ of the set $P$. Stability means that $R$ is close to $P$ provided $h$ is small enough. Below we consider families of operators

$$D_h : \zeta \rightarrow R = D_h \zeta$$

(we call them algorithms) and select families with appropriate stability properties. One more property important to us is physical realizability of algorithms. A physically realizable algorithm forms set $R$ in real time. Its output at time $t$ is the "part" of the set $R$ "realized" up to time $t$ and consequently it can not depend on future values of measurement results $\zeta$. Formally such an algorithm $D_h$ can be defined as an operator which assigns to a pair $(\zeta, t)$ the set $R(t)$ of functions $(\zeta, u)$ defined on the time interval $[t_0, t]$:

$$D_h : (t, \zeta) \rightarrow R(t) = R(\zeta, t) =$$
$$= \{(\xi, u) : \xi = \xi(\tau) \in R^n, u = u(\tau) \in V, t_0 \leq \tau \leq t\}.$$

For these sets the following conditions are fulfilled.

1) The set corresponding to the final time instant $R(\vartheta)$ is the final output of the algorithm $R = D_h \zeta$.

2) If $t_2 > t_1$ then functions $(\xi, u)$ belonging to $R(t_2)$ being restricted to the interval $[t_0, t_1]$ lie in $R(t_1)$. (See Fig. 1).

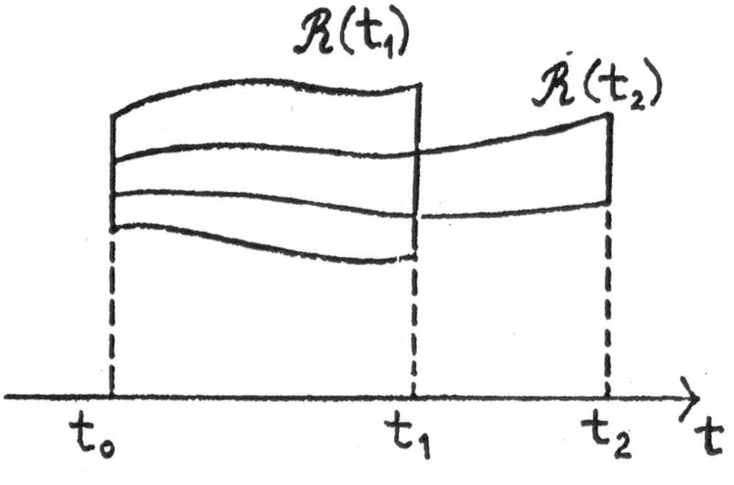


Fig. 1

3) If measurements $\zeta_1$ ans $\zeta_2$ coincide up to time $t$, then the sets corresponding to them at time t coincide too $R(\zeta_1, t)$. (See Fig. 2.).

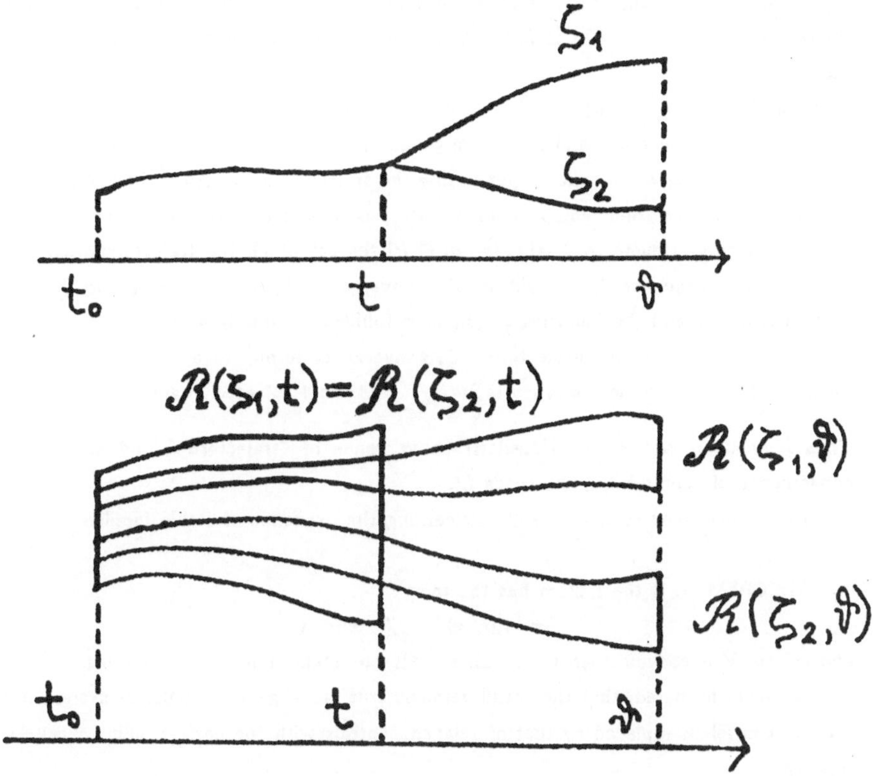


Fig. 2

Below we deal with physically realizable algorithms only.

Now let us specify the stability property. We introduce three types of stability basing on three definitions of convergence of sets. Let $C=C(T, R^r)$ and $L^p=L^p(T, R^r)$, $p>1$, be standerd spaces of functions. We call elements of the product $C\times L^p$ quasiprocesses and consider algorithms whose outputs are quasiprocesses. Let $Q_i$ be a sequence of sets of quasiprocesses. We need the following slightly modified definition of convergence of sets introduced by Mosco [7]. Sequence of sets $Q_i$ converges to the set $P$ in the sense of Mosco if

1) for any pair $(x, v)\in P$ there exists a sequence $(\xi_i, u_i)$ of elements of $Q_i$ wich converges to $(x, v)$ in the strong topology of the space of guasiprocesses, and

2) for any sequence $(\xi_i, u_i)$ of elements of $Q_i$ there exists an element $(x, v)\in P$ and a subsequence $(\xi_{ij}, u_{ij})$ such that its first components $\xi_{ij}$ converge in $C$ to $x$ and its second components $u_{ij}$ converge to $v$ in the weak—star topology of $L^p$.

Mosco convergence may be used for appoximation of minimal value $\varphi=\varphi(x, v)$ on $P$. Indeed if functional $\varphi$ is continuous (in the space of quasiprocesses) and convex in $v$, then minimal values of $\varphi$ over the sets $Q_i$ converge to $\varphi_* = \min \varphi\ (x, v)$; this value may serve an estimation of the "worst quality" process compatible with measurement $z$.

A family $D_h$ of algorithms is called Mosco-stable if for any sequences $h_i$, $\zeta_i$ such that $h_i \to 0$, $|\zeta_i(t) - z(t)| \leq h_i$ the corrisponding outputs converge to $P$ in the sense of Mosco.

The Hausdorff stability and $X$-stability are introduced in the similar way basing on the Hausdorff convergence and $X$-convergence of sets. The latter is defined as follows. Let $V(x)$ be the set of all controls generating the trajectory $x$ and $\Xi_i = [\xi : (\xi, u) \in Q_i]$ be the set of the first components of all quasiprocesses from $Q_i$; we call them quasitrajectories. Denote by $X=[x: (x, v) \in P]$ the set of all the trajectories compatible with $z$. The sequence $Q_i$ is said to $X$-converge to $P$ if $\Xi_i$ converge to $X$ in the Hausdorff metric and the following condition is fulfilled. If quasiprocesses $(\xi_i, u_i)$ lie in $Q_i$ and the corresponding quasitrajectories $\xi_i$ converge uniformly to a trajectory $x$ from $X$ them the $L^p$-half distance $d(u_i, v(x)) = \inf_{v \in V(x)} |u_i - v|_{L^p}$ goes to zero.

Thus $X$-convergence means Hausdorff convergence of "trajectories" and non-uniform convergence of "controls" in the space $L^p$.

Let us pass now to some results concerning the existence of stable families.

THEOREM 1. If the system has the form

$$\dot{x} = f_1(t, x) + f_2(t, x)v$$

and the set $V$ is convex, then there exists a Mosco-stable family of algorithms.

It should be noted, that the result remains true for a general nonlinear system if the class of controls is replaced by that of relaxed controls with the corresponding weak-star topology.

For the case of $X$-stability some extra conditions are to be imposed. First of all suppose that the set $X$ of all trajectories compatible with $z$ is closed in C. Denote by $V_* = V_*(x)$ the set of all controls $v$ such that $z = Gf\ (t, x\ (t), v\ (t))$ for almost all $t \in T$. It is clear that for any trajectory $x$ compatible with $z$ we have $V(x) \subset V_*(x)$. We introduce the following

Condition A: for any trajectory $x$ compatible with $z$ the two sets coicide $V_*(x) = V(x)$.

THEOREM 2. If condition A is fulfilled, then there exists an $X$-stable family of algorithms.

A simple counterexample shows that condition A is essential:

$$x = u,\ x(0) = 0,$$
$$x, u \in R,\ |u| \leq 1,\ T = [0, 1],$$
$$G \equiv 0.$$

Here $z = 0 \cdot x = 0$. We have $V_* \neq V = [x]$, so $A$ is not true. There is no $X$-stable family.

THEOREM 3. If for any trajectory $x$ compatible with $z$ all the controls from $V_*$ are equivalent with respect to the Lebesque measure, then there exists a Hausdorff stable family of algorithms.

The proofs of the three existence theorems are based on the positional control principle of N. N. Krasovskii [8; 9] and some ideas of the theory of ill-posed problems. We give here a brief outline of the method. Consider a finite partition of time interval $T$

$$t_0 < t_1 < \cdots < t_N = \vartheta, \quad t_{i+1} - t_i = \delta,$$

and introduce an auxiliary control system which we call a model

$$\dot{w} = F_w(t, w_i, \xi_i, \zeta_i, u_i, u_i),$$
$$\dot{\xi} = F_\xi(t, w_i, \xi_i, \zeta_i, u_i, u_i),$$
$$t_i \le t < t_{i+1}, \; w_i = w(t_i), \; \xi_i = \xi(t_i), \; \zeta_i = \zeta(t_i),$$

where $\dim \xi = \dim x$. The controls are given by the closed-loop control law

$$u_i = u_i(t_i, w_i, \zeta_i, \xi_i),$$
$$u_i = u_i(t_i, w_i, \zeta_i, \xi_i).$$

We call it a strategy. Note that the model equations and the values of strategy depend on the current states of model and the current measurement results.

The model and the strategy should be chosen so that pairs $(\xi, u)$ are close to the processes from the set $P$. It can be shown that this way leads to the desired results.

Algorithms approximate the set of processes $P$. That is why algorithms form sets of motions of the model. However in some particular cases it suffices to form a single motion. This makes algorithms much more convenient for practical computation. Now we shall concentrate on the algorithms of this kind.

Let us consider a system with the right-hand side affine in $v$

$$\dot{y} = f_1(t, y, z) + f_2(t, y, z)\, v,$$
$$\dot{z} = g_1(t, y, z) + g_2(t, y, z)\, v,$$

where $f_i$, $g_i$ are Lipschitz, $x = (y, z)$, coordinates $z$ are measured. The initial state of the system is supposed to be given

$$y(t_0) = y_0, \; z(t_0) = z_0.$$

Suppose that the range of matrix $g_2$ is equal to the dimension of the control vector $v$, which we denote by $r$. The last assumption implies the uniqueness of process $(x, v)$ compatible with the measurement $z$. As above the measurement results $\zeta$ are not precise

$$|\zeta(t) - z(t)| \le h.$$

Let us construct a model and a strategy ensuring stable approximation of $(x, v)$ in the product $C \times L^2$. Thus we consider a Hausdorff-stable family of algorithms. The model is of the form

$$\dot{\xi}_z = \bar{u}_i, \; \xi_{z0} = z_0,$$
$$\dot{\xi}_y = f_1(i) + f_2(i) g_i, \; \xi_{y0} = y_0,$$
$$\dot{w} = g_1(i) + g_2(i) u_i, \; w_0 = z_0,$$
$$t_i \le t < t_{i+1}.$$

Here the following notations are used

$$f_1(i)=f_1(t_i,\ \xi_{1i},\ \zeta_i),$$

$$g_1(i)=g_1(t_i,\ \xi_{1i},\ \zeta_i),$$

$$\bar{g}_i=g_2^{+}(i)(u_i - g_1(i)),$$

$g_2^{+}(i)$ is the pseudoinverse matrix for $g_2(i)$.

A few words on the sense of the components of the model. The second component $\xi_2$ approximates the component $y$ of the system. Here the unknown control $v$ is replaced by the term $\bar{g}_i$. This term as one can easily see, coincides with $v$ at each moment $t_i$ if vector $\xi_2$ follows exactly along $y$ and the measurements are precise $\zeta(t)=z(t)$. Actually these conditions are not true and $\bar{g}_i$ differs from $v$. Component $\xi_1$ is an approximation of $z$. Component $w$ (whuich is also an approximation of $z$) is used for forming $L^2$ approximation $u$ of control $v$.

The required properties of the model are guarantied by the strategy

$$u_i^{(1)}=-K_1\ sign\ (\xi_{1i}^{(1)}-\zeta_i^{(1)}),$$

$$(K_1=const\ \geq\ |\ \dot{z}^{(1)}(t)\ |\ )$$

$u_i$ − minimum point or

$$\beta(u)=2(w_i\ -\ \xi_i)'\ g_2(i)\ u+\alpha\ |\ u\ |^2,\quad u\in V.$$

The parameter $\alpha$ and the step $\delta$ of time partition are connected with $h$ by relations

$$\alpha=\alpha_h\rightarrow 0,\ h/\alpha_h\rightarrow 0\ (h\rightarrow 0),\ \delta=\delta_h\leq const\cdot h.$$

Finally we can formulate

THEOREM 4. Let $P_h\ (w_h,\ \xi_h,\ u_h,\ u_h)$ be control processes for the model generated by the above given strategy and corresponding to measurement results $\xi_h$ satisfying $|\ \zeta_h(t) - z(t)\ |\leq h$. Then paies $(\xi_h,\ u_h)$ converge to the process $(x,\ v)$ in the product $C\times L^2$.

Let us illustrate the result by two examples.

**Example 1** System is a platform $\Pi$ moving in a horizontal plane without rotation, see Fig. 3.

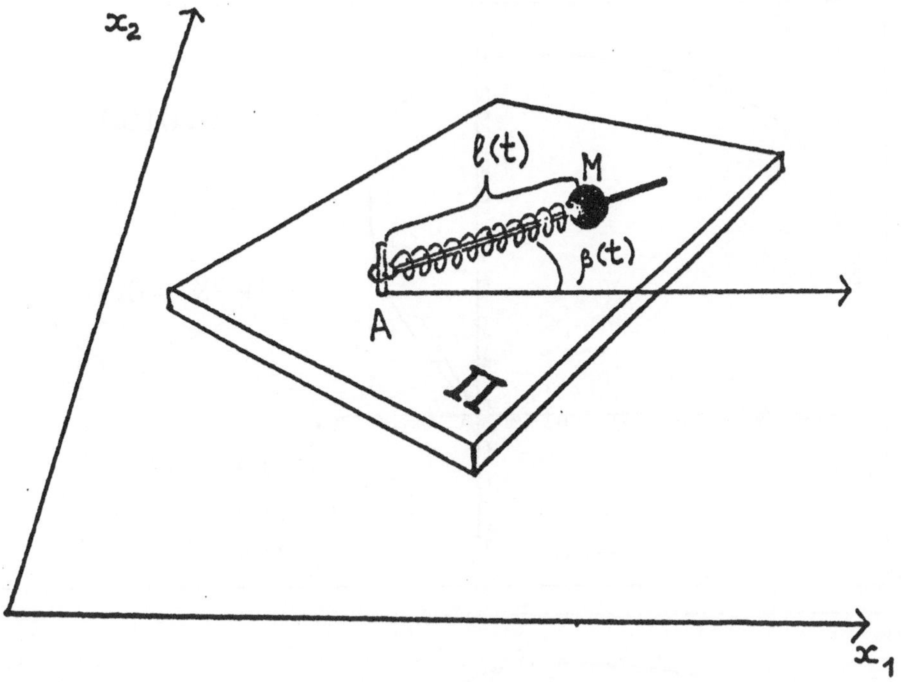


Fig. 3

A weightless bar $B$ is fastened at point $A$ on the platform. It may rotate freely around the point $A$. A material point $M$ connected with point $A$ by a spring is sliding along the bar. The measured parameters of the system are the length $l=l(t)$ of the spring, the angle $\beta = \beta(t)$ between the axis $x_1$ and the bar, and the velocities of these parameters $\lambda(t)=\dot{l}(t)$, $\omega(t)=\dot{\beta}(t)$. Measurements $l(t), \beta(t), \lambda(t), \omega(t)$ are not precise.

The problem is to approximate in real time current states $X(t)=(X_1(t), X_2(t))$, velocities $W(t)=(W_1(t), W_2(t))$ and accelerations $v(t)=(v_1(t), v_2(t))$ of the platform.. We consider acceleration $v$ as a control and approximate it in space $L^2$. State $X$ and velocity $W$ are approximated uniformly. Denote by m, $l^*$, $c$, $g$ mass of point $M$, length of the spring, rigidity of the spring, velocity of $M$ respectively. The system equations have the form

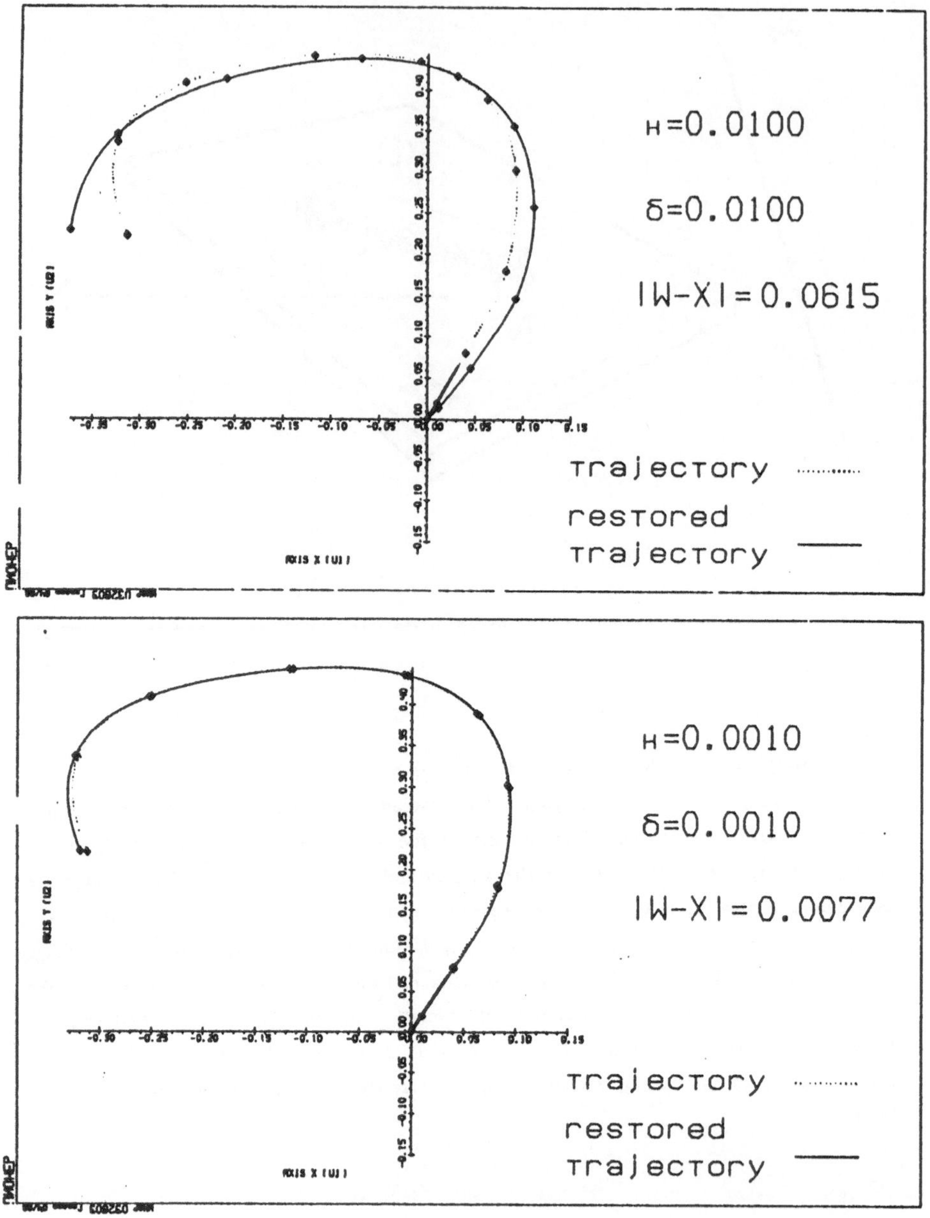
н=0.0100
δ=0.0100
|W-X|=0.0615
Trajectory
restored trajectory
н=0.0010
δ=0.0010
|W-X|=0.0077
Trajectory
restored trajectory

Fig. 4

$$\dot{X}_1=W_1,\ \dot{X}_2=W_2,\ \dot{W}_1=v_1,\ \dot{W}_2=v,$$

$$g_1=-\frac{c}{m}(l-l^*)\cos\beta,\quad g_2=-\frac{c}{m}(l-l^*)\sin\beta,$$
$$l=\lambda,\quad \beta=\omega,$$

$$\dot{\lambda}=-\frac{c}{m}(l-l^*)-v_1\cos\beta-v_2\sin\beta$$

$$\omega=\frac{a_1}{l}-\frac{a_2}{l}+\frac{\sin\beta}{l}v_1+\frac{\cos\beta}{l}v_2,$$

$$a_1=[(W_1-q_1)\cos\beta+(W_2-q_2)\sin\beta]\,\omega,$$

$$a_2=[(W_1-q_1)\sin\beta-(W_2-q_2)\cos\beta]\,\lambda,$$

We suppose that control vector $v$ lies within the fixed ball $(v_1^2+v_2^2)\leq\mu$. The paramenters $z=(l,\beta,\lambda,\omega)$ are measured, and $y=(X_1, X_2, W_1, W_2, q_1, q_2)$ are not measured.

The approximation algorithm was simulated on a computer for the following data:

$$T=[0,1],\ m=10,\ c=0.05,\ \mu=10,$$

$$X_{1,0}=X_{2,0}=W_{1,0}=W_{2,0}=q_{1,0}=q_{2,0}=\beta_0=\lambda_0=\omega_0=0,$$

$$l_0=l^*=0.5,$$

| | |
|---|---|
| $v_1(t)=2,$ | $0\leq t<0.25,$ |
| $v_1(t)=-4,$ | $0.25\leq t<0.75,$ |
| $v_1(t)=8,$ | $0.75\leq t\leq 1,$ |
| $v_2(t)=4,$ | $0\leq t<0.3332,$ |
| $v_2(t)=-4,$ | $0.3332<t\leq 1.$ |

Two variants were considered

1) $h=0.01,\ \alpha=0.1\,h^{1/2},$

2) $h=0.001,\ \alpha=0.075\,h^{1/2},$

where $h=t_{i+1}-t_i=\delta$ is the time step. The results of approximation of the states and controls are shown on Fig. 4–6.

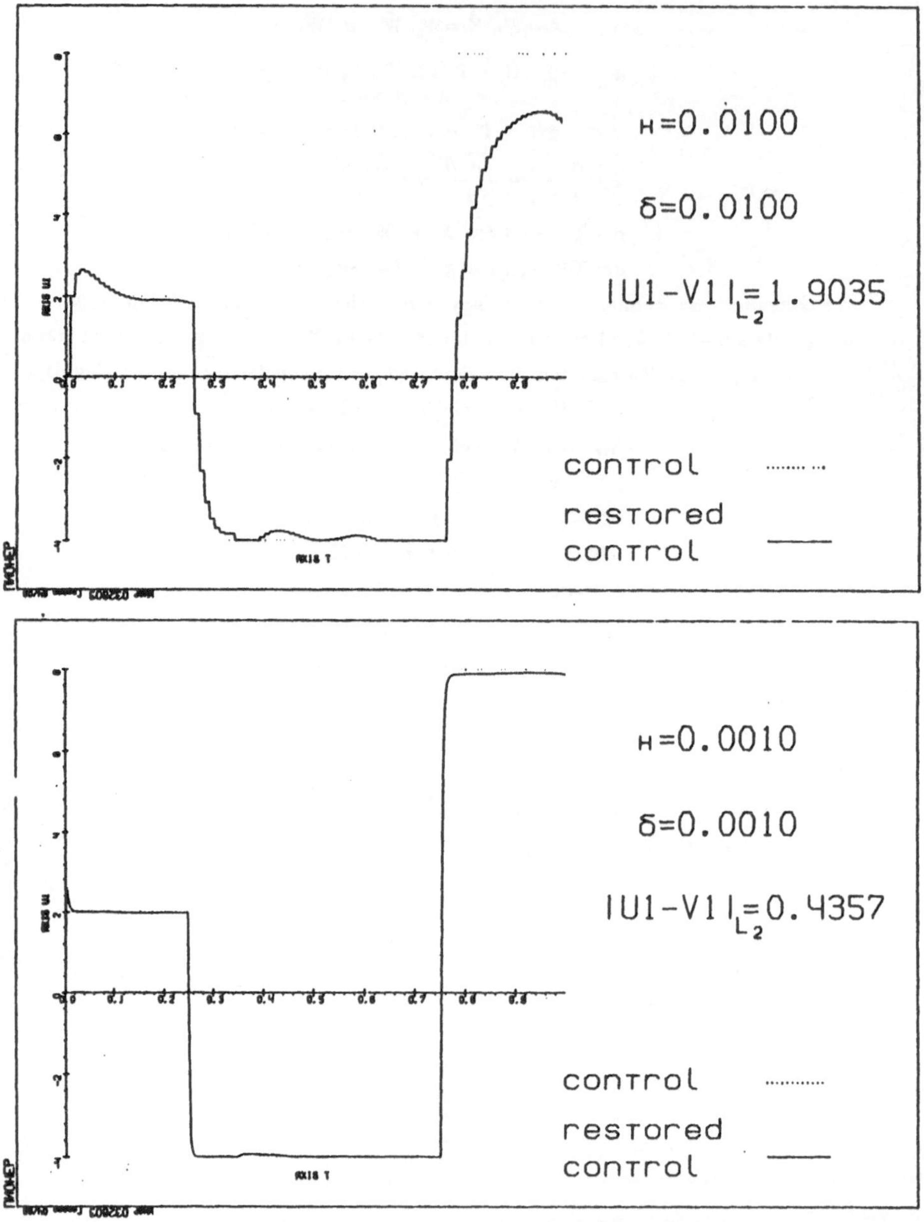
H=0.0100
δ=0.0100
‖U1-V1‖$_{L_2}$=1.9035
control
restored control
AXIS T
H=0.0010
δ=0.0010
‖U1-V1‖$_{L_2}$=0.4357
control
restored control
AXIS T

Fig. 5

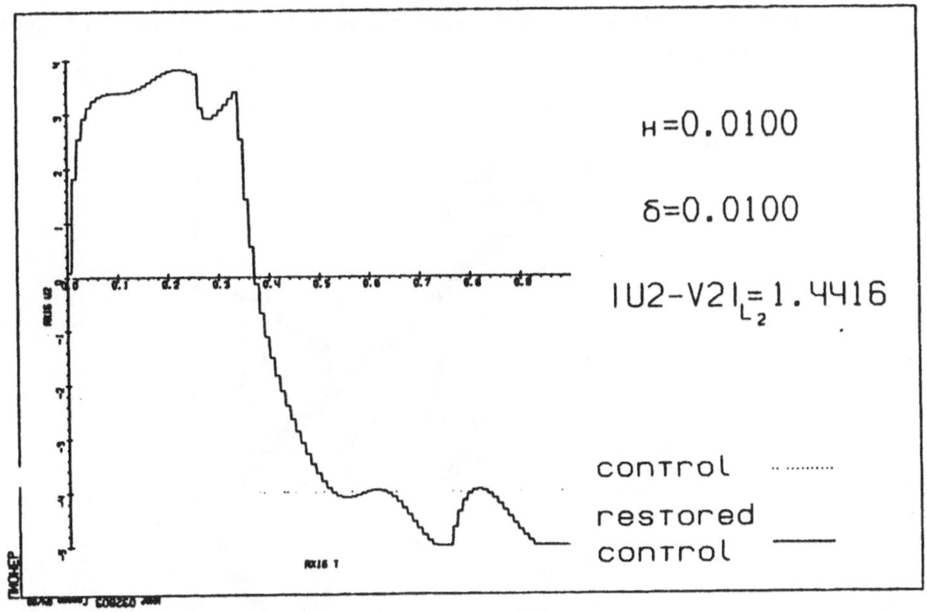
h=0.0100
δ=0.0100
|U2-V2|$_{L_2}$=1.4416
control
restored control

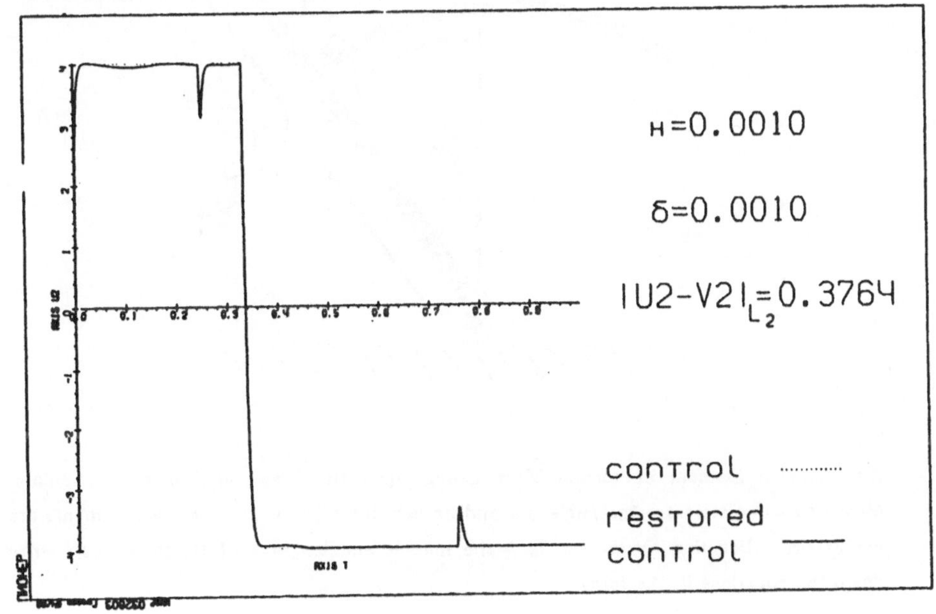
h=0.0010
δ=0.0010
|U2-V2|$_{L_2}$=0.3764
control
restored control

Fig. 6

**Example 2.** Consider a double pendulum moving within a vicinity of its upper equilibrium, see Fig. 7.

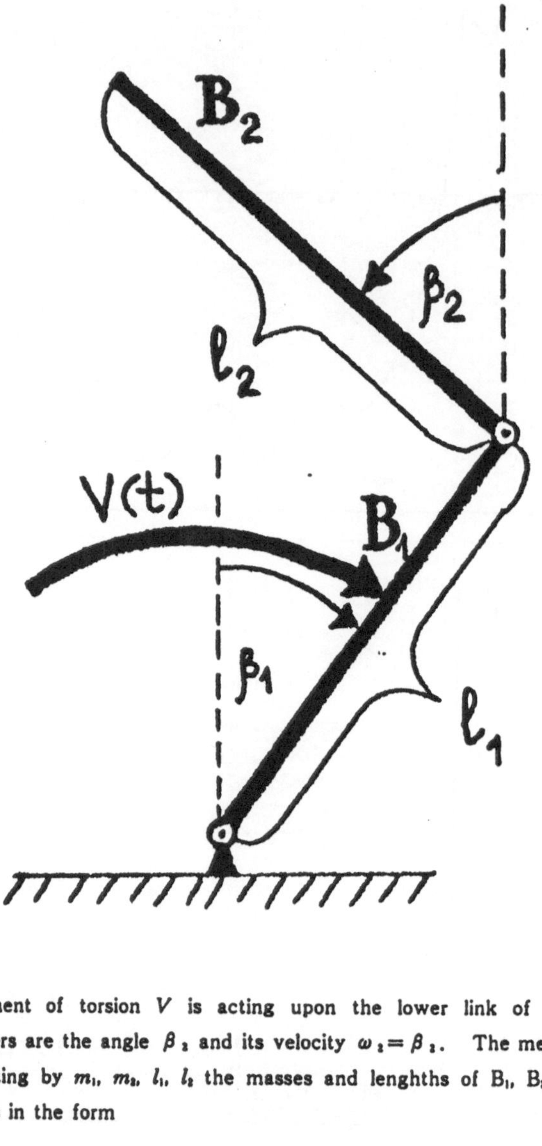


Fig. 7

An unknown moment of torsion $V$ is acting upon the lower link of the pendulum. Measured parameters are the angle $\beta_2$ and its velocity $\omega_2 = \dot{\beta}_2$. The measurements are not precise. Denoting by $m_1$, $m_2$, $l_1$, $l_2$ the masses and lenghths of $B_1$, $B_2$ we can write down the equations in the form

$$\dot{\beta}_1=\omega_1,\quad \dot{\omega}=\frac{1}{l_1(4_1m+3_2m)}[6m+2m)d\,\beta_1-9m_2g\,\beta_2+\frac{12}{l}\,v],$$

$$\dot{\beta}=\omega_2,\quad \dot{\omega}_2=\frac{1}{l_1(4m_1+3m_2)}[-9(m_1+2m_2)g\,\beta_1+6(m_1+3m_2)g\,\beta_2-\frac{18}{l_1}\,v].$$

**We suppose that $|v|\le\mu$ where $\mu$ is a given constant. The measured parameters are $z=(\beta_2,\ \omega_2)$. Parameters $y=(\beta_2,\ \omega_2)$ are not available. For simulation the following data were chosen**

$$T=[0,1],\ m_1=m_2=1,\ l_1=l_2=3,\ \mu=12$$

$$\beta_{1,0}=0.075,\ \omega_{1,0}=-0.0001,\ \beta_{2,0}=-0.2,\ \omega_{2,0}=0.01$$

$$v(t)=10s(t),\ \beta_2(t)>0.01$$

$$v(t)=-10s(t),\ \beta_2(t)<-0.01,$$

$$v(t)=-3s(t),\ |\beta_2(t)|\le 0.01,\ \beta_1(t)>0.05,$$

$$v(t)=3s(t),\ |\beta_2(t)|\le 0.01,\ \beta_1(t)<-0.05,$$

$$s(t)=1-0.5\sin\pi t,\ \alpha=0.006h^{1/2},$$

**where $h=t_{i+1}-t_i=\delta$ is the time step. Two variants were considered: 1) $h=0.001$, 2) $h=0.0001$. The results of simulation are presented on Fig. 8−9.**

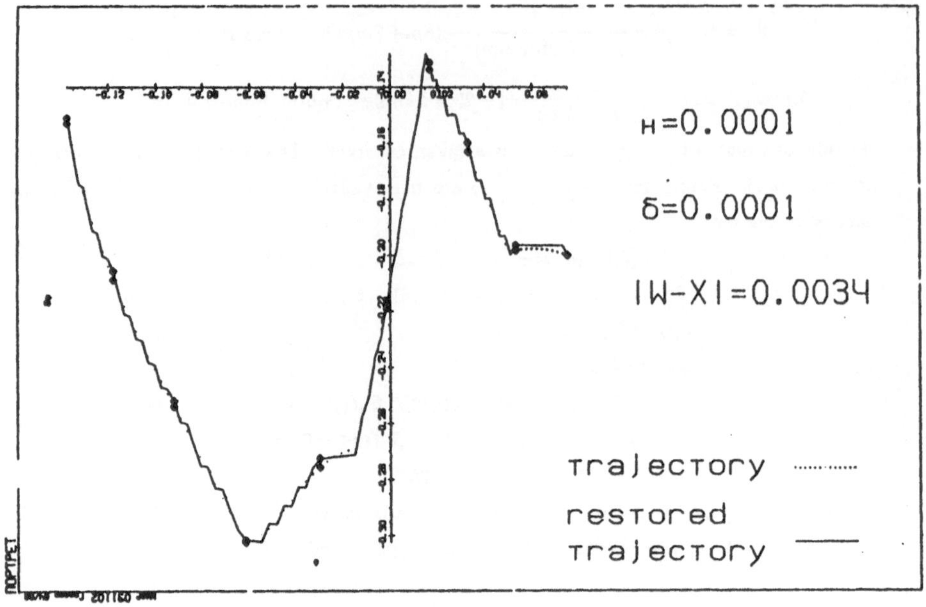
H=0.0001
δ=0.0001
|W-X|=0.0034
Trajectory
restored
Trajectory

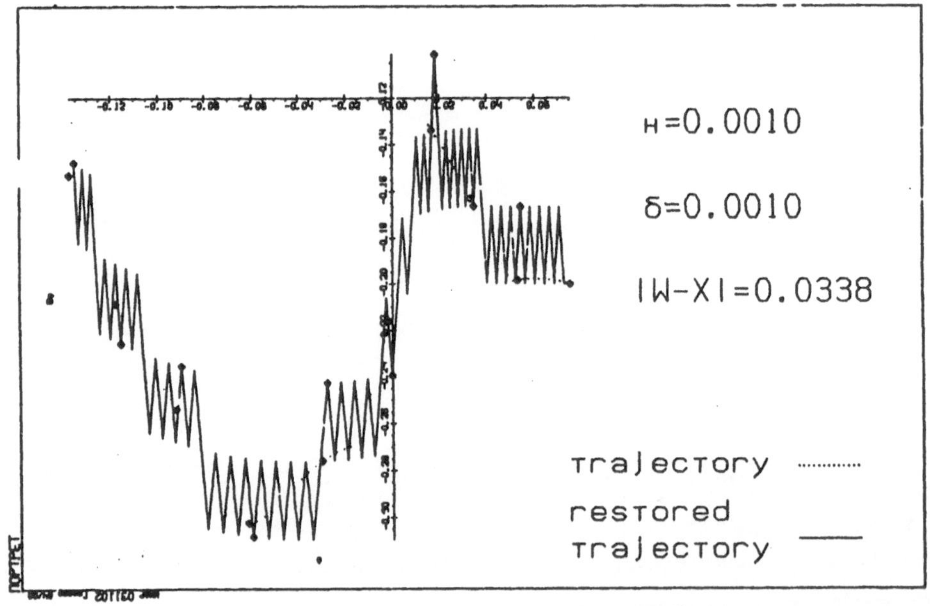
H=0.0010
δ=0.0010
|W-X|=0.0338
Trajectory
restored
Trajectory

Fig. 8

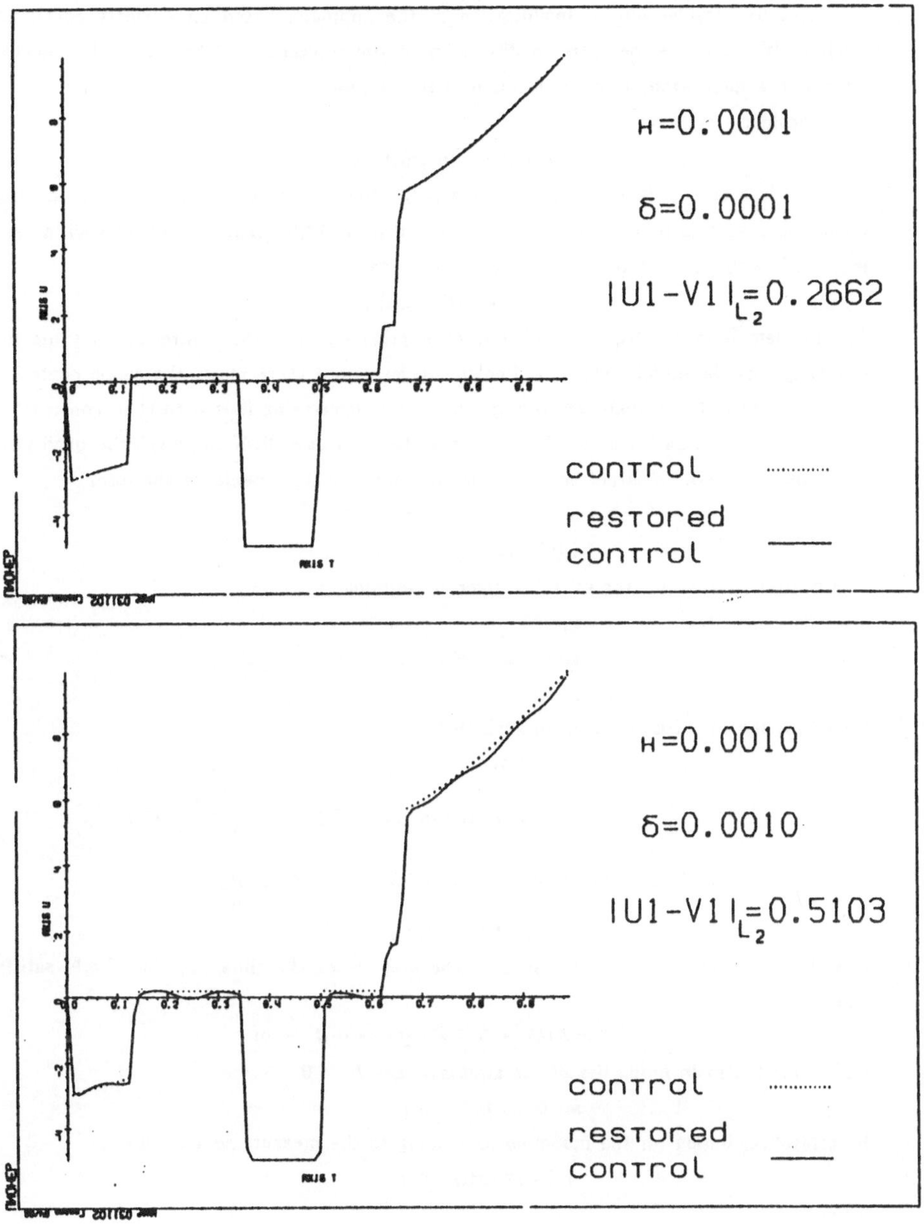
H=0.0001
δ=0.0001
|U1-V1|L2=0.2662
control
restored control
H=0.0010
δ=0.0010
|U1-V1|L2=0.5103
control
restored control

Fig. 9

Let us consider now a modification of the approach based on auxiliary controlled models. We impose some extra conditions on control realizations. These conditions enable us to find a more accurate approximation of the control.

Consider a system

$$\dot{x}=f(t,\ x,\ v),\ x(t_0)=x_0,$$

where $v$ is scalar, $f$ is Lepschitz, $x_0$ is given. Suppose it is known that the control realization $v=v(t)$ is monotonically decreasing and $a\leq v(t)\leq b$. (An example of such a case is given below). The state vector $z=x$ is measured

$$|\ \zeta(t)\ -x(t)\ |\ \leq h.$$

The problem is to construct in real time an approximation of the control $v$. Assume for simplicity that the initial state $x_0$ is known precisely, and there exists the single control $v$ corresponding to the actual trajectory $x$. Since $v$ is decreasing it is natural to construct a decreasing approximation of $v$. We shall see that this essentially improves the quality of approximation. We construct the approximation introducing a model of the form

$$\dot{w}=f(t,\ \zeta_{i-1},\ u_i),\ t_i\leq t\leq t_{i+1},$$

$$w(t_0)=x_0.$$

The control strategy for the model is given by formula

$$u_0=b$$

$$u_i=maxV_i,\ V_i\neq\phi,$$

$$u_i=u_{i-1},\ V_i=\phi.$$

Here $V_i$ is the set of all $u\in[a,\ b]$ such that

$$a\leq u\leq u_{i+1},$$

$$(w_i-\zeta_i)\cdot(f(t_{i-1},\ \zeta_{i-1},\ u)\ -\ \frac{\zeta_i-\zeta_{i-1}}{\delta})\leq\mu,$$

$$\mu=2[\ |\ x_0\ |\ +(K+1)(\vartheta-t_0)][c(\delta+h)+2\frac{h}{\delta}+(K+1)h],$$

$$K=const\geq\ |\ x(t)\ |\ ,$$

$c$ is the Lipschitz constant for $f$ in $x$. The step $\delta$ of the time partition shorld satisfy conditions

$$\delta=\delta(h)\rightarrow0,\ h/\delta(h)\rightarrow0\ (h\rightarrow0).$$

Let us list the main properties of the strategy. Let $h\rightarrow0$,

$$u_h(t)=u_i,\ t_i\leq t<t_{i+1}$$

be control realization for the model corresponding to the measurement results $\zeta_h$,

$$|\ \zeta_h(t)-x(t)\ |\ \leq h.$$

Then

1) $u_h(t)$ are monotonically decreasing,

2) $u_h(t)\geq v(t)$ almost everywhere,

3) $u_h$ converges to $v$ in $L^2$,

4) if $v$ is continuous then for any $\varepsilon>0$ we have that $u_h$ converges to $v$ uniformly on $[t_0+\varepsilon,\ \vartheta]$.

Let us give an illustration to these statements.

Example 3. Consider the simplest oscillating system

$$\ddot{x}=-v(x-l_0)/m$$

(a material point and a spring). We suppose that the coefficient $v=v(t)$ (the rigidity of the spring) is a monotonically decreasing function of time, $0\leq v\leq c$. (For instance, there might be a number of springs breaking one by one in the course of oscillation). (See Fig. 10).

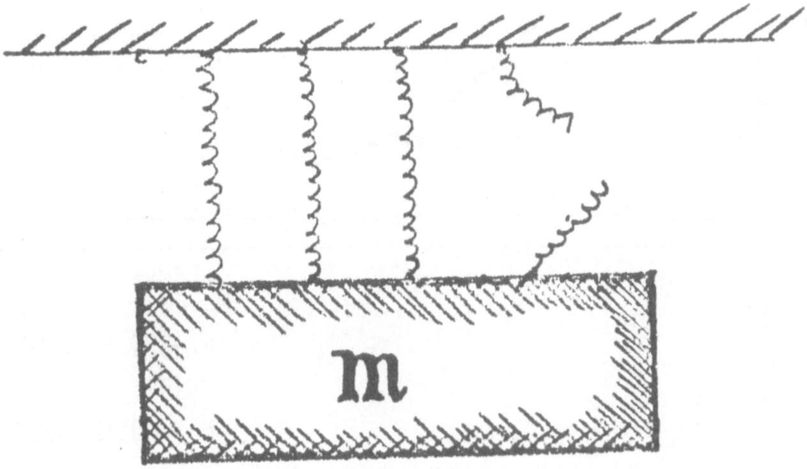


Fig. 10

Current states and velocities of the oscillating point are measured

$$|\zeta_1(t)-x(t)|\leq h,\quad |\zeta_2(t)-\dot{x}(t)|\leq h.$$

Our goal is to approximate $v$ in real time. The simulations were carried out using the above given algorithm for the data $T=[0,1]$, $l_0=1$, $m=0.5$ and two variants of control choice. The first variant corresponds to $v(t)=3(1-t^2)$ and the second one to

$$v(t)=3,\quad 0\leq t<\tfrac{1}{4},$$
$$v(t)=\tfrac{9}{4},\quad \tfrac{1}{4}\leq t<\tfrac{1}{2},$$
$$v(t)=\tfrac{3}{4},\quad \tfrac{1}{2}\leq t<\tfrac{3}{4},$$
$$v(t)=0,\quad \tfrac{3}{4}\leq t\leq 1.$$

The results are presented on Fig. 11–12.

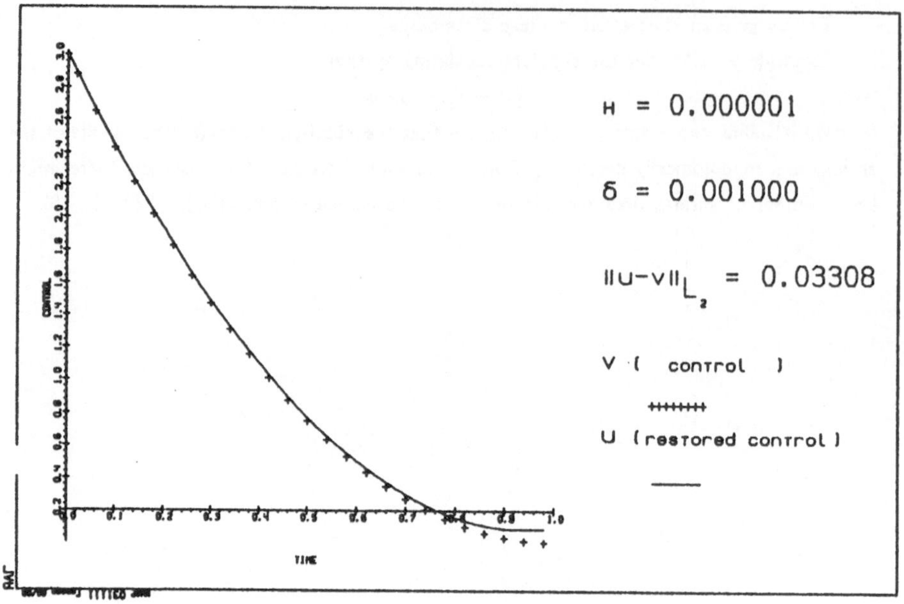
H = 0.000001
δ = 0.001000
||U−V||L2 = 0.03308
V ( control )
U (restored control)
TIME
CONTROL

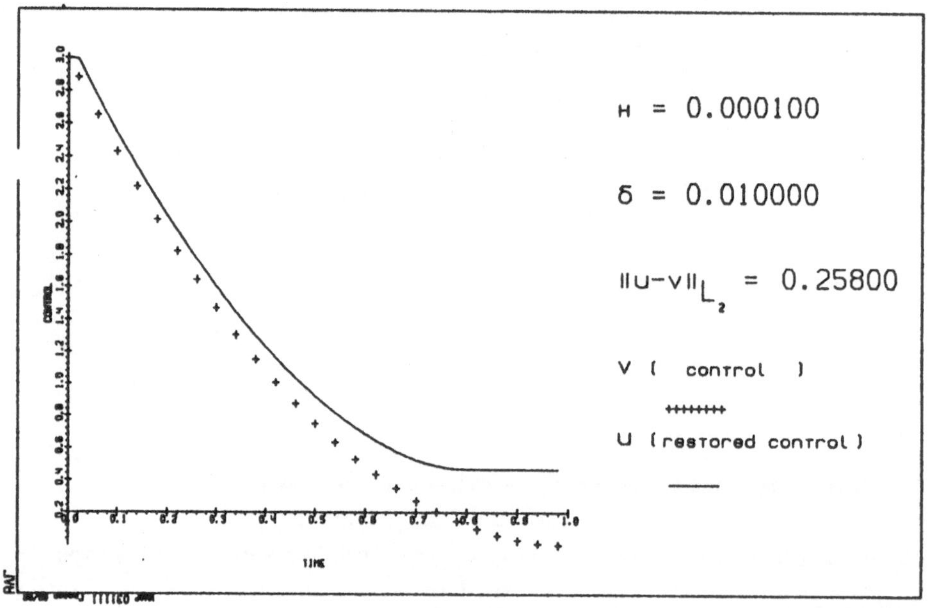
H = 0.000100
δ = 0.010000
||U−V||L2 = 0.25800
V ( control )
U (restored control)
TIME
CONTROL

Fig. 11

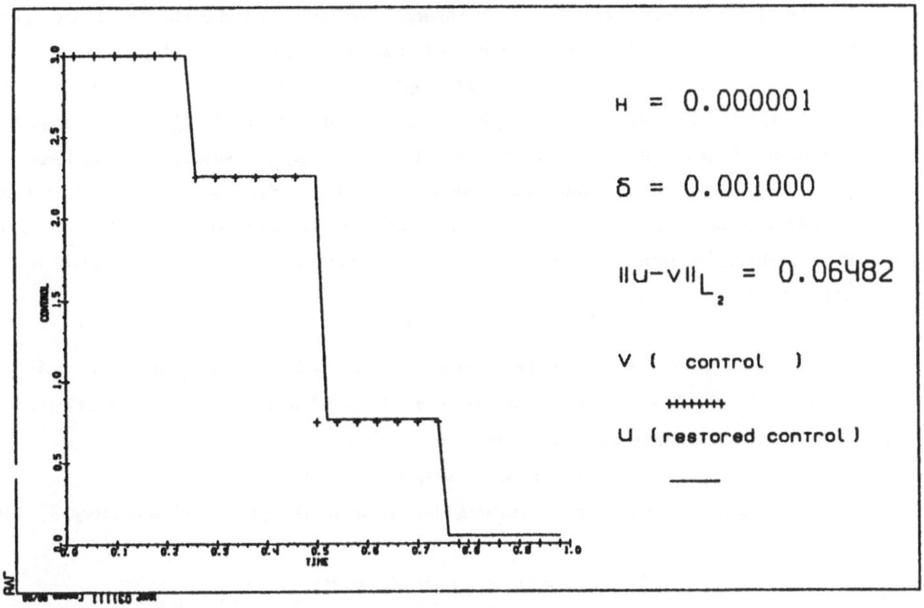
H = 0.000001
δ = 0.001000
‖u−v‖L2 = 0.06482
V ( control )
U (restored control )
CONTROL
TIME

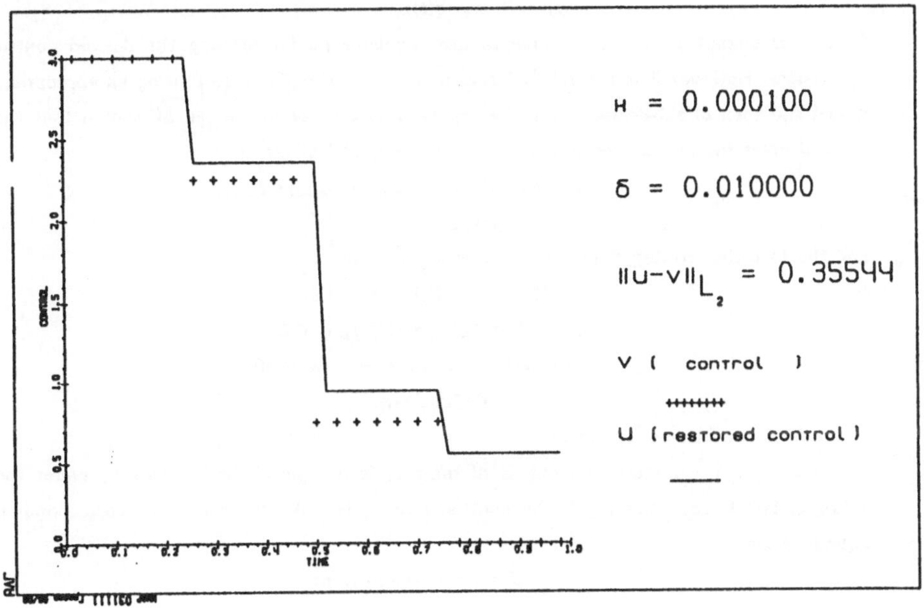
H = 0.000100
δ = 0.010000
‖u−v‖L2 = 0.35544
V ( control )
U (restored control )
CONTROL
TIME

Fig. 12

The approach under consideration is applicable to some control problems. Let us give one of the applications of this kind. The control system is given by the equation

$$\dot{x}=f(t, x, p) + g(t, x, q), x(t_0)=x.$$

Here $f$, $g$ are Lipschitz, $p=p(t) \in P$ is the control, $q=q(t) \in Q$ is the unknown disturbance, $P$, $Q$ are given compacta. Functions $p(t)$, $q(t)$ are supposed to be Lebesge measurable. Denote by $x_*=x_*(t)$ the prescribed trajectory. Suppose that one has to lead the system along the prescribed trajectory not knowing the disturbance realization $q$. The control $p$ should be formed on the basis of current measurements of $x$. In general these measurements $\zeta$ are not precise

$$| \zeta(t)-x(t) | \leq h.$$

The control $p$ should be stable in the following sense. If $h$ is small enough then $v = | x-x_* |_C + | \dot{x}-\dot{x}_* |_{L^2}$ is also small. Actually the problem is to find a stable control $p$. Let us assume that there exists a regulator of the form

$$p_*=p_*(t, v), v=g(t, x_*(t), q)$$

($p_*$ is Lipschitz). This regulator ensures motion along the prescribed trajectory $x_*$ for arbitrary $q(t)$

$$\dot{x}_*=f(t, x_*, p_*(t, v(t))+v(t),$$
$$v(t)=g(t, x_*, q(t)).$$

Since $v$ is unknown we are not able to use regulator $p_*$ for forming the desired control realization. However it is natural to form an $L^2$-approximation $u$ to $v$ using an appropriate model and then to substitute $u$ into the regulator instead of $v$. It can be shown that the method gives the desired result. We can use the model of the form

$$w=f(t_i, \zeta_i, p_*(t_i, u_i))+u_i, t_i \leq t < t_{i+1},$$
$$w(t_0)=x_0$$

and the following strategy: $u_i$ is the minimum point of

$$\beta(u)=(w_i-\zeta_i)' u+\alpha | u_i|^2,$$
$$u \in G_i= \{g(t_i, x_*(t_i), q): q \in Q\} ,$$
$$\alpha = \alpha(h) \to 0, h/\alpha \to 0 (h \to 0),$$
$$\delta \leq const\cdot h.$$

Let us consider an illustrative

**Example 4.** A material point $M$ of mass $m$ is moving along the axis $x_1$ under the action of two forces. Force $p$ is the control, force $q$ is a disturbance. The corresponding equations are

$$\dot{x}_1=\dot{x}_2, x_2=(p+q)/m.$$

We suppose that

$$| p | \leq \mu, | q | \leq \nu, \mu > \nu.$$

The prescribed motion is given by relations

$$\dot{x}_{1*}(t)=\dot{x}_{2*}(t), x_{2*}(t)=s_*(t).$$

Here $s_*$ is the prescribed time-dependent acceleration, $| s_*(t) | \leq (\mu - \nu) /m$.

The problem is to form a stable control force ensuring $L^2$-closeness of the actual acceleration to the prescribed one. The above presented algorithm solves the problem. The data tor simulation were

$T=[0,2]$, $x_1(0)=x_2(0)=0$, $m=1$, $\mu=3$, $\nu=1$,

$s_*(t)=\sin \pi t$, $0 \leq t \leq 1$ (starting),

$s_*(t)=0$, $1<t<1.5$ (free motion),

$$s_*(t)=-2\sin\frac{\pi-1.5}{0.5},\quad 1.5\leq t\leq 2 \text{ (braking)}.$$

$$q(t)=\left(\frac{t}{1.5}\right)^2,\quad 0\leq t<1.5,$$

$q(t)=-0.5$, $1.5\leq t\leq 2$.

time step $\delta=t_{i+1}-t_i$ is equal to $h$ taking one of the values 1) $h=0.2$, 2) $h=0.02$. For simulation results see Fig. 13.

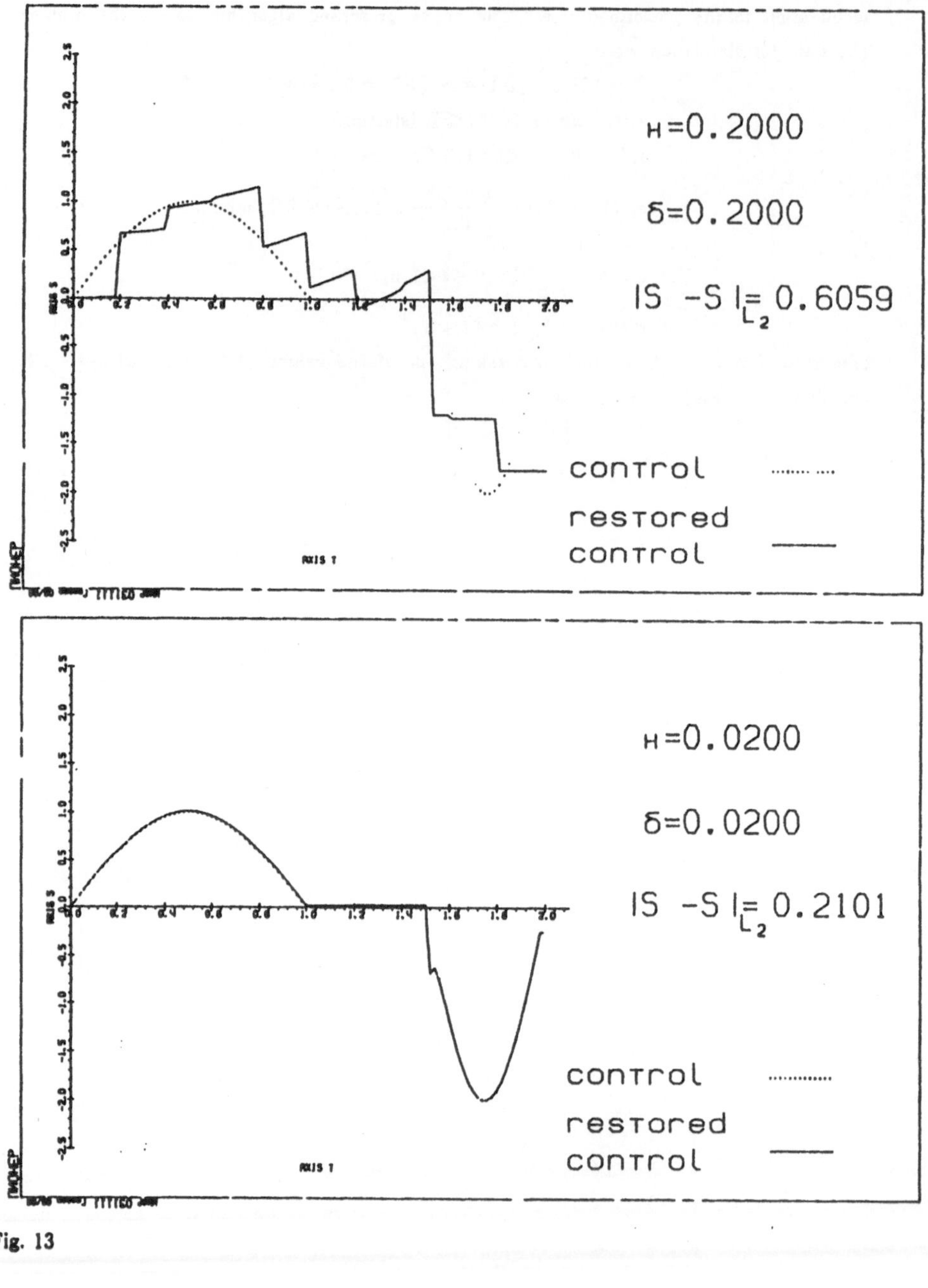
H=0.2000
δ=0.2000
|S −S|= 0.6059
L2
control
restored
control
AXIS 1
H=0.0200
δ=0.0200
|S −S|= 0.2101
L2
control
restored
control
AXIS 1

Fig. 13

Let us mention in conclusion that the method of models and strategies can be applied to other classes of control systems. Some of these clsses are

1) systems with noncontinuous right—hand side,

2) systems with aftereffect

$$x=f(t,\ x_t,\ v),$$

$$x_t=x(t+s),\ -\tau\leq s\leq 0\ ,\ \ \tau=const>0,$$

3) systems with varying time lags as controls

$$x=f(t,\ x(t-v)),\ \ 0\leq v\leq\tau\ ,$$

4) systems with time lags in coordinates and controls

$$x=f(t,\ x,\ x(t-\tau),\ v(t),\ v(t-\tau)),$$

5) systems governed by a Couchy problem or standard initial boundary value problem for parabolic and hyperbolic equations, as well as some other types of partial differential equations.

## References.


1. Osipov Yu. S., Kryazhimskii A. V. On dynamical solving of operator equations. Dokl. Akad. Nauk USSR, 1983, vol. 269, no. 3, p. 552—556 (in Russian).
2. Osipov Yu. S., Kryazhimskii A. V. Method of Lyapunov functions for the problem of motion modelling. Stability of a motion. Novosibirsk, Nauka, 1985, p. 53—56 (in Russian).
3. Kryazhimskii A.V., Osipov Yu. S. On control modelling in a dynamical system. Jzv. Akad. Nauk USSR. Tekhn. Kibernet., 1983, no. 2, p. 51—60 (in Russian).
4. Osipov Yu. S., Kryazhimskii A. V. Stable solutions to inverse problems of control systems dynamics. Trudy Mat. Inst. Steklov. (in Russian, to appear).
5. Kryazhimskii A. V. Optimization of ensured result for the dynamical systems. Proceedings of ICM—86 (to appear).
6. Osipov Yu. S. Problems of dynamical reconstruction. Chislo and mysl. Moscow, Znanie, 1987, no. 10, p. 7—27 (in Russian).
7. Mosco U. Convergence of convex sets and of solutions of variational inequalities. Adv. Math., 1969, vol. 3, no. 4, p. 510—585.
8. Krasovskii N. N. Controlling of a dynamical system. Moscow, Nauka, 1985 (in Russian).
9. Krasovskii N. N., Subbotin A. I. Positional differential games. Moscow, Nauka, 1974 (in Russian).

# APPLICATION OF OPTIMIZATION-BASED METHODS IN CONTROL SYSTEM DESIGN

W. Y. NG
University Engineering Department,
Cambridge, CB2 1PZ,
United Kingdom.

## 1. Introduction

In this paper, we discuss the application of the class of optimization-based design methods in practical design of control systems. In the introduction, we seek to clarify our area of concern.

Firstly, it is practical designs that we are addressing. By practical, we mean they are real life problems. The design specifications are often vague, even inexact, contrary to text-book problems which specifications are clear and fit the problem statements of the design methods being illustrated. For a text-book problem, it is all a matter of executing the particular design method and finding a solution, often with optimality, while for a practical design, the major effort is often in structuring the problem, clarifying the specifications as more and more about limitations and trade-offs are known, choosing suitable design methods and selecting among alternative candidate designs. Finding any candidate solution is by no means a major part of the problem. Also, with respect to the specifications which are imprecise, the search is for a satisficing design, rather than an optimal one. The final design is a product of much inevitable compromise and trade-offs executed to a satisfactory degree by the designer. We also note that computer aids are indispensible for practical designs of controllers and computer-aided design (CAD) is an equivalent terminology in our context.

Zakian and Al-Naib (1) first proposed the formulation of a control system design problem as a general set of inequalities. It marks the beginning of the development of a whole class of control system design methods which we shall call optimization-based methods (2 - 7). The main strength of such methods lies in that unlike the control-theoretic design methods, no restriction is placed on the class of specifications. Both control-theoretic and application-oriented objectives are treated uniformly. This property together with the use of numerical optimizations as the means to search for solutions shall characterise them. Employing such methods in design involves selection of a set of controllers to be searched, represented by a parametrization with a set of design variables, selection of an initial design, and the execution of optimization to search for solutions among the set. There are two other advantages of such methods. Firstly, the optimization search is continuous in the controller set (or parameter space). This makes any resultant controller an open design, one which neighborhood with similar performance is also known. Such open design is valuable as further selection is possible. Secondly, much about the limitations and trade-offs can be obtained by tracing the controllers along the paths tracked by the optimization search.

## 2. A Design Strategy

In this section, we propose a strategy to apply the class of optimization-based methods to a practical controller design problem seen as a multi-objective programming problem (MOP) of the form :

$$\min_{x} J(x) \qquad x \in X \subset R^{p}, \quad J \in R^{q}$$

where Ji, $i \in Q = \{1,2,..,q\}$ are the quantitative performance indices measuring the objectives and x is a parametrization representing a selected set of controllers from which solutions are sought. The existence of an initial set of design objectives is assumed which has been translated into the form of satisfactory levels of the indices, i.e. we seek a solution x such that :

$$J_i(x) \leq C_i^{(0)} \qquad \text{for all } i \in Q$$

which have not been satisfied by an initial design :

$$x^{(0)}$$

The initial design objectives may be vague and inexact. Their clarification to achieve compatibility and trade-offs is a major task of the strategy, which does it by guiding the designer to change the values of the satisfactory levels.

The emphasis of this strategy is in the interaction support of the following three aspects of design : management of the design process, designer's comprehension of the problem, and decision support, to refining specifications for trade-offs, compatibility and removal of redundancy.

### Management of the Design Process

To manage the design process, we suggest the following scheme to split up the different stages :

(a) Exploratory :
exploring a simplified version of the optimization search

for an empirical understanding of the design problem ;

(b) Formulation / Re-formulation :
formulating a substitute problem by scalarizing the MOP design problem to exercise search and trade-offs with appreciation of compatibility ;

(c) Search :
for generating one or more candidate controller(s) ;

(d) Evaluation :
analysing the present candidate and its place in the design history to update problem understanding ;

A design iteration ends with the evaluation of a candidate design from stage (d). If the design is not satisfactory, the designer may choose to start either from (a) or (b) or (c).

**Designer's Comprehension**

This is achieved by creating a database of PO solutions

$$X^{*} = \{ x^{*}(i) \}$$

which interrogation should reveal relationships among the design variables and the indices. This database is initiated with entries from the initial exploratory stage described as an interactive algorithm below :

STEP 0 : Select/construct a set of information-rich indices Jei's (IRI's) to approximate the original MOP to the following form

$$\min_{x} \ Je(x) \qquad Je \in R^{m} , \ m < q$$

Set $x^{0} = x^{(0)}$

STEP 1 : Assign bad and good values for the IRI's, favouring one of them, $Je_{j}$, say.

STEP 2 : PO Surface Sampling

conduct iterative minimax optimization with initial design $x^{0}$

$$\min_{x} \max_{i} \frac{Je_i(x) - Je_{i(good)}}{Je_{i(bad)} - Je_{i(good)}}$$

If this is a first execution
stop when solution has converged and store final solution in database ;
else sample and store points from the optimization trajectory and stop when
either i) convergence starts
or ii) $Je_j$ attains satisfactory value ;

Final point obtained is $x^f$.

STEP 3 : If more samples are wanted,
set $x^0 = x^f$, go to STEP 2 ;
else
continue.

STEP 4 : Evaluate original full index vector for the sample points obtained as

$$J(x^{(i)})$$

Remove non-PO points, i.e.

$$x^{(j)} \quad \text{such that} \quad J(x^{(j)}) \geq J(x^{(i)}) \text{ for some } i$$

Save others in database.

STEP 5 : <u>Index Grouping</u>

Assign screening vector

$$b \in R^s, \; s \leq q, \quad b = \{ b_i : i \in S \subset Q \}$$

Obtain subset of database

$$\tilde{X}^* = \{ x^*(i) : J_k(x^*(i)) \leq b_k \quad \text{for all } k \in S \}$$

Assign minimum correlation e, $0 < e < 1$ to partition the set of indices Q into

$$G = \{ Q_l : l \quad (1,2,\ldots,g) = L \}$$

such that

$$Q_1 \cup Q_2 \cup \ldots \cup Q_g = Q$$

and

$$Q_l \cap Q_k = \emptyset \qquad \text{for all } l,k \in L$$

and

$$cor(J_i, J_j) \geq e \qquad \text{for all } i,j \in Q_l, \text{ for all } l \in L$$

where cor(Ji,Jj) is the correlation of the two indices in the database subset.

If satisfied with G, then stop ;
else select another e and re-do index grouping.

This algorithm requires much interaction from the designer which will give him much control over the PO surface being sampled by selecting the good and bad values. However, this is possible only if a small set of representative IRI is chosen ( m 5, say), otherwise the interaction will be too complicated. In the index grouping step, assignment of b allows the designer to define an area of his concern. The choice of a large e in step 5 will result in many small groups if such conservatism is desired. Choosing a small e will result in fewer groups which gives a more crude approximation to the relationships among the groups. Applications of this grouping procedure on the database shall enhance the designer's understanding of the problem. However, a careful control of step 2 for a reasonably large sample with uniformity is vital. The Simplex Direct Search Method of Nelder and Mead (8) is suitable for this purpose. Unlike other optimization routines which step size is often optimized to some extent, that of Simplex is inherited across iterations making the generated series of points relatively more uniform. Also, the well-known robustness of this method is invaluable since we are constantly changing the overall objective function to be minimized in step 2.

## Decision Support

A given set of bounds of the performance indices can be tested against the database with the support of an index grouping G so that compatibility and trade-offs can be appreciated. If necessary, the bounds will be changed to achieve them. Ensuring compatibility is important. According to our experience, a set of incompatible bounds results in the subsequent optimization converging to points which performance index values will suffer unplanned trade-offs. Subsequent search will have to undo this before desirable designs can be found.

The formulation stage achieving this is described as an interactive algorithm as follows :

STEP 0 : A given set of index bounds, $c^{(0)}$ is to be modified to achieve compatibility and express desirable trade-offs.

STEP 1 : Set $k = 1$ and

$$C^{(k)} = C^{(0)}$$

STEP 2 : Identify active index set

$$A = A_1 \cup A_2 \cup .. \cup A_1$$

where

$$A_1 \subset Q_1$$

and

$i \in Q_1$ and $i \in A_1$ implies $J_i$ is less active than $J_j$ for some $j \in Q_1$ as decided by the designer with pairwise comparisons of members of $Q_1$ ;

STEP 3 : Determine if $c_i^{(k)}$'s are compatible.

Terminate if compatibility is satisfactory

otherwise relax some $c_i^{(k)}$'s to trade-off.

Set $c^{(k+1)} = c^{(k)}$, $k = k+1$, go to STEP 2 .

Proper graphics support for the interaction points of the above two algorithms will be invaluable, for example, plots of the overall objective function in step 2 of the exploratory stage for convergence check and plots of performance index values for pairwise comparison in step 2 of the formulation stage.

## Conclusion

The two interactive algorithms are useful tools supporting the design strategy. The region of interest in the PO surface can be mapped out in the exploratory stage by collecting a set of designs, uniformly sampled by the Simplex optimization routine with designer's interaction guided by the first interactive algorithm. Analysis of the resultant PO solution database, using the second algorithm for the formulation stage, revealed the trade-offs required to enhance the compatibility of specifications in a progressive manner. In this way, the designer could prescribe specifications with good chance of being achieved in the search stage. The design direction will be under good control with such specifications. In practical designs when a primary concern is in clarifying the goals achievable, such controllability is invaluable. It should be noted that the interactive algorithms depend much on the decisions and choices made by the designer. However, by making their effect transparent, the strategy rightly stays as a support to the designer, who remains the ultimate decision-maker of the process.

## Acknowledgement

This research is supported by grants from St. John's College, Cambridge and the University Engineering Department. Thanks are due to my supervisor Dr. Maciewjowski, and Dr. Steinhauser during a visit to DFVLR, West Germany for many helpful discussions.

## REFERENCES

1. Zakian, V. and Al-Naib, L., 1973, "Design of Dynamical and Control Systems by the Method of Inequalities", Proc. IEE, Vol. 120, No 11.

2. Becker, R.G., Heunis, A.J. and Mayne, D.Q., "Computer-aided Design of Control Systems via Optimisation", 1979, Proc. IEE, Vol. 126, No.6.

3. Polak, E., Mayne, D.Q. and Stimler, D.M., "Control System Design via Semi-Infinite Optimization : A Review", 1984, Proc. IEEE, Vol. 72, No. 12.

4. Nye, W.T. and Tits, A.L., "An Application-oriented Optimization--based Methodology for Interactive Design of Engineering Systems", 1986, I. J. Ctl., Vol. 43, No. 6.

5. Kreisselmeier, G. and Steinhauser, R., "Application of Vector Performance Optimization to a Robust Control Loop Design for a Fighter Aircraft", 1979, I. J. Ctl., Vol. 37, No. 2.

6. Tabak, D., Schy., A.A. Giesy, D.P. and Johnson, K.G., "Application of Multiobjective Optimization in Aircraft Control Systems Design", 1979, Automatica, Vol. 15.

7. Fleming, P.J. and Pashkevich, A.P., "Application of Multi-Objective Optimization to Compensator Design for SISO Control Systems", 1986, Elect. Let., Vol. 22, No. 5.

8. Nelder, J.A. and Mead, R., "A Simplex Method for Function Minimization", 1965, Computer Journal, Vol. 7.

MODELLING SUPPORT SYSTEM FOR SYSTEM DYNAMICS

KISHI Mitsuo*, NODA Keitaro**, YAMASHITA Yuji*, and TAGUCHI Katashi*
* Department of Naval Architecture, College of Engineering, University of Osaka Prefecture, Sakai, Osaka 591, JAPAN
** Nippon Kokan, Kawasaki, Kanagawa 210, JAPAN

## 1. INTRODUCTION

The process of system dynamics inquiry consists of two phases as shown in Figure 1 [1]: i) Conceptual phase, which addresses the problem definition and system conceptualization. ii) Technical phase, which addresses the system representation, simulation of system behaviour, evaluation of model validity, and system analysis. In system dynamics, it is possible to simulate system behaviour quantitatively using DYNAMO [2] after the conceptual phase. The conceptual phase (system modelling) is critically important; however, it is close to being an art. The system modelling is presented by specialists basing on their own knowledge and experience about both the system and system dynamics, that is to say not a systematic way. Therefore, some heuristic computer aided techniques for system/structural modelling, such as ISM [3], Cognitive Map [4], etc., have been proposed, but they are not so precise for system dynamics modelling.

In this study, a prototype of modelling support system for system dynamics is designed by organizing the knowledge structures of generic patternized expectations and the rules on how to construct system dynamics models. The proposed system covers the system conceptualization and the model representation. The system is a production-rule-oriented consultation system encoded in PROLOG [5]. PROLOG, which is a programming language based on predicate logic, is good for processing causal relations in system dynamics models [6],[7].

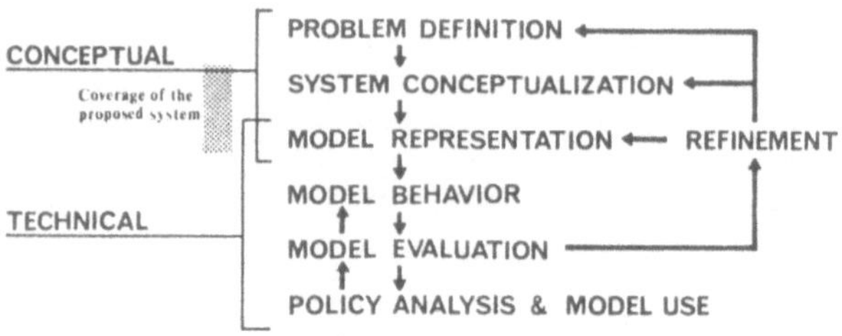


Figure 1 Process of system dynamics inquiry [1]

Brief executing processes of the proposed system are: 1) Extracting concepts (elements) within a system by perceiving action/decision making and by inferring the causal relations. 2) Preparing a causal-loop diagram of the system automatically by integrating the causal relations and by eliminating inappropriate relations so as to be precise for system dynamics models. 3) Transforming the causal-loop diagram into a flow-diagram by identifying system levels, rates, auxiliaries and parameters automatically. 4) Generating a simulation program semiautomatically by defining equations on the causal relations and by specifying initial values for variables, etc. To facilitate future modelling work about a system related to the ones dealt with in the past, the proposed system has a knowledge base of facts acquired in the systems modelling.

## 2. SYSTEM MODELLING

The first phase in system dynamics modelling is the problem definition which defines a model purpose, system boundary, and the level of concepts aggregation [1], [8],[9]. This phase is critically important; however, it is close to being an art. In the following, the subsequent phases after the problem definition are investigated.

### 2.1. Extracting Causal Relations

After the problem definition, there lies the phase of system conceptualization which involves, for example, listing the concepts of a system, extracting the causal relations between them, and identifying the feedback structures in conjunction with time delays. "System dynamics deals with change" [8]. Any change is induced by some action (including phenomena). Therefore, in this study, the system conceptualization is pursued by perceiving action/decision making and by inferring the relations between them. The concepts extracted from a system are expected to be measurable or potentially measurable.

In the system conceptualization phase, a causal-loop diagram is to be prepared. Concepts of a system and causal relations between concepts correspond to the nodes and the links in the diagram, respectively. The nodes (concepts) can be classified into "action" or "state" node [6]. The process of action making is expressed schematically in Figure 2. Action occurs owing to some information/input, and the action changes the state of the object. The action making process is considered as a fundamental unit of causal relations in systems. The links between nodes can be classified into, so to call it, "action making (information → action)", "action resulting (action → state)", or "simple relation (i.e. irrelevant to action making)" links. The whole structure of a system is depicted by selecting the action making units and by interconnecting them directly or through the simple links among them. The process for selecting action making units is, for example, as follows: i) extract the action making unit with explicit decision maker, ii) do that with explicit action maker, iii) do that without explicit decision and action makers (i.e. phenomenon).

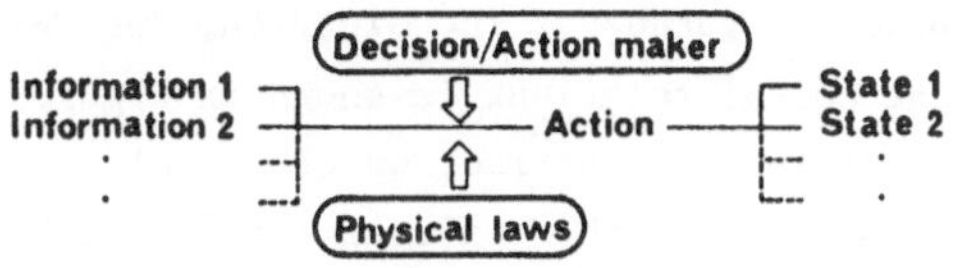


Figure 2 Process of action making

It can be said that there is a great variety of data structures to represent causal relations [7]. In this paper, the data structures of action, state, and information nodes are given in the form of predicate logic formula as follows:

Action(AN, OBJ, AM, DM, SYS) (a)

State(SN, SBJ) (b)

Inform(IN) (c)

where AN is the name of an action, OBJ the object of the action, AM/DM the action/decision maker of the action, SYS the (sub-)system which contains the action. SN is the name of a state, SBJ the subject (substance) of the state, and IN the name of information. The data structure to represent a link between nodes is:

Link(INN, TNN, LK, SIGN) (d)

where INN is the name of the initial node, TNN the name of the terminal node, LK the kind of the link (e.g. action making or simple relation), and SIGN the sign of the effect of INN on TNN (positive or negative). It matters little if there are null inputs to AM, DM, SYS in (a), LK, SIGN in (b).

In the next place, extraction of causal relations in the upper stream of the information nodes is pursued. The algorithm for extracting the causal relations is shown in Figure 3.

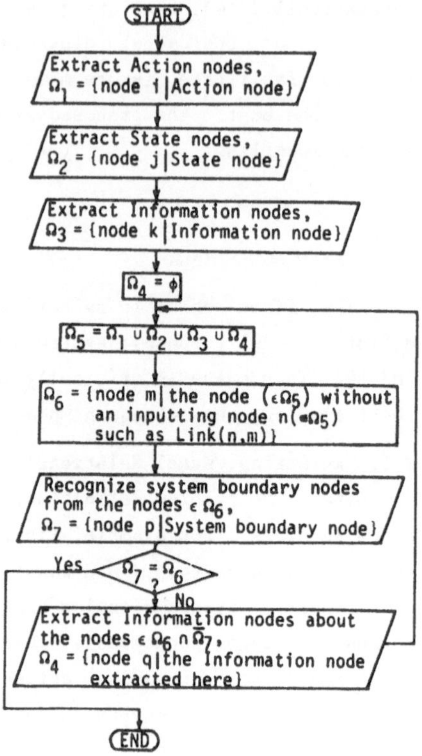


Figure 3 Algorithm for extracting causal relations

After the extraction of causal relations, all the information nodes are classified into action or state nodes, giving null input to SBJ for the state nodes not influenced by action nodes directly.

The causal-loop diagram is prepared by interconnecting the causal relations obtained above. The process can be automated by using PROLOG or LISP, because those programming languages have such a function inherently. In causal-loop diagrams, the links between the nodes related indirectly should be eliminated, otherwise the number of links will increase, resulting in a complicated diagram. In Figure 4, some of the knowledge required for eliminating such inappropriate links are

```
Rule 1: If Link(A,B,_,_),
        and Path(A,B),
        and Action(A,_,_,_,_),
        and Action(B,_,_,_,_),
        and Path(A,B) does not contain Action node
            except A and B,
      then Link(A,B,_,_) is eliminated.
Rule 2: If Link(A,B,_,_),
        and Path(A,B),
        and State(A,_),
        and Action(B,_,_,_,_),
        and Path(A,B) does not contain Action node
            except B,
      then Link(A,B,_,_) is eliminated.
Rule 3: If Link(A,B,_,_),
        and Path(A,B),
        and State(A,_),
        and State(B,_),
        and Path(A,B) does not contain Action node,
      then Link(A,B,_,_) is eliminated.
Rule 4: If Link(A,B,_,_),
        and Path(A,B),
        and Action(A,OBJa,_,_,_),
        and State(B,SBJb),
        and OBJa is not SBJb,
        and Path(A,B) does not contain Action node
            except A,
      then Link(A,B,_,_) is eliminated.
```

Figure 4 Rules for eliminating inappropriate links

represented as the heuristic rules, where, Path(A, B) is a sequence of links (excluding Link(A,B)), and the each link in the path is directed toward node B and away from node A. Those rules, of course, are not absolute; however, the causal-loop diagram resulting from applying these rules illustrates a clearcut structure of a system.

For large-scale and/or complex systems, there are difficulties in visualizing a causal-loop diagram and in understanding the overall structure of a system. One of the efficient attempts in dealing with large-scale systems is to decompose the system into smaller subsystems. There are several decomposition methods [10]. Introduced in this study is a heuristic method considering causal relations about action making rather than mathematical method based on graph theory. Figure 5 shows the algorithm of decomposition by the heuristic approach (see Figure 10, for reference).

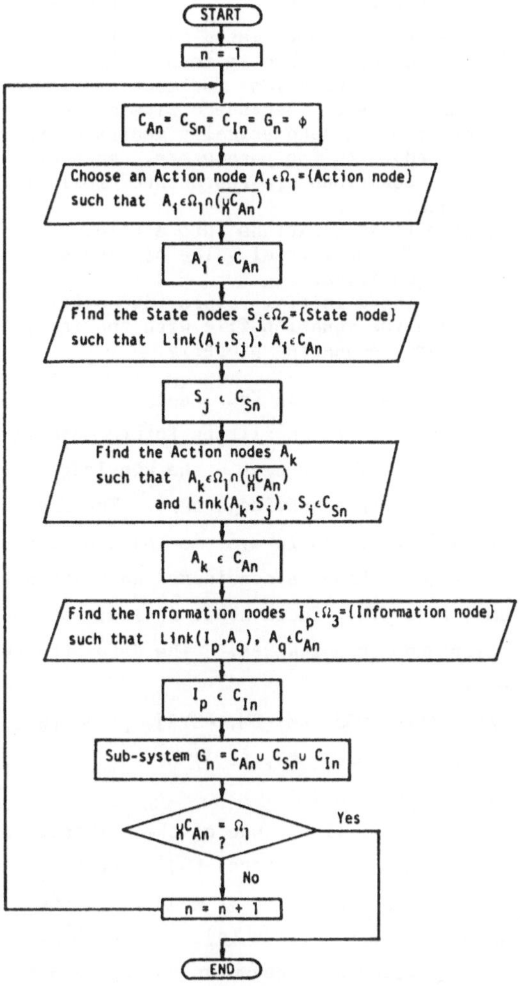


Figure 5 Algorithm for decomposition

## 2.2. Developing Computer Model

In developing a computer model to simulate the system behaviour, it is generally helpful to prepare a flow-diagram by refining the causal-loop diagram. The necessary step in refining the causal-loop diagram into the flow diagram is the identification of system levels, rates, auxiliaries, and parameters (in this paper, "parameter" means the system boundary node). The rules required for the identification are represented in Figure 6.

```
Rule 5: If Action(A,_,_,_),
        then Node A is Rate.
Rule 6: If Link(A,B,_,_) does not exist,
         and Node B is not Rate,
         then Node B is Parameter.
Rule 7: If Link(A,B,_,_),
         and Action(A,OBJa,_,_,_),
         and State(B,SBJb),
         and OBJa is SBJb,
        then Node B is Level.
Rule 8: If Node A is not Parameter,
         and State(A,_),
         and Node A is not Level,
        then Node A is Auxiliary.
```

Figure 6 Rules for identifying system levels, etc.

The computer model is developed by formulating the flow-diagram. In the following, the process of generating a computer program is described:

1) Create an abbreviated name for each variable (node) provided that one-to-one correspondence is found between them.
2) Write the equation for each variable as a function of the variables in the upper stream. Because of its conventional/standardized format, the rate-level equation is automatically generated referring to "SIGN" (positive or negative) in (d). As for parameters, each of them is expressed as a function of time. The equations are represented following a computer language statement, e.g. BASIC.
3) Save these equations into a file.
4) Specify the initial value of each system level. They are expressed in the form of equalities.
5) Save these initial values into a file.
6) Merge the equation file with the initial value file, and complete a computer program adjusting properly.

### 2.3. Organizing Acquired Knowledge

In order to to facilitate future modelling work about a system related to the ones dealt with in the past, the knowledge acquired in the systems modelling should be accumulated in a knowledge base. The facts to be stored in the knowledge base are the causal relations in systems and the mathematical equations defined on variables. The knowledge base is considered as a library/catalogue of dynamic structures of systems [8]. The facts about action, state, and information nodes are codified according to some indexes. The data structures of the stored facts are given as follows:

Action(AN, OBJ, AM, DM, [SYS], [IN], [SN], [EQ]) (e)

State(SN, SBJ, [AN], [EQ]) (f)

Inform(IN, [AN]) (g)

where [ * ] is a list, one of the symbolic expression, that means a set of atoms (i.e. arbitrary characters) [5], and EQ the equation for the node defined in the phase of developing computer model. In addition, the facts about links are stored: the data structure follows (d). In the phase of extracting causal relations, "AN" and "SYS" are employed as keywords to retrieve the information from the knowledge base.

## 3. STRUCTURE OF THE PROPOSED SYSTEM

Fundamental specifications for the modelling support system are:

1) The proposed system is intended for the users without/with technical knowledge about system dynamics modelling.
2) Initially, the system has the rules and knowledge only about system dynamics modelling procedure.
3) Knowledge/facts acquired in systems modelling are accumulated to facilitate future modelling work.

A skeleton of the system is shown in Figure 7. The system is composed of:

A) Knowledge Base (KB) - This contains the heuristic rules and procedure knowledge about system dynamics modelling and the facts acquired in systems modelling in the past. The rules and procedure knowledge are stored in Procedure Knowledge Base (PKB), and the facts are stored in Fact Base (FB).

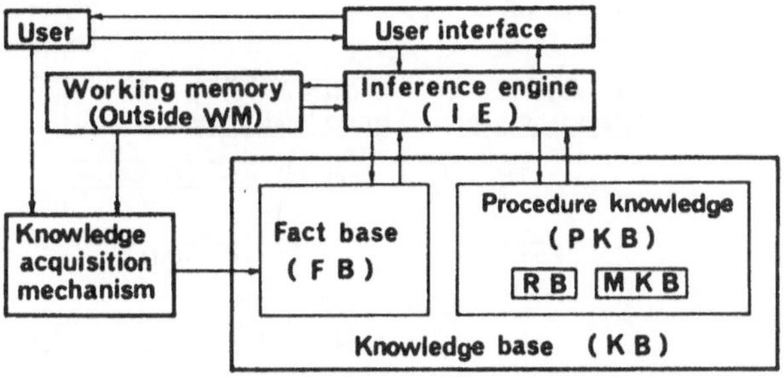


Figure 7 Skeleton of the proposed system

B) Fact Base (FB) - The facts about causal relations in systems and mathematical equations on variables are stored. The data structures of the facts are given by (d)-(g).
C) Procedure Knowledge Base (PKB) - The rules to operate the facts are stored in Rule Base (RB). The procedure knowledge about system dynamics modelling are stored in Meta-Knowledge Base (MKB). PKB is the core program of the system encoded in PROLOG.
D) Rule Base (RB) - The rules, that would be applied to the facts, are stored. Each rule has a precondition, and it can be applied if the precondition is satisfied. Such rules are called "production rules".
E) Meta-Knowledge Base (MKB) - The procedure knowledge about system dynamics modelling, including the meta-knowledge/meta-rules, i.e. knowledge/rules about how to use other knowledge/rules, are stored.
F) Inference Engine (IE) - This draws new conclusions from given facts applying the rules and procedure knowledge which are loaded from PKB at the system starting. In addition, the facts in FB are provided to IE according to demand. Facts are operated and stored in Inside Working Memory (Inside WM). In this system, PROLOG interpreter plays the role of IE.
G) Working Memory (WM) - This is a storage area used for the facts and other short-term information. In this system, because of the limitation of the memory capacity, Outside WM is equipped using an external memory device in addition to the Inside WM of IE.
H) Knowledge Acquisition Mechanism - This extracts the knowledge about causal relations in systems from the facts in WM, and codifys them to be stored in FB. Sometimes the knowledge is directly inputted by the user.

Both Inference Engine and the proposed system are based on "production system" [11].

Using the rules and procedure knowledge described in chapter 2, the proposed system generates a system dynamics model. To facilitate writing equations in the phase of formulating causal relations, the proposed system supports some built-in functions such as TABLE, PULSE, etc. in DYNAMO.

## 4. APPLICATION EXAMPLES

### 4.1. Operation Test

An operation test is given to examine the validity and applicability of the proposed system. Three model builders (i.e. Mr. A: a student without knowledge about system dynamics, Mr. B: a student with a little knowledge about system dynamics, but he is inexperienced in this expert system, and Mr. C: one of the authors) get an exercise to develop an ecosystem model of an island on where hare and fox are inhabiting. Modelling of the ecosystem structure is carried out using the proposed system. Figure 8 shows the results illustrating the system's output with causal-loop diagrams. Though model A and model B are fairly analogous, there are many difference

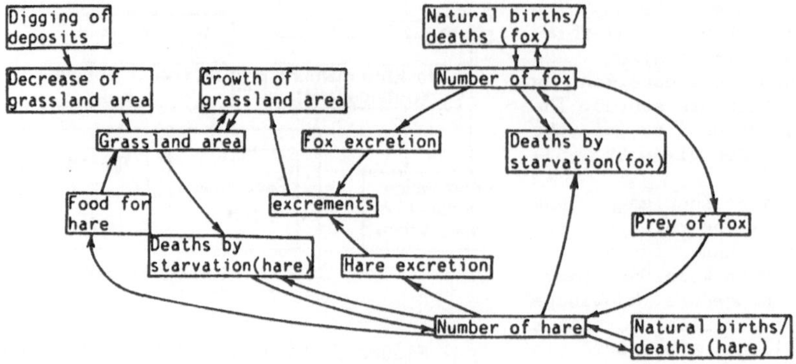


(A) MODEL A

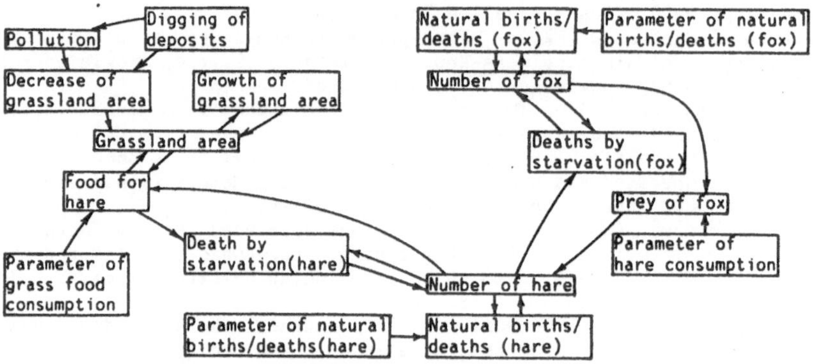


(B) MODEL B

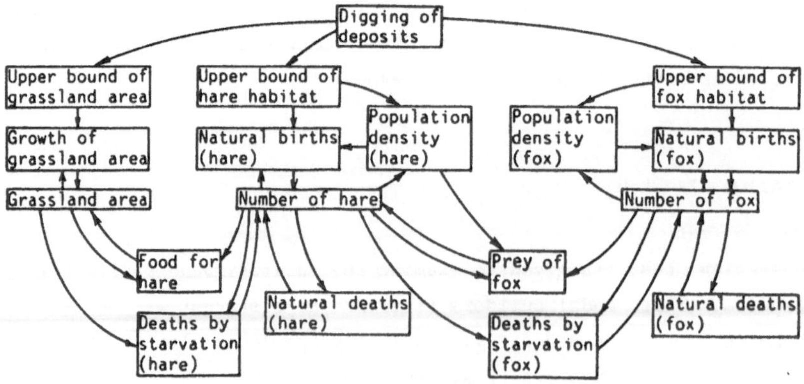


(C) MODEL C

Figure 8 Causal-loop diagrams for an ecosystem

in the three models with regard to the level of the problem recognition. However, the each model succeeded in extracting the same critical concepts and feedback loops in the system. If the model builders enter on the phase of system formulation, then those models will be refined fairly well.

**4.2. Modelling of Maritime Industries**

On account of complicated international and economical circumstances, shipbuilding and shipping industries in the developed countries are going through their serious recession. In this section, a system dynamics model for maritime industries is presented using the proposed system. Extraction of the causal relations of the system are pursued by the authors. Maritime industries are composed of shipping, shipbuilding, and port & harbour. And further, for example, shipping is divided into specialized markets (tanker, bulker, container, etc.). Figure 9 shows a causal-loop diagram for the tanker fleets system obtained using the proposed system, and the identified system levels, rates, etc. Figure 10 shows an example of the proposed system's output for the model. Japan Maritime Research Institute has already provided a similar model [12]. Although the two models differ in their elaborateness, the central structures of the causal relations in the models are equivalent.

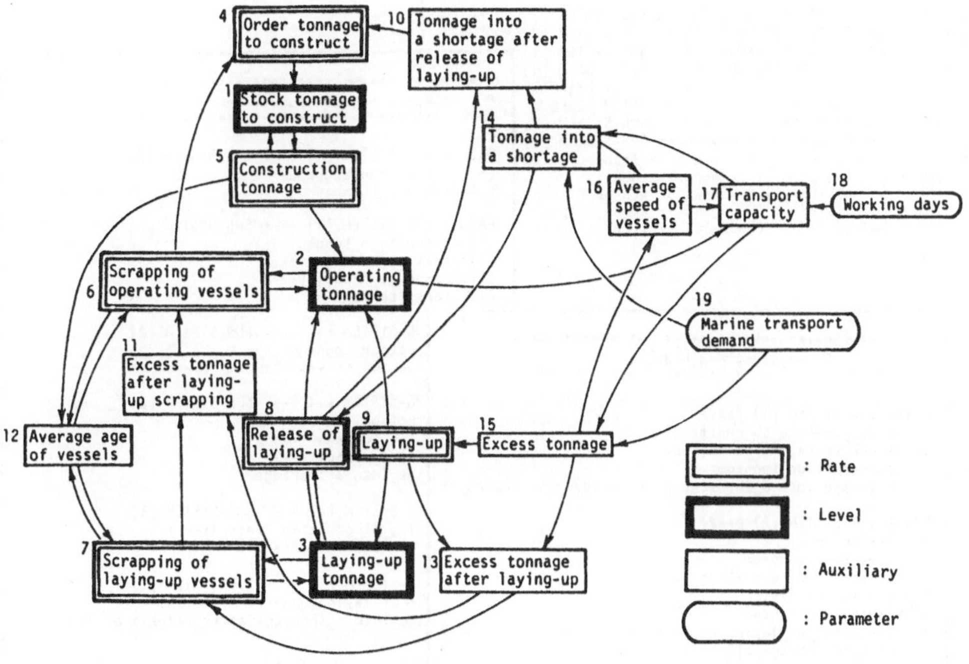


Figure 9 Causal-loop diagram for tanker fleets system

* Causal Links (Sub-System) *
##### SUB-SYSTEM [1] #####
NODE = order_tonnage_to_construct
NODE = stock_tonnage_to_construct
NODE = construction_tonnage
NODE = tonnage_into_a_shortage_after_release_of_laying_up
LINK = order_tonnage_to_construct -> stock_tonnage_to_construct
LINK = stock_tonnage_to_construct -> construction_tonnage
LINK = construction_tonnage -> stock_tonnage_to_construct
LINK = tonnage_into_a_shortage_after_release_of_laying_up -> order_tonnage_to_construct
##### SUB-SYSTEM [2] #####
NODE = laying_up
NODE = operating_tonnage
NODE = scrapping_of_operating_vessels
NODE = release_of_laying_up
NODE = laying_up_tonnage
NODE = scrapping_of_laying_up_vessels
NODE = excess tonnage after laying up
* Causal Links (Total-System) *
NODE = SUB-SYSTEM [1]
NODE = SUB-SYSTEM [2]
NODE = (16)average_speed_of_vessels
NODE = (17)transport_capacity
NODE = (18)working_days
NODE = (19)marine_transport_demand
LINK = SUB-SYSTEM [1](NODE= 5) -> SUB-SYSTEM [2](NODE= 2)
LINK = SUB-SYSTEM [1](NODE= 5) -> SUB-SYSTEM [2](NODE=12)
LINK = SUB-SYSTEM [2](NODE= 6) -> SUB-SYSTEM [1]
LINK = SUB-SYSTEM [2](NODE= 8) -> SUB-SYSTEM [1]
LINK = transport_capacity -> SUB-SYSTEM [2](NODE
LINK = transport_capacity -> SUB-SYSTEM [2](NODE
LINK = marine_transport_demand -> SUB-SYSTEM [2]
LINK = marine_transport_demand -> SUB-SYSTEM [2]
LINK = average_speed_of_vessels -> transport_cap
LINK = working_days -> transport_capacity
LINK = SUB-SYSTEM [2](NODE= 2) -> transport_capa
LINK = SUB-SYSTEM [2](NODE=14) -> average_speed_
LINK = SUB-SYSTEM [2](NODE=15) -> average_speed_
* Paths between Sub-System *
PATH = SUB-SYSTEM [1] ---> SUB-SYSTEM [2]
(5)construction_tonnage--->
(2)operating_tonnage
PATH = SUB-SYSTEM [1] ---> SUB-SYSTEM [2]
(5)construction_tonnage--->
(2)operating_tonnage--->
(17)transport_capacity--->
(15)excess_tonnage
* Sub-systems and Nodes *
##### SUB-SYSTEM [1] #####
4---order_tonnage_to_construct
1---stock_tonnage_to_construct
5---construction_tonnage
10---tonnage_into_a_shortage_after_release_of_laying_up
##### SUB-SYSTEM [2] #####
9---laying_up
2---operating_tonnage
6---scrapping_of_operating_vessels
8---release_of_laying_up
3---laying_up_tonnage
7---scrapping_of_laying_up_vessels
13---excess_tonnage_after_laying_up
15---excess_tonnage
11---excess_tonnage_after_laying_up_scrapping
12---average_age_of_vessels
14---tonnage_into_a_shortage

Figure 10 Example of the system's output

## 5. CONCLUSIONS

This paper is concerned with a design and implementation of modelling support system for system dynamics. The application examples are provided to verify the applicability of the proposed system. The results are summarized as follows:

1) The process for system dynamics modelling is investigated. Some rules and procedure knowledge for preparing system dynamics models are presented, and the data structures of causal relations are given in the form of predicate logic formula.
2) A modelling support system for system dynamics is designed including the knowledge base about causal relations.
3) The results of the application examples are in the following: i) The proposed system effectively supports the system dynamics modelling not only by the user with little technical knowledge about system dynamics but also by experienced user. ii) In eliminating inappropriate links in the causal-loop diagrams for large-scale systems, the proposed system needs some heuristic rules to avoid the problem of infinite number of path combinations.

Future work planned includes the development of more sophisticated knowledge acquisition mechanism by means of hierarchical conceptual classification and conceptual information processing.

## REFERENCES


[1] Roberts, N., Andersen, D.F., Deal, R.M., Garet, M.S., and Shaffer, W.A., Introduction to Computer Simulation; A System Dynamics Modeling Approach, Addison-Wesley, 1983.

[2] Pugh, A.L., DYNAMO User's Manual, Fifth Edition, MIT Press, 1976.

[3] Sage, A.P., Methodology for Large-Scale Systems, McGraw-Hill, 1977.

[4] Axelrod R. ed.), Structure of Decision: The Cognitive Maps on Political Elites, Princeton University Press, 1976.

[5] Clocksin, W.F., and Mellish, C.S., Programming in Prolog, Springer-Verlag, 1983.

[6] Korn, J, Cumbers, J.D., Huss, F., "Computer Aided Systems Modelling", in: Sriram D., and Adey R. (eds.), Applications of Artificial Intelligence in Engineering Problems, Vol. 1, Springer-Verlag, 1986, pp 201-213.

[7] Nolan, P.J., and McCarthy, M.A., "AI Frame-Based Simulation in System Dynamics", in: Sriram D., and Adey R. (eds.), Applications of Artificial Intelligence in Engineering Problems, Vol. 1, Springer-Verlag, 1986, pp 527-538.

[8] Forrester, J.W., "System Dynamics - Future Opportunities", TIMS Studies in the Management Science, No. 14, North-Holland, 1980, pp 7-21.

[9] Starr, P.J., "Modeling Issues and Decisions in System Dynamics", TIMS Studies in the Management Science, No. 14, North-Holland, 1980, pp 45-59.

[10] Vemuri, V., Modeling of Complex Systems: An Introduction, Academic Press, 1978.

[11] Nilsson, N.J., Principles of Artificial Intelligence, Springer-Verlag, 1980.

[12] SD Research Society in JMRI, "SD Model for Shipping and Shipbuilding - Tanker Market" (in Japanese), Journal of the Japan Maritime Research Institute, No. 130, 1977, pp 5-21.

## STABILITY AND CONTROL OF LINEAR PERIODIC SYSTEMS

Akira Ichikawa
Faculty of Engineering, Shizuoka University
Hamamatsu, 432 Japan

### 1. Introduction.

Recently periodic systems have attracted much attention of researchers [3], [6], [10], [17]. In this paper we take infinite dimensional linear periodic systems and consider the tracking problem. We first recall some sufficient conditions for stability of homogeneous equations and for the existence of periodic solutions [10], [11]. We consider both deterministic and stochastic systems.

The problem of tracking periodic signals has been studied by Artstein and Leizarowitz[1] for a finite dimensional time invariant linear system. They have introduced the notion of overtaking optimality and obtained a unique optimal control under the controllability and observability condition. It is given by the feedback law involving the solution of an algebraic Riccati equation.

We have recently considered the average quadratic control problem for infinite dimensional linear periodic systems with periodic inputs [10]. Under the stabilizability and detectability condition one can show the existence of a unique nonnegative periodic solution to the usual Riccati equation. We have shown that the optimal control is a feedback law given by this periodic solution.

These two problems are closely related but we cannot extend the results in [1] directly to our infinite dimensional systems due to the lack of controllability. So we take the average cost criterion rather than seeking overtaking optimality. We show that the periodic solution of the same Riccati equation yields an optimal control for our tracking problem. For a time invariant system the optimal law coincides with the one in [1]. We also take a stochastic system and consider a tracking problem. Similar results are obtained.

Possible applications to systems described by partial differential equations and delay differential equations are demonstrated by simple examples.

### 2. Stability and periodic solutions of linear systems.

Let $Y$ be a real separable Hilbert space and let $A(t)$ be a possibly unbounded linear operator on $Y$ with $A(t+T)=A(t)$, $t \in R$. We assume that $A$ generates a strongly continuous evolution operator $U(t,s)$, $t \geq s$ [18], [20]. We also assume that there exists a family of linear operators $A_n(t)$ generating evolution operators $U_n(t,s)$ differentiable on $Y_0$ dense in $Y$ and converges strongly in $Y$ to $U(t,s)$ uniformly on $0 \leq t \leq T$.

First we consider the equation

$$\begin{aligned} &(i) \quad y' = A(t) + f(t), \\ &(ii) \quad y(0) = y_0 , \end{aligned} \tag{2.1}$$

where $f$ is a continuous T-periodic function in $Y$. We define the mild solution of (2.1) by

$$y(t) = U(t,0)y_0 + \int_0^t U(t,s)f(s)ds . \tag{2.2}$$

The operator $A(t)$(or $U(t,s)$) is called exponentially stable if

$$|U(t,s)| \leq Ce^{-a(t-s)}, \quad t \geq s \quad \text{for some} \quad C,a > 0$$

The following result is known [11]:

Proposition 2.1 $A(t)$ is exponentially stable if and only if there exists a strongly continuous T-periodic nonnegative solution in $L(Y)$ to the Liapunov equation

$$P' + A^*P + PA + I = 0 . \tag{2.3}$$

If $A(t)$ is exponentially stable, then (2.1i) has a unique T-periodic solution given by

$$y(t) = \int_{-\infty}^t U(t,s)f(s)ds . \tag{2.4}$$

Thus it corresponds to the initial value

$$y_0 = \int_{-\infty}^0 U(0,s)f(s)ds . \tag{2.5}$$

Next we consider a stochastic evolution equation. Let $(\Omega,F,F_t,P)$ be a stochastic basis and let $(W_i)$ and $W(t)$ be a k-dimensional and an H-valued independent Wiener process respectively where $H$ is a real separable Hilbert space and $\mathrm{cov}[W(t)] = tW$, $W$ a nuclear operator in $H$ [4]. Consider

$$\begin{aligned} &(i) \quad dy = [A(t)y+f(t)]dt + G_i(t)ydw_i + G(t)dw , \\ &(ii) \quad y(0) = y_0 , \end{aligned} \tag{2.6}$$

where $G_i \in L(Y)$, $G \in L(H,Y)$ are strongly continuous and T-periodic. For each $F_0$-measurable $y_0$ with $E|y_0|^2 < \infty$ we define the mild solution of (2.6) by the unique solution in $C([0,L];L_2(\Omega,Y))$ of

$$y(t) = U(t,0)y_0 + \int_0^t U(t,s)f(s)ds + \int_0^t U(t,s)G(s)dw(s) + \int_0^t U(t,s)G_i(s)y(s)dw_i(s) , \tag{2.7}$$

where the repeated $i$ denotes the summation from $i = 1$ to $k$. The existence and uniqueness of a mild solution on an arbitrary interval $[0,L]$ is well-known [7], [13]. We assume that the mild solution can be approximated by strong solutions of approximating systems of (2.6) involving $A_n(t)$. So the Ito's formula for (2.6) is formally applicable [7], [10]. Consider now the homogeneous part of (2.6)

$$dy = A(t)ydt + G_i(t)ydw_i(t) ,$$
$$y(s) = y_s . \qquad (2.8)$$

Let $V(t.s)$ be the stochastic fundamental solution [7], [15] of (2.8) so that $y(t,s;y_s) = V(t,s)y_s$. The system (2.8)(or $(A;G_i)$) is said to be exponentially stable if

$$E|V(t,s)y_s|^2 \leq Ce^{-a(t-s)}E|y_s|^2, \quad t \geq s \quad \text{for some } C,a > 0.$$

Proposition 2.2 $(A;G_i)$ is stable if and only if there exists a strongly continuous T-periodic nonnegative solution in $L(Y)$ to the Liapunov equation

$$P' + A^*P + PA + G_i^*PG_i + I = 0 \qquad (2.9)$$

We recall that a stochastic process $y(t)$ is T-periodic if the joint distribution of $y(t_1+T), y(t_2+T),\cdots, y(t_n+T)$ for any $t_1,\cdots, t_n$ is independent of T [17]. If $(A,G_i)$ is exponentially stable, then (2.6) has a unique T-periodic mild solution given by [10]

$$y(t) = \int_{-\infty}^{t} V(t,s)f(s)ds + \int_{-\infty}^{t} V(t,s)G(s)dw(s) . \qquad (2.10)$$

Weaker conditions for the existence of periodic solutions can be found in [10].

3. Tracking of periodic signals.

First we consider the deterministic problem :

$$y' = A(t)y + B(t)u, \quad y(0) = y_0, \qquad (3.1)$$

$$J(u) = \overline{\lim_{L\to\infty}}(1/L)\int_0^L [|M(y-h)|^2 + \langle Nu,u\rangle ]dt, \qquad (3.2)$$

where $u$ is a control in a real reparable Hilbert space $U$, $B \in L(U,Y)$, $N \in L(U)$, $M \in L(Y,Y_1)$, $Y_1$ a real Hilbert space, are T-periodic and strongly continuous and $N(t) \geq cI$ for some $c > 0$.
We wish to minimize $J(u)$ over

$$U_{ad} = \{u : \overline{\lim_{L\to\infty}}(1/L)\int_0^L |u(t)|^2 dt < \infty \text{ such that } \sup_{t\geq 0}|y(t)| < \infty\} \qquad (3.3)$$

To make our problem nontrivial we need some assumptions. Recall [6] that
(i) $(A,B)$ is stabilizable if there exists $K : (-\infty,\infty) \to L(Y)$ bounded strongly continuous such that $A - BK$ is exponentially stable.
(ii) $(A,M)$ is detectable if there exists $J : (-\infty,\infty) \to L(Y)$ bounded strongly continuous such that $A - JM$ is exponentially stable.
If $(A,B)$ is stabilizable, then the control

$$u = -K(t)y + f(t) \qquad (3.4)$$

is admissible for each T-periodic strongly continuous function f. To solve our control problem we need to consider the Riccati equation

$$Q' + A^*Q + QA + M^*M - QBN^{-1}B^*Q = 0 . \tag{3.5}$$

The following is known [6] :

Proposition 3.1 (i) Suppose $(A,B)$ is stabilizable. Then there exists a strongly continuous T-periodic nonnegative solution to (3.5).
(ii) If $(A,M)$ is detectable, then there exists at most one solution which is strongly continuous T-periodic and nonnegative. Moreover, if $Q$ is a solution of (3.5), then $A - BN^{-1}B^*Q$ is exponentially stable.

Now we are ready to state our main result.

Theorem 3.1 Suppose $(A,B)$ is stabilizable and $(A,M)$ detectable. Then the feedback control

$$\bar{u} = -N^{-1}B^*[Qy + r] \tag{3.6}$$

is optimal and

$$J(\bar{u}) = (1/T)\int_0^T [|Mh|^2 - \langle BN^{-1}B^*r, r\rangle]dt, \tag{3.7}$$

where $r$ is the unique T-periodic solution of

$$r' + (A^* - QBN^{-1}B^*)r - M^*Mh = 0 \tag{3.8}$$

given by

$$r(t) = -\int_t^\infty U_0^*(s,t)M^*(s)M(s)h(s)ds \tag{3.9}$$

and $U_0$ is the evolution operator generated by $A - BN^{-1}B^*Q$.

Proof. Let $y$ be the mild solution of (3.1) corresponding to an admissible control $u$. Differentiate $\langle Qy,y\rangle + 2\langle r,y\rangle$ formally and remove terms involving $A$, $A^*$ using (3.5), (3.6). Then integrate the resulting equality from $0$ to $L$ we obtain

$$\langle Q(L)y(L),y(L)\rangle + 2\langle r(L),y(L)\rangle - \langle Q(0)y(0),y(0)\rangle - 2\langle r(0),y(0)\rangle$$

$$= -\int_0^L [|M(y-h)|^2 + \langle Nu,u\rangle]dt + \int_0^L [|Mh|^2 - \langle BN^{-1}b^*r,r\rangle]dt$$

$$+ \int_0^L |N^{1/2}[u + N^{-1}B^*(Qy+r)]|^2 dt .$$

Note that $y$ is bounded and that there exists a unique T-periodic solution of (3.8) since $A - BN^{-1}B^*Q$ is exponentially stable. Now dividing this by $L$ and taking superior limit we obtain

$$J(u) = (1/T)\int_0^T [|Mh|^2 - < BN^{-1}B^*r,r >]dt + \overline{\lim_{L\to\infty}}(1/L)\int_0^L |N^{1/2}[u + N^{-1}B^*(Qy+r)]|^2 dt \quad (3.10)$$

This procedure can be justified using approximating systems involving $A_n(t)$ [2], [5], [6]. The optimality of $\bar{u}$ and (3.7) now follow easily.

The closed loop system corresponding to $\bar{u}$ is

$$y' = (A - BN^{-1}B^*Q)y - BN^{-1}B^*r \,. \quad (3.11)$$

Note that $A - BN^{-1}B^*Q$ is exponentially stable by Proposition 3.1 and that $BN^{-1}B^*r$ is T-periodic. Thus as in (2.4) there exists a unique T-periodic solution of (3.11) and it is given by

$$y_p(t) = -\int_{-\infty}^t U_0(t,s)B(s)N^{-1}(s)B^*(s)r(s)ds \quad (3.12)$$

Let $y(t;y_0)$ be the solution of (3.11) with $y(0) = y_0$. Then it is easy to see that $|y_p(t) - y(t,y_0)| \to 0$ exponentially. Hence $y_p(t)$ is globally exponentially asymptotically stable.

As in [10] we can show that three cases below are included in our results. First we take a different class of controls :

$$U_{ad} = \{u \in L_2([0,T],U) : \ u \text{ is T-periodic and the mild solution of (3.1) has a T-periodic solution for some } y_0\}.$$

We define the cost functional by

$$J_p(u) = (1/T)\int_0^T [|M(y-h)|^2 + < Nu,u >]dt \,. \quad (3.13)$$

Then we have :

<u>Corollary 3.1.</u> The feedback control (3.6) is optimal and $J_p(\bar{u}) = J(\bar{u})$.

Now we consider the time invariant case. We assume that operators A, B, M, N and h are constant. So A is the infinitesimal generator of a strongly continuous semigroup U(t). We assume that (A,B) is stabilizable and (A,M) detectable in the usual sense [4], [21]. Thus there exists a unique nonnegative solution to the algebraic Riccati equation

$$A^*Q + QA + M^*M - QBN^{-1}B^*Q = 0 \,. \quad (3.14)$$

Furthermore, $A - BN^{-1}B^*Q$ is exponentially stable. Hence we have :

Corollary 3.2. Suppose (A,B) is stabilizable and (A,M) detectable. Then the feedback control

$$\bar{u} = -N^{-1}B^*[Qy + r] \tag{3.15}$$

is optimal and

$$J(\bar{u}) = |Mh|^2 - \langle BN^{-1}B^*r, r\rangle, \tag{3.16}$$

where $r = (A^* - QBN^{-1}B^*)^{-1}M^*Mh$ and $Q$ is the unique nonnegative solution of (3.14).

An equivalent formulation of the above problem is the following :

$$\text{minimize } J_s(u) = |M(y-h)|^2 - \langle Nu, u\rangle,$$
$$\text{subject to } Ay + Bu = 0 . \tag{3.17}$$

Note that the control (3.15) is admissible.

Corollary 3.3. Under the hypothesis of Corollary 3.2 the feedback law (3.15) is optimal and $J_s(\bar{u}) = J(\bar{u})$.

Next we consider the stochastic problem given by

$$dy = [A(t)y + B(t)u]dt + G_i(t)y dw_i + G(t)dw, \quad y(0) = y_0 \tag{3.18}$$

$$J(u) = \lim_{L\to\infty}(1/L)E\int_0^L [|M(y-h)|^2 + \langle Nu, u\rangle]dt . \tag{3.19}$$

We now wish to minimize $J(u)$ over

$$U_{ad} = \{u : u \text{ is } F_t\text{-measurable, with } \overline{\lim_{L\to\infty}}(1/L)E\int_0^L |u(t)|^2 dt < \infty$$
$$\text{such that } \sup_{t\geq 0} E|y(t)|^2 < \infty\} .$$

We assume that $(A,B;G_i)$ is stabilizable and $(A,M:G_i)$ detectable i.e., there exists $K : (-\infty,\infty) \to L(Y,U)$ and $J : (-\infty,\infty) \to L(Y)$ bounded and strongly continuous such that $(A - BK;G_i)$ and $(A - JM;G_i)$ are exponentially stable. Then we have [10] :

Proposition 3.2. (i) Suppose $(A,B;G_i)$ is stabilizable. Then there exists a strongly continuous T-periodic nonnegative solution to the Riccati equation

$$Q' + A^*Q + QA + M^*M + G_i^*QG_i - QBN^{-1}B^*Q = 0 . \tag{3.20}$$

(ii) If $(A,M;G_i)$ is detectable, then there exists at most one such solution. If $Q$ is such a solution, then $(A - BN^{-1}B^*Q;G_i)$ is exponentially stable.

We have results similar to the deterministic case.

Theorem 3.2. Suppose $(A,B;G_i)$ is stabilizable and $(A,M;G_i)$ detectable. Then the feedback law

$$\bar{u} = -N^{-1}B^*[Qy + r] \tag{3.21}$$

is optimal and

$$J(\bar{u}) = (1/T)\int_0^T [|Mh|^2 - \langle BN^{-1}B^*r, r\rangle + tr.GWG^*Q]dt \tag{3.22}$$

where $Q$ is the unique T-periodic strongly continuous nonnegative solution of (3.20) and $r$ is the unique T-periodic strongly continuous solution of

$$r' + (A^* - QBN^{-1}B^*)r - M^*Mh = 0 \ . \tag{3.23}$$

Proof. We calculate $d[\langle Qy,y\rangle + 2\langle r,y\rangle]$ along the solution corresponding to an admissible control $u$ and proceed as in the deterministic case.

The closed loop system correspondig to the optimal control $u$ is

$$dy = [(A - BN^{-1}B^*Q)y - BN^{-1}B^*r]dt + G_i(t)ydw_i + G(t)dw \ . \tag{3.24}$$

Since $(A - BN^{-1}B^*Q;G_i)$ is exponentially stable, (3.24) has a unique T-periodic solution

$$y_p(t) = -\int_{-\infty}^t V_0(t,s)B(s)N^{-1}(s)B^*(s)r(s)ds + \int_{-\infty}^t V_0(t,s)G(s)dw(s), \tag{3.25}$$

where $V_0(t,s)$ is the fundamental solution of homogeneous system obtained from (3.24). Moreover, $y_p(t)$ is exponentially asymptotically stable in mean square.

## 4. Examples

Example 4.1. Consider the beam equation

$$\begin{aligned} &\partial^2y/\partial t^2 + 2b\partial y/\partial t + \partial^4y/\partial x^4 = \sin\pi^2 t\,\sin\pi x \qquad 0 < x < 1,\ 1 > b > 0, \\ &y(t,0) = y(t,1) = y''(t,0) = y''(t,1) = 0 \ . \end{aligned} \tag{4.1}$$

We take $Y = D(A_0^{1/2}) \times L_2(0,1)$, $A_0 = d^4/dx^4$,

$$D(A_0) = \{y \in L_2(0,1) : y^{(i)} \in L_2(0,1),\ i=1,2,3,4,\ \ y(0) = y(1) = y''(0) = y''(1) = 0\},$$

and

$$A = \begin{bmatrix} 0 & I \\ -A_0 & -2bI \end{bmatrix}, \quad D(A) = D(A_0) \times D(A_0^{1/2}) \ .$$

Then $A$ generates a $C_0$-semigroup

$$S(t) = e^{-bt}\begin{bmatrix} \cos Dt + bD^{-1}\sin Dt \ , & D^{-1}\sin Dt \\ -(b^2D^{-1}+D)\sin Dt \ , & \cos Dt - bD^{-1}\sin Dt \end{bmatrix} \tag{4.2}$$

where $D = \sqrt{A_0 - b^2}$ and

$$(\cos D\, t)y = \sum_{n=1}^{\infty} 2\cos\sqrt{n^4\pi^4 - b^2}t < y, \sin \pi x > \sin n\pi x$$

and $\sin D\, t$ is defined in a similar manner. The unique $(2/\pi)$-periodic solution of (4.1) is given by

$$\begin{bmatrix} y(t) \\ \partial y/\partial t \end{bmatrix} = \int_{-\infty}^{t} S(t-s) \begin{bmatrix} 0 \\ \sin \pi^2 s \sin \pi x \end{bmatrix} ds$$

Hence

$$y(t) = (-1/2\pi^2 b)\cos \pi^2 t \sin \pi x .$$

Now consider the tracking problem

$$\partial^2 y/\partial t^2 + 2a\partial y/\partial t + \partial^4 y/\partial x^4 = u, \quad 0 < x < 1, \ 1 > a > 0, \ y(t,0) = y(t,1) = y''(t,0) = y''(t,1) = 0,$$

$$J(u) = \overline{\lim_{L\to\infty}}(1/L)\int_0^L [4|\partial y/\partial t - \sin \pi^2 t \sin \pi x|^2 + |u|^2]\, dt .$$

We take $Y$ as before but replace $A$ by

$$\overline{A} = \begin{bmatrix} 0 & I \\ -A_0 & -2aI \end{bmatrix}, \quad D(\overline{A}) = D(A) .$$

We take $U = L_2(0,1)$, $Bu = \begin{bmatrix} 0 \\ u \end{bmatrix}$ and $N = I$. Then

$$M^*M = 4\begin{bmatrix} 0 & 0 \\ 0 & I \end{bmatrix} = 4M_0, \quad B^*B = M_0 \quad \text{and} \quad A^* = \begin{bmatrix} 0 & -I \\ A_0 & -2aI \end{bmatrix} .$$

Thus the algebraic Riccati equation (3.14) has a nonnegative solution $Q = (-2a + 2\sqrt{a^2+1})I$. The generator of the optimal closed system is $\overline{A} - BN^{-1}B^*Q = A$ with $b = \sqrt{a^2+1}$ .
Note that

$$S^*(t) = e^{-bt}\begin{bmatrix} \cos D\, t + b\, D^{-1}\sin D\, t , & A_0^{-1}\, D^{-1}\sin D\, t \\ -A_0(b^2 D^{-1} + D)\sin D\, t , & \cos D\, t - b\, D^{-1}\sin D\, t \end{bmatrix}$$

Then optimal law is

$$\overline{u} = -B^*\left[Q\begin{bmatrix} y \\ \partial y/\partial t \end{bmatrix} + \begin{bmatrix} r_1 \\ r_2 \end{bmatrix}\right] = -2(\sqrt{a^2+1})\partial y/\partial t - r_2 ,$$

where

$$r(t) = \begin{bmatrix} r_1 \\ \\ r_2 \end{bmatrix} = -\int_t^{\infty} S^*(s-t) \begin{bmatrix} 0 \\ \cdot \\ 4\sin \pi^2 s \sin \pi x \end{bmatrix} ds .$$

Hence

$$r_2(t) = -(2/b)\sin \pi^2 t \sin \pi x = -(2/\sqrt{a^2+1})\sin \pi^2 t \sin \pi x .$$

The minimal cost (3.7) is

$$J(u) = (\pi/2)\int_0^{2/\pi} (2\sin^2 \pi^2 t - (2/b^2)\sin^2 \pi^2 t )dt = 1 - (1/b^2) = 1 - 1/(a^2+1) .$$

The periodic solution of the optimal system is

$$y_p(t) = -1/(\pi^2(a^2+1))\cos \pi^2 t \sin \pi x .$$

Example 4.2. Consider the delay system [12], [19],

$$x_1'(t) = -(\pi/2)x_1(t-1) - x_2(t) ,$$

$$x_2'(t) = u(t) .$$

In this case we take $Y = M_2(-1,0) \times R$, $M_2(-1,0) = R \times L_2(-1,0)$ and

$$D(A) = \{\begin{bmatrix} (x_1,x) \\ x_2 \end{bmatrix} : x_1 \in R,\ x \in H^1(-1,0),\ x_2 \in R,\ x(0) = x_1 \} ,$$

$$A\begin{bmatrix} (x_1,x) \\ x_2 \end{bmatrix} = \begin{bmatrix} (-(\pi/2)x_1(-1) - x_2, \quad dx/d\theta) \\ 0 \end{bmatrix} .$$

Then A has eigenvalues $\pm(\pi/2)i$ and real and imaginary parts of eigenvectors are given respectively by

$$\phi_1 = \begin{bmatrix} (1, \cos(\pi/2)\theta) \\ 0 \end{bmatrix}, \quad \phi_2 = \begin{bmatrix} (0, \sin(\pi/2)\theta) \\ 0 \end{bmatrix} .$$

Hence

$$S(t)\phi_1 = \cos(\pi/2)t\phi_1 - \sin(\pi/2)t\phi_2$$

and the uncontrolled system has a 4-periodic solution

$$y(t) = \begin{bmatrix} (\cos(\pi/2)t, \quad \cos(\pi/2)(t+\theta)) \\ 0 \end{bmatrix} .$$

This solution is not asymptotically stable but we can control the system in such a way that $x_1(t)$ stays close to $\cos(\pi/2)t$. For this purpose we consider the tracking problem

$$J(u) = \overline{\lim_{L\to\infty}}(1/L)\int_0^L [[x_1(t)-\cos(\pi/2)t]^2 + u^2(t)]dt.$$

Define

$$Bu = \begin{bmatrix} (0,0) \\ u \end{bmatrix}, \quad M\begin{bmatrix} (x_1,x) \\ x_2 \end{bmatrix} = x_1,$$

then it is known that (A,B) is stabilizable and (M,A) detectable [19]. Hence the optimal feedback control yields the best 4-periodic solution approximating $\cos(\pi/2)t$.

References.


[1] Z. Artstein and A. Leizarowitz, Tracking periodic signals with the overtaking criterion, IEEE Trans. Automat. Contr., Vol. AC-30, 1123-1126, 1985.

[2] V. Barbu and G. Da Prato, Hamilton-Jacobi Equations in Hilbert Spaces, Pitman, London, 1983.

[3] F. Colonius, Optimal periodic control, Report n. 140, Forschungsschwerpunkt Dynamische Systeme, Universität Bremen, Oct. 1985.

[4] R.F. Curtain and A.J. Pritchard, Infinite Dimensional Linear Systems Theory, LN in Control Inf Sci., 8(1978), Springer-Verlag, New York.

[5] G. Da Prato, Quelques résultats d'existence, unicité et regularité pour un problème de la théorie du contrôle, J. Math. Pures Appl., 52(1973), 353-375.

[6] G. Da Prato, Synthesis of optimal control for an infinite dimensional problem, SIAM J. Control Optimiz., 25(1987), 706-714.

[7] G. Da Prato, Equations aux derivées partielles stochastiques et applications, R.I. N.148, Centre de Mathématiques Appliquées, Ecole Polytechnique, France, 1986.

[8] G. Da Prato and A. Ichikawa, Stability and quadratic control for linear stochastic equations with unbounded coefficients, Bollettino U.M.I.,(6) 4-B(1985), 987-1001.

[9] G. Da Prato and A. Ichikawa, Optimal control of linear systems with almost periodic inputs, SIAM J. Control Optimiz., 25(1987), 1007-1019.

[10] G. Da Prato and A. Ichikawa, Optimal control for linear periodic systems, Appl. Math. Optim., to appear.

[11] G. Da Prato and A. Ichikawa, Liapunov equations for time-varying linear systems, Systems Control Letters, 9(1987), 165-172.

[12] J. Hale, Theory of Functional Differential Equations, Springer-Verlag, New York, 1977.

[13] A. Ichikawa, Optimal control of a linear stochastic evolution equation with state and control dependent noise, Proc. IMA Conference "Recent Theoretical Developments in Control", Leicester, U.K., Academic Press, New York, 1978, 383-401.

[14] A. Ichikawa, Dynamic programming approach to stochastic evolution equations, SIAM J. Control Optimiz., 17(1979), 162-174.

[15] A. Ichikawa, Bounded solutions and periodic solutions of a linear stochastic evolution equation, 5th Japan-USSR Symposium on Probability, Kyoto, Japan, 1986, LN in Math., to appear.

[16] A. Ichikawa, Stability of Flexible systems, IMACS International Symposium on modelling and Simulation of Distributed Parameter Systems, Hiroshima, Japan, 1987.

[17] T. Morozan, Periodic solutions of affine stochastic differential equations, Stoch. Anal. Appl., 4(1986), 87-110.

[18] A. Pazy, Semigroups of Linear Operators and Applications to Partial Differential Equations, Springer-Verlag, New York, 1983.

[19] J.M. Schumacher, A direct approach to compensator design for distributed parameter systems, SIAM J. Control Optim., 21(1983), 823-836.

[20] H. Tanabe, Equations of Evolution, Pitman, London, 1979.

[21] J. Zabczyk, Remarks on the algebraic Riccati equation in Hilbert space, Appl. Math. Optim., 2(1976), 251-258.

# ON STOCHASTIC CONTROL PROBLEMS : AN ALGORITHM FOR THE VALUE FUNCTION AND THE OPTIMAL POLICY - SOME APPLICATIONS.

Roberto GONZALEZ and Edmundo ROFMAN

Facultad de Ciencias Exactas e Ingeniería. Av Pellegrini 250 - (2000) Rosario, ARGENTINA.

INRIA - Domaine de Voluceau, 78153 Le Chesnay Cedex, FRANCE

## § 0 - INTRODUCTION

The aim of this paper is to propose an approximation procedure to compute the value function V and its optimal feedback policy $(\hat{\tau}, \hat{u}, \hat{\xi})$ related to stochastic control problems having stopping time, continuous and impulse controls in each strategy.

As we did in [4], [5] for deterministic problems we will employ here as basic tool of analysis, the characterization of V as the maximum element of a suitable set $\mathcal{W}$ of functions w. In [4] the definition of $\mathcal{W}$ requires that w be subsolution of the Hamilton-Jacobi equation associated to the deterministic control problem, i.e.

$$(0.1) \qquad \frac{\partial w(x)}{\partial x} \cdot f(x,u) + l(x,u) - \alpha\, w(x) \geq 0, \qquad \forall u \in U.$$

Now, for the stochastic problem, we deal-instead of (0.1) - with

$$(0.2) \qquad A(u)\, w \leq \gamma(u) \quad \text{in } \mathcal{D}'(\Omega),\ \Omega \subset \mathbb{R}^n, \qquad \forall u \in U,$$

where A is a second order differential operator.

§ 4 - REFERENCES

[1] BENSOUSSAN A. and LIONS J.L., "Contrôle impulsionnel et inéquations quasi-variationnelles", DUNOD, Paris (1982).

[2] CIARLET P.G. and RAVIART P.A. "Maximum principle and uniform convergence for the finite elements method", Computer Methods in Applied Mechanics and Engineering, 2 (1973), 17-31.

[3] CRANDALL M.G. and LIONS P.L., "Viscosity solutions of Hamilton-Jacobi equations" Trans. AMS t, 282, (1984), 487-502.

[4] GONZALEZ R. and ROFMAN E., "On deterministic control problems : An approximation procedure for the optimal cost" Part I : The stationary case, SIAM J. on CONTROL and OPT. 23, 2 (1985), 242-266.

[5] GONZALEZ R. and ROFMAN E., "On deterministic control problems : An approximation procedure for the optimal cost" Part II : The non stationary case, SIAM J. on CONTROL and OPT. 23, 2 (1985), 267-285.

[6] GONZALEZ R. and ROFMAN E., "On the optimisation of a short-run model of energy production system" Lect Notes in Control and Information Sc. 84, Proceedings of the 12th IFIP Conf. Budapest (1985). Sp Verlag 1986, 757-765.

[7] GONZALEZ R. and ROFMAN E., "On the computation of optimal control policies of energiy production systems with random perturbations" accepted for presentation to the 26th C.D.C. IEEE, Los Angeles.Dec.87.

[8] MENALDI J.L., "Sur les problèmes de temps d'arrêt, contrôle impulsionnel et continu correspondant à des opérateurs dégénérés", Thèse d'Etat, Univ. Paris IX Dauphine. Dec.(1980).

The numerical data have been provided by the National French Company EDF ( Electricity of France) : they describe a forecast of the French system for the year 2000.

The demand function D(.) is a stochastic step function in [0,T] (in our numerical exemple T : a week, $i_T$ = 167) i.e.

(3.1) $$D(t) = (t_i) \quad . \quad \forall t \in [t_i, t_{i+1}), \; i = 0,1,\dots,i_T$$

where $D(t_i)$ are independant gaussien variables of mean $D_i$ and variance $\sigma_i$ . We have supposed that at the beginning of each interval $[t_i, t_{i+1})$, $D(t_i)$ can be measured and it was possible to decide, without delay, the new power levels needed to satisfy the new demand.

For solving our discrete problem in the stochastic approach we have computed 368.640 values of the function V (x,t) and we have obtained the associated optimal feed-back policies. The time of computation was 51' on a DPS 8 (Bull) Multics Computer. Then we have made several simulations of the operation of the system. ( Each simulation employed less than 2").

It is interesting to remark than the cost of the optimal policy obtained at each simulation was always about 2,5 % less than that obtained for the some stochastic demand D (t) using the table of V(x,t) given by the deterministic approach presented in [6].

switchs within that set of values is, at most, n. If $V_n$ is the solution of $\mathcal{P}_n$ we can show

(2.13) $$V_1 \geq \cdots \geq V_n \geq V_{n+1} \geq V$$

$$\lim_{n\to\infty} V_n = V$$

On the other hand we consider the discretized problem $\mathcal{P}_n^h$ for which we prove

(2.14) $$\lim_{\|h\|\to 0} \bar{w}_n^h = V_n$$

(2.15) $$\bar{w}_n^h \geq \bar{w}_{n+1}^h \geq \cdots \geq \bar{w}^h \quad , \forall_n .$$

So, $\overline{\lim}_{\|h\|\to 0} \bar{w}^h \leq V_n$ ; then, using (2.13) we obtain (2.12). Finally (2.9) and (2.12) give (2.8).

## § 3 - THE NUMERICAL SOLUTION OF A STOCHASTIC PROBLEM. SOME COMMENTS.

We have studied a short-run model of energy production management with random pertubations in the demand D(t). More precisely we deal with the optimization of an electricity production system which comprises three hydraulic plants (two of pumped type) and seven thermic plants (one nuclear, two of coal, two of fuel, one gas powered and one external). (See [7])

If $x \in \mathbb{R}^3$ is the stock of energy in the dams, our purpose was to obtain the optimal cost V (x,t) and the optimal production policy $P(.) \in \mathbb{R}^{10}$ of the system, hour by hour, during a week.

where $\|h\|$ is the maximum of the diameters of the triangles of $\Omega^h$, is achieved in two steps. We will briefly give here the main ideas.

In the first part we show

(2.9) $$\underline{\lim}_{\|h\|\to 0} \bar{w}^h \geq V.$$

For that we regularize the elements of (1.1) by means of a convolution with a function of $C^\infty(\mathbb{R}^2)$ having a parameter $\rho > 0$. These functions $w_\rho$ can be approximate by functions $w_{\rho,\alpha}$ with this property : the linear finite element $w^h_{\rho,\alpha}$, taking the same values of $w_{\rho,\alpha}$ in the vertex of the triangulation $\Omega^h$, belongs to $\mathscr{W}^h$. So,

(2.10) $$\bar{w}^h \geq w^h_{\rho,\alpha}$$

If we consider in (2.10) the lower limits for $\|h\|\to 0$, then the limits for $(\rho,\alpha) \to (0,0)$, we obtain

(2.11) $$\underline{\lim}_{\|h\|\to 0} \bar{w}^h \geq w.$$

Finally, as w is an arbitrary element of W, (2.9) is proved.

The second part is devoted to show

(2.12) $$\overline{\lim}_{\|h\|\to 0} \bar{w}^h \leq V.$$

We consider a sequence of auxiliar problems $\mathscr{P}_n$ for which the controls $u_n$ can take in (1.1) a finite number of values and the number of or, in an equivalent form,

$$(2.6)\qquad w^h(x_i^h) \le \beta_i^h \left(\gamma(u,x_i^h) + T_i^h \sum_j w^h(a_{ij}^h) \frac{l_j^h}{2p_j^h} + w^h(b_i^h) \frac{\|f(x_i^h)\|}{\|x_i^h - b_i^h\|}\right)$$

with $$\beta_i^h = \left(\alpha + \frac{\|f(x_i^h)\|}{\|x_i^h - b_i^h\|} + T_i^h \sum_j \frac{l_j^h}{2p_j^h}\right)^{-1}, \quad T_i^h = \left(\frac{1}{6} \sum_j l_{ij}^h\, p_{ij}^h\right)^{-1}.$$

As $a_{ij}^h$ and $b_{ij}^h$ are convex combinations of the neighbouring vertex of $x_i^h$ it is possible to show, using the linearity of the function $w^h$, the following Discrete Maximum Principale (DMP) for the operator $A^h$ :

(2.7) (DMP) : If C is a subset of vertex of $\Omega^h$ satisfying $A^h w^h(x_i^h) \le 0 \quad \forall x_i^h \in C$, there exists $\Gamma$, $0 < \Gamma < 1$ such that

$$w^h(x_i^h) \le \Gamma \left( \max_{x_i^h \in C} (w^h(x_k^h)) \vee 0\right).$$

A first consequence of this DMP is that a unique solution $\bar{w}^h$ exists for the problem $(\mathscr{P}^h)$ when $A^h$ is given by (2.5). Moreover, being (2.6) similar to the inequalities that appear in [4] we can compute $\bar{w}^h$ with iterative algorithms of the same type than those used in [4].

Finally we can prove the uniform convergence of $\bar{w}^h$ to V, being V the solution of problem $(\mathscr{P})$ with A(u) given by (2.1). The proof of

$$(2.8)\qquad \lim_{\|h\|\to 0} |V(x) - \bar{w}^h(x)| = 0 \qquad \forall x \in \Omega$$

$$(2.3) \qquad \Delta^h w^h (x_i^h) = \frac{\sum_j \left(\frac{w^h (a_{ij}^h) - w^h(x_i^h)}{p_{ij}^h}\right) \frac{l_j^h}{2}}{\frac{1}{6} \sum_j l_{ij}^h \, p_{ij}^h}$$

$\Sigma_j$ considering all triangles having $x_i^h$ es vertex.

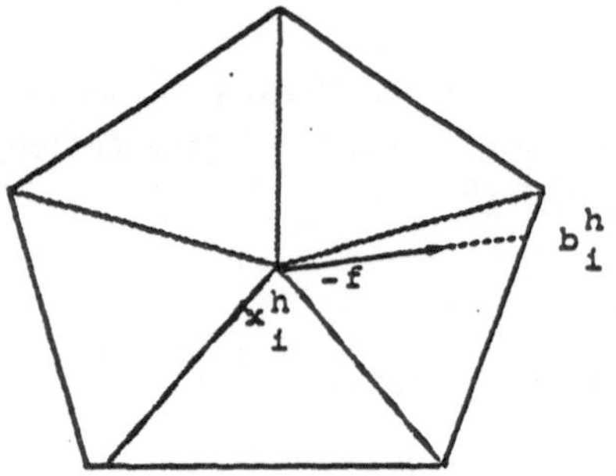


Figure 2

On the other hand $f.\nabla$ is discretized as it was done in [4], i.e. we consider $\nabla$ in the direction $-f$ :

$$(2.4) \qquad -f \, . \, \nabla w^h(x_i^h) = \frac{-w^h (b_i^h) + w^h (x_i^h)}{\|x_i^h - b_i^h\|} \, \|f \, (x_i^h)\|$$

After (2.3) and (2.4) the inequality $A^h(u) \, w^h \leq \gamma(u)$ can be written as

$$(2.5) \qquad \frac{\sum_j \left(\frac{w^h (x_i^h) - w^h (a_{ij}^h)}{p_{ij}^h}\right) \frac{l_j^h}{2}}{\frac{1}{6} \sum_j l_{ij}^h \, p_{ij}^h} + \frac{w^h(x_i^h) - w^h(b_i^h)}{\|x_i^h - b_i^h\|} \, \|f(x_i^h)\| +$$

$$+ \alpha \, w^h(x_i^h) \leq \gamma(u, x_i^h)$$

and a trigulation $\Omega^h$ as it is shown in the Figure 1, i.e. each normal from the center $x_i^h$ to the opposite side of the j-triangle define a point ($a_{ij}^h$) of that side.

$l_{ij}^h$ = lenght of the opposite side of the j-triangle ;

$p_{ij}^h$ = height of the j-triangle

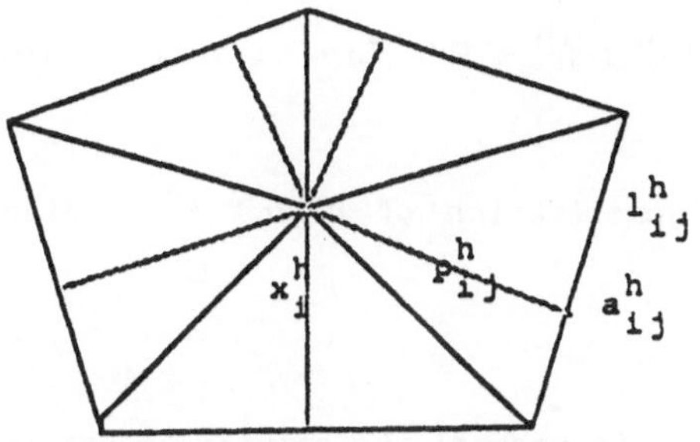


Figure 1

Using the classic Green's formule as reference we define the following discretization $\Delta^h$ of the Laplace operator $\Delta$ :

(2.2) $$\Delta^h w^h(x_i^h) \,.\, \int \psi(x)\, dx = \int - \nabla^h w^h . \nabla\psi \, dx$$

where $\psi$ is the linear finite element with value 1 in $x_i^h$ and 0 in the remaining vertex.

From Figure 1 we obtain for (2.2) :

(1.3) $\qquad w_1^h \le w_2^h \qquad w_1^h(x_i^h) \le w_2^h(x_i^h), \ \forall x_i^h$ vertex of $\Omega^h$.

and ($\mathcal{P}^h$) discretized version of ($\mathcal{P}$) :

($\mathcal{P}^h$) : Find the maximum element $\bar{w}^h$ of the set $\mathcal{W}^h$ with respect to the partial order (1.3) being

(1.4) $$\mathcal{W}^h = \{w^h : \Omega^h \to \mathbb{R} / A^h(u)w^h \le \gamma(u), \quad \forall u \in U^h\}$$

where $U^h$ is a finite discretization of U and $A^h$ a discretization of the operator A.

**REMARK** : It is not possible, in general, to ensure the exixtence of $\bar{w}^h$. In fact we can show that ($\mathcal{P}^h$) has a unique solution if the choice of the discretization allows $A^h$ to verify a Discrete Maximum Principle (DMP), (cfr.[2]).

In what follows we present

## § 2 - AN EXEMPLE OF A SUITABLE DISCRETIZATION

Let us suppose $\Omega \subset \mathbb{R}^2$,

(2.1) $$A(u) = -\Delta + f.\nabla + \alpha = -\left(\frac{\partial}{\partial x_1^2} + \frac{\partial}{\partial x_2^2}\right) + \left(f_1 \frac{\partial}{\partial x_1} + f_2 \frac{\partial}{\partial x_2}\right) + \alpha, \ \alpha > 0$$

The choice of a suitable discretisation scheme for A represents the main difficulty for the definition of our new algorithm. Because of that we will center the discussion on this point and this paper will be devoted to introduce in the algorithm presented in [4] the necessary changes to take into account (0.2) instead of (0.1).

## § 1 - THE PROBLEM $\mathscr{P}$

Let us consider the set

$$\mathscr{W} = \{w \in W_0^{1,\infty}(\Omega) \;/\; A(u)\, w \geq \gamma(u) \text{ in } \mathscr{D}'(\Omega), \quad \Omega \subset \mathbb{R}^n,\ \forall u \in U\}, \tag{1.1}$$

where $\Omega$ is an open bounded set, u(.) is a measurable function of time in a compact set $U \subset \mathbb{R}^m$ and $\gamma(u,x)$ a continuous function in $U \times \Omega$, Lipschitz in x.

We consider the following problem :

$(\mathscr{P})$ : Find the maximum element V of the set $\mathscr{W}$ defined by (1.1), i.e. find $V(x) \in \mathscr{W}$ such that

$$V(x) \geq w(x), \qquad \forall x \in \Omega,\ \forall w \in \mathscr{W}. \tag{1.2}$$

Questions concerning existence and unicity of V can be seen in [1], [3], [8]. Here we will compute V as the limit of the solutions of a sequence of approximate problems $(\mathscr{P}^h)$. To simplify the presentation we will suppose that $\Omega$ is polyhedric.

Being $w^h$ a linear finite element over a triangulation $\Omega^h$ of $\Omega$ we introduce the natural partial order

# SOLVING SLP RECOURSE PROBLEMS: THE CASE OF STOCHASTIC TECHNOLOGY MATRIX, RHS & OBJECTIVE

K. Frauendorfer
Institut für Operations Research, Universität Zürich
Moussonstrasse 15, CH-8044 Zurich

ABSTRACT: In this paper an approach is presented for approximating the recourse function of a SLP recourse problem incorporating stochastic technology matrix, right-hand side and objective, using supporting functions. It will turn out that this approach is equivalent to the one discretizing the distributions by applying Jensen's and (generalized) Edmundson-Madansky's inequality - having been first outlined in [13]. Monotonicity and convergence will be proved. In the final part, computational aspects will be discussed, followed by application of the new approach to a small stochastic LP for illustration.

## 1. Introduction

We consider linear stochastic programming problems (SLP) with recourse of the form

$$\min_{x \in X} \Psi(x,F,G) := c'x + \iint_{\Xi\Theta} Q(x,\xi,\eta)\,dG(\eta)\,dF(\xi) \qquad (1.1)$$

$$\text{where } Q(x,\xi,\eta) := \min q'(\eta)\cdot y \qquad (1.2)$$
$$W\cdot y = h(\xi) - T(\xi)\cdot x$$
$$y \geq 0 .$$

$X \subset R^n$ denotes a convex polyhedral set, $x \in R^n$ the first stage decision with $c \in R^n$ the according cost vector. $\xi = (\xi_1,\ldots,\xi_K)$, $\eta = (\eta_1,\ldots,\eta_L)$ is a K-dimensional, L-dimensional random vector with distribution function F, G. $\Xi$, $\Theta$ represent half-open K-dimensional, L-dimensional rectangles containing the support of $\xi$, $\eta$. We assume that $\xi$ and $\eta$ are stochastically independent ($\xi$ and $\eta$ may, however, consist each of stochastically dependent components).

The function $Q(x,\xi,\eta)$ is called the recourse function as it refers to the recourse problem (1.2) which has decision x as a parameter and involves the random coefficients $q(\eta)$, $h(\xi)$ and $T(\xi)$. T and h, q respectively, are assumed to be linear affine in $\xi$, $\eta$ respectively - i.e.

$$T(\xi) = T_0+\sum_k T_k\xi_k,\ h(\xi) = h_0+\sum_k h_k\xi_k;\ q(\eta) = q_0+\sum_l q_l\eta_l\ -$$

with given (mxn) matrices $T_i$, m-vectors $h_i$ and $n_1$-vectors $q_i$. W is called the recourse matrix of order $(mxn_1)$; $y\in R^{n_1}$ denotes the second stage decision or recourse decision.

Ensuring that problem (1.1) is well defined, we assume complete recourse, $\{u|W'u\leq q(\eta)\}\neq\emptyset$ $\forall\eta\in cl\Theta$ (i.e.: finiteness of $Q(x,\xi,\eta)$ $\forall(\xi,\eta)\in cl(\Xi x\Theta)$) and $cl\Xi\subset R^K$, $cl\Theta\subset R^L$ being compact.

A detailed discussion of properties of SLP problems with recourse can be found in [11] and [18]. The basic properties we will use are the piecewise linearity of $Q(x,\xi,\eta)$ and further the convexity of $Q(x,\xi,\eta)$ in x and $\xi$ for fixed $\eta\in\Theta$, the concavity of $Q(x,\xi,\eta)$ in $\eta$ for fixed x and $\xi$ and the finiteness and convexity of $\bar{Q}(x):=\int\int_{\Xi\Theta}Q(x,\xi,\eta)dG(\eta)dF(\xi)$ on X.

In the last decade several approximation schemes have been worked out for solving SLP recourse problems (see e.g. [2], [3], [6], [7], [12], [13], [14], [18]) concentrating on the cases: h stochastic, h and T stochastic or q stochastic. These approximation schemes may be classified as follows: some approximate the recourse function, others the distribution function.

In this paper a new idea is proposed of approximating the recourse function $Q(x,\xi,\eta)$, using supporting functions for the case q, h and T stochastic, yielding lower and upper bounds for $\bar{Q}(x)$ (section 2); finally it will turn out that this approach is equivalent to the one which approximates the distribution by discrete distributions, having first been outlined in [13]. In section 3 we prove monotonicity and convergence of the obtained bounds. In the final part (section 4) we discuss computational aspects and demonstrate the approximation approach on a small stochastic LP.

Before we start we have to state some conventions:

$\Omega:=\Xi x\Theta$ with $\Xi:=\mathop{X}_{k=1}^{K}[a_{k0},a_{k1})$, $\Theta:=\mathop{X}_{l=1}^{L}[b_{l0},b_{l1})$ with $a_{k0}<a_{k1}$ $\forall k$, $b_{l0}<b_{l1}$ $\forall l$. $\nu:=(\nu_1,\dots,\nu_K)$, $\mu(\mu_1,\dots,\mu_L)$ with $\nu_k,\mu_l\in\{0,1\}$ identify $a_\nu:=(a_{1\nu_1},\dots,a_{K\nu_K})$, $b_\mu:=(b_{1\mu_1},\dots,b_{L\mu_L})$ the $2^K$, $2^L$ vertices of $cl\Xi$, $cl\Theta$. $\bar{\nu}_k:=1-\nu_k$, $\bar{\mu}_l:=1-\mu_l$, hence $a_{\bar{\nu}},b_{\bar{\mu}}$ represent the vertices opposite $a_\nu,b_\mu$.

A partition of $\Omega$ is denoted by $S^j$:

$$S^j := \{\Omega^r \mid \Omega^r \text{ half-open rectangles, } r=1,\dots,r^j \text{ with } \Omega^{r_1} \cap \Omega^{r_2} = \emptyset \ \forall r_1 \neq r_2, \bigcup_r \Omega^r = \Omega\}. \tag{1.3}$$

The half-open subrectangles of $\Xi$ and $\Theta$ that define the cells $\Omega$ are denoted $\Xi^{s_r}$ and $\Theta^{t_r}$; $a_\nu^{s_r}$ and $b_\mu^{t_r}$ identify the vertices of $cl\Xi^{s_r}$ and $cl\Theta^{t_r}$.

$$p(\Xi^{s_r}) := P(\xi \in \Xi^{s_r}),\ p(\Theta^{t_r}) := P(\eta \in \Theta^{t_r}),\ p^r := P((\xi,\eta) \in \Omega^r) \tag{1.4}$$

$$\left.\begin{array}{l}(\hat{\xi},\hat{\eta}) := E(\xi,\eta),\ \hat{\xi}^{s_r} := E(\xi \mid \xi \in \Xi^{s_r}) \text{ for } p(\Xi^{s_r})>0,\ \hat{\eta}^{t_r} := E(\xi \mid \xi \in \Theta^{t_r}) \\ \text{for } p(\Theta^{t_r})>0;\ \hat{\xi}_k^{s_r} := \frac{1}{2}(a_{k0}+a_{k1}) \quad \forall k \text{ for } p(\Xi^{s_r})=0 \text{ and} \\ \hat{\eta}_l^{t_r} := \frac{1}{2}(b_{l0}+b_{l1}) \quad \forall l \text{ for } p(\Theta^{t_r})=0 .\end{array}\right\} \tag{1.5}$$

$\hat{G}, \hat{G}^{t_r}, \hat{F}, \hat{F}^{s_r}$ denote the one-point distributions on $\hat{\xi}, \hat{\xi}^{s_r}, \hat{\eta}, \hat{\eta}^{t_r}$.

(1.6)

For $\Omega^r$ with $p^r>0$: $\tilde{G}^{t_r}, \tilde{F}^{s_r}$ denote the discrete distributions of the vertices of $\Theta^{t_r}, \Xi^{s_r}$ where $\eta$ attains the value $b_\mu^{t_r}$ with probability $p^0(b_\mu^{t_r})$ and $\xi$ attains the value $a_\nu^{s_r}$ with probability $p^0(a_\nu^{s_r})$, with

$$p^0(b_\mu^{t_r}) := \frac{1}{p(\Theta^{t_r})} \int_{\Theta^{t_r}} \prod_{l=1}^{L} \frac{|\eta_l - b_{l\bar{\mu}_l}^{t_r}|}{(b_{l1}^{t_r} - b_{l0}^{t_r})} \, dG(\eta) \tag{1.7}$$

$$p^0(a_\nu^{s_r}) := \frac{1}{p(\Xi^{s_r})} \int_{\Xi^{s_r}} \prod_{k=1}^{K} \frac{|\xi_k - a_{k\bar{\nu}_k}^{s_r}|}{(a_{k1}^{s_r} - a_{k0}^{s_r})} \, dF(\xi). \tag{1.8}$$

REMARK: as outlined in [6], $p^0(b_\mu^{t_r})$ and $p^0(a_\nu^{s_r})$ are determined by

$$\int_{\Theta^{t_r}} \prod_{l\in\Lambda} \eta_l \, dG(\eta) \ \forall \Lambda \subset \{1,\dots,L\} \text{ and by } \int_{\Xi^{s_r}} \prod_{k\in\Lambda} \xi_k \, dF(\xi) \ \forall \Lambda \subset \{1,\dots,K\}.$$

Let $\tilde{G}, \tilde{F}$ be defined on $\Theta, \Xi$ according to (1.7), (1.8).

Further, subject to partition $S^j$, $\tilde{\tilde{G}}^j, \tilde{\tilde{F}}^j, \hat{\hat{G}}^j, \hat{\hat{F}}^j$ are defined by

$$\left.\begin{array}{ll}\tilde{\tilde{G}}^j := \sum_{t_r : p(\Theta^{t_r})>0} p(\Theta^{t_r}) \cdot \tilde{G}^{t_r} & \hat{\hat{G}}^j := \sum_{t_r} p(\Theta^{t_r}) \cdot \hat{G}^{t_r} \\ \tilde{\tilde{F}}^j := \sum_{s_r : p(\Xi^{s_r})>0} p(\Xi^{s_r}) \cdot \tilde{F}^{s_r} & \hat{\hat{F}}^j := \sum_{s_r} p(\Xi^{s_r}) \cdot \hat{F}^{s_r} .\end{array}\right\} \tag{1.9}$$

## 2. Approximation of the Expectation

We start with the approximation of $Q(x,\xi,\eta)$ on $cl(\Xi x\Theta)$ by linearizing the recourse function $Q(x,\cdot,b_\mu)$ at an arbitrary $\bar{\xi}\epsilon\Xi$ $\forall b_\mu$ and $Q(x,a_\nu,\cdot)$ at an arbitrary $\bar{\eta}\epsilon\Theta$ $\forall a_\nu$. Since due to the properties mentioned in the introduction the recourse function is convex in $\xi$ and concave in $\eta$ for all x we have

$$Q(x,\xi,b_\mu)\geq Q(x,\bar{\xi},b_\mu)+v(x,\bar{\xi},b_\mu)'(\xi-\bar{\xi}):=D_{\bar{\xi}}(x,\xi,b_\mu) \tag{2.1}$$

$$Q(x,a_\nu,\eta)\leq Q(x,a_\nu,\bar{\eta})+w(x,a_\nu,\bar{\eta})'\cdot(\eta-\bar{\eta}):=H_{\bar{\eta}}(x,a_\nu,\eta)\quad , \tag{2.2}$$

where $v(x,\bar{\xi},b_\mu)$, $w(x,a_\nu,\bar{\eta})$ respectively, is some subgradient of Q at $(x,\bar{\xi},b_\mu)$, $(x,a_\nu,\bar{\eta})$ respectively, available from the multipliers, the primal solution of the recourse problems at $(x,\bar{\xi},b_\mu)$, $(x,a_\nu,\bar{\eta})$, respectively.

Next we define on $cl(\Xi x\Theta)$

$$L_{\bar{\xi}}(x,\xi,\eta):=\sum_\mu \lambda_\mu(\eta)\cdot D_{\bar{\xi}}(x,\xi,b_\mu) \text{ with } \lambda_\mu(\eta):=\prod_{l=1}^{L}\frac{|\eta_l-b_{l\bar{\mu}_l}|}{(b_{l1}-b_{l0})}\quad . \tag{2.3}$$

As obviously $\sum_\mu \lambda_\mu(\eta)\cdot b_\mu=\eta$, we get together with (2.1) and (2.3)

$$L_{\bar{\xi}}(x,\xi,\eta)\leq Q(x,\xi,\eta)\quad \forall(\xi,\eta)\epsilon cl(\Xi x\Theta)\quad , \tag{2.4}$$

where by construction $L_{\bar{\xi}}(x,\cdot,\cdot)$ is linear in $\xi$ and linear in $\eta_l$, separately $\forall l$. Similarily, we obtain on $cl\Omega$

$$U_{\bar{\eta}}(x,\xi,\eta):=\sum_\nu \tau_\nu(\xi)\cdot H_{\bar{\eta}}(x,a_\nu,\eta) \text{ with } \tau_\nu(\xi):=\prod_{k=1}^{K}\frac{|\xi_k-a_{k\bar{\nu}_k}|}{(a_{k1}-a_{k0})}\quad . \tag{2.5}$$

Again, as $\sum_\nu \tau_\nu(\xi)\cdot a_\nu=\xi$, we have together with (2.2) and (2.5)

$$U_{\bar{\eta}}(x,\xi,\eta)\geq Q(x,\xi,\eta)\quad \forall(\xi,\eta)\epsilon cl\Xi x\Theta\quad , \tag{2.6}$$

where $U_{\bar{\eta}}(x,\cdot,\cdot)$ is linear in $\xi_k$, separately $\forall k$, and linear in $\eta$.

Integrating (2.4) and (2.6) yields lower and upper bounds for the expectation of $Q(x,\cdot,\cdot)$ :

$$\begin{aligned}\bar{L}_{\bar{\xi}}(x)&:=\iint_{\Xi\Theta} L_{\bar{\xi}}(x,\xi,\eta)dG(\eta)dF(\xi)\leq\iint_{\Xi\Theta} Q(x,\xi,\eta)dG(\eta)dF(\xi)\leq\\ &\leq\iint_{\Xi\Theta} U_{\bar{\eta}}(x,\xi,\eta)dG(\eta)dF(\xi):=\bar{U}_{\bar{\eta}}(x)\quad .\end{aligned} \tag{2.7}$$

Applying now (1.7), (1.8), (2.3), (2.5) and its property of linearity, we obtain for $\bar{L}_{\bar{\xi}}(x)$ and $\bar{U}_{\bar{\eta}}(x)$ :

$$\bar{L}_{\bar{\xi}}(x)=\sum_{\mu} D_{\bar{\xi}}(x,\hat{\xi},b_\mu)\cdot p^0(b_\mu)=\sum_{\mu}[Q(x,\bar{\xi},b_\mu)+v(x,\bar{\xi},b_\mu)'\cdot(\hat{\xi}-\bar{\xi})]\cdot p^0(b_\mu),$$
$$\bar{U}_{\bar{\eta}}(x)=\sum_{\nu} H_{\bar{\eta}}(x,a_\nu,\hat{\eta})\cdot p^0(a_\nu)=\sum_{\nu}[Q(x,a_\nu,\bar{\eta})+w(x,a_\nu,\bar{\eta})'\cdot(\hat{\eta}-\bar{\eta})]\cdot p^0(a_\nu). \tag{2.8}$$

As by (2.1), (2.2) $\forall\bar{\xi}\in\Xi$, $\forall\bar{\eta}\in\Theta$, the inequalities

$$D_{\bar{\xi}}(x,\hat{\xi},b_\mu)\le D_{\hat{\xi}}(x,\hat{\xi},b_\mu),\quad H_{\bar{\eta}}(x,a_\nu,\hat{\eta})\ge H_{\hat{\eta}}(x,a_\nu,\hat{\eta}) \tag{2.9}$$

hold and hence $\forall\bar{\xi}\in\Xi$, $\forall\bar{\eta}\in\Theta$

$$\bar{L}_{\bar{\xi}}(x)\le L_{\hat{\xi}}(x),\quad \bar{H}_{\bar{\eta}}(x)\ge H_{\hat{\eta}}(x), \tag{2.10}$$

we obviously have to linearize $Q(x,\cdot,b_\mu)$ at $\bar{\xi}=\hat{\xi}$ and $Q(x,a_\nu,\cdot)$ at $\bar{\eta}=\hat{\eta}$ to obtain the 'best' lower and upper bounds for the expectation of Q. In doing this the scalar products in (2.8) become zero and we realize that $\bar{L}_{\hat{\xi}}(x)$, $\bar{U}_{\hat{\eta}}(x)$ represent those bounds which one would obtain by applying the Jensen inequality and the generalization of the Edmundson-Madansky inequality due to [6], i.e.

$$\bar{L}_{\hat{\xi}}(x)=\iint_{\Xi\Theta} L_{\hat{\xi}}(x,\xi,\eta)\,d\tilde{G}(\eta)\,d\hat{F}(\xi)=\iint_{\Xi\Theta} Q(x,\xi,\eta)\,d\tilde{G}(\eta)\,d\hat{F}(\xi),$$
$$\bar{U}_{\hat{\eta}}(x)=\iint_{\Xi\Theta} U_{\hat{\eta}}(x,\xi,\eta)\,d\hat{G}(\eta)\,d\tilde{F}(\xi)=\iint_{\Xi\Theta} Q(x,\xi,\eta)\,d\hat{G}(\eta)\,d\tilde{F}(\xi). \tag{2.11}$$

We may state the following:

*Corollary:* $\bar{L}_{\hat{\xi}}(x)$, $\bar{U}_{\hat{\eta}}(x)$ are convex in x on X.

*Proof:* Convexity of $Q(\cdot,\hat{\xi},b_\mu)$, $Q(\cdot,a_\nu,\hat{\eta})$ in x and (2.11) implies convexity of $\bar{L}_{\hat{\xi}}(x)$, $\bar{U}_{\hat{\eta}}(x)$.

REMARK: Similar statements for $\bar{L}_{\bar{\xi}}(x)$, $\bar{U}_{\bar{\eta}}(x)$ with $\bar{\xi}\ne\hat{\xi}$, $\bar{\eta}\ne\hat{\eta}$ are in general not valid.

Further, is $Q(x,\xi,\eta)$ linear on $\mathrm{cl}\Omega$, then $L_{\bar{\xi}}(x,\xi,\eta)=Q(x,\xi,\eta)=U_{\bar{\eta}}(x,\xi,\eta)$ $\forall(\bar{\xi},\bar{\eta})\in$ Int $\Omega$ and hence

$$L_{\bar{\xi}}(x)=\iint_{\Xi\Theta} Q(x,\xi,\eta)=U_{\bar{\eta}}(x)\quad \forall(\bar{\xi},\bar{\eta})\in \text{Int }\Omega\,. \tag{2.12}$$

As already discussed in [5] and as becomes evident here again in paying attention to the construction of $L(x,\cdot,\cdot)$ and $U(x,\cdot,\cdot)$, the generalization of Edmundson-Madansky inequality yields the true expectation value for functions being linear in their components separately.
A similar statement for the Jensen inequality holds only in the case of stochastically independent components. For the dependent case, we have to assume joint linearity of the function in $\xi$, or $\eta$ respectively, for

obtaining the exact expectation value by applying Jensen inequality.

From now on we concentrate on the linearization - due to (2.1), (2.2) - at $\hat{\xi},\hat{\eta}$ and define

$$D(x,\xi,b_\mu):=D_{\hat{\xi}}(x,\xi,b_\mu),\ H(x,a_\nu,\eta):=H_{\hat{\eta}}(x,a_\nu,\eta),$$

$$L(x,\xi,\eta):=L_{\hat{\xi}}(x,\xi,\eta),\ U(x,\xi,\eta):=U_{\hat{\eta}}(x,\xi,\eta),\ \bar{L}(x):=\bar{L}_{\hat{\xi}}(x),\ \bar{U}(x):=\bar{U}_{\hat{\eta}}(x).$$

## 3. Improving the Approximation

We consider a partition $S^j=\{\Omega^1,\dots,\Omega^{r^j}\}$ of $\Omega$ into half-open rectangles due to (1.3). With $\Omega^r=\Xi^{s_r}\times\Theta^{t_r}$, we may linearize $Q(x,\cdot,b_\mu^{t_r})$ at $\hat{\xi}^{s_r}$, $Q(x,a_\nu^{s_r},\cdot)$ at $\hat{\eta}^{t_r}$ according to (2.1) and (2.2), but now subject to $\Omega^r$, obtaining $D^{s_r}(x,\xi,b_\mu^{t_r})$, $H^{t_r}(x,a_\nu^{s_r},\eta)$, which further define - similarly to (2.3) and (2.5) - $L^r(x,\xi,\eta)$, $U^r(x,\xi,\eta)$ on $\mathrm{cl}\Omega^r$.

So $L^r(x,\xi,\eta)$, $U^r(x,\xi,\eta)$ respectively, represent the functions that approximate $Q(x,\xi,\eta)$ from below, above respectively, on $\mathrm{cl}(\Omega^r)$ for all $r=1,\dots,r^j$. Hence we have

$$\begin{aligned}\bar{L}^j(x) &:= \sum_r \int_{\Omega^r} L^r(x,\xi,\eta)dG(\eta)dF(\xi) \le \\ &\le \int_\Omega Q(x,\xi,\eta)dG(\eta)dF(\xi) \le \\ &\le \sum_r \int_{\Omega^r} U^r(x,\xi,\eta)dG(\eta)dF(\xi) := \bar{U}^j(x)\ .\end{aligned} \tag{3.1}$$

We have to stress that by construction $\forall r$ :

$$\begin{aligned} L^r(x,\hat{\xi}^{s_r},b_\mu^{t_r}) &= D^{s_r}(x,\hat{\xi}^{s_r},b_\mu^{t_r}) = Q(x,\hat{\xi}^{s_r},b_\mu^{t_r}) \\ U^r(x,a_\nu^{s_r},\hat{\eta}^{t_r}) &= H^{t_r}(x,a_\nu^{s_r},\hat{\eta}^{t_r}) = Q(x,a_\nu^{s_r},\hat{\eta}^{t_r})\ ,\end{aligned} \tag{3.2}$$

and therefore due to the definition of $L^r$, $U^r$ and (1.5), (1.6), (1.7),

$$\begin{aligned}\bar{L}^j(x)&=\sum_r p^r\int_{\Omega^r} Q(x,\xi,\eta)d\tilde{G}^{t_r}(\eta)d\hat{F}^{s_r}(\xi)\ ,\\ \bar{U}^j(x)&=\sum_r p^r\int_{\Omega^r} Q(x,\xi,\eta)d\hat{G}^{t_r}(\eta)d\tilde{F}^{s_r}(\xi)\ .\end{aligned} \tag{3.3}$$

By (1.4), (1.9) we may rewrite (3.3) as

$$\bar{L}^j(x)=\int_\Omega Q(x,\xi,\eta)d\tilde{\tilde{G}}^j(\eta)d\hat{\hat{F}}^j(\xi),\ \bar{U}^j(x)=\int_\Omega Q(x,\xi,\eta)d\hat{\hat{G}}^j(\eta)d\tilde{\tilde{F}}^j(\xi)\ , \tag{3.4}$$

illustrating again the equivalence to the application of the Jensen and the generalized Edmundson-Madansky inequality on partition $S^j$, see [6], [7].

MONOTONICITY:

*Theorem:* Let $S^1$ and $S^2$ be partitions of $\Omega$ according to (1.3) such that $S^2$ results from $S^1$ by partitioning at least one cell $\Omega^r \in S^1$ by a cutting hyperplane orthogonal to one coordinate axis. For the lower and upper bounds, $\bar{L}^j(x)$ and $\bar{U}^j(x)$ j=1,2 due to (3.1) then follows:

$$\bar{L}^1(x) \le \bar{L}^2(x), \; \bar{U}^1(x) \ge \bar{U}^2(x) \; . \tag{3.5}$$

*Proof:* (We shortly outline the proof for $\bar{L}^1(x) \le \bar{L}^2(x)$): It suffices to concentrate on the case of subdividing $\Omega$ into two half-open rectangles $\Omega^1$ and $\Omega^2$, assuming $\Omega^1,\Omega^2 \ne \emptyset$ (i.e. $\Omega^1,\Omega^2 \subsetneq \Omega$), as otherwise (3.12) obviously holds.

We consider two different cases:

CASE i)

$\Omega^1 = \Xi^1 x \Theta$, $\Omega^2 = \Xi^2 x \Theta$; i.e. $\Omega$ has been partitioned orthogonal to a $\xi_k$

CASE ii)

$\Omega^1 = \Xi x \Theta^1$, $\Omega^2 = \Xi x \Theta^2$; i.e. $\Omega$ has been partitioned orthogonal to a $\eta_l$.

<u>CASE i)</u>: By linearity of $D(x,\cdot,b_\mu)$ in $\xi$ and by convexity of $Q(x,\cdot,b_\mu)$ in $\xi$ we get $\forall b_\mu$

$$D(x,\hat{\xi},b_\mu) = \sum_{s_r=1}^{2} p(\Xi^{s_r}) \cdot D(x,\hat{\xi}^{s_r},b_\mu) \le \sum_{s_r=1}^{2} p(\Xi^{s_r}) \cdot D^{s_r}(x,\hat{\xi}^{s_r},b_\mu) \; , \tag{3.6}$$

where $D^1(x,\xi,b_\mu)$, $D^2(x,\xi,b_\mu)$ denotes the linearization at $\hat{\xi}^1,\hat{\xi}^2$, the conditional expectation of $\Xi^1,\Xi^2$ according to (1.5).

Applying linearity of $D,D^{s_r}$ in $\xi$ and integrating (3.6) subject to $dF(\xi)$, we get

$$L(x,\hat{\xi},\eta) \le p(\Xi^1) \cdot L^1(x,\hat{\xi}^1,\eta) + p(\Xi^2) \cdot L^2(x,\hat{\xi}^2,\eta) \tag{3.7}$$

for all $\eta \in \Theta$ and hence integrating (3.7) subject to $dG(\eta)$ yields $\bar{L}^1(x) \le \bar{L}^2(x)$.

<u>CASE ii)</u>: Let $b^j_{\mu 0}$ denote the vertices of $\Theta^j$, j=1,2, and $\lambda^j_{\mu 0}(\eta)$ be defined according to (2.3) $\forall \eta \in \text{cl}\; \Theta^j$, j=1,2. $D(x,\xi,b^j_{\mu 0})$ is defined according to (2.1) for j=1,2, subject to $\text{cl}(\Xi x \Theta^j)$.

As D is linear in $\xi$ for fixed $x,b^j_{\mu 0}$, we may concentrate on $D(x,\hat{\xi},b^j_{\mu 0})$.

By (3.2) and the concavity of $Q(x,\xi,\cdot)$ in $\eta$ we obtain $\sum_\mu \lambda_\mu(b^j_{\mu_0})\cdot D(x,\hat{\xi},b_\mu) \le D(x,\hat{\xi},b^j_{\mu_0})$. Hence $\forall \eta \in cl(\Theta^j)$ we have

$$\sum_{\mu_0} \lambda^j_{\mu_0}(\eta)\cdot\sum_\mu \lambda_\mu(b^j_{\mu_0})\cdot D(x,\hat{\xi},b_\mu) \le \sum_{\mu_0}\lambda^j_{\mu_0}(\eta)\cdot D(x,\hat{\xi},b^j_{\mu_0}) \quad . \tag{3.8}$$

As further

$$\sum_{\mu_0}\lambda^j_{\mu_0}(\eta)\cdot\lambda_\mu(b^j_{\mu_0}) = \lambda_\mu(\eta) \;\; \forall \eta \in cl\Theta^j,$$

we get by (3.8) for $\eta \in cl\Theta^j$ $L(x,\hat{\xi},\eta) \le L^j(x,\hat{\xi},\eta)$, yielding $\bar{L}^1(x) \le \bar{L}^2(x)$ by integration with respect to $dG(\eta)$.

To prove $\bar{U}^1(x) \ge \bar{U}^2(x)$ we simply have to apply the same arguments as above for $\hat{\eta}$ on $-H(x,\xi,\hat{\eta})$ yielding finally $-\bar{U}^1(x) \le \bar{U}^2(x)$. //

CONVERGENCE: Assuming that each subinterval $\Omega^r$ becomes arbitrarily small with respect to $\max[(a^{s_r}_{k1}-a^{s_r}_{k0}),(b^{t_r}_{l1}-b^{t_r}_{l0})]$ within a sequence of partitions $S^j$, we enforce - by continuity of $Q(x,\xi,\eta)$ - $\forall \varepsilon > 0$ $U^r(x,\xi,\eta)-L^r(x,\xi,\eta) < \varepsilon$ $\forall (\xi,\eta) \in \Omega^r \in S^j$ with $j \ge J_\varepsilon$ and hence obtain convergence of $\bar{L}^j(x)$, $\bar{U}^j(x)$ to $\bar{Q}(x)$ for $j \to \infty$.

Of course, in practice it is impossible to let j run to infinity. For some cells $\Omega^r$ it would not even be necessary to become arbitrarily small: for instance if $Q(x,\xi,\eta)$ is linear on $\Omega^r$; in this case $L^r=Q=U^r$ on $\Omega^r$, implying that it does not make any sense to partition $\Omega^r$. This property together with the monotonicity allows to develop an efficient refining strategy (see e.g. [6],[7]) for gaining significant improvements of the approximation.

## 4. Computational Aspects

For evaluating $\bar{L}^j(x),\bar{U}^j(x)$ subject to a certain partition $S^j$, one has to compute conditional probabilities and conditional expectations which might be a difficult task for arbitrary distributions. As already outlined in [6],[7], we therefore replace the given distributions G,F by discrete ones resulting from sampling in $\Omega$ to reduce the computational effort for determining $p^r,\hat{\xi}^{s_r},\hat{\eta}^{t_r}$ of $\Omega^r \in S^j$, due to (1.4)-(1.8). In doing so, a statistical error occurs which has to be kept negligible compared to a tolerance level - say $\varepsilon$ - we prescribe for the optimal value of the SLP problem with recourse. For this purpose, one may apply the idea of [6] to evaluate an upper bound for the sample variances $\hat{\sigma}^2(x)$ of the recourse function $Q(x,\xi,\eta)$ being determined by

- $\max_{\nu,\mu} [U^r(x,a_\nu^{s_r},b_\mu^{t_r})-L^r(x,a_\nu^{s_r},b_\mu^{t_r})]$ ,
- the expectation of $U^r(x,\xi,\eta)$, $L^r(x,\xi,\eta)$ ,
- an upper bound for $\int_{\Omega^r}[L^r(x,\xi,\eta)]^2 dG(\eta)dF(\xi)$, available by applying the generalized Edmundson-Madansky inequality (note that $L^r(x,\cdot,\cdot)$ being linear in the components $\xi_k,\eta_l$ and hence $[L^r(x,\cdot,\cdot)]^2$ being convex in the components $\xi_k,\eta_l$ justify the application of the above-mentioned inequality).

Next, by applying the Central Limit Theorem (see e.g. [4], [9], [10]), one may determine a sufficiently small confidence interval for the expectation of $Q(x,\xi,\eta)$ - subject to a certain sampling size N - and therefore keep control on the statistical error; see [6], [7].

For determining a sufficiently accurate approximation for the optimal first-stage decision x* that solves the original SLP problem (1.1), we have to solve $\min\{\bar{L}^j(x)|x\epsilon X\}$ - a LP with dual blockangular structure - yielding $\hat{x}^j$. With $\bar{L}^j(\hat{x}^j),\bar{U}^j(\hat{x}^j)$ we have lower and upper bounds of the optimal objective value, telling us whether we are sufficiently accurate and may accept $\hat{x}^j$ as approximation of x* or whether we have to improve the approximation. In the latter case we have to perform a refining step, i.e. to evaluate $\bar{L}^{j,r}(\hat{x}^j):=\int_{\Omega^r}L^r(x^j,\xi,\eta)dG(\eta)dF(\xi)$ and $\bar{U}^{j,r}(\hat{x}^j):=\int_{\Omega^r}U^r(\hat{x}^j,\xi,\eta)dG(\eta)dF(\xi)$, lower and upper bounds for the objective subject to $\Omega^r$ and determine the set of cells $\Omega^r$ for which $\bar{U}^{j,r}(\hat{x}^j)-\bar{L}^{j,r}(\hat{x}^j)$ exceeds our tolerance; these cells will be further partitioned orthogonal to a certain coordinate. As the computational effort for solving SLP problems with recourse increases with the number of cells and the number of vertices that build up a partition, we are strongly motivated to keep the number of refining steps as small as possible. In [6], a refining strategy - extensively tested in [6], [7] - has been developed for the convex case (i.e. stochasticity only in the right-hand side of the recourse problem), taking advantage of the fact that partitioning orthogonal to the coordinate subject to which the recourse function is 'mostly nonlinear' promises to be 'most efficient'.

This refining strategy may be generalized to the convex-concave case (i.e. stochasticity in the right-hand side as well as in the objective of the recourse problem) and may be further applied to SLP problems of the form (1.1).

The implementation of an algorithm incorporating the above-mentioned properties is being worked out and will finally illustrate the effort for solving recourse problems in the case of stochasticity in T, h and q.

REMARK: For solving $\min\{\bar{L}^j(x)\,|\,x\in X\}$, one should use an algorithm that takes advantage of the special structure. In fact there exist some algorithms that may be expected to be efficient in solving dual-blockangular-structured LPs; i.e. the L-shape algorithms [1],[8],[17], the basis-reduction method of [16] or the regularized decomposition method described in [15]; so far, to the author's knowledge, efficient implementation of these algorithms has been verified only for SLP recourse problems with deterministic objectives (i.e. $q=q(\eta)\ \forall\eta\in\Theta$); hence an extension of one of these, for problems with q, h and T stochastic, would be of great importance.

We conclude with illustrating the approximation idea of the recourse function described in section 2 and consider the following stochastic LP:

EXAMPLE: $Q(\xi,\eta):=\min\{\langle q(\eta)\cdot y\rangle\,|\,Wy\le h(\xi),y\ge 0\}$ ,

with $q(\eta)=\begin{pmatrix}-2\\1\end{pmatrix}+\begin{pmatrix}4\\-4\end{pmatrix}\eta$, $h(\xi)=\begin{pmatrix}12\\6\end{pmatrix}+\begin{pmatrix}2\\4\end{pmatrix}\xi$, $W=\begin{pmatrix}1&4\\1&1\end{pmatrix}$

and $\eta\sim U[0,1]$, $\xi\sim U[0,1]$; $\Xi=[0,1]$, $\Theta=[0,1]$ .

We get $v(\hat{\xi},b_\mu)=-8$ for $\hat{\xi}=\frac{1}{2}$, $b_\mu=0$ and $v(\hat{\xi},b_\mu)=-\frac{3}{2}$ for $\hat{\xi}=\frac{1}{2}$, $b_\mu=1$, yielding $D(\xi,b_\mu)=-12-8\xi$ for $b_\mu=0$ and $D(\xi,b_\mu)=-9-\frac{3}{2}\xi$ for $b_\mu=1$, and further $L(\xi,\eta)=$ $=(1-\eta)D(\xi,0)+\eta\cdot D(\xi,1)=-12-8\xi+3\eta+\frac{13}{2}\eta\cdot\xi$ .

Similarly we get $w(a_\nu,\hat{\eta})=-12$ for $a_\nu=0,\hat{\eta}=\frac{1}{2}$ and $w(a_\nu,\hat{\eta})=-14$ for $a_\nu=1,\hat{\eta}=\frac{1}{2}$, yielding $H(a_\nu,\eta)=3-12\eta$ for $a_\nu=0$ and $H(a_\nu,\eta)=\frac{7}{2}-14\eta$ for $a_\nu=1$; and further $U(\xi,\eta)=(1-\xi)\cdot H(0,\eta)+\xi H(1,\eta)=3+\frac{1}{2}\xi-12\eta-2\eta\xi$.

We have $-12-8\xi+3\eta+\frac{13}{2}\eta\xi\le Q(\xi,\eta)\le 3+\frac{1}{2}\xi-12\eta-2\eta\xi$, and by integrating the bounds $\bar{L}:=-\frac{103}{8}\le\bar{Q}\le-\frac{13}{4}:=\bar{U}$.

## References


[1] J. Birge: Decomposition and Partitioning Methods for Multi-Stage Stochastic Linear Programs (Operations Research 33, 989-1007)

[2] J. Birge, R. J.-B. Wets: Designing Approximation Schemes for Stochastic Optimization Problems, in Particular for Stochastic Programs with Recourse (Mathematical Programming Study 27 (1986) 54-102)

[3] J. Birge, R. J-B. Wets: A Sublinear Approximation Method for Stochastic Programming (Technical Report, Department of Industrial and Operations Engineering, University of Michigan, 1986)

[4] W. Feller: An Introduction to Probability Theory and Its Applications, Vol.II (Wiley & Sons, New York, 1966)

[5] K. Frauendorfer: Bounding the Expectation of a Function of a Multivariate Random Variable - with Application to Stochastic Programming (Proceedings of the 1986 Symposium on Operations Research; in: Methods of Operations Research 57, 1987, Athenäum, 13-25)

[6] K. Frauendorfer: Solving SLP Recourse Problems with Arbitrary Multivariate Distributions - The Dependent Case (to appear in Mathematics of Operations Research)

[7] K. Frauendorfer, P. Kall: A Solution Method for SLP Recourse Problems with Arbitrary Distributions - The Independent Case (Manuscript, Institut für Operations Research, University of Zurich, 1986)

[8] H. Gassmann: Multi-Period Stochastic Programming (Ph.D., University of British Columbia, 1987)

[9] B.W. Gnedenko: Lehrbuch der Wahrscheinlichkeitsrechnung (Kurs Teorii Verayatuostei) 5. erweiterte Auflage (Akademie-Verlag, Berlin, 1968, English translation published as "Theory of Probability", Chelsea N.Y., 1962)

[10] P. Hall: Rates of Convergence in the Central Limit Theorem (Research Notes in Methematics 62, 1982)

[11] P. Kall: Stochastic Programming (Springer-Verlag, Berlin, 1976)

[12] P. Kall, K. Frauendorfer, A. Ruszczynski: Approximations in Stochastic Programming with Recourse (to appear in Numerical Methods for Stochastic Optimization, Y. Ermoliev and R. J.-B. Wets eds., Springer-Verlag)

[13] P. Kall, D. Stoyan: Solving Stochastic Programming Problems with Recourse, Including Error Bounds (Mathematische Operationsforschung, Statistik, Ser. Optimization 13, 1982, 341-447

[14] J.L. Nazareth, R. J.-B. Wets: Algorithms for Stochastic Programs: The Case of Nonstochastic Tenders (Mathematical Programming Study 28, 1986, 1-28)

[15] A. Ruszczynski: A Regularized Decomposition Method for Minimizing a Sum of Polyhedral Functions (Mathematical Programming 35, 1986, 309-333)

[16] B. Strazicky: Computational Experience with an Algorithm for Discrete Recourse Problems (Stochastic Programming, Proceedings of the 1974 Oxford International Conference, M. Dempster ed., Academic Press, London, 263-274

[17] R. Van Slyke, R. Wets: L-Shaped Linear Program with Applications to Optimal Control and Stochastic Linear Programs (SIAM Journal on Applied Mathematics 17, 1969, 638-663)

[18] R. J.-B. Wets: Stochastic Programming: Solution Techniques and Approximation Schemes (A. Bachem, M. Grötschel, B. Korte, eds., Mathematical Programming, The State of the Art, Springer-Verlag, Berlin, 1983, 566-603)

# Optimal Input for Autoregressive Model Discrimination Based on the Kullback's Divergence

Katsuji Uosaki and Toshiharu Hatanaka

Department of Applied Physics, Osaka University
Suita, Osaka 565, Japan

Keywords. Identification; input design; model discrimination; Kullback's divergence; autoregressive model.

Abstract. Experiment design for dynamic systems identification and modeling has attracted considerable attention to obtain the maximal information from the observation data. Most studies in this area have been devoted to accurate parameter estimation within a specified model structure. However, the problem of primary importance in system identification might be to determine the model structure itself. From this point of view, we consider in this paper the optimal input design problem for discriminating efficiency two rival autoregressive models. An optimal input is derived which maximizes the time increment of the Kullback's divergence. It is exemplified the applicability of the proposed input for efficient autoregressive model discrimination by simulation studies.

## 1. Introduction

Experiment design for dynamic system modeling and identification has attracted considerable attention to obtain the maximal information from the observation data. Different aspect of experiment design problems in system identification, e.g., input signal synthesis, sampling instants, feedback setting, etc., have been discussed and summarized in Åström and Eykhoff (1970), Mehra (1974), Goodwin and Payne (1977), Zarrop (1979), and Söderström and Stoica (1983). Most studies in the optimal input design problem are for accurate parameter estimation within a specified model structure with some constraints on input/output (see, for example, Aoki and Staly (1970), Ng and Qureshi (1981), Stoica and Söderström (1982), and Królikowski and Eykhoff (1985)). The problem of primary importance in system identification and modeling, however, might be determination of the model structure itself. From this point of view, Uosaki et al. (1984a, 1984b) presented an optimal input to determine one model among two rival autoregressive models with different order based on the Ds-criterion, which maximizes the power of likelihood ratio test, when the system output is constrained in some stochastic sense. While, Królikowski and Eykhoff (1984) considered the same problem based on the idea of alternative instrumental determinant ratio (AIDR) test and discussed the influence of input design on discrimination power of order test for autoregressive models.

In this paper, we consider the same input design problem for autoregressive model discrimination and derive an optimal input maximizing the time increment of the Kullback's divergence which expresses the distance of models.

## 2. Models and Divergence

Consider the autoregressive model with a controllable input signal $u_t$

$$y_t=a_1y_{t-1}+\cdots+a_ny_{t-n}+bu_{t-1}+\varepsilon_t \tag{1}$$

where $\varepsilon_t$ is independently normally distributed with mean zero and constant variance $\sigma^2$. We assume the system is stable, i.e., the polynomial $A(z^{-1})=1-a_1z^{-1}-\cdots-a_nz^{-n}$ has its all zeros strictly inside the unit circle. We consider the problem to find an input $u_{t-1}$ which determines efficiently the order of the autoregressive model is order n or n-1 under the input constraint,

$$|u_t| \leq C \tag{2}$$

where C is a given constant.

The order determination problem can be expressed as a model discrimination problem, that is, discriminating one autoregressive model among the following two rival autoregressive models:

$$\begin{aligned} M_1 \;:\; y_t&=a_1^{(1)}y_{t-1}+a_2^{(1)}y_{t-2}+\cdots+a_n^{(1)}y_{t-n}+b^{(1)}u_{t-1}+\varepsilon_t^{(1)} \\ &=\theta^{(1)T}y_{t-n}^{t-1}+b^{(1)}u_{t-1}+\varepsilon_t^{(1)} \end{aligned} \tag{3a}$$

$$\begin{aligned} M_2 \;:\; y_t&=a_1^{(2)}y_{t-1}+a_2^{(2)}y_{t-2}+\cdots+a_{n-1}^{(2)}y_{t-(n-1)}+b^{(2)}u_{t-1}+\varepsilon_t^{(2)} \\ &=\theta^{(2)T}y_{t-n}^{t-1}+b^{(2)}u_{t-1}+\varepsilon_t^{(2)} \end{aligned} \tag{3b}$$

where $\varepsilon_t^{(j)}$ is an independent normal random variable with mean zero and variance $\sigma_{(j)}^2$, (j=1,2), respectively, and

$\theta^{(1)} = (a_1^{(1)}, a_2^{(1)}, \cdots, a_{n-1}^{(1)}, a_n^{(1)})^T,$

$\theta^{(2)} = (a_1^{(2)}, a_2^{(2)}, \cdots, a_{n-1}^{(2)}, 0\ )^T,$

$y_k^t = (y_t, y_{t-1}, \cdots, y_k)^T, \ k \leq t.$

As a measure of distance between these two models, we employ the Kullback's divergence. The Kullback's divergence for discriminating between $M_1$ and $M_2$ may be defined by

$$J_t[1:2;y^t] = I_t[1:2;y^t] + I_t[2:1;y^t] \tag{4}$$

where $I_t[1:2;y^t]$ is the Kullback-Leibler discrimination information measure (KLDI) (Kullback (1959)) defined by

$$I_t[i:j;y^t] = \int p_i(y^t|u^{t-1}) \log \frac{p_i(y^t|u^{t-1})}{p_j(y^t|u^{t-1})} dy^t \tag{5}$$

where $p_j(y^t|u^{t-1})$ is the probability density function of $y^t=(y_t,y_{t-1},\cdots,y_1)^T$ given $u^{t-1}=(u_{t-1},u_{t-2},\cdots,u_1)^T$ under the model $M_j$, $(j=1,2)$, respectively. It is known that the divergence has the following properties as the KLDI:

(i) $J_t[1:2;y^t] \geq 0$ (6a)

(ii) $J_t[1:2;y^t] = 0$ if and only if $p_1(y^t|u^{t-1}) = p_2(y^t|u^{t-1})$ (6b)

(models are identical)

These facts suggest that it becomes easier to determine the models as the divergence $J_t[1:2;y^t]$ is larger since the distance between the models is larger. Therefore, it is natural to find an input which maximizes $J_t[1:2;y^t]$, or maximizes the time increment of the divergence defined by

$$\begin{aligned} \Delta J_t[1:2;y^t] &= J_t[1:2;y^t] - J_{t-1}[1:2;y^{t-1}] \\ &= \Delta I_t[1:2;y^t] + \Delta I_t[2:1;y^t] \end{aligned} \tag{7}$$

where

$$\Delta I_t[i:j;y^t] = I_t[i:j;y^t] - I_{t-1}[i:j;y^{t-1}], \tag{8}$$

in order to maximize the distance between the models and then discriminate the models efficiently.

Application of the chain rule,

$$p_j(y^t|u^{t-1}) = p_j(y_t|y^{t-1},u^{t-1})p_j(y^{t-1}|u^{t-1}) \tag{9}$$

to (5) yields

$$\Delta I_t[i:j;y^t] = \int p_i(y^{t-1}|u^{t-1})dy^{t-1}$$

$$\cdot \int p_i(y_t\ y^{t-1},u^{t-1})\log \frac{p_i(y_t|y^{t-1},u^{t-1})}{p_j(y_t|y^{t-1},u^{t-1})}\,dy_t$$

$$=E_i[\tilde{I}_t[i:j;y_t]], \tag{10}$$

where

$$\tilde{I}_t[i:j;y_t] = \int p_i(y_t|y^{t-1},u^{t-1})\log \frac{p_i(y_t|y^{t-1},u^{t-1})}{p_j(y_t|y^{t-1},u^{t-1})}\,dy_t \tag{11}$$

and $E_i[\cdot]$ means the expectation with respect to the probablity density function of $y^t$ given $u^{t-1}$ under the model $M_i$. Then, the time increment of the divergence can be expressed by

$$\Delta J_t[1:2;y^t] = E_1[\tilde{I}_t[1:2;y_t]] + E_2[\tilde{I}_t[2:1;y_t]] \tag{12}$$

Here, we used the fact $u_{t-1}$ does not effect on $y^{t-1}$.

In the following, we will seek for an input $u_{t-1}$ maximizing the time increment of the Kullback's divergence $\Delta J_t[1:2;y^t]$ under the input constraint (2).

## 3. Input Design for Model Discrimination

By the assumptions of linearity of the system (1) and normality of $\varepsilon_t$, the probability distributions of $y^t$ given $u^{t-1}$ and $y_t$ given $y^{t-1}$ and $u^{t-1}$ under the model $M_j$ (j=1,2) are also normal, i.e.,

$$p_j(y^t|u^{t-1}) = (2\pi)^{-t/2}|\Sigma_{(j)}|^{-1}\exp(-\tfrac{1}{2}(y^t-\mu^t_{(j)})^T\Sigma^{-1}_{(j)}(y^t-\mu^t_{(j)})), \tag{13a}$$

$$p_j(y_t|y^{t-1},u^{t-1}) = \frac{1}{\sqrt{2\pi}\sigma_{(j)}}\exp(-\frac{1}{2\sigma^2_{(j)}}(y_t-m_{t(j)})^2), \quad (j=1,2) \tag{13b}$$

where

$$\mu^t_{(j)} = E_j[y^t|u^{t-1}],$$

$$\Sigma_{(j)} = E_j[(y^t-\mu^t_{(j)})(y^t-\mu^t_{(j)})^T|u^{t-1}],$$

$$m_{t(j)} = E_j[y_t|y^{t-1},u^{t-1}].$$

Using (13b), we have

$$I_t[i:j;y_t] = -\frac{1}{2}\log\frac{\sigma^2_{(i)}}{\sigma^2_{(j)}}+\frac{\sigma^2_{(i)}-\sigma^2_{(j)}}{2\sigma^2_{(j)}}+\frac{(m_{t(i)}-m_{t(j)})^2}{2\sigma^2_{(j)}} \tag{14}$$

The expectation of the last term in (14) with respect to the probability density function of $y^t$ given $u^{t-1}$ under the model $M_i$ is given by

$$E_i[\frac{(m_{t(i)}-m_{t(j)})^2}{2\sigma^2_{(j)}}]$$

$$= \frac{1}{2\sigma^2_{(j)}} E_i[((\theta^{(i)T}y^{t-1}_{t-n}+b^{(i)}u_{t-1})-(\theta^{(j)T}y^{t-1}_{t-n}+b^{(j)}u_{t-1}))^2]$$

$$= \frac{1}{2\sigma^2_{(j)}} [tr\{(\Sigma^{t-1}_{t-n(i)}+(\mu^{t-1}_{t-n(i)})(\mu^{t-1}_{t-n(i)})^T)\delta\theta\delta\theta^T\}$$

$$+2\delta\theta^T\mu^{t-1}_{t-n(i)}\delta b u_{t-1}+(\delta b)^2u^2_{t-1}] \tag{15}$$

where

$$\mu^{t-1}_{t-n(i)} = E_i[y^{t-1}_{t-n}|u^{t-1}],$$

$$\Sigma^{t-1}_{t-n(i)} = E_i[(y^{t-1}_{t-n}-\mu^{t-1}_{t-n(i)})(y^{t-1}_{t-n}-\mu^{t-1}_{t-n(i)})^T|u^{t-1}],$$

$$\delta\theta = (a^{(1)}_1-a^{(2)}_1, a^{(1)}_2-a^{(2)}_2,\cdots,a^{(1)}_{n-1}-a^{(2)}_{n-1},a^{(1)}_n)^T,$$

$$\delta b = b^{(1)}-b^{(2)}.$$

Here, we should note that $\mu^{t-1}_{t-n(i)}$ and $\Sigma^{t-1}_{t-n(i)}$ are independent of $u^{t-1}$.

Thus, we can express $\Delta J_t[1:2;y^t]$ as

$$\Delta J_t[1:2;y^t] = \alpha u^2_{t-1} + \beta u_{t-1} + \gamma \tag{16}$$

where

$$\alpha = \frac{(\delta b)^2}{2\sigma^2_{(1)}} + \frac{(\delta b)^2}{2\sigma^2_{(2)}},$$

$$\beta = \frac{\delta\theta^T \mu_{t-n(2)}^{t-1} \delta b}{2\sigma_{(1)}^2} + \frac{\delta\theta^T \mu_{t-n(1)}^{t-1} \delta b}{2\sigma_{(2)}^2},$$

$$\gamma = \frac{\sigma_{(2)}^2 - \sigma_{(1)}^2 + tr[\Sigma_{t-n(2)}^{t-1} + (\mu_{t-n(2)}^{t-1})(\mu_{t-n(2)}^{t-1})^T)\delta\theta\delta\theta^T]}{2\sigma_{(1)}^2}$$

$$+ \frac{\sigma_{(1)}^2 - \sigma_{(2)}^2 + tr[\Sigma_{t-n(1)}^{t-1} + (\mu_{t-n(1)}^{t-1})(\mu_{t-n(1)}^{t-1})^T)\delta\theta\delta\theta^T]}{2\sigma_{(2)}^2}$$

It is easy to see that under the input constraint (2) the time increment of the divergence $\Delta J_t[1:2;y^t]$, which is a quadratic function of $u_{t-1}$, attains its maximum if we select $u_{t-1}$ as

$$u_{t-1}^o = C \operatorname{sgn}(\frac{\beta}{\alpha}) \quad (17)$$

In practice, we are of course unaware of the true values of $\delta\theta$, $\delta b$ and $\mu_{t-n(j)}^{t-1}$, so the optimal input $u_{t-1}^o$ has to be constructed by using their estimates computed recursively with the observation.

4. Numerical Example

Consider the problem to determine whether the order of given autoregressive model is 1 or 2 under the input constraint $|u_t| \leq 1$. The true system is of order 2 and given by

$$y_t = 0.5y_{t-1} - 0.1y_{t-2} + u_{t-1} + \varepsilon_t,$$

where $\varepsilon_t$ is independently normally distributed with zero mean and unit variance.

Here we use the criterion the Akaike's Information Criterion (AIC) (Akaike (1974)) defined by

$$AIC(p) = \log(\frac{RSS(p)}{t}) + \frac{2p+2}{t},$$

where RSS(p) denotes the residual sum of squares for the model of order p, for determination of the order of autoregressive models. It is shown that the AIC is an unbiased estimator of the (average) mean log-likelihood

$$S(g:f(\cdot|\theta)) = g(x)\log f(x|\theta)dx,$$

and that the KLDI appeared in definition of the divergence can be expressed by using (average) mean log-likelihood $S(\cdot:\cdot)$ as

$$I(g:f) = S(g:g) - S(g:f(\cdot|\theta)).$$

Thus, the divergence criterion discussed in this paper is found to be closely related with the AIC.

Figure 1 shows the difference of the AIC defined by

$\Delta AIC(2) = AIC(1)-AIC(2)$.

It is seen that $\Delta AIC(2)$ by the proposed input $u^o_{t-1}$ becomes to take positive value as number of observation increases while $\Delta AIC(2)$ by random input is negative. Since the AIC should take the minimum for true order 2, the proposed input makes discrimination of the true model of order 2 clearer and easier compared to random input. This indicates the applicability of the proposed input for autoregressive model discrimination.

5. Conclusions

Optimal input design problem under the input constraint is discussed in this paper for efficient discrimination of autoregressive models. An optimal input is derived, which maximizes the time increment of Kullback's divergence and may yield the big value of the divergence to make difference of the models clearer. The input can be obtained by using the recursive estimates of the model parameters. Simulation study indicates the proposed input makes discrimination of the true model against the

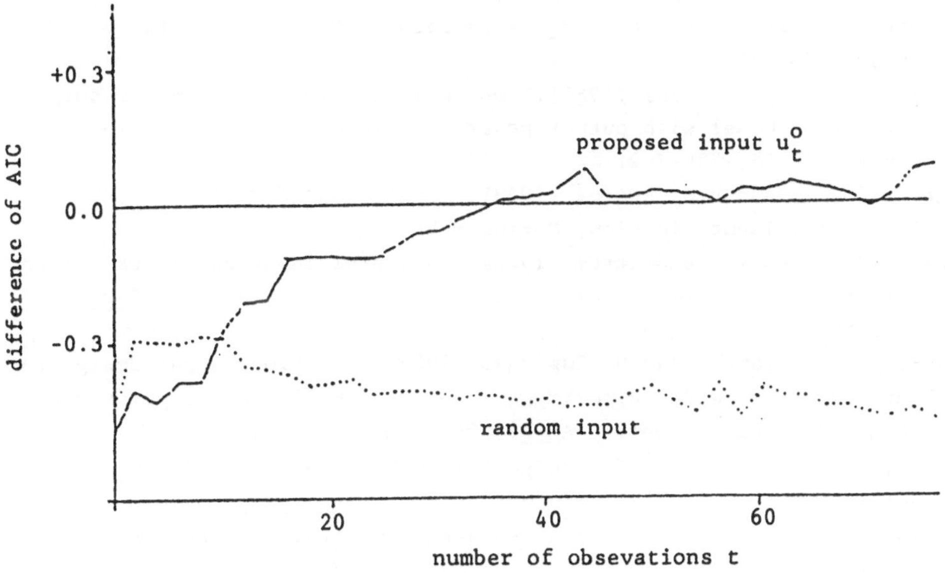


Fig.1 The difference of AIC

rival model easier compared to random input. This approach can be applicable to discrimination of more general models such as autoregressive moving average models, and it is now under investigation.

References

Akaike, H. (1974). A new look at the statistical model identification, IEEE Trans. Auto. Contr., AC-19, 716-723.

Aoki, M. and R.M. Staley (1970). On input synthesis in parameter identification, Automatica, 6, 431-440.

Åström, K.J. and P. Eykhoff (1970). System identification - a survey, Automatica, 7, 123-162.

Goodwin, G.C. and R.L. Payne (1977). Dynamic System Identification, Academic Press.

Królikowski, A. and P. Eykhoff (1984). Aspects of input signal design for model order and parameter estimation in linear dynamical systems, Identification, Adaptive and Stochastic Control, Pergamon Press, 746-751.

Królikowski, A. and P. Eykhoff (1985). Input signal design for system identification: A comparative analysis. IFAC Identification and System Parameter Estimation 1985, Pergamon Press. 915-920.

Kullback, S. (1959). Information Theory and Statistics, J.Wiley.

Mehra, R.K. (1974). Optimal input signals for parameter estimation in dynamic systems -survey and new results-, IEEE Trans. Auto. Contr., AC-19, 753-768.

Ng, T.S. and Z.H. Qureshi (1981). Optimal experiment design for autoregressive model with output power constraint, IEEE Trans. Auto. Contr., AC-26, 739-742.

Söderström, T. and P.G. Stoica (1983). Instrumental Variable Methods for System Identification, Springer-Verlag.

Stoica, P.G. and T. Söderström (1982). A useful input parameterization for optimal experiment design, IEEE Trans. Auto. Contr., AC-27, 986-989.

Uosaki, K., I. Tanaka and H. Sugiyama (1984a). Optimal input design for autoregressive model discrimination with constrained output variance, IEEE Trans. Auto. Contr., AC-29, 348-350.

Uosaki, K., I. Tanaka and H. Sugiyama (1984b). Optimal input design for autoregressive model discrimination with output amplitude constraints, Identification, Adaptive and Stochastic Control, Pergamon Press, 741-745.

Zarrop, M.B. (1979). Optimal Experiment Design for Dynamic System Identification, Springer-Verlag.

# IDENTIFICATION OF LINEAR DISCRETE-TIME STOCHASTIC SYSTEM USING COVARIANCE INFORMATION

Seiichi Nakamori* and Hidekatsu Tokumaru**

* Department of Technics, Faculty of Education, Kagoshima University, 20-6, 1, Kourimoto, 890, Japan

**Department of Applied Mathematics and Physics, Faculty of Engineering, Kyoto University, Kyoto, 606, Japan

This paper presents the new identification procedure of the discrete-time stochastic model in linear time-variant system. The model to be identified has the random input with the white Gaussian property. The variance of the white Gaussian input and the auto-covariance function of the signal are assumed to be known. The linear time-variant system is identified by these covariance information.

## 1. Introduction

The identification problems are solved mainly by the following four methods(1). 1)Off-line identification methods. 2)Recursive identification as nonlinear filtering(Bayesian approach). 3) Stochastic approximation. 4)Model reference techniques and pseudo-linear regressions. In addition to these methods, R. Mehra(2) presents the correlation method which identifies the discrete-time constant parameter system in its caninical form. The discrete-time dynamic system is excited by the white Gaussian input and its variance is known. The measurement noise is also white Gaussian. The problem is to identify the state transition matrix, input matrix of the white Gaussian input to the system, and the output matrix in the canonical form.

This paper treats the similar identification problem to that by R. Mehra(2). However, the objective system to be identified is extended to the linear time-variant discrete-time system. The dynamic system has the white Gaussian input and its variance is known in advance. The signal is observed with the additive white Gaussian noise to the signal. The new identification algorithm here identifies the state transition matrix and the input matrix. The characteristic of the current procedure is that the autocovariance function of the signal is expressed in the semi-degenerate kernel form. The presented identification algorithm is applicable to identification problems of general kinds of signals. The reason for this assertion is based on the property that the semi-degenerate kernel(3) is the function which can express the autocovariance functions of the nonstationary or stationary stochastic processes.

The digital simulation results show that the presented algorithm identifies the state transition matrix and the input matrix well.

## 2. Problem statement of estimation problems

This section introduces the basic estimation problems. The current identification procedure is deeply related to the estimation problems. Therefore, the estimation problem is explained for a start.

The time-variant stochastic dynamic system can be represented as

$$\left.\begin{aligned} z(k+1) &= \Phi(k+1,k)z(k)+\Gamma(k)u(k) \\ y(k) &= z(k)+v(k) \end{aligned}\right\} \qquad (1)$$

in the linear discrete-time system, where $z(k)$:$n\times 1$ signal vector. Although $z(k)$ is called the state vector, we use the word "signal vector" because we mainly treat the signal estimation and modeling problems. $\Phi(k+1,k)$:$n\times n$ state transition matrix. $\Gamma(k)$:$n\times 1$ input vector. $u(k)$:Scalar input forcing function of white noise. $y(k)$:$n\times 1$ vector of observed value. $v(k)$:$n\times 1$ vector of white Gaussian observation noise. We assume that $E(z(k))=0$, $E(v(k))=0$, $E(u(k))=0$, $E(z(k)v^T(s))=0$, $E(u(k)v^T(s))=0$, $E(v(K)v^T(s))=R(k)\delta(k-s)$, and $E(u(k)u^T(s))=Q(k)\delta(k-s)$, for $1\le k,s<\infty$. Here, $\delta(k-s)$ is the Dirac delta function(4).

It is the specific characteristic that the autocovariance function of the signal is expressed by the semi-degenerate kernel of

$$K(k,s)=\begin{cases} \sum_{i=1}^{M} A_i(k)B_i^T(s), & 0\le s\le k, \\ \sum_{i=1}^{M} B_i(k)A_i^T(s), & 0\le k\le s. \end{cases} \qquad (2)$$

Here, $A_i(k)$ and $B_i(s)$ are $n\times \ell$ bounded deterministic matrix functions. The semi-degenerate kernel is appropriate for expressing general autocovariance function of the nonstationary or stationary stochastic processes(3).

The filtering estimate $\hat{z}(k,k)$ is written as a linear combination of the observation set $\{y(i),\ 1\le i\le k\}$.

$$\hat{z}(k,k)=\sum_{i=1}^{k} h(k,i)y(i) \qquad (3)$$

The optimal filtering estimate is calculated by the optimal impulse response function. In this paper, the linear least-squares filtering problem, which minimizes the mean-square value of the filtering error, is considered.

$$J=E((z(k)-\hat{z}(k,k))^T(z(k)-\hat{z}(k,k)) \tag{4}$$

From the orthogonal projection lemma(4) that the filtering error is orthogonal to the observed value,

$$z(k)-\sum_{i=1}^{k} h(k,i)y(i)\perp y(s),\ 1\le s<k, \tag{5}$$

the Wiener-Hopf equation

$$E(z(k)y^T(s))=\sum_{i=1}^{k} h(k,i)E(y(i)y^T(s)) \tag{6}$$

is obtained. In Eq.(5), $\perp$ denotes the notation of orthogonality. Eq. (6) is transformed into

$$h(k,s)R(k)=K(k,s)-\sum_{i=1}^{k} h(k,i)K(i,s) \tag{7}$$

by using preliminary statistical properties for the signal and noise. Eq.(7) is the basic equation which plays an important role in the filtering problem. The Cauchy system for the optimal impulse response function is derived by the invariant imbedding method in the discrete-time system. The detailed derivation process is reported in S. Nakamori(5) and the sequential algorithm for the discrete-time prediction estimate is developed. In the next section, the filtering algorithm by S. Nakamori(5) is summarized in (Theorem 1).

## 3. Filtering algorithm by covariance information

(Theorem 1)

If the autocovariance function of the signal is expressed by Eq. (2), the filtering estimate is calculated sequentially by the following Eqs.(8)-(11).

$$\hat{z}(k,k)=\sum_{i=1}^{M} A_i(k)e_i(k) \tag{8}$$

$$e_i(k)=e_i(k-1)+J_i(k,k)(y(k)-\sum_{j=1}^{M} A_j(k)e_j(k-1)) \tag{9}$$

$$J_i(k,k)=(B_i^T(k)-\sum_{m=1}^{M} r_{im}(k-1)A_m^T(k))\{R(k)+\sum_{j=1}^{M}(B_j(k)-\sum_{n=1}^{M}A_n(k)r_{nj}(k-1))$$
$$A_j^T(k)\}^{-1} \tag{10}$$

$$r_{ij}(k)=r_{ij}(k-1)+J_i(k,k)(B_j(k)-\sum_{n=1}^{M}A_n(k)r_{nj}(k-1)) \tag{11}$$

Initial conditions of the difference Eqs.(9) and (11), at k=0, are as follows.

$$e_i(0)=0,\ r_{ij}(0)=0,\ i,\ j=1,\ 2,\ldots,\ M. \tag{12}$$

## 4. Kalman filter equations

The main objective of this paper lies in the development of new identification algorithm using covariance information. The construction of the state-space model for the signal has been studied as indicated in the introduction. The direction of this paper to the identification problem is somewhat different from the previous researches in the point that we make use of the Kalman filter equation with the filter presented in (Theorem 1). These two kinds of filters are used to identify the state-space model in terms of the covariance information.

Now, let us describe the Kalman filter equation(6) for the purpose of the design of new identification algorithm in section 6.

$$\hat{z}(k,k)=\hat{z}(k,k-1)+G(k)(y(k)-\hat{z}(k,k-1)) \tag{13}$$

$$G(k)=M(k,k-1)(M(k,k-1)+R(k))^{-1}$$
$$=P(k,k)R^{-1}(k) \tag{14}$$

$$M(k,k-1)=\Phi(k,k-1)P(k-1,k-1)\Phi^T(k,k-1)+\Gamma(k-1)Q(k-1)\Gamma^T(k-1) \tag{15}$$

$$P(k,k)=(I-G(k))M(k,k-1) \tag{16}$$

$G(k)$ is called the Kalman gain. $P(k,k)$ is the filtering error covariance matrix. $M(k,k-1)$ represents the covariance matrix of $z(k)-\hat{z}(k,k-1)$.

## 5. Filtering error covariance matrix

The filter of (Theorem 1) and the Kalman filter are designed based on the minimum mean-square criterion. Therefore, it is expected that both filters have the equivalent estimation properties statistically. A point of contact from these two filters is considered to be the filtering error covariance matrix $P(k,k)$. In this section, it is shown that the filtering error covariance matrix of the filter of (Theorem 1) is represented in terms of the covariance information only.

(Theorem 2)

The filtering error covariance matrix is expressed by Eq.(17), provided that the autocovariance function of the signal is given by Eq.(2).

$$P(k,k)=K(k,k)-\sum_{i=1}^{M} A_i(k)\sum_{j=1}^{M}\sum_{n=1}^{M} r_{in}(k)A_n^T(k) \tag{17}$$

Proof. The filtering error covariance matrix $P(k,k)$ of the filtering estimate $\hat{z}(k,k)$ is defined by

$$P(k,k)=E((z(k)-\hat{z}(k,k))(z(k)-\hat{z}(k,k))^T) \tag{18}$$

at time k. $z(k)-\hat{z}(k,k)$ is orthogonal to $\hat{z}(k,k)$. Namely, it follows that $E((z(k)-\hat{z}(k,k))\hat{z}^T(k,k))=0$. Then Eq.(18) becomes

$$P(k,k)=E((z(k)-\hat{z}(k,k))z^T(k))$$

$$=K(k,k)-\sum_{i=1}^{M} A_i(k)E(e_i(k)z^T(k)). \tag{19}$$

In S. Nakamori(5), the function $e_i(k)$ is given by

$$e_i(k)=\sum_{j=1}^{k} J_i(k,j)y(j),\quad i=1,2,\ldots,M. \tag{20}$$

Substituting Eq.(20) into Eq.(19), one obtains

$$P(k,k)=K(k,k)-\sum_{i=1}^{M} A_i(k)\sum_{j=1}^{k} J_i(k,j)E(y(j)z^T(k)). \tag{21}$$

The quantity $E(y(j)z^T(k))$ is developed as follows by taking into account of $K(j,k)=\sum_{i=1}^{M} B_i(j)A_i^T(k)$, for $0\le j\le k$, from Eq.(2). Also, $r_{in}(k)$ is given by

$$r_{in}(k)=\sum_{j=1}^{k} J_i(k,j)B_n(j) \tag{22}$$

(5). Using Eq.(22), one can rewrite Eq.(21) as

$$P(k,k)=K(k,k)-\sum_{i=1}^{M} A_i(k)\sum_{n=1}^{M} r_{in}(k)A_n^T(k) \tag{23}$$

(Q.E.D.).

6. New identification algorithm

This section presents new identification algorithm by referring to the results of section 3, 4 and 5. The identification algorithm is summarized in (Theorem 3) for the time-variant nonstationary system.

(Theorem 3)

If the autocovariance function of the signal is given by Eq.(2) in the semi-degenerate kernel form, the state transition matrix and the input vector are identified by the following equations sequentially.

$$J_i(k,k)=(B_i^T(k)-\sum_{m=1}^{M} r_{im}(k-1)A_m^T(k))\{R(k)+\sum_{j=1}^{M}(B_j(k)-\sum_{n=1}^{M} A_n(k)r_{nj}(k-1))\, A_j^T(k)\}^{-1} \tag{24}$$

$$r_{ij}(k)=r_{ij}(k-1)+J_i(k,k)(B_j(k)-\sum_{n=1}^{M} A_n(k)r_{nj}(k-1)) \tag{25}$$

$$M(k+1,k)=K(k+1,k+1)-\sum_{i=1}^{M} A_i(k+1)\sum_{n=1}^{M} r_{in}(k)A_n^T(k+1) \tag{26}$$

$$P(k,k)=K(k,k)-\sum_{i=1}^{M} A_i(k)\sum_{n=1}^{M} r_{in}(k)A_n^T(k) \tag{27}$$

The state-transition matrix $\Phi(k+1,k)$ is identified by

$$K(k+1,k+1)-M(k+1,k)=\Phi(k+1,k)(K(k,k)-P(k,k))\Phi^T(k+1,k), \tag{28}$$

where the filtering error covariance matrix $P(k,k)$ is calculated by Eqs.(24), (25) and (27) sequentially. Since the state transition matrix $\Phi(k+1,k)$ is already identified by Eq.(28) and $P(k,k)$ is computed by Eqs.(24), (25) and (27), the input vector $\Gamma(k)$ is identified by

$$\Gamma(k)Q(k)\Gamma^T(k)=M(k+1,k)-\Phi(k+1,k)P(k,k)\Phi^T(k+1,k) \tag{29}$$

Initial conditions of the difference Eqs.(25), at k=0, are

$$r_{ij}(0)=0,\ i,\ j=1,\ 2,\ldots,\ M. \tag{30}$$

Proof. Eqs.(24) and (25) are equal to Eqs.(10) and (11). Eq.(27) is same with Eq.(23). Eq.(29) is obtained by replacing k with k+1 in Eq.(15). Eqs.(26) and (28) have to be proved. Let us prove Eq.(26) at first. M(k+1,k) is the function which is defined by

$$M(k+1,k)=E((z(k+1)-\hat{z}(k+1,k))(z(k+1)-\hat{z}(k+1,k))^T) \tag{31}$$

(4). Since $z(k+1)-\hat{z}(k+1,k)$ is orthogonal to $\hat{z}(k+1,k)$, Eq.(31) is transformed into

$$M(k+1,k)=E((z(k+1)-\hat{z}(k+1,k))z^T(k+1)). \tag{32}$$

From Eqs.(8) and (9), it is straightforward to show that

$$\hat{z}(k,k)=\sum_{i=1}^{M} A_i(k)e_i(k-1)+\sum_{i=1}^{M} A_i(k)J_i(k,k)(y(k)-\sum_{j=1}^{M} A_j(k)e_j(k-1)). \tag{33}$$

If one compares Eq.(33) with Eq.(13), one finds that $\hat{z}(k,k-1)$ is given by

$$\hat{z}(k,k-1)=\sum_{i=1}^{M} A_i(k)e_i(k-1) \tag{34}$$

and the Kalman gain is formulated as

$$G(k)=\sum_{i=1}^{M} A_i(k)J_i(k,k). \tag{35}$$

From Eq.(34), M(k+1,k) in Eq.(32) is rewritten as

$$M(k+1,k)=K(k+1,k+1)-\sum_{i=1}^{M} A_i(k+1)E(e_i(k)z^T(k+1)). \tag{36}$$

Substitution of Eq.(20) into Eq.(36) yields

$$M(k+1,k)=K(k+1,k+1)-\sum_{i=1}^{M} A_i(k+1)\sum_{j=1}^{k} J_i(k,j)E(y(j)z^T(k+1)). \tag{37}$$

The quantity $E(y(j)z^T(k+1))$ is calculated as follows by considering that

$$K(j,k+1)=\sum_{i=1}^{M} B_i(j)A_i^T(k+1),\ \text{for } 0\le j\le k+1, \tag{38}$$

from Eq.(2). One obtains Eq.(26) by substituting Eq.(38) into Eq.(37) and using Eq.(22).

Let us derive Eq.(28) next. It is found that the autocovariance of the signal satisfies

$$K(k+1,k+1)=\bar{\Phi}(k+1,k)K(k,k)\Phi^{T}(k+1,k)+\Gamma(k)Q(k)\Gamma^{T}(k). \tag{39}$$

From Eq.(39) and the equation, which is obtained by replacing k with k+1 in Eq.(15), Eq.(28) is derived(Q.E.D.).

## 7. Digital simulation example

In this example, the identification problem of the time-variant nonstationary system is considered. The scalar state-space model for the signal z(k) is denoted by

$$z(k+1)=a(k)z(k)+b(k)u(k). \tag{40}$$

a(k) and b(k) are generated by the following equations.

$$\left.\begin{aligned} a(k+1)&=0.9a(k)+u_1(k),\quad E(u_1^2(k))=0.1^2 \\ b(k+1)&=0.8b(k)+u_2(k),\quad E(u_2^2(k))=0.1^2. \end{aligned}\right\} \tag{41}$$

The variances of u(k) and v(k) are Q and R respectively. The problem is to identify the parameters a(k) and b(k) by (Theorem 3). The functions $A_1(k)$ and $B_1(k)$, which appeared in the semi-degenerate kernel of Eq.(2), are specified as follows, for M=1.

$$\left.\begin{aligned} A_1(k)&=a(k-1)a(k-2)a(k-3)\cdots\cdots\cdots\cdots a(2)a(1)a(0) \\ B_1(s)&=A^{-1}(s)K(s,s) \end{aligned}\right\} \tag{42}$$

The autocovariance function K(k,k) is updated by

$$K(k+1,k+1)=a^2(k)K(k,k)+b^2(k)Q. \tag{43}$$

The value of Q is 1.0. As in example 1, the values of R are 0.1, 1.0 and 5.0. Table 1 shows the identification results of the parameters a(k) and b(k) for Q=1 and $R=0.1^2$. First five samples of the identification results are shown. Table 2 shows the identification results of the parameters a(k) and b(k) for Q=1 and R=1.0. Table 3 shows the identification results of the parameters a(k) and b(k) for Q=1 and R= $5.0^2$.

Table 1 Identification results for Q=1 and R=$0.1^2$.

| Time k | Estimate of a | Estimate of b |
|---|---|---|
| 1 | 0.980000 | 0.200000 |
| 2 | 0.980000 | 0.200000 |
| 3 | 0.980000 | 0.200000 |
| 4 | 0.980000 | 0.200000 |
| 5 | 0.980000 | 0.200000 |

Table 2 Identification results for Q=1 and R=1.0.

| Time k | Estimate of a | Estimate of b |
|---|---|---|
| 1 | 0.980000 | 0.200000 |
| 2 | 0.980000 | 0.200000 |
| 3 | 0.980000 | 0.200000 |
| 4 | 0.980000 | 0.200000 |
| 5 | 0.980000 | 0.200000 |

Table 3 Identification results for Q=1 and R=$5^2$.

| Time k | Estimate of a | Estimate of b |
|---|---|---|
| 1 | 0.980000 | 0.200000 |
| 2 | 0.980000 | 0.200000 |
| 3 | 0.980000 | 0.200000 |
| 4 | 0.980000 | 0.200000 |
| 5 | 0.980000 | 0.200000 |

## 8. Conclusions

This paper designed the new identification algorithm. The characteristics of this algorithm are summarized as follows.
(1) The new identification algorithm of the time-variant or time-invariant stochastic systems is developed in the discrete-time system.
(2) The algorithm is applied to the modeling of the stochastic system with the white Gaussian input.
(3) The variances of the white Gaussian observation noise and the white Gaussian input noise, and the functions $A_i(k)$ and $B_i(k)$ of Eq. (2) are needed in the algorithm.

The digital simulation results in section 7 reveal that the identification algorithm of (Theorem 3) has the much accurate identification properties.

## References

(1) Ljung, L. and T. Söderström(1983). Theory and practice of recursive identification. MIT Press.

(2) Mehra, R.(1971). On-line identification of linear dynamic systems with applications to Kalman filtering, IEEE Trans. Autom. Contr., vol. AC-16, No.1, pp.12-21.

(3) Nakamori, S. and A. Hataji(1981). New prediction algorithms by covariance information based on innovation theory, Automatica, vol. 17, No.2, pp.379-386.

(4) Sage, A. and J. Melsa(1971). Estimation theory with applications to communications and control. McGraw-Hill.

(5) Nakamori, S.(1983). New design of linear discrete-time predictor using covariance information suitable for predictions of air pollution levels, Proc. of the Fourth International Symposium on the Use of Computers for Environmental Engineering Related to Buildings, pp.246-249.

(6) Kalman, R. and R. Bucy(1961). New results in linear filtering and prediction theory, ASME, J. Basic Eng., vol.83D, pp.95-108.

# OPTIMUM STATE ESTIMATION USING A NEW CLASS OF ROBUST DOUBLY OPTIMIZED RECURSIVE ESTIMATORS

Pero J. Radonja
"RUDI CAJAVEC" Res. and Dev. Dept.
11000 Beograd ,YUGOSLAVIA

ABSTRACT:A new class of robust doubly optimized recursive (DOPTIR) estimators minimize total real estimator output noise. Total output noise consists both of noise due to input observation noise and roundoff noise due to rounding after arithmetic operations. The doubly optimized estimators obtained show very favourable performances in cases of decreased levels of input observation noise, compared with those which we adopted during estimator design. Furthermore, it can be seen that the DOPTIR estimator shows significantly greater robustness to the finite accuracy of the coefficients than the standard recursive estimator.

## 1 INTRODUCTION

In many cases of digital signal processing it is impossible to avoid error due to quantization effects during performing of algorithms.The quantization of coefficients in algorithms introduces a deterministic error whereas the quantization in arithmetic introduces a roundoff noise.The analysis of the roundoff noise in estimators is very complex because of time variable coefficients and complex statistical properties of input samples.Consequently, we shall first define the basic statistical properties of input signals.

## 2 BASIC STATISTICAL PROPERTIES OF ESTIMATOR INPUT SIGNAL

Assume that we have a set of N independent input signals $y_i(k)$ ,$i=1,2,3,....,N$ ,which form column matrix $\mathbf{y}(k)$.This observation vector,that is,the estimator input signal $\mathbf{y}(k)$ can be described by [1]

$$\mathbf{y}(k)=C(k)\mathbf{x}(k) + \mathbf{n}(k) \tag{1}$$

The matrix $C(k)$ is a diagonal measurement matrix with terms $c_i(k)$,$i=1,2,...,N$ , and represents filtering which is introduced by a real physical system through which signals are transmitted - Fig.1. $\mathbf{n}(k)$ is the column observation noise vector with terms $n_i(k)$,$i=1,2,...,N$.The observation noises $n_i(k)$ are zero mean stationary white noise processes and uncorrelated with each other and also with respect to the signal. The corresponding diagonal covariance matrix is $R_o = E[n(k)n(k)^T]$ with terms $\sigma_{ni}$ $i=1,2,...,N$. The column vector of the original signal $x(k)$ with terms $x_i(k), i=1,2,..,N$ can be represented by

$$x(k) = \Phi(k)x(k-1) + w(k) \qquad (2)$$

A signal dynamics matrix $\Phi(k)$ is a diagonal matrix with terms $\varphi_i(k)$, $i=1,2,....,N$ and represents the correlation between two successive samples of signal. The zero mean unpredictable white noise component of the signal is represented by column vector $w(k)$ with terms $w_i(k)$, $i=1,2,....,N$. A corresponding diagonal covariance matrix is $Q_o = E[w(k)w(k)^T]$.

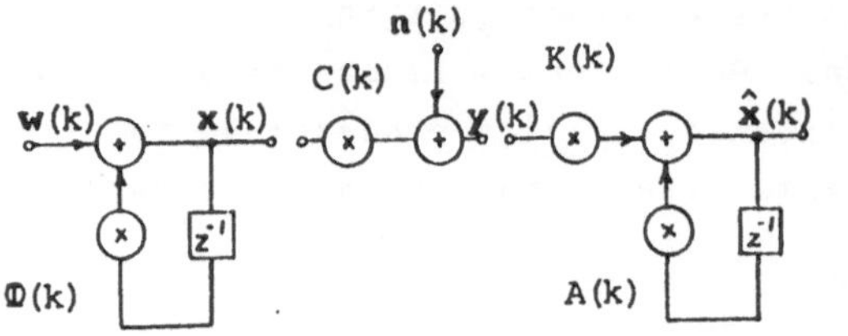


Fig.1 State-space model & estimator

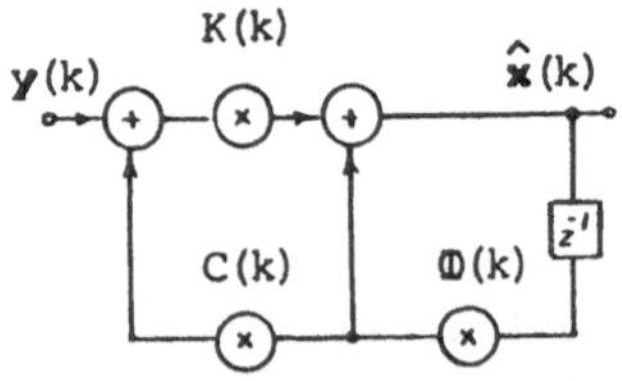


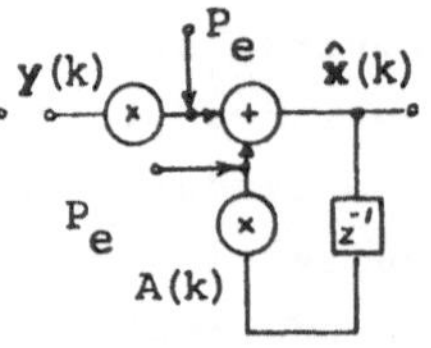


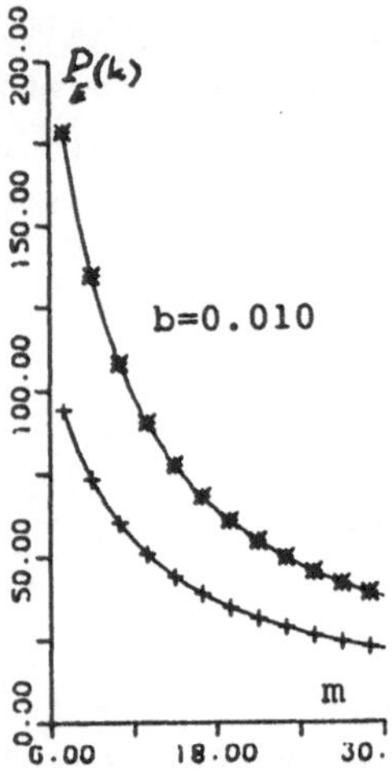


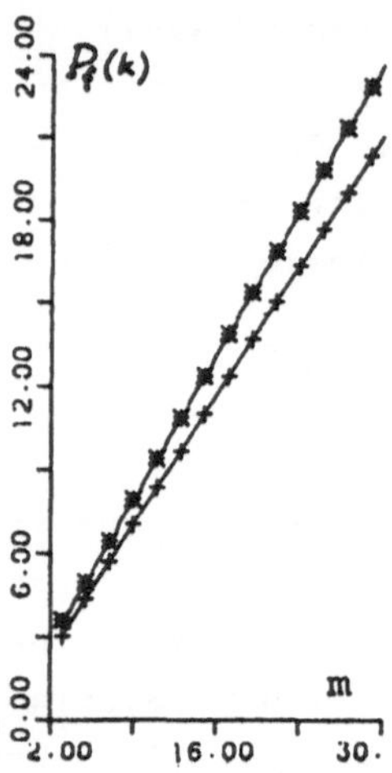


Fig.2 Alternate version

Fig.3 Equivalent net.

Fig.4 Estimation noise

Fig.5 Roundoff noise

## 3 COVARIANCE OF THE ESTIMATION ERROR

The covariance matrix of the estimation error, that is, the noise of the estimation can be obtained using [3]

$$P_E(k) = E\{[x(k)-\hat{x}(k)][x(k)-\hat{x}(k)]^T\} \qquad (3)$$

where $\hat{x}(k)$ denotes the estimated value of $x(k)$. Using an assumption that the estimator has a structure as in Fig.1 the estimated value $\hat{x}(k)$ can be expressed as

$$\hat{x}(k) = A(k)\hat{x}(k-1) + K(k)y(k) \qquad (4)$$

Gain matrix $K(k)$ and transition matrix $A(k)$ which ensure minimum variance of the estimation error will be obtained by solving the following set of matrix equations,[2],(orthogonality principle)

$$E\{[x(k)-\hat{x}(k)]y(k)^T\} = 0 \qquad (5)$$

$$E\{[x(k)-\hat{x}(k)]\hat{x}(k-1)^T\} = 0 \qquad (6)$$

Now we shall define an a priori estimate $\hat{x}(k|k-1)$ and the corresponding covariance matrix $P_E(k|k-1)$ by equations (7) and(8)

$$\hat{x}(k|k-1) = \Phi(k)\hat{x}(k-1) \qquad (7)$$

$$P_E(k|k-1)=E\{[x(k)-\hat{x}(k|k-1)]x(k)^T\} \qquad (8)$$

Using Eqs.(8),(7) and (2) we obtain

$$P_E(k|k-1)=\Phi(k)P_E(k-1)\Phi(k)^T + Q_0 \qquad (9)$$

Using Eqs.(5) and (6) ,equation (3) become

$$P_E(k) = E\{[x(k)-\hat{x}(k)]x(k)^T\} \qquad (10)$$

On the basis of Eqs.(5),(1),(10) and (4) it is easy to obtain

$$K(k) = P_E(k)C^T(k)R_0^{-1} \qquad (11)$$

The known relation between A(k) and K(k) in Kalman filtering

$$A(k) = [I_N - K(k)C(k)]\Phi(k) \qquad (12)$$

we can get using (5),(6),(2),(4) and (1).Based on Eq.(12) we can draw the alternative structure of the estimator as in Fig.2.From Eqs.(10),(4),(12),(1) and (8) we get

$$P_E(k) =[I_N - K(k)C(k)]P_E(k|k-1) \qquad (13)$$

With the aid of Eqs.(13) and (11) we finally get

$$K(k) = P_E(k|k-1)C^T(k)[C(k)P_E(k|k-1)C^T(k)+R_0]^{-1} \qquad (14)$$

Now we have a set of three matrix equations (9),(14) and (13) for recursive computing covariance of the estimation noise $P_E$ (k),and coefficients matrices K(k) and A(k),[3] .

Normalized variances of the estimation noise (13),versus the number of input samples m,in cases of matched filtering (curve denoted by *) and estimation of one unvaried signal (curve denoted by +) are shown in Fig.4.The noise to signal ratio has the value 0.010.The coefficients in matrices K(k) and A(k) have 12 bit accuracy.The accuracy of arithmetic operations is 8 bits.

## 4 ROUNDOFF NOISE COVARIANCE MATRIX

The effect of the quantization constraint manifests itself in several different ways.One of them is the appearance of error due to rounding or truncation in the arithmetic.We shall model this effect by means of an additive equivalent noise source at each point in the estimator where roundoff occurs. Let us now define the basic statistical properties of the equivalent roundoff noise source for the most frequent case, which occure when results of arithmetical operations are rounded.Note that, for rounding, the roundoff noise is zero mean.If we assume that fixed point numbers are represented by $b_o$+1 bit binary words,the roundoff noise variance,for fixed point arithmetic [5,6] is

$$\sigma_e^2 = \frac{1}{12} 2^{-2b_o} \tag{15}$$

that is, $P_e = \mathrm{diag}[\sigma_{ei}^2]$, i=1,2,......,N for the matrix case.

A statistical model , the equivalent network of a real recursive estimator implemented in fixed point arithmetic is shown in Fig.3. The equivalent sources of roundoff noise $P_e$ are introduced only due to multiplications.The roundoff noise covariance matrix of the considered estimator can be computed using [4]

$$P_f(k) = 2P_e + P_f(k-1)A^2(k) \tag{16}$$

The lines denoted by * and + in Fig.5 represent the normalized roundoff noise variance Eq.(16) in the case of the simultaneous estimation of two signals.The line denoted by * refers to the Gaussian pulse while the line denoted by + refers to the unvaried signal.The normalization is performed using $\sigma_e^2$ for $b_o$+1=8 bits.

## 5 OPTIMIZATION OF REAL ESTIMATORS IN VIEW OF TOTAL ESTIMATOR OUTPUT NOISE

In the standard optimization procedure the estimator is optimized only taking account of the error of estimation.In other words in the standard procedure the effect of roundoff noise is neglected.Now using the relations from preceding sections we can compute the variance both of the noise of estimation Eq.(13) and the roundoff noise Eq.(16).On the basis of introduced notations and uncorrelation between estimation and roundoff noise,the variance of the total real estimator output noise is given by

$$P_t(m)=P_E(m) + P_f(m) \tag{17}$$

The question now arises: How much can we decrease the total estimator output noise if we optimize the coefficients of the estimator in view of both the noise of estimation and the roundoff noise.

Optimum gain matrix K(k) and state transition matrix A(k) can be obtained solving

$$E[(x(k) - \hat{x}(k))\hat{x}(k-1)^T] - \frac{1}{2}\cdot\frac{\partial P_f(k)}{\partial A(k)} = 0 \tag{18}$$

$$E[(x(k) - \hat{x}(k))y(k)^T] = 0 \tag{19}$$

In Eq.(18) we use notation

$$\frac{\partial P_f(k)}{\partial A(k)} = \frac{\partial P_f(k)}{\partial a_1(k)} + \frac{P_f(k)}{a_2(k)} + \cdots\cdots + \frac{\partial P_f(k)}{\partial a_N(k)}$$

For simplicity we introduce $P_0(k)$ with its defining relation

$$P_0(k) = E[(x(k) - \hat{x}(k))x(k)^T] \tag{20}$$

Using Eqs.(3),(18) and (20) we can write

$$P_E(k) = P_0(k) - \frac{1}{2}\cdot\frac{\partial P_f(k)}{\partial A(k)}\, A^T(k) \tag{21}$$

With the aid of Eqs.(19),(1),(20) and (4) it is easy to find

$$K(k) = P_0(k)C^T(k)R_0^{-1} \tag{22}$$

Assume further that $\hat{x}(0)=0$ and using Eq.(4) we can derive

$$\hat{x}(k-1)=\sum_{j=2}^{k-1}\Big(\prod_{\ell=k-1}^{j}A(\ell)\Big)K(j-1)y(j-1)+K(k-1)y(k-1) \tag{23}$$

Let us introduce $J_{k-1}, J_{k-2}, \ldots\ldots, J_1$ with their defining relations

$$J_j = E[(x(k) - \hat{x}(k))y(j)^T] \qquad j=1,2,\ldots,k-1 \tag{24}$$

We can obtain a more convenient form of Eq.(18) using Eqs.(23) and (24) If we denote $E[x(k)x(k)^T]$ by $P_x(k)$ and compute all Eqs.(24) then we can group the terms multiplied by A(k) and terms multiplied by $[I_N - K(k)C(k)]\Phi(k)$. In this way we obtain two resulting matrices $RM_1(k)$ and $RM_2(k)$.

$$RM_1(k)= P_x(k-1)C^T(k-1)K^T(k-1)+ \cdots + \frac{1}{2}\, A^{-1}(k)\,\frac{\partial P_f(k)}{\partial A(k)} \tag{25}$$

$$RM_2(k)=P_x(k-1)C^T(k-1)K^T(k-1)+A(k-1)P_x(k-2)C^T(k-2)K^T(k-2)A^T(k-1)+\cdots \tag{26}$$

Now the state transition matrix A(k) can be written in the known form

$$A(k)=[I_N - K(k)C(k)]\Phi_R(k) \tag{27}$$

where we have used following notation

$$\Phi_R(k)= \Phi(k)RM_2(k)RM_1(k)^{-1} \tag{28}$$

The form of the derived Eq.(27) is a very interesting result. Eq.(27) shows the invariance of a form of relationship between the matrix A(k) and the matrix expression $I_N$- K(k)C(k),see Eq.(12). The relation (28) is the fundamental relation of the new class of doubly optimized recursive (DOPTIR) estimators. Letting $\hat{x}(k \mid k-1)$ denote an a priori estimate of x(k)

$$\hat{x}(k|k-1) = \Phi_R(k)\hat{x}(k-1) \tag{29}$$

$$P_0(k|k-1) = E[(x(k) - \hat{x}(k|k-1))x(k)^T] \tag{30}$$

Using Eqs.(20),(29),(30),(4),(1) and (27) it is easy to find

$$P_0(k) = [I_N - K(k)C(k)]P_0(k|k-1) \tag{31}$$

On the other hand using (30) as the starting relation we get

$$P_0(k|k-1)=\Phi_R(k)P_0(k-1)\Phi^T(k)+[\Phi(k)-\Phi_R(k)]P_x(k-1)\Phi^T(k)+Q_0 \tag{32}$$

To express K(k) as the function of $P_0(k|k-1)$ we use Eqs.(22) and (31)

$$K(k)=P_0(k|k-1)C^T(k)[R_0 + C(k)P_0(k|k-1)C^T(k)]^{-1} \tag{33}$$

Now we have all the necessary relations which define the new class of doubly optimized recursive (DOPTIR) estimators.In the first step we shall calculate $\Phi_R(k)$ using (28) and then we shall use Eqs.(32),(33) and (31).A new value of $\Phi_R(k+1)$ can then be calculated and the process repeated.

## 6 BASIC PERFORMANCES AND ADVANTAGES OF DOUBLY OPTIMIZED ESTIMATORS

In this section we shall consider the reduction of the total estimator output noise and other effects due to the optimal procedures outlined in the preceding section.

Let us consider first the reduction of variance of the total real estimator output noise in the case of simultaneous estimation of two sets of input samples with previously defined statistical properties -

Fig.4.The reduction of variance of the total estimator output noise in the case of matched filtering is represented by the curve denoted by *in Fig.6.The improvement is about 5% for m = 30.In the case of estimation of an unvaried signal,the improvement is greater (about 7%) as we can see in the same figure,the curve denoted by +.

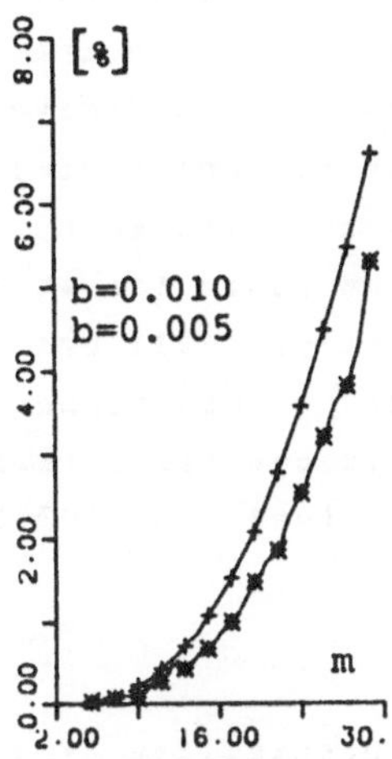


Fig.6
Reduction of $P_t$

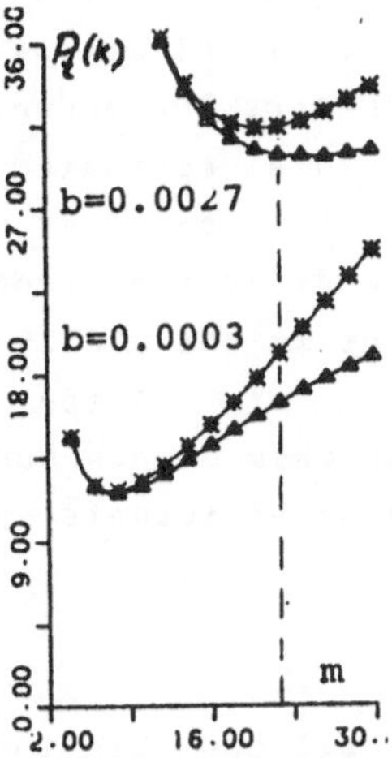


Fig.7 Total output noise

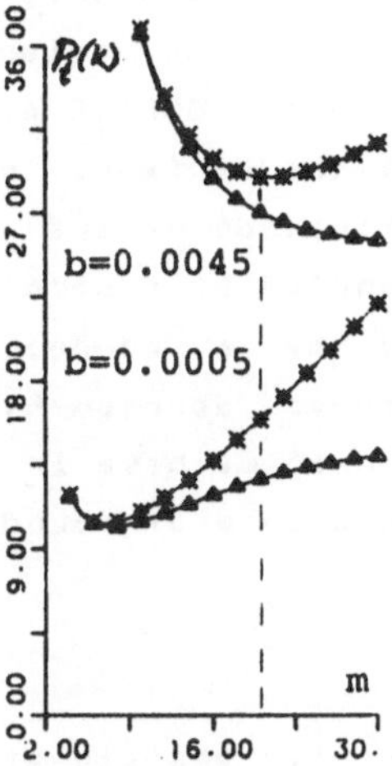


Fig.8 Total output noise

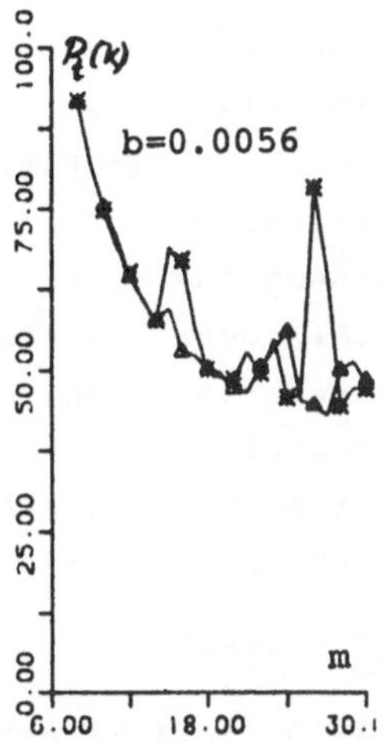


Fig.9
Robustness

A necessary condition for designing the estimator is the known observation noise-to-signal ratio,that is, parameter b.Considering the fact that we always need the minimum value of the estimator output noise it is evident that the estimator must display the best performance for the maximum observation noise. Consequently,in practice,the estimator relatively frequently estimates the signal with a lower observation noise than we have adopted during the design of the estimator.This means that the behaviour of the estimator in the case of decreased observation noise is sometimes important.

We shall now consider the behaviour of both the standard and DOPTIR estimators in the case of decreased observation noise. Let as assume that the estimation procedure will be the same as previously.

For example, assume that parameter b is 0.0003 instead 0.0027 as a consequence of decreased observation noise variance and that all coefficients both of the standard and DOPTIR estimators are unchanged.In the case of matched filtering, the variance of total noise on the output of the standard estimator is 15% greater than the corresponding variance on the output of the DOPTIR estimator as we can see in Fig.7,curves denoted by * and Δ ,respectively. In Fig.8 the case of estimation of an unvaried parameter is shown. The variance of total noise on the output of the standard estimator is 29% greater than the corresponding variance on the output of the DOPTIR estimator,(b=0.0045 and b=0.0005).Evidently the DOPTIR estimator shows very favourable behaviour when input observation noise is decreased.

Let us now consider the robustnees of estimators,that is ,the effect of rounded coefficients on the variance of total estimator output noise in the case of standard and DOPTIR estimators.The accuracy of coefficients is 8 bits.In Fig.9,the variance of total estimator output noise versus the number of estimation steps in the case of matched filtering,is shown.The curve denoted by * corresponds to the standard estimator while the curve denoted by Δ corresponds to the DOPTIR estimator.It can be seen that the curve denoted by Δ is closer to a monotonously decreasing curve which we have in the case of nonrounded coefficients.In other words the influence of rounded coefficients is less in DOPTIR estimators.

7 EXAMPLES

As a first illustrative example we shall consider the possibility of the application of racursive estimators in matched filtering.Assume that we have 9 independent data sets to be estimated simultaneously.The standard approach is to use 9 nonrecursive (transversal) filters with coefficients matched to the shape of the envelope of input samples to be expected.Assume further that noise-to-signal ratio,that is, parameter b,has the values 0.0008, 0.0016, 0.0024,....,0.0072 and that input samples correspond to the sampled Gaussian pulses. In Fig.10 the variance of the total output noise versus the input samples number,in the case of application of the standard nonrecursive filters,is shown.

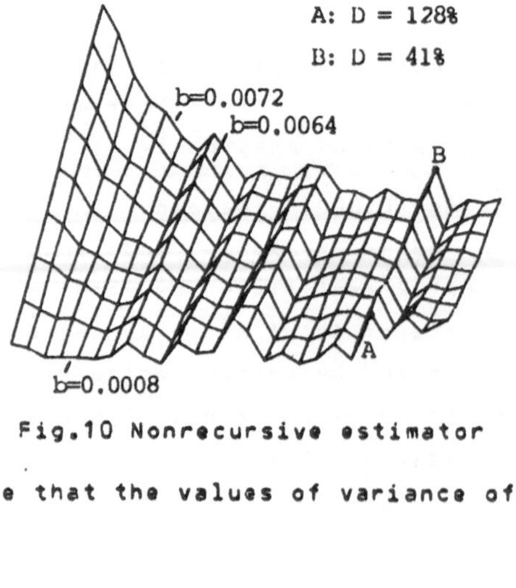


Fig.10 Nonrecursive estimator

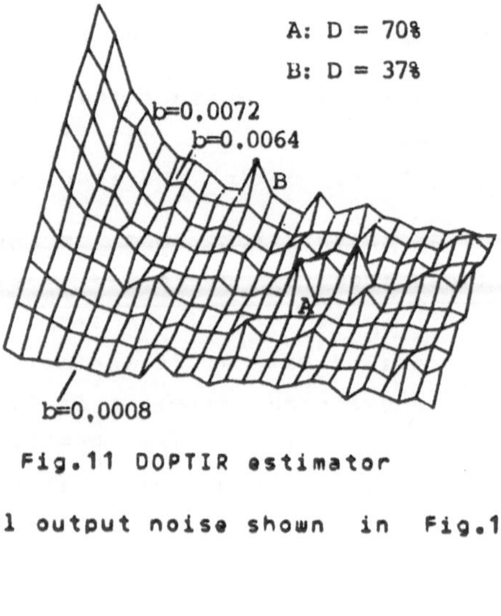


Fig.11 DOPTIR estimator

Note that the values of variance of total output noise shown in Fig.10

represents the results of simulation. This is performed using 300 uncorrelated input sequences for each m, that is, for each number of input samples. In Fig.10 it can be seen that the influence of rounded coefficients is very large. (The accuracy of coefficients is 8 bits.) Practically, it is impossible to perform correct matched filtering in cases when we have 16,19 or 26 input samples. The largest irregularity or degradation of performances D, defined by $D=(P_t - P_{treg})/P_{treg}$, is 128% for b=0.0008. ($P_{treg}$ is regular value of $P_t$) For b=0.0072 the irregularity is 41%.

In the case of application of DOPTIR estimators we shall get Fig.11, for the same input signals and the same procedure of simulation. Now we can see that we have only two points with greater irregularity and that the greatest irregularity is 70%. This is a significant improvement compared with nonrecursive filters.

In Fig.12, the variances of total output noise both of the standard nonrecursive matched filter (results denoted by *) and DOPTIR estimator (results denoted by Δ) are again presented. It can be seen that the DOPTIR estimator has a significantly lower variance of total output noise, particularly for m > 8. Furthermore, the results in Fig.12 verify the conclusion derived on the basis of Figs.10 and 11 in view of the lower sensitivity to the finite accuracy of the coefficients.

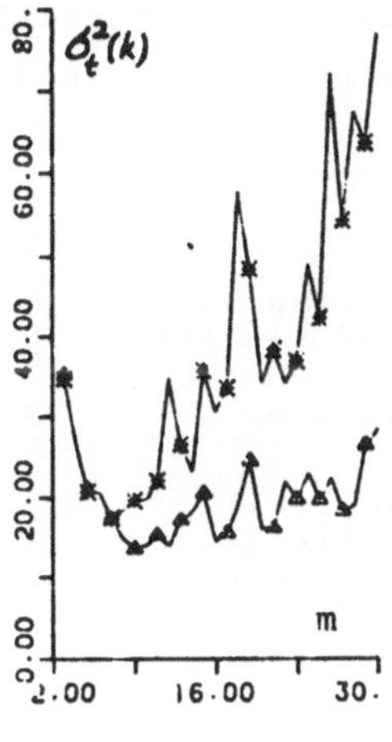

Fig.12 Total output noise

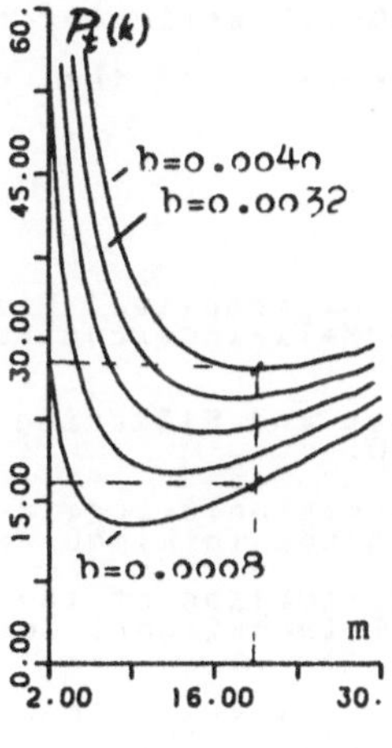

Fig.13 Standard recursive estim.

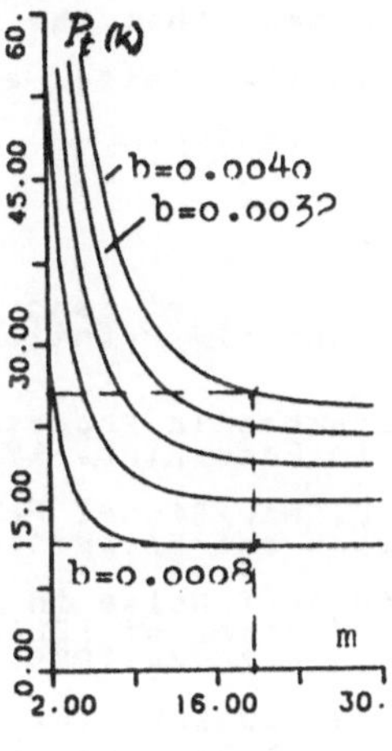

Fig.14 DOPTIR estimator

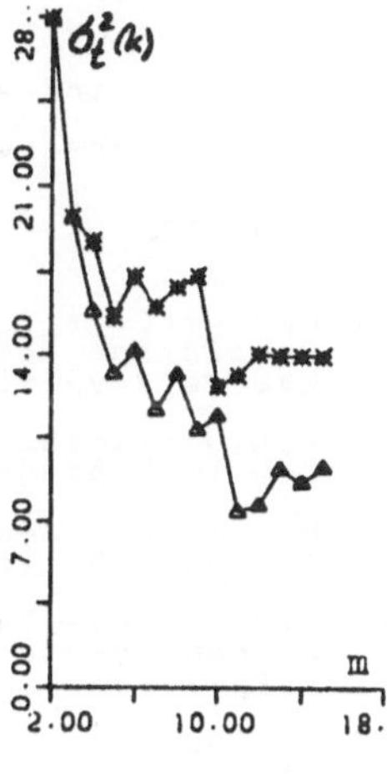

Fig.15 Total output noise

As another example we shall consider simultaneous estimation of 5 uncorrelated unvaried parameters. This problem can be treated as 5 simultaneous cases of scalar tracking (output regulation). In the case of application of the standard recursive estimator (for b=0.0040) the unknown unvaried state is estimated with minimum error ($P_{tmin}$) after 19 input samples/Fig.13. As we can see in Fig.13 $P_t(k)$ has value 27.18 $\sigma_e^2$ and 16.25 $\sigma_e^2$ for b=0.0040 and 0.0008/ respectively.However if we take DOPTIR estimators we shall get results as in Fig.14.Now the $P_t(k)$ has value 25.50 $\sigma_e^2$ and 11.25 $\sigma_e^2$ respectively.Evidently the application of DOPTIR estimators ensure lower total output noise/that is/ less error on the output /10-50%.

The results of the simulation are presented in Fig.15.It is seen that actually the DOPTIR estimator (Δ) has lower total output noise than the standard estimator (*).

## 8 CONCLUSION

The proposed new classes of doubly optimized recursive estimators ensure lower total output noise/which was the goal of the optimization.

The doubly optimized estimators obtained show very favourable performances in cases of decreased levels of input observation noise/ compared with those which we adopted during estimator design.In other words/in this case/the DOPTIR estimator has significantly lower total output noise (15 - 30%) compared to the noise which we had in the case of application of the standard estimator.The cases of matched filtering and estimation of an unvaried signal are both valid.

Furthermore/ we have seen that the DOPTIR estimator shows significantly greater robustness to the finite accuracy of the coefficients than the standard recursive estimator.

## 9 REFERENCES


[1] Graham C.Goodwin and Kwai Sang Sin/ Adaptive Filtering Prediction and Control/ Prentice-Hall/Inc./Englewood Cliffs/ New Jersey/1984.

[2] A.H.Jazwinski/Stochastic Processes and Filtering Theory/ New York:Academic Press/Inc./1970.

[3] M.Schwartz and L.Shaw/Signal Processing:Discrete Spectral Analysis/Detection and Estimation/New York:McGraw-Hill/1975.

[4] P.J.Radonja/"Roundoff Noise in Estimation of the Signal Random Parameter"/Proc. of IEEE International Conference on ASSP/Tampa/FL/March 26-29/1985.

[5] A.V.Oppenheim and R.W.Schafer/Digital Signal Processing/ Prentice-Hall/Inc./Englewood Cliffs/N.J./1975.

[6] L.R.Rabiner and B.Gold/Theory and Application of Digital Signal Processing/Prentice-Hall/Inc./Englewood Cliffs/ N.J./1975.

# CHANDRASEKAHR FILTERING AND SMOOTHING METHODS GIVEN COVARIANCE MODEL AND PROPERTIES OF X AND Y FUNCTIONS

Masanori Sugisaka
Department of Electrical Engineering
Oita University
Oaza Dannoharu, Oita 870-11, Japan

Abstract-This paper is concerned with the Chandrasekhar filtering and smoothing methods given the covariance model of continous stationary stochastic signals in the presence of white Gaussian noises. Two types of Riccati differential equations systems are presented. One is for calculating the Chandrasekhar's X-function and the other is for the Chandrasekhar's Y-function. The explicit relationships between the two solutions to the Riccati differential equations systems and the Chandrasekhar's X- and Y-functions are shown. Also derived are the relations between the Chandrasekhar's X-function and Y-function. In addition, another type of Chandrasekhar's X- and Y-functions that is different from the original ones in terms of numerical stability and accuracy and the corresponding Riccati differential equations systems are presented. Similar formulas are also given for this type of Chandrasekhar's X- and Y-functions.

## 1. Introduction

Casti-Kalaba-Murthy have presented[1] the filtering equations based on the Chandrasekhar's X- and Y-functions assuming that the covariance model of the signals is represented as the infinite weighted sum of exponential functions with different damping ratios. The filtering equations are refered to as the Chandrasekhar filter that is superior to the Kalman filter in terms of the computational time and storage if the state space model is given and the dimension of the outputs is less than that of the states[2]-[3].

However, the structure of this filter and the relationships between the Chandrasekhar's X- and Y-functions have not been clear until now. This paper addresses to clarify the structure of the Chandrasekhar filter given the covariance model and then presents the explicit relationships between the Chandrasekhar's X- and Y-functions introducing two types of Riccati differential equations systems. In addition, another type of Chandrasekhar's X- and Y-functions is proposed and the corresponding Riccati differential equations systems are presented. In section 2 problem statement is given. Section 3 gives the Chandrasekhar filtering and smoothing formulas using the covariance model. In section 4 presents, at first, two types of Riccati differential equations systems and then the relationships between the Chandrasekhar's X- and Y-functions and the solutions to the Riccati systems. Also given are the relationships between the Chandrasekhar's X-function and Y-function. Section 5 introduces another type of Chandrasekhar's X- and Y-functions that is different from the original ones by Casti-Kalaba-Murthy in terms of stability and accuracy in the numerical computations and the similar formulas to the original ones are derived for this X- and Y-functions. Conclusion follows in section 6.

## 2. Problem Statement

The linear least-squares estimate (LLSE) of continuous stationary signals in the presence of white Gaussian noises is specified by the convolution integral of the measured output with the impulse response function (IRF)h, which obeys the Fredholm integral equation of the second kind(FIESK) and its resolvent identity, i.e.,

$$\hat{x}(t/C)=\int_0^C h(t,s;C)y(s)ds(\text{LLSE}), \tag{2.1}$$

$$h(t,s;C)\bar{R}=\begin{cases} k(t-s)-\int_0^C h(t,u;C)k(u-s)du, & (2.2a)\\ k(t-s)-\int_0^C k(t-u)h(u,s;C)du, & (2.2b)\end{cases}$$

(Resolvent Identity)

$$y(t)=x(t)+v(t),\quad t\varepsilon\ [0,C], \tag{2.3}$$

where y(t) is the measured output, x is the zero-mean stationary stochastic signal with the covariance given by

$$k(t-s)=E[x(t)x(s)]=\sum_{i=1}^{n} a_i\exp(-\lambda_i|t-s|) \tag{2.4}$$

uncorrelated with the measurement noise v(t), v(t) is the zero-mean white Gaussian noise with the covariance $\bar{R}$. The notation E denotes the expectation. For the filtering problem (t=C in (2.1)), Casti-Kalaba-

Murthy [1] presented the initial-value solution (IVS) of (2.2) based on the Chandrasekhar's X- and Y-functions and designed the filter using the covariance information called the Chandrasekhar filter (CF) by substituting the IVS into (2.1). Later, Sugisaka[5] presented the Chandrasekhar smoothers(CS's).

In the present paper, we focus attention on the structure of the filter and the relationships between the Chandrasekhar's X- and Y-functions. In other words, our objectives are to clarify the relationships between the solutions to Riccati systems and the Chandrasekhar's X- and Y-functions introducing two types of Riccati differential equations systems and to show how the Chandrasekhar's X-function is related to the Y-function. Also, introduced is the another type of Chandrasekhar's X- and Y-functions that is different from the original ones in stability and accuracy for evaluating the numerical solutions. Our next objective is to show both the structure of the X- and Y-functions introduced and the relationships between them applying the same procedure as that used for the original ones.

## 3. Chandrasekhar Filter and Smoothers [1],[4]-[5]

The filtering problem is to obtain the estimate x(t/C) at t=C , namely,

$$\hat{x}(C)=\hat{x}(C/C)=\int_0^C A(t,C)y(t)dt, \qquad (3.1)$$

where the IRF of the filter A(s,C)=h(C,s;C) is the solution of the FIESK such that

$$A(t,C)\bar{R}=K(C-t)-\int_0^C A(u,C)K(u-t)du, \quad t\in[0,C]. \qquad (3.2)$$

Introduce the auxiliary function $J_i$ which is the solution of the FIESK given by

$$J_i(t,C)\bar{R}=\exp[-\lambda_i(C-t)]-\int_0^C J_i(u,C)K(u-t)du, \quad i=1,\cdots,n, \quad t\in[0,C], \qquad (3.3)$$

to express

$$A(t,C)=\sum_{i=1}^n a_i J_i(t,C), \quad t\in[0,C]. \qquad (3.4)$$

Casti-Kalaba-Murthy [1] showed that the CF is specified by the following IVS.

CF:

$$\hat{x}(C)=\sum_{i=1}^{n} a_i L_i(C), \tag{3.5}$$

$$\dot{L}_i(C)=-\lambda_i L_i(C)+X_i(C)(y(C)-\hat{x}(C)), \tag{3.6a}$$

$$L_i(0)=0(\text{I.C.}),\ i=1,\cdots,n, \tag{3.6b}$$

$$\dot{X}_i(C)=-Y_i(C)\sum_{j=1}^{n} a_j Y_j(C), \tag{3.7a}$$

$$X_i(0)=1/\bar{R}(\text{I.C.}),\ i=1,\cdots,n, \tag{3.7b}$$

$$\dot{Y}_i(C)=-\lambda_i Y_i(C)-X_i(C)\sum_{j=1}^{n} a_j Y_j(C), \tag{3.8a}$$

$$Y_i(0)=1/\bar{R}(\text{I.C.}),\ i=1,\ldots,n. \tag{3.8b}$$

In addition, the Chandrasekhar fixed-point and initial-point smoothers are given by the following IVS's[5].

Fixed-Point Smoother:

$$\hat{x}_C(t/C)=A(t,C)(y(C)-\hat{x}(C)), \tag{3.9a}$$

$$=\sum a_i J_i(t,C)(y(C)-\hat{x}(C)), \tag{3.9b}$$

$$\hat{x}(t/C)\big|_{C=t}=\hat{x}(t)(\text{I.C.}), \tag{3.9c}$$

$$J_{i,C}(t,C)=-\lambda_i J_i(t,C)-X_i(C)\sum_{j=1}^{n} a_j J_j(t,C), \tag{3.10a}$$

$$J_i(t,C)\big|_{C=t}=X_i(t)(\text{I.C.}), \tag{3.10b}$$

Initial-Point Smoother:

$$\dot{\hat{x}}(0/C)=\Sigma\, a_j Y_j(C)(y(C)-\hat{x}(C)), \qquad (3.11a)$$

$$\hat{x}(0/C)\big|_{C=0}=x(0)=0(\text{I.C.}), \qquad (3.11b)$$

where the notation '.' means the ordinary derivative, the notation 'I.C.' means initial condition, and the notations $\hat{x}_C(t/C)$ and $J_{i,C}$ mean the partial derivatives of $\hat{x}(t/C)$ and $J_i$ with respect to C, respectively.

We note that the Chandrasekhar's X- and Y-function, which obeys the set of ordinary differential equations with known initial conditions given by (3.7)-(3.8), were defined by

$$X_i(C)=J_i(C,C), \qquad (3.12a)$$
$$Y_i(C)=J_i(0,C), \qquad (3.12b)$$

and that these functions were related to the auxiliary function $J_i$ by

$$X_i(C)\bar{R}=1-\int J_i(u,C)K(u-C)du, \qquad (3.13a)$$
$$=1-\int J_i(C-u,C)K(u)du, \qquad (3.13b)$$
$$Y_i(C)\bar{R}=\exp[-\lambda_i C]-\int J_i(u,C)K(u)du, \qquad (3.13c)$$

where we omitted the upper and lower limits C and 0 of the integrals above(in the succeedings both limits will be omitted). The function $J_i(C-u,C)$ in (3.13b) which is the backward function of $J_i(u,C)$ satisfies the FIESK given by

$$J_i(C-u,C)\bar{R}=\exp[-\lambda_i u]-\int J_i(C-\tau,C)K(\tau-u)d\tau\ ,\ u\in[0,C].(3.14)$$

The physical meanings of the Chandrasekhar's X- and Y-func- tions are the followings.

$$A(C,C)=\Sigma\, a_i X_i(C)=(1/\bar{R})E(x(C)-\hat{x}(C))^2, \qquad (3.15a)$$
$$A(0,C)=\Sigma\, a_i Y_i(C)=(1/\bar{R})E(x(C)-\hat{x}(C))(x(0)-\hat{x}(0/C)). \qquad (3.15b)$$

In addition, the physical meaning of the IRF or resolvent kernel $h(t,s;C)$ is such that

$$h(t,s;C)=(1/\bar{R})E(x(t)-\hat{x}(t/C))(x(s)-\hat{x}(s/C)). \qquad (3.16)$$

## 4. Chandrasekher's X- and Y-Functions and Riccati Systems

We shall show two Riccati systems in order to evaluate the Chandrasekhar's X- and Y-functions. One is for the X-function and is given by

$$\dot{R}_{ij}^{+}(C)=-\lambda_i R_{ij}^{+}(C)+X_i(C)X_j(C)\exp[\lambda_j C]\bar{R},\quad i,j=1,\ldots,n, \tag{4.1a}$$

$$R_{ij}^{+}(0)=0(I.C.). \tag{4.1b}$$

The other is for the Y-function. It is described as

$$\dot{R}_{ij}^{-}(C)=-\lambda_i R_{ij}^{-}(C)+X_i(C)Y_j(C)\bar{R}, \tag{4.2a}$$

$$R_{ij}^{-}(0)=0(I.C.). \tag{4.2b}$$

Utilizing the solutions $R_{ij}^{+}$ and $R_{ij}^{-}$ to the above Riccati systems, the Chadrasekhar's X and Y-functions can be expressed by

$$X_i(C)=(1/\bar{R})(1-\Sigma\, a_j\exp[-\lambda_j C]R_{ij}^{+}(C)), \tag{4.3}$$

$$Y_i(C)=(1/\bar{R})(\exp[-\lambda_i C]-\Sigma\, a_j R_{ij}^{-}(C)), \tag{4.4}$$

where the functions $R_{ij}^{+}$ and $R_{ij}^{-}$ were difined by

$$R_{ij}^{+}(C)=\int J_i(u,C)\exp[\lambda_j u]du, \tag{4.5}$$

$$R_{ij}^{-}(C)=\int J_i(u,C)\exp[-\lambda_j u]du, \tag{4.6}$$

respectively(The derivations of (4.1)-(4.4) are omitted for reason of space).

Conversely, we can represent the functions $R_{ij}^{+}$ and $R_{ij}^{-}$ defined above in terms of the Chandrasekhar's X- and Y-functions. It is seen that the functions $R_{ij}^{+}$ and $R_{ij}^{-}$ are expressed by

$$R_{ij}^{+}(C)=(\lambda_i+\lambda_j)^{-1}(X_i(C)X_j(C)-Y_i(C)Y_j(C))\exp[\lambda_j C]\bar{R}, \tag{4.7}$$

$$R_{ij}^{-}(C)=(\lambda_i-\lambda_j)^{-1}(X_i(C)Y_j(C)-Y_i(C)X_j(C))\bar{R}, \tag{4.8}$$

(The derivations of (4.7)-(4.8) are also omitted). Then interchanging the order of the subscripts we notice that

$$R_{ji}^{+}(C)=\exp[(\lambda_i-\lambda_j)C]R_{ij}^{+}(C), \tag{4.9}$$

$$R_{ji}^{-}(C)=R_{ij}^{-}(C). \tag{4.10}$$

Namely, while the function $R_{ij}^{+}$ is not symmetric, the function $R_{ij}^{-}$ is symmetric. Substituting (4.7) and (4.8) into (4.3) and (4.4), respectively, we obtain

$$X_i(C)\bar{R}=1-\sum_{j=1}^{n} a_j(\lambda_i+\lambda_j)^{-1}(X_i(C)X_j(C)-Y_i(C)Y_j(C)), \tag{4.11}$$

$$Y_i(C)\bar{R}=\exp[-\lambda_i C]-\sum_{j=1}^{n} a_j(\lambda_i-\lambda_j)^{-1}(X_i(C)Y_j(C)-Y_i(C)X_j(C)). \tag{4.12}$$

The above algebraic equations shows important relationships between the Chandrasekhar's X- and Y- functions. Similiarly, substituting (4.7) and (4.8) into (4.1) and (4.2), respectively, we have alternative representations for the Riccati systems as follows.

$$\dot{R}_{ij}^{+}(C)=\lambda_j R_{ij}^{+}(C)+Y_i(C)Y_j(C)\exp[\lambda_j C]\bar{R}, \tag{4.13}$$

$$\dot{R}_{ij}^{-}(C)=-\lambda_j R_{ij}^{-}(C)+Y_i(C)X_j(C)\bar{R}. \tag{4.14}$$

If we assume the solution for (3.2) to be given by

$$A(t,C)=\sum a_i\exp[-\lambda_i C]J_i^{*}(t,C),\ t\in[0,C], \tag{4.15}$$

then, we have another type of CF and CS's which are similar to the CF and CS's in section 3 [4]-[5]. Refer [5] for the filtering and smoothing equations. In the next section, we shall present the similiar formulas for this case as shown in the previous sections.

## 5. Another Type of Chandrasekhar X- and Y-Functions and Riccati Systems

We see that the solution $J^{*}(t,C)$ defined by (4.15) satisfies the FIESK given by

$$J_i^{*}(t,C)\bar{R}=\exp[\lambda_i C]-\int J_i^{*}(u,C)K(u-t)du,\ i=1,\dots,n,\ t\in[0,C] \tag{5.1}$$

We shall define another type of Chandrasekhar's X- and Y- function in terms of the function introduced above, viz.,

$$X_i^*(C)=J_i^*(C,C)=\exp[\lambda_i C]X_i(C), \quad (5.2a)$$

$$Y_i^*(C)=J_i^*(0,C)=\exp[\lambda_i C]Y_i(C). \quad (5.2b)$$

Then we have the following Riccati systems,

$$\dot{R}_{ij}^{*+}(C)=X_i^*(C)X_j^*(C)\bar{R}, \quad (5.3a)$$

$$R_{ij}^{*+}(0)=0(I.C.), \quad (5.3b)$$

$$\dot{R}_{ij}^{*-}(C)=X_i^*(C)Y_j^*(C)\exp[-\lambda_j C]\bar{R}, \quad (5.4a)$$

$$R_{ij}^{*-}(0)=0(I.C.), \quad (5.4b)$$

where the functions $R_{ij}^{*+}$ and $R_{ij}^{*-}$ were defined by

$$R_{ij}^{*+}(C)=\int J_i^*(u,C)\exp[\lambda_j u]du=\exp[\lambda_i C]R_{ij}^{+}(C), \quad (5.5a)$$
$$R_{ij}^{*-}(C)=\int J_i^*(u,C)\exp[-\lambda_j u]du=\exp[\lambda_i C]R_{ij}^{-}(C). \quad (5.5b)$$

The functions $X_i^*$ and $Y_i^*$ are given by

$$X_i^*(C)=(1/\bar{R})(\exp[\lambda_i C]-\Sigma\, a_j\exp[-\lambda_j C]R_{ij}^{*+}(C)), \quad (5.6)$$

$$Y_i^*(C)=(1/\bar{R})(1-\Sigma\, a_j R_{ij}^{*-}(C)). \quad (5.7)$$

Then, utilizing the relationships between $X^*$, $Y^*$, $R_{ij}^{*+}$, $R_{ij}^{*-}$ and X, Y, $R_{ij}^{+}$, $R_{ij}^{-}$ shown in (5.2) and (5.5), respectively, we easily obtain the following results.

$$R_{ij}^{*+}(C)=(\lambda_i+\lambda_j)^{-1}(X_i^*(C)X_j^*(C)-Y_i^*(C)Y_j^*(C))\bar{R}, \quad (5.8)$$

$$R_{ij}^{*-}(c)=(\lambda_i-\lambda_j)-1(X_i^*(C)Y_j^*(C)-Y_i^*(C)X_j^*(C))\bar{R}\exp[-\lambda_j C], \quad (5.9)$$

$$R_{ji}^{*+}(C)=R_{ij}^{*+}(C), \quad (5.10)$$

$$R_{ji}^{*-}(C)=\exp[-(\lambda_i-\lambda_j)C]R_{ij}^{*-}(C), \quad (5.11)$$

$$X_i^*(C)\bar{R}=\exp[\lambda_i C]-\sum_{j=1}^{n} a_j\exp[-\lambda_j C](\lambda_i+\lambda_j)^{-1}$$
$$\times (X_i^*(C)X_j^*(C)-Y_i^*(C)Y_j^*(C))\bar{R}, \qquad (5.12)$$

$$Y_i^*(C)\bar{R}=1-\sum_{j=1}^{n} a_j\exp[-\lambda_j C](\lambda_i-\lambda_j)^{-1}$$
$$\times (X_i^*(C)Y_j^*(C)-Y_i^*(C)X_j^*(C))\bar{R}, \qquad (5.13)$$

$$\dot{R}_{ij}^{*+}(C)=(\lambda_i+\lambda_j)R_{ij}^{*+}(C)+Y_i^*(C)Y_j^*(C)\bar{R}, \qquad (5.14)$$

$$\dot{R}_{ij}^{*-}(C)=(\lambda_i-\lambda_j)R_{ij}^{*-}(C)+Y_i^*(C)X_j^*(C)\exp[-\lambda_j C]\bar{R}. \qquad (5.15)$$

It is obvious that utilizing the results shown in section 4 and the relationships given by (5.2) and (5.5) yields easily the above equations.

## 6. Conclusion

This paper showed two types of Riccati differential equations with zero initial conditions where one is for calculating the Chandrasekhar's X-function and the other is for the Chandrasekhar's Y-function. Furthermore, the explicit relationships between the two solutions to the Riccati differential equations and the Chandrasekhar's X- and Y-functions were presented. Also derived were the relations between the Chandrasekhar's X-function and Y-function.

In addition, another type of Chandrasekhar's X- and Y-functions, which is slightly different from the original Chandrasekhar's X- and Y-functions by Casti-Kalaba-Murthy, was introduced and the similar formulas for another type of Chandrasekhar's X- and Y-functions were presented. It was noticed that the new X- and Y-functions is different from the original X- and Y-functions in accuracy and stability for calculating the numerical solutions.

References

[1] J. Casti, R. Kalaba, and V. K. Murthy: A New Initial-Value Method for On-Line Filtering and Estimation, IEEE Trans. Information Theory, IT-18, pp.515-518(1972).

[2] T. Kailath, L. Ljung, and M. Morf: Recursive Input-Output and State Space Solutions for Continuous-Time Linear Stationary Signals, IEEE Trans. Automatic Control, AC-28, pp.897-906(1983).

[3] J. Casti: Dynamical Systems and Their Applications: Linear Theory, Academic Press, New York(1977).

[4] M. Sugisaka: Initial-Value Solutions for On-Line Smoothing and Filtering of Linear Stationary Signals, IEEE Trans. Automatic Control, AC-29, pp.174-177(1984).

[5] M. Sugisaka: The Design of Chandrasekhar-Type Filters and Smoothers, J. Soc. Instrum. & Control Eng., Vol 23, No-4, pp.347-352(1988).

# NEW APPROACH TO IMAGE MODELLING WITH APPLICATION TO RESTORATION PROBLEMS

**A. Germani(1) and L. Jetto(2)**

**(1) Dip. di Sistemi Universitá della Calabria-87036 Arcavacata (Cosenza),Italy and IASI-CNR V.le Manzoni 30 00185 Rome, Italy.**

**(2) Dip. di Elettronica ed Automatica Università di Ancona, 60131 Ancona, Italy.**

## 1. INTRODUCTION

In the last years much attention has been devoted to the Kalman filtering approach to restoration problems of images degraded by additive gaussian noise. As a consequence many efforts have been made towards the construction of image models suitable to recursive restoration techniques. In [1-9] are proposed methods based on the knowledge of image autocorrelation function. Of course in practical situations noise-free image is not available and therefore it is not possible to estimate the autocorrelation function with a given precision. For this reason several authors [10-14] proposed adaptive filters based on identification-estimation algorithms.

In this work a new technique to image restoration problem is proposed. Here it is developed a $2-D$ Kalman filtering approach where the computational burden for the identification of image parameters is greatly reduced. To this purpose the following hypotheses are assumed:

*i) Regularity assumption*. The image is modelled by an ensamble of $2-D$ differentiable surfaces. In this case the signal which represents the gray level and its spatial derivatives up to a suitably chosen order $\bar{n}$ can be assumed as the state vector.

*ii) Stochastic assumption*. All the derivatives of order $\bar{n}+1$ of the $2-D$ signal are modelled by means of zero-mean independent gaussian random fields.

These hypotheses are based on the fact that most of images are constituted by the union of open disjoint subregions whose interior is enough regular to be well described by a smooth $2-D$ gaussian process. The boundaries of each subregion are the image edges which represent sharp discontinuities in the distributions of the gray level. Of course the image is not differentiable on these dicontinuities so that the filter will have to be suitably adapted to their presence as it is specified in [15]. An image is defined to be *homogeneous* if it does not contain any discontinuity.

## 2. HOMOGENEOUS IMAGE EQUATION

Let us indicate with $x(r,s)$ the value of the original image at spatial coordinates $(r,s)$ where the continuous variables r and s denote the vertical and horizontal position respectively. Here we assume $(r,s) \in [0,1]^2$.

Because of the smoothness assumption let us define as the state:

$$X(r,s) = [\frac{\partial^n x(r,s)}{\partial r^{n-\alpha} \partial s^{\alpha}}, n = 0, \dots, \bar{n}, \alpha = 0, \dots, n]^T$$

If $\bar{n}$ is the maximum order of derivatives taken into account the dimension of $X(r,s)$ is $N = \frac{(\bar{n}-1)(\bar{n}+2)}{2}$. Now if we put $r = r(u) = r_0 + \gamma u; s = s(u) = s_0 + \beta u$ the following equation can be written:

$$\dot{X}(r(u),s(u)) = \frac{\partial}{\partial r}X(r(u),s(u))\dot{r}(u) + \frac{\partial}{\partial s}X(r(u),s(u))\dot{s}(u) \tag{2.1}$$

Let us introduce the following notations:

$$\frac{\partial}{\partial r}X(r(u),s(u)) = AX(r(u),s(u)) + BW_r(r(u),s(u)) \tag{2.2}$$

$$\frac{\partial}{\partial s}X(r(u),s(u)) = A'X(r(u),s(u)) + BW_s(r(u),s(u)) \tag{2.3}$$

where $A$ and $A'$ are $(N \times N)$ matrices whose elements $a_{l,m}$ and $a'_{l,m}$ are such that:

$$a_{l,m} = \begin{cases} 1 & \text{if } m = l + 1 + \left[\frac{\sqrt{8l-7}-1}{2}\right] \\ 0 & \text{otherwise} \end{cases} \tag{2.4}$$

$$a'_{l,m} = \begin{cases} 1 & \text{if } m = l + 2 + \left[\frac{\sqrt{8l-7}-1}{2}\right] \\ 0 & \text{otherwise} \end{cases} \tag{2.5}$$

where the vectors $W_r(r(u),s(u))$ and $W_s(r(u),s(u))$ have dimension $\bar{n}+1$ and are given by

$$W_r(r(u),s(u)) = [\frac{\partial^{\bar{n}+1}x(r,s)}{\partial r^{\bar{n}-\alpha+1}\partial s^{\alpha}};\ \alpha = 0,1,...,\bar{n}]^T$$

$$W_s(r(u),s(u)) = [\frac{\partial^{\bar{n}+1}x(r,s)}{\partial r^{\bar{n}-\alpha}\partial s^{\alpha+1}};\ \alpha = 0,1,...,\bar{n}]^T$$

and where the $(N \times (\bar{n}+1))$ matrix $B$ has the form

$$B = \begin{bmatrix} O \\ I \end{bmatrix}$$

The dimensions of the null block and of the identity matrix are $(N - (\bar{n}+1)) \times (\bar{n}+1)$ and $(\bar{n}+1) \times (\bar{n}+1)$ respectively. It is possible to show that $A$ and $A'$ commute [15]. Using the (2.2) and (2.3) the equation (2.1) can be rewritten in the following form :

$$\dot{X}(r(u),s(u)) = \dot{r}(u)AX(r(u),s(u)) + \dot{s}A'X(r(u),s(u)) + \dot{r}(u)BW_r(r(u),s(u)) + \dot{s}(u)BW_s(r(u),s(u)) \tag{2.6}$$

which is defined *homogeneous image equation.*

## 3. STATE SPACE REPRESENTATION OF THE DISCRETIZED IMAGE.

Let us denote with

$$x_{i,j} = x(i\Delta_r, j\Delta_s),\ i,j = 0,1,...,M$$

the value of the sampled image at the pixel with vertical coordinate $i\Delta_r$ and horizontal coordinate $j\Delta_s$, where $\Delta_r$ and $\Delta_s$ denote the distance between two adjacent pixels on a same column or row

respectively. If the image is sampled with an equal number of pixels $M+1$ on each column and on each row, the normalized value of $\Delta_r$ and $\Delta_s$ are both equal to $\frac{1}{M}$.
We consider the situation where the image is observed under additive white gaussian noise $v_{i,j} \sim N(0, \sigma_v^2)$:

$$y_{i,j} = x_{i,j} + v_{i,j} \tag{3.1}$$

In order to obtain a state space representation for the discretized image let us consider the following semicausal model:

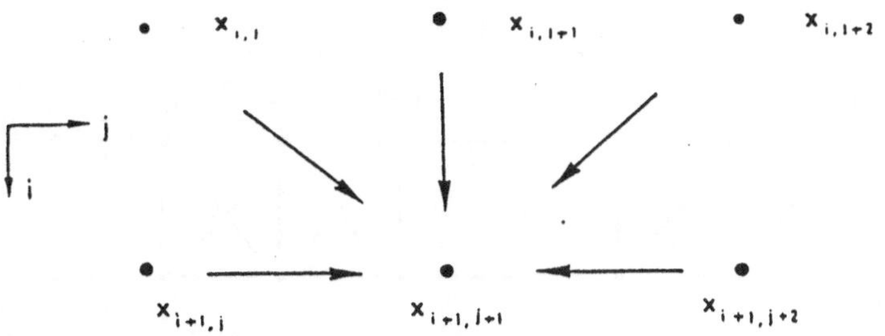


*Fig. 1*

By integrating the eqn.(2.6) along the directions indicated in Fig. 1 the following equations are obtained:

$$X_{i+1,j+1} = H_s(\Delta_s)X_{i+1,j} + W^{(1)}_{i+1,j+1} \tag{3.2}$$

$$X_{i+1,j+1} = H(\Delta_s, \Delta_r)X_{i,j} + W^{(2)}_{i+1,j+1} \tag{3.3}$$

$$X_{i+1,j+1} = H_r(\Delta_r)X_{i,j+1} + W^{(3)}_{i+1,j+1} \tag{3.4}$$

$$X_{i+1,j+1} = H(-\Delta_s, \Delta_r)X_{i,j+2} + W^{(4)}_{i+1,j+1} \tag{3.5}$$

$$X_{i+1,j+1} = H_s(-\Delta_s)X_{i+1,j+2} + W^{(5)}_{i+1,j+1} \tag{3.6}$$

$$H_s(\Delta_s) = e^{A'\Delta_s};\ H_r(\Delta_r) = e^{A\Delta_r};\ H(\Delta_s, \Delta_r) = e^{A'\Delta_s + A\Delta_r}$$

and

$$W^{(1)}_{i+1,j+1} = \int_0^1 H_s(\Delta_s(1-\tau))\Delta_s BW_s((i+1)\Delta_r, j\Delta_s + \Delta_s\tau)d\tau \tag{3.7}$$

$$W^{(2)}_{i+1,j+1} = \int_0^1 H(\Delta_s(1-\tau), \Delta_r(1-\tau))B(W_r(i\Delta_r + \Delta_r\tau, j\Delta_s + \Delta_s\tau) + \Delta_s W_s(i\Delta_r + \Delta_r\tau, j\Delta_s + \Delta_s\tau))d\tau \tag{3.8}$$

$$W^{(3)}_{i+1,j+1} = \int_0^1 H_r(\Delta_r(1-\tau))\Delta_r BW_r(i\Delta_r + \tau\Delta_r, (j+1)\Delta_s)d\tau \tag{3.9}$$

$$W^{(4)}_{i+1,j+1} = \int_0^1 H(-\Delta_s(1-\tau), \Delta_r(1-\tau))B(\Delta_r W_r(i\Delta_r + \tau\Delta_r, (j+2)\Delta_s - \Delta_s\tau) -$$

$$-\Delta_s W_s(i\Delta_r + \Delta_r\tau, (j+2)\Delta_s - \Delta_s\tau))d\tau \tag{3.10}$$

$$W^{(5)}_{i+1,j+1} = -\int_0^1 H_s(-\Delta_s(1-\tau))\Delta_s B W_s((i+1)\Delta_r, (j+2)\Delta_s - \Delta_s\tau)d\tau \tag{3.11}$$

The eqns. (3.3) and (3.5) have been obtained exploiting the commutativity of the two semigroups $e^{A'\Delta_s}$ and $e^{A\Delta_r}$ which derives from the commutativity of their generators. A way to fast compute the matrix exponentials is reported in [15].

Now let us consider the following ensamble of pixels:

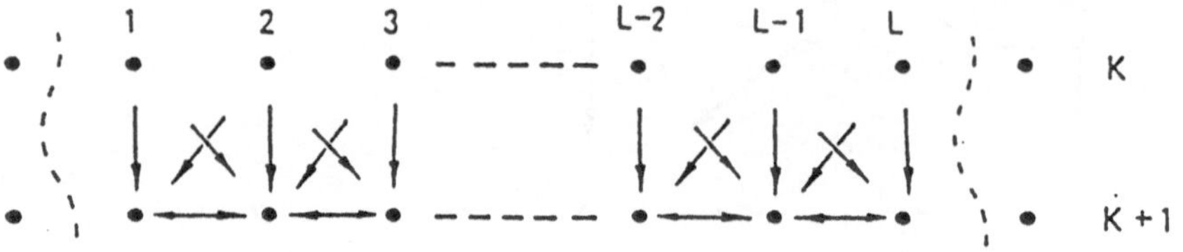


*Fig. 2*

From (3.2)-(3.6), taking into account the scheme of Fig.2 and defining the following vectors and matrices:

$$\tilde{X}_{i+1,j+1} = \begin{pmatrix} X_{i+1,1} \\ \cdot \\ \cdot \\ \cdot \\ X_{i+1,L} \end{pmatrix} ; \quad Z_i = \begin{pmatrix} \frac{1}{3}(W^{(3)}_{i+1,1} + W^{(4)}_{i+1,1} + W^{(5)}_{i+1,1}) \\ \frac{1}{5}(W^{(1)}_{i+1,2} + W^{(2)}_{i+1,2} + W^{(3)}_{i+1,2} + W^{(4)}_{i+1,2} + W^{(5)}_{i+1,2}) \\ \vdots \\ \frac{1}{5}(W^{(1)}_{i+1,L-1} + W^{(2)}_{i+1,L-1} + W^{(3)}_{i+1,L-1} + W^{(4)}_{i+1,L-1} + W^{(5)}_{i+1,L-1}) \\ \frac{1}{3}(W^{(1)}_{i+1,L} + W^{(2)}_{i+1,L} + W^{(3)}_{i+1,L}) \end{pmatrix} \tag{3.12}$$

$$F = \begin{pmatrix} 0 & \frac{H_s(-\Delta_s)}{3} & 0 & 0 & \dots & 0 \\ \frac{H_s(\Delta_s)}{5} & 0 & \frac{H_s(-\Delta_s)}{5} & 0 & \dots & 0 \\ 0 & \frac{H_s(\Delta_s)}{5} & 0 & \frac{H_s(-\Delta_s)}{5} & \dots & 0 \\ \vdots & \ddots & \ddots & \ddots & \ddots & \vdots \\ 0 & \dots & 0 & \frac{H_s(\Delta_s)}{5} & 0 & \frac{H_s(-\Delta_s)}{5} \\ 0 & \dots & 0 & 0 & \frac{H_s(\Delta_s)}{3} & 0 \end{pmatrix}$$

$$E = \begin{pmatrix} \frac{H_r(\Delta_r)}{3} & \frac{H(-\Delta_s,\Delta_r)}{3} & 0 & 0 & \dots & 0 \\ \frac{H(\Delta_s,\Delta_r)}{5} & \frac{H_r(\Delta_r)}{5} & \frac{H(-\Delta_s,\Delta_r)}{5} & 0 & \dots & 0 \\ \vdots & \ddots & \ddots & \ddots & \ddots & \vdots \\ 0 & \dots & 0 & \frac{H(\Delta_s,\Delta_r)}{5} & \frac{H_r(\Delta_r)}{5} & \frac{H(-\Delta_s,\Delta_r)}{5} \\ 0 & \dots & 0 & 0 & \frac{H(\Delta_s,\Delta_r)}{3} & \frac{H_r(\Delta_r)}{3} \end{pmatrix}$$

we get:

$$\tilde{X}_{i+1} = F\tilde{X}_{i+1} + E\tilde{X}_i + Z_i$$

or equivalently:

$$\tilde{X}_{i+1} = [I - F]^{-1}E\tilde{X}_i + [I - F]^{-1}Z_i \tag{3.13}$$

The particular structure of the band matrix $[I-F]$ has allowed to define a fast procedure for computing its exact inverse [15].
Taking into account the (3.1) the following measure equation can be associated to the eqn. (3.13)

$$\tilde{Y}_{i+1} = C\tilde{X}_{i+1} + V_{i+1} \tag{3.14}$$

where

$$\tilde{Y}_{i+1} = \begin{pmatrix} Y_{i+1,1} \\ \vdots \\ Y_{i+1,L} \end{pmatrix} ; C = \begin{pmatrix} C' & 0 & \dots & 0 \\ 0 & C' & \dots & 0 \\ \vdots & & & \\ 0 & 0 & \dots & C' \end{pmatrix} ; V_{i+1} = \begin{pmatrix} v_{i+1,1} \\ \vdots \\ v_{i+1,L} \end{pmatrix}$$

$$C' = (1 \quad 0 \quad \dots \quad 0)$$

## 4.MODELLING OF THE INPUT NOISE.

The eqns. (3.13) and (3.14) have a form amenable to the Kalman filter implementation as soon as it is proved that $Z_i$ is a white noise sequence . To this purpose we observe that owing to the stochastic assumption the two vectors $W_r(r,s)$ and $W_s(r,s)$ can be modelled as gaussian random fields with the following properties:

$$E[W_s(r,s)] = 0 \; ; \; E[W_s(r,s)W_s^T(\bar{r},\bar{s})] = \psi_s\delta(\|(r,s) - (\bar{r},\bar{s})\|) \tag{4.1}$$

$$E[W_r(r,s)] = 0 \; ; \; E[W_r(r,s)W_r^T(\bar{r},\bar{s})] = \psi_r\delta(\|(r,s) - (\bar{r},\bar{s})\|) \tag{4.2}$$

$$E[W_r(r,s)W_s^T(\bar{r}.\bar{s})] = \psi_{r,s}\delta(\|(r,s) - (\bar{r},\bar{s})\|) \tag{4.3}$$

where $\psi_s$ and $\psi_r$ are diagonal matrices.
from (3.2)-(3.6) it follows :

$$W_{i,j}^{(5)} = -H_s(-\Delta_s)W_{i,j+1}^{(1)} \tag{4.4}$$

$$W_{i,j}^{(2)} = H_s(\Delta_s)W_{i,j-1}^{(3)} + W_{i,j}^{(1)} \tag{4.5}$$

$$W_{i,j}^{(4)} = H_s(-\Delta_s)W_{i,j+1}^{(3)} + H_s(-\Delta_s)W_{i,j+1}^{(1)} \tag{4.6}$$

Substituting the (4.4) (4.6) into (3.12) and taking into account (4.1)-(4.3) it can be verified that

$$E[Z_iZ_j^T] = Q_z\delta_{i,j}$$

so that $Z_i$ is a white sequence distributed as $\sim \mathcal{N}(0,Q_z)$ . Moreover from (3.7) and (4.1) it results :

$$E[W_{i,j}^{(1)}W_{l,m}^{(1)T}] = \int_0^{\Delta_s}\int_0^{\Delta_s} H_s(\Delta_s - \tau)B\psi_sB^TH_s^T(\Delta_s - \theta)\delta(|i-l|\Delta_r + |(j-m)\Delta_s + (\tau - \theta)|)d\theta d\tau$$

$$= \delta_{i,l}\delta_{j,m}Q_s$$

where

$$Q_s = \int_0^{\Delta_s} H_s(\Delta_s - \tau)B\psi_s B^T H_s^T(\Delta_s - \tau)d\tau$$

In the same way from (3.9) and (4.2) it results:

$$E[W_{i,j}^{(3)} W_{l,m}^{(3)^T}] = \delta_{i,l}\delta_{j,m}Q_r$$

where

$$Q_r = \int_0^{\Delta_r} H_r(\Delta_r - \tau)B\psi_r B^T H_r^T(\Delta_r - \tau)d\tau$$

and finally

$$E[W_{i,j}^{(1)} W_{l,m}^{(3)^T}] =$$

$$\int_0^{\Delta_s}\int_0^{\Delta_r} H_s(\Delta_s - \tau)B\psi_{r,s}B^T H_r^T(\Delta_r - \theta)\delta(|(i-l)\Delta_r + (\Delta_r - \theta)| + |(j-m)\Delta_s + (\Delta_s - \tau)|)d\theta d\tau = 0$$

In conclusion after a simple calculation the following covariance matrix is found $Q_s =$

$$\begin{pmatrix}
\frac{1}{9}S_1 & \frac{1}{15}T & \frac{1}{15}U & 0 & 0 & 0 & \dots & 0 \\
\frac{1}{15}T^T & \frac{1}{25}(S_1+S_2) & \frac{1}{25}T & \frac{1}{25}U & 0 & 0 & \dots & 0 \\
\frac{1}{15}U^T & \frac{1}{25}T^T & \frac{1}{25}(S_1+S_2) & \frac{1}{25}T & \frac{1}{25}U & 0 & \dots & 0 \\
0 & \frac{1}{25}U^T & \frac{1}{25}T^T & \frac{1}{25}(S_1+S_2) & \frac{1}{25}T & \frac{1}{25}U & \dots & 0 \\
\vdots & \ddots & \ddots & \ddots & \ddots & \ddots & \ddots & \vdots \\
0 & \dots & 0 & \frac{1}{25}U^T & \frac{1}{25}T^T & \frac{1}{25}(S_1+S_2) & \frac{1}{25}T & \frac{1}{15}U \\
0 & \dots & 0 & 0 & \frac{1}{25}U^T & \frac{1}{25}T^T & \frac{1}{25}(S_1+S_2) & \frac{1}{15}T \\
0 & \dots & 0 & 0 & 0 & \frac{1}{15}U^T & \frac{1}{15}T^T & \frac{1}{9}S_3
\end{pmatrix}$$

where

$$S_1 = [Q_r + H_s(-\Delta_s)Q_r H_s^T(-\Delta_s) + 4H_s(-\Delta_s)QH_s^T(-\Delta_s)]$$

$$S_2 = [4Q_s + H_s(\Delta_s)Q_r H_s^T(\Delta_s)];\ S_3 = [4Q_s + H_s(\Delta_s)Q_r H^T(\Delta_s) + Q_r]$$

$$T = [Q_r H_s^T(\Delta_s) + H_s(-\Delta_s)Q_r - 4H_s(-\Delta_s)Q_s];\ U = H_s(-\Delta_s)Q_r H_s^T(\Delta_s)$$

Under the reasonable hypothesis of image with finite limited and isotropic spectrum it is possible to get precise estimates of $\psi_s$ and $\psi_r$ [15].

### 5. NUMERICAL RESULTS.

Two different kinds of images have been considered to test the performance and the applicability of the algorithm: A) homogeneous field (64 × 64) generated by:

$$x_{i,j} = A_0 + \sum_{l=1}^{4} A_l cos(l\omega_0((i-1)\Delta + (j-1)\Delta)$$

where

$$A_0 = 5;\ A_1 = 10;\ A_2 = 9;\ A_3 = 8;\ A_4 = 7;\ \omega_0 = 0.4\pi;\ \Delta = \frac{5}{63}$$

B) simulated eight-bit image of 256 × 256 pixels consisting of concentric rings as shown in Fig.3. The bidimensional field has been generated to simulate a really homogeneous image with finite ,limited and isotropic spectrum.The synthesized image has been chosen because it is provided with sharp edges.The filter is adapted to their presence by introducing a linear edge detector operator $h_{i,j}$. Information concerning the size and the location of the edges is obtained by means of a discrete convolution of $y_{i,j}$ with $h_{i,j}$ [16]. This information is used to control the input of the filter in order to improve its step response as discussed in [15].

The originals have been corrupted by a zero-mean white gaussian noise with a variance such that $S.N.R.(signal variance/noise\ variance) = 1$. According to the image representation of equations (3.13),(3.14) the Kalman filter has been implemented as a strip filtering processor [5] by partitioning the image into parallel strips of width L. The Riccati equation has been implemented off line starting from an initial value of the error covariance matrix equal to $[I - F]^{-1}Q_x[I - F^T]^{-1}$ and the steady state gain has been used to process real data.The homogeneous field has been processed with three models of increasing order corresponding to the assumption $\bar{n} = 0, 1, 2$ respectively, while $\bar{n} = 0$ has been chosen for the image B in that it can be considered a piecewise constant image. Numerical results are summarized in the following table and figures.

| | $\bar{n} = 0$ | $\bar{n} = 1$ | $\bar{n} = 2$ |
|---|---|---|---|
| S.N.R.I. | 6.45 db | 7.96 db | 8.39 db |
| iteration number for filter stabilization | 17 | 62 | 148 |

TABLE 5.1 Results relative to the homogeneous field.

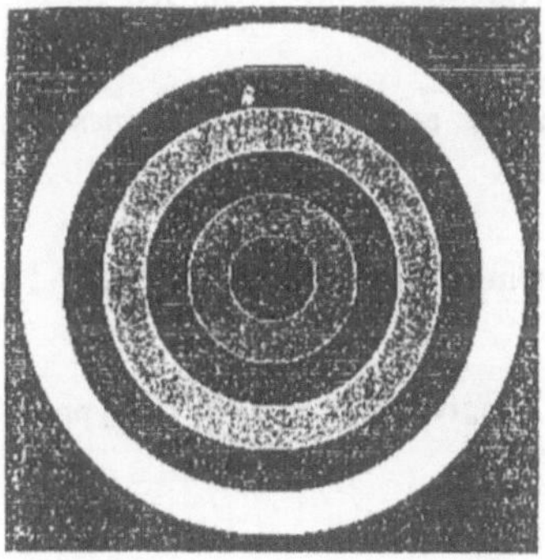

*Fig.3*
*Original Image B*

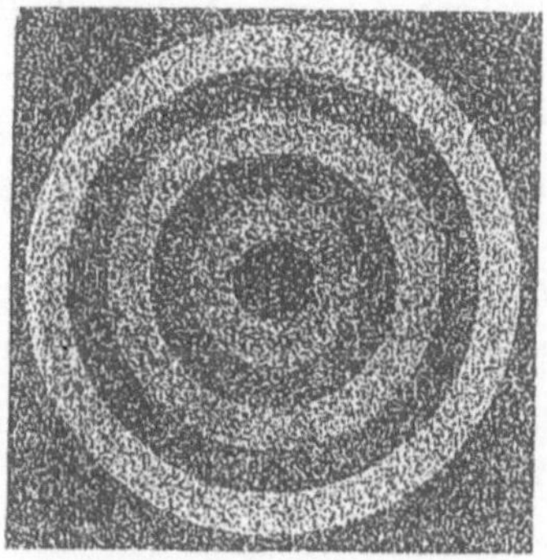

*Fig.4*
*Noisy Image B*

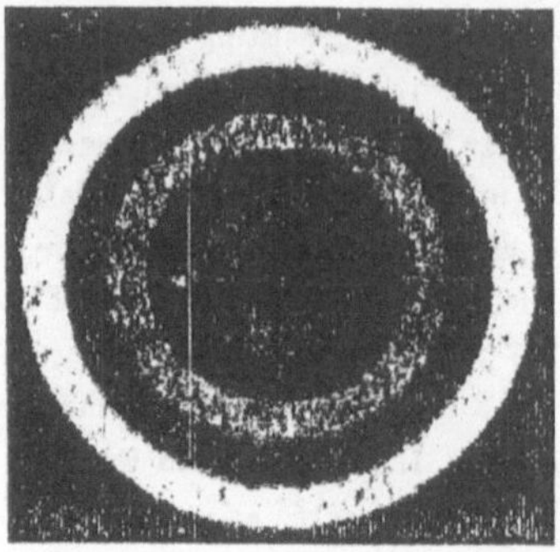

*Fig. 5*
*Filtered Image B*
*S. N. R. I. = 10.17 db*

## References


1. N.E. NAHI, Role of recursive estimation in statistical image enhancement, *Proc. IEEE* Vol. 60, pp. 872-877, July 1972.

2. N.E. NAHI and T. ASSEFI, Bayesan recursive image estimation. *IEEE Trans. on Comp.* Vol. C-21, pp. 734-738, July 1972.

3. S.R. POWELL and L.M. SILVERMAN, Modeling of two-dimensional covariance functions with applications to image enhancement, *IEEE Trans. Automat. Contr.*, Vol. AC-19, pp. 8-13, Feb. 1974.

4. A.K. JAIN and E. ANGEL, Image rstoration, modeling, and reduction of dimensionality, *IEEE Trans. Comput.* Vol. C-23, pp. 470-476, May 1974.

5. J.W. WOODS and C.H. RADEWAN, Kalman filtering in two dimensions, *IEEE Trans Inform. Theory* Vol. IT-23, pp. 473-482, July 1977.

6. A. HABIBI, Two dimensional Bayesian estimate of images, *Proc. IEEE* Vol. 60, pp. 878-883, July 1972.

7. A.K. JAIN, Partial differential equation and finite difference methods in image processing. Part I: image representation *JOTA* Vol. 23, pp. 65-91, Sept 1977.

8. T. KATAYAMA and M. KOSAKA, Smoothing algorithms for two-dimensional image processing, *IEEE Trans. on Syst. Man and Cybern.* Vol. SMC-8, pp. 62-66, Jan. 1978.

9. T. KATAYAMA and M. KOSAKA, Recursive filtering algorithm for a two-dimensional system, *IEEE Trans. Automat. Contr.* Vol. AC-24, pp. 130-132, Feb. 1979.

10. T. KATAYAMA, Restoration of noise images using a two-dimensional linear model, *IEEE on Syst. Man and Cybern.* Vol. SMC-9, pp. 711-717, Nov. 1979.

11. H.R. KESHAWAN and M.D. SRINATH, Enhancement of noisy images using an interpolative model in two dimension, *IEEE on Syst. Man and Cybern.* Vol. SMC-8, pp. 247-258, April 1978.

12. H. KAUFMAN, J.W. WOODS, S. DRAVIDA and A.M. TEKALP, Estimation and identification of two dimensional images, *IEEE Trans. on Automat. Control* Vol. AC-28, pp. 745-756, July 1983.

13. P.E. WELLSTEAD and J.R. CALDAS PINTO, Self-tuning filters and predictors for two-dimensional system. Part I: Algorithms, *Int. J. Control* Vol. 42, pp. 457-478, 1985.

14. P.E. WELLSTEAD and J.R. CALDAS PINTO, Self-tuning filters and predictors for two dimensional systems. Part II: Smoothing application.

15. A. GERMANI and L. JETTO, Image modelling and restoration: a new approach. Rpt.IASI-CNR, submitted.

16. J. BIEMOND and J.J. GERBRANDS, An edge-preserving recursive noise-smoothing algoritm for image data, *IEEE Trans on Syst. Man and Cybern.* Vol. SMC-9, pp. 622-627, Octob. 1979.

# HIGHER ORDER SENSITIVITY OF SOLUTIONS TO CONVEX PROGRAMMING PROBLEMS WITHOUT STRICT COMPLEMENTARITY *)

Kazimierz Malanowski

Systems Research Institute
Polish Academy of Sciences
ul. Newelska 6, 01 - 447 Warszawa

Abstract

We consider a family of convex programming problems that depend on a vector parameter. It is shown that the solutions of the problems and the associated Lagrange multipliers are arbitrarily many times directionally differentiable functions of the parameter, provided that the data of the problems are sufficiently regular. The characterizations of the respective derivatives are given.

Key words: Parametric Convex Programming, Sensitivity Analysis, Higher--Order Directional Differentiability of Solutions, Lipschitz Continuity of Derivatives.

## 1. Introduction

Most papers concerning sensitivity analysis of mathematical programming problems deal with the so called optimal - value function, which to every value of the parameter assigns the corresponding optimal value of the cost function. Differential properties of the optimal-value function has been analysed, e.g., in [4, 5, 11].

Differentiability of solutions to such problems was proved under much stronger assumptions, among which strict complementarity plays a crucial role (see [3]).

Without the strict complementarity assumption, only directional differentiability with respect to parameters, can be proved. The research in this direction was initiated by J.H. Bigelow and N.Z. Shapiro [1] and completed by K. Jittorntrum [7], who proved that, if mathematical programming problems depending on a parameter satisfy the strong se-

---

*) This paper was completed when the author visited the Laboratory for Flight Systems Research of UCLA and it was supported in part under NASA grant NAG 2-303.

cond order sufficient conditions, as well as some constraints regularity, then their solutions are directionally-differentiable functions and the directional derivatives are given by the solutions of some auxiliary quadratic programming problems.

The same result was obtained by A. Shapiro [11] as a corollary to his second order sensitivity analysis of the optimal value function.

One has to mention also the paper by R.S. Dembo [2], who analysed sensitivity of solutions to geometric programming problems, including higher-order differentiability.

In this paper we consider a family of strongly convex programming problems whose data depend on a vector parameter. It is assumed that the gradients of the binding constraints are linearly independent.

We give a characterization of the first-order directional derivatives w.r.t. the parameter of the solutions and associated Lagrange multipliers of these problems. This characterization is slightly different than that of Jittorntrum [7].

This characterization is more complicated than that in [7]. However, it enables us to show that, for a fixed direction, the solutions and the associated Lagrange multipliers are arbitrarily many times directionally differentiable functions of the parameter, provided that the data are sufficiently regular.

The directional derivatives are given as the solution and the associated multipliers of an auxiliary quadratic problem of optimization subject to linear, equality-type constraints.

The explicit form of this quadratic problem is derived in the case of the second-order derivative and can be easily constructed for higher-order ones.

As far as the author knows the results similar to those presented here are contained only in [10], where a slightly different approach was used.

## 2. Problem statement

Let $H \subset R^q$ be an open and convex set of parameters. We consider a family $\{P_h\}$ of the following convex programming problems $(P_h)$ depending on $h \in H$:

$(P_h)$ find $u(h) \in \Phi_h$ such that

$$f(u(h),h) = \min_{u \in \Phi_h} f(u,h), \tag{2.1}$$

where

$$\Phi_h = \{u \in R^n \mid \phi^i(u,h) \leqslant 0,\ i \in I\}, \tag{2.2}$$
$$I := \{1,2,\ldots,m\}.$$

We denote

$$\phi(u,h) = [\phi^1(u,h),\phi^2(u,h),\ldots,\phi^m(u,h)]^T.$$

Assume that for some integer $p \geqslant 1$, the following conditions hold:

(i) $f(.,.)$ is a $(p+1)$-times continuously differentiable function on $R^n \times H$,

(ii) $f(.,h)$ is strongly convex uniformly with respect to $h \in H$, i.e. there exists a constant $\beta > 0$, independent of $h$ and such that

$$\langle v, D^2_{uu} f(u,h) v \rangle \geqslant \beta \|v\|^2, \quad \forall u,v \in R^n, \quad \forall h \in H, \tag{2.3}$$

(iii) $\phi^i(.,.)$, $i \in I$, are $(p+1)$-times continuously differentiable functions on $R^n \times H$,

(iv) for each $h \in H$, $\phi^i(.,h)$, $i \in I$, are convex functions of $u$,

(v) for each $h \in H$, the admissible set $\Phi_h$ is non-empty:

$$\Phi_h \neq \emptyset. \tag{2.4}$$

It is well known that under (ii), (iv) and (v) Problem $(P_h)$ has a unique solution $u(h)$.

Let us denote by

$$I^o_h = \{i \in I \mid \phi^i(u(h),h) = 0\} \tag{2.5}$$

the set of indices of all the constraint functions binding at $u(h)$.

In addition to (i) through (v) we assume that at the points $u(h)$ the following constraints regularity conditions hold:

(vi) there exists a constant $\gamma > 0$ such that

$$\| [D_u\phi^i(u(h),h)]_{I^o_h} v \| > \gamma \|v\| \tag{2.6}$$

for every $h \in H$ and every $v$ of appropriate dimension. Here $[D_u\phi^i(u(h),h)]_{I^o_h}$ denotes the matrix whose columns are the gradients of all constraint functions binding at $u(h)$.

Let us define the usual Lagrangian associated with $(P_h)$:

$$\mathcal{L}(.,.,.) : R^n \times R^m \times H \to R^1,$$

$$\mathcal{L}(u,\lambda,h):=f(u,h)+\sum_{i=1}^{m}\lambda^i\phi^i(u,h). \tag{2.7}$$

Note that by (vi) the set $\Phi_h$ has a nonempty interior and there exists a uniquely defined Lagrange multiplier $\lambda(h)$ such that $u(h)$ is characterized by Kuhn-Tucker conditions:

$$D_u\mathcal{L}(u(h),\lambda(h),h)=D_uf(u(h),h)+\sum_{i\in I}\lambda^i(h)D_u\phi^i(u(h),h)=0, \tag{2.8}$$

$$\lambda^i(h)\phi^i(u(h),h)=0,\quad \lambda^i(h)\geqslant 0,\quad i\in I. \tag{2.8a}$$

The following important result was obtained by W.W. Hager [6]:

Lemma 2.1

If conditions (i) through (vi) are satisfied with p=1, then for every compact and convex set $\mathcal{H}\subset H$ there exists a constant $c>0$ such that for every $h_1,h_2\in\mathcal{H}$,

$$\|u(h_2)-u(h_1)\|\leqslant c\|h_2-h_1\|, \tag{2.9a}$$

$$\|\lambda(h_2)-\lambda(h_1)\|\leqslant c\|h_2-h_1\|. \tag{2.9b}$$

We intend to show that, for any fixed $h\in H$ and any fixed direction $g\in R^q$, $u(h+\alpha g)$ and $\lambda(h+\alpha g)$ are p-times directionally differentiable functions of $\alpha$, i.e., for any $\alpha>0$

$$u(h+\alpha g)=u(h)+\alpha u^{(1)}(h,g)+\frac{\alpha^2}{2}u^{(2)}(h,g)+\dots\dots+\frac{\alpha^p}{p!}u^{(p)}(h,g)+o(\alpha^p), \tag{2.10a}$$

$$\lambda(h+\alpha g)=\lambda(h)+\alpha\lambda^{(1)}(h,g)+\frac{\alpha^2}{2}\lambda^{(2)}(h,g)+\dots\dots+\frac{\alpha^p}{p!}\lambda^{(p)}(h,g)+o(\alpha^p) \tag{2.10b}$$

where $\frac{o(\alpha^p)}{\alpha^p}\xrightarrow[\alpha\downarrow 0]{}0.$

The directional derivatives $u^{(j)}(h,g)$ and $\lambda^{(j)}(h,g)$, $j=1,\dots,p$ are characterized as the solutions and the associated Lagrange multipliers of some auxiliary quadratic programming problems. Our considerations are restricted to the first and the second order derivatives. Higher-order differentiability can be analysed in exactly the same way.

## 3. First-order directional differentiability

It was proved by K. Jittorntrum [7] that under the assumptions (i) through (vi), with p=1, for any fixed $h\in H$ and $g\in R^q$, $u(h+\alpha g)$ and $\lambda(h+\alpha g)$ are directionally differentiable functions of $\alpha$.

The directional derivatives

$$u^{(1)}(h,g) := \lim_{\alpha\downarrow 0} \{ \frac{1}{\alpha}[u(h+\alpha g) - u(h)]\} \tag{3.1a}$$

$$\lambda^{(1)}(h,g) := \lim_{\alpha\downarrow 0} \{ \frac{1}{\alpha}[\lambda(h+\alpha g) - \lambda(h)]\} \tag{3.1b}$$

are characterized in [7] (see also [8]) respectively as the solution and the associated Lagrange multiplier of an auxiliary quadratic programming problem.

Below we are going to give an equivalent but slightly different characterization. To this end, for a given h, define the following sets of indices :

$$J_h^o = \{i \in I_h^o \mid \lambda^i(h) > 0\}, \tag{3.2a}$$

$$K_h^o = I \setminus I_h^o \tag{3.2b}$$

as well as the sets

$$L_{h,g}^o = \{i \in I_h^o \setminus J_h^o \mid \exists \alpha^i > 0,\ \forall \alpha \in (0,\alpha^i),\ \lambda^i(h+\alpha g) > 0\}, \tag{3.3a}$$

$$M_{h,g}^o = \{i \in I_h^o \setminus J_h^o \mid \exists \beta^i > 0,\ \forall \beta \in (0,\beta^i),\ \phi^i(u(h+\beta g), h+\beta g) < 0\}, \tag{3.3b}$$

$$N_{h,g}^o = (I_h^o \setminus J_h^o) \setminus (L_{h,g}^o \cup M_{h,g}^o) \tag{3.3c}$$

depending both on the point h and the direction g.

We need the following auxiliary results:

<u>Lemma 3.1</u>

If $i \in N_{h,g}^o$, then there exists a sequence $\{\bar{\alpha}_n\}\downarrow 0$ such that

$$\phi^i(u(h+\bar{\alpha}_n g), h+\bar{\alpha}_n g) = 0,\ \lambda^i(h+\bar{\alpha}_n g) = 0 \quad \forall \bar{\alpha}_n \in \{\bar{\alpha}_n\}$$

<u>Proof</u>

By definition (3.3c) if $i \in N_{h,g}^o$, then for any $\alpha_n > 0$ there exist $\alpha_n', \alpha_n'' \in (0,\alpha_n)$ such that $\phi^i(u(h+\alpha_n' g), h+\alpha_n' g) = 0$ and $\lambda^i(h+\alpha_n'' g) = 0$. Without any loss of generality, we can assume that $\alpha_n'' \leq \alpha_n'$. If $\alpha_n' = \alpha_n''$, then we put $\bar{\alpha}_n = \alpha_n'$. Suppose that $\phi^i(u(h+\alpha_n'' g), h+\alpha_n'' g) < 0$, i.e. $\alpha_n'' < \alpha_n'$. By continuity of $u(h+\alpha g)$ with respect to $\alpha$ (Lemma 2.1), as well as by (iii) and in view of $\phi^i(u(h+\alpha_n' g), h+\alpha_n' g) = 0$, we find that there exists

$$\tilde{\alpha}_n = \min\{\alpha \in (\alpha_n'', \alpha_n'] \mid \phi^i(u(h+\alpha g), h+\alpha g) = 0\}$$

Since by (2.8a) we have $\lambda^i(h+\alpha g)=0$ for all $\alpha \in (\alpha_n'' \tilde{\alpha}_n)$, then by continuity of $\lambda^i(h+\alpha g)$ with respect to $\alpha$, we get $\lambda^i(h+\tilde{\alpha}_n g)=0$. Hence we can choose $\bar{\alpha}_n = \tilde{\alpha}_n$.

□

Lemma 3.2

Let $\ell(\cdot)$ be a real-valued Lipschitz continuous function. If there exist two different numbers $k_1$ and $k_2$ such that, for some sequences $\{\alpha_n\}\downarrow 0$ and $\{\beta_n\}\downarrow 0$,

$$\ell(\alpha + \alpha_n) = \ell(\alpha) + \alpha_n k_1 + o(\alpha_n),$$

$$\ell(\alpha + \beta_n) = \ell(\alpha) + \beta_n k_2 + o(\beta_n),$$

then, for any $k \in (k_1, k_2)$, there exists a sequence $\{\gamma_n\}\downarrow 0$ such that

$$\ell(\alpha + \gamma_n) = \ell(\alpha) + \gamma_n k. \tag{3.4}$$

Proof

Suppose that $k_1 > k_2$. Then there exists an index $\bar{n}$ such that

$$k_1 + \frac{o(\alpha_n)}{\alpha_n} > k$$

and

$$k_2 + \frac{o(\beta_n)}{\beta_n} < k \qquad \text{for all} \quad n > \bar{n}.$$

Hence

$$\ell(\alpha + \alpha_n) > \ell(\alpha) + \alpha_n k$$

$$\ell(\alpha + \beta_n) < \ell(\alpha) + \beta_n k$$

i.e.

$$\frac{1}{\beta_n}[\ell(\alpha+\beta_n)-\ell(\alpha)] < k < \frac{1}{\alpha_n}[\ell(\alpha+\alpha_n)-\ell(\alpha)] \quad \text{for all} \quad n > \bar{n}.$$

By continuity of the function $\frac{1}{\beta}[\ell(\alpha+\beta)-\ell(\alpha)]$ with respect to $\beta$ on any interval $[\beta_n, \alpha_n]$, where $\beta_n, \alpha_n > 0$, we find that there exists $\gamma_n \in [\beta_n, \alpha_n]$ such that (3.4) holds. We complete the proof by noting that $\{\alpha_n\}\downarrow 0$, $\{\beta_n\}\downarrow 0$.

□

We are now in a position to prove the following theorem:

Theorem 3.1

If conditions (i)-(vi) are satisfied with $p=1$, then, for any fixed $h \in H$ and $g \in R^q$, $u(h+\alpha g)$ and $\lambda(h+\alpha g)$ are directionally differentiable functions of $\alpha$ and the directional derivatives $u^{(1)}(h,g)=v(h,g)$, $\lambda^{(1)}(h,g)=\mu(h,g)$ are given as the solution and the associated multiplier of the following quadratic problem of optimization, subject to linear, equality-type constraints:

$(QP^1_{h,g})$ find $v(h,g)\in X_{h,g}$ such that

$$k^1(v(h,g),h,g) = \min_{v\in X_{h,g}} k^1(v,h,g) \tag{3.5}$$

where

$$k^1(v,h,g) = \frac{1}{2} < v, Q(h)v > + < q^1(h,g), v >, \tag{3.6}$$

$$Q(h) = D^2_{uu} f(u(h),h) + \sum_{i\in I} \lambda^i(h) D^2_{uu}\phi^i(u(h),h), \tag{3.6a}$$

$$q^1(h,g) = [D^2_{uh} f(u(h),h) + \sum_{i\in I} \lambda^i(h) D^2_{uh}\phi^i(u(h),h)]g, \tag{3.6b}$$

$$X_{h,g} = \{v\in R^n \mid \chi^i(v,h,g) := < D_u\phi^i(u(h),h), v > + \\ + < D_h\phi^i(u(h),h), g > = 0, \quad i\in J^o_h\cup L^o_{h,g}\}. \tag{3.7}$$

Moreover

$$\mu^i(h,g) = 0 \quad \text{for} \quad i\in I\setminus(J^o_h\cup L^o_{h,g}) = K^o_h\cup M^o_{h,g}\cup N^o_{h,g}. \tag{3.8}$$

Proof

Note that by (ii), (iv) and (2.8a) the matrix $Q(h)$ is positive definite, while by (vi) the set $X_{h,g}$ is non-empty. Hence the problem $(QP^1_{h,g})$ has a unique solution $v(h,g)$, while by (vi) the associated Lagrange multipliers $\mu^i(h,g)$, $i\in J^o_h\cup L^o_{h,g}$, are defined uniquely.

In view of (3.6) and (3.8), a necessary and sufficient condition of optimality can be written

$$\begin{aligned} &Q(h)v(h,g) + q^1(h,g) + \sum_{i\in I}\mu^i(h,g) D_u\phi^i(u(h),h) = \\ &= [D^2_{uu} f(u(h),h) + \sum_{i\in I}\lambda^i(h) D^2_{uu}\phi^i(u(h),h)]v(h,g) + \\ &+ [D^2_{uh} f(u(h),h) + \sum_{i\in I}\lambda^i(h) D^2_{uh}\phi^i(u(h),h)]g + \\ &+ \sum_{i\in I}\mu^i(h,g) D_u\phi^i(u(h),h) = 0. \end{aligned} \tag{3.9}$$

Let $\{\alpha_n\}\downarrow 0$ be an arbitrary sequence. By Lemma 2.1,

$$\left\|\frac{1}{\alpha_n}[u(h+\alpha_n g) - u(h)]\right\| \leqslant c,$$

$$\|\frac{1}{\alpha_n}[\lambda(h+\alpha_n g)-\lambda(h)]\| \leq c.$$

Hence we can extract a subsequence $\{\alpha'_n\} \subset \{\alpha_n\}$ such that

$$\frac{1}{\alpha'_n}[u(h+\alpha'_n g)-u(h)] \to v, \tag{3.10a}$$

$$\frac{1}{\alpha'_n}[\lambda(h+\alpha'_n g)-\lambda(h)] \to \mu, \tag{3.10b}$$

for some $v \in R^n$, $\mu \in R^m$.

We are going to show that the pair $(v,\mu)$ satisfies (3.9) along with (3.7) and (3.8). Hence $v$ is a solution and $\mu$ is an associated Lagrange multiplier of $(QP^1_h)$. Therefore they are unique and independent of the choice of the sequences $\{\alpha_n\}$ and $\{\alpha'_n\}$. This will prove the theorem.

In order to show that $v$ and $\mu$ satisfy (3.9), we take the difference quotient of (2.8) at $h+\alpha'_n g$ and at $h$. Passing to the limit and using (i), (iii) and (3.10), we arrive at (3.9).

To show that $v \in X_{h,g}$ note that, by the definition (3.2b) and by continuity of $\lambda(\cdot)$, for $\alpha_n > 0$ sufficiently small we have $\lambda^i(h+\alpha_n g)>0$ $i \in J^o_h$. Hence by the complementarity slackness (2.8a) we obtain

$$\phi^i(u(h+\alpha_n g), h+\alpha_n g)=0 \qquad i \in J^o_h \tag{3.11}$$

for $\alpha_n > 0$ sufficiently small. This shows that $\chi^i(v,h,g)=0$ for $i \in J^o_h$.

Taking advantage of the definition (3.3a) and using the same argument as above, we find that (3.11) holds also for $i \in L^o_{h,g}$, which completes the proof that $v \in X_{h,g}$.

It remains to prove (3.8). By the definition (3.2b) as well as by (iii) and by the continuity of $u(\cdot)$,

$$\phi^i(u(h+\alpha_n g), h+\alpha_n g) < 0 \qquad i \in K^o_h,$$

for $\alpha_n > 0$ sufficiently small, i.e., we have by the complementarity slackness (2.8a) that

$$\lambda^i(h+\alpha_n g) = 0 \qquad i \in K^o_h \tag{3.12}$$

for $\alpha_n \geq 0$ sufficiently small. Hence (3.8) is satisfied on $K^o_h$.

Taking advantage of the definition (3.3b) and using the same argument as above, we show that (3.12) holds also for $i \in M^o_{h,g}$. We now proceed to consider the set $N^o_{h,g}$. It is easy to see that it is enough to consider the following two cases of sequences $\{\alpha_n\}\downarrow 0$:

(1) $\quad \phi^i(u(h+\alpha_n g), h+\alpha_n g) < 0, \quad \lambda^i(h+\alpha_n g) = 0,$

(2) $\quad \phi^i(u(h+\alpha_n g), h+\alpha_n g) = 0, \quad \lambda^i(h+\alpha_n g) > 0.$

In the case (1), the condition (3.8) is satisfied, while in the case (2) we have $\mu^i \geqslant 0$. If

$$\mu^i = 0, \tag{3.13}$$

then (3.8) holds. However, if

$$\mu^i > 0, \tag{3.14}$$

then the condition (3.8) is violated.

We shall show that (3.14) can not be satisfied for any $i \in N^o_{h,g}$. Assume the oposite, namely that, for some $j \in N^o_{h,g}$, there exist two different sequences $\{\check{\alpha}_n\}\downarrow 0$ and $\{\hat{\alpha}_n\}\downarrow 0$ for which (3.13) and (3.14) are satisfied respectively.

In the case of (3.13) v is given by the solution of $(QP^1_{h,g})$, while, it is easy to see, that in the case of (3.14) v is given by the unique solution of the problem $(\widehat{QP}^1_{h,g})$, which is formulated as $(QP^1_{h,g})$ with the exception that the equality type constraints in (3.7) hold for the set of indices $J^o_h \cup L^o_{h,g} \cup \{j\}$. By (vi) the admissible set for $(\widehat{QP}^1_{h,g})$ is non-empty, hence this problem has a unique solution $\hat{v}$. The associated Lagrange multipliers are unique by (vi).

Note that for all sequences $\{\alpha_n\}\downarrow 0$ any cluster point of $\{\frac{1}{\alpha_n}[\lambda^i(h+\alpha_n g) - \lambda^i(h)]\}$ must satisfy either (3.13) or (3.14). Hence, if there exist sequences $\{\check{\alpha}_n\}$ and $\{\hat{\alpha}_n\}$, then the sequences $\{\frac{1}{\alpha_n}[\lambda^i(h+\alpha_n g) - \lambda^i(h)]\}$ would have two different isolated cluster points. This is impossible by Lemma 3.2, since $\lambda^i(h+\alpha g)$ is a Lipschitz continuous function of $\alpha$. Hence, the limits in (3.10) must be the same for all sequences $\{\alpha_n\}\downarrow 0$.

By Lemma 3.1, there is a sequence $\{\bar{\alpha}_n\}$ for which (3.13) holds. This completes the proof of the theorem.

□

Note that in view of (3.8) the optimality conditions for $(Q^1_{h,g})$ can be expressed, in terms of the Lagrangian (2.7), as follows:

$$D^2_{uu}\mathcal{L}(u(h),\lambda(h),h)v(h,g) + D^2_{u\lambda}\mathcal{L}(u(h),\lambda(h),h)\mu(h,g) + D^2_{uh}\mathcal{L}(u(h),\lambda(h),h)g = 0,$$

$$(D^2_{\lambda u}\mathcal{L}(u(h),\lambda(h),h)v(h,g) + D^2_{\lambda h}\mathcal{L}(u(h),\lambda(h),h)g,\mu(h,g)) = 0. \quad (3.15)$$

From the proof of Theorem 3.1 we obtain immediately:

Corollary 3.1

We have:

$$\chi^i(v(h,g),h,g) \leqslant 0 \quad \text{for} \quad i \in I^o_h \setminus J^o_h, \quad (3.16a)$$

$$\mu^i(h,g) \geqslant 0 \quad \text{for} \quad i \in I^o_h \setminus J^o_h, \quad (3.16b)$$

$$\chi^i(v(h,g),h,g) = 0 \quad \text{for} \quad i \in N^o_{h,g}. \quad (3.16c)$$

Remark 3.1

In view of (3.16a) and (3.16b), $v(h,g)$ can be interpreted as the solution to the following quadratic programming problem:

$(\widetilde{QP}^1_{h,g})$ find $v(h,g) \in \tilde{X}_{h,g}$ such that

$$k^1(v(h,g),h,g) = \min_{v \in \tilde{X}_{h,g}} k^1(v,h,g),$$

where

$$\tilde{X}_{h,g} = \{v \in R^n \mid \chi^i(v,h,g) \begin{cases} = 0 & \text{for} \quad i \in J^o_h \\ \leqslant 0 & \text{for} \quad i \in I^o_h \setminus J^o_h \end{cases}\}$$

and $k^1(v,h,g)$, $\chi^i(v,h,g)$ are defined as in $(QP^1_{h,g})$. Characterization of $u^{(1)}(h,g)$ as the solution to $(\widetilde{QP}^1_{h,g})$ was given by Jittorntrum [7]. The characterization by $(\widetilde{QP}^1_{h,g})$ is simpler since it requires that the sets $I^o_{h+\alpha g}$ and $J^o_{h+\alpha g}$ be known only at the point $\alpha = 0$, while in the case of $(QP^1_{h,g})$ the set must be known on some interval $[0,\bar{\alpha})$. On the other hand Theorem 3.1 gives an explicit characterization of the set of those indices $i \in I$ for which the constraint functions $\chi^i(v,h,g)$ are binding at $v(h,g)$. This last property enables us to obtain, in the next section, certain forms of the higher-order directional derivatives.

## 4. Higher-order directional differentiability

In this section we show that the results of Theorem 3.1 can be extended to higher-order directional differentiability of the solutions to convex programming problems, provided that the data of these problems are sufficiently regular.

We restrict ourselves to second – order directional derivatives, since the higher-order differentiability can be analysed in exactly the same way.

We are going to prove the existence and to find the form of

$$w(h,g) := u^{(2)}(h,g) = \lim_{\alpha \downarrow 0} \frac{1}{\alpha} [\, v(h+\alpha g,g) - v(h,g)] , \tag{4.1a}$$

$$\nu(h,g) := \lambda^{(2)}(h,g) = \lim_{\alpha \downarrow 0} \frac{1}{\alpha} [\, \mu(h+\alpha g,g) - \mu(h,g)] . \tag{4.1b}$$

We shall repeat the outline of the proof of Theorem 3.1.

<u>Proposition 4.1</u>

If conditions (i)-(vi) are satisfied with $p=2$, then for every compact and convex set $\mathcal{H} \subset H$ there exists a constant $c > 0$ such that for every $h, h+\alpha g \in \mathcal{H}$, with $\alpha > 0$,

$$\| v(h+\alpha g,g) - v(h,g) \| < c\alpha , \tag{4.2a}$$

$$\| \mu(h+\alpha g,g) - \mu(h,g) \| < c\alpha . \tag{4.2b}$$

<u>Proof</u>

We apply Hager's abstract result concerning Lipschitz continuity of solutions to constrained processes. By Theorem 2.1 in [6], in order to prove (4.2) it is enough to show that:

1) $v(h+\alpha g,g)$ and $\mu(h+\alpha g,g)$ are continuous functions of $\alpha$,
2) these functions are Lipschitz continuous, with the same Lipschitz modulus, for all pairs $(h,h+\alpha g) \in \mathcal{H} \times \mathcal{H}$ such that

$$\{i \in I_h^o \mid \chi^i(v(h,g),h,g)=0\} = \{i \in I_{h+\alpha g}^o \mid \chi^i(v(h+\alpha g,g),h+\alpha g,g)=0\}.$$

We shall prove that $v(h+\alpha g,g)$ and $\mu(h+\alpha g,g)$ are locally Lipschitz continuous functions of $\alpha$, which of course implies 1).

Indeed, from the definitions (3.2) and (3.3) it follows that, for $\alpha > 0$ sufficiently small,

$$J_h^o \cup L_{h,g}^o \subset J_{h+\alpha g}^o \subset I_{h+\alpha g}^o \subset J_h^o \cup L_{h,g}^o \cup N_{h,g}^o .$$

Hence also

$$J_h^o \cup L_{h,g}^o \subset J_{h+\alpha g}^o \cup L_{h+\alpha g,g}^o \subset J_h^o \cup L_{h,g}^o \cup N_{h,g}^o . \tag{4.3}$$

On the other hand it follows from (3.8) and (3.16c) that the solution to $(QP^1_{h,g})$ does not change if we modify the set of constraints $X_{h,g}$ putting in (3.7) $i \in J^o_h \cup L^o_{h,g} \cup N$ instead of $i \in J^o_h \cup L^o_{h,g}$, where $N$ is any subset of $N^o_{h,g}$. Hence by (4.3) we can put in (3.7) $i \in J^o_{h+\alpha g} \cup L^o_{h+\alpha g,g}$, i.e. we can treat both problems $(QP^1_{h,g})$ and $(QP^1_{h+\alpha g,g})$ as having the same set of binding constraints. By the results of [5], this implies that (4.2) hold for an $\alpha > 0$ sufficiently small and a constant $c > 0$ which can be chosen independent of $\alpha$.

To complete the proof of the proposition it remains to show 2). This is done in exactly the same way as in the proof of Theorem 3.1 in [6]. □

In a way similar to (3.3), we defin e the sets

$$L^1_{h,g}=\{i \in N^o_{h,g} \mid \exists\, \alpha^i>0,\ \forall \alpha \in (0,\alpha^i),\ \mu^i(h+\alpha g,g)>0\}, \tag{4.4a}$$

$$M^1_{h,g}=\{i \in N^o_{h,g} \mid \exists\, \beta^i>0,\ \forall \beta \in (0,\beta^i),\ \chi^i(v(h+\beta g,g),h+\beta g,g)<0\}, \tag{4.4b}$$

$$N^1_{h,g}=N^o_{h,g} \setminus (L^1_{h,g} \cup M^1_{h,g}). \tag{4.4c}$$

Repeating the proof of Lemma 3.1 we obtain:

<u>Lemma 4.1</u>

If $i \in N^1_{h,g}$, then there exists a sequence $\{\bar{\alpha}_n\} \downarrow 0$ such that

$$\chi^i(v(h+\bar{\alpha}_n g,g),h+\bar{\alpha}_n g,g)=0,\ \mu^i(h+\bar{\alpha}_n g),g)=0 \quad \forall \bar{\alpha}_n \in \{\bar{\alpha}_n\}. \tag{4.5}$$

Now we can prove the result analogous to Theorem 3.1:

<u>Theorem 4.1</u>

If conditions (i)-(vi) are satisfied with $p=2$, then, for any fixed $h \in H$ and $g \in R^q$, there exist second directional derivatives of $u(h+\alpha g)$ and $\lambda(h+\alpha g)$ defined by (4.1).

They are characterized respectively as the unique solution and the associated multiplier of the following quadratic problem of optimization:

$(QP^2_{h,g})$ find $w(h,g) \in \Psi_{h,g}$ such that

$$k^2(w(h,g),h,g) = \min_{w \in \Psi_{h,g}} k^2(w,h,g) \tag{4.6}$$

where

$$k^2(v,h,g) = \frac{1}{2} < v,Q(h)v > + < q^2(h,g),v > , \tag{4.7}$$

$$Q(h)=D^2_{uu}f(u(h),h)+\sum_{i\in I}\lambda^i(h)D^2_{uu}\phi^i(u(h),h), \tag{4.7a}$$

$$q^2(h,g)=[D_hQ(h)g]v(h,g)+D_hq^1(h,g)g+\sum_{i\in I}\mu^i(h,g)[D^2_{uu}\phi^i(u(h),h)v(h,g)+$$

$$+D^2_{uh}\phi^i(u(h),h)g]=[D^3_{uuu}f(u(h),h)v(h,g)]\,v(h,g)+2[D^3_{uuh}f(u(h),h)v(h,g)]g+$$

$$+[D^3_{uhh}f(u(h),h)g]g+\sum_{i\in I}\{\lambda^i(h)([D^3_{uuu}\phi^i(u(h),h)v(h,g)]v(h,g)+$$

$$+2[D^3_{uuh}\phi^i(u(h),h)v(h,g)]g+[D^3_{uhh}\phi^i(u(h),h)g]g)+$$

$$+2\mu^i(h,g)(D^2_{uu}\phi^i(u(h),h)v(h,g)+D^2_{uh}\phi^i(u(h),h)g)\}, \tag{4.7b}$$

$$\Psi_{h,g}=\{w\in R^n \mid \psi^i(w,h,g)=0,\quad i\in J^o_h\cup L^o_{h,g}\cup L^1_{h,g}\}, \tag{4.8}$$

$$\psi^i(w,h,g):=\langle D_u\phi^i(u(h),h),w\rangle+[\langle D^2_{uu}\phi^i(u(h),h)v(h,g)+$$

$$+D^2_{uh}\phi^i(u(h),h)g,v(h,g)\rangle+\langle D^2_{hu}\phi^i(u(h),h)v(h,g)+D^2_{hh}\phi^i(u(h),h)g,g\rangle]. \tag{4.8a}$$

Moreover

$$\nu^i(h,g)=0 \quad \text{for} \quad i\in K^o_h\cup M^o_{h,g}\cup M^1_{h,g}\cup N^1_{h,g}. \tag{4.9}$$

Proof

The proof of the theorem will be very similar to that of Theorem 3.1. It is easy to see that both the solution and associated Lagrange multipliers of $(QP^2_{h,g})$ are defined uniquely. Taking into account (4.9), the optimality condition for $(QP^2_{h,g})$ can be write as

$$Q(h)w(h,g)+q^2(h,g)+\sum_{i\in I}\nu^i(h,g)D_u\phi^i(u(h),h)=0. \tag{4.10}$$

By Proposition 4.1 from any sequences $\{\alpha_n\}\downarrow 0$ we can extract a subsequence $\{\alpha'_n\}\subset\{\alpha_n\}$ such that

$$\frac{1}{\alpha'_n}[v(h+\alpha'_n g,g)-v(h,g)]\to w, \tag{4.11a}$$

$$\frac{1}{\alpha'_n}[\mu(h+\alpha'_n g,g)-\mu(h,g)]\to \nu. \tag{4.11b}$$

We shall show that the pair $(w,\nu)$ satisfies (4.10) together with (4.8) and (4.9), i.e. it is the solution and the Lagrange multiplier of $(QP^2_{h,g})$.

Taking difference quotient of (3.9) at $(h+\alpha_n' g)$ and at h, and passing to the limit we arrive at (4.10).

The proof of (4.8) is a simple repetition of the proof of (3.7) in Theorem 3.1. Also (4.9) for $i \in K_h^o \cup M_{h,g}^o \cup M_{h,g}^1$ is proved in the same way as (3.8) for $i \subset K_h^o \cup M_{h,g}^o$.

Hence it remains to prove (4.9) for $i \in N_{h,g}^1$. Like in the proof of the analogous result in Theorem 3.1 it is enough to consider the following two cases of sequences $\{\alpha_n\} \downarrow 0$:

$$(1) \qquad \chi^i(v(h+\alpha_n g,g),h+\alpha_n g,g) \neq 0, \ \mu^i(h+\alpha_n g,g) = 0,$$

$$(2) \qquad \chi^i(v(h+\alpha_n g,g),h+\alpha_n g,g) = 0.$$

Note that, in contrast to the case of the first-order derivatives, here the signs of $\chi^i(v(h+\alpha_n g,g),h+\alpha_n g,g)$ and $\mu^i(v(h+\alpha_n g,g)$ may be either positive or negative.

In the case (1), the condition (4.9) is satisfied, but it might be violated in the case (2), if

$$\nu^i(h,g) \neq 0. \tag{4.12}$$

However taking advantage of Lemma 3.2 as well as Proposition 4.1 and using the same argument as in the proof of Theorem 3.1 we find that (4.12) is excluded.

Hence (4.9) holds. Moreover, we obtain as in Corollary 3.1

$$\psi^i(w(h,g),h,g) = 0 \quad \text{for} \quad i \in N_{h,g}^1.$$

□

<u>Remark 4.1</u>

In contrast to the case of the first order derivatives the inequalities analogous to (3.16a) and (3.16b) may not be satisfied by $\psi^i(w(h,g),h,g)$ and $\nu^i(h,g)$. Therefore we are unable to characterize the second order derivatives as the solutions to a convex programming problem analogous to $(\widetilde{QP}_{h,q}^1)$.

By Theorems 3.1 and 4.1 we obtain in a standard way:

<u>Corollary 4.1</u>

If conditions (i)-(vi) are satisfied with p=2, then, for any fixed $h \in H$, $g \in R^q$ and any $\alpha > 0$,

$$u(h+\alpha g) = u(h) + \alpha u^{(1)}(h,g) + \frac{\alpha^2}{2} u^{(2)}(h,g) + o(\alpha^2), \tag{4.13a}$$

$$\lambda(h+\alpha g) = \lambda(h) + \alpha\lambda^{(1)}(h,g) + \frac{\alpha^2}{2}\lambda^{(2)}(h,g) + o(\alpha^2), \tag{4.13b}$$

where $u^{(1)}(h,g)$, $\lambda^{(1)}(h,g)$ and $u^{(2)}(h,g)$, $\lambda^{(2)}(h,g)$ are characterized in Theorems 3.1 and 4.1.

Provided that the data of Problem $(P_h)$ are sufficiently regular we can repeat recursively the argument of the proof of Theorem 4.1 to obtain:

Corollary 4.2

If conditions (i)-(vi) hold, then for any fixed $h \in H$, $g \in R^q$ and any $\alpha > 0$,

$$u(h+\alpha g) = u(h) + \alpha u^{(1)}(h,g) + \frac{\alpha^2}{2} u^{(2)}(h,g) + \ldots + \frac{\alpha^p}{p!} u^{(p)}(h,g) + o(\alpha^p), \tag{4.14a}$$

$$\lambda(h+\alpha g) = \lambda(h) + \alpha\lambda^{(1)}(h,g) + \frac{\alpha^2}{2} \lambda^{(2)}(h,g) + \ldots + \frac{\alpha^p}{p!} \lambda^{(p)}(h,g) + o(\alpha^p), \tag{4.14b}$$

where $u^{(i)}(h,g)$ and $\lambda^{(i)}(h,g)$ are given as the solutions and the associated Lagrange multipliers of some auxiliary quadratic problems of optimization, subject to linear, equality-type constraints, which can be derived recursively in the same way as $(QP^2_{h,g})$.

Now let us consider the so called optimal-value function which, to every value of the parameter, assigns the corresponding optimal value of the cost function in $(P_h)$:

$$F(\cdot) : H \longrightarrow R^1,$$

$$F(h) := f(u(h),h). \tag{4.15}$$

Corollary 4.3

If conditions (i)-(vi) hold, then, for any fixed $h \in H$, $g \in R^q$ and any $\alpha > 0$,

$$F(w+\alpha g) = F(h) + \alpha F^{(1)}(h,g) + \frac{\alpha^2}{2} F^{(2)}(h,g) + \ldots + \frac{\alpha^{p+1}}{(p+1)!} F^{(p+1)}(h,g) + o(\alpha^{p+1}) \tag{4.16}$$

where

$$F^{(i)}(h,g) = \frac{d^{i-1}}{d\alpha^{i-1}} \langle D_h\ (u(h+\alpha g), \lambda(h+\alpha g), h+\alpha g)\Big|_{\alpha=0}, g \rangle \tag{4.16a}$$

Proof

It follows from (2.7) and (2.8a) that

$$F(h)=\mathcal{L}(u(h),\lambda(h),h).$$

Hence Corollary 4.2 yields (4.16a) for $i=1,\dots,p$. On the other hand, (2.8), (3.7) and (3.8) yield

$$F^{(1)}(h,g)=\langle D_u\mathcal{L}(u(h),\lambda(h),h),u^{(1)}(h,g)\rangle+\langle D_\lambda\mathcal{L}(u(h),\lambda(h),h),\lambda^{(1)}(h,g)\rangle+$$

$$+\langle D_h\mathcal{L}(u(h),\lambda(h),h),g\rangle=\langle D_h\mathcal{L}(u(h),\lambda(h),h),g\rangle, \quad (4.17)$$

i.e., $F^{(1)}(h,g)$ does not depend on $u^{(1)}(h,g)$, $\lambda^{(1)}(h,g)$. Therefore, $F^{(i)}(h,g)$ depends only on $u^{(j)}(h,g)$, $\lambda^{(j)}(h,g)$ with $j\le i-1$. Thus (4.16) holds since $u^{(p)}(h,g)$ and $\lambda^{(p)}(h,g)$ exist. □

Remark 4.2

The second order directional derivative $F^{(2)}(h,g)$ can be expressed in a simple form:

$$F^{(2)}(h,g)=\langle D^2_{uu}\mathcal{L}(u(h),\lambda(h),h)u^{(1)}(h,g),u^{(1)}(h,g)\rangle+$$

$$+2\langle D^2_{hu}\mathcal{L}(u(h),\lambda(h),h)u^{(1)}(h,g),g\rangle+$$

$$+\langle D^2_{hh}\mathcal{L}(u(h),\lambda(h),h)g,g\rangle \quad (4.18)$$

Indeed, differentiating (4.17) with respect to $\alpha$ we obtain

$$F^{(2)}(h,g)=\langle D^2_{hu}\mathcal{L}(u(h),\lambda(h),h)u^{(1)}(h,g)+D^2_{h\lambda}\mathcal{L}(u(h),\lambda(h),h),\lambda^{(1)}(h,g)+$$

$$+D^2_{hh}\mathcal{L}(u(h),\lambda(h),h)g,g\rangle. \quad (4.19)$$

Multiplying (3.15) by $u^{(1)}(h,g)$, adding to (4.19) and using (3.15a) we arrive at (4.18).

Note that (4.18) was first derived by A. Shapiro [11] who used a completely different approach.

## References


[1] J.H. BIGELOW and N.Z. SHAPIRO: Implicit Function Theorem for Mathematical Programming and for Systems of Inequalities, Mathematical Programming, 6 (1974), 141-156.

[2] R.S. DEMBO: Sensitivity Analysis in Geometric Programming, Journal of Optimization. Theory and Applications 37 (1982), 1-21.

[3] A.V. FIACCO: Introduction to Sensitivity and Stability Analysis in Nonlinear Programming, Academic Press, New York, 1983.

[4] J. GAUVIN and A. DUBEAU: Differential Properties of the Marginal Functions in Mathematical Programming, Mathematical Programming Study 19 (1982), 101-119.

[5] B. GOLLAN: On the Marginal Function in Nonlinear Programming, Mathematics of Operations Research 9 (1984), 208-221.

[6] W.W. HAGER: Lipschitz Continuity for Constrained Processes, SIAM Journal on Control and Optimization 17 (1979), 321-338.

[7] K. JITTORNTRUM: Solution Point Differentiability without Strict Complementarity in Nonlinear Programming, Mathematical Programming Study 21 (1984), 127-138.

[8] K. MALANOWSKI: Differential Sensitivity of Solutions to Convex Programming Problems without Strict Complementarity Assumption. Technical Report No. ZTS 3-4/83, Systems Research Institute of the Polish Academy of Sciences (Warsaw, Poland, 1983).

[9] K. MALANOWSKI: Differentiability with Respect to Parameters of Solutions to Convex Programming Problems, Mathematical Programming 33 (1985), 352-361.

[10] K. MALANOWSKI: Stability of Solutions to Convex Problems of Optimization. Lecture Notes in Control and Information Sciences, vol. 93, Springer-Verlag, Berlin, 1987.

[11] A. SHAPIRO: Second Order Sensitivity Analysis and Asymptotic Theory of Parametrized Nonlinear Programs, Mathematical Programming 33 (1985), 280-299.

[12] R.T.ROCKAFELLAR: Marginal Values and Second Order Necessary Conditions for Optimality, Mathematical Programming 26 (1983), 245--286.

# DISTRIBUTED NONLINEAR OPTIMIZATION ALGORITHMS FOR GENERAL NETWORK PROBLEMS

D.K. Subramanian
Department of Computer Science and Automation
Indian Institute of Science, Bangalore-560 012, India

Summary

There are a number of large networks which occur in many problems dealing with the flow of power, communication signals, water, gas, transportable goods, etc. Both design and planning of these networks involve optimization problems. The first part of this paper introduces the common characteristics of a nonlinear network (the network may be linear, the objective function may be non linear, or both may be nonlinear). The second part develops a mathematical model trying to put together some important constraints based on the abstraction for a general network. The third part deals with solution procedures; it converts the network to a matrix based system of equations, gives the characteristics of the matrix and suggests two solution procedures, one of them being a new one. The fourth part handles spatially distributed networks and evolves a number of decomposition techniques so that we can solve the problem with the help of a distributed computer system. Algorithms for parallel processors and spatially distributed systems have been described.

There are a number of common features that pertain to networks. A network consists of a set of nodes and arcs. In addition at every node, there is a possibility of an input (like power, water, message, goods etc) or an output or none. Normally, the network equations describe the flows amoungst nodes through the arcs. These network equations couple variables associated with nodes. Invariably, variables pertaining to arcs are constants; the result required will be flows through the arcs. To solve the normal base problem, we are given input flows at nodes, output flows at nodes and certain physical constraints on other variables at nodes and we should find out the flows through the network (variables at nodes will be referred to as across variables).

The optimization problem involves in selecting inputs at nodes so as to optimise an objective function; the objective may be a cost function based on the inputs to be minimised or a loss function or an efficiency function.

The above mathematical model can be solved using Lagrange Multiplier technique since the equalities are strong compared to inequalities. The Lagrange multiplier technique divides the solution procedure into two stages per iteration. Stage one calculates the problem variables $x$ and stage two the multipliers $\lambda$ . It is shown that the Jacobian matrix used in stage one (for solving a nonlinear system of necessary conditions) occurs in the stage two also.

A second solution procedure has also been imbedded into the first one. This is called total residue approach. It changes the equality constraints so that we can get faster convergence of the iterations.

Both solution procedures are found to converge in 3 to 7 iterations for a sample network.

The availability of distributed computer systems - both LAN and WAN - suggest the need for algorithms to solve the optimization problems. Two types of algorithms have been proposed - one based on the physics of the network and the other on the property of the Jacobian matrix. Three algorithms have been deviced, one of them for the local area case. These algorithms are called as regional distributed algorithm, hierarchical regional distributed algorithm (both using the physics properties of the network), and locally distributed algorithm (a multiprocessor based approach with a local area network configuration). The approach used was to define an algorithm that is faster and uses minimum communications. These algorithms are found to converge at the same rate as the non distributed (unitary) case.

## 1. Introduction

There are many networks in existence today. Most of them are conceived as transportation systems - transporting oil, water, power, gas, communication signals, data amoungst offices, information, mail, goods etc. The advent of inexpensive computer communications systems has changed the style of working. The networks can be grouped into two groups - linear and nonlinear. One of the important problems associated with networks is planning the network topology amoungst a number of centres - called nodes - so as to meet the traffic/flow capacities from sources to networks via various paths. Another important problem is that given a network topology and sources and destinations, it is intended to find an optimal quanta of flows in the lines so that overall cost is minimum. Whereas many techniques have been developed to handle linear networks, generalisation of modelling and solution algorithms for nonlinear networks is not being taken up in a major way so that we can abstract general features of many such networks and build faster algorithms to handle planning and operational problems. Many models definitely exist to study water, power, gas, communications systems. (1, -7) the main objective of this paper is to derive a generalisation for these networks and develop solution algorithms which lead to convergence at a faster rate.

The paper does not consider the stochastic nature of a network nor the dynamics - time variations of a network. The model devised is for the steady state operations and optimization. It can be used for topology planning, steady state analysis, location of sources, location of destination nodes, line capacity determination etc.

## 2. General Abstraction of a Network

Some of the abstractable features of a nonlinear network are:

(i) The network consists of a set of sources which feed the input. Let us define this set as S. Let $s_i$ be a member of this set S. i will be the index for nodes.

(ii) A network consists of a set of destinations which absorb and consume. Let us define this set as D. Let $d_j$ be a member node of this set D. Again j is a node.

(iii) There may be intermediate junction nodes which may be neither sources or destinations. Let us call their set as F (floating nodes). Let $f_k$ be a member node of this set F.

(iv) There is always one node - which is a source node - but the input supplied by this node is not known. This node normally handles the fluctuations. This is known as the reference node. Let us denote this node by the index N.

(v) Flows are through lines - arcs. Network topology gives the connection graph amongst these nodes. Capacity of flow through an arc is known. Arcs are bidirectional.

(vi) Network variables are of two kind: one is the across variable at each node - this can be capacity, voltage, pressure, etc. Another is the through variable for each arc. Let us call by

$x_i$ the across variable at ith node and $y_{ij}$ the through variable for the arc connecting nodes i and j. Each arc can be characterised by a function as follows:

$$Y_{ij} = h(x_i, x_j) \tag{1}$$

There is a parameter-which gives the dependence of y on $\Delta x$ ($dy/d\,\Delta x$) - associated with each arc ($\Delta x = x_i - x_j$). We include this parameter also in the function explicitly since it is required to be evaluated in capacity planning problems.
So equation (1) becomes

$$Y_{ij} = h_{ij}(x_i, x_j, a_{ij}) \tag{2}$$

Also $$a_{ij} = a_{ji} \tag{3}$$

(vii) At each source node, there is an input. Let us call this $P_i$ and the output at each destination node be denoted as $O_i$.

(viii) Let the total number of nodes be M. Then we cannot calculate $x_1, x_2, \ldots x_M$. One of them depends on the others. As will be seen later, there will be one redundant equation.

(ix) There will be a loss in the arc in some networks and none in some other cases. Equations (1)/(2) can take care of this situation.

Now we can write the equations governing the network. It is possible to write a flow balance equation for each node as

$$P_i - \sum_{j=1}^{M} Y_{ij} = 0 \tag{4}$$

$i \in S$

$$\sum_{j=1}^{M} Y_{ij} - R_i = 0 \tag{5}$$

$i \in D$

$$\sum_{j=1}^{M} Y_{ij} = 0 \tag{6}$$

$i \in F$

The equation for reference node is also similar, but it can be obtained from (4), (5), (6) also as a redundant equation.

The most important aspect of this model is that it can be generalised in the following form

$$f_i(x_i, P_i) + \sum_{j=1}^{M} a_{ij}x_j = 0 \quad (i \in S) \tag{7}$$

$$f_i(x_i, R_i) + \sum_{j=1}^{M} a_{ij}x_j = 0 \quad (i \in D) \tag{8}$$

$$\sum_{j=1}^{M} a_{ij}x_j = 0 \quad (i \in F) \tag{9}$$

In general this can be written as

$$G(X,P) + AX = 0 \tag{10}$$

where G is a vector of functions $f_i$

$$G = [f_1, f_2, \ldots\ldots \quad ]^T \tag{11}$$

A is a matrix of order n x n

X is a vector of across variables

$$X = [x_1, x_2, \ldots\ldots\ldots x_n]^T \tag{12}$$

(n is the number of equations = M-1)

Some of the abstractable characteristics of equation (10) are

(i) The functions $f_i$ depend only $x_i$ and not on other x's.

(ii) The functions can be inverse or quadratic or cubic on x.

(iii) A is a symmetric matrix.

(iv) A is a sparse matrix, if the network is sparse as is usually the case.

(v) Consider the equation for ith node. It will have nonzero A terms only for these across variables of nodes connected to i by arcs. All other nodes do not directly influence node i. Hence the structure of A will be similar to the network topology. This is useful for parallel algorithms solving the network problem in a wide area network of computers.

(vi) Losses in arcs can be imbedded into the equation for the Nth node (reference node)

We will introduce an extra equation for node N which is called the residue equation. The residue equation is a function of all inputs, outputs and losses in lines. It can be written as

$$Z(P_1, P_2, \ldots P_M, O_1, O_2, \ldots \quad x_1, x_2, \ldots x_n) = 0 \qquad (13)$$

The actual equation structure depends on the type of network under consideration. This equation is extremely useful in convergence studies for optimization.

Equations (10) and (13) represent the steady state model for a network. The next section describes an algorithm for steady state analysis. The algorithm is well known in the study of networks.

## 3. Algorithm for Steady State Analysis

Since most of these equations are very well behaved we can use Newton-Raphson technique to solve them. But first, the steady state problem is defined as follows:

Given output requirements $O_i$ $(i \in D)$
input flows $P_i$ $(i \in S)$
and $x_N$ (variable for reference node N)

it is intended to find out the values of $x_i$
$(i = 1, 2, \ldots$ M but not N) and as a consequence $Y_{ij}$. s - flows in arcs.

The algorithm is given as follows:

(i) Assume some initial values for all $x_i$ s (value for $x_N$ is given)

(ii) Let us get the derivative of network equation (10) as

$$G'(X) + A = 0$$

G' is a diagonal matrix

Then Newton Raphson equation becomes

$$\Delta X = -(G'(X) + A)^{-1} (G(X, P) + AX) \qquad (14)$$

Solve this for $\Delta X$ (increments for X)

(iii) Update X

Repeat steps (II) and (III) until solution is got.

There are several ways of solving (ii) one of the methods is explained below:

$$\text{Let } J_0 = G(X_0, P) + A \qquad (15)$$

Then we can modify the equation (14) as

Start with a value $X_1$

$$J_0 \Delta X = -(G(X_1, P) + AX) \qquad (16)$$

$$X_2 = X_1 + \Delta X$$

we can calculate the next improvement as

$$J_0 \; \Delta X = - \left[(G(X_1,P) + AX_1 + G(X_2,P) + AX_2\right] \qquad (17)$$

This procedure gives a cubic convergence. Reference (8) indicates the use of this procedure for a power network.

4. <u>Network Optimization</u>

Normally, we would like to minimize losses or total cost of inputs; the problem of optimization obtains the inputs to be fed at different sources such that the total cost is a minimum. The total cost can be cost of inputs, or cost of pumping the inputs in the case of water or gas or cost due to losses or penalties due to the environmental improvements needed due to the inputs (this is considered in the location optimization algorithm). The objective function can be written as

$$\min \sum_{i \in S \cup N} c_i(P_i) \qquad (18)$$

This is a constrained optimization problem, the constraints are given by equations (10) and (13).

The above is a nonlinear programming problem. There are inequalities imbedded into the model to consider situations like actual input should be less than the capacity available in a source node. These can be easily handled in the algorithm separately and hence are not discussed here.

For our discussion purposes, we redefine our model as follows:

$$\min \quad C = \sum c_i(p_i) \qquad (19)$$

subject to $\hat{G}_i(X,P) = 0$ (20)

$i=1,2,..M \ (i \neq N)$ (21)

$Z = 0 \quad (i = N)$

$\hat{G}$ is the function defined as

$$\hat{G}_i = G_i \ (X,P) + \sum_{j=1}^{M} a_{ij} \, x_j \tag{21}$$

Let us write the algorithms for solution of the above minimization problem.

4.1 Lagrange Multiplier Approach with total Residue equation

We can write the unconstrained function as

$$L = C + \sum_{i=1}^{M} \lambda_i \, G_i + \lambda_N \, Z \tag{22}$$

To minimize L, we get the following necessary conditions

(I) $\frac{\partial L}{\partial x_i} = 0$ gives is

$$J^T K_1 \Lambda = K_2 \, \lambda_N \tag{23}$$

$\Lambda$ is a vector of $\lambda_i$ (excluding $\lambda_N$) and J is defined in (15); $K_1$ is another matrix. Normally diagonal one; $K_2$ is a vector.

II $\frac{\partial L}{\partial p_i} = 0$ gives

$$-\frac{dc_i}{dp_i} = \lambda_i + \lambda_N \frac{dZ}{dp_i} \tag{24}$$

III $\frac{\partial L}{\partial \lambda_i} = 0$ gives the steady state equations

$$\hat{G}_i = 0 \tag{25}$$

The solution procedure consists of

(i) Assume initial values for X and P.

(ii) Calculate new set of values for X by solving the steady state equation (25). Store the Jacobian matrix obtained.

(iii) Calculate $\Lambda$ (excluding $\lambda_N$) using equation (23) and the Jacobian matrix stored in the previous step.

(iv) Calculate $P_N$ from $\hat{G}_N = 0$ or $Z = 0$ (26)

(v) Calculate $\lambda_N$ from $\frac{\partial L}{\partial P_N} = 0$ (27)

$$\text{i.e.} \quad \frac{dC_N}{dp_N} + \lambda_N \frac{dZ}{dP_N} = 0 \qquad (28)$$

(vi) Calculate $P_i$ $(i \in S)$ from (24)

(vii) Goto step (ii) and repeat until convergence is got.

The above method has been applied to water and power networks and converges to solution in three to seven iterations.

## 5. Distributed Optimization Algorithms

The optimization problem discussed in the previous section deals with operational aspects and hence has to be solved very frequently in a day - as and when there is a change in the destination requirements. Since monitoring computers measure the inputs, variables and outputs, we can get fast parallel algorithms to solve this problem and minimise not only time, but also data communications required between computers. Some distributed algorithms are discussed below.

### 5.1 Regional Distributed Algorithm

Let us first define a regional tendency. If we can identify a subnetwork which has a few connections(paths/arcs) to the remaining network, then we can say that the network exhibits regional tendencies.

- That is several subnetworks are loosely connected/coupled together to form a big network. (Graph theory terminologies are avoided here for the purposes of readability).

Whenever we have a large network with several regional networks in it, we can decouple the optimization algorithm into a number of region based algorithms.

The solution algorithm then becomes:

(i) Assume initial values for X and P

(ii) For each region do

Assume inflows/outflows from/to neighbouring regions;
Treat the arcs from neighbouring regions as sources or destinations depending on whether the flow is input or output;
Solve the regional optimization problem as a local, decoupled problem
end

(iii) Update values of P and flows in arcs which interconnect regions.

(iv) Repeat steps (ii) and (iii) until convergence is obtained.

The advantages of this algorithm are

(i) In some networks, a region may have a geopolitical administrative representation and inter regional flows may be fixed. Hence decoupling is simpler and each region behaves as an independent network.

(ii) Data transmissions amoungst the computers in different regions are minimum.

(iii) Convergence is quite good, since the problem is decomposed into problems of smaller sizes and each subnetwork for regions has got a residue equation.

We can modify the above algorithm by incorporating a sensitivity model so that we get a fast, feasible solution first and we move to an optimal solution quickly from this as the solution space becomes small. The approach is given below.

We have the steady state equation as

$$\hat{G}(X,P,0) = 0 \qquad (29)$$

Since we move from one operating point to another; let us call the first operating point as $(X_o,P_o,0_o)$ and the second as $(X_1,P_1,0_1)$

Then we know that

$$\hat{G}(X_o,P_o,0_o) = 0 \qquad (30)$$

Also
$$\hat{G}(X_1,P_1,0_1) = 0 \qquad (31)$$

Let $X_1 = X_0 + \Delta X$ and $P_1 = P_0 + \Delta P$.

If $\Delta P$ is distributed amoungst the various inputs and the small variation can be adjusted by the reference node input, then we can calculate $\Delta X$ from

$$G(X_0,P_1,0_1) + H \ \Delta X = 0 \qquad (32)$$

H is the Jacobian matrix.

Equation (32) replaces Newton-Raphson solution procedure in step (ii) of the optimization algorithm described in Section 4.1.

## 5.2. Hierarchical Distributed Algorithm

In the regional algorithm (See 5.1), the inter region flows are to be assumed. There was no basis for making this assumption. This difficulty is solved by invoking a two stage process.

We normally develop the network topology for a large system with recognizable regions. We also derive a minimal equivalent network. This removes many source/destination nodes as well as intermediate nodes by network reduction transformations. So we have now two networks one is a minimal network with a few important nodes and arcs including interconnection arcs and another the original one. As described in the previous section, let us decompose the original network into a number of regional networks. Now let us call the minimal network as the global min network.

We can write the solution procedure for optimization:

(i) Establish initial values - may be the previous operating values-for P and X (0 is given).

(ii) Solve the optimization for the global min. network. This gives the flows in interconnecting arcs. Use these as inputs or outputs in the corresponding regional networks. Now regional networks are totally decoupled and independent.

(iii) Solve the regional optimizations in parallel.

(iv) Goto step (ii) and repeat until convergence is obtained.

This algorithm avoids assumptions for inter regional flows - when not scheduled -, but data communications time increases compared to the previous distributed algorithm problem.

## 6. Conclusion

This paper has developed a general model for network optimization and a number of algorithms have also been developed. The first algorithm uses the property of a zero net residue in the flows in the network. This property is used to obtain guaranteed convergence of the optimization algorithm.

The availability of a number of processors at several places (both LAN and WAN) gave the impetus to develop distributed algorithms. Two algorithms have been developed. The first one uses a physical regionalism that can be found in the network topology and with the aid of this, we can reduce the number of data transfers and time - one of the important bottlenecks in data communications - The second algorithm is an improvement over the regional algorithm. It provides a global minimal network to obtain inter regional flows. With the help of these flows, the regional networks can be decoupled.

## 7. References

1. D.K. Subramanian, 'Exact Economic Dispatch Without Dual Variables', IEEE Summer Power Meeting, 1970.

2. D.K. Subramanian, 'A Residual Current Method for Exact Economic Dispatch', Symposium on Power Systems Operations and Control, I.I.T. Delhi, 1972.

3. S.R. Kshirsagar and D.K. Subramanian 'Computer Aided Analysis of Large Scale Water Distribution Networks', Annual Conference, American Water Works Association, Chicago Ill., 1978.

4. D.K. Subramanian, 'A Static Data Structure for Steady State Analysis of Large Networks', Second Midwest Symposium on Circuits and Networks, Philadelphia, 1979.

5. D.K. Subramanian, 'A Total Residue Optimization Model for Power and Urban Water Distribution Systems', Third International Symposium on Large Engg. Systems, Newfoundland, Canada, 1980.

6. D.K. Subramanian 'Optimal Design and Operation of Large Urban Water Distribution Systems', Proc National Seminar on Water Supply Systems, Bangalore, 1983.

7. A.K.N. Reddy and D.K. Subramanian 'Design of Rural Energy Centres', Proc. Acad. of Sciences, Vol. 2, pt 3, pp 395-416, 1979.

8. D.K. Subramanian 'Fast Decoupled Load Flow Solution by a Higher Order Recursive Algorithm', IFAC Symposium on Computer Applications in Large Power Systems, Delhi, 1979.

# A MULTIPLEX ALGORITHM FOR LINEAR PROGRAMMING PROBLEMS

J.A. Snyman and M. van Rooyen
Department of Mathematics and Applied Mathematics,
University of Pretoria, Pretoria 0002, Republic of South Africa

Scope and Purpose - Since the relatively recent publications of polynomial-time iterative interior methods by Khachian [3] and Karmarkar [2] there has been a revival of interest in non-simplex methods for solving linear programming problems. One of the important reasons for the renewed interest is the emergence of parallel computers which may make older and discarded non-simplex methods, that can exploit the parallelism, more attractive. Other reasons are given by Zeleny [4]. Here the purpose is to develop and investigate a non-simplex (multiplex) method which follows a path through the interior and along higher dimensional hypersurfaces bounding the feasible polytope, rather than only along the one-dimensional edges as is the case with the simplex method. The strategy employed on deciding on the path is the product of computer experimentation and allows for maximum freedom to revert to steps in higher dimensional hypersurfaces which one may expect to be more economic. In this and other respects the present method appears to be different from some older methods which it resembles.

Abstract - In the multiplex method proposed here the solution is found by following a gradient path through the interior of the feasible region and through subspaces of reduced dimension corresponding to the bounding hypersurfaces of the feasible region. The path moves from any initial feasible point through a sequence of linear steps to a vertex of the polytope defined by the constraints. Although similar the current method differs fundamentally from Rosen's [5] gradient projection method and the strategy adopted in adding and dropping constraints from the enforcing set is different from that of the reduced gradient method of Wolfe [6]. Once the path reaches a vertex the algorithm determines whether or not it is optimal by applying a simple perturbation procedure for which the perturbed points are generated as a by-product of the computed path to the vertex. A theoretical convergence argument is put forward and using the computer the method has successfully been applied to a large number of test problems.

## 1. INTRODUCTION

Since Khachian's [3] and Karmarkar's [2] publications of polynomial-time projection algorithms for linear programming (LP) problems there has been a marked revival of interest in non-simplex ways of approaching the old LP-problem. For a brief summary see, for example, Mitra et al [4]. In this paper a multiplex and geometric method, is proposed and implemented. The proposed method is called a multiplex method because the path generated from an initial feasible point is not a simple path along the one-dimensional edges of the feasible polytope, as is the case with the simplex method, but may be through the interior and along multi-dimensional hypersurfaces bounding the feasible region.

Although similar the current method differs fundamentally from Rosen's gradient projection method [5] in that the ascent directions are obtained from the gradients of reduced problems of lesser dimension. These directions, when translated to the original space, do not necessarily correspond to Rosen's gradient projection directions. In fact, at any stage of the path, the required reduced problem may be chosen in a non-unique fashion that in practice results in a non-unique multiplex path through and along the hypersurfaces and edges of the feasible polytope. The present method also exhibits similarities to the reduced gradient method of Wolfe [6]. The special strategy adopted here and which is the product of experiment, allows for maximum freedom to revert back to steps along higher dimensional hypersurfaces and makes the current formulation and implementation to appear different from that of Wolfe. In short the method presented here is conceptually a simple geometric one, requiring no slack variables in its formulation, no Lagrangian multipliers in its interpretation and has successfully been applied to test problems.

The basic idea of the multiplex method is to take a step from an initial feasible point in the direction of the gradient vector until a constraint is met. The constraint is then used to eliminate a variable from the problem thus giving rise to a reduced problem of dimension one less. Considering the reduced problem in its own right a further step is computed in the direction of its gradient vector until the next constraint is met and a further variable may then be eliminated from the problem. Rules are determined and applied according to which constraints are either added or dropped from the active enforcing set. For example, if the substitution of the new constraint implies the activity of <u>all</u> the previous constraints then it is added to the active set. If not then only the latest constraint is retained and we restart building up the active enforcing set of constraints from scratch. Theoretical considerations indicate and numerical experiments confirm that, under normal conditions, the procedure adopted terminates at a vertex of the feasible polytope defined by the constraints. Thus a path is generated from any initial feasible point through a sequence of linear steps to a vertex of the polytope.

Once a vertex has been reached the new algorithm determines whether or not it is optimal by applying a simple perturbation procedure for which the perturbed points are generated as a by-product of the computed path to the vertex. If not optimal the algorithm proceeds by restarting from a perturbed point (on a suitable edge) with increased function value and the path is continued until the next vertex is reached.

## 2. THE MULTIPLEX ALGORITHM

### 2.1 Problem statement

We consider the following formulation of the LP-problem:

$$\begin{aligned}&\text{maximize} \quad f(x) = c^t x\\ &\text{subject to the linear constraints}\\ &g_i(x) = a_i^t x - b_i \leq 0, \quad i=1,2,\dots,m\end{aligned} \qquad (1)$$

where $c$, $a_i$, $x$ are column vectors in $R^n$, $b_i \in R$. In matrix notation the constraints may also be written as $Ax - b \leq 0$ where $a_i^t$ corresponds to the i-th row of $A \in R^{m \times n}$ and $b \in R^m$. The traditional nonnegativity constraints, if specified, may be included in the above constraint formulation in which case $m \geq n$. For the purpose of this paper we assume that the constraints are non-degenerate and that a feasible starting point, say $x^1$, is known.

### 2.2 Initial step

The multiplex (MP) algorithm takes a step in the direction of the gradient vector $c$, such that the next point $x^2$ is given by

$$x^2 = x^1 + hc$$

where $h = \text{maximum}\ \{\lambda \,|\, A(x^1+\lambda c) - b \leq 0\}$. If $x^1$ is feasible then $h \geq 0$ and the above is equivalent to defining

$$h = \text{minimum}\ \{h_i \,|\, h_i \geq 0\}$$

where

$$h_i = \begin{cases} -(c^t a_i)^{-1} g_i(x^1) & \text{if } c^t a_i > 0 \\ -1 & \text{otherwise.} \end{cases} \qquad (2)$$

Notice that if $c^t a_i < 0$ then $c$ points away from the constraint boundary $g_i(x) = 0$ and we can therefore choose to ignore it. We do this effectively by assigning the value $-1$ to the corresponding $h_i$. If $c^t a_i = 0$, $c$ lies in the affine space $\{x \,|\, g_i(x) = 0\}$ so that this constraint places no restriction on $h$. Thus even in this case we choose to ignore the constraint and in this way we allow for maximum freedom in constructing the ascent path.

If $h \geq 0$ and finite there exists at least one active constraint at $x^2$, say

$g_k$. We now propose that the next step taken by the MP-algorithm be in the affine space defined by the active constraint $g_k(x) = 0$. We prefer to call this an enforcing constraint to distinguish it from other constraints which may also be active but with negative $h_i$ values. To obtain a feasible direction in the mentioned affine space we eliminate a variable from our problem by means of the enforcing constraint. This leads us to the reduced LP-problem.

## 2.3 Reduced LP-problem

For greater generality assume that there are $p$ enforcing constraints identified by the index set $I = \{w_i\}_{i=1}^{p}$. Clearly the system of active constraints $A_I x = b_I$, where $A_I^t = [a_{w_1}, b_{w_2}, \dots, b_{w_p}]$ can be written as

$$Px_a + Qx_o = d \tag{3}$$

where $P$ is a non-singular matrix in $R^{p \times p}$, $Q \in R^{p \times (n-p)}$, $x_a \in R^p$, $x_o \in R^{n-p}$, $d \in R^p$ and where the relationship between $x_a$, $x_o$ and the original $x$ is given by $x = E\begin{bmatrix} x_a \\ x_o \end{bmatrix}$ for some appropriate permutation matrix E. We may now solve for the $p$ dependent variables $x_a$ i.t.o. the $n-p$ independent variables $x_o$ :

$$x_a = P^{-1}d - P^{-1}Qx_o \tag{4}$$

Substituting the above into the original objective function and the constraints of LP-problem (1), we obtain the reduced LP-problem:

$$\begin{aligned} &\text{maximize } f^*(x_o) = c^{*t}x_o + K \\ &\text{such that} \\ &\quad g_j(x_o) = a_j^{*t}x_o - b_j^* \le 0, \quad j \notin I \end{aligned} \tag{5}$$

where

$$\begin{aligned} c^{*t} &= (c_o^t - c_a^t P^{-1}Q), \quad K = c_a^t P^{-1}d \\ a_j^{*t} &= (a_{jo}^t - c_{ja}^t P^{-1}Q) \quad \text{and} \quad b_j^* = a_{ja}^t P^{-1}d - b_j; \end{aligned} \tag{6}$$

and where $c_o$, $a_{jo}$ and $c_a$, $a_{ja}$ respectively correspond to the coefficients of $x_o$ and $x_a$ in the original objective function and constraints. We note that the reduced problem has dimension $n-p$ with $m-p$ constraints.

The next step in $R^{n-p}$ may now be constructed as in section 2.2 for the initial step but now for the reduced problem. This gives another enforcing constraint which may now be added to our index set $I$ resulting in a further reduction process followed by the computation of yet another step. Such an iterative procedure of just adding each new enforcing constraint soon ends up with $p = n$, i.e. at a vertex that in general is not necessarily optimal. We now propose a strategy, the product of experimention, that increases the probability of non-zero steps along higher dimensional hyperplanes and hypersurfaces, thus delaying early termination at a vertex. We also add a perturbation procedure that assures eventual termination at an optimal vertex.

The basic strategy is as follows. If the computed stepsize is zero, the new enforcing constraint is added for the next iteration as before. If, however, the stepsize is greater than zero, then to begin with only one of the new constraints is enforced in the next iteration. This allows for the attractive possibility of a non-zero step being taken in the (n-1)-dimensional hyperplane corresponding to the new constraint. If however the stepsize (with only one constraint enforced) is zero then we allow for two situations that may arise. In the first case if all the previous enforcing constraints give $h_i$'s that are zero we add them all to the current new constraint in the next iteration. Otherwise if one or more of the previous constraints yield a negative $h_i$ (i.e. $a_i^{*t}c^* \leq 0$) then we add only one of the previous constraints with an $h_i = 0$ to the current constraint for the next iteration. This strategy is formally set our in Step 6 of Section 2.4. Once a vertex is finally reached the new algorithm tests whether or not it is optimal by applying a simple perturbation procedure (step 9 of 2.4) for which the perturbed points are generated as by-products of the computed path to the vertex and so that the solution of a new system of equations is not required (see section 3). The convergence of the algorithm is discussed in section 4. A formal statement of the algorithm now follows.

## 2.4 Formal MP-algorithm

Step 1. Assume a feasible starting point $x^1$ is known. Set $x^2 \leftarrow x^1 + hc$ where $h$ is calculated according to eqs. (2). We may now choose an enforcing constraint active at $x^2$, say $g_k$. Set $w_1 \leftarrow k$, $p \leftarrow 1$, $i \leftarrow 1$ and flag $v \leftarrow 0$.

Step 2. Set $i \leftarrow i + 1$.

Step 3. Assume a set of enforcing constraints $W = \{w_i\}_{i=1}^{p}$ active at the feasible point $x^i$.

Step 4. Eliminate $p$ variables from LP-problem (1) by means of the set of equations representing the $p$ enforcing constraints. This implies the construction of an invertible matrix $P$ and a permutation matrix $E$ corresponding to eqs. (3). Calculate the coefficients $a_j^*$, $c^*$ and $b_j^*$ as defined by (6) to yield the reduced LP-problem (5). If $c^* = 0$ STOP.

Step 5. Calculate the next feasible solution by taking a step in the direction of the reduced gradient $c^*$ to give

$$x_o^{i+1} = x_o^i + hc^*, \qquad\qquad x_a^{i+1} = P^{-1}d - P^{-1}Qx_o^{i+1}$$

where $h$ is determined by eqs (2) with the coefficients replaced by those corresponding to the reduced problem. This yield one new enforcing constraint, say $g_s$.

Step 6. Cases I and II: If $h > 0$ and $v = 0$ or $1$, then set $v \leftarrow 1$ and store the "old" set of enforcing constraints $W_{old} \leftarrow W$ and set $p_{old} \leftarrow p$. Use only the new enforcing constraint in the next iteration, i.e. set $p \leftarrow 1$, $w_1 \leftarrow s$ and go to step 2.
Case III: If $h = 0$ and $v = 0$ go to step 7.
Case IV: If $h = 0$ and $v = 1$ (i.e. $p = 1$) then (a) if $h_w = 0$ for all $w \in W_{old}$ then set $p \leftarrow p_{old} + 1$, $W \leftarrow W_{old} \cup W$, $v \leftarrow 0$ and go to Step 2; otherwise (b)(i.e. $\exists\, w \in W_{old}$ such that $h_w = -1$) set $v \leftarrow 0$ and go to step 7.

Step 7. Set $p \leftarrow p + 1$ and $w_p \leftarrow s$.

Step 8. If $p = n$ then $x^{i+1}$ is a vertex and go to step 9; otherwise go to step 2.

Step 9. (Perturbation procedure) Compute the perturbation points $y^j$, $j=1,2,\ldots,n$, each lying on a separate edge emanating from the vertex $x^{i+1}$ and given by the solution of the systems

$$Py^j = d - \epsilon^j \tag{7}$$

where $\epsilon^j = [0,0,\ldots,0,\epsilon,0,\ldots,0]^t$, for some suitably small $\epsilon > 0$ appearing in the j-th position. If $f(y^j) \le f(x^{i+1})$ for all $j$ then $x^{i+1}$ is optimal and STOP; otherwise choose the first $y^j$ for which $f(y^j) > f(x^{i+1})$ and set the new starting point $x' \leftarrow y^i$ and go to step 1.

## 3. IMPLEMENTATION

Step 4 in the MP-algorithm requires the construction of eqs. (3) and the computation of the solution (4). This is done economically by exploiting the fact that in the application of the MP-algorithm we either have only one enforcing constraint or one enforcing constraint is added to a set (the previous set or the "old" set) for which the solution has already been computed. In the case of only one constraint $(a_k^t - b_k = 0)$ the construction is trivial. Set $x_a \leftarrow x_i$ where $a_{ki} \neq 0$, then $P = a_{ki}$. In the second case when the p-th constraint $(p \ge 2)$ is added we already have available the solution for the previous $p-1$ enforcing constraints in the form

$$Ix_a + P^{-1}Qx_o = P^{-1}d$$

where $I$ is the identity matrix. Adding the new (p-th) constraint,

$a_a^t x_a + a_o^t x_o = d_p$, the solution for the new $x_a$ is obtained by applying Gauss-Jordan elimination to the following system:

$$\left[\begin{array}{ccc} I & P^{-1}Q & P^{-1}d \\ \hline a_a^t & a_o^t & d_p \end{array}\right] \tag{8}$$

Adding appropriate multiples of the upper rows to the p-th row yields the modified system:

$$\left[\begin{array}{c:c:c:c} I & \vdots & B & P^{-1}d \\ \hline 0\ 0 \cdots 0\ 0 & \alpha & \bar{a} & \bar{d} \end{array}\right]$$

where $\alpha$ may be zero. If $\alpha$ is zero we interchange the p-th column with a column to the right (the k-th) for which $\bar{a}_k \neq 0$. The diagonal entry is then normalized by dividing the p-th row by $\alpha$. We now add appropriate multiples of the p-th row to the upper rows to ensure zero entries in the p-th column above the diagonal and thus obtain the pxn solution matrix:

$$[\ I \ \vdots\ P^{-1}Q \ \vdots\ P^{-1}d\ ]^{(p)}$$

which gives the new dependent vector $x_a$ i.t.o. the new reduced independent vector $x_o$. The superscript denotes that the result is for p enforcing constraints.

The calculation of the perturbed points required for the perturbation procedure may be incorporated in the above elimination procedure by noticing that we need only extend system (8) at each iteration by $[0,0,\ldots,\epsilon]^t$ on the right hand side to obtain the perturbation vectors (see Step 9, Section 3.4):

$$[\ I \ \vdots\ P^{-1}d \ \vdots\ P^{-1}\epsilon I]^{(n)}$$

and thus the perturbed points are computed as a by-product of the computed path to the vertex. $\epsilon$ must of course be chosen small enough to ensure that, in the absence of degeneracy, the perturbed points remain feasible.

Because in practice the computer calculations are done in finite arithmetic we introduce in the implementation a tolerance parameter, $\delta > 0$, that is used when testing for zero. Thus, for example we set $h \leftarrow 0$ if $|h| < \delta$ (or preferably if $|h|\|c^*\| < \delta$ for a more scale free test).

In the application of the MP-algorithm a feasible starting point may be obtained, if necessary, in the usual way of first applying the method to the auxiliary problem.

## 4. CONVERGENCE OF THE METHOD

To prove convergence of the method to an optimal vertex we require the following two lemmas.

Lemma 1. If the LP-problem (1) has a bounded solution then $h^{(k)} \rightarrow 0$ where $h^{(k)}$ denotes the steplength taken at the k-th iteration.

Proof. It can easily be shown that the increase in the objective function is $\Delta f = \|c^*_{(k)}\|^2 h^{(k)}$ where $c^*_{(k)}$ is the reduced grandient at the k-th iteration and for this algorithm $h^{(k)} \geq 0$. Moreover, and less trivially, we can show that in the case where $c^*_{(k)} = 0$ and $h^{(k)}$ is thus not defined, that we have termination at an optimal hypersurface defined by the enforcing constraints. On the other hand if $\|c^*_{(k)}\| \neq 0$ throughout then, since we assume a finite number of constraints bounding the feasible region, there exists a $\gamma > 0$ such that $\|c^*_{(k)}\|^2 \geq \gamma$ for all possible k. Now the boundedness of the solution implies that $\sum_k \|c^*_{(k)}\|^2 h^{(k)}$ be convergent and since $\|c^*_{(k)}\|^2 \geq \gamma > 0$ for all k it follows that $h^{(k)} \rightarrow 0$. □

Lemma 2. The perturbation procedure in step 9 of the MP-algorithm determines optimality. This statement is obvious. □

We are now in a position to formulate a convergence theorem.

Theorem. If the solution of LP-problem (1) is bounded then the MP-algorithm terminates in a finite number of steps at an optimal hypersurface or vertex.

Proof. From step 6 of the MP-algorithm we deduce the following possible relationships between iterations:

I $v = 0$ & $h > 0$ leads to $v = 1$ & $p = 1$ in next iteration;
II $v = 1$ & $h > 0$ leads to $v = 1$ & $p = 1$ in next iteration;
III $v = 0$ & $h = 0$ leads to $p \leftarrow p+1$ & $v = 0$ in next iteration;
IV $v = 1$ & $h = 0$ leads to either:
(a) $p \leftarrow p_{old}+1$ & $v = 0$ in next iteration;
(b) $p \leftarrow p+1 = 2$ & $v = 0$ in next iteration.

If the algorithm does not terminate we must have an infinite incidence of cases I, II or IV(b) which are the only cases in which the number of enforcing constraints are reduced, $p \longleftarrow 1$ in cases I and II and $p \longleftarrow 2$ in case IV(b). Since IV(b) can only follow on steps I and II non-termination due to the infinite occurrence of case

IV(b) implies an infinite number of steps $h^{(k)} > 0$. In fact an infinite number of steps with $h^{(k)} \geq \delta$ (tolerance) must occur, which contradicts Lemma 1. Thus case IV(b), similarly to cases I and II, can only occur a finite number of times and since in all the other possible cases namely III and IV(a), the number of enforcing constraints is increased, this implies that termination at a vertex with $p = n$ (or at an optimal hypersurface) must follow in a finite number of iterations. If the vertex is optimal (Lemma 2) we stop, otherwise restart to yield next vertex (with increased function value) in a finite number of steps. Since the feasible polytope has only a finite number of vertices termination at the optimal vertex is obtained in a finite number of steps. □

## 5. RESULTS AND CONCLUSION

Figures 1(a) and (b) give a three-dimensional representation of the working of the MP-algorithm when applied respectively to the simple problem

(a) maximize $f = x_1 + 2x_2 + 3x_3$ ; such that

$$0 \leq x_i \leq 1, \quad i=1,2,3$$

and the "pathological" Klee-Minty problem

(b) maximize $f = 100x_1 + 10x_2 + x_3$; such that

$$x_1 \leq 1, \quad 20x_1 + x_2 \leq 100, \quad 200x_1 + 20x_2 + x_3 \leq 10000, \quad x_i \geq 0, \ i=1,2,3.$$

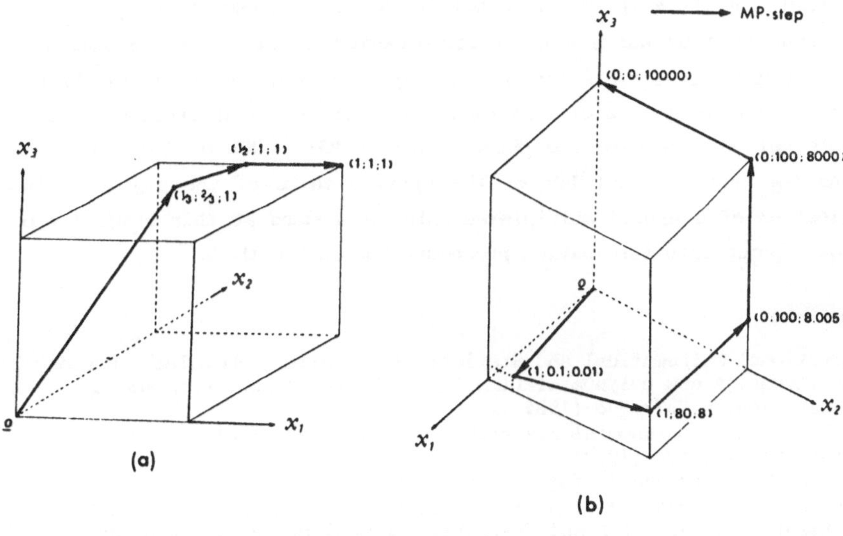


FIGURE 1 Geometrical representation of the MP-steps required for the solution of problems (a) and (b). For the sake of clarity the figure for (b) is not drawn to scale.

The figures clearly illustrate how the MP-algorithm not only takes steps along the

1-dimensional edges of the feasible polytope but also through the internal space and along higher dimensional planes.

The MP-algorithm has been implemented in a FORTRAN program that also allows for the solution of the auxiliary problem if necessary. The program was successfully used in the solution of many problems with $n \leq 20$ and $m \leq 60$. For small values of $n$ ($\leq 10$) the performance of the MP-algorithm appeared reasonably competitive with that of the IMSL subroutine ZX3LP [1], but for larger problems the comparison was poor and the program, as it stands, is probably not suitable for large problems. One must however bear in mind that the prototype program developed here was not designed with optimum efficiency in mind, but only to demonstrate that the algorithm may successfully be applied to LP-problems. The program, for example, uses the standard linear equations solver LEQT1F [1] of the IMSL subroutine library. In large LP-problems which occur in practice the coefficient matrix is usually very sparse and one may expect that if we adapt the program to exploit this sparseness in the Gauss-Jordan elimination procedure, that the performance of the MP-algorithm would be greatly improved. It also remains, of course, to take advantage of the considerable parallelism inherent in the new algorithm.

It is significant that in many cases the first termination occurs at an optimal or near optimal vertex. This indicates that the current algorithm, in agreement with the work of Mitra et al [4], may be used in a hybrid scheme that uses both the new method and the simplex method. Not only can the MP-algorithm provide a better starting point for the simplex method but it can also be used to provide a so-called "purification" step by which a non-basic internal solution may be improved to an extreme point basic feasible solution. As reported by Mitra et al [4] the execution of this purification step is of importance where iterative internal methods, such as Karmarkar's method, are used and where in about 25% - 30% of the total number of iterations one reaches about 80% of the optimum value of the objective function. The application of a hybrid multiplex-simplex procedure at this stage may possibly be the most appropriate termination procedure for such methods.

## 6. REFERENCES

1. International Mathematical and Statistical Libraries, IMSL Inc. Houson, Texas.
2. N. Karmarkar, A new polynomial-time algorithm for linear programming. *Combinatorica*, 4, 373-395 (1984).
3. L.G. Khachian, A polynomial algorithm in linear programming. *Soviet Mathematics Doklady*, 20, 191-194 (1979)
4. G. Mitra, M. Tamiz and J. Yadegar, *Experimental investigation of an interior search method within the simplex framework*. Technical Report TR/06/86, Department of Mathematics and Statistics, Brunel University Uxbridge. (1986).
5. J.B. Rosen, The gradient projection method for nonlinear programming Part I, Linear constraints. *Journal of the Society for Industrial and Applied Mathematics*, 8, 181-217 (1960)
6. P. Wolfe, Methods of nonlinear programming, in J. Abadie (Red). *Nonlinear Programming*, 97-131, North Holland, Amsterdam.
7. M. Zeleny, An external reconstruction approach to linear programming. *Computers and Operations Research*, 13, 95-100 (1986).

# NEWTON-TYPE ALGORITHMS WITH NONMONOTONE LINE SEARCH FOR LARGE-SCALE UNCONSTRAINED OPTIMIZATION

L. Grippo, F. Lampariello and S. Lucidi
Istituto di Analisi dei Sistemi ed Informatica del CNR
Viale Manzoni 30, 00185 Roma, Italy

## ABSTRACT

In this paper we define globally convergent algorithms for the solution of large dimensional unconstrained minimization problems. The algorithms proposed employ a nonmonotone steplength selection rule along the search direction which is determined by means of a Truncated-Newton algorithm. Numerical results obtained for a set of test problems are reported.

## 1. INTRODUCTION

Conjugate gradient methods are currently considered to be the only methods which are applicable to truly large dimensional unconstrained minimization problems, that is problems with hundreds or thousands of variables. This is due to the fact that, in spite of their relatively low efficiency, these methods do not require matrix operations for computing the search directions.

Quite recently, the Truncated-Newton method proposed in [1] has proved to be valuable in the solution of large-scale problems [2], [3], [4], since it allows to reduce the computational cost of determining the Newton's direction while retaining a fast asymptotic convergence rate. A remarkable feature of some implementations of these methods is the possibility of computing an approximate solution to the Newton equation by means of a conjugate gradient algorithm, without the need for storing the Hessian matrix. Thus, it is possible to exploit the information on second order derivatives with similar storage requirements as the nonlinear conjugate gradient methods. When the Hessian matrix is not available the information on second order derivatives can be replaced by finite difference approximations at the expense of an additional gradient evaluation for each minor iteration of the conjugate gradient algorithm.

As regards the need for ensuring global convergence, the same strategies developed for modifying Newton's method can be applied for stabilizing Truncated-Newton algorithms. In particular, it has been shown in [5], in connection with Newton's method, that considerable computational savings can be obtained by adopting a "nonmonotone line search technique", that is by relaxing some acceptability criterion for the stepsize in a way that allows an increase of the function values without affecting the global convergence properties. The application of this technique to the stabilization of a Truncated-

Newton algorithm has already been considered in [4], where preliminary computational experience has been reported.

In this paper we present new results on the application of a Truncated-Newton algorithm with a nonmonotone line search to the solution of large dimensional unconstrained minimization problems. In particular we show that the global convergence requirement can be made compatible with an ultimate quadratic convergence rate even in the absence of an a priori estimate of the smallest eigenvalue of the Hessian matrix.Moreover we consider an extension of this algorithm in which finite difference approximations for the computation of the second order derivatives are used. Numerical results for a set of large dimensional problems are reported.

## 2. STEPLENGTH SELECTION FOR NEWTON-TYPE METHODS

Consider the problem:

$$\min_{x \in \mathcal{R}^n} f(x) \tag{1}$$

where $f : \mathcal{R}^n \to \mathcal{R}$.

We assume that, for a given $x_o \in \mathcal{R}^n$, the level set

$$\Omega_o = \{x \in \mathcal{R}^n : f(x) \leq f(x_o)\}$$

is compact and that both the gradient $g(x)$ and the Hessian matrix $H(x)$ of $f$ exist and are continuous on $\Omega_o$.

In order to determine a local solution of problem (1) we consider an algorithm of the form:

$$x_{k+1} = x_k + \alpha_k d_k, \qquad k = 0, 1, \ldots \tag{2}$$

where $x_o \in \mathcal{R}^n$ is a given starting point, $d_k \neq 0$ is an "approximation" to the Newton's direction $-H^{-1}(x_k)g(x_k)$ and $\alpha_k$ is the stepsize. (In the sequel we adopt the notation $f_k = f(x_k)$, $g_k = g(x_k)$, $H_k = H(x_k)$).

For the computation of $\alpha_k$ we use the nonmonotone line search technique proposed in [5], by imposing that the function value at each new iterate satisfies an Armijo's condition with respect to the maximum of the function values computed in a prefixed number of previous iterates. More specifically, we consider the following algorithm:

*Nonmonotone Line search algorithm* (NL)

*Data:* $x_k$, $g_k$, $d_k$, $\sigma \in (0,1)$, $\gamma \in (0,1)$, $f_k^* = \max_{0 \leq j \leq m(k)} [f(x_{k-j})]$, with $m(k) \leq k$.

*Step 1:* Set $\alpha = 1$.

*Step 2:* Compute $f_\alpha = f(x_k + \alpha d_k)$. If $f_\alpha \leq f_k^* + \gamma \alpha g_k^T d_k$, set $\alpha_k = \alpha$ and stop.

*Step 3:* Set $\alpha = \sigma\alpha$ and go to Step 2.

△

Under the assumption that $m(k)$ is bounded by a prefixed integer for all $k$, we can state the following theorem, whose proof can be derived, with minor modifications, from that given in [5].

THEOREM 1.*Let $\{x_k\}$ be the sequence defined by the iterative scheme (2). Assume that:*

*(i) there exist positive scalars $c_1$, $c_2$, $\nu_1$ and $\nu_2$ such that:*

$$g_k^T d_k \leq -c_1 \| g_k \|^{\nu_1}, \quad \| d_k \|^{\nu_2} \leq c_2 \| g_k \|;$$

*(ii) the stepsize $\alpha_k$ is computed by means of Algorithm NL above.*

*Then:*

*(a) the sequence $\{x_k\}$ remains in $\Omega_o$ and every limit point $\hat{x}$ satisfies $g(\hat{x}) = 0$;*

*(b) no limit point of $\{x_k\}$ is a local maximum of $f$;*

*(c) if the number of the stationary points of $f$ in $\Omega_o$ is finite, the sequence $\{x_k\}$ converges.*

$\triangle$

## 3. TRUNCATED-NEWTON ALGORITHMS WITH NONMONOTONE LINE SEARCH

The computation of $d_k$ is based on the use of a truncated conjugate gradient algorithm for the minimization of the quadratic model

$$\phi(d) = f_k + g_k^T d + \frac{1}{2} d^T H_k d \tag{3}$$

which approximates $f(x_k + d)$. More specifically, we consider the following algorithm, which is a modified version of the algorithm proposed in [2].

*Truncated-Newton algorithm* (TN)

*Data:* $g_k$, $H_k$, $\eta_k > 0$, $\epsilon \in (0,1)$, $\sigma_1 > 0$, $\sigma_2 > 0$.

*Step 1:* Set $p_o = 0$, $r_o = g_k$, $s_o = -r_o$, $i = 0$.

*Step 2:* Set $q_i = H_k s_i$. If $| s_i^T q_i | \geq \min\left[\epsilon, \| g_k \|\right] \| s_i \|^2$, go to Step 3; otherwise set $d_k = -g_k$ and stop.

*Step 3:* Compute: $\lambda_i = -s_i^T r_i / s_i^T q_i$, $\quad p_{i+1} = p_i + \lambda_i s_i$ and $r_{i+1} = r_i + \lambda_i q_i$.
If $\| r_{i+1} \| / \| g_k \| > \eta_k$, compute: $\beta_i = r_{i+1}^T q_i / s_i^T q_i$, $\quad s_{i+1} = -r_{i+1} + \beta_i s_i$, set $i = i + 1$ and go to Step 2.

*Step 4:* Set $i_k = i + 1$ and:

$$d = \begin{cases} p_{i+1} & \text{if } g_k^T p_{i+1} \leq 0; \\ -p_{i+1} & \text{if } g_k^T p_{i+1} > 0. \end{cases}$$

If $| g_k^T d | \geq \sigma_1 \| g_k \|^3$ and $\| d \|^2 \leq \sigma_2 \| g_k \|$, set $d_k = d$ and stop; otherwise set $d_k = -g_k$ and stop.

$\triangle$

We remark that Algorithm TN differs from the Truncated-Newton Algorithm of [2] for the tests at Step 2 and 4. In particular, in Algorithm TN the conjugate gradient iteration is not terminated when a direction of negative curvature (i.e. $s_i^T H_k s_i < 0$) is encountered, provided that $| s_i^T H_k s_i |$ is not too small. In other words, the algorithm attempts to determine a good approximation of Newton's direction, even if $H_k$ is not positive definite. However, if $s_j^T H_k s_j < 0$ for some $j$, it is no longer true, in general, that the vectors $p_i$, $i \geq j$ are descent directions for $f$. The instructions at Steps 2 and 4 in Algorithm TN ensure that $d_k$ is a descent direction satisfying an angle condition. In particular condition (i) of Theorem 1 is satisfied with $\nu_1 > 2$ and $\nu_2 > 1$. It will be shown, in the proof of Theorem 2, that this allows to establish both global convergence towards stationary points and an ultimate quadratic convergence rate when the Hessian matrix at the limit point is positive definite. This extends to the Truncated-Newton Method a result already known [6] for globally convergent modifications of Newton's Method.

It can be easily verified that Algorithm TN terminates in a finite number of iterations. In fact, consider the sequences $\{s_i\}$ and $\{r_i\}$ produced by Algorithm TN; by following the same reasoning employed in [7] for the case that $H_k$ is positive definite, we have that if, for some $m > 0$, $| s_i^T H_k s_i | > 0$ for $i = 0, 1, \ldots, m$, then:

$$r_i^T r_j = 0, \qquad 0 \leq i, j \leq m, \qquad i \neq j.$$

This implies that if the algorithm does not terminate at Step 2, it produces in at most $n$ iterations a point $p_{m+1}$ such that $r_{m+1} = 0$ and hence it terminates at Step 4. We can now define the following algorithm for the minimization of $f$.

*Truncated-Newton algorithm with Nonmonotone Line search* (TNNL)

*Data:* $x_o$, $\theta > 0$, integer $M \geq 0$.

*Step 1:* Set $k = 0$, $\eta_o = \theta$, $m(0) = 0$ and compute $f_o$, $g_o$.

*Step 2:* If $\| g_k \| = 0$ stop; otherwise compute the Hessian matrix $H_k$ and determine the search direction $d_k$ by means of Algorithm TN.
If $d_k = -g_k$, set $m(k) = 0$.

*Step 3:* Compute $\alpha_k$ by means of Algorithm NL. Set $x_{k+1} = x_k + \alpha_k d_k$, $k = k + 1$, compute $g_k$, set: $\eta_k = \min [\theta/k, \| g_k \|]$ , $m(k) = \min [m(k-1) + 1, M]$ and go to Step 2.

△

We note that $M \geq 0$ is the maximum number of previous iterates which are taken into account at each iteration during the minimization process. Obviously, for $M = 0$ the line search algorithm reduces to Armijo's method.

The properties of the preceding algorithm are summarized in the next theorem.

THEOREM 2. *Let $\{x_k\}$ be the sequence produced by the Algorithm TNNL. Then either the algorithm terminates at some $x_\mu$ such that $g(x_\mu) = 0$ or it produces an infinite sequence such that:*

*(a) every limit point $\hat{x}$ of $\{x_k\}$ satisfies $g(\hat{x}) = 0$;*

*(b) no limit point of $\{x_k\}$ is a local maximum of $f$;*

*(c) if the number of the stationary points of $f$ in $\Omega_o$ is finite, the sequence $\{x_k\}$ converges.*

*(d) If $x_k \to x^*$, where $H(x^*)$ is positive definite and Lipschitz continuous, the sequence $\{x_k\}$ converges at a quadratic convergence rate.*

PROOF. If the algorithm terminates the assertion follows from Step 2. Assume that the algorithm produces an infinite sequence $\{x_k\}$. Algorithm NL ensures that $\{x_k\}$ remains in the compact set $\Omega_o$; therefore, by the continuity assumptions, there exists $w$ such that $\|g_k\| \leq w$ for all $k$. Now by the instructions at Step 2 and 4 of Algorithm TN we have:

$$g_k^T d_k \leq -c_1 \| g_k \|^3, \quad \| d_k \|^2 \leq c_2 \| g_k \|$$

where $c_1 = \min\left[\sigma_1, w^{-1}\right]$ and $c_2 = \max\left[\sigma_2, w\right]$. This implies that conditions (i) and (ii) of Theorem 1 are satisfied so that (a), (b) and (c) follow from Theorem 1. In order to prove (d) we show first that, if $x_k \to x^*$ where $H(x^*)$ is positive definite, both the test at Step 2 and the test at Step 4 of Algorithm TN are satisfied for $k$ sufficiently large. In fact, since $H(x^*)$ is positive definite, there must exist an index $k_1$ and numbers $\mu_2 \geq \mu_1 > 0$ such that, for all $k \geq k_1$:

$$\mu_1 \|x\|^2 \leq x^T H_k x \leq \mu_2 \|x\|^2, \qquad x \in \mathcal{R}^n,\ x \neq 0.$$

Since $\|g_k\| \to 0$ there exists a $k_2 \geq k_1$ such that for all $k \geq k_2$ we have:

$$\|g_k\| \|s_i\|^2 \leq \mu_1 \|s_i\|^2 \leq s_i^T H_k s_i$$

and hence the test at Step 2 is satisfied.

Let now $i_k$ be the iteration index produced by Algorithm TN at iteration $k$ of Algorithm TNNL; then we can write:

$$H_k p_{i_k} + g_k = r_{i_k}, \quad \text{where} \quad \|r_{i_k}\| \leq \eta_k \|g_k\| \leq \|g_k\|^2,$$

so that, for $k \geq k_2$, we obtain:

$$g_k^T p_{i_k} = -g_k^T H_k^{-1} g_k + g_k^T H_k^{-1} r_{i_k} \leq \left[ -\frac{1}{\mu_2} + \frac{1}{\mu_1} \|g_k\| \right] \|g_k\|^2.$$

Hence, for $k$ sufficiently large, say $k \geq k_3 \geq k_2$, we have, as $\|g_k\| \to 0$, that $g_k^T p_{i_k} < 0$ and that, for any given $\sigma_1 > 0$:

$$|g_k^T p_{i_k}| \geq \left[ \frac{1}{\mu_2} - \frac{1}{\mu_1} \|g_k\| \right] \|g_k\|^2 \geq \sigma_1 \|g_k\|^3.$$

Moreover, for $k \geq k_3$ we have

$$\|p_{i_k}\| \leq \frac{1}{\mu_1} \left[ \|g_k\| + \|r_{i_k}\| \right] \leq \frac{2}{\mu_1} \|g_k\|$$

and hence, for any given $\sigma_2 > 0$:

$$\|p_{i_k}\|^2 \leq \frac{4}{\mu_1^2} \|g_k\|^2 \leq \sigma_2 \|g_k\|.$$

We can conclude that, for $k$ sufficiently large, the test at Step 4 of Algorithm TN is satisfied by the direction $d_k = p_{i_k}$. We observe now that, for $k$ sufficiently large, as the matrix $H_k$ is positive definite and the test at Step 2 of Algorithm TN is satisfied, the assumptions of Theorem A.4 of [2] hold and hence there exists an index $k_4 \geq k_3$ such that the stepsize $\alpha_k = 1$ is acceptable in Algorithm NL for $k \geq k_4$. Thus, for $k \geq k_4$, we have $x_{k+1} = x_k + d_k$,where $H_k d_k = -g_k + r_k$, with $\|r_k\| \leq \|g_k\|^2$, so that the quadratic convergence rate follows from the assumptions made and the results of [1].

△

Algorithm TNNL can be modified in order to avoid the need for evaluating second order derivatives. In particular it is possible, as proposed in [2] and [8] to replace the matrix vector product $q_i = H_k s_i$ computed in Algorithm TN with the finite difference approximation:

$$H_k s_i \simeq \frac{1}{\tau_i}\Big[g(x_k + \tau_i s_i) - g(x_k)\Big] \tag{4}$$

where, denoting by $\rho$ the "machine precision", the stepsize $\tau_i$ can be chosen as:

$$\tau_i = \frac{\sqrt{\rho}}{\|s_i\|}.$$

The discretized version of Algorithm TNNL, obtained by employing the finite difference formula (4), will be coded as TNNLD.

## 4. NUMERICAL RESULTS

Algorithms TNNL and TNNLD defined in the preceding section have been tested on a set of large dimensional problems obtained from standard test functions.

The parameters which appear in the algorithms have been chosen as follows:

$$M = 10,\ \sigma = 0.5,\ \gamma = 10^{-3},\ \theta = 10^{-1},$$

$$\varepsilon = 10^{-8},\ \sigma_1 = 10^{-12},\ \sigma_2 = 10^{12},\ \rho = 10^{-16}$$

For each problem the computational results are reported by specifying the number $n_l$ of line searches required to attain convergence, the number $n_f$ of function evaluations, the number $n_g$ of gradient evaluations (note that for Algorithm TNNL we have obviously $n_g = n_l + 1$) and the largest number $i_{max} = \max_k i_k$ of conjugate gradient iterations required for approximating the Newton's direction with the prescribed accuracy.

When the sequence $\{f_k\}$ obtained with $M = 10$ is not monotonic, we add also, for comparison, the results obtained for $M = 0$ which correspond to the use of Armijo's line search rule.

For all problems the termination criterion was $\| g_k \| \leq 10^{-5}$, although in most of cases, the last step achieves a considerably higher accuracy.

Problem 1. Separated Rosenbrock function [8]

$$f(x) = \sum_{i=1}^{n/2} [100(x_{2i} - x_{2i-1}^2)^2 + (1 - x_{2i-1})^2]$$

$$x_o = [-1.2, 1, \ldots, -1.2, 1]^T \ , \ x^* = [1, \ldots, 1]^T \ , \ f(x^*) = 0.$$

This function is an $n$-dimensional extension of Rosenbrock's function with a block-diagonal Hessian matrix. It can be easily seen that, because of the problem structure, the behaviour of a Newton-type method does not depend on the problem dimension. In Table 1 we report, as an example, the results obtained for $n = 1000$. We note that the use of the nonmonotone line search is beneficial both in terms of line searches and in terms of function evaluations and that Algorithm TNNLD has the same behaviour of Algorithm TNNL; moreover, as expected, the search direction is computed in at most two conjugate gradient iterations. In particular we remark that the Truncated-Newton method has a built-in capability of exploiting the problem structure; in fact, in this problem, Algorithm TNNLD outperforms any unstructured Quasi-Newton method.

TABLE 1. Results for Problem 1.

| Code | $n$ | $n_l$ | $n_f$ | $n_g$ | $i_{max}$ |
|---|---|---|---|---|---|
| TNNL ($M = 10$) | 1000 | 14 | 21 | 15 | 2 |
| TNNL ($M = 0$) | 1000 | 22 | 31 | 23 | 2 |
| TNNLD ($M = 10$) | 1000 | 14 | 21 | 40 | 2 |
| TNNLD ($M = 0$) | 1000 | 22 | 31 | 63 | 2 |

Problem 2. Extended Rosenbrock function [8]

$$f(x) = \sum_{i=1}^{n-1} [100(x_{i+1} - x_i^2)^2 + (1 - x_i)^2]$$

$$x_o = [-1.2, 1, \ldots, -1.2, 1]^T \ , \ \hat{x}_o = [2, \ldots, 2]^T, \quad x^* = [1, \ldots, 1]^T \ , \ f(x^*) = 0.$$

This function, which is still an $n$-dimensional extension of Rosenbrock's function, has a tridiagonal Hessian matrix. Two initial points, among those proposed in the literature, have been considered. The results obtained starting from $x_o$ and $\hat{x}_o$ are shown in Tables 2 and 3 respectively. From Tables 2 and 3 we may note that also in this case the use of the nonmonotone line search technique is advantageous; this is more evident in correspondence to the starting point $x_o$ and for increasing values of the problem dimension. It is interesting to observe that $i_{max}$ remains approximately constant at a value which is well beyond the value of $n$ and that Algorithm TNNLD has the same behaviour of Algorithm TNNL.

TABLE 2. Results for Problem 2 starting from $x_o$.

| Code | $n$ | $n_l$ | $n_f$ | $n_g$ | $i_{max}$ |
|---|---|---|---|---|---|
| TNNL ($M = 10$) | 50 | 86 | 87 | 87 | 20 |
| | 100 | 150 | 153 | 151 | 24 |
| | 1000 | 1335 | 1341 | 1336 | 26 |
| TNNL ($M = 0$) | 50(*) | 90 | 102 | 91 | 21 |
| | 100 | 168 | 196 | 169 | 21 |
| | 1000 | 1494 | 1698 | 1495 | 25 |
| TNNLD ($M = 10$) | 50 | 86 | 87 | 1226 | 20 |
| | 100 | 150 | 153 | 2455 | 24 |
| | 1000 | 1335 | 1341 | 27664 | 26 |
| TNNLD ($M = 0$) | 50(*) | 91 | 107 | 1414 | 24 |
| | 100 | 168 | 196 | 2756 | 21 |
| | 1000 | 1496 | 1711 | 30991 | 25 |

(*) the termination criterion is satisfied at a point near $[-1, 1, \ldots, 1]^T$.

TABLE 3. Results for Problem 2 starting from $\hat{x}_o$.

| Code | $n$ | $n_l$ | $n_f$ | $n_g$ | $i_{max}$ |
|---|---|---|---|---|---|
| TNNL ($M = 10$) | 100 | 14 | 15 | 15 | 24 |
| | 1000 | 13 | 14 | 14 | 19 |
| | 10000 | 13 | 14 | 14 | 27 |
| TNNL ($M = 0$) | 100 | 14 | 18 | 15 | 22 |
| | 1000 | 14 | 18 | 15 | 21 |
| | 10000 | 14 | 17 | 15 | 23 |
| TNNLD ($M = 10$) | 100 | 14 | 15 | 158 | 24 |
| | 1000 | 13 | 14 | 144 | 19 |
| | 10000 | 13 | 14 | 148 | 27 |
| TNNLD ($M = 0$) | 100 | 14 | 18 | 183 | 22 |
| | 1000 | 14 | 18 | 180 | 21 |
| | 10000 | 14 | 17 | 159 | 23 |

In Table 4 we show the effects of varying the parameter $\theta$ (which appears in the truncation criterion), in correspondence to the starting point $x_o$ and for $n = 100$. In the last column we report the cumulative number $i_c$ of conjugate gradient iterations, that is: $i_c = \sum_{k=1}^{n_l} i_k$.

It can be observed that for decreasing values of $\theta$ both the number of line searches and the number of function evaluations remain approximately constant, whereas the number of conjugate gradient iterations increases considerably. This evidentiates the benefits of the Truncated-Newton approach, especially in the case of Algorithm TNNLD.

Another interesting observation is that the nonmonotone line search rule is beneficial in all cases, even when an inaccurate approximation of the Newton's direction is employed in the initial stages of the minimization process.

TABLE 4. Effects of varying $\theta$ in Problem 2 ($n = 100$)

| Code | $\theta$ | $n_l$ | $n_f$ | $n_g$ | $i_c$ |
|---|---|---|---|---|---|
| TNNL ($M = 10$) | 1 | 149 | 150 | 150 | 1862 |
| | $10^{-1}$ | 150 | 153 | 151 | 2304 |
| | $10^{-3}$ | 147 | 148 | 148 | 3103 |
| | $10^{-5}$ | 147 | 148 | 148 | 3880 |
| TNNL ($M = 0$) | 1 | 166 | 193 | 167 | 2123 |
| | $10^{-1}$ | 168 | 196 | 169 | 2587 |
| | $10^{-3}$ | 165 | 189 | 166 | 3491 |
| | $10^{-5}$ | 165 | 190 | 166 | 4325 |
| TNNLD ($M = 10$) | 1 | 149 | 150 | 2012 | 1862 |
| | $10^{-1}$ | 150 | 153 | 2455 | 2304 |
| | $10^{-3}$ | 147 | 148 | 3250 | 3103 |
| | $10^{-5}$ | 147 | 148 | 4050 | 3902 |
| TNNLD ($M = 0$) | 1 | 166 | 192 | 2278 | 2111 |
| | $10^{-1}$ | 168 | 196 | 2756 | 2587 |
| | $10^{-3}$ | 164 | 185 | 3630 | 3465 |
| | $10^{-5}$ | 165 | 192 | 4553 | 4387 |

Problem 3. Dixon function [3]

$$f(x) = (x_1 - 1)^2 + \sum_{i=2}^{n} i(2x_i^2 - x_{i-1})^2, \quad x_o = [1, \ldots, 1]^T \ , \ f(x^*) = 0.$$

This function has a tridiagonal Hessian matrix with a full set of distinct eigenvalues. From Table 5 we note that in this case the maximum number $i_{max}$ of conjugate gradient iterations increases with $n$. This is due to the fact that the condition number of the Hessian matrix increases with the problem dimension.

TABLE 5. Results for Problem 3

| Code | $n$ | $n_l$ | $n_f$ | $n_g$ | $i_{max}$ |
|---|---|---|---|---|---|
| TNNL ($M = 10$) | 50 | 8 | 9 | 9 | 39 |
| | 100 | 8 | 9 | 9 | 51 |
| | 1000 | 10 | 11 | 11 | 204 |
| | 10000 | 12 | 13 | 13 | 599 |
| TNNLD ($M = 10$) | 50 | 8 | 9 | 155 | 41 |
| | 100 | 8 | 9 | 198 | 54 |
| | 1000 | 10 | 11 | 756 | 210 |
| | 10000 | 12 | 13 | 2270 | 601 |

## 4. CONCLUSIONS

On the basis of the computational experience the following indications can be given.

(i) A nonmonotone line search technique may allow a considerable saving, also when it is employed in association with a Truncated-Newton algorithm. In particular, a reduction of both the number of line searches and of the number of function evaluations can be obtained even if inaccurate approximations of the Newton directions are employed in the initial stages of the minimization process. It was experienced that the advantages of the nonmonotone line search increase especially in the solution of ill-conditioned problems.

(ii) The Truncated-Newton method appears to be a valuable tool for the solution of large dimensional problems. In particular, as already observed in [2], it may significantly outperform the nonlinear conjugate gradient method, when the Hessian matrix has some specific sparsity structure, or when accurate solutions are sought.

(iii) When the problem dimension is very large and the computation of the matrix vector product $H_k s_i$ is time-consuming, Algorithm TNNLD appears to be much more convenient than Algorithm TNNL. In fact, if the Hessian matrix cannot be stored, it may be necessary, in Algorithm TNNL, to evaluate the second order derivatives at each minor iteration of the conjugate gradient algorithm.

## REFERENCES


[1] R.S. Dembo, S.C. Eisenstat and T. Steihaug, *Inexact Newton methods*, SIAM J. Numer. Anal., 19 (1982), pp.400-408.

[2] R.S. Dembo and T. Steihaug, *Truncated-Newton algorithms for large-scale unconstrained optimization*, Math. Prog., 26 (1983), pp.190-212.

[3] L.C.W. Dixon and R.C. Price, *Numerical experience with the Truncated-Newton method*, Tech. Rep. No. 169, Numerical Optimization Centre, The Hatfield Polytechnic, Hatfield, UK, 1986.

[4] L. Grippo, F. Lampariello and S. Lucidi, *A Truncated-Newton method with nonmonotone line search for unconstrained optimization*, J. Optim. Theory Appl., (to appear).

[5] L. Grippo, F. Lampariello and S. Lucidi, *A nonmonotone line search technique for Newton's method*, SIAM J. Numer. Anal., 23 (1986), pp.707-716.

[6] D.P. Bertsekas, *Constrained optimization and Lagrange multiplier methods*, Academic Press, New York, 1982.

[7] M.R. Hestenes, *Conjugate direction methods in optimization*, Springer-Verlag, New York, 1980.

[8] N.K. Garg and R.A. Tapia, *QDN: A variable storage algorithm for unconstrained optimization*, Tech. Rep., Department of Mathematical Sciences, Rice University, Houston, TX, 1977.

[9] J.P. Bulteau and J.P. Vial, *A restricted trust region algorithm for unconstrained optimization*, J. Optim. Theory Appl., 47 (1985), pp.413-435.

# CENTERED NEWTON METHOD FOR MATHEMATICAL PROGRAMMING

Kunio TANABE
The Institute of Statistical Mathematics
4-6-7 Minamiazabu, Minatoku, Tokyo, Japan 106

ABSTRACT

The purpose of this paper is to introduce a generic class of algorithms for solving a system of nonlinear equations, Linear Programming problems, Quadratic Programming problems, Nonlinear Programming problems and general complementarity problems. The algorithms were obtained by modifying the standard Newton-Raphson method applied to a system of nonlinear equations in the complementarity conditions so that it is biased towards 'center curve' passing through the solutions. The search direction of the methods is a positive combination of the Newton direction and a 'centering' direction which is also given by applying the Newton method to a projected system of the complementarity equations. These two directions guide the generated sequence of the approximations towards the solution and the center variety respectively. A class of 'penalized norms' and 'guiding cones' is also introduced for choosing step lengths in bivariate search.

## 1. CENTERED NEWTON METHOD FOR SOLVING A SYSTEM OF NONLINEAR EQUATIONS

We consider the problem of solving a system of nonlinear equations,

$$(1.1)\qquad r_1(z) = 0, \quad r_2(z) = 0, \ldots, \quad r_k(z) = 0,$$

where $r_i(z)$'s are twice continuously differentiable function of an k-dimensional column vector variable $z$ of real entries. The set of the solutions of the system (1.1) will be denoted by $V$. The set of points which satisfies at least one of the equations in (1.1) will be denoted by $V'$. The k-dimensional column vector whose i-th entry is $r_i(z)$ will be denoted by $r(z)$, which is a differentiable mapping from the k-dimensional Euclidean space $R^k$ to itself. The Jacobian matrix of $r(z)$ will be denoted by $J_r(z)$. The k-dimensional column vector whose i-th entry is $\mathrm{sign}(r_i(z))$ will be denoted by $s(z)$, where sign(y) = -1(if y<0), 0(if y=0) and 1(if y>0). Let the set $S$ of singular points of the system (1.1) be defined by $S = \{z \in R^k: \text{ rank } J_r(z) < k\}$, and let $S^c$ be the complement of the set $S$.

Given an d-dimensional vector $v$, let $v_i$ denote its i-th entry, let $v^t$ denote the transpose of $v$, let $|v|$ denote the vector whose i-th entry is $|v_i|$, let $[v]$ denote the diagonal matrix of order d, whose i-th diagonal entry is $v_i$, i.e., $[v] = diag(v_1, v_2, \dots, v_d)$, let $\|v\|_1$ and $\|v\|_2$ denote the 1-norm $\sum|v_i|$ and the Euclidean norm of v respectively, and let $\pi(v)$ denote the product of all the elements of v, i.e., $\pi(v) = v_1 v_2 \dots v_d$. Given two vectors u and v, let $(u,v)$ denote the inner product $v^t u$ of the two vectors, and let $[u]/[v]$ denote the diagonal matrix, $[u][v]^{-1} = [v]^{-1}[u]$. Let $1_d$ denote the d-dimensional column vector whose entries are all one. Given two column vectors x and y, let $(x; y)$ denote the column vector $(x^t, y^t)^t$.

We will describe a generic class of iterative methods of the form,

(1.2) $$z^{i+1} = z^i + d(\alpha, \beta; z^i), \quad (i=0,1,2,\dots).$$

The displacement vector $d(\alpha, \beta; z) \in R^k$ of our methods at the current point z is given by the following linear equation,

(1.3) $$\begin{aligned} J_r(z)d(\alpha, \beta; z) &= -(\alpha I + \beta P(z))r(z) \\ &= -(\alpha+\beta)r(z) + \beta(\sum|r_i(z)|/\|s(z)\|_2^2)s(z), \end{aligned}$$

where $P(z) = I - s(z)(s(z))^t/\|s(z)\|_2^2$ is an orthogonal projector, and $\alpha$ and $\beta$ are nonnegative numbers.

The vector $dn(z) = d(1, 0; z)$ is the Newton-Raphson direction for solving Eq.(1.1). The second term in the right hand side of Eq.(1.3) is introduced to bias the Newton direction towards the 'center variety' defined by

(1.4) $$C = \{z \in R^k: |r_1(z)| = |r_2(z)| = \dots = |r_k(z)|\}.$$

The set C is typically a union of curves passing through all the points in V. Specifically, the vector $dc(z) = d(0, 1; z)$, which will be called 'centering direction', is obtained by applying the Newton-Raphson method to the projected system of nonlinear equations, $P(z)r(z) = 0 \in R^k$. (See [7,8] for an analysis of continuous analogues of the Newton-Raphson method applied to an underdetermined system of nonlinear equations.) The vector dc(z), which vanishes on the center variety C, is a descent direction of the function,

(1.5) $$\rho(z) = (\|r(z)\|_1/k)/\pi(|r(z)|)^{1/k} = (\sum|r_i(z)|/k)/(\pi|r_i(z)|)^{1/k},$$

which is the ratio of the arithmetic mean to the geometric mean of all the elements of $|r(z)|$. dc(z) is tangential to the variety defined by $\|r(z)\|_1$ = const., i.e.,

(1.6) $$(\nabla\rho(z), dc(z)) \le 0, \text{ and } (\nabla\|r(z)\|_1, dc(z)) = 0, \text{ on } (V')^c.$$

The function $\rho(z)$ takes its minimum value 1 at the points in the center variety C, and serves as a measure of deviation from the center variety. On the other hand, the Newton direction $dn(z)$ is a descent direction of the norm $\|r(z)\|_1$ and tangential to the variety defined by $\rho(z)$ = const., i.e.,

(1.7) $\quad (\nabla\rho(z), dn(z)) = 0$, and $(\nabla\|r(z)\|_1, dn(z)) \leq 0$, on $(V')^c$.

See Figure. The search direction defined by (1.3) is a positive combination of the Newton direction $dn(z)$ and the centering direction $dc(z)$, i.e., $d(\alpha, \beta; z) = \alpha dn(z) + \beta dc(z)$, which will be called the 'complementary decomposition' with respect to $\|r(z)\|_1$ and $\rho(z)$. The method which uses these directions will be called 'Centered Newton' method.

The centering variety and the centering direction are affected by bad scaling of the nonlinear equations in (1.1). Hence, in practice, we scale each of the equations in (1.1) by

(1.8) $\quad r_i(z) := r_i(z)/\|\nabla r_i(z)\|_2, \quad i=1,2,\ldots,k,$

and apply the centered Newton method.

## 2. PLANE SEARCH: PENALIZED NORM AND GUIDING CONE

In this section we discuss how to determine the bi-variate step lengths', $\alpha$ and $\beta$ in the generic class of (modified) centered Newton methods,

(2.1) $\quad z^{i+1} = z^i + q(\alpha, \beta; z^i) = z^i + \alpha dn(z^i) + \beta dc(z^i), (i=0,1,2,\ldots).$

We introduce two kinds of methods for this purpose, which will be called 'penalized norm' method and 'guiding cone' method respectively.

The first method is used in the general centered Newton methods.

PENALIZED NORM METHOD: The centered Newton direction $\alpha dn(z) + \beta dc(z)$ is always a descent direction of the penalized norm,

(2.2) $\quad \mu(\omega, z) = (\rho(z))^{\omega}\|r(z)\|_1 = k(\|r(z)\|_1/k)^{1+\omega}/\Pi(|r(z)|)^{\omega/k}$

$\quad = k(\Sigma\, |r_i(z)|/k)^{1+\omega}/(\Pi|r_i(z)|)^{\omega/k},$

for any positive numbers $\alpha$, $\beta$ and $\omega$. We have the inequality,

(2.3) $\quad \mu(\omega, z) = (\rho(z))^{\omega}\|r(z)\|_1 \geq \|r(z)\|_1,$

which implies that if $\mu(z)$ is reduced to zero, so is $\|r(z)\|_1$. Hence, we can determine $\alpha$ and $\beta$ by referring to the function, $\mu(\omega, z^i + \alpha dn(z^i) + \beta dc(z^i))$. Specifically, we determine the step lengths $\alpha$ and $\beta$ so that the step length $\alpha$ in the Newton direction is as large as possible under the conditions,

(2.4) $\quad 0 < \alpha \leq 1,\ 0 \leq \beta$, and

(2.5) $\quad \mu(\omega,\ z^i + \alpha dn(z^i) + \beta dc(z^i)) < (1-\delta)\mu(\omega,\ z^i)$,

where $\delta$ is a small positive number which is appropriately chosen. Note here that $\omega$ can be smaller than 1.

The second method is used in the centered Newton methods for solving mathematical programming problems.

GUIDING CONE METHOD: For a positive number $\lambda \geq 1$, we define a 'guiding cone' by

(2.6) $\quad \mathrm{Cone}(r,\ \lambda) = \{z \in R^k : \|r(z)\|_1/k \leq \lambda(\Pi(|r(z)|))^{1/k}\}.$

With an appropriate choice of the value $\lambda > 1$, we determine the step lengths $\alpha$ and $\beta$ so that the step length $\alpha$ in the Newton direction is as large as possible under the conditions, (2.4),

(2.8) $\quad z^{i+1} = z^i + \alpha dn(z^i) + \beta dc(z^i) \in \mathrm{Cone}(r,\ \lambda)$, and

(2.9) $\quad \|r(z^{i+1})\|_1 < (1-\delta)\|r(z^i)\|_1.$

Taking into account of the inequality constraints, we specify further the cone in the following sections.

## 3. CENTERED NEWTON METHOD FOR LINEAR AND QUADRATIC PROGRAMMING.

In this section we consider the problem of solving the quadratic (and linear) programming problem,

(3.1) Maximize $c^t x - x^t Dx/2$ subject to $Ax \leq b$, and $x \geq 0$,

where A is an m x n real matrix, D is an n x n nonnegative definite matrix, $y,\ b \in R^m$ and $x,\ c \in R^n$ are column vectors. The problem reduces to a linear programming problem when D is the zero matrix. The dual problem of (3.1) is

(3.2) Minimize $b^t y + x^t Dx/2$ subject to $A^t y + Dx \geq c$, and $y \geq 0$.

These problems are equivalent to one of finding vectors x and y which satisfy the complementarity conditions,

(3.3) $\quad r(z) = \begin{pmatrix} [x](A^t y+Dx-c) \\ [y](b-Ax) \end{pmatrix} = \begin{pmatrix} [A^t y+Dx-c]x \\ [b-Ax]y \end{pmatrix} = 0 \in R^{m+n}$, and

(3.4) $\quad x \geq 0,\ y \geq 0,\ b - Ax \geq 0$, and $A^t y + Dx - c \geq 0$,

where $z = (x;\ y)$ is an (m+n) dimensional column vector. We assume that an initial interior feasible point $z^0 = (x^0,\ y^0)$ is known, which satisfies the interior feasibility conditions,

(3.5) $\quad x > 0,\ y > 0,\ b - Ax > 0$, and $A^t y + Dx - c > 0.$

If the centered Newton method given in Sections 1 and 2 is applied to (3.3) in such a way that the interior feasibility (3.5) is maintained throughout the iteration, we obtains the search directions, dn = (xn; yn) and dc = (xc; yc) by solving the linear equations,

$$(3.6)\quad \begin{bmatrix} [A^t y+Dx-c]/[x]+D & A^t \\ -A & [b-Ax]/[y] \end{bmatrix} \begin{bmatrix} xn \\ yn \end{bmatrix} = - \begin{bmatrix} A^t y+Dx-c \\ b-Ax \end{bmatrix},$$

$$(3.7)\quad \begin{bmatrix} [A^t y+Dx-c]/[x]+D & A^t \\ -A & [b-Ax]/[y] \end{bmatrix} \begin{bmatrix} xc \\ yc \end{bmatrix} = - \begin{bmatrix} A^t y+Dx-c \\ b-Ax \end{bmatrix} + \gamma \begin{bmatrix} [x]^{-1} 1 \\ [y]^{-1} 1 \end{bmatrix},$$

where the scalar function $\gamma$ is defined by,

$$(3.8)\qquad \gamma = \|r(z)\|_1/(m+n) = (b^t y - c^t x + x^t Dx)/(m+n) \geq 0.$$

Recall here that the inequalities (1.6-7) hold.

We generate a sequence of feasible points $\{z^i\}$ by the iteration,

$$(3.9)\qquad z^{i+1} = (x^{i+1}; y^{i+1}) = (x^i; y^i) + \alpha(xn; yn) + \beta(xc; yc),$$

where the step lengths $\alpha$ and $\beta$ are determined by the methods suggested in Section 2. Since the relative interior feasibility conditions (3.5) are easily satisfied throughout the iterations, we use the following feasible center variety and feasible cone in (2.8), which are defined respectively by

$$(3.10)\qquad C+ = \{z \in R^{m+n}: [x](A^t y+Dx-c) = \gamma 1, \text{ and } [y](b-Ax) = \gamma 1, \text{ where } \gamma \geq 0, \text{ and } x \text{ and } y \text{ satisfy } (3.4)\}, \text{ and}$$

$$(3.11)\qquad \mathrm{Cone}+(r, \lambda) = \{z \in R^{m+n}: (b^t y - c^t x + x^t Dx)/(m+n) \leq \lambda[\pi(x)\pi(y)\pi(b-Ax)\pi(A^t y+Dx-c)]^{1/(m+n)}, \text{ where } x \text{ and } y \text{ satisfy } (3.4)\},$$

where $\lambda \geq 1$. The penalized norm is also re-defined in this case by

$$(3.12)\qquad \mu+(\omega, z) = (m+n)\{(b^t y-c^t x+x^t Dx)/(m+n)\}^{1+\omega} / \{\pi(x)\pi(y)\pi(b-Ax)\pi(A^t y+Dx-c)\}^{\omega/(m+n)}.$$

In the rest of this section we consider the linear programming case where $D = 0$. Then, the equations (3.6-7) give the solutions,

$$\begin{aligned} xn &= \{[A^t y-c]/[x] + A^t([y]/[b- Ax])A\}^{-1} c, \\ yn &= -\{[b- Ax]/[y] + A([x]/[A^t y-c])A^t\}^{-1} b, \\ (3.13)\qquad xc &= \{[A^t y-c]/[x] + A^t([y]/[b- Ax])A\}^{-1} \\ &\qquad \{c + \gamma([x]^{-1} 1 - A^t [b-Ax]^{-1} 1)\}, \end{aligned}$$

$$yc = -\{[b-Ax]/[y] + A([x]/[A^ty-c])A^t\}^{-1} \{b - y([y]^{-1}1 + A[A^ty-c]^{-1}1)\}.$$

If the condition (3.5) is satisfied, we have

$$(3.14) \qquad c^txn \geq 0, \text{ and } b^tyn \leq 0.$$

We have also

$$(3.15) \qquad b^tyn - c^txn = -(b^ty - c^tx), \text{ and } b^tyc - c^txc = 0.$$

Hence, we have

$$(3.16) \quad \|r(z + \alpha dn + \beta dc)\|_1 = b^t(y + \alpha yn + \beta yc) - c^t(x + \alpha xn + \beta xc) = (1-\alpha)(b^ty - c^tx) = (1-\alpha)\|r(z)\|_1.$$

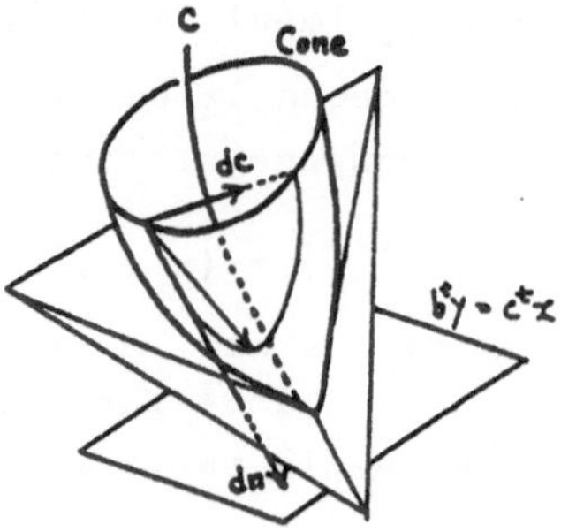


Figure: Newton Direction, Centering Direction, Center Curve, and Guiding Cone.

4. CENTERED NEWTON METHOD FOR LINEAR PROGRAMMING: NONSYMMETRIC COMPLEMENTARITY ENFORCING APPROACH

In this section we consider the problem of solving the linear programming problem,

$$(4.1) \qquad \text{Maximize } c^tx \quad \text{subject to } Ax = b, \text{ and } x \geq 0,$$

where $A$ is an $m \times n$ real matrix, $y, b \in R^m$ and $x, c \in R^n$ are column vectors. The dual problem of (4.1) is

$$(4.2) \qquad \text{Minimize } b^ty \quad \text{subject to } A^ty \geq c.$$

These problems are equivalent to one of finding vectors $x$ and $y$ which satisfy the complementarity conditions,

$$(4.3) \quad r(z) = \begin{pmatrix} [x](A^ty-c) \\ b-Ax \end{pmatrix} = \begin{pmatrix} [A^ty-c]x \\ b-Ax \end{pmatrix} = 0 \in R^{m+n}, \text{ and}$$

$$(4.4) \qquad x \geq 0, \quad b - Ax = 0, \text{ and } A^ty - c \geq 0,$$

where $z = (x; y)$ is an $(m+n)$-dimensional column vector. We assume that an initial interior feasible point $z^0 = (x^0, y^0)$ is known, which satisfies the relative interior feasibility conditions,

(4.5) $\quad x > 0, \quad b - Ax = 0, \quad \text{and} \quad A^t y - c > 0.$

If the centered Newton method given in Sections 1 and 2 is applied to Eq.(4.3) in such a way that the relative interior feasibility (4.5) is maintained throughout the iteration, we obtains the search directions, $dn = (xn; yn)$ and $dc = (xc; yc)$ by solving the linear equations,

$$(4.6) \quad \begin{bmatrix} [A^t y-c]/[x] & A^t \\ -A & 0 \end{bmatrix} \begin{bmatrix} xn \\ yn \end{bmatrix} = -\begin{bmatrix} A^t y-c \\ b- Ax \end{bmatrix},$$

$$(4.7) \quad \begin{bmatrix} [A^t y-c]/[x] & A^t \\ -A & 0 \end{bmatrix} \begin{bmatrix} xc \\ yc \end{bmatrix} = -\begin{bmatrix} A^t y-c \\ b- Ax \end{bmatrix} + \begin{bmatrix} \gamma [x]^{-1} 1 \\ 0 \end{bmatrix}.$$

where the scalar function $\gamma$ is defined by

$$(4.8) \quad \gamma = \|r(z)\|_1/n = (b^t y - c^t x)/n \geq 0.$$

Note that the second equality in (4.8) holds if the equality in (4.5) is satisfied. The equations (4.6-7) give the solutions

$$(4.9) \quad \begin{aligned} xn &= -x + ([x]/[A^t y-c])A^t \{A([x]/[A^t y-c])A^t\}^{-1} b, \\ yn &= -\{A([x]/[A^t y-c])A^t\}^{-1} b, \\ xc &= -x + ([x]/[A^t y-c])A^t \{A([x]/[A^t y-c])A^t\}^{-1} b \\ &\quad + \gamma [I - ([x]/[A^t y-c])A^t \{A([x]/[A^t y-c])A^t\}^{-1} A][A^t y-c]^{-1} 1, \\ yc &= -\{A([x]/[A^t y-c])A^t\}^{-1} (b - \gamma A[A^t y-c]^{-1} 1). \end{aligned}$$

Eq.(4.7) is slightly different from the one which would be obtained by applying directly the centered Newton method described in Section 1. It has been modified so as to take advantage of the fact that the condition (4.5) is easily kept satisfied, since a single step of the Newton method enforces the equality constraint, i.e., $A(x + xn) = b$.

We generate a sequence of feasible points $\{z^i\}$ by the iteration, (3.9), where the step lengths $\alpha$ and $\beta$ are determined by the methods suggested in Section 2. However, the relative interior feasibility conditions (4.5) are easily satisfied throughout the iterations, we use the following 'relative' feasible center variety and 'relative' feasible cone in (2.8), which are defined respectively by

$$(4.10) \quad C+ = \{z \in R^{m+n}: [x](A^t y-c) = \gamma 1, \text{ and } b-Ax = 0, \text{ where } \gamma = (b^t y-c^t x)/n \geq 0, \text{ and } x \text{ and } y \text{ satisfies } (\bar{4}.4)\}, \text{ and}$$

$$(4.11) \quad Cone+(r, \lambda) = \{z \in R^{m+n}: (b^t y-c^t x)/n \leq \lambda [\Pi(x)\Pi(A^t y-c)]^{1/n}, \text{ and } x \text{ and } y \text{ satisfies } (4.4)\},$$

where $\lambda \geq 1$. The penalized norm is also re-defined in this case by

(4.12) $\quad \mu+(w, z) = n\{(b^t y - c^t x)/n\}^{1+w}/\{\Pi(x)\Pi(A^t y - c)\}^{w/n}$.

It is interesting to note that when the point (x; y) is in the Center variety C+, the Newton direction yn in (5.13) is given by

(4.13) $\quad yn = -\gamma(A[A^t y - c]^{-2} A^t)^{-1} b,$

which is a component of the Karmarkar's direction analyzed in [1], and that the centering direction yc is the sum of two components, yn and

(4.14) $\quad \gamma^2 (A[A^t y - c]^{-2} A^t)^{-1} A[A^t y - c]^{-1} 1$

which are contained in the search directions used in [2,14].

## 5. CENTERED NEWTON METHOD FOR NONLINEAR PROGRAMMING

In this section we consider the problem of solving a nonlinear programming problem,

(5.1) $\quad$ Minimize $f(x)$ subject to $g(x) \geq 0 \quad \varepsilon\ R^m$,

where $f(x)$ and $g_i(x)(i=1,2,\dots,m)$ are twice continuously differentiable functions of an n-dimensional column vector x. Under the mild conditions there exist a multiplier y* corresponding to the optimal solution x* such that z* = (x*; y*) is a solution of the Kuhn-Tucker condition,

(5.2) $$r(z) = \begin{pmatrix} \nabla f(x) - J_g^t(x) y \\ \theta[y]g(x) \end{pmatrix} = 0 \quad \varepsilon\ R^{m+n},$$

(5.3) $\quad g(x) \geq 0 \quad$ and $\quad y \geq 0,$

where z = (x; y) is an (m+n)-dimensional column vector and $\theta$ is a positive number which is introduced to balance the residuals of the two sets of components of r(z). We assume that an initial interior feasible point $z^0 = (x^0, y^0)$ is known, which satisfies the interior feasibility conditions,

(5.4) $\quad g(x) > 0 \quad$ and $\quad y > 0.$

If the centered Newton method given in Sections 1 and 2 is applied to (5.2) in such a way that the interior feasibility (5.4) is maintained throughout the iteration, we obtains the search directions, dn = (xn; yn) and dc = (xc; yc) by solving the linear equations,

(5.5) $$\begin{bmatrix} H(x, y) & -J_g^t(x) \\ J_g^t(x) & [g(x)]/[y] \end{bmatrix} \begin{bmatrix} xn \\ yn \end{bmatrix} = - \begin{bmatrix} \nabla f(x) - J_g^t(x) y \\ g(x) \end{bmatrix},$$

$$(5.6)\quad \begin{bmatrix} H(x,y) & -J_g^t(x) \\ J_g(x) & [g(x)]/[y] \end{bmatrix}\begin{bmatrix} xc \\ yc \end{bmatrix} = -\begin{bmatrix} \nabla f(x)-J_g^t(x)y \\ g(x) \end{bmatrix} + \begin{bmatrix} \gamma 1 \\ (\gamma/\theta)[y]^{-1}1 \end{bmatrix}.$$

where $H(x, y) = \nabla^2 f(x) - \sum y_i \nabla^2 g_i(x)$ is the Hessian matrix of the Lagrangean function, $L(x, y) = f(x) - (g(x), y)$, and the scalar function $\gamma$ is defined by

$$(5.7)\quad \gamma = \|r(z)\|_1/(m+n) = \{\|\nabla f(x)-J_g^t(x)y\|_1 + \theta(g(x), y)\}/(m+n) \geq 0.$$

If we put $y^\dagger = y + yn$, then Eq.(5.5) is equivalent to

$$(5.8)\quad \begin{bmatrix} H(x,y) & -J_g^t(x) \\ J_g^t(x) & [g(x)]/[y] \end{bmatrix}\begin{bmatrix} xn \\ y^\dagger \end{bmatrix} = -\begin{bmatrix} \nabla f(x) \\ 0 \end{bmatrix}.$$

Similarly, if we put $y^{\dagger\dagger} = y + yc$, then Eq.(5.6) is equivallent to

$$(5.9)\quad \begin{bmatrix} H(x,y) & -J_g^t(x) \\ J_g(x) & [g(x)]/[y] \end{bmatrix}\begin{bmatrix} xc \\ y^{\dagger\dagger} \end{bmatrix} = -\begin{bmatrix} \nabla f(x) \\ 0 \end{bmatrix} + \begin{bmatrix} \gamma 1 \\ (\gamma/\theta)[y]^{-1}1 \end{bmatrix}.$$

To avoid the costly computation and the singulatity of the Hessian matrix $H(x, y)$, we can replace it by some positive definite approximation. Note that as long as the interior feasibility condition (5.4) is satisfied, the coefficient matrix is a nonsingular matrix in this case.

We generate a sequence of feasible points $\{z^i\}$ by the iteration, (3.9), where the step lengths $\alpha$ and $\beta$ are determined by the methods suggested in Section 2. We use also the following feasible center variety and feasible cone in (2.8), which are defined respectively by

$$(5.10)\quad C+ = \{z \in R^{m+n}:\ \nabla f(x)-J_g^t(x)y = \gamma 1,\ [y]g(x) = (\gamma/\theta)1,$$
$$\text{where } \gamma = \|r(z)\|_1/(m+n) \geq 0, \text{ and } x \text{ and } y \text{ satisfy (5.3)}\}, \text{ and}$$

$$(5.11)\quad Cone+(r, \lambda) = \{z \in R^{m+n}:\ [\|\nabla f(x)-J_g^t(x)y\|_1 + \theta(g(x), y)]/(m+n)$$
$$\leq \lambda\{\theta^m\pi(|\nabla f(x)-J_g^t(x)y|)\pi(g(x))\pi(y))\}^{1/(m+n)},$$
$$\text{where } x \text{ and } y \text{ satisfy (5.3)}\},$$

where $\lambda \geq 1$. The penalized norm is also redefined by

$$(5.12)\quad \mu+(\omega, z) = (m+n)[\{\|\nabla f(x)-J_g^t(x)y\|_1 + \theta(g(x), y)\}/(m+n)]^{1+\omega}$$
$$/\{\theta^m\pi(|\nabla f(x)-J_g^t(x)y|)\pi(g(x))\pi(y))\}^{\omega/(m+n)}.$$

APPENDIX

The centered Newton method can be applied to the problem of solving a general complementarity problem,

$$(6.1)\quad r(z) = [z]h(z) = [h(z)]z = 0,\quad z \geq 0,\quad h(z) \geq 0,$$

where $h(z)$ is a twice continuously differentiable mapping from the k-dimensional Euclidean space $R^k$ to itself. The Newton direction $dn(z)$ and the centering direction $dc(z)$ are obtained by solving the linear equations,

$$(6.2)\quad \{[h(z)]/[z] + J_h(z)\}dn(z) = -h(z),$$

$$(6.3)\quad \{[h(z)]/[z] + J_h(z)\}dc(z) = -(I - [z]^{-1}11^t[z]/k)h(z).$$

For more detail, see [13].

REFERENCES


[1] I. Adler, N. Karmarker and G. Veiga, An Implementation of Karmarker's algorithm for linear programming, Working Paper, Operations Research Center, Univ. of California, Berkeley, 1986.

[2] M. Iri and H. Imai, A multiplicative Barrier function method for linear programming, Algorithmica 1 (1986), 455-482.

[3] H. Imai, Extension of the multiplicative penalty function method for linear programming, Manuscript, (1986).

[4] N. Karmarkar, A new polynomial time algorithm for linear programming, Combinatorica 4 (1984), 373-395.

[5] N. Megiddo, Pathways to the optimal set in linear programming, Proceedings of the 7th Mathematical Programming Symposium, Japan, 1986, 1-35

[6] G. Sonnevend, An "analytical center" for polyhedrons and new classes of global algorithms for linear(smooth, convex) programming, Proceedings of the 12th IFIP Conference on System Modelling and Optimization; Lecture Notes in Control and Information Sciences, Springer, 1986.

[7] K. Tanabe, Continuous Newton-Raphson method for solving an underdetermined system of nonlinear equations, Journal of nonlinear analysis 3 (1979), 495-503.

[8] K. Tanabe, A geometric method in nonlinear programming, Journal of Optimization Theory and Applications 30 (1980), 181-210.

[9] K. Tanabe, A unified method for designing constrained optimization algorithms, Proceedings of the 2nd Mathematical Programming Symposium, Japan, 1981, 47-69.

[10] K. Tanabe, Global analysis of continuous analogues of the Levenberg-Marquardt and the Newton-Raphson methods for solving nonlinear equations, Annals of the Institute of Statistical Mathematics 37 (1985), 189-203.

[11] K.Tanabe, Complementarity-enforcing centered Newton method for linear programming, Manuscript at the Symposium, "New methods for Linear Programming", The Institute of Statistical Mathematics, (Feb., 1987).

[12] K. Tanabe, Centered Newton method for solving a system of nonlinear equations: Global method, Forthcoming.

[13] K. Tanabe, Complementality-enforcing centered Newton method for Mathematical Programming: Global Method, The Institute of Statistical Mathematics Cooperative Research Report 5, "New Methods for Linear Programming", (1987).

[14] G-de-Ghellinck and J-P Vial, A polynomial Newton method for linear programming, CORE Discussion Paper 8614 (1986).

[15] H. Yamashita, Polynomially and quadratically convergent method for linear programming, Mathematical Systems Institute Inc. Technical Report (1986).

# APPLICATION OF GENERALIZED RAIFFA SOLUTION TO MULTICRITERIA BARGAINING SUPPORT

Piotr Bronisz, Lech Krus
Systems Research Institute, Polish Academy of Sciences
6 Newelska st., 01-447 Warsaw, Poland

## 1. Introduction.

The growing interest in development of decision support systems (DSS) and their practical application is observed, especially in the domain of multicriteria decision making. In the DSS approach explicitely given utility functions are not assumed, therefore the problem can not be directly boiled down to one-shot unicriterial optimization. In that place, an interactive learning process is constructed supporting the decision maker in finding Pareto optimal solution according to his preferences. In this paper, the similar approach is proposed in the case of a bargaining problem.

A multicriteria bargaining problem is considered. The set of players, $N=\{1,2,...,n\}$, is assumed, each player has m criteria. The problem is formulated by a pair $(S,d)$ where $S \subset R^{n*m}$ is an agreement set and d is a disagreement point. The agreement set consists of feasible outcomes, any one of which will be the result if it is specified by unanimous agreement of all the players. In the event that no unanimous agreement is reached, the disagreement point is the result.

The problem is a generalization of the bargaining problem formulated and considered by Nash [1950], Raiffa [1953], Roth [1979], and others where preferences of particular player are expressed by explicitely given utility function.

In the paper, we assume that no utility functions of the players are given explicitely. An interactive negotiation scheme is considered. The goal of the scheme is to facilitate the players in finding Pareto outcomes in agreement set, according to their preferences and under some "fairness" rules. According to the scheme a sequence of nondominated cooperative solutions is generated. At each iteration, particular player, $i\in N$, expresses his preferences by means of an improvement direction vector, $\lambda_i\in\mathbb{R}^m_+$, which indicates the direction of improvement from the disagreement point. The most preferable improvement direction vectors selected by the players are basis to determinate a solution. It is shown (Bronisz, Krus [1987]) that under four axioms that describe the "fairness" rules: weak Pareto optimality, symmetry, invariance under positive affine transformations, and restricted monotonicity , there is the unique solution generalizing the Raiffa solution concept to multicriteria bargaining problem.

## 2. Problem formulation and solution.

For every point $x=(x_1,\dots,x_n)\in\mathbb{R}^{n*m}$, $x_i\in\mathbb{R}^m$, $x_i=(x_{i1},\dots,x_{im})$, let $x_{ij}$ denote the value of the j-th criterion for the i-th player. We employ a convention that for $x,y\in\mathbb{R}^k$, $x\geq y$ implies $x_i\geq y_i$ for all $i=1,\dots,k$ , $x>y$ implies $x\geq y$, $x\neq y$, $x\gg y$ implies $x_i>y_i$ for all $i=1,\dots,k$ .

We confine our consideration to the class, $\mathfrak{B}$, of all multicriteria bargaining problems $(S,d)$ satisfying the following conditions:

(i) S is compact and there is $x\in S$ such that $x\gg d$,

(ii) S is comprehensive, i.e. for $x\in S$ if $d\leq y\leq x$ then $y\in S$.

The proposed assumptions are analogous as those by Thomson [1980] in unicriteral case except convexity which is not assumed here. Condition (i) says that the agreement set is closed and bounded as well as that the problem is non-degenerate, i.e. there is an outcome in S which is better for all the players than the disagreement point (For simpicity, we assume that each player $i\in N$ maximizes values of his criteria). Condition (ii) states that the objectives are disposable, i.e. that if the players can reach the outcome x then they can reach any outcome worse than x.

Let $\lambda_i\in\mathbb{R}^m$ be an improvement direction vector of the i-th player and let $\lambda\in\mathbb{R}^{n*m}$, $\lambda=(\lambda_1,\dots,\lambda_n)$.

We define an **utopia point**, $u(S,d,\lambda)\in\mathbb{R}^{n*m}$ as

$$u(S,d,\lambda)=(u_1(S,d,\lambda_1),\dots,u_n(S,d,\lambda_n)),$$

$$u_i(S,d,\lambda_i)=\max_{\geq}\{x_i\in\mathbb{R}^m:\ x\in S,\ x\geq d,\ x_i=d_i+a\lambda_i\ \text{ for some } a\in\mathbb{R}_+\}.$$

Intuitively, the i-th component of an utopia point, $u_i(S,d,\lambda_i) \in \mathbb{R}^m$, is the maximal outcome in S for the i-th player according to his improvement direction vector $\lambda_i$.

A solution for the multicriteria bargaining problems is a function

$$f : \mathfrak{B} \times \mathbb{R}_+^{n*m} \longrightarrow \mathbb{R}^{n*m}$$

which associates to each bargaining problem $(S,d) \in \mathfrak{B}$ and each improvement direction vector $\lambda$, $\lambda \gg 0$, a point of S, denoted $f(S,d,\lambda)$.

We impose on a solution the following four axioms:

**A1.** Weak Pareto optimality. There is no $x \in S$ such that $x \gg f(S,d,\lambda)$.

**A2.** Invariance under positive affine transformations of criteria. Let $Tx = (T_1x_1,\dots,T_nx_n)$ be an arbitrary affine transformation such that $T_ix_i=(a_{ij}x_{ij}+b_{ij})_{j=1,\dots,m}$, $a_{ij}>0$, $i \in N$, and let $Lx = (L_1x_1,\dots,L_nx_n)$ be a linear transformation connected with T, i.e. $L_ix_i=(a_{ij}x_{ij})_{j=1,\dots,m}$, $i \in N$.

Then $f(TS,Td,L\lambda) = T\,f(S,d,\lambda)$.

**A3.** Symmetry. For any point $x \in R^{n*m}$ and for any permutation on N, $\pi$, let $\pi^*x=(x_{\pi(1)},\dots,x_{\pi(n)})$. We say that $(S,d)$ is a symmetric problem if $d_1=d_2=\dots=d_n$ and if $x \in S$ then for every permutation on N, $\pi$, $\pi^*x \in S$.

If $(S,d)$ is a symmetric problem and $\lambda_1=\lambda_2=\dots=\lambda_n$ then $f_1(S,d,\lambda) = f_2(S,d,\lambda) = \dots = f_n(S,d,\lambda)$.

**A4.** Restricted Monotonicity. If $(S,d)$, $(T,d)$, $\lambda \in \mathbb{R}^{n*m}$ are such that $u(S,d,\lambda) = u(T,d,\lambda)$ and $S \subset T$ then $f(S,d,\lambda) \le f(T,d,\lambda)$

The first three axioms are usually imposed on the solutions of axiomatical bargaining models. The restricted monotonicity axiom assures that all the players benefit (or at least not lose) from any enlargement of the set of feasible outcomes if the utopia point does not change. It is shown (Bronisz, Krus [1987]) that the four axioms stated above define the unique solution to the multicriteria bargaining problem, what is presented in the following theorem.

**Theorem.** There is the unique function, G, satisfying the axioms A1 - A4. It is the function defined by

$$G(S,d,\lambda) = \max_{\geq}\{x\in S:\ x=d + h[u(S,d,\lambda)-d] \text{ for some } h\in\mathbb{R}_+\}.$$

The solution is a generalization of the solution proposed by Raiffa [1953] and axiomatically characterized by Kalai, Smorodinsky [1975] (in two-person case) and by Thomson [1980] (in n-person case) for the multicriterial case. It is easy to notice that in the unicriterial case, i.e. when $m=1$, each game $(S,d)$ has a unique utopia point which coincides with the ideal point and the solution coincides with the Raiffa solution (see Raiffa [1953], Roth [1979]). Properties of the solution have been detaily studied in Bronisz, Krus [1987].

**3. Interactive negotiation scheme.**

The solution has been used in an interactive system supporting negotiation process over the bargaining problem.

The interactive negotiatioin scheme consists of two phases.

In the first one, each player $i\in N$, acts separately. He explores his maximal outcomes, $u_i(S,d,\lambda_i)$, according to his successive improvement direction vectors, $\lambda_i$. This phase allows each player to estimate the ranges of his criteria and enables the initial approximation of his preferable improvement direction vector. Let $\lambda^0_i$, $i\in N$, be the improvement direction vectors selected by the players in the first phase.

The second phase is played in several rounds. At each round t, $t=1,2,...$ , the outcome $G(S,d,\lambda^{t-1})$ is calculated, where $\lambda^{t-1}$, $\lambda^{t-1}=(\lambda^{t-1}_1,...,\lambda^{t-1}_n)$, is the direction selected by the players in round $(t-1)$. If the players agree on the outcome $G(S,d,\lambda^{t-1})$, the process ends, otherwise they modify their improvement direction vectors. For that, each player $i\in N$ can independently test different directions (related solutions are calculated assuming that the actions of the other players do not change) and then he selects a new improvement direction vector $\lambda^t_i$.

The presented scheme can support achievement of an agreement outcome. The scheme based on the presented solution concept seems to be reasonable because the properties resulting from the axioms assure "fairness" of the solution. Moreover, it is remarkable that the outcome $G(S,d,\lambda)$ is compatible with the players' improvement direction vectors, i.e. for each $i\in N$ there is a number $\beta$ such that

$G_i(S,d,\lambda)-d_i=\beta\lambda_i$, what prevents direction manipulations by the players. The interactive scheme has been applied and tested in negotiation support system on joint development program.

**References**

Bronisz P., L. Krus, [1987], "The Raiffa Solution for Multicriteria Bargaining Problems", Working Paper ZTSW-1-17/87, Systems Research Institute, Polish Academy of Sciences.

Kalai E., M. Smorodinsky, [1975], "Other Solutions to Nash's Bargaining Problem", Econometrica, Vol. 43, pp. 513-518.

Nash J.F., [1950] "The Bargaining Problem", Econometrica, Vol. 18, pp. 155-162.

Raiffa H., [1953], "Arbitration Schemes for Generalized Two-Person Games", Annals of Mathematics Studies, No. 28 pp. 361-387, Princeton.

Roth A.E., [1979], "Axiomatic Model of Bargaining", Lecture Notes in Economics and Mathematical Systems, Vol. 170, Springer-Verlag, Berlin.

Thomson W., [1980], "Two Characterization of the Raiffa Solution", Economics Letters, Vol. 6, pp.225-231.

# RECENT RESULTS ON NONDIFFERENTIABLE EXACT PENALTY FUNCTIONS

G. Di Pillo
Dipartimento di Informatica e Sistemistica, Università di Roma "La Sapienza"
Via Eudossiana 18, 00184 Roma, Italy

L. Grippo
Istituto di Analisi dei Sistemi ed Informatica del CNR
Viale Manzoni 30, 00185 Roma, Italy

## ABSTRACT

In this paper we introduce a quite natural definition of exactness for penalty functions and we show that the best known class of nondifferentiable penalty functions is exact according to this definition. Moreover, we introduce a new class of nondifferentiable exact penalty functions containing a barrier term which causes the unconstrained minimizers to be located in the interior of a compact set; this allows the construction of an unconstrained algorithm which can be shown to be globally convergent towards K-T points of the constrained problem.

## 1. INTRODUCTION

In recent years a considerable attention has been devoted to nondifferentiable exact penalty functions for the solution of nonlinear programming problems (see, e. g. [1], [2], [5-7]). However, most of the literature on this subject is mainly concerned with conditions which ensure that the penalty function has a local (global) minimum point at a local (global) minimum point of the constrained problem, for all sufficiently small positive values of the penalty parameter. On the other hand, the main motivation for the use of penalty methods is that of solving the constrained problem by employing some unconstrained minimization algorithm. Hence, for non convex problems, it appears to be of interest the study of converse properties which ensure that local (global) minimizers of the penalty functions are local (global) solutions of the constrained problem.

In this paper we adopt a quite natural definition of exactness which accounts for this requirement and we show that an important class of nondifferentiable exact penalty functions is exact according to this definition.

Another point which has to be taken into account is the fact that the properties of exactness can be established under the assumption that the penalty parameter is smaller than a threshold value depending on a compact set $\mathcal{D}$ which contains the problem solutions of interest. Thus, in the non convex case, it may happen that the level set of the penalty function corresponding to a given value of the penalty parameter and to some fixed initial point, even if compact, need not be contained in $\mathcal{D}$. Although this may be irrelevant from the conceptual point of view in connection with the notion of exactness, it may assume a considerable interest from the computational point of view, when unconstrained descent methods are employed for the minimization of the penalty function. In fact, it may happen that the sequence produced by an unconstrained algorithm may be attracted towards critical points of the penalty function which are out of the set $\mathcal{D}$ where exactness is established.

Therefore, we consider a new class of exact nondifferentiable penalty functions which allows to overcome this difficulty under suitable regularity and compactness assumptions on the feasible set. More specifically, the functions considered incorporate a barrier term which causes the unconstrained minima to be located in the interior of a compact perturbation of the feasible set. This device allows to define an algorithm model, based on an automatic adjustment rule for the penalty coefficient, which is globally convergent towards Kuhn - Tucker points of the problem.

In the sequel we shall consider the general nonlinear programming problem:

$$\begin{aligned} &\text{minimize } f(x) \\ &\text{subject to } g(x) \leq 0, \ h(x) = 0, \end{aligned} \tag{P}$$

where $f : \mathbb{R}^n \to \mathbb{R}$, $g : \mathbb{R}^n \to \mathbb{R}^m$, $h : \mathbb{R}^n \to \mathbb{R}^p$, $p \leq n$ are continuously differentiable functions and the feasible set $\mathcal{F} := \{x \in \mathbb{R}^n : \ g(x) \leq 0, \ h(x) = 0\}$ is assumed to be non empty.

We shall also make reference to the problem:

$$\text{minimize } f(x), \ x \in \mathcal{F} \cap \mathcal{D}, \tag{$\tilde{\mathrm{P}}$}$$

where $\mathcal{D}$ is a compact subset of $\mathbb{R}^n$ such that $\mathcal{F} \cap \mathring{\mathcal{D}} \neq \emptyset$.

For any $x \in \mathbb{R}^n$ we define the index sets:

$$\begin{aligned} I_0(x) &:= \{i : g_i(x) = 0\}, \\ I_+(x) &:= \{i : g_i(x) \geq 0\}. \end{aligned}$$

Everywhere below we suppose that the following assumptions are satisfied.

ASSUMPTION (A1). *Any global solution of Problem* ($\tilde{\mathrm{P}}$) *belongs to the set* $\mathring{\mathcal{D}}$.

ASSUMPTION (A2). *For any* $x \in \mathcal{D}$ *the gradients* $\nabla h_j(x), j = 1, \ldots, p$ *are linearly independent and there exist a* $z \in \mathbb{R}^n$ *such that:*

$$\begin{aligned} &\nabla g_i(x)'z < 0, \quad i \in I_+(x), \\ &\nabla h_j(x)'z = 0, \quad j = 1, \ldots, p. \ \triangleleft \end{aligned}$$

We note that Assumption (A1) concerns the selection of the set $\mathcal{D}$ ; it can be satisfied, in particular, by a proper choice of $\mathcal{D}$, whenever the global solutions of Problem

(P) belong to a bounded subset of $\mathcal{F}$. We note also that if $x \in \mathcal{F}$, Assumption (A2) reduces to the *Mangasarian-Fromovitz* constraint qualification.

2. DEFINITIONS OF EXACTNESS

In essence, an *exact penalty function* for Problem (P) is a function $F(x;\varepsilon)$, where $\varepsilon > 0$ is a *penalty parameter*, with the the property that there is an appropriate parameter choice such that a single unconstrained minimization of $F(x;\varepsilon)$ yields a solution to Problem (P) . In particular, we require that there is an *easy* way for finding correct parameter values, by imposing that exactness is retained for all $\varepsilon$ ranging on some set of nonzero measure. More specifically, we take $\varepsilon \in (0, \varepsilon^*]$ where $\varepsilon^* > 0$ is a suitable *threshold value.*

For a given $\varepsilon > 0$, let $F(x;\varepsilon)$ be a continuous real function defined on a set $\mathcal{E}$, such that $\mathring{\mathcal{D}} \subseteq \mathcal{E} \subseteq \mathcal{D}$ and consider the problem:

$$\text{minimize } F(x;\varepsilon), \; x \in \mathring{\mathcal{D}} \, . \tag{Q}$$

Since $\mathring{\mathcal{D}}$ is an open set, any local solution of Problem (Q), provided it exists, is unconstrained and thus Problem (Q) can be considered as an essentially unconstrained problem.

We introduce now the following definition.

DEFINITION 1. *The function $F(x;\varepsilon)$ is said to be an exact penalty function for Problem* (P) *with respect to the set $\mathcal{D}$ if there exists an $\varepsilon^* > 0$ such that, for all $\varepsilon \in (0, \varepsilon^*]$:*

(i) *Problem* (Q) *admits a global solution and any global minimum point of Problem* (Q) *is a global minimum point of Problem* ($\tilde{\text{P}}$);

(ii) *any global minimum point of Problem* ($\tilde{\text{P}}$) *is a global minimum point of Problem* (Q);

(iii) *any local unconstrained minimizer of Problem* (Q) *is a local solution of Problem* (P). ◃

The properties stated in the preceding definition guarantee that the constrained problem can actually be solved over $\mathcal{D}$ by means of the unconstrained minimization of $F(x;\varepsilon)$ for sufficiently small values of the parameter $\varepsilon$.

Properties (i) and (ii) jointly establish the equivalence between Problem (Q) and problem ($\tilde{\text{P}}$) with respect to global minimizers; property (iii) ensures that local unconstrained minimizers of the penalty function have a meaning with reference to the original problem. We remark that if all global solutions of Problem (P) are contained in $\mathring{\mathcal{D}}$, then Problem (P) and Problem ($\tilde{\text{P}}$) possess the same global solutions. In this case, properties (i) and (ii) of Definition 1 imply that global solutions of Problem (P) and global minimizers of Problem (Q) are the same.

The properties considered in the Definition 1 do not characterize the behaviour of $F(x;\varepsilon)$ on the boundary of $\mathring{D}$. Thus, in the general case, it may happen that the sequence of points produced by an unconstrained algorithm is attracted towards a stationary point of $F(x;\varepsilon)$ out of $D$ or that it does not admit any limit point. Therefore, it may be difficult to construct minimizing sequences for $F(x;\varepsilon)$ which are globally convergent on $\mathring{D}$ towards the solutions of the constrained problem.

In order to avoid this difficulty, it is necessary to impose further conditions on $F(x;\varepsilon)$ and we are led to introduce the notion of *global exactness* of a penalty function.

DEFINITION 2. *The function $F(x;\varepsilon)$ is said to be a globally exact penalty function for Problem (P) with respect to the set $D$ if it is exact in the sense of Definition 1 and, moreover, for any $\varepsilon > 0$ and for any $\hat{x} \in \partial D$ there exists a neighbourhood $B(\hat{x};\rho)$ such that if $\{x_k\} \subseteq \mathring{D}$ and $\lim_{k\to\infty} x_k = \hat{x}$, we have:*

$$\liminf_{k\to\infty} F(x_k;\varepsilon) > F(x;\varepsilon),$$

*for all $x \in B(\hat{x};\rho) \cap \mathring{D}$.* ◃

The condition given above excludes the existence of minimizing sequences for $F(x;\varepsilon)$ originating in $\mathring{D}$ which have limit points on the boundary.

## 3. EXACTNESS OF NONDIFFERENTIABLE PENALTY FUNCTIONS

The best known class of nondifferentiable penalty functions is defined by:

$$J_q(x;\varepsilon) := f(x) + \frac{1}{\varepsilon}\left\| [g^+(x)'\ h(x)']' \right\|_q,$$

where $\varepsilon > 0$, $\|\cdot\|$ denotes the $\ell_q$ norm over $\mathbb{R}^{m+p}$ for $1 \le q \le \infty$ and $g^+(x)$ denotes the vector with components:

$$g_i^+(x) := \max[0, g_i(x)].$$

In particular, we have:

$$J_q(x;\varepsilon) = f(x) + \frac{1}{\varepsilon}\left[\sum_{i=1}^{m} (g_i^+(x))^q + \sum_{j=1}^{p} |h_j(x)|^q\right]^{1/q},$$

for $1 \le q < \infty$, and:

$$J_\infty(x;\varepsilon) = f(x) + \frac{1}{\varepsilon}\max\left[g_1^+(x),\dots,g_m^+(x), |h_1(x)|,\dots,|h_p(x)|\right].$$

The next proposition establishes the correspondence between critical points of $J_q(x;\varepsilon)$ and K-T triples for Problem (P) and extends results already given in the literature in the case $q = 1$ and $q = \infty$ (see, for instance, [1]); a detailed proof can be found in [3].

PROPOSITION 1. *There exist an $\varepsilon^* > 0$ such that, for all $\varepsilon \in (0, \varepsilon^*]$ :*

(a) *if $x_\varepsilon$ is a critical point of $J_q(x;\varepsilon)$, there exist $\lambda_\varepsilon, \mu_\varepsilon$ such that $(x_\varepsilon, \lambda_\varepsilon, \mu_\varepsilon)$ is a K-T triple for Problem* (P);

(b) *If $(\bar{x}, \bar{\lambda}, \bar{\mu})$ is a K-T triple for Problem* (P) *such that $\bar{x} \in \mathring{\mathcal{D}}$, the point $\bar{x}$ is a critical point for $J_q(x;\varepsilon)$.* ◃

We can now prove the following theorem on the exactness of $J_q(x;\varepsilon)$.

THEOREM 1. *The function $J_q(x;\varepsilon)$ is an exact penalty function for Problem* (P) *with respect to the set $\mathcal{D}$ in the sense of Definition 1.*

PROOF. We must show that conditions (i),(ii) and (iii) of Definition 1 are satisfied.

Since $\mathcal{D}$ is compact and $J_q(x;\varepsilon)$ is continuous on $\mathcal{D}$, for any $\varepsilon > 0$ there exists a point $x_\varepsilon{}^* \in \mathcal{D}$ such that:

$$J_q(x_\varepsilon{}^*;\varepsilon) = \min_{x \in \mathcal{D}} J_q(x;\varepsilon).$$

We prove first, by contradiction, that there exists an $\varepsilon_1^* > 0$ such that, for all $\varepsilon \in (0, \varepsilon_1^*]$ the point $x_\varepsilon^*$ is a global solution to Problem $(\tilde{\mathrm{P}})$.

Suppose that this assertion is false. Then, for any integer $k$ there must exist an $\varepsilon_k \leq 1/k$ and a global minimizer $x_k$ of $J_q(x;\varepsilon_k)$ on $\mathcal{D}$ such that $x_k$ is not a global solution of Problem $(\tilde{\mathrm{P}})$.

Let $\tilde{x}$ be a global minimizer of Problem $(\tilde{\mathrm{P}})$; then, we have $J_q(\tilde{x};\varepsilon_k) = f(\tilde{x})$, and hence we can write:

$$J_q(x_k;\varepsilon_k) = \min_{x \in \mathcal{D}} J_q(x;\varepsilon_k) \leq J_q(\tilde{x};\varepsilon_k) = f(\tilde{x}). \tag{1}$$

Since $\mathcal{D}$ is compact, there exists a convergent subsequence, which we relabel $\{x_k\}$, such that $\lim_{k\to\infty} x_k = \hat{x} \in \mathcal{D}$.

By (1) we have:

$$\limsup_{k\to\infty} J_q(x_k;\varepsilon_k) \leq f(\tilde{x}),$$

which implies, by construction of $J_q(x;\varepsilon)$ that $\hat{x} \in \mathcal{F} \cap \mathcal{D}$ and $f(\hat{x}) \leq f(\tilde{x})$, whence it follows that $\hat{x}$ is a global minimum point of Problem $(\tilde{\mathrm{P}})$.

Recalling Assumption (A1), we have $\hat{x} \in \mathring{\mathcal{D}}$ and therefore, since $\lim_{k\to\infty} x_k = \hat{x}$, it follows that for $k$ large enough , the points $x_k$ belong to $\mathring{\mathcal{D}}$ and are critical points of $J_q(x;\varepsilon)$. Then, by Proposition 1 we have, for sufficiently large values of $k$:

$$x_k \in \mathcal{F} \text{ and } J_q(x_k;\varepsilon_k) = f(x_k),$$

so that, by (1), $x_k \in \mathcal{F} \cap \mathring{\mathcal{D}}$ is both a global minimum point of $J_q(x;\varepsilon_k)$ on $\mathcal{D}$ and a global minimum point of Problem $(\tilde{\mathrm{P}})$ and this contradicts our original assumption.

It can be concluded that there exists an $\varepsilon_1^* > 0$ such that for all $\varepsilon \in (0, \varepsilon_1^*]$ any global minimizer $x_\varepsilon^*$ of $J_q(x;\varepsilon)$ on $\mathcal{D}$ is a global solution to Problem $(\tilde{\mathrm{P}})$. On the other hand, by Assumption (A1), the global solutions of Problem ($\tilde{P}$) are in $\mathring{D}$ and hence, for all $\varepsilon \in (0, \varepsilon_1^*]$, we have that $x_\varepsilon^* \in \mathring{D}$ is also a global minimizer for Problem (Q).

Thus we have proved that for $\varepsilon \in (0, \varepsilon_1^*]$, Problem (Q) admits a global solution . Now, if $\varepsilon \in (0, \varepsilon_1^*]$ and $\tilde{x}_\varepsilon$ is another global solution of Problem (Q) it must hold that:

$$J_q(\tilde{x}_\varepsilon; \varepsilon) = J_q(x_\varepsilon^*; \varepsilon)$$

so that $\tilde{x}_\varepsilon$ is a global minimizer of $J_q(x; \varepsilon)$ on the whole set $D$ and hence a global solution to Problem ($\tilde{P}$). Thus, property (i) of Definition 1 is satisfied.

Let now $\varepsilon \in (0, \varepsilon_1^*]$ and let $x_\varepsilon$ be any global minimizer of Problem (Q). By condition (i) we have that $x_\varepsilon$ is a global solution to Problem ($\tilde{P}$) so that, by definition of $J_q(x; \varepsilon)$ we have:

$$f(x_\varepsilon) = J_q(x_\varepsilon; \varepsilon). \tag{2}$$

On the other hand, if $\bar{x}$ is another global minimizer of Problem ($\tilde{P}$), we have

$$f(\bar{x}) = J_q(\bar{x}; \varepsilon). \tag{3}$$

Therefore, as $f(x_\varepsilon) = f(\bar{x})$, (2) and (3) imply that $J_q(\bar{x}; \varepsilon) = J_q(x_\varepsilon; \varepsilon)$ and this proves that $\bar{x}$ is a global solution to Problem (Q). Thus, also property (ii) of Definition 1 is established.

Let now $\varepsilon_2^*$ be the threshold value for $\varepsilon$ considered in Proposition 1. It follows that for $\varepsilon \in (0, \varepsilon_2^*]$, if $x_\varepsilon \in \mathring{D}$ is a local minimum point (and hence a critical point) of $J_q(x; \varepsilon)$ we have, in particular, $x_\varepsilon \in \mathcal{F}$. Therefore, for $\varepsilon \in (0, \varepsilon_2^*]$ we have: $J(x_\varepsilon; \varepsilon) = f(x_\varepsilon)$ so that there exists a neighborhood $B(x_\varepsilon; \rho)$ such that

$$f(x_\varepsilon) \le J_q(x; \varepsilon) \quad \text{for all} \quad x \in B(x_\varepsilon; \rho),$$

whence

$$f(x_\varepsilon) \le f(x) \text{ for all } \quad x \in B(x_\varepsilon; \rho) \cap \mathcal{F},$$

which proves that $x_\varepsilon$ is a local minimizer of Problem (P). Finally, letting $\varepsilon^* = \min[\varepsilon_1^*, \varepsilon_2^*]$ the proof is complete. ◁

We consider now a nondifferentiable penalty function with global exactness properties, under suitable compactness assumptions on the feasible set $\mathcal{F}$.

Let $\beta = (\alpha_0, \ \alpha')'$ with $\alpha_0 \in \mathbb{R}$, $\alpha \in \mathbb{R}^m, \beta > 0$, and consider the set

$$S_\beta := \left\{ x \in \mathbb{R}^n : g(x) \le \alpha, \left\| h(x) \right\|_2^2 \le \alpha_0 \right\}.$$

Suppose that the following assumption is satisfied.

ASSUMPTION (A3). *The set $S_\beta$ is compact.* ◁

Obviously, this assumption implies that the feasible set is compact. We can take $D = S_\beta$, so that $\mathcal{F} \subset \mathring{D}$, Problem ($\tilde{P}$) reduces to the original problem (P) and Assumption (A1) of Section 1 is satisfied.

Let us introduce the functions:

$$a_0(x) := \alpha_0 - \left\| h(x) \right\|_2^2,$$
$$a_i(x) := \alpha_i - g_i(x), \quad i = 1, \dots, m$$

and denote by $A(x)$ the diagonal matrix:

$$A(x) := \mathrm{diag}\big(a_i(x)\big), \quad i = 1, \dots, m.$$

We have, obviously, that $a_i(x) > 0, \quad i = 0, 1, \dots, m$, for all $x \in \mathring{D}$.

Then, we consider the following function

$$Z_q(x;\varepsilon) := f(x) + \frac{1}{\varepsilon} \left\| \left[ \left(A^{-1}(x) g^+(x)\right)' \ \frac{h(x)'}{a_0(x)} \right]' \right\|_q,$$

where $\varepsilon > 0$ and $1 \le q \le \infty$. In particular,we have:

$$Z_q(x;\varepsilon) = f(x) + \frac{1}{\varepsilon} \left[ \sum_{i=1}^{m} \left( \frac{g_i^+(x)}{a_i(x)} \right)^q + \frac{1}{a_0(x)^q} \sum_{j=1}^{p} \left| h_j(x) \right|^q \right]^{1/q},$$

for $1 \le q < \infty$, and:

$$Z_\infty(x;\varepsilon) = f(x) + \frac{1}{\varepsilon} \max \left[ \frac{g_1^+(x)}{a_1(x)}, \dots, \frac{g_m^+(x)}{a_m(x)}, \frac{|h_1(x)|}{a_0(x)}, \dots, \frac{|h_p(x)|}{a_0(x)} \right].$$

The following proposition holds (for the proof see [4]).

PROPOSITION 2. *There exist an $\varepsilon^* > 0$ such that, for all $\varepsilon \in (0, \varepsilon^*]$ :*

(a) *if $x_\varepsilon$ is a critical point of $Z_q(x;\varepsilon)$, there exist $\lambda_\varepsilon, \mu_\varepsilon$ such that $(x_\varepsilon, \lambda_\varepsilon, \mu_\varepsilon)$ is a K-T triple for Problem* (P)*;*

(b) *if $(\bar{x}, \bar{\lambda}, \bar{\mu})$ is a K-T triple for Problem* (P) *such that $\bar{x} \in \mathring{D}$, the point $\bar{x}$ is a critical point for $Z_q(x;\varepsilon)$.* ◁

Using the preceding result, it is possible to establish the properties of exactness of $Z_q(x;\varepsilon)$, which are stated in the following theorem.

THEOREM 2. *The function $Z_q(x;\varepsilon)$ is a globally exact penalty function for Problem* (P) *with respect to the set $D$ in the sense of Definition 2.*

PROOF. By construction, we have $\lim_{k \to \infty} Z_q(x_k;\varepsilon) = +\infty$ for any sequence $\{x_k\} \subset \mathring{D}$ such that $x_k \to y \in \partial D$. Hence, by Definition 2 we have that $Z_q(x;\varepsilon)$ is globally exact if it is exact in the sense of Definition 1. This can be shown as in the proof of Theorem 1, making use of Proposition 2 in place of Proposition 1. ◁

## 4. A GLOBALLY CONVERGENT ALGORITHM

On the basis of the preceding results we can define an algorithm model which can be proved to be globally convergent towards K-T triples of the problem.

We assume that it is available an unconstrained minimization algorithm defined by an iteration map $\mathcal{A} : \mathring{D} \to 2^{\mathring{D}}$, such that, for any given value of $\varepsilon > 0$, the limit points of the sequence produced by $\mathcal{A}$ are critical points of $Z_q(x;\varepsilon)$.

It is also assumed that, at each step, the steepest descent direction is computed by solving the problem:

$$\min_{\|d\| \leq 1} DZ_q(x_k, d; \varepsilon_j),$$

where $DZ_q(x_k, d; \varepsilon_j)$ is the directional derivative of $Z_q(x;\varepsilon_j)$ at $x_k$ along the direction $d$.

The algorithm model described below makes use of a preselected sequence $\{\varepsilon_j\}$, with $\varepsilon_{j+1} < \varepsilon_j$, $j = 0, 1, \ldots$ and $\varepsilon_j \to 0$ as $j \to \infty$.

ALGORITHM MODEL

Initial guess: $z_o = x_o \in \mathring{D}$

*Step 0* : set $j = 0$;

*Step 1* : set $k = 0$ and $x_o = z_j$;

*Step 2* : if $\min_{\|d\| \leq 1} DZ_q(x_k, d; \varepsilon_j) = 0$ go to Step 3; else go to Step 4;

*Step 3* : if $\left\| \left[ g^+(x_k)' \; h(x_k)' \right]' \right\|_q = 0$ stop; else go to step 6;

*Step 4* : if $\left| \min_{\|d\| \leq 1} DZ_q(x_k, d; \varepsilon_j) \right| \geq \left\| \left[ g^+(x_k)' \; h(x_k)' \right]' \right\|_q$ go to Step 5; else go to Step 6;

*Step 5* : compute $x_{k+1} \in \mathcal{A}(x_k)$ set k = k+1 and go Step 2;

*Step 6* : set $z_{j+1} = x_k$, $j = j + 1$ and go to Step 1.

The convergence properties of the algorithm are stated in the following theorem, whose proof can be found in [4].

THEOREM 3. *Suppose that for every $\varepsilon > 0$ and every $x_o \in \mathring{D}$ any limit point of the sequence $\{x_k\}$ generated by the iteration map $\mathcal{A}$ is a critical point of $Z_q(x;\varepsilon)$ belonging to $\mathring{D}$.*

*Then, either the algorithm terminates at some $x_\nu \in \mathring{D}$ and there exists $\lambda_\nu, \mu_\nu$ such that $(x_\nu, \lambda_\nu, \mu_\nu)$ is a K-T triple for Problem* (P) *, or the algorithm produces an infinite sequence $\{x_k\} \subset \mathring{D}$ such that for each limit point $\bar{x}$ there exist $\bar{\lambda}$ and $\bar{\mu}$ which yield a K-T triple $(\bar{x}, \bar{\lambda}, \bar{\mu})$ for Problem* (P). ◁

It must be remarked that, because of the barrier terms in the expression of $Z_q(x;\varepsilon)$, the level set $\{x \in \mathring{D}: Z_q(x;\varepsilon) \leq Z_q(x_o;\varepsilon)\}$ is compact for any $\varepsilon > 0$ and $x_0 \in \mathring{D}$. Therefore, the assumption of Theorem 3 can be easily satisfied by employing any globally convergent unconstrained minimization method for nondifferentiable functions.

## REFERENCES


[1] Bertsekas, D.P. (1982). *Constrained Optimization and Lagrange Multiplier Methods*, Academic Press, New York.

[2] Conn, A.R. (1982). Penalty Function Method. Powell, M.J.D. (Ed.), *Nonlinear Optimization 1981*. Academic Press, New York.

[3] Di Pillo, G. and Grippo, L. (1985). On the Exactness of a Class of Nondifferentiable Penalty Functions. Report R-127, IASI-CNR, Rome; *J. Optim. Theory Appl.*, (to appear).

[4] Di Pillo, G. and Grippo, L. (1987). Globally Exact Nondifferentiable Penalty Functions. Report 10.87, Dipartimento di Informatica e Sistemistica, University of Rome "La Sapienza", Rome.

[5] Evans, J.P., Gould, F.J. and Tolle, J.W. (1973). Exact Penalty Functions in Nonlinear Programming. *Math. Programming* **4** 72-97.

[6] Fletcher, R.E. (1983). Penalty Functions. Bachem, A., Grötschel, M. and Korte, B. (Eds.) *Mathematical Programming , The State of the Art*, Springer-Verlag, Berlin, 87-114.

[7] Han, S.P. and Mangasarian, O.L. (1979). Exact Penalty Functions in Nonlinear Programming. *Math.Programming* **17** 251-269.

# A PENALTY FUNCTION FORMULATION FOR INTERACTIVE MULTIOBJECTIVE PROGRAMMING PROBLEMS

Teruo Sunaga, Mohammed Abdul Mazeed
and Eiji Kondo
Department of Mechanical Engineering
Faculty of Engineering, Kyushu
University, Fukuoka-812

## 1. INTRODUCTION

Many of the complex engineering design problems consist of several conflicting and noncommensurable criteria which must be optimized simultaneously. A natural approach for such problems is a multicriterion (multiobjective) optimization, but not a scalar optimization. This approach allows the designer to look for alternative optimal solutions. Also it enables the designer to appreciate more fully the range of possibilities in the feasible region.

Recently many papers have dealth with the computation of Pareto optima, and among them $l_p$ - norm methods and minimax methods are widely recognized. However it can be said that solution methods for nonlinear multiobjective programming problems are not yet well established. In this paper we propose an efficient and pragmatic mathematical model for nonlinear multiobjective programming problems. This model by the use of Morrison's penalty function [ 6 ] transforms the minimax formulation of the constrained multiobjective programming problem into a series of smooth unconstrained problems of scalar objectives. For the optimization of these scalar objectives, we propose a derivative free Box's Complex Method [ 1 ] with some simple modifications. Finally we formulate a beam design problem as a multiobjective programming problem and solve it by our model.

## 2. A MINIMAX FORMULATION OF MULTIOBJECTIVE PROGRAMMING PROBLEMS

Among the various solution techniques of multiobjective programming problems, compromise solutions based on $l_p$ - norm methods and weighting methods based on the linear sum of objective functions have received considerable attention [ 2, 7 ]. However these methods have certain drawbacks. Norm methods do not generate all Pareto optima; besides it is quite possible that the designer is never satisfied with any solution obtained by these methods. The ability of the linear weighting method depends on the form of Pareto optimum set. If the form of the Pareto optimum set is non-convex, the linear weighting method may not determine a great part of the Pareto optimum set. In view of these points, it can be said that the minimax method proposed here is an efficient method of locating any desirable point on the Pareto optimum set.

Next we describe a minimax formulation for the general nonlinear mulitobjective programming problem. The general nonlinear multiobjective programming problem can be expressed as :

$$\text{Min.}\{f_1(x),\ldots\ldots,f_i(x),\ldots\ldots,f_M(x)\} \tag{1}$$

s.t.

$$g_l(x) \le 0, \qquad l=1,\ldots,L \tag{2}$$

$$h_j(x) = 0, \qquad j=1,\ldots,J \tag{3}$$

where $x\in R^n$, $J<n$ and $f_i, g_l, h_j$ are assumed to be nonlinear and smooth.

Problem (1)-(3) can be transformed into an equivalent minimax problem as

$$\min_X \max_i \ \{\rho_i[f_i(x) - f_i^0]\} \tag{4}$$

where

$$X = \{x: g(x)\le 0, h(x)=0\},$$

$f_i^0$ are aspiration levels supplied by the decision maker and $\rho_i$ are scaling factors. The scaling factors $\rho_i$ can be expressed as $\rho_i=1/R_i$ where $R_i$ is the range of the objective function $f_i$ in the objective space [ 8 ]. One advantage of this scaling is that additional scaling of the objectives will not be necessary.

It is reasonable to say that the minimax formulation (4) always finds a Pareto optimal solution. Now the following theorem always holds.

**THEOREM**: Minimax formulation (4) does not lose any Pareto optima of the original problem, or conversely the solution of the formulation (4) gives only Pareto optimal solutions.

A similar proposition and a proof for it can be found in Kosi and Silvennoinen [ 2 ].

Let $x^*$ and $\bar{f}^*$ represent the solution and optimal value respectively for problem (4). Then

$$\max_i \{\rho_i[f_i(x^*) - f_i^0]\} \le \bar{f}^*, \qquad i=1,\ldots,M \tag{5}$$

or

$$\rho_i\{f_i(x^*)-f_i^0\}\le \bar{f}^*, \qquad i=1,\ldots,M \tag{6}$$

or

$$f_i(x^*) \le f_i^0 + \bar{f}^*/\rho_i, \qquad i=1,\ldots,M \tag{7}$$

Criterion values obtained in (7) are the best possible values with respect to the aspiration levels proposed by the decision maker; hence they represent a Pareto solution in the objective space. If the decision maker is not satisfied with this solution, he is asked to set the new aspiration levels; once again problem (4) is reattempted. This process continues till the decision maker is satisfied with one solution.

## 3. PENALTY FUNCTION FORMULATION

In this section we propose an efficient penalty function formulation for the minimax problem (4). Let problem (4) be written as

$$\min_{X} \max_{i} \{\bar{f}_i(x)\} \tag{8}$$

where

$$\bar{f}_i(x)=\rho_i\{f_i(x)-f_i^q\}, \qquad i=1,\dots,M \tag{9}$$

Here the functions $\bar{f}_i(x)$ are assumed to be smooth; however the main difficulty in solving (8) is related to the kinks in the objective

$$F(x) = \max_{i} \bar{f}_i(x), \qquad i=1,\dots,M$$

and the constraints. These kinks are points at which $F(x)$ and the constraints are not differentiable, and the solution might occur at such a kink. Numerical methods for solving minimax problems consist of methods of approximating $F(x)$ and the constraints by a close enough function in which the kinks are smoothed out.

For the minimax formulation (4), here we propose the following least pth function.

$$F(x,\bar{f}^{(k)},w)= \sum_{i=1}^{M} \{\bar{f}_i(x)-\bar{f}^{(k)}\}_+^p + w\sum_{l=1}^{L} \{g_l(x)\}_+^p + w\sum_{j=1}^{J} |h_j(x)|^p \tag{10}$$

Here $\bar{f}^{(k)}$ is the estimate of $\bar{f}^*$ at any iteration $k$ such that $\bar{f}^{(k)}\leq\bar{f}^*$, $w$ is a positive weighting parameter that controls constraint violations and

$$a_+ = \begin{cases} a & \text{if } a\geq 0 \\ 0 & \text{otherwise} \end{cases} \tag{11}$$

For $p=1$, function (10) is an exact penalty function and for $p=\infty$, it is a minimax formulation of problem (8). This function has discontinuous first partial derivatives for p=1 and p=∞, and is differentiable for $1<p<\infty$. For p=1 and p=∞ many of the smooth methods can not be applied for the minimization of function (10). But for building algorithms with differentiability in mind, any value $1<p<\infty$ can be considered. In this paper we are interested particularly when p=2. When $M=1$ and p=2, that is for a single criterion problem, formulation (10) is the same as the formulation proposed by Morrision [ 6 ].

The following theorem guarantees that for a constant penalty parameter $w$, the successive optimization and updating of (10) will give the optimal solution of problem (8).

**CONVERGENCE THEOREM**: Let $\mathcal{F}^{(1)}$ be the initial estimate of $\mathcal{F}(x)$ such that

$$\mathcal{F}^{(1)} \leq \mathcal{F}^* \tag{12}$$

Let

$$\min_x F(x,\mathcal{F}^{(k)},w) = F(x^{(k)},\mathcal{F}^{(k)},w) = F^{(k)} \tag{13}$$

If $\mathcal{F}^{(k)}$ is updated as

$$\mathcal{F}^{(k+1)} = \mathcal{F}^{(k)} + \sqrt{F^{(k)}/M} \tag{14}$$

then $x^{(k)}\to x^*$, $\mathcal{F}^{(k)}\to\mathcal{F}^*$ and $F^{(k)}\to 0$.

**Proof**: Since $x^{(k)}$ and $x^*$ are solutions to (10) and (8) respectively, the following relation holds.

$$F(x^{(k)},\mathcal{F}^{(k)},w) \leq F(x^*,\mathcal{F}^{(k)},w) = \sum_{i=1}^{M}\{\mathcal{F}_i(x^*) - \mathcal{F}^{(k)}\}^2_+ \tag{15}$$

On the other hand, based on the facts that $\mathcal{F}^*$ is the optimal value to problem (8) and that if $a \leq b$ then $a_+ \leq b_+$, the following relations hold.

$$\mathcal{F}_i(x^*) - \mathcal{F}^{(k)} \leq \mathcal{F}^* - \mathcal{F}^{(k)}, \qquad i=1,\ldots,M \tag{16}$$

$$\{\mathcal{F}_i(x^*) - \mathcal{F}^{(k)}\}_+ \leq \{\mathcal{F}^* - \mathcal{F}^{(k)}\}_+, \qquad i=1,\ldots,M \tag{17}$$

From relations (15) and (17), we have

$$F^{(k)} = F(x^{(k)},\mathcal{F}^{(k)},w) \leq F(x^*,\mathcal{F}^{(k)},w) = \sum_{i=1}^{M}\{\mathcal{F}_i(x^*) - \mathcal{F}^{(k)}\}^2_+ \tag{18}$$

$$\leq M\{\mathcal{F}^* - \mathcal{F}^{(k)}\}^2_+$$

If

$$\mathcal{F}^{(k)} \leq \mathcal{F}^* \tag{19}$$

then, from relation (18) we have

$$\mathcal{F}^{(k)} + \sqrt{F^{(k)}/M} = \mathcal{F}^{(k+1)} \leq \mathcal{F}^*. \tag{20}$$

By induction, (20) means that

$$\bar{J}^{(k)} \le \bar{J}^{*} \qquad \text{for all } k. \tag{21}$$

Hence there exists an $\bar{J}^{(\infty)}$ such that

$$\bar{J}^{(\infty)} \le \bar{J}^{*} \tag{22}$$

Also (20) can be written as

$$\bar{J}^{(1)} + \sum_{k=1}^{\infty}\sqrt{F^{(k)}/M} \le \bar{J}^{*}. \tag{23}$$

Since the left hand side of (23) is a nonnegative infinite series, $F^{(k)}$ converges to zero, and there exists an $x^{(\infty)}$ such that

$$F^{(\infty)} = F(x^{(\infty)},\bar{J}^{(\infty)},w) = 0 \tag{24}$$

Hence it follows that $x^{(\infty)}$ satisfies the constraints (2) and (3) and

$$\{\bar{J}_i(x^{(\infty)}) - \bar{J}^{(\infty)}\}_+^2 = 0, \qquad i=1,\ldots,M$$

or

$$\bar{J}_i(x^{(\infty)}) \le \bar{J}^{(\infty)}, \qquad i=1,\ldots,M \tag{25}$$

From (25) it follows that

$$\bar{J}^{*} \le \text{Max } \bar{J}_i(x^{(\infty)}) \le \bar{J}^{(\infty)} \tag{26}$$

From (22) and (26) it is concluded that $\bar{J}^{*}=\bar{J}^{(\infty)}$, that is as $x^{(k)}\to x^{*}$, then $\bar{J}^{(k)}\to\bar{J}^{*}$ and $F^{(k)}\to 0$. Q.E.D.

## 4. COMPLEX METHOD

The Quasi-Newton Method, a derivative based method, is perhaps the best known method for the optimization of unconstrained problems. However, there exist cases in which a function is not differentiable or it is difficult to compute the derivatives. Here we consider the derivative free Box's Complex Method [ 1 ] for constrained problems. The validity of the complex method for constrained problems was already investigated by Mazeed et. al in [ 3,4 ] and there it was reported that the complex method should be used together with a penalty function for the convergence to the precise solution.

Here we will explain Box's complex method briefly. Let $N$ ($N \ge n+1$, where $n$ is the dimension of the problem) feasible points, be set as the vertices of the complex (polyhedron). Now the point having the highest objective value is rejected, and the centroid for the rest of the points is determined. Then the new vertex is located along the line joining the rejected point and the centroid at a distance equal to or greater than the distance from the rejected point to the centroid. If the new vertex gives a value higher than that of the rejected point, it is replaced by another vertex located half the distance from the new vertex to the centroid. If a constraint is violated, the new vertex is also moved halfway in towards the centroid. Repetition of this procedure for the new complex is known as the complex method.

The above procedure performs well when the centroid is a feasible point; this happens when all of the functions are convex. We propose the following procedure for the general case. That is, when the centroid is infeasible or when the objective value at the centroid is higher than the highest value corresponding to the rejected point of the complex, the new vertex is sought on a line joining the rejected point and the point that gives the minimum objective value.

A detailed algorithm for the modified complex method described above can be found in references [ 3, 4 ].

## 5. MULTIOBJECTIVE BEAM DESIGN PROBLEM

In a beam design problem several objectives such as minimization of the volume of the beam and minimization of the static compliance of the beam could be thought of. The objectives which are conflicting in nature can be handled simultaneously by a multiobjective mathematical model. In this section we will consider a problem from Osyczka [ 7 ] which consists of two objectives: minimization of the volume of the beam (VOL) and minimization of the static compliance of the beam (STATC). The beam design problem is as shown in Fig. 1.

The variables in this problem are $x_1$, length of the part 1 of the beam and $x_2$, interior diameter of the beam. The constraints for this problem are as follows.

Bounds imposed on the length of part 1 of the beam are:

$$0.0 \le x_1 \le 1000 \text{ mm}$$

Bounds imposed on the interior diameter of the beam are:

$$40.0 \le x_2 \le 75.2 \text{ mm}$$

It is assumed that the beam in Fig. 1 should resist the maximum force $F_{max}=3820$ N and the permissible bending stress of the beam material $\sigma_g=180$ N/mm$^2$. Thus the bending strength constraints are:

For part 1

$$\frac{F_{max} x_1}{\frac{D_2^4 - x_2^4}{32 D_2}} \le \sigma_g$$

For part 2

$$\frac{F_{max}l}{\frac{D_1^4-x_2^4}{32D_1}} \leq \sigma_g$$

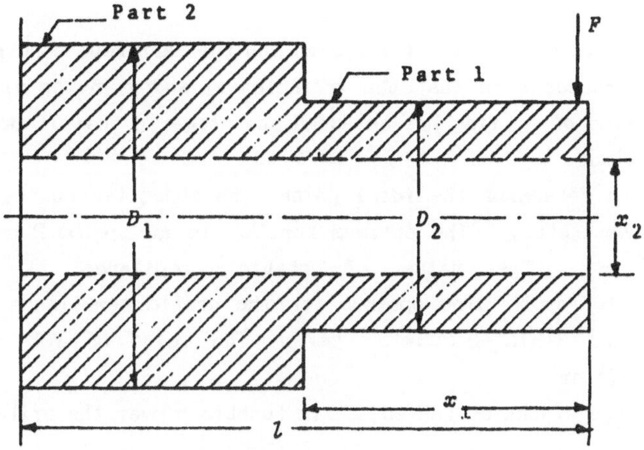


Fig. 1 Drawing of the beam

The objective volume of the beam (VOL) is given as:

$$VOL = \frac{\pi}{4}\{x_1(D_2^2-x_2^2)+(l-x_1)(D_1^2-x_2^2)\}$$

The static compliance of the beam (STATC) for the displacement under the force F is:

$$STATC = \frac{64}{3\pi E}\left[\left\{\frac{1}{D_2^2-x_2^2} - \frac{1}{D_1^4-x_2^4}\right\}x_1^3 + \frac{l^3}{D_1^4-x_2^4}\right]$$

The parameters are assumed as $l=1000$ mm, $D_1$ = 100 mm, $D_2$ = 80 mm and $E=2.06\times10^5$ N/mm$^2$.

Now the beam design problem can be written as:

$$\text{Min.}VOL = \frac{\pi}{4}\{x_1(D_2^2-x_2^2)+(l-x_1)(D_1^2-x_2^2)\}$$

$$\text{Min.}STATC = \frac{64}{3\pi E}\left[\left\{\frac{1}{D_2^2-x_2^2} - \frac{1}{D_1^4-x_2^4}\right\}x_1^3 + \frac{l^3}{D_1^4-x_2^4}\right]$$

s.t.

$1-x_2/40 \leq 0$

$x_2/75.2 -1 \leq 0$

$$1 - 180(D^4 - x_2^4)/(9.78\times10^6 x_1) \leq 0$$
$$x_1 \geq 0$$
$$x_1 \leq 1000$$
$$x_2 \geq 0$$
$$x_1 \leq 80$$

Here the first three constraints are scaled and the last four constraints are imposed to the search process such that the variables $x_1$ and $x_2$ never violate the imposed bounds. This kind of scaling has an added advantage in that it does not modify the objective functions.

In order to determine the ideal point, the objective functions VOL and STATC are optimized separately. The optimum for VOL is at $x_1 = 165.2$ mm, $x_2 = 75.2$ mm and $VOL^* = 2.945\times10^6$ mm$^3$. The value of static compliance at this point is $STATC_{VOL^*} = 4.990\times10^{-4}$ mm/N. The optimum for the static compliance STATC is at $x_1 = 0$ mm, $x_2 = 40$ mm and $STATC^* = 3.383\times10^{-4}$ mm/N. The value of VOL at this point is $VOL_{STATC^*} = 6.597\times10^6$ mm$^3$.

Now the range widths of the objective functions over the criterion space can be determined as

$$R_1 = VOL^* - VOL_{STATC^*} = 3.652\times10^6 \text{ mm}^3$$
$$R_2 = STATC^* - STATC_{VOL^*} = 1.607\times10^{-4} \text{ mm/N}$$

As a first step to the multicriteria analysis, let us suppose that the first aspiration level is $VOL = 3.0\times10^6$ mm$^3$ and $STATC = 3.5\times10^{-4}$ mm/N.

Now the normalized objective functions would become as

$$\bar{f}_1 = \{VOL - 3.0\times10^6\}/3.652\times10^6$$
$$\bar{f}_2 = \{STATC - 3.0\times10^{-4}\}/1.607\times10^{-4}$$

These two equations are substituted in equation (10) and the weighting parameter of the penalty function (10) is set as $w = 1000$. The resulting penalty function is optimized by the complex method and the Pareto optimum obtained is $x_1 = 232.5$ mm, $x_2 = 67.3$ mm, $VOL_P = 4.134\times10^6$ mm$^3$ and $STATC_P = 3.999\times10^{-4}$ mm/N.

Supposing that the decision maker is not satisfied with this solution and considering that he wants to decrease the volume of the beam, we assume the new aspiration levels as $VOL = 4\times10^6$ mm$^3$ and $STATC = 4.5\times10^{-4}$ mm/N. The Pareto optimum obtained for this case is $x_1 = 232.2$ mm, $x_2 = 67.5$ mm, $VOL_P = 3.601\times10^6$ mm$^3$ and $STATC_P = 4.324\times10^{-4}$ mm/N. Supposing that the decision maker is not yet satisfied with the beam volume obtained in this solution and assuming that he still wants to decrease the volume of the beam, we assume the next aspiration level as $VOL = 3.5\times10^6$ mm$^3$ and $STATC = 4.5\times10^{-4}$ mm/N. The pareto optimum obtained for this case is $x_1 = 236.5$ mm, $x_2 = 69.2$ mm, $VOL_P = 3.421\times10^6$ mm$^3$, and $STATC_P = 4.465\times10^{-4}$ mm/N. We assume that the decision maker is satisfied with this solution, and hence terminate the algorithm. The results for this problem are given in Table 1.

*Table* 1. Computational Results for Beam Design Problem

| | Variables | | Objectives | |
|---|---|---|---|---|
| | $x_1$ mm | $x_2$ mm | *VOL* $mm^3$ | *STATC* mm/N |
| Optimum of VOL | 165.2 | 75.2 | $2.945\times10^6$ | $4.990\times10^{-4}$ |
| Optimum of STATC | 0.0 | 40.0 | $6.597\times10^6$ | $3.383\times10^{-4}$ |
| Pareto Optimum for *VOL*=$3.0\times10^6$ *STATC*=$3.5\times10^{-4}$ | 232.5 | 62.4 | $4.134\times10^6$ | $3.999\times10^{-4}$ |
| Pareto Optimum for *VOL*=$4.0\times10^6$ *STATC*=$4.5\times10^{-4}$ | 237.2 | 67.5 | $3.601\times10^6$ | $4.324\times10^{-4}$ |
| Pareto Optimum for *VOL*=$3.5\times10^6$ *STATC*=$4.5\times10^{-4}$ | 236.5 | 69.2 | $3.421\times10^6$ | $4.465\times10^{-4}$ |

Another engineering problem that can be treated as a multicriterion optimization problem is the metal cutting problem. In a problem involving single-pass, single-tool operation, several objectives such as minimization of cost, minimization of cutting time, maximization of production rate, maximization of metal removal rate and maximization of tool life could be thought of. A similar application of our algorithm to the metal cutting problem can be found in [ 5, 9 ].

## 6. CONCLUSIONS

In this paper we treated the minimax formulation of the general nonlinear multiobjective programming problem. The minimax formulation of the multiobjective programming problem is superior to $l_p$- norm methods and linear weighting method in the sense that it has the ability to determine the desired Pareto optimal solution. And we proposed a penalty function formulation for this minimax formulation. For the optimization of this penalty function, we proposed a modified version of the complex method. Also we have shown that a solution determined by our algorithm is always a Pareto optimal solution. An advantage of our algorithm is that any point in the space of objectives chosen by the decision maker can be considered as the

**referenc objective. Finally we have formulated the beam design problem as a multicriterion optimization problem and solved it by our algorithm.**

## REFERENCES

[ 1 ] Box, M. J.: A New Method of Constrained Optimization and Comparison with Other Methods, *Computer Journal*, 1965, 8(42), pp. 42-52.

[ 2 ] Koski, J. and Silvennoinen, R.: Norm Methods and Partial Weighting in Multicriterion Optimization of Structures, *International Journal for Numerical Methods in Engineering*, Vol.24, 1987, pp. 1101-1121.

[ 3 ] Mazeed, M. A., Sunaga, T. and Kondo, E.: Optimization of Nonlinear Programming Problems using Penalty Functions and Complex Method, *Journal of the Operations Research Society of Japan*, Vol. 30, NO.4, 1987.

[ 4 ] Mazeed, M. A., Sunaga, T. and Kondo, E.: Nonlinear Optimization Using the Least Square Function and the Complex Method, *Memoirs of the Faculty of Engineering, Kyushu University, Fukuoka*, Vol. 47, NO.3, 1987.

[ 5 ] Mazeed, M. A., Sunaga, T. and Kondo, E.: A Mathematical Model for Interactive Multiobjective Programming Problems, *Proceedings of the International Conference on Modelling and Simulation, Melbourne, Australia*, 1987.

[ 6 ] Morrison, D. D.: Optimization by Least Squares, *SIAM J. Numer., Anal.*, 5(1), 1968, pp. 83-88.

[ 7 ] Osyczka, A.: *Multicriterion Optimization in Engineering*, Ellis Horwood Limited, 1984.

[ 8 ] Steur, R. E.: *Multiple Criteria Optimization Theory, Computation, and Application*, John Wiley & Sons, Inc., 1986.

[ 9 ] Sunaga, T., Mazeed, M. A., Kondo, E. and Kiyota, T.: A Practical Approach to Multiobjective Programming Problems, *Transactions of the Japan Society of Mechanical Engineers*, (In Japanese), 1987.

# MULTICRITERIA OPTIMIZATION IN THE DISO SYSTEM

Evtushenko Y., Mazourik V., Ratkin V.
Computing Center of the USSR Academy of Sciences
Moscow B-333, Street Vavilova 40, U.S.S.R.

## 1. INTRODUCTION

From the mathematical point of view *multicriteria optimization* (MCO) is a natural generalization of optimization problems. The need of decision making in contradictory situations makes MCO methods so interesting for us. MCO deals with one of the most sophisticated aspects of human activity which is to achieve several goals by the single act of decision making. MCO models and ordinary optimization are not very much different in task definition, giving us hope to use the similar numerical methods.

This paper gives the overview of the MCO package as one of the main parts of the DISO - *dialogue system for optimization problem solving* which was developped in the Computing Center of the USSR Academy of Sciences. Two MCO methods are described in this paper. Both methods are based on the idea of nonuniform covering technique and inclusion function approach, which was initially developped for global extremum search [1-5]. The complexity of MCO tasks makes it necessary to create effective numerical methods to find both a single point of *Pareto set* and an approximation of this set also. The paper describes two MCO algorithms which differ in the interpretation of the solution and as a consequence in the complexity of numerical calculations. The main features of the MCO package are described also.

## 2. OVERVIEW OF THE DISO SYSTEM

The basic feature of the DISO system is the integration principle. Unlike other dialogue systems of this class the DISO system includes several interconnected packages for the solution of the following tasks:

- unconstrained minimization;
- nonlinear programming;
- optimal control;

- linear programming;
- global optimization;
- multicriteria optimization;
- linear algebra;
- nonlinear algebraic equations.

For all these tasks the DISO system delivers for the user the unified set of dialogue capabilities, which includes:

- task definition and analysis in text mode;
- automatic optimization class recognition and correspondent dialogue package initiation;
- local analysis of task definition functions in a given point;
- changing of optimization method and its control parameters in the dialogue session;
- asynchronous control of the solution process which makes it possible to stop the process at any moment;
- control of the hierarchical solution process with the automatic or manual creation of the subordinate tasks from the list mentioned above;
- choice of the numerical or analytical differentiation schemes;
- control of the numerical and graphical interaction in the process of solution.

From the implementation point of view dialogue capabilities of the DISO system are based on the new approach of the multiwindow technique which includes the capability to access the values of the variables using special fields in the windows. The methods in each optimization class have different forms of graphical output in accordance with their basic mathematical schemes.

The integrated mode of the DISO system makes it really powerfull instrument for the solution of different application tasks. It is essential, that the system makes it possible to change the task definition in the process of the solution and to transfer the task from one mathematical model to another. The typical example of such transformation takes place if you add a restrictions to the initial unconstrained optimization problem thus creating nonlinear programming problem. Another example: transforming all the criteria but one to the restrictions will change the MCO problem to the more simple class of nonlinear programming. It is important to note that all the numerical results achieved so far will remain accessible after these transformations.

Most optimization models have hierarchical structure. The nonlinear programming methods, for example, may reduce the task to unconstrained minimization on every iteration. Optimal control methods usually create subordinate tasks of nonlinear programming and so on. In all these cases the DISO system provides the dialogue capabilities of the correspondent class for the solution of the subordinate problem. Finishing the solution will move the user again to the level of the initial problem. This feature of the DISO system proved to be really valuable for the solution of several difficult applied problems saving time and increasing the accuracy of the solution.

The MCO dialogue package plays a central role in the DISO system, because its mathematical model can be obviously treated as a generalization of the other optimization models listed above.

## 3. STATEMENT OF MCO PROBLEM

The multicriteria problem that we consider has the following form:

$$\min_{x \in X} F(x), \tag{1}$$

where $x \in R^n$ is a vector of decision variables, $X$ is the feasible decision set, $F = [F^1, F^2, \dots, F^m]$ is the objective vector, $F$: $R^n \to R^m$, vector-function $F$ is continuous on $X$. The decision set $X$ is assumed to be a closed and bounded (therefore compact). The goal is to find the efficient (Pareto) set $X_*$ of $X$ with respect to $F$, that is

$$X_* = \left\{x \in X: \text{ if } F(w) \leqslant F(x) \text{ for some } w \in X \text{ then } F(w) = F(x)\right\}. \tag{2}$$

We shall use the following convention: if $a, b \in R^s$ then $a \leqslant b$ if and only if $a^i \leqslant b^i$ for all $1 \leqslant i \leqslant s$.

We propose two extentions of ε-optimality concept which were developed for scalar optimization problem to vector case. In the first extention we introduce ε-efficient set as follows:

$$X_*^\varepsilon = \left\{x \in X: F(x) \leqslant F(x_*) + \varepsilon, \text{ where } x_* \in X_*\right\}, \tag{3}$$

where vector $\varepsilon \in R^m$ has all positive components and is named accurasy vector.

The set $W_*^\varepsilon$ is called ε-net of the Pareto set iff:

1) for any point $x \in X_*$ there exists a point $z \in W_*^\varepsilon$ such that $F(z) \leqslant F(x) + \varepsilon$;
2) there are no two different points $x$ and $z$ in $W_*^\varepsilon$ such that $F(x) \leqslant F(z)$.

Let $X$ and $P_l$ be a compact right parallelepipeds parallel to the coordinate axis (abbreviated as a box in the sequel):

$$X = \left\{x \in R^n: a \leqslant x \leqslant b\right\},$$

$$P_i = \left\{x \in R^n:\ a_i \leq x \leq b_i\right\},\ P_i \subset X,\ a_i \in R^n,\ b_i \in R^n,\ i = 1, 2, \ldots.$$

The main diagonal of the box $P_i$ we denote as $d_i = b_i - a_i$, the midpoint of the box is $c_i = \frac{1}{2}(b_i + a_i)$.

Let's introduce the $m$-dimentional vector-function $Q(P)$, for which every $j$-th component is defined by the condition:

$$Q^j(P) = \min_{x \in P} F^j(x),\ P \subset X.$$

We assume that for $Q(P)$ it is possible to find vector-function $G(P)$ which is the lower estimation of $Q(P)$ on the box $P$. This function must satisfy two conditions:

$$G(P) \leq Q(P), \tag{4}$$

$$\lim_{\|d_i\|_\infty \to 0} (G(P_i) - Q(P_i)) = 0. \tag{5}$$

Here we introduced the sequence of the boxes which satisfies the following conditions:

$$P_{i+1} \in P_i,\ \lim_{\|d_i\|_\infty \to 0} P_i = P_\infty \in X,\ i=1, 2, \ldots,$$

where $P_\infty$ is accumulating point.

For the given vector-function $F(x)$ the vector-function $G(P)$ can be found either on the basis of interval analysis [6] or by introducing some additional hypothesis. For example, supposing that all the components of $F(x)$ on the $X$ set satisfy the Lipschitz condition with constants $L^j = \sum_{i=1}^{n} \max_{x \in X} \left|\frac{\partial F^j(x)}{\partial x^i}\right|$, $1 \leq j \leq m$, we have:

$$G^j(P_i) = F^j(c_i) - \frac{1}{2} \cdot L^j \cdot \|d_i\|_\infty. \tag{6}$$

If in addition $F(x)$ satisfies Lipschitz condition with constants $M^j = \sum_{k=1}^{n} \sum_{i=1}^{n} \max_{x \in X} \left|\frac{\partial^2 F^j(x)}{\partial x^i \partial x^k}\right|$, $1 \leq j \leq m$, then:

$$G^j(P_i) = F^j(c_i) - \frac{1}{2} \cdot \|d_i\|_\infty \cdot \min\left\{L^j,\ \|F^j_x(c_i)\|_1 + \frac{1}{4} \cdot M^j \cdot \|d_i\|_\infty\right\}. \tag{7}$$

It is obvious that these functions $G(P)$ satisfy conditions (4), (5).

## 4. DESCRIPTION OF MCO METHODS

Now we are going to describe two algorythms for the approximate solution of the problem (1). During the computation process the algorythms will generate the sequence of the boxes

$$B_k = \{P_1, P_2, \ldots, P_k\}, \text{ all } P_i \subset X,$$

and the corresponding sequence of these boxes midpoints

$$N_k = \{c_1, c_2, \ldots, c_k\}.$$

Let each box $P_i$ be linked with the structure $S_i=(c_i, d_i, G_i)$, where $G_i=G(P_i)$. We call the set $S$ for the sequence $B_k$ the structure list

$$S = \{S_1, S_2, \ldots, S_k\}.$$

Algorythms differ in the interpretation of the solution and time consuming. The $W_*^\varepsilon$ set is obtained by one of them (second algorythm) and a single point $x_r$ from the $X_*^\varepsilon$ set - by another (first algorythm). The general scheme of both algorythms is described below.

**Algorithm.**

Initial actions:

1). Let $P_1 = X$, calculate $c_1$, $d_1$, $F(c_1)$, $G_1 = G(P_1)$ and set $B_1 = \{P_1\}$, $S_1=(c_1, d_1, G_1)$, $S = \{S_1\}$.

Main cycle:

2). Choose the box $P_s$ from $B_k$, for which $\min_{1\leqslant i\leqslant k} \max_{1\leqslant j\leqslant m} G_i^j$ is achieved.

3). Choose a coordinate direction $t$ in box $P_s$, parallel to which $P_s$ has an edge of maximum length, i.e. $d_s^t = \max_{1\leqslant i\leqslant n} d_s^i$. Bisect $P_s$ in the direction $t$, getting boxes $P_\alpha$, $P_\beta$ with midpoints $c_\alpha$,$c_\beta$ and diagonals $d_\alpha$,$d_\beta$ respectively.

4). Calculate $F(c_\alpha)$, $F(c_\beta)$, modify $W_k$ (or $x_r$) and define the vector values $G_\alpha = G(P_\alpha)$, $G_\beta = G(P_\beta)$.

5). Remove the box $P_s$ from the sequense $B_k$, i.e. remove the structure $S_s$ from the list $S$. Add into the list $S$ two structures: $S_\alpha=(c_\alpha, d_\alpha, G_\alpha)$ and $S_\beta=(c_\beta, d_\beta, G_\beta)$, assuming $S_s = S_\alpha$ and $S_{k+1} = S_\beta$.

6). For all $S_i$ from the list $S$ check the following condition: if $F(x_j) \leqslant G(P_i)+\varepsilon$ for any $x_j$ from $W_k$ (or for $x_r$) then remove $S_i$ from the list $S$ and remove $P_i$ from $B_k$. Order new list $\{S_{i_1}, S_{i_2}, \ldots, S_{i_p}\}$ and give it the name $S = \{S_i\}_{1\leqslant i\leqslant p}$.

7). Let $k = p$. If $k \neq 0$, i.e. $S$ is not empty then goto 2.

Concluding operations:

8). Invoke output procedure. Stop computation.

**The constraction rule for $W_*^{\varepsilon}$.**

Let $W_1 = \{c_1\}$.

Let we have $N_k$, $N_{k+1}$, $W_k$. Then

i) if there exists $x_i \in W_k$ such that $F(x_i) \leqslant F(x_{k+1})$, then $W_{k+1} = W_k$;

ii) otherwise $W_{k+1} = (W_k \setminus V) \cup \{x_{k+1}\}$, where $V = \left\{x_i \in W_k:\ F(x_{k+1}) \leqslant F(x_i)\right\}$.

The above (second) algorithm defines as a rezult the $\varepsilon$-net of the Pareto set (i.e. $W_q = W_*^{\varepsilon}$) using the finite number of $F$ evaluations.

**The rule for finding a single point from $X_*^{\varepsilon}$.**

Let $x_r = c_1$.

Let $c_k$ was obtained. Then if $F(c_k) \leqslant F(x_r)$ then $x_r = c_k$.

The above (first) algorithm defines the point $x_r \in X_*^{\varepsilon}$ using the finite number of $F$ evaluations.

5. PROBLEM SOLVING IN MCO PACKAGE

The task below was solved in order to investigate the quality of MCO algorithms. Full definition of this task is given in [7]. The solution based on the sequence of minimization subtasks was found there. This task includes two criteria, five parameters and has the following form:

$$\min_{x \in X} F(x), \tag{8}$$

where $F^1(x) = 1 - \prod_{i=1}^{n} \left[1 - \left(1 - r^i\right)^{x^i + 1}\right]$,

$$F^2(x) = \sum_{i=1}^{n} c^i \cdot x^i,$$

$X = \{x^i \in \mathrm{N}:\ 0 \leqslant x^i \leqslant 10,\ 1 \leqslant i \leqslant n\}$, $n=5$, vectors $c$ and $r$ are given in the Table 1.

Two solutions were obtained using both algorithms with different accuracy vectors $\varepsilon$. $G$ vector function was defined in accordance to (6). Lipshits constants vector was chosen to be $L = [0.7,\ 0.7]$. This value is good higher estimation for the given vector function. The results for the first algorithm are given in Table 2 for different accuracies. Table 3 contains the results for the second algorithm with the accuracy vector $\varepsilon = [0.1,\ 0.35]$.

| $i$ | 1 | 2 | 3 | 4 | 5 |
|---|---|---|---|---|---|
| $c^i$ | 0.13 | 0.13 | 0.15 | 0.14 | 0.15 |
| $r^i$ | 0.90 | 0.75 | 0.65 | 0.80 | 0.85 |

Table 1. Coefficients for task (8).

| accurasy | | criteria | | parameters | | | | | crit. |
|---|---|---|---|---|---|---|---|---|---|
| $\varepsilon^1$ | $\varepsilon^2$ | $F^1(x)$ | $F^2(x)$ | $x^1$ | $x^2$ | $x^3$ | $x^4$ | $x^5$ | evaluations |
| 0.20 | 0.50 | 0.575 | 0.43 | 0 | 0 | 0 | 2 | 1 | 163 |
| 0.15 | 0.40 | 0.444 | 0.44 | 0 | 0 | 1 | 1 | 1 | 335 |
| 0.10 | 0.35 | 0.336 | 0.55 | 1 | 1 | 1 | 1 | 0 | 591 |

Table 2. Single points from the $X_*^\varepsilon$ set for different accuracies.

| No. | $F^1(x)$ | $F^2(x)$ | $x^1$ | $x^2$ | $x^3$ | $x^4$ | $x^5$ |
|---|---|---|---|---|---|---|---|
| 1 | 0.096 | 1.12 | 1 | 2 | 2 | 2 | 1 |
| 2 | 0.125 | 0.98 | 1 | 2 | 2 | 1 | 1 |
| 3 | 0.166 | 0.85 | 1 | 1 | 2 | 1 | 1 |
| 4 | 0.198 | 0.83 | 1 | 2 | 1 | 1 | 1 |
| 5 | 0.236 | 0.70 | 1 | 1 | 1 | 1 | 1 |
| 6 | 0.305 | 0.57 | 0 | 1 | 1 | 1 | 1 |
| 7 | 0.336 | 0.55 | 1 | 1 | 1 | 1 | 0 |
| 8 | 0.396 | 0.42 | 0 | 1 | 1 | 1 | 0 |
| 9 | 0.446 | 0.41 | 1 | 1 | 1 | 0 | 0 |
| 10 | 0.497 | 0.28 | 0 | 1 | 1 | 0 | 0 |

Table 3. $\varepsilon$-net of the Pareto set $W_*^\varepsilon$ for $\varepsilon = [0.1, 0.35]$.

## 6. MCO PACKAGE DIALOGUE CAPABILITIES

As was already mentioned above, the DISO system provides the user the unified set of dialogue capabilities to control every step of the solution process from the task definition up to the analysis of the numerical results.Let's overview briefly these capabilities.

The MCO task definition can be done using text processor which is part of the DISO system. A special language DIFALG is used for task definition. This language is very similar to the ALGOL-60. The inner form of the function representation is created as a result of the task definition compilation process. The calculation of the numerical values of the functions in a given point is based then on the interpretation of this inner form. The peculiarity of the DIFALG language is that its semantics includes the notion of the differentiation. Function evaluation may be done in parallel with the first and second derivatives of this function in a given point by the user request. The important point is that it gives the user not the numerical approximation but the exact value of the derivatives, which corresponds to the analytically evaluated value. The differentiation algorithms are based on a special highly effective approach which qualitatively increases the calculation speed.

Task definition includes comment lines. The system automatically defines the optimization class and passes the control to the correspondent dialogue monitor analysing these comments. For example, to define MCO problem, the user marks with the special comment those functions in the DIFALG listing which he wants to be included to the set of the criteria. The other comments mark the functions to be included to the equality and nonequality restriction sets. If any, the problem will be classified as nonlinear (single or multicriteria) programming model. Special comments define the mode of optimization: local or global, give initial point value, parallelepipedal restrictions in the parameter space etc. The user may return to the task definition step at any moment of the solution, make some modifications and continue the solution process.

The user initiates the dialogue session finishing the task definition. The set of control windows become available to him at this moment. Each window provides the user with some resources to control the solution process. The MCO dialogue monitor opens the access to the three windows.

The first one provides the capabilities for manual analysis of the functions local properties in a given point. It also controls the MCO method choice, initiation and termination. As was mentioned above, the DISO windows are structured in a sence that they include the different sets of fields to access the variables and control their values. The first MCO window contains the fields for the decision vector, criteria values vector, list of methods available etc. The user may move the cursor to the parameter vector and give this vector some initial value. The cursor movement to the criteria field automatically yields in the recalculation of the criteria functions in a given point with immediate output of the numerical results in this field. The field which is connected with the list of available methods plays the role of the menu. Each method has its own list of control parameters. The choice of the method automatically provides the user with the new list of fields for control parameters which appear in the window and become accessible by the ordinary routine of the cursor movement. Another field is responsible for the type of the solution: the user can choose $\varepsilon$-approximation of the Pareto set or single-point solution as it was mentioned above. Finally, the window has the menu field to run the method, stop it and switch to the task modification mode.

The second window provides the user with the view of the Pareto set in the process of its creation. With the help of the fields in this window the user can point out two coordinate axis to define the two-dimentional plain to create the projection of the pareto set. The solution process will show each new Pareto point in the special graphical field of this window. Finishing the solution process the user may enter the mode of Pareto set analysis. The cursor will take the form of the arrow in the Pareto field. The user can move this arrow from one Pareto point to another, vizualising correspondent numerical values of criteria and parameter vectors. The user can store the Pareto set in a file with a given name or retrieve previously defined Pareto set and continue the solution to achieve more accurate Pareto approximation.

The third window gives the possibility to vizualise the covering technique in the parameter space. Each method has its own covering strategy. The user can vizually estimate the efficiency of the chosen covering scheme for the given task to be solved.

The system provides the possibility of asynchronous control which includes, for example, switching from one window to another while the optimization process continues it's progress.

## 7. CONCLUSION

The MCO methods described in this paper were tested using several tasks and proved to be rather effective. There exist several reasons of the methods success. First, unlike the known methods, these methods are strongly oriented on the effective use of the computer memory which decreases the amount of function evaluation. This is most valuable feature if function evaluation takes long time. Second, the interval analysis technique makes it possible to eliminate the initial Lipshitz constant estimation. Third, there exists a simple and natural way to organize parallel calculations by the feasible domain division between several processors. Fourth, the described methods permit the inclusion of local search algorithms, which may speed the calculations enormously. Fifth, the second of the proposed methods creates the Pareto set using nonunified one-way covering technique, instead of the common approach, which is based on the manifold solution of the auxiliary tasks on the feasible domain.

Dialogue MCO package, which is part of the DISO system, is a promising numerical basis for the implementation of numerous decision support systems.

## 8. REFERENCES

[1] Evtushenko, Y.G.: *Numerical Method for the Global Search*. Zhurnal Vychislitelnoi Matematiki i Matematicheskoi Fiziki, Vol. 11, No 6, 1971, pp. 1390-1403.

[2] Evtushenko, Y.G.: *Numerical Optimization Technique*. Springer-Verlag, New-York, 1985.

[3] Evtushenko, Y.G., Potapov, M.A., *A Nondifferentiable Approach to Multicriteria Optimization*. Lecture Notes in Economics and Mathematical Systems, Springer-Verlag. 1984, 255: pp. 97-102.

[4] Ratschek, H., 1985. *Inclusion Functions and Global Optimization*. Mathematical Programming. 33 (3): 300-317.

[5] Evtushenko, Y.G., Ratkin, V.A.: *Bisection Method for Global Optimization* (in Russian). Izvestija Akademii Nayk SSSR, Tehnicheskaya Kibernetika, No. 1, 1987, pp. 119-127.

[6] Moore, R.E.: *Interval Analysis*. Prentice-Hall, Englewood Clifs, 1966.

[7] Saaty T.L. *Optimization in Integers and Related Extremal Problems*. McGraw-Hill, New York, 1970.

**Sensitivity Analysis of A Descriptor Distributed Parameter System and Its Application to Shape Optimization**

Masato Koda
IBM Research, Tokyo Research Laboratory,
5-19 Sanban-cho, Chiyoda-ku, Tokyo 102, Japan

## *1. Introduction*

There exists a wealth of results on the sensitivity analysis of lumped constant parameters where the parameters and their variations are constant in space and time; see, for example, Frank (1978) and Cukier et al. (1978). The familiar "elementary" sensitivities are partial derivative sensitivity coefficients (or system Jacobian) which relate the input and output variables of the system that is usually defined using the state-space formulation. This sensitivity information can generally be used for further analysis for studies in model development, model validatation, experimental design, etc.

In distributed parameter systems described by partial differential equations, there usually exist, in addition to constant parameters, parameters and their variations that are spatially and/or temporally varying. Recently, a new variational approach to compute most general sensitivity measures for distributed parameter systems has been proposed by Koda (1982a), Koda and Seinfeld (1982), and Koda et al. (1979). The method is based on the computation of the functional derivative, a distributed analogue of the sensitivity coefficient.

The functional derivative sensitivities can be defined utilizing the system outputs (state variables) and simultaneous solutions for the adjoint (costate variable) equations. In other words, we can compute in real time the sensitvity or infinite-dimensionalgradient based on the variational approach. Naturally, the gradient information can be used in the numerical optimization, and this capability for the fine tuning of the shape design or domain optimization in fluid dynamics has been shown and demonstrated by Koda (1984).

In this paper, we wish to show that the same variational approach can be extended to distributed parameter systems in descriptor form (Luenberger 1973). In particular, we consider unsteady fluid dynamical systems for incopmressible flows which can be either viscous or inviscid. Contrast to the time-varying momentum (Navier-Stokes) equations, the continuity equation for incompressible flow has no explicit time derivative terms; this makes it difficult to adequately characterize the system in conventional state-space framework. Hence, a descriptor variable characterization is adopted in the present study.

Considering the structural similarity, we choose to model the descriptor system as the limit of the singularly perturbed system. For lumped systems, Koda (1982b) has shown that the variational approach can be effectively applicable to the singularly perturbed systems. In this paper, a descriptor distributed parameter system is defined as the limiting case of Chorin's "artificial compressibility" formulation (Chorin 1967) for incompressible fluid flows.

Utilizing the variational approach, it is shown that the functional derivative sensitivities for the optimal shape design problem can be obtained by solving both the descriptor equation and the linear codescriptor (adjoint) equation with similar structure. This paper may be viwed in several ways as a continuation of the general results derived in Koda (1984) to the descriptor distributed parameter system. The present analysis, however, is more realistic and general to allow descriptor variable formulations and provides several important reasons for interest in modelling and optimization of descriptor distributed parameter systems.

## *2. Descriptor Distributed Parameter Systems*

In this section, we wish to illustrate our rationals and motivations for modelling distributed parameter systems in descriptor form. We consider equations of motion for two-dimensional incompressible viscous fluids

$$\frac{\partial u}{\partial t} = -u\frac{\partial u}{\partial x} - v\frac{\partial u}{\partial y} - \frac{\partial p}{\partial x} + \nu\left(\frac{\partial^2 u}{\partial x^2} + \frac{\partial^2 u}{\partial y^2}\right) \tag{1a}$$

$$\frac{\partial v}{\partial t} = -u\frac{\partial v}{\partial x} - v\frac{\partial v}{\partial y} - \frac{\partial p}{\partial y} + \nu\left(\frac{\partial^2 v}{\partial x^2} + \frac{\partial^2 v}{\partial y^2}\right) \tag{1b}$$

with the continuity equation

$$\frac{\partial u}{\partial x} + \frac{\partial v}{\partial y} = 0 \tag{2}$$

where u and v are the velocity components in x- and y-coordinate directions, respectively, p is the pressure, $\nu$ is the kinematic viscosity, and t is the time. It is important to note that the system cannot be adequately characterized by the state-space formulation since the continuity equation (2) has no explicit time derivative terms.

The solution of these equations presents major difficulties which in part are inherent in the continuity equation (2) and in part due to the special role of the pressure in the equations (1). Thus, it usually requires the large amount of computer time to overcome these difficulties and to numerically obtain the physically meaningful solution.

If the goal is to obtain the steady solution only, Chorin (1967) proposed an efficient method by adding a time derivative of the pressure to the continuity equation (2). This term is multiplied by an "artificial compressibility," which is a small quantity to be determined that the steady limit is reached as fast as possible. Then, we can apply the standard compressible flow techniques for the incompressible equations.

We thus introduce the auxiliary system of equations

$$\frac{\partial u}{\partial t} = -\frac{\partial}{\partial x}(u^2) - \frac{\partial}{\partial y}(uv) - \frac{\partial p}{\partial x} + \nu\Delta u \equiv f_1 \tag{3a}$$

$$\frac{\partial v}{\partial t} = -\frac{\partial}{\partial x}(uv) - \frac{\partial}{\partial y}(v^2) - \frac{\partial p}{\partial y} + \nu\Delta v \equiv f_2 \tag{3b}$$

$$\varepsilon\frac{\partial p}{\partial t} = -\frac{\partial u}{\partial x} - \frac{\partial v}{\partial y} \equiv f_3 \tag{3c}$$

where $\varepsilon$ is the artificial compressibility and $\Delta$ denotes the Laplacian operator. It is important to note that the conservation-law formulation is employed in the above auxiliary system (3) since the divergence free constraint of the original continuity equation (2) has been replaced by a time evolution equation for the pressure, (3c). In the artificial compressibility method, however, the auxiliary system has no physical meaning until steady state is reached. This technique has been extensively applied to fluid flow and heat transfer problems, and an excellent survey of the method has been made by Peyret and Taylor (1983).

We rewrite the auxiliary system (3) in matrix-vector form

$$E_\varepsilon\frac{\partial}{\partial t}\begin{bmatrix} u \\ v \\ p \end{bmatrix} = \begin{bmatrix} f_1 \\ f_2 \\ f_3 \end{bmatrix} \equiv f \tag{4}$$

where $E_\varepsilon$ is the 3 x 3-dimensional matrix defined by

$$E_\varepsilon = \begin{bmatrix} 1 & 0 & 0 \\ 0 & 1 & 0 \\ 0 & 0 & \varepsilon \end{bmatrix} \tag{5}$$

This formulation naturally falls into the framework of singularly perturbed systems since $\varepsilon > 0$ is a small number.

In the inviscid limit, one-dimensional version of this singularly perturbed system can be written as follows:

$$\frac{\partial}{\partial t}\begin{bmatrix} u \\ p \end{bmatrix} = -\begin{bmatrix} 2u & 1 \\ 1/\varepsilon & 0 \end{bmatrix}\frac{\partial}{\partial x}\begin{bmatrix} u \\ p \end{bmatrix} \tag{6}$$

The eigenvalues of this system are $u + (u^2 + 1/\varepsilon)^{1/2}$ and $u - (u^2 + 1/\varepsilon)^{1/2}$, which are real and distinct, therefore, the system is hyperbolic. For any positive values of $\varepsilon$, the entire flow field is always subsonic since the eigenvalues are opposite in sign. This guarantees smooth solution without discontinuity, which typically represents itself as a shock in transonic regions.

The limiting system ($\varepsilon \to 0$) of (4) is the descriptor distributed parameter system

$$E\frac{\partial z}{\partial t} = f(z, z_x, z_y, z_{xx}, z_{yy}) \tag{7}$$

where the descriptor vector is denoted by z(x,y,t), whose components are primitive variables u, v, and p, i.e., $z^T = [u \ v \ p]$. In (7), partial derivatives are formally denoted by subscripts, e.g., $z_x = \partial z / \partial x$ and $z_{xx} = \partial^2 z / \partial x^2$, etc., and E denotes the singular matrix defined by

$$E = \begin{bmatrix} 1 & 0 & 0 \\ 0 & 1 & 0 \\ 0 & 0 & 0 \end{bmatrix} \tag{8}$$

Note that the time evolution equation (7) has the reduced order equal to rank(E) = 2, hence the initial and boundary conditions are required for $Ez(x,y,t)$ or only for u(x,y,t) and v(x,y,t).

The descriptor system model (7) is consistent with the standard primitive variable formulation of incompressible flows, and it is numerically advantageous and more straightforward to use descriptor or primitive variables in the analysis of fluid dynamical systems, especially, in three-dimensional flows. It is also known that the primitive (descriptor) variable formulation is more accurate on the boundaries (Roache 1976), which is particularly advantageous in the numerical shape design problem that will be treated in the subsequent sections.

## 3. *Optimal Shape Design Problem*

We now proceed with the formal problem formulation. We use the optimal shape design or domain optimization problem as a basis for the analysis. Such problems have been considered in Koda (1984), Pironneau (1984), and Fujii (1986, 1987). Our main concern here is to derive the functional derivative sensitivities of the descriptor distributed parameter system with respect to boundary variations.

We consider the following general descriptor system

$$E\frac{\partial z}{\partial t} = f(z, z_x, z_y, z_{xx}, z_{yy}) \tag{9}$$

where z(x,y,t) is the n-dimensional descriptor vector defined on the two-dimensional spatial domain $\Omega$, and E denotes the n x n-dimensional singular matrix with rank(E) < n. The initial and boundary conditions which are consistent with (9) are assumed to be given as follows:

$$Ez(x,y,0) = Ez_0 \tag{10}$$

$$g(z) = 0 \qquad on\ \partial\Omega \tag{11}$$

where $\partial\Omega$ denotes the boundary of $\Omega$. We further assume that $g_z = \partial g / \partial z$ is invertible. The system evolves from t = 0 to t = T, the period in which we are interested.

In our problem, the domain $\Omega$ is bounded by two boundary surfaces, $\Gamma$ and S; the inner surface S is unknown and the outer surface $\Gamma$ is known (see Fig. 1). This is a free boundary problem in which a part of the boundary, S, is unknown and must be determined. Let V denote an area surrounded by S. Then the optimal shape design problem is to determine the coordinate location of each point of the boundary S according to some appropriate design criteria.

We consider the following design performance functional

$$J = \int_0^T \int_\Omega D(z, z_x, z_y, z_{xx}, z_{yy})dxdydt \tag{12}$$

which is to be minimized subject to the constraint

$$I = \int_V h(x, y)dxdy = const. \tag{13}$$

where $D(z, z_x, z_y, z_{xx}, z_{yy})$ and h(x,y) are to be defined appropriately. In particular, we wish to derive a design sensitivity of J with respect to variations in the boundary coordinates by utilizing the functional derivative sensitivity arguments.

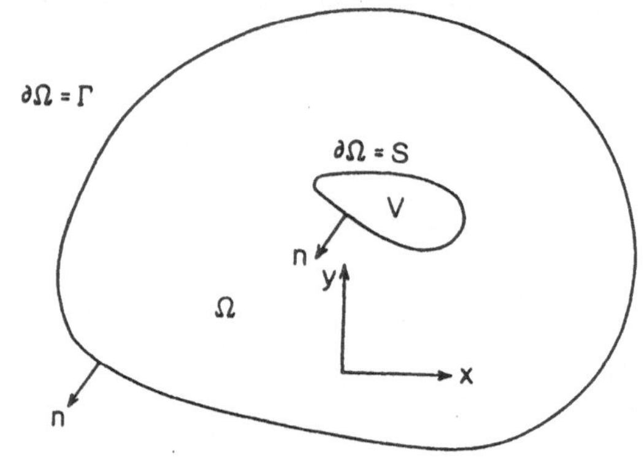


Fig. 1. Problem geometry.

## 4. Variational Approach

In this section, we explore a mathematical formalism of descriptor distributed parameter systems and derive a codescriptor variable equation using the variational approach, then the optimal shape design problem is briefly discussed guided by the (functional derivative) design sensitivy consideration.

The present optimal design problem can be treated in an optimal control framework for the desciptor distributed parameter systems, where the control is the location of the boundary. We transform the constrained optimal control problem defined in Section 3 into an unconstrained problem through the introduction of adjoint (codescriptor) functions. We thus define the augmented design performance functional by

$$J_A = J + \int_0^T \int_\Omega \psi^T(x, y, t)\left[f(z, z_x, z_y, z_{xx}, z_{yy}) - E\frac{\partial z}{\partial t}\right]dxdydt \tag{14}$$

where $\psi$(x,y,t) is the n-dimensional codescriptor variable function. Then if S* is the boundary that minimizes J, it also clearly minimizes the augmented design performance functional (14).

### *4.1. Transformation of Domain Variations*

Consider a small perturbation in the boundary coordinate location, $\delta(\partial\Omega) = \delta S$, which transforms the initial boundary $\partial\Omega$ to $\partial\tilde{\Omega} = \partial\Omega + \delta(\partial\Omega)$ with a consequent mapping of the domain $\Omega$ into $\tilde{\Omega} = \Omega + \delta\Omega$, the perturbed domain corresponding to $\partial\tilde{\Omega}$. This perturbation of the boundary transforms the local coordinate system in the initial unperturbed domain to the global coordinate system in the perturbed domain $\tilde{\Omega}$ whose coordinates are denoted by $\xi$ and $\eta$.

We suppose that there exists one parameter family transformations that map the initial domain $\Omega$ onto $\tilde{\Omega}$ as follows:

$$\xi = \varphi_1(x, y, z, z_x, z_y, z_{xx}, z_{yy}; \sigma) \tag{15a}$$

$$\eta = \varphi_2(x, y, z, z_x, z_y, z_{xx}, z_{yy}; \sigma) \tag{15b}$$

where $\sigma$ is a parameter of the transformations, and the new value of the descriptor vector, $\tilde{z} = \tilde{z}(\xi, \eta, t)$ is mapped onto the original descriptor function z(x,y,t) by the transformation

$$\tilde{z} = \Phi(x, y, t, z, z_x, z_y, z_{xx}, z_{yy}; \sigma) \tag{16}$$

We assume that the above transformations (15) and (16) are one-to-one, continuously differentiable and invertible with $\sigma = 0$ corresponds to the identity map. One can note that the idea of mapping the perturbed domain onto the unperturbed domain is natural and straightforward, and it goes back to the early work of Hadamard (1910).

The first variations of x, y, and z are defined as

$$\delta x \equiv \xi - x = \sigma \frac{\partial}{\partial\sigma} \varphi_1(x, y, z, z_x, z_y, z_{xx}, z_{yy}; 0) \tag{17a}$$

$$\delta y \equiv \eta - y = \sigma \frac{\partial}{\partial\sigma} \varphi_2(x, y, z, z_x, z_y, z_{xx}, z_{yy}; 0) \tag{17b}$$

and

$$\delta z \equiv \tilde{z}(\xi, \eta, t) - z(x, y, t) = \sigma \frac{\partial}{\partial\sigma} \Phi(x, y, t, z, z_x, z_y, z_{xx}, z_{yy}; 0) \tag{18}$$

where we have employed the Taylor expansions assuming that $\sigma$ is a small quantity. The total variation of the descriptor function can be computed as:

$$\delta z = \delta\tilde{z} + z_x\delta x + z_y\delta y \tag{19}$$

where $\delta\tilde{z} = \tilde{z}(x, y, t) - z(x, y, t)$ is a perturbation of the descriptor function in usual variational sense. This implies that the total variation of z(x,y,t) can be decomposed into a contribution due to the usual variation of the descriptor function, $\delta\tilde{z}$, and into a contribution due to the coordinate transformations, $z_x\delta x + z_y\delta y$. The transformation Jacobian is approximated (order one relative to $\sigma$) as

$$\left|\frac{\partial(\xi, \eta)}{\partial(x, y)}\right| = 1 + \frac{\partial}{\partial x}\delta x + \frac{\partial}{\partial y}\delta y \tag{20}$$

The first variation of the augmented design performance functional is defined as the principal linear part (relative to $\sigma$) of the difference

$$\delta J_A \equiv \int_0^T \int_{\tilde{\Omega}} \Big(D(\tilde{z}, \tilde{z}_\xi, \tilde{z}_\eta, \tilde{z}_{\xi\xi}, \tilde{z}_{\eta\eta})$$

$$+ \psi^T(\xi, \eta, t)\Big[f(\tilde{z}, \tilde{z}_\xi, \tilde{z}_\eta, \tilde{z}_{\xi\xi}, \tilde{z}_{\eta\eta}) - E\frac{\partial\tilde{z}}{\partial t}\Big]\Big) d\xi d\eta dt$$

$$- \int_0^T \int_\Omega \Big( D(z, z_x, z_y, z_{xx}, z_{yy})$$

$$+ \psi^T(x,y,t)\Big[f(z, z_x, z_y, z_{xx}, z_{yy}) - E\frac{\partial z}{\partial t}\Big]\Big) dxdydt \tag{21}$$

Utilizing the Jacobian of transformations, (20), the domain of integration $\tilde{\Omega}$ of the first integral in the RHS of (21) can be transformed into the unperturbed domain $\Omega$. Thus, the present formulation needs not follow the unknown perturbed domain $\tilde{\Omega}$.

### *4.2. Codescriptor Variable Equation*

After a rather lengthy calculation of (21), using the Taylor's theorem and integration by parts, and retaining terms of order one relative to $\sigma$, we obtain

$$\delta J_A = \int_0^T \int_\Omega \left[ \frac{\partial \psi^T}{\partial t} E + H_z - \left(H_{z_x}\right)_x - \left(H_{z_y}\right)_y + \left(H_{z_{xx}}\right)_{xx} + \left(H_{z_{yy}}\right)_{yy} \right] \delta \tilde{z}\, dxdydt$$

$$+ \int_0^T \int_\Omega \frac{\partial}{\partial x}\left( D\delta x + \left[ H_{z_x} - \left(H_{z_{xx}}\right)_{xx} \right] \delta \tilde{z} + H_{z_{xx}} \delta \tilde{z}_x \right) dxdydt$$

$$+ \int_0^T \int_\Omega \frac{\partial}{\partial y}\left( D\delta y + \left[ H_{z_y} - \left(H_{z_{yy}}\right)_{yy} \right] \delta \tilde{z} + H_{z_{yy}} \delta \tilde{z}_y \right) dxdydt$$

$$- \int_\Omega \psi^T(x,y,T) E \delta \tilde{z}(x,y,T) dxdy \tag{22}$$

where we have used $E\delta\tilde{z}_0 = 0$, and where the Hamiltonian functional H is defined as follows:

$$H(x,y,t; z, z_x, z_y, z_{xx}, z_{yy}) = D(z, z_x, z_y, z_{xx}, z_{yy}) + \psi^T(x,y,t) f(z_x, z_y, z_{xx}, z_{yy}) \tag{23}$$

In (22), partial derivatives of H are indicated by subscripts, i.e., $H_{z_x} = \partial H / \partial z_x$, $H_{z_{xx}} = \partial H / \partial z_{xx}$, etc.

From the fist integral of the RHS of (22) and utilizing the arbitrariness of the variations $\delta\tilde{z}$ in $\Omega$, we can obtain the following codescriptor variable equation:

$$E^T \frac{\partial \psi}{\partial t} = -f_z^T \psi + \left(f_{z_x}{}^T \psi\right)_x + \left(f_{z_y}{}^T \psi\right)_y - \left(f_{z_{xx}}{}^T \psi\right)_{xx} - \left(f_{z_{yy}}{}^T \psi\right)_{yy}$$

$$- D_z{}^T + \left(D_{z_x}{}^T\right)_x + \left(D_{z_y}{}^T\right)_y - \left(D_{z_{xx}}{}^T\right)_{xx} - \left(D_{z_{yy}}{}^T\right)_{yy} \tag{24}$$

Since the present variational approach is based on the linear (infinitesimal) perturbation theory, the codescriptor system (24) is expreesed as a system of linear equations.

Several remarks are warranted here for the above codescriptor system. It is important to note that the linear system (24) is described by using the singular matrix E of the original descriptor system. This implies that the codescriptor system involves the time evolution equation whose order is equal to rank(E), which is less than full rank. Thus, the descriptor and codescriptor systems are structurally similar, and the conventional state-space formulation is inadequate to characterize neither of the systems.

From the last integral in (22), the codescriptor function $\psi(x,y,t)$ must satisfy the terminal condition

$$E^T\psi(x,y,T) = 0 \tag{25}$$

This indicates that the codescriptor system (24) should be solved backward in time while the descriptor system (9) must be solved separately forward in time.

Considering the variations in the boundary conditions (11), i.e. $g_z\delta z = 0$ *on* $\partial\Omega$, the appropriate boundary conditions can be obtained for the codescriptor system, and this will be treated in Subsection 4.3. Solution of this codescriptor variable equation, in conjunction with the descriptor variable equation, yields the functional derivative sensitivities.

### 4.3. Functional Derivative Design Sensitivity

We are now in a position to follow some geometric and computational aspects of the problem. The change of the design performance functional due to a shift of a boundary tangential to S is certainly smaller than the change due to a normal shift. Thus, we can restrict our sensitivity arguments only to the boundary perturbation which is normal to the boundary surface S. We therefore specify the following geometric relationship for the variations in the boundary coordinates:

$$\delta x(s) = \delta n(s)\cos\theta \tag{26a}$$

$$\delta y(s) = \delta n(s)\sin\theta \tag{26b}$$

where θ is the angle between the positive x-axis and the outward normal, n(s), at a point s on the boundary, and where s denotes the arc length along S (see Fig. 2). Then, the functional derivative design sensitivity $\delta J_A / \delta n(s)$ can be derived by extending the analysis of Koda (1984), but for reasons of brevity, details of the derivation is omitted here.

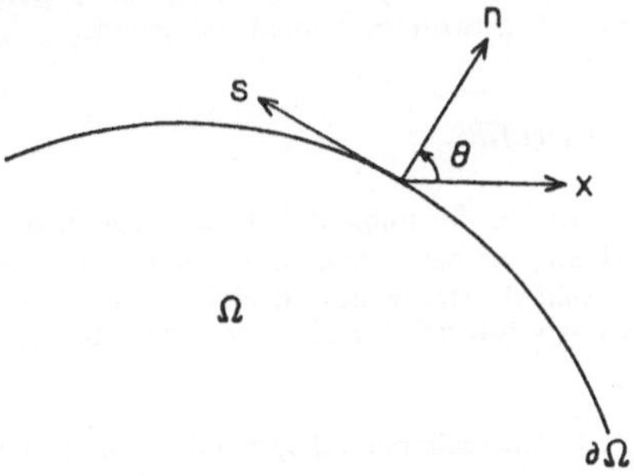


Fig. 2. Coordinates at the boundary.

Employing the Green's theorem to transform (22) into the boundary integral over $\partial\Omega = S$, and approximating $\delta\tilde{z}$ by $\delta z$, we obtain the following relationship between $\delta J_A$ and the boundary variation $\delta n(s)$:

$$\delta J_A = \int_S\int_0^T\Big[-D + (P\cos\theta + Q\sin\theta)z_n + \left(H_{z_{xx}}\cos^2\theta + H_{z_{yy}}\sin^2\theta\right)z_{nn}$$

$$-\left(H_{z_{xx}} - H_{z_{yy}}\right)\sin\theta\cos\theta z_{sn}\Big]dt\delta n(s)ds$$

$$-\int_S\int_0^T\left(H_{z_{xx}}\cos^2\theta + H_{z_{yy}}\sin^2\theta\right)\delta z_n dt ds \tag{27}$$

where the subscript indicates partial derivatives with respect to the boundary coordinates n and s, i.e., $z_n = \partial z / \partial n$, etc., and where we have defined

$$P \equiv H_{z_x} - \left(H_{z_{xx}}\right)_x \tag{28a}$$

$$Q \equiv H_{z_y} - \left(H_{z_{yy}}\right)_y \tag{28b}$$

In the above formulation, we have used (26) and the condition, $\delta z = 0$ on S.

From the second integral in the RHS of (27), we specify the boundary conditions for the codescriptor system (24) as

$$H_{z_{xx}}\cos^2\theta + H_{z_{yy}}\sin^2\theta = 0 \qquad on\ S \tag{29}$$

Using (27) and (28), we obtain the desired functional derivative design sensitivity as follows:

$$\frac{\delta J_A}{\delta n(s)} = \int_0^T \left[-D + (P\cos\theta + Q\sin\theta)z_n - \left(H_{z_{xx}} - H_{z_{yy}}\right)\sin\theta\cos\theta z_{sn}\right]dt \tag{30}$$

We have the following remarks on the derived design sensitivity:

1) The design sensitivity functional thus derived are defined through the descriptor and codescriptor variable functions. Thus results are more general to include the state-space systems as a special case with E = I (where I denotes the identity matrix).

2) In the above development, the descriptor variable formulation is essentialy guided by the computational considerations of Chorin (1967) for incompressible flows using the time marching solution techniques. Therefore, the numerical implementation of the results is self-evident for steady solutions.

3) The functional derivative sensitivity can be interpreted as the "infinite-dimensional" gradient, and hence a modification of the standard gradient method can be implemented as an iterative unconstrained optimization technique.

4) In the analysis, the original two-dimensional boundary adaptation is treated as the one-directional operation aligned with the normal to the boundary surface. Thus, the present approach is considerably simpler and numerically advantageous than two-dimensional adaptations.

## 5. *Example - Mininmum Drag Profile*

We consider here a minimum drag problem for numerical shape design in the unsteady two-dimensional fluid that is treated in Section 2. Hence, the basic equation is the descriptor system (7) and (8), with the descriptor vector $z^T = [u\ v\ p]$. A suitable steady flow field is specified as the initial condition. The boundary conditions are given as $u = v = 0$ on S, and, at the outer boundary far from S, the flow is assumed to be uniform.

For the design performance functional of this minimum drag problem, we shall determine that the integrand in (12) takes the following form:

$$D(z_x, z_y) = \frac{\nu}{2}\left[2\left(u_x^2 + v_y^2\right) + \left(u_y + v_x\right)^2\right] \tag{31}$$

which is the rate of energy dissipated in the fluid. Then, applying the sensitivity analysis techniques developed in Section 4, we can derive the design sensitivity as

$$\frac{\delta J_A}{\delta n(s)} = \nu\int_0^T z_n{}^T E\left[z_n + 2\psi_n\right]dt \tag{32}$$

where the codescriptor vector $\psi(x,y,t)$ is governed by the following linear partial differential equation:

$$E\frac{\partial\psi}{\partial t} = \begin{bmatrix} u_x & v_x & 0 \\ u_y & v_y & 0 \\ 0 & 0 & 0 \end{bmatrix}\psi - \begin{bmatrix} u & 0 & 1 \\ 0 & u & 0 \\ 1 & 0 & 0 \end{bmatrix}\frac{\partial\psi}{\partial x} - \begin{bmatrix} v & 0 & 0 \\ 0 & v & 1 \\ 0 & 1 & 0 \end{bmatrix}\frac{\partial\psi}{\partial y}$$

$$-\nu E\left(\frac{\partial^2\psi}{\partial x^2} + \frac{\partial^2\psi}{\partial y^2}\right) - \nu\begin{bmatrix} \Delta u \\ \Delta v \\ 0 \end{bmatrix} \tag{33}$$

where we have used (24). Utilizing (25) and (29), the terminal and boundary conditions for $E\psi(x,y,t)$ are given as $\psi_1(x,y,T) = \psi_2(x,y,T) = 0$, and $\psi_1 = \psi_2 = 0$ on S, respectively.

The codescriptor system (33) yields the "continuity-like" equation which has no explicit time derivative term of $\psi_3$, i.e.

$$\frac{\partial \psi_1}{\partial x} + \frac{\partial \psi_2}{\partial y} = 0 \tag{34}$$

Thus, the codescriptor variable $\psi_3$ plays the similar role as pressure does in the original equations (1) and (2). Note that the Chorin's artificial compressibility method may be effectively applied to (33) for the steady solutions. In such a case, the boundary layer properties which are inherent in the descriptor or singularly perturbed systems may be important for the numerical implementation.

We now derive the necessary condition for optimality of this minimum drag problem. We set $h(x,y) \equiv 1$ in the constraint (13); this is the constant volume condition for the optimal shape. Then, applying the present variational approach, we can obtain the first variation

$$\delta I = \int_S \delta n(s) ds = 0 \tag{35}$$

Combining (32) and (35), and employing the Lagrangean multiplier rule, we can derive the necessary condition for optimality as follows:

$$\frac{\delta J_A}{\delta n(s)} = \nu \int_0^T z_n{}^T E \left[ z_n + 2\psi_n \right] dt = \lambda = const. \qquad on\ S \tag{36}$$

where $\lambda$ denotes the constant multiplier associated with the constraint (13).

Using this necessary condition, iterative numerical optimization algorithms can be implemented for the minimum drag problem. It is easily shown that the present results for the optimal shape design reduce to those previously derived by Pironneau (1973, 1974) for the purely steady-state cases and by Koda (1984) for unsteady case.

## *6. Conclusions*

A new method for the sensitivity analysis of the descriptor distributed parameter systems has been illustrated based on the functional derivative sensitivity analysis and variational approach. It has been shown that the functional derivative can be described by simultaneous solutions for the descriptor and codescriptor variable equations which are structurally similar. An application of the present variational approach to the optimal shape design problem in the fluid dynamics is briefly discussed and the results are parallel in several ways to those obtained in Koda (1984).

In motivating the descriptor variable formulation, we suggested a structural similarity between the descriptor distributed parameter systems and the singularly perturbed systems. A descriptor system is obviously the limiting case of a singularly perturbed system, but the full exploitation of this structural relationship for modelling and optimization remains as a topic for further study. The boundary layer properties of both the descriptor and codescriptor variable equations seem to be important for the numerical implementation.

### *References*

Chorin, A.J. (1967). A numerical method for solving incompressible viscous flow problems. J. Comp. Phys., 1, 12-26.

Cukier, R.I., H.B. Levine, and K.E. Shuler (1978). Nonlinear sensitivity analysis of multi-parameter model systems. J. Comp. Phys., 26, 1-42.

Frank, P.M. (1978). Introduction to System Sensitivity Theory. Academic Press, New York.

Fujii, N. (1986). Necessary conditions for a domain optimization problem in elliptic boundary value problems. SIAM J. Control and Optimization, 24, 346-360.

Fujii, N. (1987). Domain optimization problems with a boundary value problem as a constraint. Proc. 4th IFAC Symposium on Control of Distributed Parameter Systems, 5-9, Pergamon Press, London.

Hadarmard, J. (1910). Lessons on the Calculus of Variation (in French). Gauthier-Villards, Paris.

Koda, M. (1982a). Sensitivity analysis of atmospheric diffusion equations. Atmos. Envir., 16, 2595-2601.

Koda, M. (1982b). Sensitivity analysis of singularly perturbed systems. Int. J. Syst. Sci., 13, 909-919.

Koda, M. (1984). Optimum design in fluid mechanical distributed-parameter systems. Large Scale Syst., 6, 279-291.

Koda, M., A.H. Dogru, and J.H. Seinfeld (1979). Sensitivity analysis of partial differential equations with application to reaction and diffusion processes. J. Comp. Phys., 30, 259-282.

Koda, M., and J.H. Seinfeld (1982). Sensitivity analysis of distributed parameter systems. IEEE Trans. Automat. Control, AC-27, 951-955.

Luenberger, D.G. (1973). Dynamic equations in descriptor form. IEEE Trans. Automat. Control, AC-22, 312-321.

Peyret, R. and T.D. Taylor (1983). Computational Methods for Fluid Flow. Springer-Verlag, New York.

Pironneau, O. (1973). On optimum profiles in Stokes flow. J. Fluid Mech., 59, 117-128.

Pironneau, O. (1974). On optimum design in fluid mechanics. J. Fluid Mech., 64, 97-110.

Pironneau, O. (1984). Optimum Shape Design for Elliptic Systems. Springer-Verlag, New York.

Roache, P.J. (1976). Computational Fluid Dynamics. Hermosa Publishers, Albuquerque, New Mexico.

# EXISTENCE OF AN OPTIMAL DOMAIN IN A DOMAIN OPTIMIZATION PROBLEM

Nobuo Fujii

Department of Control Engineering, Faculty of Engineering Science,

Osaka University, Toyonaka, Osaka 560, JAPAN

_abstract_. Existence of an optimal domain in a domain optimization problem is studied. By a domain optimization problem we denote an optimization problem in which an object function depending on the domain through the solution of the boundary value problem defined on the domain should be optimized. A new class of domains and a new notion of convergence of domains are introduced. With these new notions the lower semicontinuity results show the existence of an optimal domain for a wide class of optimization problems; the solution of the Dirichlet problem corresponding to the domain is a genuine (classical) solution.

Keywords. Domain optimization, Boundary value problem, Lower semicontinuity, Genuine solution, Convergence of domains.

## 1. Introduction

Domain optimization problems have received attentions in these years. Cea [1] enumerated various domain optimization problems which arise in various areas of engineering and applied science such as structural mechanics, acoustics, electric field and fluid flow.

One of the typical domain optimization problems is as follows: Let $g(x,u,p)$ be a function of $(x,u,p) \in R^2 \times R \times R^2$, where $R$ denotes the one dimensional Euclidean space. Let $\Omega$ be a bounded domain (open set) in $R^2$. Let us define

$$J(\Omega;u) \equiv \int_\Omega g(x,u(x),\nabla u(x))\,dx \qquad (1.1)$$

where $\nabla u$ stands for the gradient of $u$ and $u(x)$ is the solution of the following boundary value problem:

$$\Delta u(x) - k(x)u(x) = f(x) \qquad (x \in \Omega), \qquad (1.2)$$

$$u(x) = 0 \qquad (x \in \partial\Omega), \qquad (1.3)$$

$k(x)$ and $f(x)$ being functions defined on $R^2$. Under the appropriate conditions, this boundary value problem admits a unique soltion, which depends on the domain $\Omega$. Thus, $J(\Omega;u)$ depends on $\Omega$ through not only domain of integration but also the solution $u$ of the boundary value problem. We can formulate the following optimization problem:

$$\underset{\Omega}{\text{Minimize}}\ J(\Omega;u) \qquad (1.4)$$

subject to

$$\int_\Omega h(x)\,dx = c \text{ (const.)}, \qquad (1.5)$$

$h(x)$ being a given function. Constraint (1.5) is a generalization of the requisition that the area of $\Omega$ should be left constant. Throughout the paper, we shall confine ourselves to domains which are contained in a sufficiently large domain, say $\Omega_0$.

We shall call a domain which attains the minimum of $J(\Omega;u)$ the optimal domain. There are few investigations on the existence of the optimal domain. Pironneau [2] gave results on the existence for rather special cases; he treated the generalized (or weak) solution for the corresponding boundary value problem. Chenais [3] investigated the existence of the optimal domain introducing an important class of domains. The present author [4] derived some lower semicontinuity results and tried to apply them to the existence of the optimal domain for both Neumann and Dirichlet problems; his result for Dirichlet problems was not complete in that the generalized solution did not necessarily satisfy the boundary condition.

In this paper, we shall concentrate on some classes of domains narrower than those by the above authors and shall introduce a stronger notion of convergence of domains to show the existence of the optimal domain in the classical sense; i.e., we shall treat the genuine (classical) solution of the boundary value problem. In section 2, we shall introduce a new notion of convergence of domains. In section 3, we shall give an existence result.

Throughout the paper, the following notations will be commonly used.

$R$: Euclidean 1-space.

$R^2$: Euclidean 2-space.

$]a,b[$: open interval bounded by a and b.

$\langle\cdot,\cdot\rangle$: inner product in $R^2$.

$|\cdot|$: Euclidean norm in $R^2$.

$\partial\Omega$: boundary of $\Omega$.

$L^2(\Omega)$: Hilbert space of functions square integrable on $\Omega$.

$\Delta$: Laplacian operator in $R^2$.

$\nabla$: gradient operator in $R^2$.

$H^m(\Omega)$: Sobolev space of order m.

$H^m{}_0(\Omega)$: closure with respect to the norm in $H^m(\Omega)$ of infinitely differentiable functions with compact support.

$C^{m,\alpha}$: space of m-times differentiable functions whose mth derivatives are Holder continuous with exponent $\alpha$.

$\|\cdot\|$: norm in $C^{m,\alpha}$.

## 2. Convergence of domains

In order to prove the existence of the optimal domain, we introduce a new notion of convergence of domains. We can consider various kinds of convergence, however. Pironneau [2] made use of the Hausdorff distance to show the existence of the optimal domain in a class of domain optimization problems. Chenais [3] introduced interesting classes of domains, which will be stated in the sequel, to argue the existence of a solution of a domain identification problem transforming the problem into a domain optimization problem. In these approaches, however, the solution of the boundary value problems were the generalized (weak) solutions. The present author [4] tried to show the existence of the optimal domain for a wider class of domain optimization problems. His attempt was not complete in the case of Dirichlet problems; the solution of the boundary value problem corresponding to an "optimal" domain was not shown to satisfy the boundary condition.

In order to give a classical result in which the solution of the boundary value problem corresponding to the optimal domain is a genuine (classical) solution, we shall introduce a concept of convergence of domains which is a little stronger than usual.

Let us begin with introducing a cone $C_x$ of angle $\theta$ and of height h by

$$C_x = C_x(\xi,\theta,h) \equiv \{x \varepsilon R^2 | <x,\xi_x> > |x| \cos\theta,\ |x| < h\},$$

where $\xi_x$ denotes a unit vector in $R^2$.

<u>Definition</u> 2.1 [3]

Let $\theta \ \varepsilon\ ]0,\pi/2[$, $h > 0$, $r > 0$ $(2r \leq h)$ be three given numbers. A domain $\Omega \subset \Omega_0$ is said to satisfy "cone property" iff for any $x\ \varepsilon\ \partial\Omega$, there exists a cone $C_x = C_x(\xi_x,\theta,h)$ such that

$$\text{for any } y\ \varepsilon\ B(x,r) \cap \Omega,\quad y + C_x\ \varepsilon\ \Omega,$$

where $B(x,r)$ stands for the open disc in $R^2$ with center at x and radius r. ■

<u>Definition</u> 2.2 [3]

for $\theta\ \varepsilon\ ]0,\pi/2[$, $h > 0$, $r > 0$ $(2r \leq h)$, $\Pi(\theta,h,r)$ is the set of all subdomains of $\Omega_0$ satifying the cone property of definition 2.1. We shall say that $\{\Omega_n\}$ $\Pi(\theta,h,r)$ converges to $\Omega\ \varepsilon\ \Pi(\theta,h,r)$ strongly in $L^2(\Omega_0)$ and write as $\Omega_n \overset{s}{\rightarrow} \Omega$ in $L^2(\Omega_0)$ when the characteristic functions $\chi_n$ of $\Omega_n$ converge to that $\chi$ of $\Omega$ strongly in $L^2(\Omega_0)$. Thus, the strong $L^2(\Omega_0)$ topology is introduced in $\Pi(\theta,h,r)$. Likewise, we shall say $\Omega n \rightarrow \Omega$ a.e. in $\Omega_0$, where $\chi_n \rightarrow \chi$ a.e. in $\Omega_0$. ■

The following lemma will turn out to be important.

<u>Lemma</u> 2.1 [3]

The class $\Pi(\theta,h,r)$ of domains is relatively compact and is closed with respect to the strong $L^2(\Omega_0)$ topology. More definitely, for any infinite subset $\{\Omega_\nu\}$ of $\Pi(\theta,h,r)$, there exist a domain $\Omega\ \varepsilon\ \Pi(\theta,h,r)$ and a subsequence $\{\Omega_n\} \subset \{\Omega_\nu\}$ such that

$$\Omega_n \overset{s}{\rightarrow} \Omega \quad \text{in } L^2(\Omega_0). \quad ■ \tag{2.1}$$

Let us introduce another class of domains due to Chenais.

<u>Definition</u> 2.3 [3]

Let k and $\delta$ be two given positive numbers. We denote by $Lip(k,\delta)$ the set of all subdomains $\Omega$ of $\Omega_0$ each of which satisfies the following conditions: For any $x\ \varepsilon\ \partial\Omega$ there exist a local coordinate system $(y_1,y_2)$ and a function $\phi_x$: $\hat{P}(x) \rightarrow R$, which is Lipschitz continuous with constant k, such that

$$y\ \varepsilon\ P_{\delta\delta'}(x) \cap \Omega \Leftrightarrow \{y\ \varepsilon\ P_{\delta\delta'}(x),\ y_2 > \phi_x(y_1)\}$$

with $\delta` = k\delta$, where $\hat{P}_\delta(x)$ and $P_{\delta\delta`}(x)$ is defined by

$$\hat{P}_\delta(x) \equiv \{y_1 |\ |y_1 - y_2| < \delta\}, \tag{2.2}$$

$$P_{\delta\delta`}(x) \equiv \{y \in R^2 |\ |y_1 - x_1| < \delta,\ |y_2 - x_2| < \delta`\}. \quad \blacksquare \tag{2.3}$$

Proposition 2.1 [3]

For any k, $\delta > 0$, there exists $(\theta,h,r)$ such that

$$\mathrm{Lip}(k,\delta) \subset \Pi(\theta,h,r). \quad \blacksquare$$

Let us introduce a narrower class of domains

Definition 2.4

Let m be a nonnegative integer and k, $\delta$, $\alpha < 1$ be three given numbers. We denote by $C^{m,\alpha}(\delta,k)$ the set of subdomains $\Omega$ of $\Omega_0$ such that: For any $x \in \partial\Omega$ there exist a local coordinate $(y_1,y_2)$ and a function $\phi_x$: $\hat{P}_\delta(x) \to R$, which is a function of $C^{m,\alpha}$class with $\|\phi_x\|_{m,\alpha} \le k$, such that

$$y \in P_{\delta\delta`}(x) \cap \Omega \Leftrightarrow \{y \in P_{\delta\delta`}(x),\ y_2 > \phi_x(y_1)\},$$

where $\delta` = k\delta$ and $\hat{P}_\delta(x)$, $P_{\delta\delta`}(x)$ are defined by (2.2), (2.3). ■

Since $\Omega_0$ is a bounded domain, the boundary $\partial\Omega$ of $\Omega$ is closed and bounded; i.e., $\partial\Omega$ is compact in $R^2$ then there exist a finite number of points $x_i \in \partial\Omega$ such that

$$\partial\Omega \subset \bigcup_i o_i, \qquad o_i \subset P_{\delta\delta`}(x_i) \tag{2.4}$$

Hence, we can define $\|\partial\Omega\|_{m,\alpha}$for $\Omega \in C^{m,\alpha}(\delta,k)$ by

$$\|\partial\Omega\|_{m,\alpha} \equiv \sup \max_i \|\phi^i(\cdot)\|_{m,\alpha} \tag{2.5}$$

where $\phi^i(\cdot) \equiv \phi_{x_i}(\cdot)$ and the supremum is taken over all the coverings of the form (2.4).

We immediately obtain

Proposition 2.2

For any $\Omega \in C^{m,\alpha}(\delta,k)$, $\qquad \|\partial\Omega\|_{m,\alpha} \le k$. ■

From the definition of $C^{m,\alpha}(\delta,k)$ and proposition 2.1, we easily obtain

Proposition 2.3

If $\Omega \in C^{m,\alpha}(\delta,k)$, then $\Omega \in \mathrm{Lip}(k,\delta)$. Hence, $C^{m,\alpha}(\delta,k) \subset \Pi(\theta,h,r)$ for an appropriate triplet $(\theta,h,r)$. ■

Let us fix the triplet $(\theta,h,r)$ in all that follows.

Definition 2.5

For $\Omega \in \Pi(\theta,h,r)$ and positive $\eta$, we call the set $\{x \in \Omega_0 |\ \inf_{y \in \partial\Omega} |x - y| < \eta\}$ the $\eta$-neighborhood of $\partial\Omega$ and denote it by $N_\eta(\partial\Omega)$. ■

The following proposition is obviously valid.

Proposition 2.4

Suppose that $\{\Omega_n\} \subset C^{m,\alpha}(\delta,k)$ and $\Omega_n \overset{s}{\to} \Omega \in \Pi(\theta,h,r)$ in $L^2(\Omega_0)$. Let $\eta$ be a positive number. Then there exists a positive integer $n_0$ such that

$$\partial\Omega_n \subset N_\eta(\partial\Omega) \qquad \text{for any } n \geq n_0. \quad ■ \tag{2.6}$$

Let us consider two domains $\Omega_1$, $\Omega_2 \in C^{m,\alpha}(\delta,k)$ whose boundaries $\Gamma_1 \equiv \partial\Omega_1$, $\Gamma_2 \equiv \partial\Omega_2$ lie in $\eta$-neighborhood of each other for sufficiently small $\eta > 0$. Suppose that an open set $O(x)$ is contained in $P_{\delta\delta'}(x)$ for every $x \in \Gamma_1$. Since $\Gamma_1$ is compact, we can find a finite subcovering $\bigcup_i O_i$ from the covering $\bigcup_x O(x)$ of $\Gamma_1$. From the definition of $C^{m,\alpha}(\delta,k)$ we observe that $\Gamma_1$ and $\Gamma_2$ are represented by $y_2 = \phi_1{}^i(y_1)$ and $y_2 = \phi_2{}^i(y_1)$, respectively in $O_i$, where $(y_1,y_2)$ is the local coordinate. Let $\rho_\alpha(\Gamma_2,\Gamma_1)$ be defined by

$$\rho_\alpha(\Gamma_2,\Gamma_1) \equiv \sup \max_i \|\phi_2^i - \phi_1{}^i\|_{m,\alpha} \tag{2.7}$$

where the supremum is taken over all the finite coverings. In the same manner, we can define $\rho_\alpha(\Gamma_1,\Gamma_2)$.

Definition 2.6

We define $d_\alpha(\Omega_1,\Omega_2)$ by

$$d_\alpha(\Omega_1,\Omega_2) \equiv \max\,(\rho_\alpha(\Gamma_2,\Gamma_1),\ \rho_\alpha(\Gamma_1,\Gamma_2)),$$

where $\Gamma_1 = \partial\Omega_1$, $\Gamma_2 = \partial\Omega_2$. ■

Proposition 2.5

The following properties are valid for $d_\alpha(\cdot,\cdot)$:

(i) $d_\alpha(\Omega_1,\Omega_2) \geq 0$, $\quad d_\alpha(\Omega_1,\Omega_2) = 0$ implies $\Omega_1 \equiv \Omega_2$;

(ii) $d_\alpha(\Omega_1,\Omega_2) = d_\alpha(\Omega_2,\Omega_1)$;

(iii) $d_\alpha(\Omega_1,\Omega_2) \leq d_\alpha(\Omega_1,\Omega_3) + d_\alpha(\Omega_3,\Omega_2)$. ■

We see from this proposition that $d_\alpha(\cdot,\cdot)$ behaves like a distance in $C^{m,\alpha}(\delta,k)$ provided that the domains are close to each other. The following proposition is also easily proved.

Proposition 2.6

Suppose that $\{\Omega_n\} \subset C^{m,\alpha}(\delta,k)$ and $\Omega \in C^{m,\alpha}(\delta,k)$. If $d_\alpha(\Omega_n,\Omega) \to 0$, then

$$\Omega_n \overset{s}{\to} \Omega \quad \text{in } L^2(\Omega_0). \quad ■$$

The following lemma is a corollary to proposition 2.2 and will play an essential role in showing the existence of the optimal domain; the proof is omitted here, however.

Lemma 2.2

Suppose that $\{\Omega_n\} \subset C^{m,\alpha}(\delta,k)$ lies in the $\eta$-neighborhood of $\Omega' \in \Pi(\theta,h,r)$ for some $\eta > 0$. Then there exist a domain $\Omega \in C^{m,\alpha}(\delta,k)$ and a subsequence of $\{\Omega_n\}$

still denoted by $\{\Omega_n\}$, such that

$$d_{\alpha'}(\Omega_n,\Omega) \to 0, \tag{2.8}$$

where $0 < \alpha' < \alpha$. ■

3. Existence of the optimal domain

Lower semicontinuity of the object function in optimization problems plays one of the essential roles in proving the existence of the optimal solution. We shall have the same situation in the case of the domain optimization problems. Hence, let us recall the lower semicontinuity results by the present author [4].

Let us consider a sequene of pairs $(\Omega_n,u_n)$, where $\Omega_n \in \Pi(\theta,h,r)$ and $u_n \in H_0^1(\Omega)$. For a function $u \in H_0^1(\Omega)$, we denote by $\bar{u}$ the extention of $u$ to $\Omega_0$ obtained by setting $\bar{u}(x) = 0$ for $\Omega_0 - \Omega_n$.

Assumption 3.1 [4]

Suppose that there exist $\Omega \in \Pi(\theta,h,r)$ and $u \in H_0(\Omega)$ such that

(i) $\Omega_n \to \Omega$ a.e. in $\Omega_0$,

(ii) $\bar{u}_n \to \bar{u}$ weakly in $H_0^1(\Omega_0)$. ■

Let $g(x,u,p)$ be a function of $(x,u,p) \in R^2\times R\times R^2$. We give two lower semicontinuity results.

Lemma 3.1 [4]

Let $g = g(x,u,p)$ be a function depending only on $p \in R^2$; i.e., $g = g(p)$. Assume that $g(p)$ be nonnegative, continuous and convex. Then, on assumption 3.1,

$$\int g(\nabla u(x))\, dx \le \liminf_{n\to\infty} \int_{\Omega_n} g(\nabla u_n(x))\, dx \tag{3.1}$$

holds. ■

Lemma 3.2 [4]

Assume that $g(x,u,p)$ be nonnegative and continuous with respect to $(x,u,p)$ and that it satisfies any one of the following conditions:

(i) $g(x,u,p) \to \infty$ as $|p| \to \infty$ for each $(x,u)$;

(ii) $g(x,u,p)$ is strictly convex in $p$;

(iii) the derivatives $g_x$, $g_p$ and $g_{px}$ exist and are continuous.

Then, on assumption 3.1,

$$\int_\Omega g(x,u(x),\nabla u(x))\, dx \le \liminf_{n\to\infty} \int_{\Omega_n} g(x,u_n(x),\nabla u_n(x))\, dx \tag{3.2}$$

holds. ■

Let $\{\Omega_n\} \subset C^{3,\alpha}(\delta,k)$ converge to $\Omega \in C^{3,\alpha}(\delta,k)$; more precisely, $d_\alpha(\Omega_n,\Omega) \to 0$ $(n \to \infty)$. Let $u_n$ $(u)$ be the solution of the boundary value problem (1.2), (1.3) for $\Omega_n$ $(\Omega)$. We have already shown [5] the following proposition.

<u>Proposition</u> 3.1 [5]

Suppose that $d_\alpha(\Omega_n,\Omega) \to 0$. Then the corresponding solutions $u_n$ and their derivatives up to the second order converge to u and its corresponding derivatives uniformly on every compact subdomain $G \subset \Omega$. ■

Let us define an admissible set $A_d^\alpha$ of domains by

$$A_d^\alpha \equiv \{\Omega \in C^{3,\alpha}(\delta,k) \mid \int_\Omega h(x)\, dx = c\}. \tag{3.3}$$

Let $L_\alpha$ be defined by

$$L_\alpha \equiv \inf_{\Omega \in A_d^\alpha} J(\Omega;u), \tag{3.4}$$

where $J(\Omega;u)$ is given by (1.1). Since $A_d^{\alpha'} \subset A_d^\alpha$ when $\alpha \le \alpha' < 1$, we immediately see that

$$L_\alpha \le L_{\alpha'} \quad (0 < \alpha \le \alpha' < 1). \tag{3.5}$$

But we can show the following

<u>Proposition</u> 3.2

If we define $L_\alpha$ by (3.4), then $L_\alpha$ does not depend on $\alpha$; i.e.,

$$L_\alpha = L \quad \text{for every } \alpha \ (0 < \alpha < 1)$$

Moreover, $0 \le L < +\infty$. ■

We are now in a position to prove the following theorem which is the main result of this paper.

<u>Theorem</u> 3.1

For every $\alpha$ $(0 < \alpha < 1)$, There exists a domain $\Omega^*$ in $A_d^\alpha$ which attains the infimum L, provided that $g(x,u,p)$ satisfies one of the conditions of lemma 3.1 or lemma 3.2. ■

<u>Proof</u>

Let $\alpha'$ be such that $0 < \alpha < \alpha' < 1$. Let $\{\Omega_n\} \subset A_d^{\alpha'}$ be a minimizing sequence i.e.,

$$\lim_{n \to \infty} J(\Omega_n;u_n) = L.$$

In view of lemma 2.1, there exist $\Omega' \in \Pi(\theta,h,r)$ and a subsequence, still called $\{\Omega_n\}$, such that

$$\Omega_n \overset{s}{\to} \Omega' \text{ in } L^2(\Omega_0).$$

From proposition 2.4, we observe that for sufficiently small $\eta > 0$ there exists a positive integer $n_0$ such that

$$\Gamma_n \equiv \partial\Omega_n \subset N_\eta(\partial\Omega'), \qquad n \ge n_0.$$

Due to lemma 2.2, there exist a domain $\Omega^* \in C^{3,\alpha}(\delta,k)$ and a subsequence of $\{\Omega_n\}$ still denoted by $\{\Omega_n\}$, such that

$$d_\alpha(\Omega_n,\Omega^*) \to 0$$

From proposition 2.6, it follows that

$$\Omega_n \rightharpoonup \Omega^* \quad \text{in } L^2(\Omega_0).$$

Hence by choosing a subsequence, we observe that

$$\Omega_n \to \Omega^* \quad \text{a.e. in } \Omega_0.$$

This means that the subsequence $\{\Omega_n\}$ finally obtained satsfies assumption 3.1,(i).

Let $u^*$ be the solution corresponding to $\Omega^*$. Obviously, $u^* \in C^{3,\alpha}(\Omega)$; hence, $u^* \in H_0^1(\Omega)$. Proposition 3.1 tells us that $u_n$ and their derivatives up to the second order converge to $u^*$ and its corresponding derivatives uniformly on every compact subdomain $G$ of $\Omega$. Hence, we can easily see that $\bar{u}_n$ converges to $\bar{u}^*$ strongly in $H_0\ (\Omega_0)$. Therefore,

$$\bar{u}_n \to \bar{u}^* \quad \text{weakly in } H^1(\Omega_0).$$

This means that $u_n$ and $u$ satisfy assumption 3.1, (ii).

Thus, from lemma 3.1 or 3.2, we immediately observe that

$$L \leq J(\Omega^*;u^*) \leq \liminf_{n \to \infty} J(\Omega_n;u_n) = L,$$

i.e.,

$$J(\Omega^*;u^*) = L.$$

This completes the proof. ■

Thus, we have shown the existence of the optimal domain such that the solution of the corresponding boundary value problem is a genuine (classical) one.

## 4. Concluding remarks

We have shown that there exists an optimal domain in some classes of smooth domains which ensure the genuine solution of the boundary value problem on each of them. In a class of less smooth domains, there may happen to be a domain which attains a value of the object function smaller than L. The problem to find the widest class of domains in which a domain assumes the infimum of the object function and the corresponding solution is a genuine one is still left to be investigated.

## References


[1] Cea, J., Problems of shape optimal design, Optimization of Distributed Parameter Structures, Edited by E. J. Haug and J. Cea, Sijthoff and Noordhoff, Alphen aan den Rijn, Holland, vol.2, 1005-1048(1981).

[2] Pironneau, O., Optimal Shape Design for Elliptic Systems, Springer-Verlag, New York(1984)

[3] Chenais, D., On the existence of a solution in a domain identification problem, J. Math. Anal. Appl. 52,189-219(1975).

[4] Fujii, N., Lower-semicontinuity in domain optimization problems, J. Optm. Theor. Appl., in press.

[5] Fujii, N., Second variation and its application in a domain optimization problem, Proceedings of the 4th IFAC symposium on Control of Distributed Parameter Systems, Edited by H. E. Rauch, 431-436, Pergamon Press, Oxford(1986).

# Second Order Necessary Optimality Conditions for Domain Optimization Problem with a Neumann Problem

Yoshito Goto, Nobuo Fujii and Yasuyuki Muramatsu
Department of Control Engineering, Faculty of Engineering Science, Osaka University, Toyonaka, Osaka 560, Japan.

Abstract. In this paper, we treat a domain optimization problem in which the boundary value problem is a Neumann problem. In the case where the domain $\Omega$ is in three-dimensional Euclidean space, the first and the second order necessary conditions which the optimal domain must satisfy are derived under the constraint which is the generalization of the requisition of constant volume.

## 1. Introduction

The domain optimization ploblem is a kind of shape optimization problems. In this problem we want to find the optimal domain minimizing (or maximizing) an object function depending on a domain through the solution of a boundary value problem defined on the domain.

For example, the problem of maximum torsional rigidity solved by Polya [1] is one of such optimization problem. For another example, we know the minimum-drag flow problem investigated by Pironneau [2-3]. Other various examples are enumerated by Cea [4]. For these domain optimization problems, Zolesio [5] developed his "material derivative method" to derive the first variation of object function. Rousselet [6-7] and Koda [8-9] approached to the problems through their sensitivity analysis.

In contrast with these approachs, one of the present authors [10-11] defined unambiguously the first and the second variations of the solution of boundary value problems with a Dirichlet condition and derived the first and the second order necessary conditions for optimality.

In the problem with Neumann condition, it is more complex to derive the condition for optimality than the problem with Dirichlet condition. Simon [12] gave the first order necessary condition.

In this paper, firstly we characterize the first and the second variation of the solution assuming that they exist. Secondly we calculate the first and the second variations of the object function in term of the variations of the solution. Finally we derive the first and the second order necessary conditions for optimality in this problem using the variations of the object function.

## 2. Problem statement and notation

There may be various kinds of domain optimization problem with a Neumann problem. We shall here treat the typical one as follows: Minimize the object function

$$J(\Omega;u) \equiv \int_\Omega g(x,u(x))\,dx \tag{2.1}$$

where $u(x)$ is the solution of the boundary value problem

$$\Delta u(x) - k(x)u(x) = f(x), \qquad x \in \Omega, \tag{2.2}$$

$$\frac{\partial u}{\partial n}(x) = \kappa, \qquad x \in \Gamma, \tag{2.3}$$

subject to the constraint

$$I(\Omega) \equiv \int_\Omega h(x)\,dx = \text{const.} \tag{2.4}$$

(2.4) is a generalization of the requisition of constant volume.

For the sake of simplicity, we assume that the functions $k(x)$, $f(x)$ and $h(x)$ are defined in $R^3$ and are sufficiently smooth, that $k(x)$ is nonnegative, and that $\Gamma$ and the given function $g(x,u)$ of $x \in R^3$ and $u \in R$ are smooth enough.

Throughout the paper, the following notations will be commonly used.

$R^3$ : the three-dimensional Euclidean space.

$\Omega$ : a domain in $R^3$ with sufficiently smooth boundary $\Gamma$.

$\Gamma$ : boundary of $\Omega$.

$\Delta$ : Laplacian operator in $R^3$.

$\mathrm{grad}_\Gamma$ : gradient operator in the tangent plane of $\Gamma$.

$div_\Gamma$ : divergence operator in the tangent plane of $\Gamma$.

$\vec{n}$ : normal on $\Gamma$.

$\frac{\partial u}{\partial n}$ : normal derivative on $\Gamma$.

3. First and second variation of the solution

In order to calculate the variation of the object function, we shall clarify the first and the second variation of the solution to the boundary value problem. In this section, on the assumption that the first and the second variation of the solution exist, we shall characterize them.

First of all, we shall define the variation of domain. We suppose that bounded domain $\Omega$ has sufficiently smooth boundary $\Gamma$. Let $\rho(x)$ and $\sigma(x)$ be arbitrary smooth functions defined on $\Gamma$. We shift every boundary point $x \in \Gamma$ along the normal of $\Gamma$ at $x$ by an amount $\varepsilon\rho(x)+\varepsilon^2\sigma(x)$; this shift is considered to be positive if it has the direction of the outward normal, negative if it goes inward. We denote this shift by

$$\delta n = \varepsilon\rho(x) + \varepsilon^2\sigma(x). \tag{3.1}$$

For $\varepsilon$ small enough, $\Gamma$ is transformed into a new closed surface $\Gamma_\varepsilon$, which enclose a new domain $\Omega_\varepsilon$.

Let us consider the following boundary value problem as well as (2.2),(2.3).

$$\Delta u_\varepsilon(x) - k(x)u_\varepsilon(x) = f(x), \qquad x \in \Omega_\varepsilon, \tag{3.2}$$

$$\frac{\partial u_\varepsilon}{\partial n}(x) = \kappa, \qquad x \in \Gamma_\varepsilon. \tag{3.3}$$

Here we suppose that $k(x)$, $f(x)$ are smooth enough and $\kappa$ is constant everywhere.

Throughout this paper, we shall place the following assumption.

Assumption

We suppose that there exist functions $\phi(x)$ and $\psi(x)$ such that

$$u_\varepsilon(x) - u(x) = \varepsilon\phi(x) + \varepsilon^2\psi(x) + o(\varepsilon^2)\,, \qquad x \in \Omega, \qquad (3.4)$$

$$\frac{\partial u_\varepsilon}{\partial x_i} - \frac{\partial u}{\partial x_i} = \varepsilon\frac{\partial \phi}{\partial x_i} + \varepsilon^2\frac{\partial \psi}{\partial x_i} + o(\varepsilon^2)\,, \qquad i=1,2,3, \qquad x \in \Omega, \qquad (3.5)$$

$$\frac{\partial^2 u_\varepsilon}{\partial x_i \partial x_j} - \frac{\partial^2 u}{\partial x_i \partial x_j} = \varepsilon\frac{\partial^2 \phi}{\partial x_i \partial x_j} + \varepsilon^2\frac{\partial^2 \psi}{\partial x_i \partial x_j} + o(\varepsilon^2), \qquad i,j=1,2,3, \qquad x \in \Omega, \qquad (3.6)$$

$$\frac{\partial^3 u_\varepsilon}{\partial x_i \partial x_j \partial x_k} - \frac{\partial^3 u}{\partial x_i \partial x_j \partial x_k} = \varepsilon\frac{\partial^3 \phi}{\partial x_i \partial x_j \partial x_k} + \varepsilon^2\frac{\partial^3 \psi}{\partial x_i \partial x_j \partial x_k} + o(\varepsilon^2) \qquad i,j,k=1,2,3, \qquad x \in \Omega, \qquad (3.7)$$

hold.

We call $\phi(x)$ the first variation of the solution and $\psi(x)$ the second variation of the solution. Following two lemmas characterize these variations. The proofs of these lemma are omitted for the sake of brevity.

Lemma 1.

The first variation $\phi(x)$ is determined by the boundary value ploblem

$$\Delta\phi(x) - k(x)\phi(x) = 0, \qquad x \in \Omega, \qquad (3.8)$$

$$\frac{\partial \phi}{\partial n}(x) = \mathrm{grad}_\Gamma\, \rho(x)\ \ \mathrm{grad}_\Gamma\, u(x) - \rho(x)\frac{\partial^2 u}{\partial n^2}(x), \qquad x \in \Gamma, \qquad (3.9)$$

where $u(x)$ is the solution of (2.2),(2.3).

Lemma 2.

The second variation $\psi(x)$ is determined by the boundary value ploblem

$$\Delta\psi(x) - k(x)\psi(x) = 0, \qquad x \in \Omega, \qquad (3.10)$$

$$\frac{\partial\psi}{\partial n}(x) = \mathrm{grad}_\Gamma\ \sigma(x)\ \mathrm{grad}_\Gamma\ u(x) - \sigma(x)\frac{\partial^2 u}{\partial n^2}(x)$$
$$+ \mathrm{grad}_\Gamma\ \rho(x)\ \mathrm{grad}_\Gamma\ \phi(x) - \rho(x)\frac{\partial^2\phi}{\partial n^2}(x) - \frac{1}{2}\rho^2(x)\frac{\partial^3 u}{\partial n^3}(x)$$
$$+ 2\rho(x)\mathrm{grad}_\Gamma\ \rho(x)\ H(x)\mathrm{grad}_\Gamma\ u(x)$$
$$+ \frac{1}{2}\kappa\ \mathrm{grad}_\Gamma\ \rho(x)\ \ \mathrm{grad}_\Gamma\ \rho(x)\ , \qquad x \in \Gamma, \qquad (3.11)$$

where $\phi(x)$ is given by lemma 1.

## 4. First and second order necessary conditions

In this section, we shall derive the first and the second order necessary conditions for the optimal domain of the optimization ploblem posed in the second section.

We suppose the existence of an optimal domain. Let bounded domain $\Omega$ with sufficiently smooth boundary $\Gamma$ be an optimal domain and $u(x)$ be the corresponding solution. Let $\Omega_\varepsilon$ be the domain obtained from $\Omega$ by the boundary variation $\delta n = \varepsilon\rho + \varepsilon^2\sigma$.

The first variation $\delta^{(1)}\bar{J}$ and the second one $\delta^{(2)}\bar{J}$ of object function $J(\Omega;u)$ is defined by

$$J(\Omega_\varepsilon;u_\varepsilon) - J(\Omega;u) = \varepsilon\delta^{(1)}\bar{J} + \varepsilon^2\delta^{(2)}\bar{J} + o(\varepsilon^2), \qquad (4.1)$$

where $u_\varepsilon$ is the solution of boundary value problem for the new domain $\Omega_\varepsilon$.

After easy calculation by using (3.4), $\delta^{(1)}J$ and $\delta^{(2)}J$ are given by

$$\delta^{(1)}\bar{J} = \int_\Gamma g(x,\ u)\rho(x)\ d\Gamma + \int_\Omega \frac{\partial g}{\partial u}(x,\ u)\phi(x)\ dx, \qquad (4.2)$$

$$\delta^{(2)}\bar{J} = \int_\Gamma g(x,\ u)\sigma(x)\ d\Gamma + \int_\Gamma \frac{\partial g}{\partial u}(x,\ u)\phi(x)\rho(x)\ d\Gamma$$
$$-\frac{1}{2}\int_\Gamma \rho^2(x)g(x,\ u)\mathrm{trace}H(x)\ d\Gamma$$
$$+\frac{1}{2}\int_\Gamma \rho^2(x)\ \left(\ \frac{\partial g}{\partial x}(x,\ u) + \frac{\partial g}{\partial u}(x,\ u)\nabla u(x)\ \right)\vec{n}\ d\Gamma$$
$$+ \int_\Omega \frac{\partial g}{\partial u}(x,\ u)\psi(x)\ dx + \frac{1}{2}\int_\Omega \frac{\partial^2 g}{\partial n^2}(x,\ u)\phi^2(x)\ dx, \qquad (4.3)$$

where $\phi(x)$ and $\psi(x)$ are the first and the second variation of the solution, and $\partial g/\partial x(x,u)$ is a vector valued function defined by

$$\frac{\partial g}{\partial x}(x, u) = \left( \frac{\partial g}{\partial x_1}(x, u), \frac{\partial g}{\partial x_2}(x, u), \frac{\partial g}{\partial x_3}(x, u) \right). \tag{4.4}$$

On the other hand, the constraint (2.4) is rewritten as

$$\delta^{(1)}\bar{I} = \int_\Gamma h(x)\rho(x)\, d\Gamma = 0 , \tag{4.5}$$

$$\delta^{(2)}\bar{I} = \int_\Gamma \left( h(x)\sigma(x) + \frac{1}{2}\left( \frac{\partial h}{\partial n}(x) - h(x)\rho^2 \text{trace } H \right)\right) d\Gamma = 0. \tag{4.6}$$

Let us introduce a function $p(x)$ as the solution of

$$\Delta p(x) - k(x)p(x) = \frac{\partial g}{\partial u}(x, u), \qquad x \in \Omega, \tag{4.7}$$

$$\frac{\partial p}{\partial n}(x) = 0, \qquad x \in \Gamma. \tag{4.8}$$

Note that the boundary value problem (4.7), (4.8) has a unique solution.

Using (4.7), (4.8) and lemma 1, we obtain

$$\delta^{(1)}\bar{J} = \int_\Gamma \rho(x)\left( g(x, u) + \text{div}_\Gamma\, (p\, \text{grad}_\Gamma\, u) + p\frac{\partial^2 u}{\partial n^2} \right) d\Gamma. \tag{4.9}$$

Since domain $\Omega$ is an optimal domain, $\delta^{(1)}\bar{J}$ must be zero for every $\rho(x)$ which satisfies (4.5). Thus, in view of the well-known Lagrange multiplier rule, we have the following first order necessary condition.

Theorem 1.

Let $\Omega$ be an optimal domain and $u(x)$ be the corresponding solution. Then, there exists a constant $\lambda$ such that

$$g(x, u) + \text{div}_\Gamma\, (p\, \text{grad}_\Gamma\, u) + p\frac{\partial^2 u}{\partial n^2} = \lambda h \tag{4.10}$$

holds, where $p(x)$ is the solution of the boundary value problem (4.7), (4.8).

Since domain $\Omega$ minimize the object function $J(\Omega;u)$, the second variation $\delta^{(2)}\bar{J}$ must satisfy

$$\delta^{(2)}\bar{J} \geq 0 . \tag{4.11}$$

for every $\rho(x)$ and $\sigma(x)$ which satisfy (4.5), (4.6). We can rewrite (4.3) in the form in which $\sigma(x)$ does not appear at all. Using (4.6), (4.7), (4.8) and (4.10), we have

$$\begin{aligned}\delta^{(2)}\bar{J} = & - \frac{1}{2} \lambda \int_\Gamma ( \frac{\partial h}{\partial n} - h\rho^2 \mathrm{trace}H ) \, d\Gamma \\ & + \int_\Gamma \frac{\partial g}{\partial u} \phi\rho \, d\Gamma + \frac{1}{2} \int_\Omega \frac{\partial^2 g}{\partial u^2} \phi^2 \, dx \\ & + \frac{1}{2} \int_\Gamma ( -g \ \mathrm{trace}H + ( \frac{\partial g}{\partial x} + \frac{\partial g}{\partial u} \nabla u) \vec{n} )\rho^2 \, d\Gamma \\ & - \int_\Gamma p( \mathrm{grad}_\Gamma \rho \ \mathrm{grad}_\Gamma \phi - \rho \frac{\partial^2 \phi}{\partial n^2} - \frac{1}{2} \rho^2 \frac{\partial^3 u}{\partial n^3} \\ & \qquad + 2\rho \ \mathrm{grad}_\Gamma \rho \ H \ \mathrm{grad}_\Gamma u + \frac{1}{2} \kappa \ \mathrm{grad}_\Gamma \rho \ \mathrm{grad}_\Gamma \rho ) \, d\Gamma . \end{aligned} \tag{4.12}$$

Now, we shall consider another boundary variation $\delta n^0$ given by

$$\delta n^0 = \varepsilon\rho(x). \tag{4.13}$$

Let $\Omega_\varepsilon{}^0$ be the domain obtained from $\Omega$ by this boundary variation. Let $u_\varepsilon{}^0$, $\phi_\varepsilon{}^0$, and $\psi_\varepsilon{}^0$ be the corresponding solution, the first and the second variation of the solution, respectively. Let $\delta^{(1)}J$ and $\delta^{(2)}J$ be the first and the second variation of $J(\Omega;u)$ defined by

$$J(\Omega_\varepsilon{}^0;u_\varepsilon{}^0) - J(\Omega;u) = \varepsilon\delta^{(1)}J + \varepsilon\delta^{(2)}J + o(\varepsilon^2) . \tag{4.14}$$

Substituting $\sigma=0$ into (4.3) and (4.6) and calculating in the same manner as above, $\delta^{(2)}J$ and $\delta^{(2)}I$ are given by

$$\delta^{(2)}J = \int_\Gamma \frac{\partial g}{\partial u}\phi\rho\, d\Gamma + \frac{1}{2}\int_\Omega \frac{\partial^2 g}{\partial u^2}\phi^2\, dx$$

$$+\frac{1}{2}\int_\Gamma ( -g\ \mathrm{trace}H + ( \frac{\partial g}{\partial x} + \frac{\partial g}{\partial u}\nabla u)\vec{n}\ )\rho^2\, d\Gamma$$

$$- \int_\Gamma p(\ \mathrm{grad}_\Gamma\ \rho\ \mathrm{grad}_\Gamma\ \phi - \rho\frac{\partial^2\phi}{\partial n^2} - \frac{1}{2}\rho^2\frac{\partial^3 u}{\partial n^3}$$

$$+ 2\ \rho\mathrm{grad}_\Gamma\ \rho\ H\ \mathrm{grad}_\Gamma\ u + \frac{1}{2}\kappa\ \mathrm{grad}_\Gamma\ \rho\ \mathrm{grad}_\Gamma\ \rho\ )d\Gamma, \tag{4.15}$$

$$\delta^{(2)}I = \frac{1}{2}\int_\Gamma (\ \frac{\partial h}{\partial n} - h\ \mathrm{trace}H\ )\rho^2\, d\Gamma. \tag{4.16}$$

From (4.12), (4.15) and (4.16), we observe at once that $\delta^{(2)}\bar{J}$ is given by

$$\delta^{(2)}\bar{J} = \delta^{(2)}J - \lambda\delta^{(2)}I\ . \tag{4.17}$$

Thus, we obtain the following second order necessary condition of Kuhn-Tucker type for optimality.

Theorem 2.

Let $\Omega$ be an optimal domain and $u(x)$ be the coresponding solution. Then, there exists a constant $\lambda$ such that

$$g(x,\ u) + \mathrm{div}_\Gamma\ (\ p\ \mathrm{grad}_\Gamma\ u\ ) + p\frac{\partial^2 u}{\partial n^2} = \lambda h \tag{4.18}$$

holds, and for every $\rho(x)$ which satisfies (4.5),

$$\delta^{(2)}J - \lambda\delta^{(2)}I \geq 0 \tag{4.19}$$

holds, where $\delta^{(2)}J$ and $\delta^{(2)}I$ are given by (4.15) and (4.16) respectively.

## 5. Conclusing remarks

We have derived the first and the second order necessary conditions for optimality in a domain optimization problem (2.1), (2.2), (2.3) and (2.4) with Neumann condition.

We do not think that the optinal domain is immediately searched from those condition. Until now, numerical solutions are calculated by using first variations. But if we use the second variation $\delta^{(2)}J$ obtained here, we might reduce iterations for calculation.

In this paper, we treat the problem with the constraint (2.4) which we generalize the condition that volume of $\Omega$ is constant. As for other conditions, for example constant surface area or domain $\Omega$ including some given domain, we can also derive the second order necessary condition by using $\delta^{(2)}J$.

Also in the case where the object function is given by

$$J(\Omega;u) = \int_\Omega g(x,u(x),\nabla u(x))\, dx \ , \tag{5.1}$$

we can derive the second order necessary conditions. The calculation is more tedius,however.

In section 4, we supposed existence of $\phi$ and $\psi$ satisfying (3.4), (3.5), (3.6) and (3.7). It will have to be shown that those assmption are right.

## References


1. Polya, G.
   Torsional rigidity, principal frequency, electrostatic capacity and symmetrization.
   Quarterly Appl. Math., Vol. 6, 267-277, 1948.
2. Pironneau, O.
   On optimum profiles in Stokes flow.
   Journal of Fluid Mechanics, Vol. 59, 117-128, 1973.
3. Pironneau, O.
   On optimum design in fluid mechanics.
   Journal of Fluid Mechanics, Vol. 64, 97-110, 1974.
4. Cea, J.
   Problems of shape optimal design.
   Optimization of Distributed Parameter Structures, Vol. 2, 1049-1081, Edited by E. J. Haug and J. Cea, Sijthoff and Noordhoff, Alphen aan den Rijn, 1981.

5. Zolesio, J. P.
The material derivative (or speed) method for shape optimization. Optimization of Distributed Parameter Structures, Vol. 2, 1089-1151, Edited by E. J. Haug and J. Cea, Sijthoff and Noordhoff, Alphen aan den Rijn, 1981.
6. Rousselet, B.
Reponse dynamique et optimisation de domaine.
Preprints of IFAC 3rd symposium on Control of Distributed Parameter Systems, Edited by J. P. Babary and L. LeLetty, IFAC, Toulouse, 1982.
7. Rousselet, B.
Shape design sensitivity of a membrane.
Journal of Optimization Theory and Applications, Vol. 40, 595-623, 1983.
8. Koda, M.
Sensitivity analysis of atmospheric diffusion equation.
Atmospheric Environment, Vol. 16, 2595-2601, 1982.
9. Koda, M.
Optimum design in fluid mechanical distributed parameter systems.
Large Scale Systems, Vol. 6, 279-292, 1984.
10. Fujii, N.
Necessary conditions for a domain optimization problem in elliptic boundary value problems.
SIAM J. Control and Optimization, Vol. 24, 346-360, 1986.
11. Fujii, N.
Second order necessary conditions in a domain optimization problem.
Proceedings of the 4th IFAC Symposium on Control of Distributed Parameter Systems, Los Angeles, 1986.
12. Simon, J.
Variation with respect to domain for Neumann Condition.
Proceedings of the 4th IFAC Symposium on Control of Distributed Parameter Systems, Los Angeles, 1986.

# OPTIMAL SENSOR AND ACTUATOR LOCATIONS IN DISTRIBUTED PARAMETER SYSTEMS

Sigeru Omatu[†] and John H. Seinfeld[††]
†Department of Information Science and Systems Engineering, Faculty of Engineering, University of Tokushima, Tokushima, 770, Japan
††Department of Chemical Engineering, California Institute of Technology, Pasadena, California 91125, U.S.A.

ABSTRACT

The optimal sensor and actuator location problem in distributed parameter systems is considered. The sensor and actuator locations are chosen to minimize the performance criterion on the LQG problem with noises dependening on the actuator and sensor locations. The existence theorem for the optimal sensor and actuator locations is proved based on a property of evolution operators.Necessary and sufficient conditions are derived by using properties of operator-valued Riccati equations and evolution operators. It is shown that a duality holds between the sensor and actuator locations.

## I. INTRODUCTION

In this paper, we consider the optimal distributed parameter sensor and actuator location problem for pointwise observation and control.We first prove the existence theorem of optimal sensor and actuator locations for the distributed parameter systems with the linear quadratic cost function. Then we derive the sufficient condition for the optimal sensor andactuator locations by using the properties of evolution operators and operator-valued Riccati equations. It is shown that the sufficient condition derived here posesses the duality between sensor and actuator locations.Fianlly, we derive the necessary condition for optimality and compare it with the sufficient condition.

## II. PROBLEM STATEMENT

Let us consider the following system whose state $u(t)$ with values in a Hilbert space $\mathcal{H}$ is described by

$$du(t)=A(t)u(t)dt+B(t;c)f(t)dt$$
$$+G(t;c)d\xi(t), (1)$$
$$u(t_0)=u_0$$

where $c \in \mathcal{U}_c$ denotes a control law of actuator locations, $f(t)$ is an input with values in a Hilbert space $\mathcal{U}_f$, $\xi(t)$is a Wiener process with values in a Hilbert space $\mathcal{U}_1$. We assume that $\mathcal{U}_c$ is compact and $A(t)$ is a differential operator such that

$$A(\cdot) \in L^\infty(t_0,t_f; \mathcal{L}(\mathcal{V},\mathcal{V}')),$$

$$\langle -A(t)\phi,\phi\rangle_{\mathcal{V}} \geq \alpha||\phi||^2 ,\phi > 0, \forall\phi \in \mathcal{V}$$

where $\mathcal{V}'$ denotes the dual space of a Hilbert space $\mathcal{V}$ and $\mathcal{V} \subset \mathcal{H} \subset \mathcal{V}'$. $\mathcal{L}(\mathcal{V}, \mathcal{V}')$ shows a family of bounded linear operators from $\mathcal{V}$ into $\mathcal{V}'$ and $\langle\cdot,\cdot\rangle_{\mathcal{V}}$ and $||\cdot||_{\mathcal{V}}$ denote an inner product between $\mathcal{V}'$ and $\mathcal{V}$ and the norm in $\mathcal{V}$, respectively. Furthermore, let $B(t;c)$ and $G(t;c)$ be operators such that

$$B(\cdot;c) \in L^\infty(t_0,t_f; \mathcal{L}(\mathcal{U}_f, \mathcal{V}')),$$

$$G(\cdot) \in L^\infty(t_0,t_f; \mathcal{L}(\mathcal{U}_1, \mathcal{V}')).$$

and let the statistics of $u_0$ and $\xi(t)$ be given by

$$E[\xi(t)]=0,E[u_0]=\bar{u}_0$$

$$E[(u_0-\bar{u}_0)\circ(u_0-\bar{u}_0)]=P_0$$

$$E(\xi(t)\ \xi(s)]= \int_{t_0}^{\min(t,s)} \tilde{Q}(\tau)\, d\tau$$

where $P_0$ and $\tilde{Q}(\tau)$ are non-negative operators and "$\circ$" is defined by

$$(h_1\circ h_2)h_3 = h_1\langle h_2,h_3\rangle_{\mathcal{U}_1},\ \forall\ h_1,h_2,\ h_3 \in \mathcal{U}_1$$

Let the observation equation be given by

$$dz(t)=F(t,s)u(t)dt+d\eta(t,s),\ s \in \mathcal{U}_s$$

where $z(t) \in \mathcal{Z}$ is the observation data and $\eta(t)$ is a Wiener process independent of $\xi(t)$ and $u_0$ such that

$$E[\eta(t,s)]=0,$$

$$E[\eta(t,s)\circ\eta(\tau,s)]=\int_{t_0}^{\min(t,\tau)} R(\sigma,s)d\sigma.$$

Here, $s \in \mathcal{U}_s$ denotes a control law of sensor locations and s is assumed to be compact. We assume that $R(t;s)$ is a positive-definite nuclear operator. Since $R(t;s)$ is nuclear and invertible, note that $\mathcal{Z}$ is a finite dimensional space[1].

We consider the optimal control problem to minimize the cost performance given by

$$J(f(\cdot),s,c)=E[\langle u(t_f),M(s)u(t_f)\rangle_{\mathcal{H}}$$

$$+\int_{t_0}^{t_f}\langle u(t),M(t;s)u(t)\rangle+\langle f(t),N(t;c)f(t)\rangle_{\mathcal{U}_f}\, dt]$$

where $M(s)$ and $M(t;s)$ are non-negative operators and $N(t;c)$ is a positive-definite operator. Denoting the optimal cost performance of $J(f(\cdot),s,c)$ with respect to $f(\cdot)$ by $V(u(t_0),t_0,s,c)$, the optimal sensor and actuator location problem is to find the optimal control laws $s^0$ and $c^0$ such that $V(u(t_0),t_0,s,c)$ is minimized with respect to $s \in \mathcal{U}_s$ and $c \in \mathcal{U}_c$:

$$V(u(t_0,t_0,s^0,c^0)=\min \quad (V(u(t_0,t_0,s,c)) \quad s \in \mathcal{U}_s, c \in \mathcal{U}_c$$

## III. PRELIMINARY RESULTS

In order to derive the optimal control laws of sensor and actuator locations, we summarize the optimal control results as follows:

$$V(u(t_0),t_0,s,c)=tr[M(s)P(t_f)]$$

$$+\int_{t_0}^{t_f}tr[M(t;s)P_c(t)]\, dt+tr[N_0P(t_0)]$$

$$+\int_{t_0}^{t_f} tr[P_c(t)P(t)R(t,s)P(t)]\, dt$$

where

$$N_0=\bar{u}_0 \circ \bar{u}_0$$

$$\bar{R}(t,s)=F^*(t,s)R^{-1}(t;s)F(t,s).$$

Here, $P(t)$ and $P_c(t)$ are given by the following operator-valued Riccati equations:

$$\frac{dP(t)}{dt}=A(t)P(t)+P(t)A^*(t)+Q(t,c)$$
$$-P(t)\bar{R}(t,s)P(t)$$
$$P(t_0)=\mathrm{cov}[\tilde{u}(t_0),\tilde{u}(t_0)],$$

where

$$P(t)=E[\tilde{u}(t)\circ\tilde{u}(t)]$$

$$Q(t,c)=G^*(t;c)Q(t)G(t;c)$$

$$\tilde{u}(t)=u(t)-\hat{u}(t)$$

and

$$\frac{dP_c(t)}{dt} = A^*(t)P_c(t)A(t)$$
$$+ M(t;s)-P_c(t)\bar{N}(t,c)P_c(t)$$

where

$$P_c(t_f)=M(s),$$

$$\bar{N}(t,c)=B(t,c)N^{-1}(t;c)B^*(t,c).$$

The properties of the evolution operator $\Omega(t,\tau)$ for $A(t)$ are as follows[1]:

$$\frac{\partial\Omega(t,\tau)}{\partial t} = A(t)\Omega(t,\tau),\ \Omega(\tau,\tau) = \mathcal{I},$$

$$||\Omega(t,\tau)||_{\mathcal{L}(\mathcal{H},\mathcal{H})} \leq 1$$

where $\mathcal{J}$ denotes the identity operator.

## IV. MAIN RESULTS

In this section, we will prove the existence theorem of the optimal sensor and actuator locations and derive necessary and sufficient conditions for the optimality.

[THEOREM 1] Let $_{c}$ and $_{s}$ be compact and let $M(t;s)$ and $N(t;c)$ be continuous with respect to t. If $M(s)$, $M(t;s)$, $R(t,s)$, $N(t,c)$, and $Q(t,c)$ are continuous with respect to s and c, then there exist the optimal sensor and actuator locations such that $V(u(t_0),t_0,s,c)$ is minimized.

PROOF We can prove the continuity of $P(t)$ and $P_c(t)$ with respect to s and c, respectively by using the same way as [2]. Since there exist minimum points of the continuous function on the compact space, the proof of the theorem is complete.

[THEOREM 2] (Sufficient Condition)

If $s^0$ and $c^0$ satisfy the following inequalities:

$$R(t,s^0) \geq R(t,s),\ \forall\, s \in \mathcal{U}_s$$

$$N(t,c^0) \geq N(t,c),\ \forall\, c \in \mathcal{U}_c \qquad (8)$$

$$M(t,s \geq M(t,s^0), M(s) \geq M(s^0),\ \forall\, s \in \mathcal{U}_s$$

$$Q(t,c) \geq Q(t,c^0),\ \forall\, c \in \mathcal{U}_c,$$

then $s^0$ and $c^0$ are the optimal sensor and actuator locations, respectively.

<u>PROOF</u> Let us define

$$\Pi(t)=P(t;s,c) - P(t;s^0,c^0)$$
$$\Pi_c(t)=P_c(t;s) - P_c o(t;s^0)$$

where $P(t;s,c)$ and $P_c(t;s)$ denote $P(t)$ and $P_c(t)$ for the sensor location s and the actuator location c. Furthermore, we define $\Delta V(u(t_0),t_0,s,c)$ by

$$\Delta V \equiv V(u(t_0),t_0,s,c) - V(u(t_0),t_0,s^0,c^0).$$

Then we can obtain the following relation:

$$\Delta V = tr[(M(s)-M(s^0))P(t_f;s,c)]$$

$$+\int_{t_0}^{t_f} tr[\Delta M(t)P(t;s,c)]dt+tr[N_0\Pi_c(t_0)]$$

$$+\int_{t_0}^{t_f} tr[\Pi_c(t)P(t;s,c)R(t,s)P(t;s,c)]dt$$

$$+\int_{t_0}^{t_f} tr[P_{c^0}(t;s^0)N(t,c^0)P_{c^0}(t;s^0)\Pi(t)]dt$$

$$+\int_{t_0}^{t_f} tr[P_{c^0}(t;s^0)\Delta Q(t)]dt \qquad (9)$$

where $\Delta M(t)$ and $\Delta Q(t)$ are defined by

$$\Delta M(t)=M(t;s)-M(t;s^0)$$

$$\Delta Q(t)=Q(t;c)-M(t;c^0).$$

Under the condition of (8), we can show that $\Pi(t)$ and $\Pi_c(t)$ are positive operators. Thus, from (9) we obtain that $\Delta V \geq 0$.Therefore, the proof of the theorem is complete.

[THEOREM 3] It is necessary for $s^0$ and $c^0$ to be optimal that

$$\tilde{\alpha}(t_0,t_f) \geq 0 \text{ for } \forall\, s \neq s^0 \text{ and } \forall\, c \neq c^0$$

$$\tilde{\alpha}(t_0,t_f) = \alpha(t_0,t_f)$$

$$+\int_{t_0}^{t_f} tr[N_0\Phi_c^*(t,t_0)P_{c^0}(t)\Delta M(t)$$

$$P_{c^0}(t;s^0)\Phi_c(t,t_0)]\,dt + \Phi_c^*(t_f,t_0)$$

$$(M(s)-M(s^0))\Phi_c(t_f,t_0) + \int_{t_0}^{t_f} tr[\Phi_s(t_f,t)$$

$$P(t;s^0,c^0)\Delta Q(t)P(t;s^0,c^0)\Phi_s^*(t_f,t)]dt$$

where $\alpha(t_0,t_f)$, $\Phi_s(t,\tau)$, and $\Phi_c(t,\tau)$ are defined by

$$\alpha(t_0,t_f)=\int_{t_0}^{t_f}\mathrm{tr}[N_0\Phi_c^*(t,t_0)P_co(t;s^0)$$

$$(N(t,c^0)-N(t,c))P_c(t;s)\Phi_c(t,t_0)]dt$$

$$+\int_{t_0}^{t_f}\int_{t_0}^{t}\mathrm{tr}[\Phi_c^*(t,\tau)P_co(t;s^0)(N(t,c^0)-$$

$$N(t,c))P_co(t;s^0)\Phi_c(t,\tau)P(\tau;s,c)R(\tau,s)$$

$$P(\tau;s,c)]\,d\tau dt+\int_{t_0}^{t_f}\int_{t_0}^{t}\mathrm{tr}[P_c(t;s)N(t,c)$$

$$P_c(t;s)\Phi_s(t,\tau)P(\tau;s^0,c^0)(R(\tau,s^0)$$

$$-R(\tau,s))P(\tau;s^0,c^0)\Phi_s^*(t,\tau)]d\tau dt,$$

$$\frac{\partial\Phi_c(t,\tau)}{\partial t} = A(t,c)\Phi_c(t,\tau),$$

$$\Phi_c(\tau,\tau)=\mathcal{J}\ ,A(t,c) = A(t)-P_co(t;s^0)N(t,c)$$

$$\frac{\partial\Phi_s(t,\tau)}{\partial t} = A(t,s)\Phi_s(t,\tau),$$

$$\Phi_s(\tau,\tau)=\mathcal{J}\ ,A(t,s)=A(t)-P(t;S^0,c^0)R(t,s).$$

From (9) we can prove theorem 3 by using the similar way to that of[2].

## V. CONCLUSIONS

The problem of optimally locating a given number of sensors and actuators for a linear distributed parameter system has been considered. The existence theorem concerning the solution of the optimal sensor and actuator locations with pointwise inputs and observations has been proved by using the properties of evolution operators. Then necessary and sufficient conditions for optimal locations have been derived based on the property of partial differential equation of Riccati type.

References

[1] Omatu,S. and J.H. Seinfeld, "Existence and Comparison Theorems for Partial Differential Equations of Riccati Type", Jounal of Optimization Theorey and Applications, Vol.36,No.2, February, 1982, pp.263-275.

[2] Omatu,S. and J.H. Seinfeld,"Optimization of Sensor and Actuator Locations in a Distributed Parameter System", Journal of Franklin Institute, Vol.315, No.5,May/June, 1983, pp.407-421.

# MODELLING OF FLEXIBLE MANIPULATOR ARMS

Shozo Tsujio
Department of Mechanical Engineering
College of Engineering, Osaka Prefectural University
Mozu-Umemachi, Sakai, Osaka 591, Japan

## 1. Introduction

The kinematics and dynamics of flexible manipulator arms having one or two links have been studied extensively in plane motion. However there are few studies which regard multi-links in spatial, three dimensional motion. The models presented in previous works are classified into three groups: (a) distributed parameter model /1 - 5/, (b) lumped parameter model /6 - 8/, and (c) simplified model for special use /9 - 12/. First dynamic models classified in above (a) are more elaborate in theory than other models. However the models are limited in simple shaped link with simple boundary conditions. Second dynamic models classified in above (b) are equivalent discrete models induced by a finite element discretization of links. In this model the limitations of link shapes and link boundary conditions are not as large as the first model, but its number of system equations is larger than the other models. The last dynamic models classified in above (c) are adequate in restricted problems but are not able to be used generally.

In this paper a dynamical model of a multi-link manipulator with elastic members is proposed which is classified in (b). The model is a lumped parameter system utilizing the Finite Element Method. The equations of motion are obtained by using Lagrange's equation. These equations consist of two parts, i.e., the equations describing large articular motion and the equations describing small elastic displacement. These equations are coupled with each other by inertial mass.

The features of the model are (1) easy to grasp the kinematic and dynamic characteristics of flexible manipulator arms by intuition, (2) easy to deal with complex-shaped links, (3) easy to deal with multi-links with rotary joints in three dimensional motion, and (4) easy to make a simulation program for general use.

The model can be used for simulation purposes in designing mechanical structure of links, planning working trajectory, and designing control strategy and its evaluation.

## 2. Kinematics of Flexible Manipulators

The following assumptions are made to derive the dynamical model of flexible manipulator arms.

1. The manipulator consists of N links in the form of an open-loop chain.
2. The motion of each joint is restricted to rotate in one degree of freedom and its compliance and damping are ignored.

3. The links are represented by a concentrated parameter system having finite number of masses and springs. Mass moment of inertia of each particle is ignored.
4. The motion of a link is described by a summation of a large articular motion and a small perturbed displacement.

In order to describe the kinematics of a manipulator, coordinate frames have to be defined. All coordinate frames are defined for the links in undeformed condition. The inertial coordinate frame $O_0$-$x_0y_0z_0$ can be fixed at an arbitrary position. Local coordinate frame $O_i$-$x_iy_iz_i$ is assigned to i-th link shown in Fig. 1. The $z_i$ axis coincides with the characteristic axis of motion of the i-th joint but its orientation is arbitrary. The origin $O_i$ is taken up as the point of intersection of the $z_i$ axis and the common perpendicular between $z_{i-1}$ and $z_i$. The $x_i$ axis is formed by the line paralleled to the common perpendicular between $z_i$ and $z_{i+1}$ and the orientation is positive when the axis directs the i-th joint to the i+1-th joint. The $y_i$ axis is chosen which completes the right-hand Cartesian coordinate system $O_i$-$x_iy_iz_i$.

$\mu_i$, $\nu_i$, $\alpha_i$ and $\theta_i$ are link parameters before the links are deformed. $\mu_i$ is a distance from $z_{i-1}$ to $z_i$ measured along the common perpendicular between $z_{i-1}$ and $z_i$. $\nu_i$ is a distance from origins $O_{i-1}$ to $O_i$ measured in direction of positive $z_{i-1}$. The angle $\theta_i$ is measured counterclockwise from positive $x_{i-1}$ to positive $x_i$ about positive the $z_i$ axis. Angle $\alpha_i$ is the angle from positive $z_{i-1}$ to positive $z_i$ measured counterclockwise about positive $x_{i-1}$ axis when $\theta_i$ is zero.

The coordinate of point $P_i$ on the i-th link is represented by a column vector or position vector $d_i$ with respect to the local frame $O_i$-$x_iy_iz_i$ (Fig. 1), i.e.,

$$d_i = ( 1, x_i, y_i, z_i )^T \tag{1}$$

where $(\cdot)^T$ indicates the transpose of vector or matrix. The position vector of the

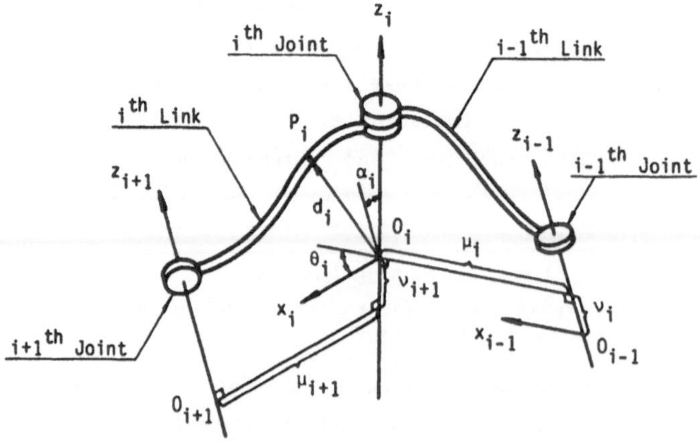


Fig. 1 Local coordinate frame and link parameters

same point $P_i$ with respect to the local frame $O_{i-1}-x_{i-1}y_{i-1}z_{i-1}$ is $d_{i-1}$ where

$$d_{i-1} = {}^{i-1}W_i \, d_i \tag{2}$$

and ${}^{i-1}W_i$ is a homogeneous transformation matrix defined by

$$ {}^{i-1}W_i = \begin{bmatrix} 1 & 0 & 0 & 0 \\ \mu_i & \cos\theta_i & -\sin\theta_i & 0 \\ 0 & \cos\alpha_i \sin\theta_i & \cos\alpha_i \cos\theta_i & -\sin\alpha_i \\ \nu_i & \sin\alpha_i \sin\theta_i & \sin\alpha_i \cos\theta_i & \cos\alpha_i \end{bmatrix} \tag{3}$$

By repeated application of Eq. (2), the position vector $d_i^0$ of the point on i-th link with respect to the inertial coordinate frame is given by

$$d_i^0 = {}^0W_1 \, {}^1W_2 \dots \dots {}^{i-1}W_i \, d_i = W_i \, d_i \tag{4}$$

To incorporate the deflection of links, elastic coordinates have to be defined. Suppose that the position vector of the g-th particle of i-th link is

$$\bar{d}_{ig} = ( 1, \bar{x}_{ig}, \bar{y}_{ig}, \bar{z}_{ig} )^T \tag{5}$$

in undeformed condition. After links have been deformed, the particle moves translationally and rotationally. Let the relative translational movements describe with

$$p_{ig} = ( p_{ig}^x, p_{ig}^y, p_{ig}^z )^T \tag{6}$$

and the relative rotational movements describe with

$$\psi_{ig} = ( \psi_{ig}^x, \psi_{ig}^y, \psi_{ig}^z )^T \tag{7}$$

The position vector after deformation is given by:

$$d_{ig} = \bar{d}_{ig} + \Gamma p_{ig} \tag{8}$$

$$\Gamma = \begin{bmatrix} 0 & 0 & 0 \\ 1 & 0 & 0 \\ 0 & 1 & 0 \\ 0 & 0 & 1 \end{bmatrix} \tag{9}$$

A position vector can be expressed in the inertial coordinate frame by multiplying the transformation matrix. The position vector $d_{ig}$ with respect to the inertial coordinate frame, $d_{ig}^0$, is given by

$$d_{ig}^{0} = W_i(\bar{d}_{ig} + \Gamma p_{ig}) \tag{10}$$

Differentiating $d_{ig}^{0}$ with respect to time t, the inertial velocity $v_{ig}^{0}$ and the inertial acceleration $a_{ig}^{0}$ are given by

$$v_{ig}^{0} = \Sigma_{j=1}^{N} U_{ij}(\bar{d}_{ig} + \Gamma p_{ig})\dot{\theta}_j + W_i\Gamma\dot{p}_{ig}$$

$$a_{ig}^{0} = \Sigma_{j=1}^{N} U_{ij}(\bar{d}_{ig} + \Gamma p_{ig})\ddot{\theta}_j + \Sigma_{j=1}^{N}\Sigma_{k=1}^{N} U_{ijk}(\bar{d}_{ig} + \Gamma p_{ig})\dot{\theta}_j\dot{\theta}_k \tag{11}$$

$$+ 2\,\Sigma_{j=1}^{N} U_{ij}\Gamma\dot{p}_{ig}\dot{\theta}_j + W_i\Gamma\ddot{p}_{ig}$$

In above equations, $\dot{\theta}_j$, etc., indicate the derivatives with respect to time, N is the number of links, and matrices $U_{ij}$ and $U_{ijk}$ are given by

$$U_{ij} = \partial W_i/\partial\theta_j \text{ for } 1 \leq j \leq i \quad \text{and} \quad U_{ij} = 0 \text{ for } j > i$$

$$U_{ijk} = \partial^2 W_i/\partial\theta_j\,\partial\theta_k \quad \text{for} \quad 1 \leq j,\ k \leq i \quad \text{and} \quad U_{ijk} = 0 \quad \text{for } j,\ k > i \tag{12}$$

## 3. Dynamics of Links

### 3.1 Kinetic energy

Let links be transformed to a finite element model. The number of particles of an n-th link is NG(n) and mass of a g-th particle is $m_{ng}$. The inertial velocity of g-th particle of n-th link being represented by $v_{ng}^{0}$ which can be given in the form of the first expression of Eq. (11), the translational kinetic energy of the particle is given by

$$2\,T_{ng} = m_{ng}\,\mathrm{tr}[v_{ng}^{0}\,(v_{ng}^{0})^T] \tag{13}$$

where tr[ · ] indicates the trace of a matrix. Summing up Eq. (13) with respect to all the particles of n-th link, the total kinetic energy for the n-th link is given by

$$2\,T_n = \Sigma_{j=1}^{N}\Sigma_{k=1}^{N} \mathrm{tr}[U_{nj}\,J_n\,U_{nk}^T]\dot{\theta}_j\dot{\theta}_k + 2\,\Sigma_{j=1}^{N} \mathrm{tr}[U_{nj}\,H_n\,W_n^T]\dot{\theta}_j$$

$$+ \Sigma_{g=1}^{NG(n)} m_{ng}\,\mathrm{tr}[\dot{p}_{ng}\,\dot{p}_{ng}^T] \tag{14}$$

where matrices $J_n$ and $H_n$ are

$$J_n = \Sigma_{g=1}^{NG(n)} m_{ng}\,(\bar{d}_{ng} + \Gamma p_{ng})(\bar{d}_{ng} + \Gamma p_{ng})^T$$

$$H_n = \Sigma_{g=1}^{NG(n)} m_{ng}\,(\bar{d}_{ng} + \Gamma p_{ng})(\Gamma\dot{p}_{ng})^T \tag{15}$$

Total kinetic energy of the manipulator is given by

$$T = \Sigma_{n=1}^{N} T_n \tag{16}$$

**3.2 Elastic potential energy**

Now we define a translational deflection vector $p_n$ and a rotational deflection vector $\psi_n$ of the n-th link as

$$\begin{aligned} p_n^T &= ( p_{n1}^T, p_{n2}^T, \dots \dots, p_{nNG(n)}^T) \\ \psi_n^T &= ( \psi_{n1}^T, \psi_{n2}^T, \dots \dots, \psi_{nNG(n)}^T) \end{aligned} \tag{17}$$

Structural stiffness matrix $K_n$ of the n-th link can be expressed with respect to these deflection vectors. Then the elastic potential energy $V_n^E$ is given by

$$2 V_n^E = ( p_n^T, \ \psi_n^T) K_n \begin{bmatrix} p_n \\ \psi_n \end{bmatrix} = ( p_n^T, \ \psi_n^T) \begin{bmatrix} K_{pp}^n & K_{p\psi}^n \\ K_{\psi p}^n & K_{\psi\psi}^n \end{bmatrix} \begin{bmatrix} p_n \\ \psi_n \end{bmatrix} \tag{18}$$

Total elastic potential energy of the manipulator is given by

$$V^E = \Sigma_{n=1}^{N} V_n^E \tag{19}$$

The structural stiffness matrix can be calculated easily by the use of the Finite Element Method. Potential energy of the system also arises from gravity. Since the potential energy due to gravity can be calculated easily, it will not be discussed here.

**3.3 Equations of motion for each link**

We choose the generalized coordinates as

$$\theta_n, \ p_{ng}, \ \psi_{ng} \qquad ( n = 1, 2, \dots \dots, N, \ g = 1, 2, \dots \dots, NG(n)) \tag{20}$$

The generalized force corresponding to joint variable $\theta_i$ is the joint torque $t_i$. For deflection variables, the generalized force corresponding to $p_{ig}$ is force $f_{ig}$ and the generalized force corresponding to $\psi_{ig}$ is moment $\tau_{ig}$. These generalized forces are due to external driving forces, gravity forces and internal link-constraint forces.

The equations of motion of the i-th link are derived by substituting the energy expressions of Eqs. (16) and (19) into the Lagrange's equations

$$\Sigma_{n=1}^{N} \Sigma_{j=1}^{N} tr[U_{ni} J_n U_{nj}^T]\ddot{\theta}_j + \Sigma_{n=1}^{N} \Sigma_{j=1}^{N} \Sigma_{k=1}^{N} tr[U_{ni} J_n U_{njk}^T]\dot{\theta}_j\dot{\theta}_k$$

$$+ 2\, \Sigma_{n=1}^{N} \Sigma_{j=1}^{N} tr[U_{ni}\, H_n\, U_{nj}^T]\dot{\theta}_j$$

$$+ \Sigma_{n=1}^{N} tr[U_{ni}\, \Sigma_{g=1}^{NG(n)} m_{ng}(\bar{a}_{ng} + \Gamma p_{ng})(\Gamma\ddot{p}_{ng})^T W_n^T] = t_i \quad (21)$$

$$m_{ig}\ddot{p}_{ig} + 2m_{ig}(W_i\Gamma)^T(\Sigma_{j=1}^{N} U_{ij}\Gamma\dot{\theta}_j)\dot{p}_{ig}$$

$$+ m_{ig}(W_i\Gamma)^T(\Sigma_{j=1}^{N} \Sigma_{k=1}^{N} U_{ijk}\, \dot{\theta}_j\dot{\theta}_k)(\bar{a}_{ig} + \Gamma p_{ig})$$

$$+ m_{ig}(W_i\Gamma)^T(\Sigma_{j=1}^{N} U_{ij}\, \ddot{\theta}_j)(\bar{a}_{ig} + \Gamma p_{ig}) + [K_{pp}{}^i]_g\, p_i + [K_{p\psi}{}^i]_g\, \psi_i = f_{ig} \quad (22)$$

$$[K_{\psi p}{}^i]_g\, p_i + [K_{\psi\psi}{}^i]_g\, \psi_i = \tau_{ig} \quad (23)$$

Notation $[\,\cdot\,]_g$ denotes the sub-matrix of matrix $[\,\cdot\,]$ corresponding to the g-th deflection variable $p_{ig}$ or $\psi_{ig}$.

It should be noticed that Eqs. (21) - (23) are the link dynamic equations having N excess equations because the coordinate system (20) has N excess degrees of freedom. Therefore N coordinates and N equations have to be taken off from the expressions in Eqs. (21) - (23) corresponding to the problems considered.

For example, if i-th link is to be shown in Fig. 2, the position of the link with respect to the rotational $z_i$ axis can be determined to ( $\theta_i + \psi_{i1}{}^z$ ), however $\theta_i$ and $\psi_{i1}{}^z$ cannot be determined individually. In other words $\theta_i$ and one of the $\psi_{ig}{}^z$ are redundant of each other in respect to the meaning of degrees of freedom in the system. In this case we can put $\psi_{i1}{}^z = 0$ and remove one expression corresponding to $\psi_{i1}{}^z$ from Eq. (23). The remaining expressions of Eqs. (21) - (23) are independent. So the problem can be solved.

Another example, nominal angles ( $\theta_i(t)$, i = 1, 2, ... ...., N) are given and the small vibrations induced by the motions should be calculated. In this problem, independent link dynamic equations are the expressions of Eqs. (22) and (23), and they are now a linear equation system with respect to $p_i$ and $\psi_i$ /6, 7/.

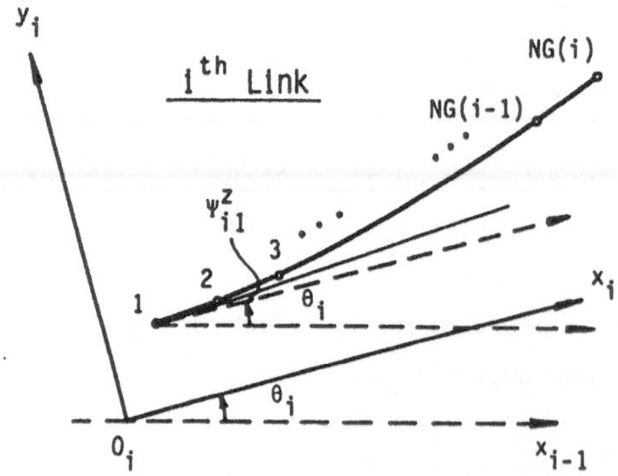


Fig. 2 i-th link and its local coordinate frame

In Eq. (23) the inertial terms are absent and these expressions are algebraic equations because the mass moments of inertia of particles are ignored due to the assumption described in Section 2. When deflection variables are small enough to be ignored in Eq. (21), this equation coincides with the dynamic equation of rigid link. By this meaning, the motion described by Eq. (21) is called articular motion. Eqs. (21) and (22) are coupled with each other with respect to inertial mass.

## 4. Coordinate Reduction

The number of dynamic equations concerned to the deflection of the i-th link is 6*NG(i). Numerical time integration of such a very large set of equations would be difficult in computational time and numerical accuracy. To solve such problems due to reducing variables some methods have been already developed for dynamical analysis in the Finite Element Method /13, 14/.

Preparing for the coordinate reduction, a complete set of deflection variables is divided into two sets:

i) a small set of variables which are connected to other links or on which external driving forces are acting (r) and

ii) a large set of remaining variables (s).

Then the deflection vector is rewritten as $u_i$, i.e.,

$$u_i^T = ((p_i^r)^T, (\psi_i^r)^T \;\vdots\; (p_i^s)^T, (\psi_i^s)^T) \tag{24}$$

The expressions in Eqs. (22) and (23) are arranged in the manner respective to the order of the elements in the deflection vector $u_i$.

Next reduced coordinate $u_i^*$ is related to deflection vector $u_i$ in the following liner expression /6, 7, 14/:

$$u_i = \begin{bmatrix} p_i^r \\ \psi_i^r \\ \cdots \\ p_i^s \\ \psi_i^s \end{bmatrix} = \begin{bmatrix} I \\ \\ \cdots \\ \Phi_I \\ \\ \end{bmatrix} \begin{bmatrix} p_i^r \\ \psi_i^r \end{bmatrix} + \begin{bmatrix} 0 \\ \\ \cdots \\ \Phi_{II} \\ \\ \end{bmatrix} \eta(t) = R_i\, u_i^* \tag{25}$$

where I is unit matrix and 0 is zero matrix. Matrix $\Phi_I$ can be determined from the deflections of the i-th link in the static equilibrium conditions concerning to the stiffness matrix $K_i$ when each particle belonging to the set (r) moves a unit length or a unit angle. Next matrix $\Phi_{II}$ can be taken as the truncated modal matrix, for example, when all of the particles belonging to the set (r) are fixed. In this case $\eta(t)$ is truncated modal coordinates.

The transformation matrix $R_i$ defined in Eq. (25) restricts the degrees of freedom of the deflection vector $u_i$. So we call the $R_i$ a restriction matrix.

When the expression of $u_i$ in Eq. (25) is substituted into Eqs. (22) and (23), and the transposed matrix of restriction matrix $R_i$ is pre-multiplied to these equations, the reduced dynamic equations of the link are obtained having variables $\theta_i$ and $u_i^*$.

## 5. Dynamic Equations of Manipulators

Link dynamic equations are derived in the previous section. In these equations displacements at a joint have to satisfy compatibility conditions: translational displacements and rotational displacements at a joint match together in two adjacent links. So these displacements are not independent in total links of a manipulator. In addition the generalized forces in these equations involve unknown internal forces which are applied to connect two adjacent links at the joints. Therefore dynamic equations of manipulator have to be assembled from the individual link dynamic equations, eliminating dependent displacements and unknown internal forces. This process can be performed by using compatibility matrix /6, 7/.

First a set of global coordinates are chosen which are independent of each other and can satisfy compatibility conditions. This global coordinates are represented to q. Next the reduced local coordinate $u_i^*$ is expressed as

$$u_i^* = C_i(\theta_i)\, q \tag{26}$$

where $C_i$ is the compatibility matrix. The matrix $C_i$ can be determined easily from the relationship between (i-1)-th local coordinate frame and i-th local coordinate frame. The compatibility matrix $C_i$ is a function of variable $\theta_i$.

Substituting Eq. (26) into Eq. (25) and substituting these resulting equations into Eqs. (21) - (23), a set of equations are obtained with respect to the variables q and $\theta = (\theta_1, \theta_2, \dots \dots, \theta_N)^T$. In this stage the generalized forces involve the unknown internal forces. Pre-multiplying the transposed compatibility matrix to these link dynamic equations and summing up with the total number of links in the manipulator, internal forces can be eliminated due to the principle of virtual work. The resulting equations are the dynamic equations of the manipulator and can be described in the following form:

$$\begin{aligned} A_{11}\ddot{\theta} + A_{12}\ddot{q} &= B_1(\theta, \dot{\theta}, q, \dot{q}) + b_1 \\ A_{21}\ddot{\theta} + A_{22}\ddot{q} &= B_2(\theta, \dot{\theta}, q, \dot{q}) + b_2 \end{aligned} \tag{27}$$

Matrices $A_{11}$ and $A_{22}$ are time-varying positive-definite matrices, and vectors $B_1$ and $B_2$ are nonlinear functions of $\theta$, $\dot{\theta}$, q and $\dot{q}$. Vectors $b_1$ and $b_2$ are generalized forces with respect to $\theta$ and q.

Eq. (27) describes the dynamic behavior of the flexible manipulator. This equation can predict the interactions between the gross articular motion and the small elastic motion. For special problems, simplified models can be introduced, evaluating the effects of the terms in Eq. (27).

## 6. Simulation Algorithm

As the dynamic equations of manipulators are given in Eq. (27), some consideration is needed in a simulation algorithm for shortening the computational time and programing for general purposes. A program has been developed and its outline is shown briefly.

Basic computational scheme of the program is as follows:

(i) starting from variables q and θ in Eq. (27), variables $p_i$ and $\psi_i$ are obtained by using Eqs. (25) and (26),

(ii) using variables $p_i$ and $\psi_i$ , terms in Eqs. (21) - (23) are calculated, and

(iii) these terms are transformed into coefficient matrices and right hand side vectors of Eq. (27) by using the transformation matrices $C_i$ and $R_i$.

The computational scheme is shown in Fig. 3. Solid line boxes indicate that the terms in the boxes are time-varying functions. On the other hand broken line boxes indicate that the terms are constants which are determined from the mechanical parameters of manipulator or are computed results of the Finite Element Analysis.

Matrices ${}^jW_i$ ( = ${}^jW_{j+1}\,{}^{j+1}W_{j+2}$ ... ... ${}^{i-1}W_i$ ), $W_i$, $U_{ij}$ and $U_{ijk}$ are calculated effectively due to recurrent formulas. In the computational process there are many numerical evaluations of inner products consisting of vectors and matrices. Trace operations effectively reduce the number of multiplications and additions in numerical evaluation for these inner products.

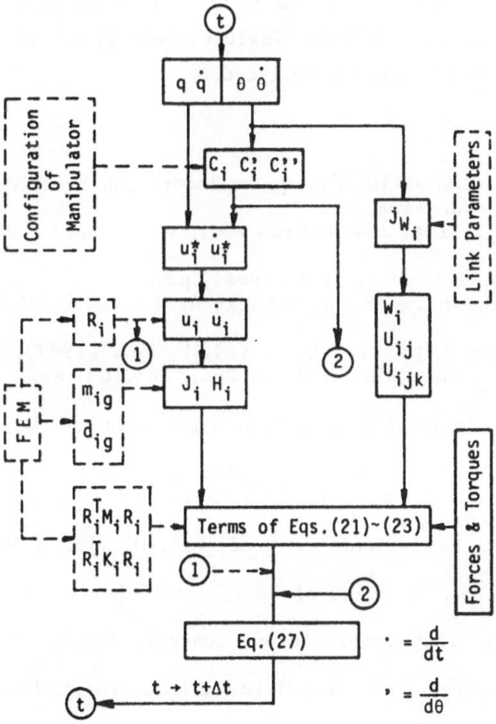


Fig. 3 Flow chart of the simulation program

As the dynamic equations of the motion are highly nonlinear, so the explicit method is more suitable than the implicit method in time integration. The forth-order Runge-Kutta-Gill method was adopted in the simulation and the time interval of integration was less than one twentieth of the shortest natural period.

## 7. Concluding Remarks

A model for dealing with dynamic behaviors of flexible manipulator arms has been presented. This model is free from an assumption of a nominal motion over time which is needed in the Sunada and Dubowsky's model /6, 7/, and does not ignore the interactions between global articular motions and small elastic deflections. The deflection curves of links used in the model can easily satisfy boundary conditions and are not excessively simplified, while the deflection curves of the Book's model /3/ are.

There are some special cases in which motions of the manipulator are constrained in a plane and elongations of links are constrained. The dynamic behaviors of such manipulators can also be evaluated systematically by the use of the restriction matrix.

The outline of the simulation program has been explained. However more detailed consideration is needed for shortening computational time when the model will be used in a real time simulator.

Flexibility of the manipulator arises from many mechanical sources. One of the most important problems is the joint compliance. If the joint is assumed to consist of two rigid links and one spring and is connected to flexible links by the use of the compatibility matrix, the problem can be simply analyzed.

## References


/1/ Truckenbrodt, A., Proc. the 4th CISM-IFToMM Symp. on Theory and Practice of Robots and Manipulators (1981, Warszawa), p.110.

/2/ Cannon, R.H.Jr. and Schmitz, E., Int. J. Robotics Res., Vol.3, No.3 (1984), p.62.

/3/ Book, W.J., Int. J. Robotics Res., Vol.3, No.3 (1984), p.87.

/4/ Sakawa, Y., Matsuno, F. and Fukushima, S., J. Robotic Syst., Vol.2, No.4 (1985), p453.

/5/ Kane, T.R. and Ryan, R.R., J. Guid. Control Dyn., Vol.10, No.2 (1987), p.139.

/6/ Sunada, W. and Dubowsky, S., Trans. ASME, J. Mech. Des., Vol.103, No.3 (1981), p.643.

/7/ Sunada, W.H. and Dubowsky, S., Trans. ASME, J. Mech. Transm. Autom. Des., Vol.105, No.1 (1983), p.42.

/8/ Chedmall, P. and Michel, G., Proc. the 15th ISIR (1985, Tokyo), p.1083.

/9/ Book, W.J., Maizza-Neto, O. and Whitney, D.E., Trans. ASME, J. Dyn. Syst. Meas. Control, Vol. 97, No.4 (1975), p.424.

/10/ Book, W.J., Trans. ASME, J. Dyn. Syst. Meas. Control, Vol.101, No.3 (1979), p.187.

/11/ Vukobratović, M. and Potkonjak, V., Dynamics of Manipulation Robots, Springer (1982), p.219.

/12/ Judd, R.P. and Falkenburg, D.R., IEEE Trans. Autom. Control, Vol.AC-30, No.5 (1985), p.499.

/13/ Ramsden, J.N. and Stoker, J.R., Int. J. Numer. Methods Eng., Vol.1, No.4 (1969), p333.

/14/ Hurty, W.C., AIAA J. Vol.3, No.4 (1965), p.678.

# COLUMN/CONSTRAINT GENERATION FOR QUADRATIC ASSIGNMENT PROBLEMS

Søren Holm and Stanisław Walukiewicz*
Institute of Management, University of Aarhus, Aarhus, Denmark

*Systems Research Institute, Polish Academy of Sciences
Warsaw, Poland

Abstract. A method using column/constraint generation techniques for solving the quadratic assignment problem (QAP) is described. First, we study the Kaufman-Broeckx linearization of QAP and next, show that the generated cuts have the same structure as the linearization constraints. We also describe an optimal solution to the separation problem. Finally, we discuss different methods for obtaining tighter equivalent formulations of the QAP.

1. Introduction. In this paper we report on a relatively new approach to the QAP using the column/constraint generation procedures. Our research was motivated in part by the encouraging results obtained by Holm [4] in studying the Kaufman-Broeckx formulation for the QAP and in part by the success of Padberg, Wolsey, and others in solving large travelling salesman problems [2,11], large pure 0-1 problems [3] and large mixed 0-1 problems [1].

The idea of our approach is as follows: First, using the Kaufman-Broeckx linearization [9] we transform the QAP into an equivalent linear mixed 0-1 problem. Next, we tighten this formulation utilizing the assignment property. Then we solve its linear programming relaxation by column generation. For choosing an initial set of columns we solve the QAP by near-optimal methods and choose columns corresponding to the variables having value one. If an optimal solution to the linear programming relaxation is not binary for the integer variables, then we construct, if possible, a cut which excludes the solution from the feasible region. If the construction of such a cut is not possible, then we apply a branch and cut procedure [11]. For surveys, see Burkard [1] and Kaki and Thompson [7].

In Section 2 we formulate the QAP and reformulate it using the Kaufman-Broeckx linearization. Next, in Section 3 we study its linear programming relaxation. A column generation procedure for solving the problem is described in Section 4 while Section 5 is devoted to a description of the constraint generation procedure. In Section 6 we show that the so called separation problem can be easily solved. In Section 7 some methods of obtaining tighter equivalent formulations are discussed.

2. Problem Formulation and Reformulation. Given two matrices $D = (d_{ij})$, $i,j \in N = \{1,\dots,n\}$ and $Q = (q_{ij})$, $i,j \in N$ the QAP can be formulated as

$$(P') \quad v(P') = \min \sum_{i=1}^{n} \sum_{j=1}^{n} \sum_{k=1}^{n} \sum_{l=1}^{n} d_{ij}\, q_{kl}\, x_{ik}\, x_{jl}$$

subject to the assignment constraints

$$\sum_{k=1}^{n} x_{ik} = 1, \qquad i \in N,$$

$$\sum_{i=1}^{n} x_{ik} = 1, \qquad k \in N,$$

$$x_{ik} = 0 \text{ or } 1, \quad (i,k) \in N^2 = N \times N,$$

where $v(P')$ is the optimal value of the problem $P'$. By $F(P')$ $(F^*(P'))$ we will denote the set of all feasible (optimal) solutions to $P'$. Without loss of generality we may assume that all data are non negative integers and, moreover, $d_{ii} = q_{ii} = 0$ for $i \in N$. Therefore $v(P') > 0$ and $F(P') \neq \emptyset$.

Kaufman and Broeckx [9] have shown that the QAP is equivalent to the following linear mixed 0-1 problem

$$(P) \qquad v(P) = \text{Min} \sum_{(i,k) \in N^2} y_{ik}$$

subject to

$$\sum_{l=1}^{n} x_{jl} = 1, \qquad j \in N,$$

$$\sum_{j=1}^{n} x_{jl} = 1, \qquad l \in N,$$

$$(1) \quad b_{ik}\, x_{ik} + \sum_{(j,l) \in N^2} d_{ij}\, q_{kl}\, x_{jl} - y_{ik} \leqslant b_{ik}, \quad (i,k) \in N^2,$$

$$x_{ik} = 0 \text{ or } 1, \quad (i,k) \in N^2,$$

$$y_{ik} \geqslant 0, \quad (i,k) \in N^2,$$

where

$$y_{ik} = x_{ik} \sum_{j=1}^{n} \sum_{l=1}^{n} d_{ij}\, q_{kl}\, x_{jl},\; b_{ik} = \sum_{j=1}^{n} \sum_{l=1}^{n} d_{ij}\, q_{kl} \,.$$

If $x^*_{ik} = 0$, then (1) reads

$$y_{ik} \geq \sum_{(j,l)\in N^2} d_{ij}\, q_{kl}\, x_{jl} - b_{ik} \leq 0, \quad (i,k)\in N^2$$

and from $y_{ik} \geq 0$ we conclude that $y^*_{ik} = 0$ in any optimal point $(x^*,y^*)\in F^*(P)$. If $x^*_{ik} = 1$, then, as we minimize the objective function,

$$y^*_{ik} = \sum_{(j,l)\in N^2} d_{ij}\, q_{kl}\, x^*_{jl} \geq 0, \qquad (i,k)\in N^2$$

in any optimal point $(x^*,y^*)\in F^*(P)$ and the corresponding linearization constraint (1) is tight at this point. As the vector $x^*$ has to satisfy the assignment constraints, then exactly $n$ constraints (1) are tight at any otpimal point $(x^*,y^*)\in F^*(P)$.

To simplify the notation we introduce

$$a_{ijkl} = \begin{cases} b_{ik} + d_{ii}q_{kk} = b_{ik} & \text{for } j=i \text{ and } l=k \\ d_{ij}q_{kl} & \text{otherwise} \end{cases}$$

and observe that $b_{ik}$ for a given $(i,k)\in N^2$ equals the sum of all elements of the $n\times n$ matrix $A_{ik} = (a_{ijkl})$. This matrix is symmetric if $D$ and $Q$ are symmetric. So we may write

(2) $$\sum_{(j,l)\in N^2} a_{ijkl}\, x_{jl} - y_{ik} \leq b_{ik}, \quad (i,k)\in N^2.$$

It is easy to check that

(3) $$\sum_{(j,l)\in N^2} a_{ijkl} = 2b_{ik} \text{ and } a_{iikk} = b_{ik}.$$

3. <u>The Linear Programming Relaxation</u>. For a given problem $P$ we denote by $\bar{P}$ its linear programming relaxation. Note, that due to the assignment constraints. $x_{jl} \leq 1$, $(j,l)\in N^2$ are redundant in $\bar{P}$.

The dual problem to $\bar{P}$ is

($\bar{D}$) $$v(\bar{D}) = \text{Max}\; \Big(\sum_{j=1}^{n} v_j + \sum_{l=1}^{n} w_l - \sum_{(i,k)\in N^2} b_{ik}\, u_{ik}\Big)$$

subject to

$$(4)\qquad v_j + w_l - \sum_{(i,k)\in N^2} a_{ijkl}u_{ik} \leqslant 0, \qquad (i,l)\in N^2,$$

$$0 \leqslant u_{ik} \leqslant 1, \qquad (i,k)\in N^2,$$

$$v_j,\ w_l \gtrless 0, \qquad j,l\in N.$$

So $\bar{D}$ has only $n^2$ constraints, $n^2$ bounded variables and 2n variables unrestricted in sign.

If $(\bar{x},\bar{y})\in F^*(\bar{P})$ and $(\bar{u},\bar{v},\bar{w})\in F^*(\bar{D})$, the complementary slackness conditions read

$$(5)\qquad \bar{u}_{ik}(b_{ik} + \bar{y}_{ik} - \sum_{(j,l)\in N^2} a_{ijkl}\,\bar{x}_{jl}) = 0, \qquad (i,k)\in N^2,$$

$$(6)\qquad \bar{v}_j(1 - \sum_{l\in N} \bar{x}_{jl}) = 0, \quad j\in N,$$

$$(7)\qquad \bar{w}_l(1 - \sum_{j\in N} \bar{x}_{jl}) = 0, \quad l\in N,$$

$$(8)\qquad \bar{x}_{jl}(\sum_{(i,k)\in N^2} a_{ijkl}\,\bar{u}_{ik} - \bar{v}_j - \bar{w}_l) = 0, \quad (j,l)\in N^2,$$

$$(9)\qquad \bar{y}_{ik}(1 - \bar{u}_{ik}) = 0, \quad (i,k)\in N^2.$$

If $\bar{x}_{ik} = 0$, then from (2) we have that $\bar{y}_{ik} = 0$.
For given $(\bar{x},\bar{y})\in F^*(P)$ we define

$$N(\bar{x}) = \{(i,k)\in N^2 : \bar{x}_{ik} > 0\} \text{ and } N(\bar{y}) = \{(i,k)\in N^2 : \bar{y}_{ik} > 0\}.$$

As it is possible to have $\bar{y}_{ik} = 0$ for $\bar{x}_{ik} > 0$, then $N(\bar{y}) \subseteq N(\bar{x})$.

If $\bar{y}_{ik} > 0$, then by (9) $\bar{u}_{ik} = 1$. If $\bar{y}_{ik} = 0$ and since $\bar{x}_{jl}$ has to satisfy the assignment constraints, then by (3) the constraint (2) is not tight at $(\bar{x},\bar{y})$ and by (5) $\bar{u}_{ik} = 0$. So, all optimal vectors $\bar{u}_{ik}$ are binary. For a given dual optimal solution $(\bar{u},\bar{v},\bar{w})$ we define $N(\bar{u}) = \{(i,k)\in N^2 : \bar{u}_{ik} = 1\}$.
So we have obtained

$$(10)\qquad N(\bar{u}) = N(\bar{y}) \subset N(\bar{x}).$$

As $\bar{D}$ has less constraints than $\bar{P}$, it will some times be reasonable to solve $\bar{D}$ instead of $\bar{P}$. We study further relations between $\bar{P}$ and $\bar{D}$

in a subsequent paper [5].

4. The Column Generation Procedure. The problem $\bar{P}$ or $\bar{D}$ is, in general, a large programming problem and therefore we suggest to solve it by the column generation procedure (see e.g. Lasdon [10]). To define an initial set of columns we solve the QAP by some heuristic methods to obtain, say, about n near-optimal solutions to the QAP and define

$$M = \{(j,l)\ N^2 : x_{jl} = 1 \text{ in some solutions to the QAP}\}.$$

It is reasonable to assume that that $|M| << n^2$, say $|M| \approx 3n$ and $M$ does not differ very much from $N(x^*)$, where $x^*$ is an optimal solution to QAP.

Having M, we obtain from $\bar{P}$ a reduced problem $\bar{P}(M)$ in the form

$$(\bar{P}(M)) \qquad v(\bar{P}(M)) = \text{Min} \sum_{(i,k)\in M} y_{ik}$$

subject to

$$(11) \qquad y_{ik} - \sum_{(j,l)\in M} a_{ijkl}\, x_{jl} - z_{ik} = -b_{ik}, \quad (i,k)\in M,$$

$$\sum_{l:(j,l)\in M} x_{jl} = 1, \quad j\in N,$$

$$\sum_{j:(j,l)\in M} x_{jl} = 1, \quad l\in N,$$

$$x_{jl} \geq 0, \quad (j,l)\in M,$$

$$y_{ik},\ z_{ik} \geq 0, \quad (i,k)\in M,$$

where $z_{ik}$ are slack variables. Using the same arguments as in the previous section, one can show that for $(i,k)\notin M$ the $(i,k)$th constraint is not tight at an optimal point, thus $\bar{y}_{ik} = 0$, and $\bar{u}_{ik} = 0$. In $\bar{P}(M)$ we have at most $|M| + 2n$ constraints and $3|M|$ variables including slacks. Obviously $F(\bar{P}(M)) \neq \emptyset$.

The relative cost coefficient for a nonbasic variable $x_{jl}$ is

$$(12) \qquad (\bar{c}_{jl})_x = (c_{jl})_x - (\bar{u}, \bar{v}, \bar{w})\ (P_{jl})_x,$$

where from the definition of $\bar{P}$, $(c_{jl})_x = 0$, $(\bar{u}, \bar{v}, \bar{w})$ is an extended optimal dual solution to $\bar{P}(M)$ with $\bar{u}_{ik} = 0$ for $(i,k) \notin M$ and $(P_{jl})_x$ is the column of $\bar{P}$ corresponding to $x_{jl}$. Thus (12) can by (10) be written as

(13) $$(\bar{c}_{ij})_x = \sum_{(i,k)\in M} a_{ijkl}\,\bar{u}_{ik} - \bar{v}_j - \bar{w}_l = \sum_{(i,k)\in N(\bar{y})} a_{ijkl} - \bar{v}_j - \bar{w}_l$$

where we may now define $N(\bar{y})$ for a given $(\bar{x},\bar{y})\in F^*(\bar{P}(M))$ as $N(\bar{y}) = \{(i,k)\in M : \bar{y} > 0\}$. If $(\bar{c}_{jl})_x < 0$, then by the simplex criterion $x_{jl}$ becomes basic. $M:=M\cup\{(j,l)\}$, and we form a new linear programming problem $\bar{P}(M)$ which in practice results in adding to the previous one the $(j,l)$-th constraint written only for $(i,k)\in M$ and adding $x_{jl}$ to the two corresponding assignment constraints.

As $\bar{u}_{ik} = 0$ for $(i,k) \notin M$, then $(\bar{c}_{ik})_y = 1 > 0$ and $(\bar{c}_{ik})_z = 0$ for any $(i,k) \notin M$. So we have to price out only the columns corresponding to the binary variables $x_{jl}$, $(j,l) \notin M$.

If $(\bar{c}_{jl})_x \geqslant 0$ for all $(j,l) \notin M$, then we have found an optimal solution $(\bar{x},\bar{y})$ to our initial problem $\bar{P}$.

5. The Constraint Generation Procedure. Assume now that we have solved the linear programming relaxation of the QAP by the column generation procedure, i.e. we have solved $\bar{P}(M)$, and its optimal solution $(\bar{x},\bar{y})$ is not integer. The idea of constraint generation consists in considering constraints (11) one by one and in constructing a cut (valid inequality) which cuts off $(\bar{x},\bar{y})$ from $F(P)$. The theory of such valid inequalities for mixed 0-1 programs was developed by Van Roy and Wolsey [12, 13]. Here we will use only so called simple generalized flow cover (GFC) inequalities.

To construct a simple GFC inequality we have to write the feasible region corresponding to a constraint of (11) in so called standard form [12].

(14) $$S = \Big\{(z,x) : \sum_{j\in M^+} z_j - \sum_{j\in M^-} z_j \leqslant b,\ l_j x_j \leqslant z_j \leqslant u_j x_j,\ u\in R^{|M^+\cup M^-|},\ x_j\in\{0,1\},\ j\in M^+\cup M^-\Big\}$$

where $0 \leqslant l_j \leqslant u_j$, $j\in M^+\cup M^-$ and $M^+$, $M^-$ are index sets. A pair of sets $(C^+,C^-)$ with $C^+\subseteq M^+$, $C^-\subseteq M^-$ is called a generalized cover if

(15) $$\sum_{j\in C^+} u_j - \sum_{j\in C^-} u_j - b = \lambda \quad \text{and} \quad \lambda > 0.$$

For given $(C^+,C^-)$ and $\lambda$ the simple GFC inequality is

(16) $$\sum_{j\in C^+} z_j + \sum_{j\in C^+} (u_j - \lambda)^+(1-x_j) \leq b + \sum_{j\in C^-} u_j + \sum_{j\in L^-} \min\{u_j,\lambda\} x_j + \sum_{j\in M^- -(C^- \cup L^-)} z_j$$

where $(u_j - \lambda)^+ = \max\{0, u_j - \lambda\}$, $L^- \subseteq M^- - C^-$ (see Corollary 4 in [13]). As we are now considering the constraints of (11) separately, we may considerably simplify notation and write a constraint in the form

(17) $$\sum_{j\in M} a_j x_j - y \leq b$$

where from the definition of problem P we know that all data are non-negative integers and moreover

$$\max_{j\in M} a_j = b, \qquad \sum_{j\in M} a_j \leq 2b \qquad \text{and} \qquad a_1 = \max_{j\in M} a_j.$$

Without loss of generality we may assume that $M = \{1,\dots,m\}$ and from the definition of problem P we know that $0 \leq y \leq bx_1$.

Now we take $M^+ = M$ and $M^- = \{m+1\}$ and write the feasible region of (17) in the form of (14)

(18) $$S = \{z \in R^{m+1}, x \in \{0,1\}^{m+1} : \sum_{j\in M^+} z_j - \sum_{j\in M^-} z_j \leq b; \quad a_j x_j \leq z_j \leq a_j x_j, \ j\in M^+; \ 0 \leq z_j \leq bx_1, \ j\in M^-\}$$

where $z_j = a_j x_j + 0y_j$ for $j\in M^+$ and $z_j = 0x_j + y_j$ for $j\in M^-$. As the set $M^-$ has in our case only one element. then we put $L^- = \emptyset$, and then either $C^- = \{m+1\}$ or $C^- = \emptyset$. If $C^- = \{m+1\}$, then the right-hand-side of (16) equals 2b, and if $C = \emptyset$, then it equals $b + y \leq 2b$. Since we are interested in tighter simple GFC inequalities, we choose $C^- = \emptyset$. So we are looking for an inequality of the form

(19) $$\sum_{j\in C^+} a_j x_j + \sum_{j\in C^+} (a_j - \lambda)^+(1 - x_j) - y \leq b$$

which cuts off $(\bar{x},\bar{y})$ from F(P). This is equivalent to finding a set $C^+ \subseteq M^+ = M$ and a positive number $\lambda$ such that

(20) $$\lambda = \sum_{j\in C^+} a_j - b.$$

The above problem is called the separation problem and can be stated in the following way:
We are asking whether for a given $(\bar{x},\bar{y}) \in F^*(\bar{P}(M))$ the maximum of the

left-hand-side of (19) over all $\lambda > 0$ and $C^+ \subseteq M^+$ is strictly greater than b. If we define

$$s_j = \begin{cases} 1, & \text{if} \quad j \in C^+ \\ 0, & \text{if} \quad j \notin C^+ , \end{cases}$$

then the problem can be formulated as an equality constrained parametric knapsack problem

$$(K(\lambda)) \qquad v(K(\lambda)) = \text{Max} \sum_{j \in M} [a_j \bar{x}_j + (a_j - \lambda)^+ (1 - \bar{x}_j)] s_j$$

subject to

$$(21) \qquad \sum_{j \in M} a_j s_j = b + \lambda ,$$

$$s_j = 0 \text{ or } 1, \quad j \in M.$$

We will study $K(\lambda)$ in the next section. To terminate our description of the constraint generation procedure we note that if $v(K(\lambda)) > b + \bar{y}$, then there exists inequality (19) which separates $(\bar{x}, \bar{y})$ from $F(P)$. If $v(K(\lambda)) \leqslant b + \bar{y}$ for all $\lambda = 1, 2, \ldots, b$, then we would consider the next constraint of (11).

6. The Separation Problem. To describe a solution of the separation problem $K(\lambda)$ we first simplify notation. Without loss of generality we may consider the restricted set $\bar{M} = \{j \in M : a_j > 0\}$ and represent it as $\bar{M} = \{1\} \cup R$.
We observe that $s_1^* = 1$ in any optimal solution $s^*$ to $K(\lambda)$, and then by (21).

$$\lambda = \sum_{j \in R} a_j \, s_j .$$

Moreover for $j \in R$

$$(a_j - \lambda)^+ s_j = (a_j - \sum_{j \in R} a_j s_j)^+ s_j = 0.$$

Therefore $K(\lambda)$ may be written as

$$v(K(\lambda)) = a_1 + \text{Max} \sum_{j \in R} [a_j \bar{x}_j - (1 - \bar{x}_1) a_j] s_j$$

$$s_j = 0 \text{ or } 1, \quad j \in R.$$

Now the optimal solution to $K(\lambda)$ is obvious. We define it in the following way for $j \in R$

(22) $$s_j^* = \begin{cases} 1, & \text{if } \bar{x}_j > 1 - \bar{x}_1, \\ 0, & \text{if } \bar{x}_j < 1 - \bar{x}_1, \\ \text{arbitrary 0 or 1,} & \text{if } \bar{x}_j = 1 - \bar{x}_1. \end{cases}$$

So we have proved the following

<u>Lemma 1</u>. Any optimal solution to the separation problem $K(\lambda)$ is given by $s_1^* = 1$ and by (22). □

Let $R^* = \{j \in R : s_j^* = 1\}$, then $C^+ = \{1\} \cup R^*$ and

$\lambda = \sum_{j \in R} a_j s_j^* = \sum_{j \in R^*} a_j$ .

The condition $v(K(\lambda)) > b + \bar{y}$ may be written as

(23) $$\sum_{j \in R^*} (\bar{x}_j + \bar{x}_1 - 1) a_j > \bar{y}$$

and the separation inequality takes the form

(24) $$\lambda x_1 + \sum_{j \in R^*} a_j x_j - y \leqslant \lambda .$$

Thus we have proved

<u>Theorem 2.</u> If (23) holds, then there exists an inequality (24) which separates $(\bar{x},\bar{y})$ from $F(P)$. □

We note that (24) has exactly the same structure as the linearization inequality (2).

Now we describe the constraint generation procedure for the (i,k)-th constraint of $P(M)$, $(i,k) \in M$.

First we form the set $\bar{M}_{ik} = \{(j,l) \in M : a_{ijkl} > 0\}$ and represent it as $M_{ik} = \{(i,k)\} \cup R_{ik}$. Next by (22) we compute

(25) $$s_{jl}^* = \begin{cases} 1, & \text{if } \bar{x}_{jl} > 1 - \bar{x}_{ik}, \\ 0, & \text{if } \bar{x}_{jl} < 1 - \bar{x}_{ik}, \\ \text{arbitrary 0 or 1,} & \text{if } \bar{x}_{jl} = 1 - \bar{x}_{ik}. \end{cases}$$

for each $(j,l) \in R_{ik}$ and let $R_{ik}^* = \{(j,l) : s_{jl}^* = 1\}$. If by (23)

$$\sum_{(j,l) \in R_{ik}^*} (\bar{x}_{jl} + \bar{x}_{ik} - 1) a_{ijkl} > \bar{y}_{ik}$$

then we add to the constraints of $P(M)$ the valid inequality (24) which takes the form

(26) $$\lambda_{ik} x_{ik} + \sum_{(j,l) \in R_{ik}^*} a_{ijkl} x_{jl} - y_{ik} \leqslant \lambda_{ik}$$

where

$$\lambda = \sum_{(j,l)\in R^*_{ik}} a_{ijkl}$$

In the next section we show that (26) can be further strengthened using so called assignment arguments.

7. <u>Tighten the Formulation of QAP</u>. In this section we are interested in methods for obtaining tighter equivalent formulations of P. We will not repeat here the well-known methods of reduction of matrix D and Q (see [4]) and only note that since all these methods try to produce as many zeros as possible in matrix D and Q, they are very much in accordance with our approach, since for sparse matrices D and Q, the matrix A will be sparse and thus the column generation, and in particular the constraint generation, will be easier as $|M|$ is small.

Consider again constraints (1) of P. As

$$0 \leqslant y_{ik} = x_{ik} \sum_{j=1} \sum_{l=1} d_{ij} q_{kl} x_{jl} \leqslant b_{ik}, \quad (i,k)\in N^2,$$

and $x_{jl}$ has to satisfy the assignment constraints, then $b_{ik}$ cannot exceed the value of the following $n\times n$ assignment problem

$$(P_{ik}) \qquad v(P_{ik}) = b_{ik} = \text{Max} \sum_{j=1}^{n} \sum_{l=1}^{n} d_{ij} q_{kl} x_{jl}$$

subject to

$$\sum_{\substack{l=1\\ l\neq k}}^{n} x_{jl} = 1, \qquad j\in N$$

$$\sum_{\substack{j=1\\ j\neq i}}^{n} x_{jl} = 1, \qquad l\in N$$

$$x_{jl} = 0 \text{ or } 1, \ (j,l)\in N^2.$$

We may reduce $b_{ik}$ further if we find all the t-best solutions to $P^t_{ik}$ for $t = 1,\ldots,T$, where $x^1 = x^*$ and $x^* \in F^*(P_{ik})$. For each $x^t$ we compute the value of the QAP, $v(x^t)$ and then

$$b_{ik} = \{v(p^q_{ik}) \ : \ v(x^q) < v(x^t), \quad t = 1,\ldots,T\}$$

The details are given in [4].

The other possibility is to rotate a given constraint of (2) in such a way that the corresponding feasible linear programming region is as small as possible. Such a procedure for a pure integer problem has been studied by Kaliszewski and Walukiewicz [8] and can easily be modified for the case when in (1) we have only one continuous variable. Obviously, rotation may be applied as cuts (19) or (24). If a rotation procedure is applied to (24), then as a result we obtain

$$x_1 + \sum_{j \in R^*} \bar{a}_j x_j - y \leqslant \lambda$$

where $a_j \leqslant \bar{a}_j \leqslant \lambda$ . This requires solving $|R^*|$ knapsack problems, and the computational load can be reduced if we solve these knapsack problems by dynamic programming (see [8] and [6]).

As usually $|R^*| < |M| - 1$ we can tighten the cut (19) by the lifting procedure. As a result we obtain

$$x_1 + \sum_{j \in M - \{1\}} \bar{a}_j x_j - y \leqslant \lambda ,$$

where $\bar{a}_j = a_j$ for $j \in R^*$ and $0 \leqslant \bar{a}_j \leqslant \lambda$ for $j \in M - (\{1\} \cup R^*)$. In general the lifting procedure requires solving, at least near-optimality, $|M| - 1 - |R^*|$ knapsack problems (see [3] and [13] for details).

Now we show that (26) and at the same time (19) can easily be strengthened by using so called assignment arguments. First we reduce $\lambda_{ik}$ to $\bar{\lambda}_{ik} \leqslant \lambda_{ik}$ by solving the following at most $(n-1) \times (n-1)$ assignment problem

$$\bar{\lambda}_{ik} = \text{Max} \sum_{(j,l)} (a_{ijkl}\, s^*_{jl}) x_{jl}$$

$$\sum_{l} s^*_{jl}\, x_{jl} = 1, \qquad j \in N - \{i\},$$

$$\sum_{j} s^*_{jl}\, x_{jl} = 1, \qquad l \in N - \{k\},$$

$$x_{jl} = 0 \text{ or } 1, \quad (j,l) \in R^*_{ik},$$

where $s^*_{jl}$ are given by (25). Next for each $(j,l) \notin R^*_{ik}$ and $j \neq i$, $l \neq k$, we compute a new coefficient $\bar{a}_{ijkl}$ as the minimal nonzero coefficient in the j-th row and the l-th column of the matrix $A_{ik} S^*_{ik} = (a_{ijkl} s^*_{jl})$. As a result we obtain

$$\text{(27)} \qquad \bar{\lambda}_{ik} x_{ik} + \sum_{(j,l) \in M^-_{ik}} \bar{a}_{ijkl}\, x_{jl} - y_{ik} \leqslant \bar{\lambda}_{ik},$$

$M^-_{ik} = \{(j,l) \in M : j \neq i,\ l \neq k\}$. Obviously, we have $1 \leqslant \bar{a}_{ijkl} \leqslant \bar{\lambda}_{ik}$ for $(j,l) \in M^-_{ik} - R^*_{ik}$ and $\bar{a}_{ijkl} = a_{ijkl}$ for $(i,j) \in R^*_{ik}$.

References


1. Burkard, R.E., Locations with Spatial Interactions-Quadratic Assignment Problem, Bericht 83-31, Technische Universität Graz, Graz, November 1983.

2. Crowder and M.V. Padberg, Solving Large-Scale Symmetric Travelling Salesman Problems to Optimality, Management Sci. 26 (1980) 495-509.

3. Crowder, H., E.L. Johnson and M.W. Padberg, Solving Large-Scale Zero-One Linear Programming Problems, Oper. Res. (1983) 803-834.

4. Holm, S., A Strengthening of the Kaufman-Broeckx Formulation of the Quadratic Assignment Problem, Odense University, January 1986.

5. Holm, S. and S. Walukiewicz, Tighter Formulation of the Quadratic Assignment Problem (in preparation).

6. Dudziński. D. and S. Walukiewicz, Exact Methods for the Knapsack Problem and Its Generalizations, European Journal of Operational Research 28 (1987) 3-21.

7. Kaki, K. and G.L. Thompson, An Exact Algorithm for the General Quadratic Assignment Problem, European Journal of Operational Research 23 (1986) 382-390.

8. Kaliszewski, I. and S. Walukiewicz, Tighter Equivalent Formulations for Integer Problems, Report of the Systems Research Institute, Warsaw 1982.

9. Kaufman, L. and F. Boreckx, An Algorithm for the Quadratic Assignment Problem Using Benders' Decomposition, European Journal of Operational Research 2 (1978) 204-211.

10. Lasdon, L.S., Optimization Theory for Large Systems, The Macmillan Co. 1970.

11. Padberg, M. and G. Rinaldi, Optimization of a 532-city Symmetric Traveling Salesman Problem by Branch-and-Cut, Oper. Res. Letters (1987) 1-7.

12. Van Roy, T.J. and L.A. Wolsey, Solving Mixed Integer Programs by Automatic Reformulation, CORE Discussion Paper 8432, Université Catholique de Louvain, Louvain-la-Neuve, June 1984.

13. Van Roy, T.J. and L.A. Wolsey, Valid Inequalities for Mixed 0-1 Programs, Discrete Applied Mathematics 14 (1986) 199-213.

# A Prallel Algorithm for the Machine Scheduling Problem

Mario Nakamori
Tokyo University of Agriculture and Technology
Nakamachi 2-24-16, Koganei, Tokyo 184
Japan

## 1. Introduction

The branch and bound method is a typical and useful technique or combinatorial optimization problems, e.g. travelling salesman problem, chromatic number problem, machine sequencing problem, etc. This method consists of two types of procedures: branching and bounding. The former generates systematically subproblems or subsolutions, whereas the latter tests whether the subproblem or subsolution under consideration is promising or not. A similar method is often used in order to obtain the minimax strategy of a game, where the branch and bound method is called "search."

In practice, the computational complexity of branch and bound methods is very steep (usually exponential) in regard to the depth of the tree generated by the branching procedure. Therefore it is important to develop techniques of acceleration for branch and bound methods. There may be many ways of acceleration, namely

(1) stronger tests for bounding;
(2) generating the more promising subproblem as early as possible in order to make tests more effective;
(3) parallel computation in multiprocessing environment.

The present paper is concerned to (3).

The branch and bound method is especially suited to parallel computation, since subproblems generated in branching can be treated by independent tasks (in computer terminology). Moreover, data communication among tasks is not so frequent and no "lock-step" synchronization is necessary. In this sense the branch and bound method is in contrast to other type of methods such as matrix multiplication, LU decomposition, etc.

The practical computational complexity of parallel branch and bound algorithm is much interesting. Let n be the number of processors available for the parallel computation. Then, we would like to ask, "Is the parallel algorithm n times quicker than sequential algorithm?" The parallel algorithm may not be n times quicker because of system overhead; it may be more than n times quicker because of some unexpected factors.

Although there have been published many reports on the construction or implementation of parallel processing systems, parallel algorithms for specified problems have not fully been discussed. The present paper is concerned to parallel computation of branch and bound algorithm for the machine sequencing problem.

At present there exist only a few experimental parallel processing systems. Therefore, most reports on parallel algorithms are based on simulation with appropriate assumption on the execution time of tasks, on the frequency of data references, and so on. From the algorithmic point of view, however, simulation is not satisfactory technique of research. The present paper is based on non-simulation method; we make no probabilistic assumption and solve "actually" several numerical examples. (Since no parallel computational environment is available, we made our experiments by pseudo-parallel program on a single processor system, but this is different from the so-called simulation.)

## 2. The machine scheduling problem

### 2.1 The problem

Suppose items $Q_1, \dots, Q_q$ are to be processed by machines $M_1, \dots, M_m$. Machines are different from one another, i.e. each operation on each item has to be processed on a specified machine and cannot be carried out on another machine. The completion time for each operation is given. The operations on each item have to be carried out in a technologically prescribed order, whereas there is freedom of choice as to the sequence of operations on each machine. Then, what is the optimal sequence of operations that minimizes the total completion time (the time that passes since the first operation begins until the last operation ends)? We are looking for this optimal sequence of operations.

The above problem, called the machine sequencing problem, reduces to that of finding a minimaximal path in a disjunctive graph. A disjunctive graph consists of

the set of vertices $V=\{v_1, \dots, v_N, s, t\}$,

the set of conjunctive arcs $A=\{a_1, \dots, a_c\}$,

the set of disjunctive arcs $B=\{b_1, \dots, b_d\}$

and is denoted by $D=(V, A, B)$.

The correspondence between the machine sequencing problem and the disjunctive network is as follows:

(1) For each operation $i$ there exists a vertex $v_i \in V$. There exist

also two dummies s (source) and t (sink).

(2) If operations i and j are on the same item and are adjacent in the technological sequence, there exists a conjunctive arc $(v_i, v_j) \in A$. There exists also a conjunctive arc $(s, v_i) \in A$ for the first operation i on each item and a conjunctive arc $(v_j, t) \in A$ for the last operation j on each item.

(3) If operations i and j are on different items but on the same machine, there exists a disjunctive pair of arcs $(v_i, v_j) \in B$ and $(v_j, v_i) \in B$. Each arc of a disjunctive pair is called the opposite of the other.

Every arc (conjunctive or disjunctive) $(v_i, v_j)$ has its own length, which is equal to the completion time of the operation i.

Let a set of arcs S be chosen such that exactly one arc of each disjunctive pair belongs to S. Such a set S is called a <u>selection</u>. There are $2^{d/2}$ of selections. Let S be a selection and G be a graph (in the ordinary sense) $G=(V, A \cup S)$. If G is circuit free, there exists the longest ath from source to sink. uch a path is called the <u>critical path</u> in G. Let us denote the length of the critical path in G by v[G]. Then the machine sequencing problem is to obtain

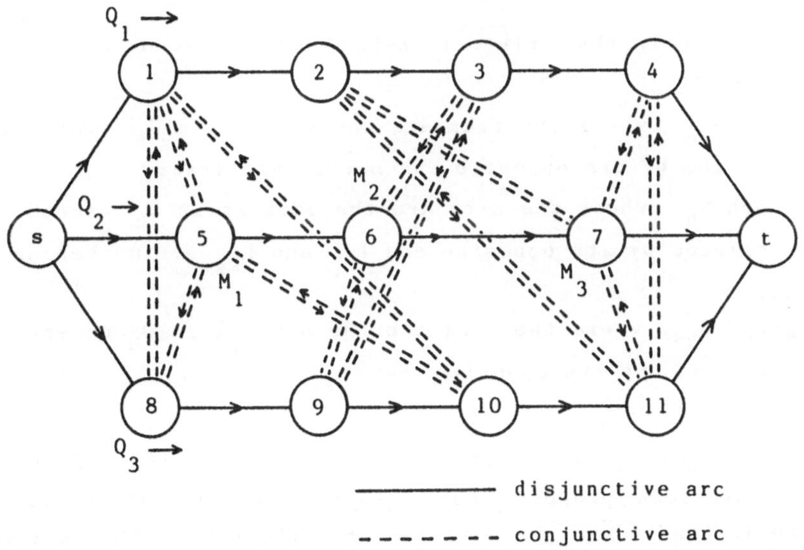


**Figure 1.** Disjunctive graph

$v[\hat{G}]=\min\{v[G] \mid G=(V, A\cup S)$, S is a selection and G is circuit free$\}$

The graph $\hat{G}$ is said to be optimal, and the critical path in G is said to be minimaximal in D.

**2.2 A branch and bound algorithm**

In the following we consider a branch and bound algorithm proposed by E. Balas [1]. This algorithm systematically generates selections (or graphs), computing the length of the critical path of each graph: first, make an arbitrary selection such that the corresponding graph is acycle; next, find the critical path; if there are disjunctive arcs in the critical path, replace them with their opposites and create new selections; repeat this procedure for each new selection. Thus a "tree" of selections is obtained.

In order to avoid duplicate generation, we mark disjunctive arcs as free or fixed. Fixed arcs are never to be reversed in the descendants. We also say that a disjunctive arc $(v_i, v_j)$ is normal if $i<j$ and reverse if $i>j$. A graph is circuit-free if its disjunctive arcs are all normal, so we are able to begin our algorithm with this graph.

The branching rule is as follows. First we make a selection consisting of only normal arcs in B and mark them as free. Let this graph be $G_0$.

Let $G_h$ be an arbitrary graph generated and $F=\{b_{k_1},\ldots, b_{k_f}\}$ be the set of free arcs in the critical path in $G_h$. We make the following graphs:

(1) A graph $G_{h'}$ where the arcs are the same as in $G_h$ except that $b_{k_1}$ is replaced by its opposite and marked as fixed;

(2) A graph $G_{h''}$ where the arcs are the same as in $G_h$ except that $b_{k_2}$ is replaced by its opposite and $b_{k_1}$ and $b_{k_2}$ are marked as fixed;

.....

(3) A graph $G_{h^{(f)}}$ where the arcs are the same as in $G_h$ except that $b_{k_f}$ is replaced by its opposite and $b_{k_1}, b_{k_2},\ldots, b_{k_f}$ are marked as fixed.

For each graph $G_{h'}, G_{h''},\ldots, G_{h^{(f)}}$ we apply the above rule again.

There have been proposed many tests. One of them is as follows: compute the length of the critical path for the graph with only conjunctive arcs and fixed disjunctive arcs. If this value is not less than the smallest length of the critical path of graphs (including free arcs) already generated, it is not necessary to generate the descendants of this graph.

Since the present paper is not mainly concerned to the way of tests, we make no further description on tests.

## 3. Making the algorithm parallel

### 3.1 The model of parallel processing system

There have been proposed many types of processor interconnection for parallel computation, and what kind of model of parallel processing system we adopt is the essential point of our discussion.

Our model is a kind of MIMD (multiple instruction stream multiple data stream) computer system. The model consists of

n processors, each having its own local memory;

a common shared memory;

a bus connecting the processing units and the common shared memory

(**Figure 2**). The common shared memory and bus connection characterizes our model. We assume that common shared memory and local memories have sufficiently large capacity.

There are other types of processor interconnection, such as network type. Network interconnection is powerful especially when data communication among processors takes place very often. By our branch and bound algorithm data communication is not so often, so we do not consider the

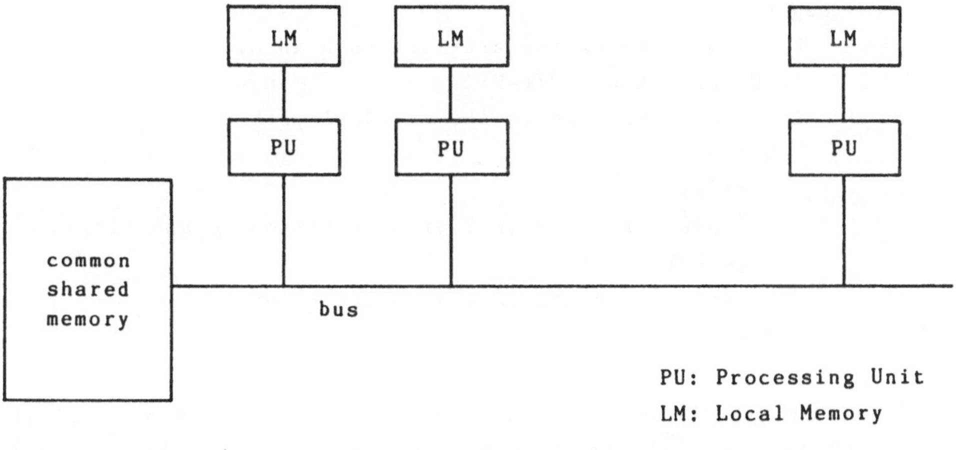


**Figure 2.** Parallel computing system

network model.

### 3.2 Parallel branch and bound algorithm

Branch and bound algorithm is often described recursivbely. Recursive algorithm, however, is not well suited to parallel computation. From the point of view of parallel computation, all subproblems generated by branching should be treated at the same level.

Thus, our algorithm is as follows:

```
main: input; VOPT←∞; S←empty;
      create a selection A such that all disjunctive arcs are
      normal;
      push A in S;
      while S is not empty do
        begin
          X←a member of S; delete X from S;
          call branch-and-bound(X) as a new task;
        end;
      wait until all tasks terminate;
      output

branch-and-bound(X):
      create a graph G using X;
      V  the lower bound of the length of the critical path;
      if V<VOPT then
        begin
          U  the length of the critical path in G;
          if U<VOPT then VOPT←U;
          for every free arc in the critical path do
            begin
              Y←X;
              replace the arc in Y with its opposite and fix;
              push Y in S;
              replace the arc in Y with its opposite
            end
        end
```

The set S (queue, stack, etc) and the variable VOPT is located in the common shared memory. The program branch-and-bound is called repeatedly. Each time this program is called, a new task is created. Hence, there are one "main" task and several "branch-and-bound" tasks in our system. All of these tasks, once created, run independently, except that simultaneous access to data in the common shared memory (such as S and VOPT) from different tasks is forbidden. The operating system is responsible for the control of tasks. Note that the operating system is assumed to satisfy the following requirements:

(1) a user program (as a task) can create a task which runs independently of the "parent" task.

(2) the "parent" task can know the status of all tasks that it created.

There is a degree of freedom on what member in S the main program takes out. We considered two strategies:

(a) the oldest member, or, "first-in first-out" (i.e., the set S is used as a queue);

(b) the member whose critical path seems to be the shortest, where the length of the critical path is predicted by the methods in [1].

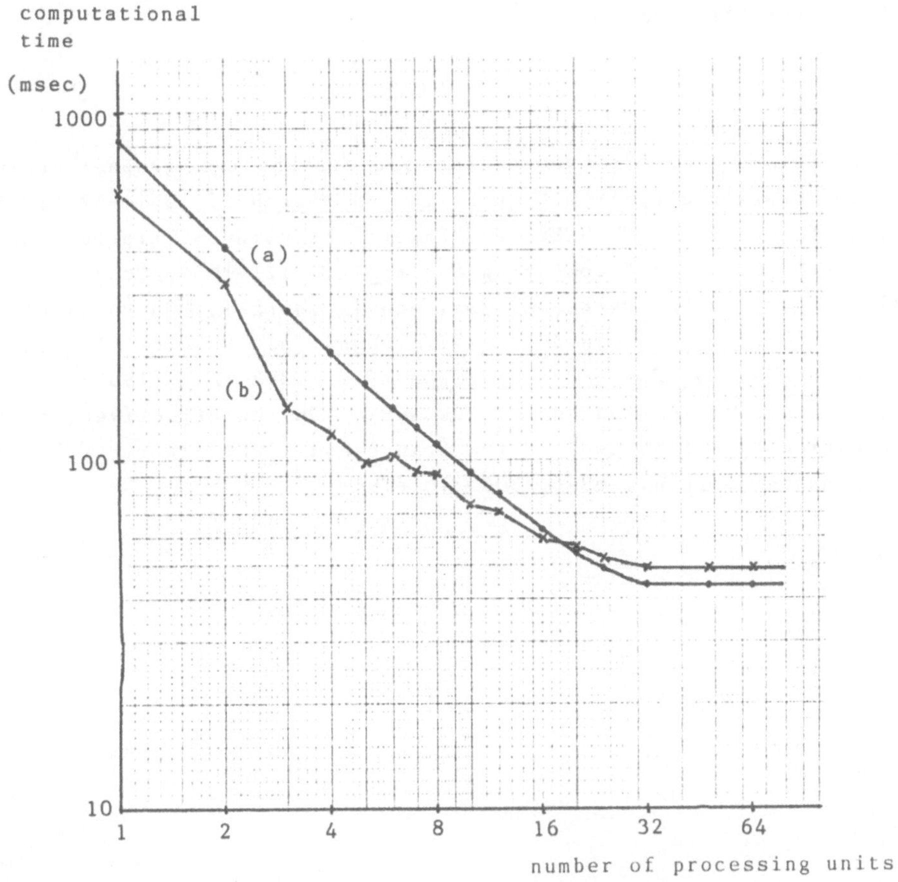


**Figure 3.** Computational time

### 3.3 Experimental result

Since we have no parallel processing system, we made our experiments through pseudo-parallel computation on a single processor. In order to do this we provided n variables as "clocks" (n is the number of processing units). Everytime access to common data takes place from a processing unit, the task with the processing unit is interrupted and the time consumed by the processing unit is measured and set in the "clock" variable. Then the n clocks are compared, and the task in the least recently used processing unit begins to run.

The computer where we made our experiment is ACOS77 TOSBAC600. Several examples of machine scheduling problems were created with three items and three machines. **Figure 3** shows typical result of an example. The curves (a) and (b) correspond to the strategies (a) and (b), respectively, in **3.2**. The curve (a) is just what we expected; (b) goes somewhat beyond our imagination, which requires further research both theoretical and experimental.

### References

[1] E. Balas, "Machine sequencing via disjunctive graphs: an implicit enumeration algorithm," Operations Research **17**, 941-957 (1969).

[2] A. K. Jones and P. Schwarz, "Experience using multiprocessor systems: a status report," Computing Surveys **12**, 121-165 (1980).

[3] H. Usui, M. Yamashita, M. Imai and T. Ibaraki, "On Parallel Searches of Game Trees ," (in Japanese), Trans. Institute of Electronics and Communication Engineers of Japan **69-D**, 1382-1393 (1986).

[4] M. Imai, Y. Yoshida and T. Fukumura, "On the effectiveness of evaluation functions used in the branch-and-bound algorithm," (in Japanese), ibid., **63-D**, 303-310 (1980).

# OPTIMAL TRADEOFF AMONG DELAY, THROUGHPUT AND FAIRNESS IN FLOW CONTROLLED NETWORKS+

Suk-gwon Chang
Dept. of Business Administration, Hanyang University
Sungdong-ku, Seoul 133, Korea

## I. INTRODUCTION

Delay, throughput and fairness are three conflicting performance criteria in flow controlled modern communication networks. This paper investigates the interrelationships of those three performance criteria and defines an "optimal tradeoff" among them. To this aim, four types of bandwidth sharing are introduced and compared each other for a variety of network configurations. The results obtained are then utilized to find the optimal operating point (OOP), which is defined to maximize the generalized power. It is shown that the maximum power can be computed via any polynomial root finding algorithm when routing is not optimized and via a variant of any multicommodity convex cost network flow algorithm when routing is to be optimized. Finally, some discussions are made on the realization of the OOP in practical flow controlled communication networks.

## II. FOUR TYPES OF RESOURCE SHARING

Consider a communication network modeled by a directed graph (N,L,C), where $N=\{1,2,...,N\}$ is the set of switching nodes, $L=\{1,2,...,L\}$ is the set of directed communication links, and $C=(C_1,C_2,...,C_L)$ is the link capacity vector of the same cardinality as L. A full-duplex link is considered as two separate directed links. All links are assumed to be error-free and all switching nodes are assumed to have unlimited buffer spaces. The network users are classified according to OD pairs so as for the users with the same OD pair to constitute a user group. Since fairness among the users belonging to the same user group is relatively straightforward, we are mainly interested in the fair allocation of network resources among user groups with different OD pairs. For simplicity, we assume without loss of generality there are W user groups and each user group has only one user, so that each user has different OD pair. Define this set of users as $W=\{1,2,...,W\}$. Note that N, L, and W are used to denote the sets as well as their cardinalities. We start with the tentative assumption that the links can be utilized up to their capacities, which will later be relaxed.

In a competing environment, each user tries to increase his throughput into the network as much as possible. If there are no other competing users or he can monopolize the entire network resources, he will increase his throughput up to the maximal flow the network allows. This maximal flow can be viewed as a good representation of his traffic requirement in any network designed appropriately on the overall traffic requirement pattern. Of course, it also measures the amount of throughput he may have to give up for the other competing users. In economical terms, it may be interpreted as his demand induced from the network topology, while the network resources, which are link bandwidths in our case, serve as network's supply to the users. If we define $m_i(p)$ as the flow on link i which constitutes the p-th user's maximal flow m(p) in the network (N,L,C), the total demand for link i is the sum of $m_i(p)$'s over all users in W.

To compute $m_i(p)$ for each user p in W, consider the following maximal flow problem with one additional objective function.

$$\text{maximize} \quad \sum_{j\in P(p)} x_j \tag{1a}$$

$$\text{minimize} \quad \sum_{j\in P(p)} h_j x_j \tag{1b}$$

$$\text{s.t.} \quad \sum_{j\in P(p)} a_{ij} x_j \le C_i, \quad i\in L \tag{1c}$$

$$x_j \ge 0, \quad j\in P(p) \tag{1d}$$

where P(p) is the set of paths for user p, $x_j$ is the flow on path j, $h_j$ is the hop count of path j, and $a_{ij}$ is the element of link-path incidence matrix satisfying the following relationship:

$$a_{ij} = \begin{cases} 1 & \text{if path } j \text{ traverses link } i \\ 0 & \text{otherwise} \end{cases} \tag{2}$$

The objective function of (1a) is the p-th user's total throughput into the network, while the one of (1b) measures the total network load imposed on the network by user p flows. It is interesting to note that this total network load is nothing but the amount of resources (bandwidths) consumed by user p. Since these two objectives are conflicting each other unless otherwise mentioned, we give absolute priority to the first objective. Then, by solving this problem we can find the maximal flow for user p with minimal network load (or equivalently with minimal consumption of resources). Using this formulation, we can compute $m_i(p)$'s, which are the lefthand-side values of the inequalities (1c) at optimum. The reasoning behind this formulation is that efficient usage of network resources is a necessary condition to determine the p-th user's true demands $m_i(p)$'s for all links in L.

The four types of bandwidth sharing schemes differ from each other in how the bandwidth of link i allocated to user p, denoted by $C_i(p)$, is computed from the demands $m_i(q)$'s for $q\in W$. Detailed descriptions follow:

Equal Link Sharing (ELS): ELS allocates the bandwidths of a certain link equally to all users (or equivalently, OD pairs) competing for that link. This is a kind of realization of Gerla's fairness definition.

$$C_i(p) = C_i \delta_i(p) / \sum_{q\in W} \delta_i(q), \quad i\in L, \ p\in W \tag{3}$$

where $\delta_i(q) = 1$ if $m_i(q) > 0$ and 0 otherwise.

Proportional Link Sharing (PLS): PLS allocates the bandwidths of a certain link to competing users in proportion to their demands $m_i(p)$'s, $p\epsilon W$. With this scheme, fairness is defined implicitly as equal achievement level among competing users.

$$C_i(p) = C_i m_i(p) / \sum_{q \epsilon W} m_i(q), \quad i \epsilon L, \; p \epsilon W \tag{4}$$

Although these two bandwidth sharing schemes are somewhat different from each other in the usage of $m_i(p)$'s when allocating link bandwidths to competing users, both of them are fair in a loose sense that for each user a certain amount of traffic is guaranteed to enter the network. However, they do not take account of the fact that a long distance user consumes more resources along his routes than a short distance user does. The following two bandwidth sharing schemes are more enhanced versions of ELS and PLS to take this point into consideration.

Equal Network Sharing (ENS): ENS is quite similar to ELS except that $\delta_i(p)$'s are normalized with respect to their weighted hop counts.

$$C_i(p) = C_i \delta_i(p) / (h_i(p) \sum_{q \epsilon W} \delta_i(q) / h_i(q)), \quad i \epsilon L, \; p \epsilon W \tag{5}$$

where $h_i(p)$ is the weighted hop count of user p flow on link i:

$$h_i(p) = \sum_{j \epsilon P(p)} a_{ij} h_j x_j / \sum_{j \epsilon P(p)} a_{ij} x_j \quad \text{at optimum of the problem (1)} \tag{6}$$

Proportional Network Sharing (PNS): PNS is quite similar to PLS except that $m_i(p)$'s are normalized with respect to their weighted hop counts.

$$C_i(p) = C_i m_i(p) / (h_i(p) \sum_{q \epsilon W} m_i(q) / h_i(q)), \quad i \epsilon L, \; p \epsilon W \tag{7}$$

The above four sharing schemes differ from each other in their implied fairness definition, so that we can not say absolute superiority of one scheme over the other. Instead, we can tradeoff these four extremes by introducing two parameters $\theta$ and $\tau$ varying on the range [0,1], which define a compromise of those four different types of fairness. $\theta$ determines the level of tradeoff between equal sharing and proportional sharing, while $\tau$ determines the level of tradeoff between link sharing and network sharing. With these parameters, the bandwidth of link i allocated to user p becomes

$$C_i(p) = C_i m_i(p)^{\theta} / (h_i(p)^{\tau} \sum_{q \epsilon W} m_i(q)^{\theta} / h_i(q)^{\tau}), \quad i \epsilon L, \; p \epsilon W \tag{8}$$

Observe that $\theta = 0$ for equal sharing and $\theta = 1$ for proportional sharing, while $\tau = 0$ for link sharing and $\tau = 1$ for network sharing.

If we define a vector C(p) of cardinality L for each user p as $C(p) = (C_1(p), C_2(p),\ldots,C_L(p))$, the link capacity vector C can be decomposed into W vectors $\{C(1),C(2),\ldots,C(W)\}$ because

$$C = \sum_{p \in W} C(p) \tag{9}$$

For each user p, we define Y(p) as the maximal flow from his source to his destination in the network (N,L,C(p)) and call it the compromising maximal flow for user p. Then, Y(p) can be computed by solving the following maximal flow problem.

$$Y(p) = \text{maximum} \sum_{j \in P(p)} x_j \tag{10a}$$

$$\text{s.t.} \sum_{j \in P(p)} a_{ij} x_j \leq C_i(p),\ i \in L \tag{10b}$$

$$x_j \geq 0,\ j \in P(p) \tag{10c}$$

If we denote the optimal link flow for link i obtained from the above problem by $Y_i(p)$ (which is the lefthand-side value of (10b) for each i at optimum), the following relations hold obviously:

$$Y_i = \sum_{p \in W} Y_i(p) \leq C_i \quad \text{for each } i \text{ in } L \tag{11}$$

Note that each Y(p), though not explicitly represented, is a function of the two parameters given the network (N,L,C).

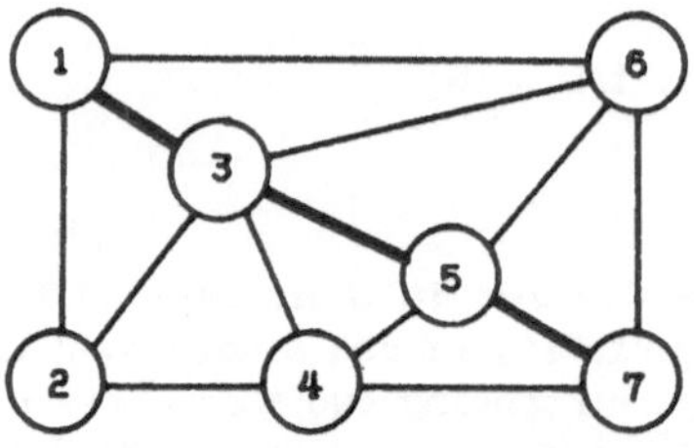


Fig. 1. Mesh A network

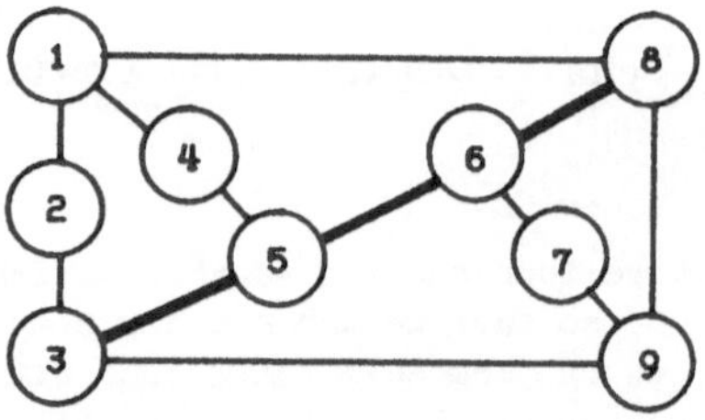


Fig. 2. Mesh B network

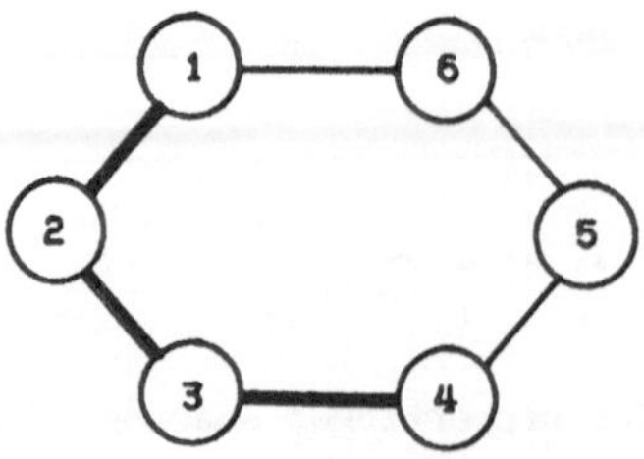


Fig. 3. Ring network

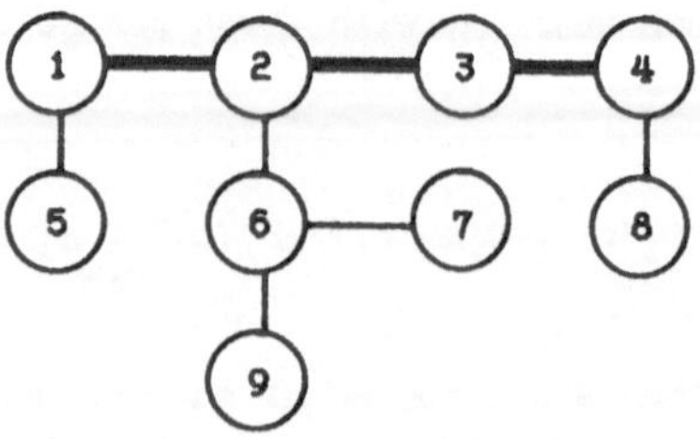


Fig. 4. Tree network

Table 1. Problem statistics

| | Mesh A | Mesh B | Ring | Tree |
|---|---|---|---|---|
| No. of nodes | 7 | 9 | 6 | 9 |
| No. of directed links | 26 | 24 | 12 | 16 |
| No. of users (OD pairs) | 7,14 | 9,18 | 6,12 | 9,18 |
| No. of paths generated | 95 (7 users) | 54 (9 users) | 12 (6 users) | 9 (9 users) |
| | 199 (14 users) | 101 (18 users) | 24 (12 users) | 18 (18 users) |

In order to evaluate our sharing schemes, practical sized networks of various configurations were constructed as shown in Fig. 1 through Fig. 4. Two of them are of mesh type with different connectivity; that is, mesh A network of Fig. 1 has more links per node than mesh B network of Fig. 2 has. The remaining two are ring and tree networks. All the links of them are assumed to be bidirectional links and to have only two kinds of capacities, 1 and 2; 1 for thin links and 2 for thick links. The experiments were performed with two sets of users (OD pairs) with different magnitude for each type of network. Detailed problem statistics are listed in Table 1.

In the experiments, one additional kind of sharing scheme was considered to quantify the cost (which is throughput degradation in our case) we should pay for fairness in different network configurations. It is the conventional multicommodity maximal flow where the total throughput is to be maximized regardless of the fairness among users. This sharing scheme, denoted by MMF, can be formally described as the following mathematical formulation.

(MMF) Maximize $\sum_{j\in P} x_j$ (12a)

s.t. $\sum_{j\in P} a_{ij}x_j \le C_i, \; i\in L$ (12b)

$x_j \ge 0, \; j\in P$ (12c)

where P is the union of P(p)'s over W; i.e., $P = \bigcup_{p\in W} P(p)$.

The results obtained from the experiments are summarized in Fig. 5 through Fig. 8, where horizontal axis measures the absolute fairness by coefficient of variation among user throughputs and vertical axis measures throughput by the throughput per user (= mean throughput). Note that the coefficient of variation, which is defined by standard deviation divided by mean throughput, is a reversed measure measuring the absolute fairness. This normalized measure enables us to compare the results for two sets of users with different magnitude.

Fig. 5 shows the summary results for mesh A network. MMF, owing to its negligence of fairness among users, achieves the highest throughput, but also the highest unfairness; i.e., the largest deviation among user throughputs. On the contrary, our sharing schemes improve fairness considerably, while achieving moderate amount of throughput for each user. It is interesting to observe that our sharing schemes have different levels of tradeoff between the two conflicting performances. Let's see the

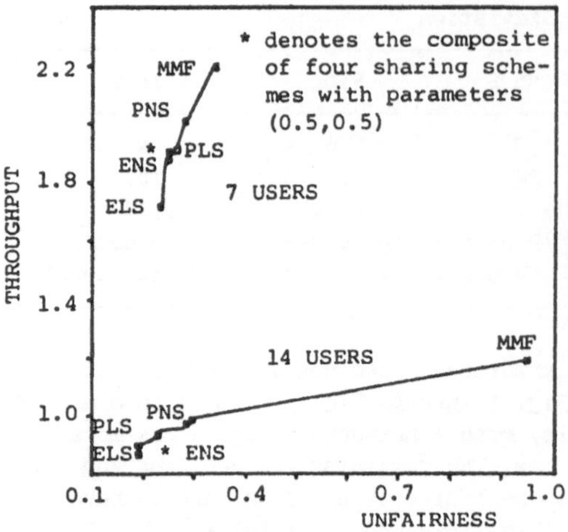


Fig. 5. Results for mesh A network

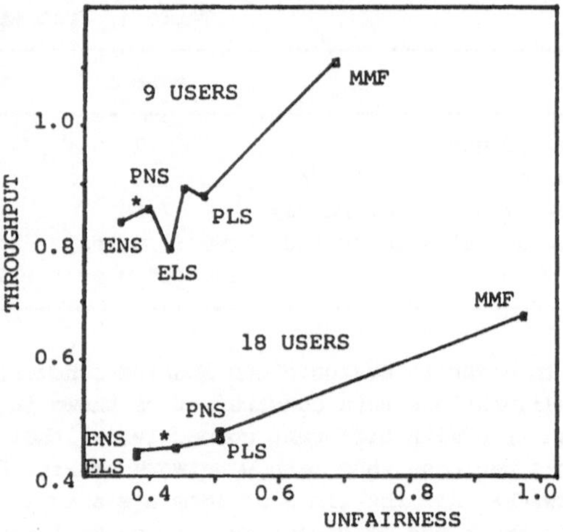


Fig. 6. Results for mesh B network

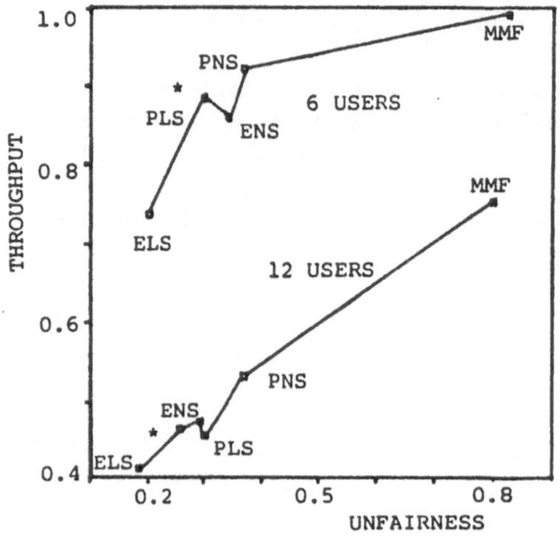


Fig. 7. Results for ring network

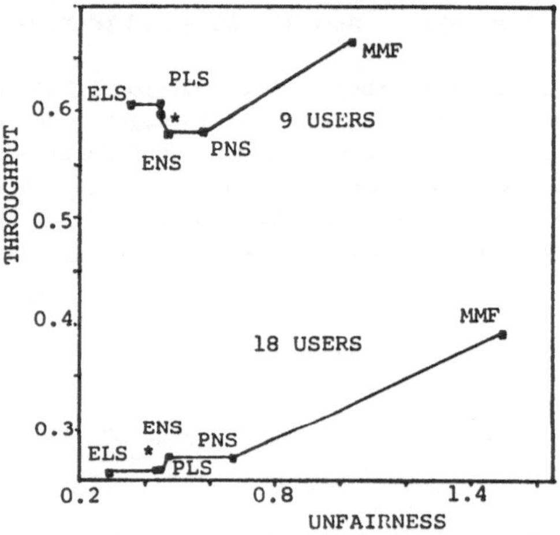


Fig. 8. Results for tree network

7 user case. Among the five sharing schemes compared, PNS achieves the largest throughput at the cost of the highest unfairness in terms of the absolute fairness, while ELS achieves the smallest throughput with the lowest unfairness. This statement also applies to 14 user case though the schemes have different orders from those for 7 user case in either performance criterion.

Now we turn our attentions to the effect of the number of communicating users on the level of tradeoff between those performances. As it can easily be seen in Fig. 5, unfairness of MMF in 14 user case appears to be much more serious than that in 7 user case. Hence, it is easily imagined that much will have to be paid for fairness particularly in 14 user case. In reality, throughput is degraded down from 2.21 of MMF to 2.02 of PNS, for instance, (which is equivalent to 9% degradation) in 7 user case, achieving 0.27 in the unfairness measure, while PNS of 14 user case shows much larger degradation ( 17% ) of throughput even for higher level ( 0.29 ) of unfairness. This observation also applies to the remaining four schemes though the corresponding values vary slightly case by case. All these observations support strongly our intuitive notion that the more are the users, the more should be paid for the same level of fairness.

The remaining three types of network configurations also show quite similar results to that of mesh A network. The only difference is that there can be found some schemes which are dominated by the other schemes in both performance criteria used. They are ELS for 9 user case of mesh B network, ENS and PLS for 6 and 12 user cases repectively of ring network, and finally ENS and PNS for 9 user case of tree network. Unfortunately, however, we can not find any systematic rule governing the dominance pattern. It seems that these are by chance or partially by the inappropriateness of the unfairness measure used in our experiments and do not necessarily mean the inferiority of one scheme over the other schemes.

As for the throughput degrading effect of the number of users, all the observations from Fig. 6 through 8 also supports strongly our previous intuitive notion, though absolute magnitudes differ from each other. With smaller set of users, PNS throughput experiences 19%, 7% and 13% degradations in mesh B, ring, and tree networks respectively. With larger set of users, the corresponding values are 28%, 29% and 31%, showing differences 9%, 22%, and 18% respectively. It is interesting to observe that the throughput degrading effect is stronger for ring and tree network configurations than for mesh type of network configurations.

## III. OPTIMAL TRADEOFF BETWEEN THROUGHPUT AND DELAY

In the previous section, we ignored delays by assuming the communication links can be utilized up to their capacities and dealt only with the relationship between fairness and throughput in a variety of network configurations. In this section, we are concerned with finding the optimal tradeoff between throughput and delay under the fairness conditions mentioned previously. We start by relaxing the assumption on link utilization and define the optimal tradeoff in terms of network power.

If we assume M/M/1 queue and the well-known Kleinrock's independence assumption, the average message delay of an open queueing network can be represented as

$$\frac{1}{\gamma} \sum_{i \in L} \frac{\lambda_i}{C_i - \lambda_i} \tag{13}$$

where $\gamma$ is total throughput in messages per second, $C_i$ and $\lambda_i$ are the i-th link capacity and the i-th link flow in messages per second repectively. Then, by definition, the generalized power with the parameter $\beta$, which determines the level of tradeoff between throughput and delay, becomes

$$\frac{\gamma^{\beta+1}}{\sum_{i \in L} \lambda_i / (C_i - \lambda_i)} \tag{14}$$

All remaining is to determine an appropriate value of $\beta$ and to optimize this unified performance measure. There may be two strategies: one is to optimize (14) without rerouting optimization and the other is with rerouting optimization. The former strategy maintains the flow pattern obtained by solving the maximal flow problems (10) in the networks $(N,L,C(p))$, $p \epsilon W$, just scaling down the overall flow pattern to optimize power. The latter strategy, on the contrary, shuffles that flow pattern aiming at smaller delay given the scaled-down user throughputs.

Strategy 1 (Routing is not optimized in terms of delay): With this strategy, we are only to find the optimal scaling factor $s^*$ which maximizes the generalized power since rerouting is not executed. If we rewrite the generalized power as a function of this scaling factor s, the problem to find the optimal operating point (OOP) reduces to the following unconstrained maximization problem:

$$\text{Maximize } p(s) = \frac{(s\gamma)^{\beta+1}}{\sum_{i \epsilon L} s\gamma_i/(C_i - s\gamma_i)} \tag{15}$$

where $\gamma_i$ is the maximal flow on link i (Refer to the eqn. (11) for precise definition).

Strategy 2 (Routing is optimized in terms of delay): Rerouting is performed in this case to minimize delay given the compromising maximal flows $\gamma(p)$'s, so that the path flows are no longer constants but decision variables. The problem to find the OOP is formulated as follows:

$$\text{Maximize} \quad \frac{(s\gamma)^{\beta+1}}{\sum_{i \epsilon L} \lambda_i/(C_i - \lambda_i)} \tag{16a}$$

$$\text{s.t.} \quad \sum_{j \epsilon P} a_{ij}x_j = \lambda_i, \; i \epsilon L \tag{16b}$$

$$\sum_{j \epsilon P(p)} x_j = s\gamma(p), \; p \epsilon W \tag{16c}$$

$$x_j \geq 0, \; j \epsilon P \tag{16d}$$

where $a_{ij}$'s are the elements of the link-path incidence matrix, $\gamma(p)$ is the compromising maximal flow for user p, which is the objective function value of the problem (10). Note that $\gamma$ is the sum of $\gamma(p)$'s over all users.

The problem (16) may be viewed as an extended version of the conventional multicommodity convex cost network flow problem and can be solved by a modification of those algorithms such as Flow Deviation method [6] by Fratta, et al., the extremal flow method [10] by Cantor and Gerla, Gradient Projection method [14] by Schwartz and Cheung, and a composite of them [1] by Bertsekas. Since algorithm development is not of our major concern in this paper, we omit the description of the algorithm.

Table 2. Comparison between two strategies[+]

| Type of network | Mesh A | | Mesh B | | Ring | | Tree | |
|---|---|---|---|---|---|---|---|---|
| | 7 users | 14 users | 9 users | 18 users | 6 users | 12 users | 9 users | 18 users |
| Strategy 1: | | | | | | | | |
| $s^*$ | 0.558 | 0.552 | 0.578 | 0.564 | 0.593 | 0.556 | 0.580 | 0.607 |
| power | 2.403 | 2.248 | 1.110 | 1.098 | 0.987 | 0.908 | 0.888 | 0.678 |
| Strategy 2: | | | | | | | | |
| $s^*$ | 0.563 | 0.572 | 0.618 | 0.644 | 0.602 | 0.696 | 0.580 | 0.607 |
| power | 2.875 | 2.829 | 1.396 | 1.622 | 1.499 | 1.598 | 0.888 | 0.678 |
| ratio[++] | 84% | 79% | 80% | 68% | 66% | 57% | 100% | 100% |

+ The problems were solved with $\beta = 1$.

$$++ \text{ ratio} = \frac{\text{power from strategy 1}}{\text{power from strategy 2}} \times 100 \ (\%)$$

In order to investigate the effect of rerouting on the performance improvement in terms of power, two problems, one for each set of users, were solved for each network configuration both with and without the routing optimization. As a resource sharing scheme, the composite of four types of sharing we analyzed in the previous section was used. Remember that both the parameter values used are 0.5.

Table 2 summarizes the results. First of all, it appears that rerouting is desired whenever possible since it decreases considerably the total message delay experienced by the users given the user throughputs allowed to the network, thus improving the unified network performance. In reality, the power achieved for mesh A network was improved from 2.403 to 2.875 and from 2.248 to 2.829 with 7 users and 14 users respectively. In other words, the powers achieved without rerouting reach only 84% and 79% of those achieved with rerouting. The results are quite similar for mesh B and ring networks except that the achievement levels decrease as we moves to mesh B and ring networks. This means that rerouting is more effective for ring type of networks than mesh type of networks. In the tree network, there can not be found any performance improvement, which is quite a natural result since each user has only one path.

Another thing worth noting here is the fact that rerouting appears more effective with larger set of users. The ratios of the powers two strategies achieve, which were represented in percentage, strongly support this statement. That is, those values for larger set of users are by far smaller than those for smaller set of users irrespective of network configuration. This implies that routing optimization contributes more to performance improvement, thereby becomes more desired in the network with larger number of users. It is also partially supported by the fact that two $s^*$ points with and without rerouting go farther from each other as the number of users becomes larger.

## IV. DISCUSSION

We've shown that the fairness among users can be achieved with only a small amount of degradation in total throughput, but the cost to pay for fairness increases as the number of users becomes larger. Furthermore, the conflict occurred when both throughput maximization and fairness improvement are persued is found to be more serious in ring and tree networks than in mesh type of networks.

Once it is completed to allocate network resources among users, there remains to be solved the problem of optimizing the tradeoff between throughput and delay. Two strategies were introduced to perform the optimization. The first is to optimize power only by scaling down the user throughputs while maintaining the flow pattern obtained during allocating link capacities among users and maximizing their flows within the allocated capacities. The second is to reroute their traffic flows, at the same time scaling down the user throughputs entering the network aiming at the maximum power. Since the first has some restrictions on routing the second does not have, it yields a suboptimal (which means "feasible, but not necessarily optimal") flow to the second. Hence it is desired to use the second strategy whenever possible since it always achieves better performance in terms of fairness, throughput and delay. Note that our fairness measure is independent of the scaling operation in both strategies. The only additional cost is the computational effort devoted to perform rerouting optimization.

Now, we turn our attention to the implementation issue of that optimal operating point (OOP). In real operating environment, network parameters including the rate of incoming traffics are slowly changing and correspondingly the control parameters adapt to those changes dynamically. As indicated by Gerla and Staskauskas [13], there may be two types of flow control. They are the rate control mode and the window control mode of operation. If we assume rate control, the implementation of the OOP is rather straightforward. But that is not the case when windows are used. We assume a network which operates Gallager-Golestaani based quasi-static routing and flow control scheme like [15], where the input rates are controlled through windows quasi-statically according to the congestion level perceived by each OD pair of users, converging to an equilibrium operating point. Then our problem is to adjust this equilibrium operating point to the OOP and to convert it to the corresponding optimal window sizes for each communicating OD pair.

Flow controlled networks are modelled more accurately by a closed queueing network. However, its applicability is yet limited to small-to-medium scale networks despite recent remarkable advances in the development of efficient algorithms based on Reiser's Mean Value Analysis. We assume for mathematical tractability an open queueing network model instead, which is a good approximation to the corresponding closed model. Then the OOP can easily be converted to the optimum window sizes of the end-to-end flow control scheme. Gallager and Golestaani [9] have suggested an equation to relate session rates to window sizes. Therefore, all remaining to do is to control the window sizes quasi-statically around those optimum values according to the congestion level perceived by each OD pair of users. A helpful discussion on this type of control mechanism can be found in Thaker and Cain [15].

References

[1] D. P. Bertsekas, "A Class of Optimal Routing Algorithms for Communication Networks," Proceedings of the Fifth International Conference on Computer Communication (ICCC), Atlanta, GA, Oct., 1980, pp. 71-75.

[2] D. P. Bertsekas, E. M. Gafni and R. G. Gallager, "Second Derivative Algorithms for Minimum Delay Distributed Routing in Networks," IEEE Trans. on Comm., Vol. COM-32, No. 8 (1984), pp. 911-919.

[3] K. Bharath-Kumar, "Optimum End-to-End Flow Control in Networks," Conference Record, International Conference on Communications, June 1980, pp. 23.3.1-23.3.6.

[4] K. Bharath-Kumar and J. M. Jaffe, "A New Approach to Performance Oriented Flow Control," IEEE Trans. on Comm., Vol. COM-29, No. 4 (1981), pp. 427-435.

[5] S. Chang and D. Tcha, "Analyzing the Effect on the Message Delay of Some System Parameter Changes in Communication Networks," Computer Networking and Performance Evaluation, North-Holland, IFIP, 1986, pp. 327-337.

[6] L. Fratta, M. Gerla and L. Kleinrock, "The Flow Deviation Method: An Approach to Store-and-Forward Communication Network Design," Networks, Vol. 3 (1973), pp. 97-133.

[7] R. Gail and L. Kleinrock, "An Invariant Property of Computer Network Power," Conference Record, International Conference on Communications, 1981, pp. 63.1.1-63.1.5

[8] R. G. Gallager, "A Minimum Delay Routing Algorithm Using Distributed Computation," IEEE Trans. on Comm., Vol. COM-25, No. 1 (1977), pp. 73-84.

[9] R. G. Gallager and S. J. Golestaani, "Flow Control and Routing Algorithms for Data Networks," Proceedings of the Fifth ICCC, Atlanta, GA, Oct., 1980, pp. 779-784.

[10] D. G. Cantor and M. Gerla, "Optimal Routing in a Packet Switched Computer Network," IEEE Trans. on Computers, Vol. C-23, No. 10 (1974), pp. 1062-1069.

[11] N. H. Gartner, "Optimal Traffic Assignment with Elastic Demands: A Review Part I. Analysis Framework," Transportation Science, Vol. 14, No. 2 (1980), pp. 174-191.

[12] N. H. Gartner, "Optimal Traffic Assignment with Elastic Demands: A Review Part II. Algorithmic Approaches," Transportation Science, Vol. 14, No. 2 (1980), pp. 192-208.

[13] M. Gerla and M. Staskauskas, "Fairness in Flow Controlled Networks," Proceedings of International Conference on Communications, 1981, pp. 63.2.1-63.2.5

[14] M. Schwarz and C. K. Cheung, "The Gradient Projection Algorithm for Multiple Routing in Message-Switched Networks," IEEE Trans. on Comm., Vol. COM-24 (1976), pp. 449-456.

[15] G. H. Thaker and J. B. Cain, "Interactions Between Routing and Flow Control Algorithms," IEEE Trans. on Comm., Vol. COM-34, No. 3 (1986), pp. 269-277.

# Distributed computing of a stochastic algorithm for combinatorial optimization problems

*ZHAO Yue*

*Takeshi FUKAO*

Department of Computer Science

Tokyo Institute of Technology

Tokyo, Japan

*ABSTRACT*

Simulated annealing method, as a general stochastic algorithm, has proven to be particularly successful for combinatorial optimization problems. But it requires a long running time for some large scale problems. This paper introduces the synchronous and partially synchronous spatial process, instead of the Metropolis procedure, in the simulated annealing method, and shows the possibility of distributed computing. We use the module partition problem as an example to show the quality of the solutions obtained by our method and point out that a parallel computing is able to be executed to shorten the running time.

## 1. Introduction

Simulated annealing method,[2] as a general stochastic algorithm, has proven to be particularly successful for combinatorial optimization problems, such as partition, placement, and wiring problems in VLSI design and traveling salesman problem and so on .[2][3] The idea comes from analogy between combinatorial optimization problems and physical disordered crystal systems such as Ising model or spin-glass, that is, they bring the objective function and minimum states in the combinatorial optimization problems into correspondence with the energy function and ground states in physical systems respectively. The global optimization is realized by controlling a parameter called temperature(degree of disorder). The knowledge well known in statistical mechanics is used for reference in this method.

A drawback of the simulated annealing method is that it requires a long running time for large scale problems. Several efforts have been made to improve the speed of this method.[8][9] This paper searches the possibility of the distributed and parallel computing to shorten the running time. Up to the present, simulated annealing method used the Metropolis procedure,[4] that is in each step of the iteration only one variable is permitted to change its state, in order to make the system reach the "equilibrium distribution" at each temperature. This means that sequential computing is needed in the general case. In this paper we consider a new stochastic process at each temperature using the concepts of the synchronous spatial process and the partially synchronous spatial process. This stochastic process differs from the traditional ones in which the transition probability depends on the change of the value of the objective function. It has a distributed and parallelizable nature in its computing. First we discuss the equilibrium distribution of the synchronous spatial process and the partially synchronous spatial process. Then we present a new simulated annealing algorithm which is based on the partially synchronous spatial process instead of the Metropolis procedure. Finally we apply this algorithm to the module partition problem and briefly discuss the results of our simulation.

## 2. Definitions

A function $g(x)$ is non-separable if it can not be separated to

$$g(x) = g_A(x^A)+g_B(x^B)$$

where $x= (x_1, \dots, x_m)$, $x^A$ and $x^B$ are sub-vectors of $x$ which do not contain the common elements. Generally a function can be written as the summation of some non-separable functions:

$$f(x) = g_1(x^1)+g_2(x^2)+\dots+g_M(x^M)$$

This is called a separation of function $f(x)$. The non-separable functions $g_1, \dots, g_M$ are called the components of $f(x)$. A separation of function $f(x)$ can be expressed by the structural graph $G$ as follows:

(1) There is a node corresponding to every variable.

(2) If $x_i$ and $x_j$ belong to the same component, there is a edge linking the corresponding nodes.

Obviously for a separation of function $f(x)$ there is a clique(complete subgraph) in graph $G$ corresponding to every component of $f(x)$. Therefore we can write $f(x)$ as

$$f(x) = \sum_{c\in C} f_c(x_c)$$

where $C$ is the set of the above-mentioned cliques.

$x(t) = (x_1(t), x_2(t), \dots, x_N(t))$ $\quad x_j \in X_j$ is the configuration of graph $G$ at time $t$. Thus the state of $x = (x_1, x_2, \dots, x_N)$ takes values in the configuration space $X = X_1 \times X_2 \times \cdots \times X_N$

A function $\pi$

$$\pi: X \rightarrow (0, 1)$$

is called a random field if

$$\sum_{x \in X} \pi(x) = 1$$

Thus a random field is just a probability distribution over the state space of $x$. Let $T_j^m$ be the operator which changes the state of the $j$th variable $x_j$ to $m$.

$$T_j^m x = (x_1, x_2, \ldots, x_{j-1}, m, x_{j+1}, \ldots, x_N)$$

Given a random field $\pi$ the conditional probability that $j$-node has state $x_j$ given that the other variable have states $x_{V-j}$ is

$$P(x_j \mid x_{V-j}) = \frac{\pi(x)}{\sum_{m \in X_j} \pi(T_j^m x)}$$

$\pi$ is a Markov random field if

$$P(x_j \mid x_{V-j}) = P(x_j \mid x_{\partial_j})$$

Where $x_{\partial_j}$ is the set of the nearest variables of $x_j$ and $V$ is the set of the nodes of the structural graph G. A Markov random field which can be written in the form

$$\pi(x) = \frac{1}{Z} \exp\{-f(x)\}$$

is a Gibbs random field.

**Definition 1**

Call a Markov process $x(t)=\{x_1(t), \ldots, x_N(t)\}$ a spatial process if:

(1) Only one node can change its state at a time.

(2) The transition rate $w(T_j^m x \mid x)$ depends on the states of $j$-node and its nearest nodes only.

(3) For any states $x$ and $T_j^m x$, it is possible to reach $T_j^m x$ from $x$ by a sequence of transitions which do not alter $x_{V-j}$.

**Definition 2**

Call a Markov process $x(t)=\{x_1(t), \ldots, x_N(t)\}$ a synchronous spatial process if:

(1) All of the nodes can change their states independently at the same time.

and (2) and (3) in definition 1.

**Definition 3**

Call a Markov process $x(t)=\{x_1(t), \ldots, x_N(t)\}$ a partially synchronous spatial process if:

(1) All of the nodes in a partitioned subset can change their states independently at the same time.

and (2) and (3) in definition 1.

### 3. The preparation in theory

In the traditional simulated annealing methods, they use the Metropolis procedure, or more generally use the transition probabilities which depend on the configurations before and after trasition, such as the change of the objective function at each temperature for example. In [5][6][7] Romeo et,al. analyzed the equilibrium distribution and convergence for this type of stochastic processes. As another type of stochastic process, we discussed the equilibrium distributions of synchronous spatial process and partially synchronous spatial process, and pointed out that, in the general case, its equilibrium distribution is not always the desirable Gibbs distribution if the transition rate is decided using the difference of the objective function values before and after transition(we called it gradient type), in Ref.[1]. In the following we show a different result for some special objective functions.

**Proposition**

The partially synchronous spatial process characterized by transition rate:

$$w(y^{\alpha} \mid x) = \prod_{j \in \Lambda^{\alpha}} w(y_j \mid x), \ \alpha=1, \ldots, M$$

$$w(y_j \mid x) = A\exp\{\frac{1}{2T} \sum_{c \in C(j)} [f_c(x_c) - f_c(T_j^{y_j} x_c)]\}$$

has the equilibrium distribution

$$\pi(x) = B\exp\{-\frac{1}{T} \sum_{c \in C} f_c(x_c)\}$$

if $C$ contains at the most 2-cliques(links or single notes).

We give the proof in the appendix

**Corollary**

The synchronous spatial process characterized by transition rate:

$$w(y \mid x) = \prod_{j} w(y_j \mid x)$$

$$w(y_j \mid x) = A\exp\{\frac{1}{2T} \sum_{c \in C(j)} [f_c(x_c) - f_c(T_j^{y_j} x_c)]\}$$

has the equilibrium distribution

$$\pi(x) = B\exp\{-\frac{1}{T} \sum_{c \in C} f_c(x_c)\}$$

if $C$ contains at the most 2-cliques(links or single nodes).

## 4. A new simulated annealing algorithm

On the basis of the proposition and corollary proved in previous section, we can give the following algorithm for the kind of combinatorial optimization problems in which the objective functions can be written in the form

$$f(x) = \sum_{c \in C} f_c(x_c)$$

where $C$ contains at the most 2-cliques.

step 1 Partition the set of variables $x_1, \ldots, x_N$ to M components $A^1, \ldots, A^\alpha, \ldots, A^M$

step 2 Give $x_i$ $(i=1, \ldots, N)$ a state randomly and set the temperature at an initial value $T_0=T_{initial}$, $k=0$, $t=0$

step 3 Choose a component $A^\alpha$ randomly . For all $j \in A^\alpha$ do step 4 synchronously.

step 4 Change the state of $x_j$ to $y_j$ $(y_j \in X_j)$ by probability

$$\frac{1}{Z}\exp\{\frac{1}{2T}\sum_{c \in C(j)}[f_c(x_c) - f_c(T_j^{y_j}x_c)]\}$$

Where

$$Z=\sum_{m \in X_j}\exp\{\frac{1}{2T}\sum_{c \in C(j)}[f_c(x_c) - f_c(T_j^{m}x_c)]\}$$

step 5 If $t<t_{end}$ then $t=t+1$ go to step 3 else go to step 6

step 6 If $T_k>T_{end}$ then $T_{k+1}=\mu T_k$ $(0<\mu<1)$; $k=k+1$; $t=0$; go to step 3
else STOP.

The difference between our algorithm and the usual simulated annealing method is in the step 4. Step 4 takes almost all of the running time. In the Metropolis procedure only one variable is permitted for state-transition, so a complete sequential computation is needed in the general case. Our algorithm has the following parallel natures

(1) It can be executed parallelly for all $x_j$ $j \in A^\alpha$.

(2) For a $j \in A^\alpha$ a parallel computing can be executed for all $m \in X_j$ to calculate Z.

If a parallel computer is used in the computations in step 4, a reduction of the running time can be expected.

## 5. Application to the module partition problem

We turn to an application of the algorithm we proposed. In the module partition problem, the modules to be placed on some chips are connected together according to a predetermined logic relation, and the chips are positioned on a two-dimension space (a plane). What we must do is to partition the set of modules and to place them on the chips.

The optimization criterion is

(i) Minimizing the total inter-chip wire length.

(ii) Making the numbers of modules on each chip balance as much as possible.

Fig.1 shows the positions of the chips, and Fig.2 shows the connecting relation of the modules. The variable $x_i$ ($i$=1, 2, . . . , $N$) expresses the state of the $i$th module, i.e. $x_i = m$ means that module $i$ will be placed on the chip $m$. The Manhattan length is used to calculate the inter-chip wire length. We introduce a function $\phi_c$ for the criterion (i)

$$\phi_c(x_c) = \phi_c(x_i, x_j) = 0 \qquad : \text{when } x_i = x_j$$

$$= a_{ij} l(x_i, x_j) \quad : \text{when } x_i \neq x_j$$

Where

$c$ : the link (2-clique) connecting module $i$ and module $j$.

$a_{ij}$ : the weight of this link. If there is no link between module $i$ and $j$ $a_{ij} = 0$.

$l(x_i, x_j)$ : the Manhattan length between chip $x_i$ and chip $x_j$.

So the criterion (i) is

$$\sum_{c \in C^*} \phi_c(x_c) \rightarrow \min$$

Where $C^*$ is the set of links shown in Fig.2. For the criterion (ii), the balance criterion, we introduce a function $\xi$ :

$$\xi(x, y) = 0 \; : \; \text{when } x \neq y$$

$$= 1 \; : \; \text{when } x = y$$

and estimate the imbalance by

$$\sum_{m=1}^{M} [\sum_i \xi(x_i, m)]^2 \rightarrow \min$$

It can be changed in form as follows

$$\sum_{m=1}^{M} [\sum_i \xi(x_i, m)]^2$$

$$= \sum_m [\sum_i \xi(x_i, m)^2 + 2\sum_{i<j} \xi(x_i, m)\xi(x_j, m)]$$

$$= \sum_i \sum_m \xi(x_i, m)^2 + 2\sum_{i<j} \sum_d \xi(x_i, m)\xi(x_j, m)$$

$$= N + 2\sum_{i<j} \xi(x_i, x_j)$$

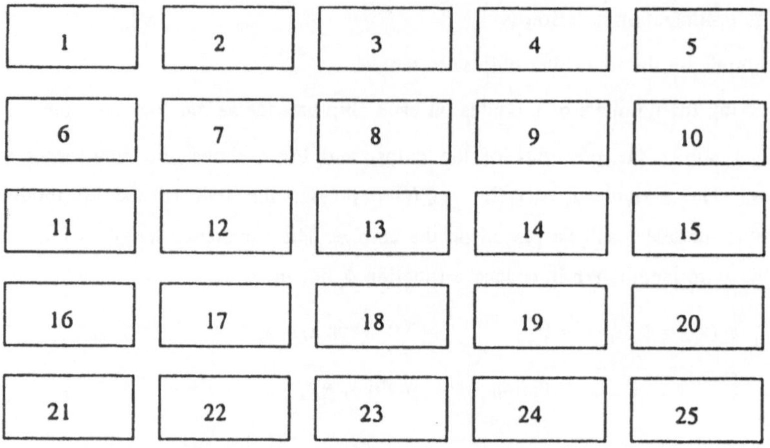


Fig.1 The positions of the chips

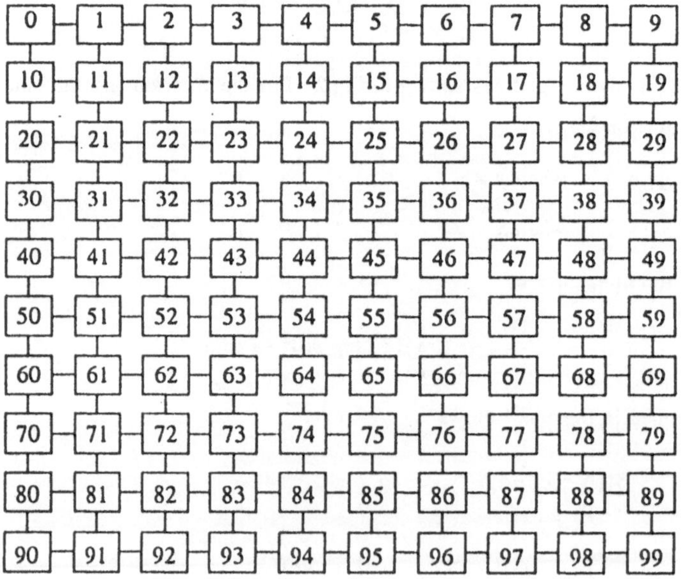


Fig.2 The connecting relation of the moduls

Finally the total objective function is

$$f(x) = \sum_{c\in C^*} \phi_c(x_c) + 2\lambda\sum_{i<j}\xi(x_i, x_j)$$

Obviously, we can write it $f(x) = \sum_{c\in C} f_c(x_c)$, where $C$ contains 2-cliques only. So the algorithm proposed in the previous section can be used to solve this problem.

In our simulation we make all the links shown in Fig.2 have the same weight 0.1 and take $\lambda = 0.03$, and in step 1 of our algorithm we make a random partition. We compare the quality and CPU-times of the results obtained by the Metropolis method with those by our algorithm. We computed this for about 10 samples(with different initial states). Table 1 shows the average values.

| the stochastic process at each temperature | | the final value of the objective function | expected time(second) if a parallel computing is executed |
|---|---|---|---|
| Metropolis procedure | | 21.25 | 816 |
| Partially synchronous spatial processes | 20-partition | 20.65 | 62.5 |
| | 10-partition | 20.82 | 26.4 |
| | 8-partition | 20.00 | 16.7 |
| | 4-partition | 20.68 | 10.1 |

Table 1 The comparison between usual simulated annealing method and our method

It is obvious from table 1 that:

(1) The results of our algorithm are frequently better than the usual one. The reason may be that, in our algorithm the system can jump out from a local minimum state more easily.

(2) There would be a big decrease in CPU-time as shown in the table if a parallel computer was used.

## 6. Conclusions

We have proposed a new simulated annealing algorithm in which we use a partially synchronous spatial process at each temperature. This partially synchronous spatial process has a node-base distributed nature, and can be executed by parallel computing. In this point our method differs from the usual simulated annealing methods. Finally we propose the following open problems:

(1) For an actual combinatorial optimization problem, how to formulate the objective function to

$$f(x) = \sum_{c \in C} f_c(x_c)$$

where $C$ contains at the most 2-cliques,

(2) Theoretical analysis of the convergence of our method.

(3) Analysis of the nature at low temperature.

(4) Developement of the parallel computing system that is suited our algorithm.

## Appendix

### Proof of the proposition

We check the detailed balance condition.

$$w(y \mid x)\pi(x) = w(x \mid y)\pi(y) \qquad (A)$$

For arbitrary states $x$ and $y$ the transition rate from $x$ to $y$ can be written

$$w(x \mid y) = \begin{cases} A & x=y \\ \prod_{j \in A^{\alpha}} w(y_j \mid x) & \text{for a } \alpha \quad y_j \neq x_j \Rightarrow j \in A^{\alpha} \\ 0 & \text{otherwise} \end{cases} \qquad (B)$$

Obviously in the first and last cases of (B) condition (A) is satisfied, so we prove about the second case only.

For the convenience of describing we introduce the following marks.

$C_1 = \{ c \mid c \subset C \text{ and } c \text{ is a 1-clique} \}$

$C_2 = \{ c \mid c \subset C \text{ and } c \text{ is a 2-clique} \}$

$C_1(j) = \{ c \mid c \subset C(j) \text{ and } c \text{ is a 1-clique} \}$

$C_2(j) = \{ c \mid c \subset C(j) \text{ and } c \text{ is a 2-clique} \}$

The left hand side of (A) is

$$w(y \mid x)\pi(x)$$

$$= \pi(x) \prod_{j \in A^{\alpha}} A\exp\{\frac{1}{2T} \sum_{c \in C(j)} [f_c(x_c) - f_c(\Gamma_j^{y_j} x_c)]\}$$

$$=\pi(x)A\exp\{\frac{1}{2T}\sum_{j\in\Lambda^a}\sum_{c\in C(j)}[f_c(x_c)-f_c(T_j^{y_j}x_c)]\}$$

$$=\pi(x)A\exp\{\frac{1}{2T}\sum_{j}\sum_{c\in C(j)}[f_c(x_c)-f_c(T_j^{y_j}x_c)]-\frac{1}{2T}\sum_{j\in\Lambda^a}\sum_{c\in C(j)}[f_c(x_c)-f_c(T_j^{y_j}x_c)]\}$$

$$=\pi(x)A\exp\{\frac{1}{2T}\sum_{j}\sum_{c\in C(j)}[f_c(x_c)-f_c(T_j^{y_j}x_c)]\}$$

$$=\pi(x)A\exp\{\frac{1}{T}\sum_{c\in C_2}f_c(x_c)+\frac{1}{2T}\sum_{c\in C_1}f_c(x_c)-\frac{1}{2T}\sum_{j}\sum_{c\in C(j)}f_c(T_j^{y_j}x_c)\}$$

$$=\pi(x)A\exp\{\frac{1}{T}\sum_{c\in C}f_c(x_c)\}\exp\{-\frac{1}{2T}\sum_{c\in C_1}f_c(x_c)-\frac{1}{2T}\sum_{j}\sum_{c\in C_1(j)}f_c(T_j^{y_j}x_c)-\frac{1}{2T}\sum_{j}\sum_{c\in C_2(j)}f_c(T_j^{y_j}x_c)\}$$

$$=AB\exp\{-\frac{1}{2T}\sum_{c\in C_1}f_c(x_c)-\frac{1}{2T}\sum_{c\in C_1}f_c(y_c)-\frac{1}{2T}\sum_{j}\sum_{c\in C_2(j)}f_c(T_j^{y_j}x_c)\}\quad (C)$$

Similarly the right hand side is

$$w(x\mid y)\pi(y)$$

$$=AB\exp\{-\frac{1}{2T}\sum_{c\in C_1}f_c(y_c)-\frac{1}{2T}\sum_{c\in C_1}f_c(x_c)-\frac{1}{2T}\sum_{j}\sum_{c\in C_2(j)}f_c(T_j^{x_j}y_c)\}\quad (D)$$

Pay attention to that the 3rd terms in (C) and in (D) are equal.

## Reference


[1] Takeshi FUKAO, ZHAO Yue: "Stochastic distributed optimization algorithm", *The Transaction of Institute of Electronics and Communication Engineers of Japan*, J-69A, 12 ,pp.1492-1501 (1986)

[2] S. Kirkpatrick, C.D. Gelatt, Jr. and M.P.Vecchi :"Optimization by simulated annealing", *Science*, 220, 4598, pp.671-680 (1983)

[3] M.Vecchi and S.Kirkpatrick :"Global wiring by simulated annealing", *IEEE Trans. On Computer Aided Design*, Vol CAD-2, No 4, Oct.1983, pp 215-222.

[4] S.Metropolis, A.Rosenbluth, A.Teller,and E.Teller :"Ewuation of state calculations by fast computing machines",*Jr.Chem.Phys.*,Vol 21, pp 1087, 1953.

[5] F.Romeo, A.Vincentelli, and C. Sechen :"Research on simulated annealing at Berkekey", *Proceedings ICCD*, oct.1984,pp 652-657

[6] F.Romeo, A.Vincentelli :"Probabilistic hill climbing algorithms: properties and applications", University of California, Berkeley, UCB/ERL M84/34,1984.

[7] D.Mitra, F.Romeo, A.Sangiovanni-Vincentelli :"Convergence and finite-time behavior of simulated annealing" University of California, Berkeley, UCB/ERL M85/23,1985.

[8] S.White :"Concepts of scale in simulated annealing", *Proceedings ICCD*, Oct 1984, pp 646-651

[9] J.W.Greene, K.J. Supowit :"Simulated annealing without rejected moves", *IEEE Trans. on Computer-Aided Design*, Vol. CAD-5 No 1, Jan.1986. pp 221-228.

# Hierarchical Simulated Annealing Optimization in Partition Problem

Yasuo Sugai and Hironori Hirata

Department of Electronics, Chiba University
1-33 Yayoi-cho, Chiba-shi 260 JAPAN

**ABSTRACT**

Simulated annealing method (SA method) is a stochastic optimization technique based on Monte Carlo method. Although SA method yields a result of high quality, it takes too much computational time to complete annealing algorithm. Then, parallelization of the SA algorithm is necessary in order to avoid the above disadvantage. A hierarchical simulated annealing method (HSA method) for partition problem is proposed for realization of parallelization to reduce computational time. The computational time of HSA method is compared with usual SA method. The HSA method hierarchically divides a given partition problem into some sub-problems applying the SA method at each stage and makes it possible to automatically choose suitable parameters in the cost functions at each stage of the hierarchical computation. Theoretical consideration and numerical experiments of particular block placement problem in VLSI layout show that the HSA method derives much more reduction of computational time and better final result than original SA method does.

## 1. INTRODUCTION

Simulated annealing method (SA method) is a stochastic optimization technique based on Monte Carlo method. Introduction of the concept of annealing into the field of combinatorial optimization by Kirkpatrick et al.[1] made much success in various field [2,3]. Although the SA method yields a result of high quality, it takes too much computational time to complete the algorithm. Therefore, parallelization of the SA algorithm is necessary in order to avoid this disadvantage.

This paper proposes a hierarchical simulated annealing method (HSA method) for the partition problem to realize a possibility of parallelization, and also evaluate the computational time compared with the original SA algorithm. The SA algorithm is an extension of Metropolis algorithm[4] to obtain the grand state of a collection of particles at any given temperature. This algorithm generates a sample of a stochastic process satisfying a stationary probability distribution at each temperature. Many iterations are necessary to put the system in a stationary configuration in stochastic sense because the iterations at initial stage makes the system in the unstationary state in computer simulation. How many times the iterations are necessary depends on the size of a given problem. Although simulation technique in physics indicates that the number of the iterations should be proportional to the number of elements of the system, it is impossible to compute such number of the iterations in practice for large scale problems, e.g., VLSI layout design.

The HSA method hierarchically divides a given problem into sub-problems whose sizes are suitable to deal with them. Because the size of the sub-problem becomes smaller along the progress of hierarchical division and parallel computation is possible from the second layer, the amount of computational time can be much reduced. Moreover, newly proposed regulation method of parameters in a cost functioncan automatically choose suitable parameters at each stage of the hierarchical computation to drastically reduce the computational time.

It is shown by the experiment of block placement problem in VLSI layout as application of partition problem that the reduction of computational time by the sequential computations of the HSA algorithm with three layer is 1/50. Moreover, the result is better than that by original SA algorithm.

## 2. THE PARTITION PROBLEM

We define the partition problem as follows.

Partition problem:

Consider a network consisting of n elements. Each element has weight $s_i$ ($i=1,\cdots,n$) and there are weighted edges $a_{ij}$ ($\geq 0$) between element i and j ($i,j=1,\cdots,n$). The weighted edges represent the strength of the relationship between two elements. The objective is to divide n elements into M groups so that the following two aims are satisfied.

Aim(i): Total strength of interactions among groups, i.e., the aggregate strength of the relationship between any two elements which belong to different groups, is to be minimized.

Aim(ii): The size of groups which is the aggregate weight of the elements belonging to a particular group is to be as same as possible.

Let $x_i$ ($=1,\cdots,M$) be a state of element i. The state means which group the element i belongs to. The configuration **X** of the network is

$$\mathbf{X}=\{x_1,x_2,\cdots,x_n\}. \quad (1)$$

A cost function f satisfying above two conditions can be defined as a function of the configuration **X**:

$$f(\mathbf{X})=\alpha[L(\mathbf{X})+\beta S^2(\mathbf{X})]. \quad (2)$$

The first and the second terms within the bracket of the right hand side of Eq.(2) represent the total strength of the relationship among groups and the size difference of the group respectively. The parameters α and β are positive constants. Details of each terms in Eq.(2) are shown in the following.

Constraint of the aim(i): $L(\mathbf{X})$

$$L(X)=\frac{1}{2}\sum_{i=1}^{n}\sum_{j=1}^{n}\phi(x_i,x_j),\quad \phi(x_i,x_j)=\begin{cases}0 & \text{for } x_i=x_j\\ a_{ij} & \text{for } x_i=x_j,\end{cases}\quad (i,j=1,\cdots,M). \tag{3}$$

Constraint of the aim(ii): S(X)

$$S(X) = \Big[\sum_{k=1}^{M}\Big\{\sum_{i=1}^{n} s_i\,\delta(x_i,k)\Big\}^2\Big]^{1/2}, \tag{4}$$

where $\delta$ is Kronecker's delta. The square in the right hand side of Eq.(4) represents the deviation of the size of the groups which is the variance of the size without constant term.

## 3. SIMULATED ANNEALING METHOD

The SA method is one of heuristic methods for combinatorial optimization problem, in which statistical thermodynamic analogy is utilized. The following concept makes it easy to understand the correspondence between physical system and stochastic method for optimization problem[5]. The probabilistic function p(**X**) for each configuration **X** is introduced. Then, optimization problem,

$$\min_{\{X\}} f(X) \tag{5}$$

can be rewritten in the form of

$$\begin{cases}\displaystyle\min_{\{p(X)\}}\ \sum_{X} p(X)f(X) & (6)\\[2ex] \text{constraints:}\quad -\displaystyle\sum_{X} p(X)\log p(X) = H,\quad \sum_{X} p(X)=1,\ p(X)>0 & (7)\end{cases}$$

A model described by Eqs.(6) and (7) can be called the stochastic model for an optimization problem. The entropy H in Eq.(7) represents the extent of the deviation of probabilistic distribution p(**X**). When the entropy is equal to zero this stochastic model is exactly identified with the original problem (5). The solution of Eqs.(6) and (7) is

$$p(X)=\frac{1}{A}\exp\{-\frac{f(X)}{T}\} \tag{8}$$

by the method of Lagrange multiplier. A constant A is a normalization factor. The parameter T is a Lagrange multiplier satisfying the first constraint in Eq.(7) that the entropy H is constant, and is equivalent to the temperature in physical system. Although the configuration **X** which gives optimum solution of the original problem can be obtained by directly searching the probabilistic distribution p(**X**), it is an unrealistic way because searching p(**X**) means looking over all possible configurations thoroughly. Therefore, we consider to simulate a stochastic process which has the probabilistic distribution (8) as a stationary one for a sample. We adopt the final result (configuration) of the sample as a solution converging the temperature T to zero. This implies the convergence of the stochastic model (6) and (7) to the original problem (5).

One of the ways to realize such a stochastic process is the Metropolis algorithm. The SA method is based on this algorithm. The SA algorithm is following:

step1: temperature T$\leftarrow$initial temperature $T_0$.

step2: loop counter N$\leftarrow$1.

step3: choose any element i (=1,•••,n) and its state $x_i$ randomly. This means to choose a new configuration, and calculate the difference of the cost function between new and old values, $\Delta f$.

step4: if $\Delta f<0$

then the new configuration is accepted and go to step 6.

step5: if $\Delta f>0$

then

if r < exp($-\Delta f/T$)

then new configuration is accepted, and the state transition is made (r=(0,1): uniform random number).

step6: N$\leftarrow$N+1

if N < (the fixed number for one temperature)

then go to step3.

step7: if the terminating criterion is true

then end

else $T_{new} \leftarrow c \cdot T_{old}$ and go to step2.

(0<c<1:temperature factor)

The Metropolis algorithm is the part from step2 to step6 in the above SA algorithm. Main characteristic of the SA algorithm is to accept the new configuration with probability exp($-\Delta f/T$) even when the state transition leads to deterioration consequentially. This property makes it possible to escape from local minima by introduction of the randomness into the system. The fixed number for one temperature in step6 is the number of iteration at that temperature to realize the stationary probabilistic distribution. The number multiplying the number of elements by the number of states of each element minus one should be considered appropriate number of iteration. In the simulation of Ising model the number of iteration is generally set at least up to the number of elements to check the states of all elements because each element has only two states, up or down. In optimization problems, since each element has several states, the huge number of iteration should be set up. The terminating criterion in step7 holds when no state transition occurs up to the prescribed number of times successively.

## 4. HIERARCHICAL SIMULATED ANNEALING METHOD

### 4.1 Hierarchical Method

The defect that the SA method needs huge computational time especially for large scale problems is due to the huge number of iteration N being proportional to the number of elements. Dividing a given problem into sub-problems in m stages reduces total number of iteration. The division number M is supposed to be written as

$$M=\prod_{i=1}^{m}\sigma_i, \quad (\sigma_i=1,\cdots,m), \tag{9}$$

where positive number $\sigma_i$ is the division number at each stage.

The HSA method is as follows:

(i) The first stage: a given problem is regarded as a sub-problem with the number of elements n and the division number $\sigma_1$. Application of the SA method divides the given problem into $M/\sigma_1$ sub-problems with the division number $\sigma_2$.

(ii) The i-th stage ($2\leq i\leq m$): the SA method is applied to each sub-problems which are given in just before stage.

### 4.2 Automatic Regulation Method of the Parameters among Stages

In general there are several objectives to be optimized in the optimization problems. A cost function is often represented in the form of weighted sum of those objectives. For practical computation, each value of the weight parameters like $\alpha$ and $\beta$ in Eq.(2) must be determined.

How to determine them plays a crucial role for getting a solution. Although the SA method itself may be gotten speeded up using parallel computation[6], the wasted computational time due to experimentally determination of parameters loses its advantage. Another regulation of parameters among stages is also necessary because of the difference of the order of the values of each parameter among stages. Hence, we propose a new method for regulation in order to automatically choose suitable values of parameters both within each stage and among stages.

Considering the fact that a probability accepting a deterioration in the Metropolis algorithm depends on the amount of change in the cost function rather than the magnitude of the cost function itself as shown in the SA algorithm, it seems that two terms $L(\mathbf{X})$ and $\beta S^2(\mathbf{X})$ in Eq.(2) equally influence a behavior of a sample when the order of the change in them is almost equal. The changes in f, L, $S^2$ are denoted $\Delta f$, $\Delta L$, $\Delta S^2$, respectively in the case that the state transition from a configuration $\mathbf{X}$ to another configuration $\mathbf{X}'$. Using $\langle\cdot\rangle$ for the average amount, the following equation is assumed:

$$\langle\Delta f\rangle = \alpha\, [\, \langle\Delta L\rangle + \beta\, \langle\Delta S^2\rangle\, ]. \tag{10}$$

Two conditions are necessary for the purpose of a suitable regulation for parameters within each stage and among stages:

[Condition I] : Regulate the value of $\beta$ so that $<\Delta L>$ and $\beta<\Delta S^2>$ are same order.

[Condition II]: Regulate the value of $\alpha$ so that $<\Delta f>$'s at all stages in hierarchical division have common order.

On the [condition I]

At first, $<\Delta L>$ and $<\Delta S^2>$ at each stage are divided by physical unity of that stage to make them no dimension. For example, when we can regard $a_{ij}$ and $s_i$ as a distance between the element i and j, and a square measurement of the element i, respectively, as a placement problem discussed later, we denote two scales at any stage i (=1,•••,m) $\hat{L}_i$ and $\hat{S}_i$, and denote the average changes by $<\Delta L_i>$, $<\Delta S_i>$, and $<\Delta f_i>$. In the case of a hierarchical division with the division number $\sigma_i$ at the stage i following relationship between the first stage and the i-th hold.

$$\hat{L}_i = \left( \prod_{k=1}^{i-1} \sigma_k^{-1/2} \right) \cdot \hat{L}_1, \qquad S_i = \left( \prod_{k=1}^{i-1} \sigma_k^{-1} \right) \cdot \hat{S}_1, \quad (i \neq 1). \tag{11}$$

Hence, Eq.(12) is rewritten using Eq.(13) as

$$<\Delta f_i> = \alpha_i \left[ \frac{<\Delta L_i>}{\hat{L}_i} + \beta_i \frac{<\Delta S_i^2>}{\hat{S}_i^2} \right], \tag{12}$$

where $\alpha_i$ and $\beta_i$ mean the parameters $\alpha$ and $\beta$ at the i-th stage. To satisfy the [condition I] for all i, it is sufficient to determine $\beta_i$ so that the order of the first term and the second term within the square bracket in the right hand side of Eq.(12) is equal. Thus, the $\beta_1$ is written as

$$\beta_1 = \frac{<\Delta L_1>}{<\Delta S_1^2>} \frac{\hat{S}_1^2}{\hat{L}_1}. \tag{13}$$

Using Eq.(11), $\beta_i$ (=2,•••,m) is written as

$$\beta_i = \left( \prod_{k=1}^{i-1} \sigma_k^{-3/2} \right) \frac{<\Delta L_i> <\Delta S_1^2>}{<\Delta L_1> <\Delta S_i^2>} \beta_1. \tag{14}$$

On the [condition II]

Assuming that the [condition I] is satisfied, we may only consider the first term $<\Delta L>$ instead of $<\Delta f>$ because the order of average change of two terms is equal. The following relation should be satisfied to regulate $\alpha$ so that the order of $<\Delta f>$'s is common among all stages.

$$\alpha_i = \left( \prod_{k=1}^{i-1} \sigma_k^{-1/2} \right) \frac{<\Delta L_1>}{<\Delta L_i>} \alpha_1, \quad (i=2,\cdots,m). \tag{15}$$

Using Eq.(13) the value of $\beta_1$ can be determined, and using Eqs.(14) and (15) the values of remaining parameters $\alpha_i$ and $\beta_i$ ($i=2,\cdots,m$) can be automatically determined if only one determines the value of the $\alpha_1$. In practical computation these values can be determined using a result of a number of random transitions before an annealing process starts.

## 5. EVALUATION OF COMPUTATIONAL TIME

As behaviors of cost functions during the annealing process can be regarded as almost similar because of the automatic regulation method for parameters, the number of the temperatures for executing simulation can be also supposed to be roughly identical for all sub-problems. Therefore, the number of iteration at one temperature can be considered as a standard measure in order to evaluate the computational time. We compare the computational time of the HSA method with that of the original SA method based on the number of iteration at one temperature in the following.

(i) The case of the original SA method

Using the number of elements n and states of each element M, the number of iteration $\tau$ is

$$\tau = nM = n\prod_{i=1}^{m} \sigma_i \ . \tag{16}$$

(ii) The case of the HSA method

The numbers of elements, states, and sub-problems at each stage are following.

| stage | element | state | sub-problem |
|---|---|---|---|
| 1 | $n$ | $\sigma_1$ | 1 |
| 2 | $n\sigma_1^{-1}$ | $\sigma_2$ | $\sigma_1$ |
| $\vdots$ | $\vdots$ | $\vdots$ | $\vdots$ |
| m | $n\prod_{j=1}^{m-1}\sigma_j^{-1}$ | $\sigma_m$ | $\sigma_{m-1}$ |

Summing up the numbers of iteration of all stages per one temperature,

$$\tau_h = n\ [\ \sigma_1 + \sigma_2 + n\sigma_1^{-1}\sigma_2^{-1}\sigma_3\sigma_2 + \sum_{i=1}^{m} (\ \sigma_i \prod_{j=1}^{i-2} \sigma_j^{-1})\ ] \tag{17}$$

is obtained in case with sequential computation.

A parallel computation is possible from the second stage. In addition each sub-problem can be independently calculated except data communication before beginning of annealing in order to accept information from the previous stage. Then, the number of iteration in parallel computation is supposed to be still a standard measure of the computational time and is shown as

$$\tau_p = n \sum_{i=1}^{m} (\sigma_i \prod_{j=1}^{i-1} \sigma_j^{-1}). \quad (18)$$

Defining relative computational time of $\tau_h$ and $\tau_p$ against $\tau$ as $\bar{\tau}_h = \tau_h/\tau$, $\bar{\tau}_p = \tau_p/\tau$, respectively, we get

$$\bar{\tau}_h = (\prod_{i=1}^{m} \sigma_i^{-1}) \sum_{i=1}^{m} (\sigma_i \prod_{j=1}^{i-2} \sigma_j^{-1}) \quad (19)$$

(sequential computation)

and

$$\bar{\tau}_p = (\prod_{i=1}^{m} \sigma_i^{-1}) \sum_{i=1}^{m} (\sigma_i \prod_{j=1}^{i-1} \sigma_j^{-1}). \quad (20)$$

(parallel computation)

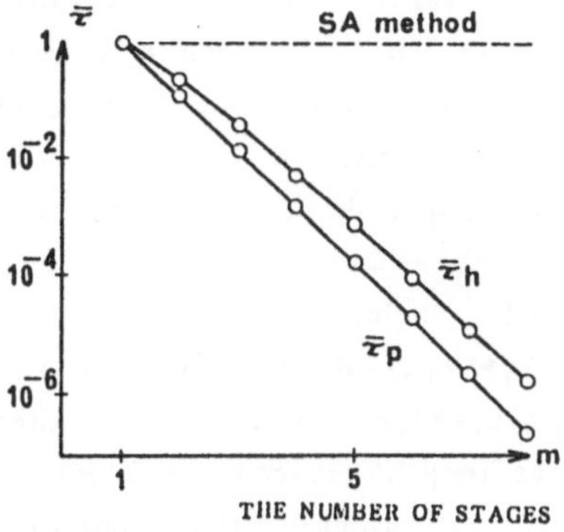


Figure 1. Comparison of computational time.

In this case relations between these ratios and the number of stages m is depicted in Fig.1 in the case with $\sigma_1=\sigma_2=9$. For example, the computational time is reduced to 22.2% (sequential) and 12.3% (parallel) in the case that $\sigma=9$, $m=2$, and $n=81$ which is one of placement problems stated in the next section.

## 6. APPLICATION TO BLOCK PLACEMENT PROBLEM IN VLSI DESIGN

We consider a simple block placement problem in VLSI layout design as an application of the partition problem: n blocks that each has a definite function are placed in a given square area so that total wire length among blocks is minimized. The HSA method for this problem is slightly modified:

(i) The first stage: a given square area in which blocks are to be placed is divided into nine sub-square-areas ($\sigma_i=\sigma=9$) having the same size in square lattice style (Fig.2). The number nine is important for this problem. The reason why nine is used as the number of division will be explained later. By the SA method, n blocks are distributed among nine sub-areas to minimize the total wire length both among sub-areas and between sub-areas and external areas which exists in the out of the original given area.

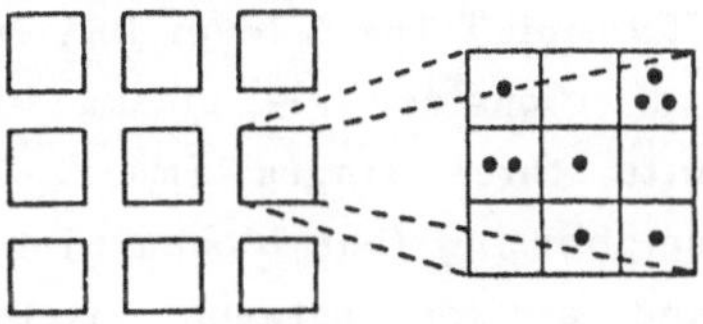

Figure 2. Hierarchical partitioning of areas( ●: block).

(ii) The i-th stage ($i=1,\cdots,m$): each given square area from the (i-1)-st stage is again divided into nine sub-square-areas so that each is the same size. An area in this stage corresponds to a sub-area in the just before stage. The blocks are distributed among sub-areas in the same manner as in the first stage.

Dividing an area into nine sub-areas is because of a rough consideration of wire length between a sub-area and the external areas. For example, it is supposed that a block having interactions with a block placed in the righthand area and with a block placed in the upper hand area should be placed in right-upper hand sub-area in a given area to shorten wire length. The element i and its state $x_i$ correspond to "block i" and "which sub-area the block i belongs to" in this placement problem, respectively. The weight $s_i$ $(i=1,\cdots,n)$ of the element i and the weighted edge $a_{ij}$ between element i and j $(i,j=1,\cdots,n)$ correspond to the square measure of the block i and the total wire length between the block i and j, respectively.

An example whose optimum placement is already known is used in this paper so as to make it possible to evaluate final results.

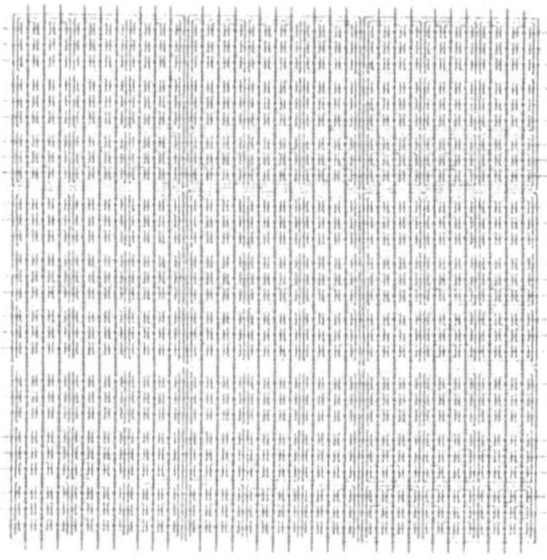

Figure 3. Optimum placement.

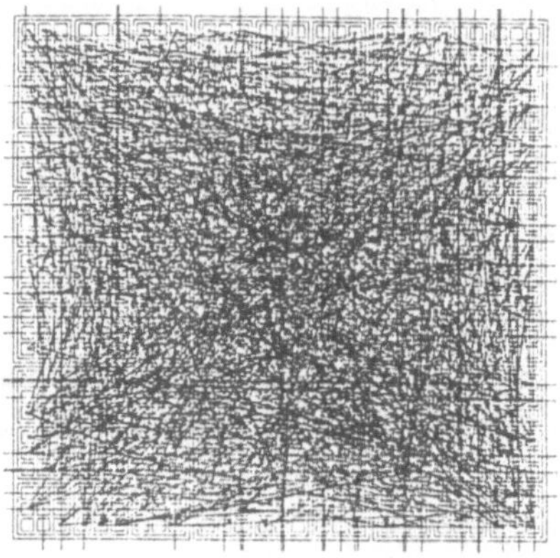

Figure 4. Random initial placement.

**[Example]** The 3-level HSA method

Consider 729 blocks which have the same square measure to be placed with three stages (m=3). Each block has only one wiring to each of neighboring four blocks in the optimum placement as shown in Fig.3. A circle and a line between circles in Fig.3 represent a block and a wiring, respectively. Random placements are used as initial placements as shown in Fig.4. The parameter $\alpha_1$ at the first stage is fixed to one.

Although only four cases succeeded in the 18 time experiments, the remainders gave near optimum solutions, e.g., Fig.5. These results are by far better than those obtained by using the original SA method, and the computational time was reduced to 56.7%. The best placement of the 18 time experiments using the original one is shown in Fig.6. These worse results are due to the fact that the calculations of the prescribed number of iteration per one temperature could not execute because of restrictive computational time of our computer. Practically, the number of iteration was only about 1/26 of the prescribed one for the original SA method. Hence, if the prescribed number of iteration is used in the original method, $\overline{\tau}_h$=0.022 is obtained, which is also better than the theoretical value $\overline{\tau}_h$=0.037. Even though the original SA method produces the results of the same quality as HSA method, the 3-level HSA method is able to have the advantage about computational time: it shows about 1/50 reduction of computational time compared with the original one.

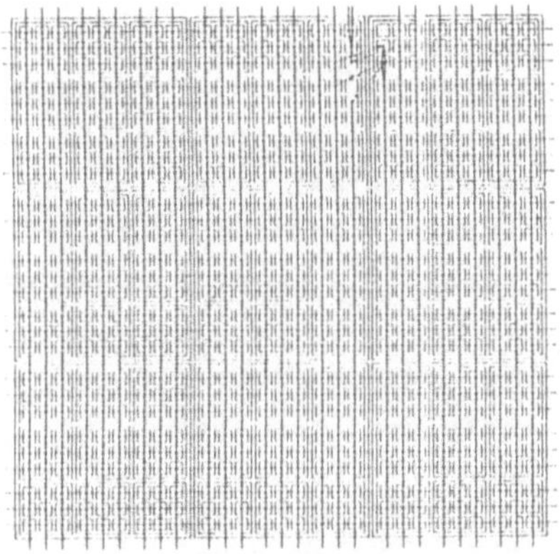

Figure 5. Near optimum placement by the HSA method.

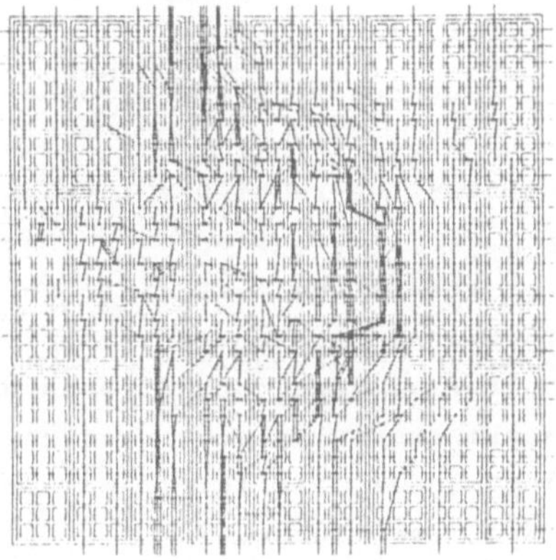

Figure 6. Best placement by the original SA method.

## 7. SUMMARY

The hierarchical simulated annealing method, i.e., the HSA method is proposed for the partition problem in order both to realize a possibility of parallelization and to reduce the computational time. New regulation method is also presented to choose suitable values of parameters both within each stage and among stages.

From our experiments of the particular block placement problem in VLSI layout design, it has been concluded that the HSA method derives much more reduction of computational time and better final result of higher quality than those of the original SA method.

## REFERENCES


1 S.Kirkpatrick, C.D.Gelatt, and M.P.Vecchi: Optimization by Simulated Annealing, Science, **220**, pp.671-680 (1983).

2 C.Sechen and A.Sangiovanni-Vincentelli: TimberWolf 3.2; A New Standard Cell Placement, Proc.IEEE 23rd DA Conf., pp.432-439(1986).

3 S.Geman and D.Geman: Stochastic Relaxation, Gibbs Distribution, and the Bayesian Restoration of Images, IEEE Trans.Pattern Anal. and Mach.Intell, **6**, pp.721-741(1984).

4 N.Metropolis, A.W.Rosenbluth, M.N.Rosenbluth, A.H.Teller, and E.Teller: Equation of State Calculations by Fast Computing Machines, J.Chem.Phys., **21**, 6, pp.1087-1092(1953).

5 T.Fukao and Z.Yue: Stochastic Distributed Optimization Algorithm (Japanese), Trans.Inst.Electron.,Inf.and Commun.Eng.,**69**, 12, pp.1492- 1501 (1986).

6 P.Banerjee and M.Jones: A Parallel Simulated Annealing Algorithm for Standard Cell Placement on a Hypercube Computer, Proc. ICCAD'86, pp.34-37(1986).

# A MIXED INTEGER 0-1 PROGRAMMING HEURISTIC FOR RESOURCE ALLOCATION IN A DECENTRALIZED SYSTEM

Probir Roy
Associate Professor, University of Missouri-Kansas City
Kansas City, Missouri 64110-2499

## INTRODUCTION

The resource allocation process in a decentralized, hierarchial system may be characterized as a system composed of a vertical arrangement of subsystems where higher level subsystems have some priority of action or right to intervention over lower level subsystems. In addition, higher level subsystems must depend upon lower level subsystems. The process of allocation, per se, is iterative in nature.

A number of mathematical programming models have been developed to represent the multilevel resource allocation process depicted above. Of These models, the multilevel, multicriterion genre' of models appear to be the most viable.

Multilevel, multicriterion resource allocation models are generally based on a method of progressive articulation that rely on an unfolding process that progressively defines the decision makers preferences and explores the criterion space. The progressive definition takes place either through a decision maker/analyst or a decision maker/machine dialogue at each iteration. At each iteration, the decision maker articulates trade-off or preference information, based on the current solution, in order to determine a new solution. Therefore, the ultimate solution depends on the accuracy of local preference choices. Herein lies a major weakness of these methods. Often times, the decision maker is incapable of providing accurate trade-off information. This incapability, among other reasons, could be due to the fact that there are too many alternatives to consider, or that the difference between alternatives is perceptually insignificant. Unfortunately, however, not nuch is known with respect to the impact of local choice inaccuracies on the ultimate solution.

Another problem faced by existing multilevel, multicriterion models is that for convergence, decision makers are theoretically required to iterate an infinite number of times. Behaviorally, such a requirement seems untenable. Finally, existing multilevel, multicriterion models do not permit integer (discrete) choices. Since a large number of real-world allocation problems involves binary choices, like accept-reject, the lack of such an integer programming capability makes current models rather restrictive and unrealistic in approach.

In light of these shortfalls in existing multilevel, multicriterion models this paper presents a mixed 0-1 programming heursitic based on simulation.

The heuristic, presented in this paper, is also an unfolding one. At each iteration in the resource allocation process, a decision maker interfaces with a computer. The computer uses the decision maker's rank ordering of the different criteria to simulate a model solution. Similar to decision support systems (DSS), the decision maker can also do sensitivity analysis if needed. Using these model solutions, the computer develops a cone of "opportunity" costs and then picks one of these for further processing. On the basis of the subordinates' plans and opportunity costs the superordinate makes reallocations. This process continues till convergence. The strength of the proposed heuristic is that once criteria are rank ordered, the computer can converge to a solution without the necessity for further decision maker intervention until the ultimate solution is obtained. The decision makers can then get together to realign their priorities if necessary. Thus, this heuristic avoids both the necessity for difficult to obtain trade-off information and the need for decision makers to iterate a large number of times. In a sense, this heuristic could be the first step towards a full fledged decentralized, multilevel resource allocation DSS.

A numerical example is provided in the paper to illustrate the heuristic.

## THE RESOURCE ALLOCATION MODEL:

Resource allocation processes are generally characterized as follows: Initially, there exists a set of inputs to the superordinate (highest) level. These inputs are derived from both external and internal stimuli and data bases. The external inputs usually represent requirements or goals which are either explicity or implicity imposed on the system by external forces ; for example, economic conditions may call for production cutbacks, competition may dictate certain pricing strategies or norms, federal or local governments might require that certain environmental pollution standards be met, etc. The internal inputs typically include information on the availability and utilization of resources such as money, labor, plant and facilities, raw materials, and proforma and actual operating and capital budgets. Internal inputs also include the superordinate's receiving and possessing information about the subordinates' requirements of shared resources, anticipated performances, likely targets, etc. Given these exogenous inputs, the information from the units below, and the superordinate's own expectations, the superordinate arrives at a utility function and a set of constraints. On the basis of this utility function and subject to the set of constraints that determine the solution space, the superordinate (or center) tentatively selects a program of resource allocations for the subordinates in the second level. This program, generally consisting of a set of resource budgets and profit plans, and additional coordination/information (in the form of binding constraints on subordinate behavior; or incentives/penalties for the attainment or non-attainment of priority targets; or information clarifying the superordinate's objectives) is passed to subordinates A subordinate at level two goes through, more or less, the same information gathering, problem solving and information communication process as the superordinate, the only difference being that the directives received by the subordinate from the superordinate are analogous to the superordinate's exogenous inputs.

The process is an iterative one, a tattonement. It begins with the highest decision unit selecting a program of resource allocations and then transmitting this program along with coordination information to each subordinate unit at the second level. Each subordinate unit at the second level then solves its allocation problem and sends appropriate information about its problem solution along with coordination information to subordinates at level three. This continues unitl the bottom levels are reached at which time the information flow reverses itself and starts flowing up through the hierarchy until it reaches the superordinate at level one. This upward flow of information is called counter planning.

During the counter planning process, the information at each decision unit is integrated and aggregated so that the superordinate, at the very top, has summary information about the ramifications of the tentative budgets or programs assigned at the beginning of the given cycle. The superordinate uses this information to arrive at new tentative budgets and the entire process repeats itself until final allocations are determined and the final plan implemented.

The allocation process, just delineated, may be mathematically formulated as a multicriterion optimization problem as follows:

$$\begin{aligned}
\max \quad & C_1X_1 + C_2X_2 + \dots + C_pX_p \\
\text{s.t.} \quad & A_1X_1 + A_2X_2 + \dots + A_pX_p = b_0 \\
& B_1X_1 \leq b_1 \\
& B_2X_2 \leq b_2 \\
& \dots \quad B_pX_p \leq b_p
\end{aligned} \qquad (P)$$

$X_i \geq 0, \ \forall i$ and some $x_i \in X_i$ integer

where

$C_i$ is an $n_i$ component vector,

$X_i$ is an $n_i$ component vector,

$A_i$ is an m x n dimensional technology matrix for the shared resources,

$B_i$ is an $m_i$ x $n_i$ dimentional technology matrix pertaining to the solely internal resources,

$b_i$ is an $m_i$ component vector of resources,

p is the number of decentralized units/subsystems.

Roy and Lakshmanan have presented an algorithm that can be used in conjunction with this model to allocate energy resources. Unfortunately, their approach faces a major problem in implementation. Their algorithm calls for an explicit enumeration of all the efficient solutions of the multicriterion problems involved at each phase, in each decentralized unit, and at each iteration. For a reasonably large real-world resource allocation situation this task may be virtually impossible. In addition to this computational problem, there is also a behavioral problem. The Roy - Lakshmanan algorithm requires decision makers to choose a "most" preferred solution from the set of efficient solutions. If the set of efficient solutions is reasonably large, there may be logical inconsistencies in the choice of the "most preferred" solution. To offset these problems and in order to make the Roy-Lakshmanan algorithm more implementable, it has been modified by the inclusion of a simulation heuristic recently developed by Roy and Basu. The Roy-Basu approach to vector max or multicriterion problems does not ask the decision maker to either choose from alternatives or to provide any trade-off information. Also the Roy-Basu heuristic is based on a finite number of solutions at each phase and hence is computationally tractable. In the next sectio.. the modified algorithm is presented.

RESOURCE ALLOCATION HEURISTIC:

The multicriterion optimization problem (P) may be broken into p subproblems, where each subproblem represents a decision making unit (for example a region or a division, or subsidiary). Assuming a linear additive utility function, the ith subproblem (for a given level of shared resources $y_i$) may be represented as follows:

$$V_i = \max_{x_i \varepsilon X_i} U(X_i) \qquad (SP_i)$$

Where,

$$U(X_i) = \sum_{j=1}^{n_i} \lambda_{ij} C_{ij} x_i$$

$$\text{s.t.} \quad A_i x_i <= y_i$$
$$B_i x_i <= b_i$$

and some elements of $x_i$ integer.

The $\lambda_{ijs}$ are known as trade-off weights that reflect the relative importance of the jth criterion to the ith decision unit. It is assumed that the weights are non-negative and normalized.

$$\Sigma\lambda_{ij} = 1, \qquad \lambda_{ij} >= 0$$

The center's problem (CP) may be represented as:

$$\max_{y_i \varepsilon F_i} \quad v \quad \sum_{i=1}^{p} V_i \,(y_i)$$

$$\text{s. t.} \quad \sum_{i=1}^{p} y_i \leq b_0 \qquad \text{(CP)}$$

where $F_j = (y_i / \quad x_i \ \varepsilon \ X_i : \ A_i x_i <= y_i)$

The heuristic proposed in this paper, solves the family of ($SP_i$) and (CP) problems iteratively.

Step 1. Each subordinate solves a relaxed problem where the constraints $A_i x_i <= y_i$ are removed. Such a solution represents an upper bound on the amount of shared resources required by the ith subunit. Let $y_i$ represent this upper bound.

Step 2. The center allocates shared resources, $y_i$, to each of the p regions.

Step 3. Given this allocation of $y_i$, each region solves its subproblem ($SP_i$) as follows:

1) The ith decision maker ranks all the criteria.
2) Using a random number generator, a weight is generated for the criterion that is the lowest ranked.
3) Using the weights generated in (2) as the base, generate all the other criteria weights in an increasing order. Empirically, we have found that random numbers generated by using a uniform distribution work very well.
4) Using the weights generated in (2) and (3) as surrogates for the $\lambda_i$s, solve $SP_i$.
5) Do (2) - (4) for m iterations
6) Pick the modal value as the "optimal" activity vector. Pick the $\lambda_i$s that have the smallest composite (objective) value for the activity vector chosen. Pass this activity vector, smallest value $V_i \,(y_i^k)$ and corresponding dual multipliers $\pi_{is}^k$ is to the center.

Step 4. The center solves a current tangential approximation

$$\max_{y,\sigma} \sum_{i=1}^{p} \sigma_i \qquad \text{(CP)}$$

s.t.

$$\sigma_i <= V_i(y_i) + \pi_i^k s \ \ (y_i - y_i^k)$$

$$\sum_{i=1}^{p} y_i = b_0$$

$$y_i \in F_i$$

For all allocations.

Let $y_i^{k+1}$, $i = 1, \ldots p$ be an optimal solution.

Step 5. Redo step 3, using $y_i^{k+1}$ as $y_i$, if $y_i^{k+1}$ is different from $y_i$. In the event that for all the subunits, $y_i^{k+1} = y_i^{k}$, than check whether $y_i^{k+1} <= \bar{y}_i$ for all i. If yes, then stop and these $y_i^{k+1}$ are to be used as the optimal resource vectors. If not, then use step 6 to determine whether $y_i^{k+1}$ is satisfactory. If not, then add p constraints $y_i <= \bar{y}_i + \varepsilon$, where $\varepsilon$ is a small perturbation, to the center's current problem and go to step 4.

Step 6. Determine if $y_i$ is a satisfactory solution. $y_i$ is a satisfactory solution if

$$\sum_{i=1}^{p} \sigma_i - \max_{1<=k < = k-1} \{ \Sigma V_i (y_i) \} < = \delta$$

where $\delta$ is an arbitrarily small number.

In the next section, the proposed heuristic is illustrated by a numerical example.

An Illustration Of The Resource Allocation Heuristic

$$\max \begin{bmatrix} 1 & 2 & -1 \\ 1 & 2 & -1 \\ 1 & -1 & 3 \end{bmatrix} \begin{bmatrix} x_{11} \\ x_{12} \\ x_{13} \end{bmatrix}$$

s.t.

$$\begin{bmatrix} 2 & 1 & 1 \\ 1 & 2 & 2 \end{bmatrix} \begin{bmatrix} x_{11} \\ x_{12} \\ x_{13} \end{bmatrix} <= \begin{bmatrix} y_{11} \\ y_{12} \end{bmatrix} \qquad (SP_1)$$

$$\begin{bmatrix} 1 & -2 & 1 \\ 1 & 3 & -1 \\ 1 & 1 & 1 \end{bmatrix} \begin{bmatrix} x_{11} \\ x_{12} \\ x_{13} \end{bmatrix} <= \begin{bmatrix} 10 \\ 20 \\ 20 \end{bmatrix}$$

$x_{11}$, $x_{12}$ >= 0, $x_{13}$ is binary

$$\max \begin{bmatrix} 1 & 4 & 1 \\ 2 & 1 & -1 \\ 1 & -1 & 3 \end{bmatrix} \begin{bmatrix} x_{21} \\ x_{22} \\ x_{23} \end{bmatrix}$$

s.t.

$$\begin{bmatrix} 1 & 1 & 2 \\ 2 & 1 & 2 \end{bmatrix} \begin{bmatrix} x_{21} \\ x_{22} \\ x_{23} \end{bmatrix} <= \begin{bmatrix} y_{21} \\ y_{22} \end{bmatrix} \qquad (SP_2)$$

$$\begin{bmatrix} 1 & -2 & 1 \\ 1 & 1 & -2 \\ 2 & 0 & 1 \end{bmatrix} \begin{bmatrix} x_{21} \\ x_{22} \\ x_{23} \end{bmatrix} <= \begin{bmatrix} 10 \\ 22 \\ 31 \end{bmatrix}$$

$x_{21}$, $x_{22}$ >=0, $x_{23}$ binary.

<u>Center</u> Max $\sigma_1 + \sigma_2$

$$\text{s.t.} \quad \sigma_i <= V_i(y_i) + \pi^k_{is} (y_i - y_i)^k \qquad \text{(CP)}$$

$$\begin{bmatrix} y_{11} \\ y_{12} \end{bmatrix} + \begin{bmatrix} y_{21} \\ y_{22} \end{bmatrix} = \begin{bmatrix} 55 \\ 45 \end{bmatrix}$$

For the purpose of comparison, it is assumed that the true weights are $\lambda_1$ = (0.5,0.3,0.2) and $\lambda_2$ = (0.6,0.3,0.1).

The true weights are however assumed to be unknown, except for ranking the criteria (when running the heuristic) and also for testing the final results of the heuristic.

<u>Solution</u>:

Step 1. Each subunit solves a relaxed problem. The implicit assumption being that that the $y_i$'s are infinite. Solving the relaxed problems we get

$x_1$ = (13.8, 2.4, 1)

$\bar{y}_1$ = (31,20.6)

also,

$x_2$ = (0,24,1)

$\bar{y}_2$ = (26,26)

Step 2. Each subunit problem was solved for the following three allocations:

$y_1$ = (30,35), $y_1$ = (60,70), $y_1$ = (0,0)

$y_2$ = (30,35), $y_2$ = (0,0), $y_2$ = (60,70)

The corresponding modal solutions were:

Subunit 1: $x_1$ = (13.2,2.6,1), $x_1$ = (13.8,2.4,1)

and $x_1$ = (0.0,0)

Subunit 2: $x_2$ = (0,2.4,1) $x_2$ = (0.0;0)

and $x_2$ = (0,24,1)

Using the smallest v values corresponding with the modal solutions and the duals that correspond to these v values - the supremal problem was set up as:

Max $\sigma_1 + \sigma_2$

s.t.

$$\begin{aligned}
y_{11} + y_{12} &= 55 \\
y_{12} + y_{22} &= 45 \\
\sigma_1 - .32_{11} &<= 6.98 \\
\sigma_1 &<= 16.9 \\
\sigma_1 - 2067y_{11} - .5667y_{12} &<= 0 \\
\sigma_2 &<= 50.05 \\
\sigma_2 - 2.55y_{21} &<= 0
\end{aligned}$$

Step 3. Solving the supremal problem we got

$y_1^1 = (30.99, 18.52)$ $y_2^1 = (24.01, 26.48)$

Step 4. Using $y_1^1$ and $y_2^1$ as the shared resource allocations, the subunit problems are re-solved. The modal solutions were

$x_1 = (14,2,0)$; $x_2 = (0,22.004,1)$

These solutions led to the following constraints that were added to the supremal problem:

$\sigma_1 \quad <= \quad 16.399$

$\sigma_2 - 1.64y_{21} \quad <= \quad -2.3904$

Going back to step 3 and re-solving the supremal problem, we got

$y_1^2 = (23.024, 16.92)$ $y_2^2 = (31.97, 28.08)$

The rest of the iterations were as follows:

Corresponding to $y_1^2$ and $y_2^2$, the modal solutions were:

$x_1 = (9.89, 3.39, 0)$; $x_2 = (0,24,1)$

These solutions led to constraints

$\sigma_1 - .32y_{11} \quad <= \quad 7.8$

$\sigma_2 \quad <= \quad 40.209$

Adding these to the supremal problem and solving:

$y_1^3 = (29.025, 18.119)$; $y_2^3 = (25.975, 26.880)$

Re-solving subunits with $y_1^3$ and $y_2^3$, the modal solutions were:

$x_1 = (13.415, 2.195, 0)$ ; $x_2 = (0,23.975,1)$,

New constraints were:

$\sigma_1 - .358y_{11} \quad <= \quad 5.28$

$\sigma_2 - 1.64y_{12} \quad < \quad -2.38$

Solving; $y_1^4 = (29.025, 17.551)$; $y_2^4 = (25.975, 27.449)$

Corresponding modal solutions were:

$x_1 = (13.4996, 2.0256, 0)$; $x_2 = (0,23.975,1)$

Only one new constraint :

$\sigma_1 - .27y_{11} \quad <= \quad 0$

Solving ; $y_1^5 = (29.025, 17.805)$; $y_2^5 = (25.975, 27.195)$

Corresponding modal solutions were:

$x_1 = (13.415, 2.1949, 0)$; $x_2 = (0,23.9749,1)$

No new constraints generated. Therefore check if $y_i^5 <= \bar{y}_i$. No! $y_{22}^5 > \bar{y}_{22}$. Therefore add the constraints $y_i <= \bar{y}_i + \varepsilon$ where $\varepsilon$ was arbitrarily chosen as 0.005.

Solving the newly created supremal problem, we got

$y_1^6 = (29.025, 20.65)$; $y_2^6 = (25.975, 24.35)$.

The corresponding modal solutions :

$x_1 = (12.615, 2.795, 1)$; $x_2 = (0,22.345,1)$

and the new constraints;

$\sigma_1 - .358\, y_{11} <= 6.125$

$\sigma_2 - 1.64\, y_{22} <= -2.38$

Solving : $y_1^7 = (29.025,\ 19.0311)$ ; $y_2^7 = (25.975,\ 25.9689)$

Corresponding solutions :

$x_1 = (13.0063,\ 2.0124,\ 1)$; $x_2 = (0, 23.9689, 1)$

One new constraint:

$\sigma_1 - .27\, y_{11} - .44\, y_{12} <= .21$

Solving ; $y_1^8 = (29.025,\ 18.995)$; $y_2^8 = (25.975, 26.005)$

Modal Solutions:

$x_1 = (12.998,\ 1.999, 1)$; $x_2 = (0, 23.975, 1)$

New Constraints :

$\sigma_1 - .7775\, y_{12} <= 1.20749$

$\sigma_2 - 1.64\, y_{21} <= -2.38$

Solving the new supremal problem ;

$y_1^9 = (29.025,\ 18.995)$ ; $y_2^9 = (25.975,\ 26.005)$

$y_i^9 = y_i^8$ and $y_i^9 <= \bar{y}_i$

Therefore stop. According to the heuristic the resource allocations should be $y_1^9$ and $y_2^9$ and the corresponding activity vectors should be $x_1 = (12.998,\ 1.999,\ 1)$ and $x_2 = (0, 23.975,\ 1)$. In order to compare the heuristic derived solution with that we would have obtained if we had known the true weights, the problem was solved as one single criterion problem. The single criterion was developed using the assumed true weights. The true solution would have been:

$x_1 = (13, 2, 1)$ ; $x_2 = (0, 24, 1)$

We see that the two solutions are similar. The slight discrepancy is due to the value of $\varepsilon$. If we had taken a smaller value, the two solutions would have been identical.

IMPLICATIONS OF THE PROPOSED DEURISTIC

It has long been recognized that resource allocation problems in decentralized systems are essentially multilevel and multicriteria in nature. Unfortunately, however, the algorithms that have been developed have been virtually impossible to implement because they have either been computationally intractable, or they have been too demanding with respect to information requirements, or they do not permit integer variables. The approach suggested in this paper would certainly enhance the implementability of such multilevel, multicriteria resource allocation models. In essence, the approach formulated in this paper could be the key to developing a Decision Support System for resource allocation in decentralized situations.

REFERENCES

1. Baum S and Carlson, R.C., "Multigoal Optimization in Managerial Science", OMEGA, vol. 2, no. 5, PP. 607-623.

2. Beckman, m. and Krelle, W. (eds), Essays and Surveys on Multiple Criteria Decision Making, Springer, Verlog, N.Y., 1983.

3. Despontin, M., Nijkamp, P., and Spronk, J. (eds), Macro-Economic Planning with Conflicting Goals, Springer - Verlog, N.Y., 1984

4. Hwang, C.L., and Masud, A.S. Md., Multiple Objective Decision Making: Methods and Applications, Springer - Verlog, N.Y. 1979

5. Nijkamp, P., and Spronk J. (eds.), Multiple Criteria Analysis, Gower Publishing, Aldershot (U.K.) 1981

6. Roy, P., and Basu, P., "A Simulation Approach to Multicriteria Decision Making with Incomplete Information", Proceedings TIMS/ORSA national Conference, Boston, April 29 - May 1, 1985.

7. Roy, P., and Lakshmanan, T.R., "A Multilevel Multicriterion Energy Resource Allocation Model", Working paper #NSF 80-2, Center for Energy and Environmental Planning, Boston University

# Symbolic Treatment of Geometric Degeneracies

*Chee-Keng Yap*[1]

Courant Institute of Mathematical Sciences
New York University
251 Mercer Street
New York, NY 10012.

Many descriptions of algorithms in computational geometry exclude degeneracies *by fiat.* Practitioners are left to their own devices for dealing with degeneracies when implementing such algorithms. Since degeneracies tend to be numerous and hard to enumerate exhaustively, this is often a reason for not implementing theoretical algorithms. This paper proposes a powerful symbolic scheme for treating degeneracies. Our method is simple to use, and is applicable for a wide variety of problems in computational geometry (in particular, whenever random perturbations are applicable). Our method is deterministic but is as efficient as probabilistic schemes. Illustrations, limitations and wider issues are discussed.

## 1 Introduction

The theoretical study of algorithm in computational geometry is an active area. Besides the inherent beauty arising from the interplay of geometrical and algorithmic properties, this area holds the promise of impact on important application areas such as robotics, graphics and VLSI. Unfortunately, the reduction of theoretical algorithms to practice has been relatively slow and this has often been commented on. In our view, two fundamental issues must be addressed in order to speed up this 'technology transfer'.

- *Fixed precision arithmetic.* Theoretical algorithms assume an exact (arbitrary precision) model of numerical computation. Its implications in the world of fixed precision computations are not fully understood.

- *Data degeneracy.* Theoretical algorithms are often described for the 'non-degenerate' cases of the inputs. Sometimes, even careful attempts at capturing all degenerate cases leave hard-to-detect gaps. This is a deep source of frustration for practitioners who often find mysterious failures in their algorithms.

We firmly believe that theoreticians are justified in making these two assumptions *as long as their goal is the understanding of the global, combinatorial structure of problems.* However, it would not justify a continuing neglect of both these issues. Both raise extremely interesting questions in their own right. Partly because of such neglect, there seems to be a credibility gap between theoreticians and implementors: the latter often view theoretical algorithms with suspicion.

Theoreticians have begun to address these questions. For instance, the recent paper [12], addresses the fixed precision issue in the context of computer graphics. It is important to realize that although fixed precision arithmetic and date degeneracy are related, they are distinct issues. In this paper we deal with the latter. The symmetry breaking rules in simplex algorithms (e.g. [3]) are the precursors of symbolic treatment of data degeneracies. In computational geometry, Edelsbrunner and his students are among the first to publish solutions to the problem of degeneracies [7], [9], [8], [10] (chapter 9.4). Our independently discovered scheme will turn out to be a generalization and simplification of their method.

Degeneracy in computational geometry is a general phenomenon. So in what sense can we justify its neglect in theoretical algorithms? One justification is that explicit handling of degeneracies obscures the centrality of the non-degenerate cases: degenerate cases normally involve an overwhelming number of cases that are disproportionate to their likelihood of occurence. But an implementor of these algorithms

[1]Supported in part by NSF grants #DCR-84-01898 and #DCR-84-01633.

must handle the degeneracies when they do arise. Most authors are correct in suggesting that a random perturbation of data will remove degeneracies with high probability. Such a suggestion is justified if the problem is *stable*, as the case generally turns out to be. However, an explicit demonstration that the problem is stable is seldom done.

In contrast to the suggestion of random perturbation, it is occasionally suggested that the algorithmic description ought to carefully work out all the degenerate cases. We take the opposite position that this is, in general, inadvisable: it is neither illuminating for a global understanding of the algorithm nor is it in the interest of the implementor who would then have to implement the numerous degenerate cases. Besides, the increase in the number of cases may lead to other programming errors or incomplete theoretical analysis. The approach advocated in this paper is to work out some general scheme to achieve suitable data perturbation. In this way, theoreticians can continue to focus on the interesting non-degenerate cases while implementors can enjoy the benefits of algorithms that have few cases and yet can handle all conceivable inputs. (Paraphrased: implementors can now join the theoretical paradise in which degeneracies are abolished.) Thus the main contribution of this paper is a practical and general scheme achieving these goals.

**Overview.** The rest of the paper is organized as follows. Section 2 explores the meaning of degeneracy and related concepts. Section 3 reviews probabilistic perturbation schemes and points out unsatisfactory properties. To illustrate the symbolic approach, we outline Edelsbrunner's method in section 4. The new scheme appears in section 5. In section 6, we explore the concept of ordered rings implied by our scheme. Section 7 investigates the computational and complexity questions arising in implementing our scheme. We conclude with some directions for future work in section 8.

## 2 What is geometric data degeneracy?

How can a general solution as proposed in the introduction be achieved? We first need an understanding of what we mean by degeneracies. In this paper, we assume that the problem input is a sequence of real numbers called the *input parameters* $\mathbf{a} = (a_1, \ldots, a_n)$. An input is *degenerate* if some polynomial $p_j(x_1, \ldots, x_{k_j})$ in a fixed set

$$D = \{p_j(x_1, \ldots, x_{k_j}) \; : \; j = 1, 2, \ldots, m \text{ and } k_j \geq 1\}, \quad (m \geq 1)$$

evaluates to zero when an *allowable substitution* for the variables $(x_1, \ldots, x_{k_j})$ is made using the $a_i$'s. Here the set $D$, called the *test polynomials*, depends only on the algorithm and not on the inputs. Although $D$ is a finite set here (the usual case in practice), our method works as well if $D$ were infinite. The 'allowable substitution' of variables in each test polynomial by input parameters is dictated by the problem. We refrain from making this more formal but this will be easy to do (for each particular case) once the following examples are understood.

**Examples of degeneracy.** When input $\mathbf{a}$ represents a set of points in the Euclidean plane, degeneracies include (i) two coincident points, (ii) three collinear points, or (iii) four cocircular points. If the input represents a set of lines, common notions of degeneracy include (iv) a vertical line, (v) two parallel lines, (vi) two perpendicular lines, or (vii) three concurrent lines. If the input represents points and lines, degeneracy may include (viii) a point lying in a line, or (ix) a line parallel to the line through two points. This list can go on. The 'allowable substitution' above may only amount to typing the variables in $D$ and the input parameters so that substitutions must respect the type distinctions. For instance, some variables and parameters correspond to the first coordinate of points and others correspond to the slope of lines, etc. More concretely, suppose the input parameters are $a_1, b_1, a_2, b_2, \ldots, a_n, b_n$ representing $n$ points in the plane, and let the set of test polynomials consists of only one polynomial,

$$\Delta(x_1, y_1, x_2, y_2, x_3, y_3)$$

which is just the 3 by 3 determinant that tests if the points $(x_1, y_1), (x_2, y_2)$ and $(x_3, y_3)$ are collinear (cf. section 4). Allowable substitution here means that (1) the substitution of the input parameters must be in pairs (i.e. for all $i, j$, $(a_i, b_i)$ must be substituted simultaneously for $(x_j, y_j)$), and that (2) the three input points to be substituted must have distinct subscripts.

**Exact model of numerical computation.** It is important to realize that this paper assumes the *exact model* of numerical computations. All numbers are represented *exactly*: for example, to represent any real algebraic number $\alpha$ exactly, it suffices to specify a polynomial $p(x)$ with integer coefficients together with an interval $I$ containing $\alpha$ but no other distinct roots of $p(x)$. The end points of $I$ are exact, say represented by rational numbers. Our development is not restricted to any particular exact representation. Note that since the integers involved (as in the coefficients of $p(x)$, or in representing $I$) can be arbitrarily long, we sometimes call this the *arbitrary precision model* (or, somewhat misleadingly, 'infinite precision model'). This contrasts with the fixed precision world of the numerical analysts. With the advent of computer algebra, the exact models are becoming more important and indeed unavoidable for some applications.

We have stated that the issues of fixed precision and data degeneracy are distinct. Indeed, any attempt to consider data degeneracy in the fixed precision models faces some very difficult problems: for instance, there are many puzzles in just trying to define what it means for three points to be collinear in the fixed precision model (actually, the world of pixels). In any case, one should begin by understanding degeneracy in exact models.

**Degree of derivation and derived degeneracies.** In the course of executing algorithms, *derived values* $\mathbf{b} = (b_1, b_2, \ldots)$ may be generated from the input parameters $\mathbf{a}$. Let us define the 'degree of derivation' of $\mathbf{b}$ be the least integer $d \geq 1$ such that for each $b_i$ in $\mathbf{b}$ there is $(d+1)$-variate polynomial $p(x, x_1, \ldots, x_d)$ with integer coefficients, whose degree in each variable is at most $d$ such that if each $x_j$ $(j = 1, \ldots, d)$ is substituted by suitable values $\gamma_j$ from $\mathbf{a}$, then $b_i$ is a root of $p(x, \gamma_1, \ldots, \gamma_d)$. For instance, if the input represents points, then $\mathbf{b}$ may be the computed distances between pairs of points, and the degree of derivation is 2. If this degree $d$ of derivation does not grow with the input size $n$, we say that the algorithm has *bounded degree of derivation*. A problem is of *bounded degree* if it has an algorithm with bounded degree of derivation. Examples of bounded degree problems include computing the convex hulls or the Voronoi diagrams. Examples of problems with unbounded degree include shortest path problems and root isolation (or more generally, cell decomposition). Our method to be described does not allow the substitution of derived values with derivation degree more than 1 (this is really the same as allowing substitution by input parameters only), but see section 8 for discussion.

Our definition of degeneracy above can be generalized thus: we say that the input parameters have *derived degeneracies* if substitution by derived values into the test polynomials of $D$ results in a zero value.

**Alternative notions of degeneracy.** In a provocative discussion, [11], Kender and Freudenstein point out that there are several, not necessarily mutually-consistent, notions of degeneracy in the literature. Kender and Freudenstein propose to unify these disparate notions by describing degeneracies to be a 'system relative' concept. Of course, this is a very general formulation and our particular notion here is 'system relative' to the extend that the particular set of test polynomials $D$ depend on the algorithm.

Degeneracies are often depicted as rare events. In computer vision, there are notions of degeneracy that belie this view, as shown by this example from [11]. Imagine a very squat pyramid and a view of the pyramid from 'below'. This view shows only the base of the pyramid and would be considered a 'degenerate view' in certain contexts of vision research. Yet, this bottom view is hardly 'rare' by any reasonable definition (in the sense of geometric measures). Perhaps it is better to call such views 'deficient' (since they give inadequate information) rather than degenerate.

Perhaps a more pertinent illustration which suggests that the 'degeneracies as rare events' view require careful interpretation is this: when dealing with highly structured scenes or robot environments, certain 'degeneracies' such as parallel lines or collinear points are features rather than accidents of the input space. For instance, in descriptions of a robot environment in a factory, we expect parallel lines to be a common feature. The explanation lies in realizing that we normally assume that the input space (for a fixed input size of $n$) is the set of all $n$-tuples of (exactly representable) real numbers. It may happen that the input space is a proper subset of all possible $n$-tuples (usually, a submanifold); 'rarity of degeneracies' must be relative to this manifold. However, our method as it stands assumes the full input space.

**Problem stability.** The fundamental assumption in this paper is that with the 'rare event' view of degeneracies, we can perturb away the degeneracy. A crucial but often implicit requirement is that the problem at hand defines a function from the input space to the output space that is 'continuous' in this sense: *a solution to a perturbed version of the input data is a reasonable approximate solution for the original data.* Then we are indeed justified in using a solution to the perturbed input as the final output. We call such problems *stable*. Since continuity arguments are involved, stability is relative to the choice of topologies on the spaces concerned.

**Example of instability/stability.** Consider the problem of constructing the Voronoi diagram of a set of planar points (sites). We would like to show that this problem is stable. Let us assume that certain four cocircular sites cause the Voronoi diagram to have a Voronoi vertex of degree 4. Any perturbation of the input that removes the cocircularity of these four sites causes the said Voronoi vertex to split into two very close Voronoi vertices of degree 3 each. There are two combinatorially distinct ways in which this split can occur. Clearly the combinatorial structure of the perturbed Voronoi diagram is different from the original Voronoi diagram. So, in the combinatorial sense, the problem would not seem stable. Another attempt at trying to show that the problem is stable is this: use the Hausdorff metric on closed point sets. Unfortunately a small perturbation may introduce an infinite Hausdorff distance between the two Voronoi diagrams. [Consider the diagram of two points $p = (-1, 0)$ and $q = (+1, 0)$ and then perturb one of the points to $p' = (\delta - 1, 0)$ for all $\delta > 0$.]

Now suppose our goal is to use this Voronoi diagram to compute an obstacle-avoiding path for a unit disc between two specified points $P$ and $Q$, viewing these sites as obstacles. (It follows from [17], that to find such a path, it is sufficient to look for one path in which the center of the disc lies in the Voronoi diagram.) Let the set of $n$ sites be represented by $\mathbf{a} \in \mathcal{E}^d$ (where $d = 2n$ and $\mathcal{E}^d$ is the Euclidean $d$-dimensional space). It is natural to measure the 'connection width' between any two points $P$ and $Q$ in the plane by the quantity $C(P, Q; \mathbf{a}) \geq 0$ defined as the maximum value attained by the clearance of some path from $P$ to $Q$ in the presence of obstacles defined by $\mathbf{a}$.[2]. Clearly, if $C(P, Q; \mathbf{a}) \leq 1$ then the unit disc has no obstacle-avoiding path connecting $P$ and $Q$. Then it is easy to show:

**Proposition 1** *For any $\delta > 0$ there is an $\epsilon > 0$ such that for all $\mathbf{a}, \mathbf{b} \in \mathcal{E}^d$, and for all points $P$ and $Q$, if $C(P, Q; \mathbf{a}) > \delta$ and $\|\mathbf{a} - \mathbf{b}\| < \epsilon$ then $|C(P, Q; \mathbf{a}) - C(P, Q; \mathbf{b})| < \delta$. Here $\|\mathbf{a}\|$ denotes the Euclidean norm.*

Such a stability or continuity property justifies the perturbation of input $\mathbf{a}$ in this application. This illustrates the kind of justification that must logically precede any application of perturbation methods.

**Induced and inherent degeneracy.** We distinguish between *inherent degeneracy* of the input data versus an *algorithm-induced degeneracy*. For example, if the input represent points to a convex hull algorithm, it is apparent that if three consecutive collinear vertices on the convex hull ought to be an inherent degeneracy of the convex hull problem. Now if the algorithm uses some kind of vertical partitioning of the input set of points, or uses the vertical sweepline paradigm, then two co-vertical points may be regarded as a degeneracy. These degeneracies are easily removed in this case, either by modifying the algorithm or by perturbing the data deterministically. In any case, they are degeneracies induced by the algorithm. This paper is concerned with algorithm-induced degeneracy. It seems that normally, induced degeneracies subsume inherent degeneracies.

## 3 Probabilistic schemes

In this paper, we assume a computational model with the following property. There is a fixed set $D$ of polynomials, independent of the input such that

> (*) The algorithm only makes decision steps based on the sign of polynomials $p(\mathbf{x}) \in D$ evaluated at $\mathbf{x} = (x_1, \ldots, x_k) := (a_{j_1}, \ldots, a_{j_k}) = \mathbf{b}$ where each $\mathbf{b}$ is an (allowable) substitution from the input parameters $\mathbf{a} = (a_1, \ldots, a_n)$. The algorithm then makes a 3-way branch depending on the sign of $p(\mathbf{b})$, with the case $p(\mathbf{b}) = 0$ considered to be degenerate.

The goal is to devise a data-perturbation method so that the algorithm never takes a degenerate branch. It is instructive to first review probabilistic methods for data perturbation often alluded to in the literature.

Perhaps the simplest solution is to arbitrarily choose either $p(\mathbf{b}) < 0$ or $p(\mathbf{b}) > 0$ whenever $p(\mathbf{b}) = 0$ is encountered. This method suffers from the problem of global consistency: how can transitivity (i.e., $p(\mathbf{b}) > q(\mathbf{b})$ and $q(\mathbf{b}) > r(\mathbf{b})$ implies $p(\mathbf{b}) > r(\mathbf{b})$) be maintained without expensive bookkeeping?

The *initial perturbation scheme* overcomes the consistency problem by preceding the entire computation with an initial random perturbation of the input data. This perturbation ought to guarantee

(1) No new degeneracies arise as a result of the perturbation.

(2) The original degeneracies are all removed.

[2]The *clearance* of a point $x$ is its distance from the closest site; the *clearance* of a path is the minimum clearance among points along the path.

Property (1) can be satisfied by computing some *a priori* upper bound on the size of the perturbation. Part (2) seems more difficult to ensure. Granted that with very high probability no degeneracy remains, the issue remains as to what the algorithm must do when a degeneracy does arise? The simplest recovery is to restart the algorithm with another random perturbation. This, in principle, can repeat indefinitely. Thus, although the probabilistic overhead complexity of the initial perturbation scheme is small, its deterministic complexity is unbounded. This is unsatisfactory from a theoretical viewpoint.

## 4 Simulation of Simplicity

We now turn to symbolic perturbation schemes. To illustrate, we briefly review a version of a scheme (called *simulation of simplicity*, or SoS for short) described in [7], [9], [8], [10]. The setting scheme is normally carried out in the setting of computing hyperplane arrangements where the test polynomials are determinants. For illustration, say a set $H$ of lines in the plane is *simple* if (i) no three lines are concurrent, (ii) no two lines are parallel, (iii) no two pairs of lines intersect on a common vertical line, and (iv) no two pairs of lines intersect on a line parallel to a line in $H$. Each of these conditions corresponds to the vanishing of a suitable determinant over the input parameters. Let a line $h_i \in H$ be given by $y = a_i x + b_i$. Now replace each input line $h_i$ by $h_i(\epsilon)$ with equation $y = a_i(\epsilon)x + b_i(\epsilon)$ where

$$a_i(\epsilon) = a_i + \epsilon^{2^{2i}}, \qquad b_i(\epsilon) = b_i + \epsilon^{2^{2i-1}}$$

It is then shown that if $\epsilon > 0$ is sufficiently small, then the new arrangement $H(\epsilon)$ is simple. However, instead of substituting actual values for $\epsilon$, we carry out the calculation symbolically. For example, to decide if the intersection of $h_i(\epsilon) \cap h_j(\epsilon)$ lies above the line $h_k(\epsilon)$, we must determine the sign of

$$\Delta(\epsilon) = det \begin{pmatrix} a_i(\epsilon) & b_i(\epsilon) & 1 \\ a_j(\epsilon) & b_j(\epsilon) & 1 \\ a_k(\epsilon) & b_k(\epsilon) & 1 \end{pmatrix}$$

It is then observed that evaluating the sign of $\Delta(\epsilon)$ amounts to evaluating a sequence of subdeterminants of the original matrix until the first non-zero entry:

$$\begin{aligned} \Delta(\epsilon) &= det \begin{pmatrix} a_i & b_i & 1 \\ a_j & b_j & 1 \\ a_k & b_k & 1 \end{pmatrix} \\ & -\epsilon^{2^{2i-1}} det \begin{pmatrix} a_j & 1 \\ a_k & 1 \end{pmatrix} + \epsilon^{2^{2i}} det \begin{pmatrix} b_j & 1 \\ b_k & 1 \end{pmatrix} + \epsilon^{2^{2j-1}} det \begin{pmatrix} a_i & 1 \\ a_k & 1 \end{pmatrix} \\ & + \cdots - \epsilon^{2^{2i-1}+2^{2j}} + \cdots \end{aligned}$$

We will show that this is actually a general phenomenon. One may regard each input parameter $a_i$ to be perturbed by an infinitesimal amount $\delta_i$ ($\delta_i = \epsilon^{2^k}$ in the preceding illustration). Furthermore there is a suitable fixed total ordering on the set of infinitesimals and their products. For instance, without loss of generality, we may assume

$$\delta_1 \ll \delta_2 \ll \cdots \ll \delta_i \ll \cdots \ll \delta_n.$$

But what, for instance, is the relation between $\delta_1\delta_3$ and $\delta_2^3$? In the next section, we will give a systematic framework for making such comparisons. (We note that [8] also uses a different infinitesimal for each variable, but they did not have to give a general rule for comparing products of infinitesimals since they restricted attention to evaluation of determinants only.)

Another remark is that it is possible to justify such uses of infinitesimals in terms of non-standard analysis.

## 5 A general scheme to avoid zeroes

We describe a procedure to evaluate any polynomial $p(\mathbf{x})$ at any value $\mathbf{x} := \mathbf{a}$. The procedure outputs the value $p(\mathbf{a})$. However, in case $p(\mathbf{x})$ is a non-zero polynomial and $p(\mathbf{a}) = 0$, then the output is one of two types of zeroes: $0-$ or $0+$. This sign information will be globally consistent in a natural sense. Such a procedure can be used as a black-box by any algorithm satisfying assumption (*) above to always avoid the degenerate branch of a decision step.

Observe that this scheme also gives us a method of comparing the values of two distinct polynomials $p, q$ at any fixed point $\mathbf{x} = \mathbf{a}$. More precisely, if the sign of the difference polynomial $p - q$ at $\mathbf{x} = \mathbf{a}$ is *positive* then we say $p(\mathbf{a}) > q(\mathbf{a})$; otherwise $p(\mathbf{a}) < q(\mathbf{a})$. Extending this, it means that we can strictly order any set of distinct polynomials $\{p_1, \ldots, p_i\}$ by their 'values' at any point $\mathbf{x} := \mathbf{a}$.

As in [16], let $\mathrm{PP} = \mathrm{PP}(x_1, \ldots, x_n)$ denote the set of all power products

$$w = \prod_{i=1}^{n} x_i^{e_i} \quad (e_i \geq 0).$$

Let $|w|$ denote $\sum_{i=1}^{n} e_i$. A total ordering $\underset{A}{\leq}$ on PP is *admissible* if for any $w, w', w'' \in \mathrm{PP}$,

1. $1 \underset{A}{\leq} w$
2. $w \underset{A}{\leq} w'$ implies $ww'' \underset{A}{\leq} w'w''$.

The two most important examples of admissible orderings are the *total degree ordering*, denoted $\underset{TOT}{\leq}$, and the *(pure) lexicographical ordering*, denoted $\underset{LEX}{\leq}$. Let $w, v$ be two power products with degree vectors $(e_1, e_2, \ldots, e_n)$ and $(d_1, d_2, \ldots, d_n)$. Then we define $w \underset{LEX}{\leq} v$ if $w = v$ or else $e_i < d_i$ at the smallest index $i$ where the $e_i$'s and $d_i$'s differ. We also define $w \underset{TOT}{\leq} v$ if $|w| < |v|$ or else $(|w| = |v|$ and) $w \underset{LEX}{\leq} v$.

Henceforth we assume some arbitrary but fixed admissible ordering $\underset{A}{\leq}$ on $PP$.

For any $w = x_1^{e_1} \cdots x_n^{e_n}$ and polynomial $p(\mathbf{x})$, $\mathbf{x} = (x_1, \ldots, x_n)$, let $p_w$ refer to the $|w|^{th}$ partial differential of $p$, where $p$ is differentiated $e_k$ times with respect to each variable $x_k$ $(k = 1, \ldots, n)$. For example, if $w = x^2 y$ then $p_w = \frac{\partial^3 p}{\partial x^2 \partial y}$. Let $S(p)$ denote the infinite list of polynomials

$$S(p) = (p_{w_0}, p_{w_1}, p_{w_2} \cdots)$$

where $w_0, w_1, w_2, \cdots$ is the list of power products in PP in increasing $\underset{A}{\leq}$-order. In particular $w_0 = 1$ and $p_{w_0} = p$. This infinite sequence has only a finite number of non-zero entries and assuming that that $p$ is not identically zero, the last non-zero entry is a constant. Our polynomial evaluation procedure proceeds as follows: given a non-zero polynomial $p(\mathbf{x})$ and a point $\mathbf{a} = (a_1, \ldots, a_n)$, we evaluate successive polynomials in $S(p)$ at $\mathbf{x} := \mathbf{a}$ until the first non-zero entry; by above remarks, termination of this procedure is guaranteed. In case $p(\mathbf{a}) \neq 0$, then we return $p(\mathbf{a})$; otherwise we return either $0+$ or $0-$, where the sign of the zero is that of the first non-zero entry evaluated in the sequence $S(p)$. It is convenient to introduce the notation

$$S(p; \mathbf{a}) = (p_{w_0}(\mathbf{a}), p_{w_1}(\mathbf{a}), p_{w_2}(\mathbf{a}) \cdots).$$

**Lemma 2**
*(a) $S(p; \mathbf{a})$ is the sequence of all zeroes iff $p$ is identically zero.*
*(b) For any $p$ and $\mathbf{a}$, the polynomial $p$ is uniquely determined by the sequence $S(p; \mathbf{a})$.*

*Proof.* (a) follows from our above remark that any non-zero polynomial has a derivative that is a non-zero constant. (b) Suppose that $p, q$ are two polynomials such that $S(p; \mathbf{a}) = S(q; \mathbf{a})$. Then $S(p - q; \mathbf{a}) = 0$ (the sequence of all zeroes). Then part (a) implies that $p - q$ is identically zero, i.e. $p = q$. **Q.E.D.**

This proof does not tell us how to reconstruct the polynomial $p$ from the sequence $S(p; \mathbf{a})$ but it is easy to give a formula to reconstruct $S(p; \mathbf{a})$ from $S(p_{x_i}; \mathbf{a})$ for all $i = 1, \ldots, n$, where $p_{x_i}$, of course, is partial differentiation with respect to $x_i$.

The evaluation procedure amounts to a function that assigns a sign (zero, positive or negative) to every polynomial, where each non-zero polynomial is either positive or negative.

We define $p(\mathbf{x}) \underset{\mathbf{a}}{>} q(\mathbf{x})$ to mean that the polynomial $p(\mathbf{x}) - q(\mathbf{x})$ has positive sign at $\mathbf{x} := \mathbf{a}$.

**Corollary 3** *For all polynomials $p$ and $q$, exactly one of the following relation holds: $p = q$ or $p \underset{\mathbf{a}}{>} q$ or $q \underset{\mathbf{a}}{>} p$.*

## 6 Ordered Rings

We now show that the relation $\underset{\sigma}{>}$ is a total ordering and has other algebraic structure as well. There is a beautiful theory of ordered fields due to Artin and Schreier [21]. One sees that the definition of ordered fields in [21] only uses the ring properties of fields. Accordingly, we may adapt that definition to rings. Our ring $R$ will be assumed to be commutative with a unit 1.

A *partially ordered ring* $R$ is a ring with an associated *sign function* $\sigma : R \to \{-1, 0, +1\}$ such that for all $a, b \in R$:

**A1** $\sigma(a) = -\sigma(-a)$. In particular, $\sigma(a) = \sigma(-a)$ iff $\sigma(a) = 0$.

**A2** $\sigma(ab) = \sigma(a)\sigma(b)$

**A3** $\sigma(a) \geq 0$ and $\sigma(b) \geq 0$ implies $\sigma(a + b) \geq 0$, with equality iff $\sigma(a) = \sigma(b) = 0$.

Call a ring element $a$ *positive*, *negative* or *nullary* according to the sign $\sigma(a)$ of $a$. Note that axiom (A1) implies that $\sigma(0) = 0$, and (A2) implies $\sigma(1) = +1$. If the sign function, in addition, satisfies the property

**A4** $\sigma(a) = 0$ if and only if $a = 0$

then we call $R$ an *ordered ring.*

We define a relation $\underset{\sigma}{>}$ on $R$ where

$$a \underset{\sigma}{>} b \text{ if and only if } \sigma(a - b) = +1.$$

Also, define $a \underset{\sigma}{\leq} b$ if either $a = b$ or $b \underset{\sigma}{>} a$. We get the expected properties as in [21]: transitivity follows from axiom (A3); $a \underset{\sigma}{>} b$ implies $a + c \underset{\sigma}{>} b + c$, and if $c$ is positive, then $ac \underset{\sigma}{>} bc$ holds as well. And $a \underset{\sigma}{>} b$ if and only if $b^{-1} \underset{\sigma}{>} a^{-1}$. Of course, $\underset{\sigma}{\leq}$ is a partial ordering on $R$, and if $R$ is an ordered ring then $\underset{\sigma}{\leq}$ becomes a total ordering. The following two consequences are less obvious:

**Lemma 4**
*1. In a partially ordered ring, $a + b \underset{\sigma}{>} a' + b'$ implies $a \underset{\sigma}{>} a'$ or $b \underset{\sigma}{>} b'$.*
*2. In an ordered ring, if $a$ and $b$ are both positive and $ab \underset{\sigma}{>} a'b'$, then $a \underset{\sigma}{>} a'$ or $b \underset{\sigma}{>} b'$.*

*Proof.* 1. $\sigma(a + b - a' - b') = +1$ implies that $\sigma(a - a') = +1$ or $\sigma(b - b') = +1$ (otherwise $\sigma(a' - a) \geq 0$ and $\sigma(b' - b) \geq 0$ and axiom (A3) implies $\sigma(a' + b' - a - b) \geq 0$, contradiction).
2. If $\sigma(a - a') = 0$ then $a = a'$ (since $R$ is an ordered ring) and $ab \underset{\sigma}{>} a'b'$ implies $ab \underset{\sigma}{>} ab'$. This means $\sigma(a(b - b')) = +1$ and since $\sigma(a) = +1$, we get $\sigma(b - b') = +1$, proving the desired result. Similarly if $\sigma(b - b') = 0$. Therefore, if we assume the result is false, we must have $\sigma(a - a') = \sigma(b - b') = -1$. We get a contradiction as follows. First, these assumptions imply $\sigma((a' - a)(b' - b)) = +1$. It follows that $\sigma((a' - a)(b' - b) + (ab - a'b')) = +1$ Rearranging terms, we get $\sigma(a(b - b') + b(a - a') = +1$. Using (A3), we infer that $\sigma(a(b - b')) = +1$ or $\sigma(b(a - a')) = +1$. If $\sigma(a(b - b')) = +1$ then applying (A2) with $\sigma(a) = +1$, we get $\sigma(b - b') = +1$ which is a contradiction. A similar contradiction arises if $\sigma(b(a - a')) = +1$. **Q.E.D.**

The main examples of ordered fields are the rational numbers and the reals. In [21] (exercises), it is pointed out that we can also order the set of univariate polynomials with coefficients over an ordered field by regarding the sign of a polynomial to be the sign of the leading coefficient. This can be generalized to multivariate polynomials once we have fixed an admissible ordering on the power products: then we may speak of the 'head monomial' and take the sign of a polynomial to be the sign of the coefficient of the head monomial. We now show another family of total orderings based on our polynomial evaluation scheme.

Let $Q$ be any ordered ring and let $R = Q[x_1, \ldots, x_n]$ be the ring of $n$-variate polynomials with coefficients from $Q$. Let $\underset{A}{\leq}$ be any fixed admissible ordering on the power products $PP = PP(x_1, \ldots, x_n)$.

**Lemma 5** *Let $u, v, u', v' \in PP$ such that either $u \neq u'$ or $v \neq v'$. If $uv \underset{A}{\leq} u'v'$ then $u \underset{A}{<} u'$ or $v \underset{A}{<} v'$.*

*Proof.* Let the exponents of $x_1, \ldots, x_n$ in $u$ and $u'$ be $\mathbf{u} = (u_1, \ldots, u_n)$ and $\mathbf{u}' = (u'_1, \ldots, u'_n)$, respectively. Similarly for let $\mathbf{v}, \mathbf{v}'$ denote the exponents with respect to $v, v'$. By a characterization of admissible orderings (see [6]), there corresponds to $\underset{A}{\leq}$ an $n$ by $n$ matrix $W$ with real entries such that $uv \underset{A}{<} u'v'$ iff

$$(\mathbf{u}+\mathbf{v})W \underset{LEX}{<} (\mathbf{u}'+\mathbf{v}')W.$$

Let $\mathbf{y} = \mathbf{u}W$, $\mathbf{z} = \mathbf{v}W$, $\mathbf{y}' = \mathbf{u}'W$ and $\mathbf{z}' = \mathbf{v}'W$. Let the $i$th component of $\mathbf{y}$ (resp. $\mathbf{y}'$, $\mathbf{z}$, $\mathbf{z}'$) be $y_i$(resp. $y'_i$, $z_i$, $z'_i$). If $i$ is the first index where either $y_i \neq y'_i$ or $z_i \neq z'_i$ then since

$$y_i + y'_i \leq z_i + z'_i$$

we must have either $y_i < y'_i$ or $z_i < z'_i$. This means either $u \underset{A}{<} u'$ or $v \underset{A}{<} v'$. **Q.E.D.**

We now extend the sign function on $Q$ to a sign function on $R = Q[x_1, \ldots, x_n]$ according to our evaluation scheme. Fix any point $\mathbf{a} \in Q^n$ and any admissible ordering $\underset{A}{\leq}$ on $PP$. For any $p \in R$, we again have the sequence $S(p)$ of its partial derivatives ordered according to $\underset{A}{\leq}$, and also the sequence $S(p; \mathbf{a})$ obtained by evaluating $S(p)$ at $\mathbf{a}$. Define the sign $\sigma(p)$ of $p$ to be the sign (since $Q$ is ordered) of the first non-zero entry in $S(p; \mathbf{a})$. If all entries in $S(p; \mathbf{a})$ are zero, then $\sigma(p)$ is defined to be 0.

**Theorem 6** *$R = Q[x_1, \ldots, x_n]$ with the sign function $\sigma$ as determined by the evaluation scheme is an ordered ring. Note that $\sigma$ depends on the choice of admissible ordering $\underset{A}{\leq}$, on the choice of $\mathbf{a} \in Q^n$, and on the sign function on $Q$.*

*Proof.* We must verify the four axioms (A1-A4). Axiom (A1) follows from the fact that $S(p; \mathbf{a}) = -S(-p; \mathbf{a})$, and (A4) comes from the fact that $S(p; \mathbf{a})$ has some non-zero constant entry unless $p$ is identically zero. (A3) comes from the fact that $S(p+q; \mathbf{a}) = S(p; \mathbf{a}) + S(q; \mathbf{a})$ where we have componentwise addition of the sequences. To show (A2), suppose that the first non-zero entry in $S(p)$ (resp. $S(q)$) is $p_u$ (resp. $q_v$) where $u, v \in PP$. We claim that the first non-zero entry in $S(pq)$ is $(pq)_{uv}$ (i.e. the partial derivative of $pq$ with respect to $uv$). We see that

$$(pq)_{uv} = \sum_{u'} p_{u'} q_{v'}$$

where $u' \in PP$ range over all divisors of $uv$ and $v'$ is given by $u'v' = uv$. By the previous lemma, unless $u = u'$ and $v = v'$, $u'v' \underset{A}{\leq} uv$ implies $u \underset{A}{>} u'$ or $v \underset{A}{>} v'$. Hence $p_{u'} = 0$ or $p_{v'} = 0$. We conclude that $(pq)_{uv} = p_u q_v$.

Suppose $uv \underset{A}{>} u'v'$. The same argument as above shows $(pq)_{u'v'} = 0$. It follows that the sign of $pq$ is equal to the sign of $p_u(\mathbf{a}) q_v(\mathbf{a})$, which is equal to $\sigma(p)\sigma(q)$. This proves (A2). **Q.E.D.**

As an example, we may obtain as a corollary that the scheme in [8] is an ordering on the ring of polynomials since it can been seen that his choice of perturbation leads to the lexicographical ordering on $PP$; this observation does not immediately come out in the original paper because the original paper were restricted to evaluating determinants.

## 7 Complexity: evaluation of sparse polynomials

Since all evaluation of polynomials are to be done through our 'black-box', it is important for this black-box to be efficient. We now consider this issue. Let $L(n, s, d)$ denote the minimum number of arithmetic operations $(+, -, \times, \div)$ sufficient to evaluate all polynomials with $s$ monomials on $n$ variables, where each variable has degree at most $d$. Hence $s \leq d + 1$ is a measure of the sparsity of the polynomial and we call this the *sparse complexity model* of polynomial evaluation. Little work has been done on this model except in special cases. Nevertheless, the model is very important especially in the multivariate setting.

To begin, we note that our algorithm need not precompute the (non-zero entries of the) sequence $S(p)$. Instead it can compute successive entries on the fly. This is because, using standard admissible orderings such as total degree or lexicographical orderings, it is relatively simple to generate successive polynomials in the sequence. Otherwise, we know that each admissible ordering is characterized by a real square matrix $M$ and if $M$ has computable elements, we can also compute successive non-zero entries of $S(p)$.

For the evaluation of univariate polynomials, we have relatively complete knowledge about the worst case complexity of the two extremes of sparsity: the dense polynomial (i.e., a polynomial of degree $d$ has $d+1$ non-zero coefficients) and the totally sparse polynomial which consists of only one monomial (i.e., the addition chain problem). From [22], [18], we may deduce that evaluating an $n$-variate polynomial with $s$ monomials, where the degree in each variables is less than $d$ has complexity at most

$$w + v\log d + H/\log H + o(H/\log H) \qquad (1)$$

where $H = ns\log d$ and $v = \min\{s,n\}$, $w = \max\{s,n\}$. This bound is tight if we insist on evaluating the monomials as separate entities. For evaluating polynomials, this is a real restriction. For instance, evaluating $(x_1+x_2)(x_3+x_4)$ as a polynomial requires strictly fewer arithmetic operations than evaluating the four component monomials separately.

In our applications, we not only want to evaluate a polynomial but *may* need to evaluate some of its derivatives as well. There is literature on evaluating a dense univariate polynomial and all of its derivatives (e.g. [1]). They are not directly suitable for our application since we need sparse multivariate polynomials, and generally do not evaluate derivatives of every order. (See also [13],[14].) The sparse model of polynomial evaluation lends itself nicely to the derivative evaluation problem since a derivative polynomial is at least as sparse as the original polynomial.

Instead of using the asymptotically optimal method of Pippenger, we use a simpler scheme based on Yao's method. The method basically says that given $x$ raised to each power of 2 less than $d$, we can evaluate $x^d$ is $\frac{\log d}{\log\log d}[1+o(1)]$ steps. Hence any $s$ term $n$-variate polynomial can be evaluated in $\frac{s\log d}{\log\log d}[1+o(1)]$ steps, (assuming an initial cost of $n\log d$ to evaluate each variable raised to powers of 2 less than $d$.) For our application, the advantage of this simple scheme comes from the potential need to evaluate derivatives: each subsequent derivative can be evaluated using the same method. The space usage is $O(n\log d)$ to store the variables raised to powers of 2.

In the univariate case, we can actually get a uniform method of polynomial evaluation whose complexity is optimal (to a multiplicative factor of $(1+o(1))$) for the entire range of $s$:

$$L(1,s,d) \le \log d + \frac{s\log(d/s)}{\log\log(d/s)}[1+o(1)]$$

To see this, we use the 'generalized Horner factorization' of a polynomial:

$$p(x) = a_1x^{d_1}(1 + a_2x^{d_2}(1 + \ldots(1 + a_sx^{d_s})\ldots))$$

where $\sum_{i=1}^{s} d_i \le d$. We first evaluate $x^k$ where $k \le d$ range over powers of 2. Now evaluate each $x^{d_i}$ in $\frac{\log(d_i)}{\log\log(d/s)}[1+o(1)]$ steps. Summing over all $i = 1,\ldots,s$, and observing that the complexity is maximized when each $d_i = d/s$, we get our result. Unfortunately, this method does not immediately generalize to multivariate polynomials.

An important observation is that one is unlikely to have to evaluate more than the first few polynomials in the sequence before turning up a non-zero value. The small probability involved here seems to be related to the small probability of residual degeneracies in probabilistic scheme.

At a small extra cost in space, we can speed up the evaluation of subsequent entries in $S(p)$. This is based on the observation that evaluating successive polynomials in the sequence $S(p)$ are closely related. The idea is that in the above method for evaluating the monomial $x_1^{d_1}x_2^{d_2}\cdots x_n^{d_n}$, we keep around the values of $x_i^{d_i}$ for each $i$. Then in the derivative (say with respect to $x_1$), we must evaluate the monomial $x_1^{d_1-1}x_2^{d_2}\cdots x_n^{d_n}$ and this can be obtained in $n + \log d_1/\log\log d_1$ multiplications (instead of $n\log d_1/\log\log d_1$). Other methods to trade off time for space can also be devised.

## 8 Directions for future work

1. The black-box we provide is not invariant under change of coordinate axes. For example, with $f(x,y) = x$ and $g(x,y) = y$ our test yields the inequality $f(0,0) > g(0,0)$ (assuming the ordering $x > y$). But if we rotate the coordinate axes of the plane by angle $\theta$, the transformed functions become $F(x,y) = x\cos\theta - y\sin\theta$ and $G(x,y) = x\sin\theta + y\cos\theta$. If $|\theta| > \pi/2$ then we get $F(0,0) < G(0,0)$, which is opposite to the unrotated case.

2. We have defined our scheme for evaluation of polynomials. It is easy to extend this to rational polynomial functions, $f(\mathbf{x}) = p(\mathbf{x})/q(\mathbf{x})$. Such rational functions might arise in, say the solution of linear systems (using Cramer's rule). We form the sequence $S(f) = (f_1, f_2, \ldots)$ as usual. It is not hard to check that again, this sequence has a finite number of non-zero entries, and there are constant entries among them. Hence the 'first non-zero element' in $S(f; \mathbf{a})$ is well-defined. Now the ordered rings become ordered fields.

3. What about evaluation of algebraic functions? By an algebraic function we understand a continuous function $r(\mathbf{x})$ such that for some polynomial $p(z; \mathbf{x})$,

$$p(r(\mathbf{x}); \mathbf{x}) = 0.$$

For each $\mathbf{x} := \mathbf{a}$, we get an algebraic number $r(\mathbf{a})$. (As discussed in section 2, such algebraic numbers are represented exactly.) Perhaps the most common instance is the square-root function $r(\mathbf{x}) = \sqrt{x_1}$, where $p(z; \mathbf{x}) = z^2 - x_1$. If we want to compare the distance $d(p, q)$ between two points $p, q$ with some value $v$, then we could clear square-roots and set up the test as $d(p, q)^2 - v^2 = 0$. One cannot do such transformation of the test polynomial with impunity because they affect the infinitesimals. For example, if $f(x) = x^2$, $F(x) = f(x)^2 = x^4$ and $g(x) = x^3$ then we have $f(0) > g(0) > F(0)$ assuming total degree admissible ordering. Hence, the comparison $f(0) : g(0)$ is not equivalent to $F(0) : g(0)$ (though it would be equivalent to $F(0) : G(0)$ where $G = g^2$). To treat algebraic functions in general, we can proceed as follows: suppose $r(\mathbf{x})$ is an algebraic function satisfying $p(r(\mathbf{x}); \mathbf{x}) = 0$. Then the partial derivatives of $r(\mathbf{x})$ is obtained by the chain rule, and we can define the sequences $S(r)$ and $S(r; \mathbf{a})$ in the usual way. For instance, in the case of the square root function, $\frac{\partial r(\mathbf{x})}{\partial x_1} = 1/2r(\mathbf{x})$ and $\frac{\partial^2 r(\mathbf{x})}{\partial x_1^2} = -1/r(\mathbf{x})^3$. The computational details of carrying out such a scheme especially if the degree of derivation is unbounded is nontrivial.

4. In some applications, there are restrictions on the perturbation. In some hidden surface removal algorithms, there is a basic test to see if two triangles $A, B$ obscure one another. Let the polygon $C$ be the intersection of the projections $A', B'$ of the triangles onto the viewplane ($xy$-plane). So $C$ has at most 6 sides. Take any vertex $P$ in $C$. By projecting $P$ back to $A$ and $B$, we get $P_A$ and $P_B$. The test reduces to comparing the $z$-coordinate of $P_A$ and of $P_B$. If their $z$-coordinates are equal, we want to do the same test on a perturbed version of $P$. We must ensure that perturbed $P$ does not go outside $C$: conceptually, we add to $P$ an infinitesimal vector directed to the interior of $C$. Unfortunately, our present perturbation scheme is not helpful here.

5. Another more far reaching direction is to to handling uncertainty in the input data. This is related to the issue of treating finite precision arithmetic. A basic question to be faced is this: what is the meaning of degeneracy in the world of imprecise input or in the world of finite precision? This is a difficult question, related, for instance, to the problem defining the digital analogues of lines and circles.

6. Computational experience and applications to actual algorithms in computational geometry ought to be carried out to expose other issues in implementing our scheme: for instance, what is a good choice of admissible ordering?

7. Other connections that could be investigated: work on numerical stability and geometric probability [4][5], singularity and catastrophe theory [19]. As pointed out by [8], the use of lexicographical rules in linear programming to break ties has been known for sometime [2],[3]. Unfortunately, the setting there is slightly different from that of the present paper.

## 9 Acknowledgement

This work was first conceived at the *Workshop on Geometric Reasoning*, Oxford University, June 30-July 3, 1986. The workshop was supported by grants from the NSF and the Science Research Council of Great Britain. I am indebted to members of the workshop (in particular, Micha Sharir and Christoph Hoffman) who commented on preliminary ideas of this scheme.

## References

[1] A.V. Aho, K. Stiglitz, and J.D. Ullman. Evaluating polynomials at fixed sets of points. *SIAM J. Computing*, 4(4):533–539, 1975.

[2] A. Charnes. Optimality and degeneracy in linear programming. *Econometrica*, 20(2):160–170, 1952.

[3] V. Chvátal. *Linear Programming*. W. H. Freeman and Company, 1983.

[4] James W. Demmel. *On condition numbers and the distance to the nearest ill-posed problem*. Technical Report 293, Dept. of Computer Science, Courant Institute, NYU, April, 1987.

[5] James W. Demmel. *The probability that a numerical analysis problem is difficult*. Technical Report 294, Dept. of Computer Science, Courant Institute, NYU, April, 1987.

[6] T. Dubé, B. Mishra, and C. K. Yap. *Admissible orderings and bounds for Gröbner bases normal form algorithm*. Report 88, NYU-Courant Robotics Lab., 1986.

[7] H. Edelsbrunner. Edge-skeletons in arrangements with applications. *Algorithmica*, 1:93–110, 1986.

[8] H. Edelsbrunner and Ernst Peter Mücke. Simulation of simplicity: a technique to cope with degenerate cases in geometric algorithms. June 1987. Manuscript.

[9] H. Edelsbrunner and R. Waupotitsch. Computing a ham-sandwich cut in two dimensions. *J. Symbolic Computation*, 171–178, 1986.

[10] Herbert Edelsbrunner. *Algorithms in Combinatorial Geometry*. Springer-Verlag, 1987.

[11] D. G. Freudenstein and J. R. Kender. What is a "degenerate" view? Manuscript, June 30-July 4, 1986. Workshop on Geometric Reasoning.

[12] D. H. Greene and F. F. Yao. Finite-resolution computational geometry. In *27th FOCS*, pages 143–152, 1986.

[13] Masao Iri. Simultaneous computation of functions, partial derivatives and estimates of rounding errors – complexity and practicality. *Japan J. of Applied Math.*, 1(2):171–178, 1986.

[14] Masao Iri and Koichi Kubota. *Methods of fast automatic differentiation and applications*. Research Memorandum RMI 87-02, Dept. of Math. Eng. and Instrumentation Physics, University of Tokyo, Japan, 1987. (extended English translation in Proceed., 7th Mathematical Programming Symp., Nagoya, Japan, November 6-7, 1986).

[15] V. J. Milenkovic. Verifiable implementation of geometric algorithms using finite precision arithmetic. Manuscript, June 30-July 4, 1986. Workshop on Geometric Reasoning.

[16] B. Mishra and C. K. Yap. *Notes on Gröbner bases*. Report 87, NYU-Courant Robotics Lab., 1986. To appear, special issue of *J. of Information Sciences*.

[17] C. Ó'Dúnlaing and C. K. Yap. A 'retraction' method for planning the motion of a disc. *J. Algorithms*, 6:104–111, 1985.

[18] Nicholas Pippenger. On the evaluation of powers and monomials. *SIAM J. Computing*, 9(2):230–250, 1980.

[19] Tim Poston and Ian Stewart. *Catastrophe Theory and its Applications*. Pitman, 1978.

[20] Jacob T. Schwartz and Micha Sharir. On the piano movers' problem: II. General techniques for computing topological properties of real algebraic manifolds. *Advances in Appl. Math.*, 4:298–351, 1983.

[21] B.L. van der Waerden. *Algebra*. Volume 1 & 2, Frederick Ungar Publishing Co., 1970.

[22] Andrew Chi-chih Yao. On the evaluation of powers. *SIAM J. Computing*, 5(1):100–103, 1976.

# A SPACE EFFICIENT ALGORITHM FOR THE GREEDY TRIANGULATION

**Andrzej Lingas**
**Department of Computer and Information Science**
**Linköping University, 581 83 Linköping, Sweden**

***Abstract*** **: An algorithm for constructing the greedy triangulation of any planar straight-line graph with $n$ vertices in time $O(n^2 \log n)$ and space $O(n)$ is presented. The upper time-space bound implied by the algorithm in particular improves Gilbert's simultaneous $O(n^2 \log n)$-time and $O(n^2)$-space bound for finding the greedy triangulation of an $n$-point planar point set. A theorem suggesting a good expected time-performance of the presented algorithm is proven.**

## 1. *Introduction*

We consider planar figures which are called *planar straight-line graphs* (PSLG for short) [PS85]. A PSLG $G$ is a is a pair $(V, E)$ such that $V$ is a set of points in the plane and $E$ is a set of non-intersecting, open straight-line segments whose endpoints are in $V$. If $G$ has no edges then it is a *planar point set*. A *diagonal* of $G$ is an open segment that neither intersects any edge of $G$ nor includes any vertex of $G$ and that has endpoints in $V$. A *triangulation* of $G$ is a maximal set of non-intersecting diagonals of $G$.

In numerical applications of triangulations, one of the proposed criteria of goodness involves the minimization of the total length of diagonals in a triangulation. A *minimum weight triangulation* (MWT for short) of a PSLG $G$ is a triangulation of $G$ that achieves the smallest possible (total diagonal) length. The complexity status of the problem of constructing MWT of a planar point set remains open [PS85]. One of the most known heuristics for MWT of a planar point set is the so called *greedy triangulation* which can be generalized to include any PSLG $G$. It inserts a diagonal $d$ of $G$ into the plane if $d$ is shortest among all diagonals of $G$ that neither intersect nor overlap with these already in the plane. A most efficient, known algorithm for the greedy triangulation due to Gilbert [Gi79,PS85] takes $O(n^2 \log n)$ time and $O(n^2)$ space.

We present a new method of constructing the greedy triangulation of any PSLG, using the so called Voronoi diagram with barriers of $G$ [Li86]. For a PSLG $G = (V, E)$, the Voronoi diagram with barriers of $G$ ($Vorb(G)$ for short) consists of regions $P(v)$, $v \in V$, such that a point $p$ is inside $P(v)$ if and only if $(p, v)$ is the shortest straight-line segment connecting $p$ with a vertex in $V$ that does not intersects any edge in $E$. Voronoi diagrams with barriers can be seen as a special case of the generalized Voronoi diagrams for shortest path problems [AA86] where the shortest paths are always straight-line segments.

Using the method of finding a shortest diagonal of $G$ on the basis of $Vorb(G)$ presented in [Li86] and the method of updating $Vorb(G)$ and the data structures supporting selection of shortest diagonal after inserting a single diagonal of $G$ presented in this paper, we obtain a new algorithm for a greedy triangulation of $G$. It runs in time $O(n^2 \log n)$ and space $O(n)$. First, the algorithm builds $Vorb(G)$. Then, it iterates the following step: insert a currently shortest diagonal of the union of $G$ with the current partial triangulation, and update the diagram with barriers and the supporting data structures. Each of these steps takes $O(n \log n)$ time and space $O(n)$. Thus, the whole algorithm for the greedy triangulation of $G$ runs in time $O(n^2 \log n)$ and space $O(n)$. In our main theorem, we give a refined upper bound on the time performance of the algorithm which suggests $O(n \log n)$ expected-time performance of the algorithm applied to a set of $n$ points uniformly distributed in a unit square.

In the next section, we present some facts of Voronoi diagrams with barriers used in further sections. In Section 3 we outline the efficient method of updating the diagrams. In Section 4 we present the algorithm for the greedy triangulation of a PSLG and the efficient method of updating the data structures used by the algorithm to select a shortest diagonal. In effect, we can derive the mentioned refined upper bound on the complexity of the algorithm.

## 2. *Properties of Voronoi Diagrams with Barriers*

Given a PSLG $G = (V, E)$, the *Voronoi diagram with barriers of G* ( $Vorb(G)$ for short, see the introduction ) has several properties analogous to those of classical Voronoi diagrams [PS85]. First of all, the regions $P(v)$, $v \in V$, comprising $Vorb(G)$, form a partition of the plane. Clearly, every edge of $Vorb(G)$ is a segment of the perpendicular bisector of a pair of vertices of $G$ from which the edge is visible. Hence, every vertex of $Vorb(G)$ is the common intersection of at least three edges of $Vorb(G)$ or residues inside an edge of $G$. It also follows that for all $v \in V$, if $m$ is the number of edges of $G$ incident to $v$ then $P(v)$ is a collection of $m$ polygonal sub-regions separated by the $m$ incident edges. We shall call the sub-regions *sectors*. Throughout the paper, we assume the standard DCEL representation (see [PS85]) of $Vorb(G)$.

Given a PSLG $G$ and $Vorb(G)$, by *the straight-line dual* of $Vorb(G)$, we mean the pair $(V, S)$ where $V$ is the set of vertices of $G$ and $S$ is the set of all open straight-line segments between pairs of non-adjacent vertices of $G$ whose regions in $Vorb(G)$ share an edge.

In [LL86], Lee and Lin have introduced the concept of a generalized Delaunay triangulation of PSLG $G$ as a triangulation $T$ of $G$ such that the interior of the circumcircle of any triangle (i.e. triangular face) in $G \cup T$ does not contain any vertex of $G$ simultaneously visible from the three vertices of the triangle. Among others, they have proved that the generalized Delaunay triangulation of $G$ is unique if no four vertices of $G$ are co-circular.

In analogy to the planar point set case, the following fact holds:

*Fact 2.1 [Li86]:* Let $G$ be a PSLG. The straight-line dual of $Vorb(G)$ is a subset of every *generalized Delaunay triangulation* of $G$.

As a corollary from Fact 2.1, we obtain the following, quantitative characterization of $Vorb(G)$.

*Fact 2.2:* Given a PSLG $G$ with $n$ vertices, there are at most $3n - 6$ edges in $Vorb(G)$.

## 3. *Updating Voronoi Diagrams with Barriers*

Clearly, if a PSLG $G = (V, E)$ has no edges then $Vorb(G)$ is equivalent to the classical Voronoi diagram of $V$, and consequently, can be constructed in time $O(n \log n)$ [PS85, F86]. Recently, Wang and Shubert have proved that $Vorb(G)$ can be constructed in time $O(n \log n)$ also in the general case [WS87].

In this section we are interested in efficiently updating the Voronoi diagram with barriers of a PSLG $G$ after inserting a new edge $e$ into $G$. In our updating procedure, besides the DCEL

**representation of $Vorb(G)$, we use another representation of $Vorb(G)$ that we call TREES.**

**For every vertex $w$ of $G$, and every sector of the region of $w$ (see Section 2), TREES contains an ordered 2-3 tree with vertices assigned to the boundary edges of the sector in clockwise order from left to right. The 2-3 tree supports the following query in logarithmic time: given a straight-line $L$, report the boundary edge or edges of the sector touched or intersected by $L$ plus the sectors (possibly of other regions) adjacent to the above edges. Note that any sector of a region in $Vorb(G)$ either is convex or can be trivially decomposed into two convex polygonal regions. Therefore, there are $O(1)$ boundary edges of the sector of the region of $w$ and $O(1)$ adjacent sectors to report. Given $Vorb(G)$, TREES can be constructed in time $O(n)$ and space $O(n)$ (see [Me84]).**

**For the new edge $e$, let $k(e)$ be the total number of vertices of the parts of the regions in $Vorb(G)$ cut off from their sites by $e$. The following theorem provides an upper bound on the cost of updating in terms of $k(e)$.**

***Theorem 3.1:* Let $G = (V, E)$ be a PSLG with $n$ vertices and let $e$ be a diagonal of $G$. Denote the PSLG $(V, E \cup \{e\})$ by $G'$. Given the DCEL and TREES representation of $Vorb(G)$, we can construct the DCEL and TREES representation of $Vorb(G')$ in time $O(k(e)\log n)$ and space $O(n)$.**

***Sketch:* Let $R$ be the set of the parts of the regions in $Vorb(G)$ cut off from their sites by $e$. $R$ can be constructed by finding the intersection of $e$ with the boundary of the region of one of its endpoints in $Vorb(G)$ and then scanning the boundaries of the parts in $R$. Using TREES, we can find the intersection in time $O(\log n)$. The scanning can be performed in time $O(k(e))$ using the DCEL representation of $Vorb(G)$. The union of elements from $R$ on a given side of $e$ forms a set of polygonal regions which can be easily found in $O(k(e)\log k(e))$ time and $O(k(e))$ space [NP82]. To prove the theorem, we shall show that the intersections of these regions with $Vorb(G')$ can be found in time $O(k(e)\log k(e))$ and space $O(k(e))$ (clearly, we have $k(e) \le n$).**

**Let $P$ be one of the above polygonal regions. Note that all sites of $G$ whose regions contribute to $P$ lie on the other side of $e$ than $P$ between the perpendicular straight-lines passing through the site endpoints of $e$. Hence, we can deduce that $P$ is convex in the direction perpendicular to $e$. It follows in particular that $P$ does not contain holes. Let $e_0$, $e_2$, ..., $e_m$ be the edges of $P$ in clockwise order. We may assume w.l.o.g that $e_m$ is a fragment of $e$. Any other edge of $P$ is a bisector boundary of a region in $Vorb(G)$ or a fragment of an edge of $G$. Now, for each of the edges $e_i$, $i = 1, ..., m-1$, that is a bisector boundary, let $v_i$ be the vertex in $M$ whose region not intersected by $e$ the boundary belongs to. It is clear that $P$ is exactly covered with the regions of the above vertices $v_i$ in $Vorb(G')$. Note that if $e_i$ is a fragment of a barrier then the neighboring edges $e_{i-1 mod\ m+1}$, $e_{i+1 mod\ m+1}$ are of bisector type and hence the vertices $v_{i-1 mod\ m+1}$ and $v_{i+1 mod\ m+1}$ are defined. Consider the polygonal chain $Q$ composed of the edges $e_i = (w_i^1, w_i^2)$ that are fragments of barriers plus all edges of the form $(w_i^2, v_{i+1\ mod\ m+1})$, $(v_{i-1\ mod\ m+1}, w_i^1)$, and $(v_j, v_{j+1})$ for $j = 0, ..., m-2$, where $v_j$ and $v_{j+1}$ are defined. Using the convexity of $P$ in the direction perpendicular to $e$ and properties of $Vorb(G)$, we can show that $Q$ is a simple polygon including $P$. Next, we claim that for each barrier $e_i$, the segment $s_i = (v_{i-1\ mod\ m+1}, v_{i+1\ mod\ m+1})$ lies within $Q$. The above claim results from the three following facts: (i) $P$ is convex in the direction perpendicular to $e$; (ii) $e_i$ cannot be perpendicular to $e$; (iii) None of the sites $v_{i-1\ mod\ m+1}$, $v_{i+1\ mod\ m+1}$ can lie strictly over the straight-line induced by $e_i$ if we assume that $e$ is placed horizontally and $e_i$ lies over $e$. Let**

$Q'$ be the simple polygon resulting from $Q$ by cutting off the quadrangles $t_i$ induced by $s_i$ and $e_i$ along the segments $s_i$. The set of the vertices of $Q'$ is equal to the set of the (defined) sites $v_i$, $0 \geq i \leq m$. By the definition of $Q'$, we can prove that $Vorb(Q') \cap Q' = Vorb(G') \cap Q'$. Thus, to find $P \cap Vorb(G')$, it is sufficient to construct the intersection $Vorb(Q') \cap P$ and the intersections $Vorb(G') \cap t_i \cap P$ where $t_i$ is defined. By [LL86,WS86], we can construct $Vorb(Q')$ in time $O(m \log m)$ and space $O(m)$. Hence, we find also $Vorb(Q') \cap P$ in time $O(m \log m)$ and space $O(m)$ [NP82]. Further, we can prove that in $Vorb(G')$, each of the quadrangles $t_i$ is covered with regions of the vertices of $Q'$ whose regions in $Vorb(Q') \cap Q'$ are adjacent to $s_i$. Now, to find $Vorb(G') \cap t_i$, we can apply a sweep-line method of constructing Voronoi diagrams due to Fortune [F86]. The sweep-line is initially placed at $s_i$ and then it moves towards $e_i$ in the direction perpendicular to $s_i$. Since the sought regions can be seen as extensions of the regions of $Q'$ adjacent to $s_i$, we can assume that they are already initiated and therefore, we do not need apply the transformation from [F86] here. In effect, we can construct all the intersections in time $O(m \log m)$ and space $O(m)$. It should be clear that also $Q'$ and the quadrangles $t_i$ can be constructed in time $O(m \log m)$ and space $O(m)$. Putting everything together, we can construct $Vorb(G') \cap P$ in time $O(m \log m)$ and consequently all the intersections of the polygonal regions induced by $R$ with $Vorb(G')$ in time $O(k(e) \log k(e))$ and space $O(k(e))$. We can assume that the above intersections are in the DCEL form since otherwise we can transform them to this form in time $O(k(e) \log k(e))$ and space $O(k(e))$. Now, we can produce the DCEL representation of $Vorb(G')$ using those of $Vorb(G)$ and of the above intersections. First, we find the difference between $Vorb(G)$ and the regions induced by $R$ by scanning the latter. Then, we merge the DCEL representation of the difference with that of the above intersections again by scanning the latter. It totally takes $O(k(e))$ time and $O(n)$ space.

It remains to produce the TREES representation of $Vorb(G')$. First of all, the new edge $e$ may split two sectors of the regions of its endpoints in $Vorb(G)$, and therefore if it is so we split the corresponding 2-3 trees in TREES accordingly. It can be done in time $O(\log n)$ [Me84]. Next, we scan the boundaries of the elements in $R$, appropriately deleting them from the affected 2-3 trees in TREES. Finally, we scan the edges of $Vorb(G')$ in the above intersections, inserting them into appropriate 2-3 trees. Totally, it takes $O(k(e) \log n)$ time. ∎

## 4. *Constructing a greedy triangulation of a PSLG*

We start this section from the following precise characterization of a shortest diagonal of a PSLG $G$ in terms of $Vorb(G)$ given in [Li86] (see the introduction for the def. of a diagonal of $G$).

*Fact 4.1 [Li86]:* A shortest diagonal of a PSLG either lies inside the union of the regions of its endpoints in $Vorb(G)$ or cuts off a triangular face in $G$ and lies within the regions of at most four vertices in $Vorb(G)$.

Using the above characterization and the mentioned algorithm of Wang and Shubert [WS86] for $Vorb(G)$, it is shown in [Li86] that a shortest diagonal of a PSLG can be found in time $O(n \log n)$ and space $O(n)$. This immediately implies a straight-forward algorithm for a greedy triangulation of a PSLG running in time $O(n^2 \log n)$ and space $O(n)$. In this paper, we present a more sophisticated algorithm. Instead of building $Vorb(G)$ from scratch every time after augmenting $G$ by a new edge, we update $Vorb(G)$ using Theorem 3.1. Then, we use Fact 4.1 also in the update-fashion. The algorithm is as follows.

1) Construct the DCEL and TREES representations of $Vorb(G)$.

2) Construct a min-heap, $HSD(G)$, of edges of the straight-line dual of $Vorb(G)$ according to the lengths of the edges.

3) Construct a set of trees, $STAR(G)$, which for each vertex $w$ of $G$ contains an ordered 2-3 tree with leaves assigned to the edges of $G$ in angular order around $w$.

4) Construct a min-heap, $HT(G)$, of edges of $Vorb(G)$ cutting off empty triangular faces from $G$ according to the lengths of the edges. For each such an edge construct two way pointers to the two other edges of $G$ of the triangular face in the tree in $STAR(G)$ for the common endpoint of the two edges.

5) Pick a shortest edge $e$ from $HSD(G)$ and $HT(G)$ and delete it from its heap. Augment $G$ by $e$.

6) Update the representations of $Vorb(G)$, $HSD(G)$ and $HT(G)$.

7) If $G$ is not a complete triangulation then go to (5).

The correctness of the above algorithm follows from Fact 4.1 and the definition of the straight-line dual of $Vorb(G)$. The first two steps of the algorithm can be performed by [WS86] in time $O(n \log n)$ and space $O(n)$. The third step can be obviously done in time $O(n \log n)$ and space $O(n)$ (see [Me84]). The fourth step can be also performed in time $O(n \log n)$ and space $O(n)$ by Lemma 3.2 in [Li86]. A single execution of the fifth step takes $O(\log n)$ time and $O(n)$ space. Since this step is performed $O(n)$ times, it totally takes $O(n \log n)$ time and $O(n)$ space. To estimate the complexity of the sixth step, we use Theorem 3.1 and the following lemma.

*Lemma 4.1:* Let $G = (V, E)$ be a PSLG with $n$ vertices and let $e$ be a diagonal of $G$. Denote the PSLG $(V, E \cup \{e\})$ by $G'$. Given the DCEL and TREES representations of $Vorb(G)$, and $HSD(G)$, $STAR(G)$, $HT(G)$, we can construct $HSD(G')$, $STAR(G')$ and $HT(G')$ in time $O(k(e) \log n)$ and space $O(n)$.

*Proof:* To obtain $HSD(G')$, we build $Vorb(G')$ analogously as in the proof of Theorem 4.1. We identify the edges of the straight-line dual of $Vorb(G)$ in one-to-one correspondence to the edges of $Vorb(G)$ totally lying on the boundaries of the parts of regions in $Vorb(G)$ cut off from their sites by $e$, deleting them from $HSD(G)$. Instead, we insert the edges of the straight-line dual of $Vorb(G')$ in one-to-one correspondence to the edges of $Vorb(G')$ that do not appear in $Vorb(G)$. Since the number of the deleted and inserted edges is $O(k(e))$, and the updating of $Vorb(G)$ to $Vorb(G')$ takes $O(k(e) \log n)$ time and $O(n)$ space, $HSD(G')$ can be also constructed in time $O(k(e) \log n)$ and $O(n)$. Further, we may assume that the DCEL and TREES representations of $Vorb(G')$ are given.

To build $STAR(G')$, we insert $e$ in the two trees in $STAR(G)$ for the endpoints of $e$. For each of the two trees, if $e$ is inserted between two edges of $G$ that previously induced an empty triangular face with a diagonal of $G$ in $HT(G)$, then we delete the diagonal from $HT(G)$ and destroy the pointers to it. Clearly, all these operations take $O(\log n)$ time and $O(n)$ space.

It remains to complete updating $HT(G)$. Augmenting $G$ by $e$ can give rise to at most two new candidates for a diagonal cutting off a triangular face from $G$ and we can find these candidates by using $STAR(G')$ in time $O(\log n)$. We test each of the candidate segments for cutting off an empty triangular face from $G$ by traversing $Vorb(G')$ along the segment and then scanning edges of $Vorb(G')$ inside the triangle. If an unexpected barrier or a fifth different region of $Vorb(G')$ is encountered during the traversal and scan, the test is failed. The method is described in more details in the proof of Lemma 3.2 in [Li86]. It takes a constant number of the operations on the representations of $Vorb(G')$, i.e. $O(\log n)$ time. ∎

Combing Theorem 3.1 with Lemma 4.1, we obtain the main result of the paper.

*Theorem 4.1:* Let $G$ be a PSLG $G$ with $n$ vertices. The algorithm (1-7) constructs the greedy triangulation of $G$ in time $O(n \log n + \log n \sum_{e \in T} k(e))$ and space $O(n)$.

Since, for all edges $e$ in $T$, $k(e) = O(n)$ holds by Fact 2.2, we obtain the following corollary.

*Corollary 4.1:* The algorithm (1-7) constructs the greedy triangulation of $G$ in time $O(n^2 \log n)$ and space $O(n)$.

Note that whenever $e$ lies inside the union of the regions of its endpoints in $Vorb(G)$ then $k(e) = 0$. Hence, by Fact 4.1, only the $e's$ passing through three or four different regions in $Vorb(G)$ can give rise to the worst-case $O(n^2 \log n)$-time performance. However, if the input PSLG $G$ is a set of $n$ points uniformly distributed in a unit square, one would expect $k(e) = O(1)$, for the overwhelming majority of edges $e$ in the greedy triangulation $T$ of $G$. Hence, Theorem 4.1 suggests the following claim.

*Claim 4.1:* The expected-time performance of the algorithm (1-7) applied to a set of $n$ points uniformly distributed in a unit square is $O(n \log n)$.

To support the above claim, one could use methods employed in exhibiting good expected behavior of the incremental algorithms for Voronoi diagrams for point sets [OIM84]. The best known upper bound on the expected-time performance of an algorithm for a greedy triangulation is $O(n^2)$ [MZ82].

## *References*

[AA86] Takao Asano and Tetsuo Asano, *Voronoi Diagram for Points in a Simple Polygon*, manuscript, Sophia University, Tokyo.

[F86] S. Fortune, *A Sweepline Algorithm for Voronoi Diagrams*, in the proceedings of the 2nd ACM Symposium on Computational Geometry, Yorktown Heights, New York.

[G79] P.D. Gilbert, *New Results in Planar Triangulations*, M.S. Thesis, Coordinated Science Laboratory, University of Illinois, Urbana, Illinois.

[Li86] A. Lingas, *Voronoi Diagrams with Barriers and the Shortest Diagonal Problem*, submitted.

[LL86] D.T. Lee and A. Lin, *Generalized Delaunay Triangulation for Planar Graphs*, to appear in Discrete and Computational Geometry, Springer Verlag.

[Me84] K. Mehlhorn, *Data Structures and Algorithms*, EATS M.T.C.S, Springer Verlag, N.Y.

[MZ82] G.K. Manacher, and A.L. Zorbrist, *The use of probabilistic methods and of heaps for fast-average-case, space-optimal greedy algorithms (extended abstract)*, manuscript.

[NP82] J. Niervergelt and F.P. Preparata , *Plane-sweeping algorithms for intersecting geometric figures*, Commun. ACM., vol. 25, no. 10. pp. 739-747.

[OIM84] T. Ohya, M.Iri, K. Murota, *Improvements of the Incremental Method for the Voronoi Diagram with Computational Comparison of Various Algorithms*, Journal of the Operations Research Society of Japan, Vol 27(4), pp. 306-336.

[PS85] F.P. Preparata and M.I. Shamos, *Computational Geometry, An Introduction*, Texts and Monographs in Computer Science, Springer Verlag, New York.

# ON THE REACHABILITY REGION

# OF A LADDER IN TWO CONVEX POLYGONS

*Minou Mansouri, Godfried Toussaint*
School of Computer Science, McGill University, Montreal

## 1 - Introduction

Two polygons are said to be *separable under translation* if one of them can be translated an arbitrary distance in some fixed direction, without intersecting with the other, [Toul]. One class of polygons that present interesting properties are *star-shaped* polygons. It is a well known result that two star-shaped n-gons are always movably separable with a single translation and that a direction for separating them can be determined in linear time [LP]. This suggests that two star shaped polygons, P and Q, can be separated by translating both of them simultaneously in some pairs of direction with respect to an arbitrary fixed point in the plane. In fact it is enough to guarantee that the *relative* motion between P and Q is correct. Let K(P) and K(Q) be the respective kernels of P and Q. Let a and b be any pair of points in the plane such that the line L(a, b) going through a and b, intersects K(P) and K(Q). Let x be any reference point in the plane, and consider the vectors xa, xb and ab, in figure 1 . We can now see that if we translate P and Q in directions xa and xb with velocities proportional to the magnitudes of xa and xb respectively, the correct relative motion between P and Q is maintained. Different pairs of points (a, b) only change the relative velocity of separation. The problem we want to solve is then :

Let there be two star shaped polygons SP and SQ with kernels K(SP) and K(SQ). And let there be two points $a \in$ Ker (SP) and $b \in$ Ker (SQ). The vector *ab* determines a *direction* of separation for SP and SQ but also a *velocity*. We are interested in finding all the pairs of points p $\in$ Ker(SP) and q $\in$ Ker(SQ) such that $\| pq \| = \| ab \|$. In other words what regions inside the two kernels determine a given velocity of separation of the two polygons ?

We first present the entire algorithm. Then we show how to compute the unreachability regions and the reachability regions in the two polygons. Finally we close with open problems.

## 2 - Preliminary results

Given two convex polygons P = $\{p_1, p_2, \ldots, p_m\}$ and Q = $\{q_1, q_2, \ldots, q_n\}$ and a line segment S = [a, b] of length r, also called a *ladder*, a *needle* or a *rod*, we want to compute the reachability regions in P and Q for the endpoints a and b respectively, with the constraint that a remains within the boundaries of P and b within those of Q. **Definition** : Given two disjoint convex polygons P and Q, let the union (set) of reachable regions PR in P and QR in Q be defined as :

PR = { p in P | for every p in PR , there exists q in Q | d(p, q) = r }
QR = { q in Q | for every q in QR , there exists p in P | d(q, p) = r }
where d(p, q) denotes the euclidean distance between p and q.

Calculating the reachable region is the same procedure for each polygon. The reachability

region(s) in a polygon is such that its boundary may contain *arcs*, it is *not* necessarily a *convex* region, it may be *disconnected* and have O(n) parts, each of which is a reachable region.
For the proof of the following lemmae refer to [Man].

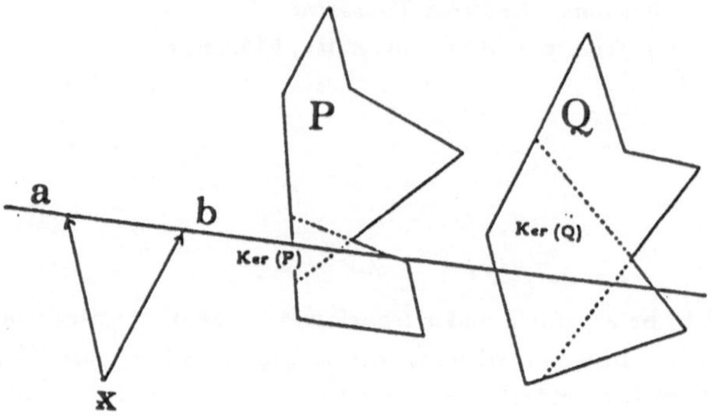


**Figure 1**

**Lemma 1 :** Let PR and QR be non empty. Then for every point x in PR, there exists point y in QR, such that d(x, y) = r and for every point s in QR, there exists point t in PR, such that d(s, t) = r.

Let dmin(P, Q) and dmax(P, Q) respectively denote the minimum and the maximum distances between P and Q, i.e.,
dmin(P, Q) = min { d(p, q) | p in P and q in Q }, over all p and q
dmax(P, Q) = max { d(p, q) | p in P and q in Q }, over all p and q.

**Lemma 2 :** A line segment S = [a, b] of length r can be placed such that $a \in P$ and $b \in Q$ *if and only if* $\text{dmin}(P, Q) \leq r \leq \text{dmax}(P, Q)$.

Let CP and CQ denote respectively the *convex hull* of circles of a given radius about each of the vertices of P and Q, UP and UQ denote respectively the *unreachable* regions in P and Q, RP and RQ denote the *reachable* regions in P and in Q.

**3 - The algorithm**

*Input :*
- two convex polygons P= $\{p_1, p_2, \dots, p_m\}$ , Q = $\{q_1, q_2, \dots, q_n\}$.
- a line segment of length r , S = [ $a$ , $b$ ].

*Output :*
Reachability regions, RP and RQ, in P and Q for point $a$ in P and point $b$ in Q.

*Algorithm REACHABILITY :*

**begin**

*step 1 :* calculate dmin (P,Q) and dmax (P,Q).

*step 2* : if dmin(P,Q) $\leq$ r $\leq$ dmax(P,Q) then continue to step 3, else stop.
*step 3* : if r = dmin(P,Q) then the reachability region is the pair of points (p, q) of P and Q that realize dmin.
Idem if r = dmax(P,Q), (we may have >1 pair of points)
*step 4* : Compute the reachability region in Q :
*1.* calculate CP with radius r.
*2.* intersect CP and Q, obtain Q'
( we know that this intersection is not empty because dmin(P,Q) $\leq$ r < dmax(P,Q) )
*3.* calculate UP
*4.* if UP = empty set then RQ = Q'
else
1. intersect UP and Q', obtain Q''
2. if Q''= empty set then RQ = Q'
else RQ = Q'- Q''
*step 5* : compute the reachability region in P, as in *step 4*.
**end.**

**Theorem 1 :** *Algorithm REACHABILITY* correctly computes PR and QR in O ( n + m ) time.
*Proof :*
The correctness follows from lemmae 1 and 2 and the results of the following sections. Let us now turn to the complexity. In step 1 , the minimum and the maximum distances can be computed in O(log m + log n) and O(n) time respectively, using Edelsbrunner [Ede] and Toussaint and Bhattacharya's algorithms [BT2]. The test in steps 2 and 3 takes constant time to perform. Step 4 and 5 have an overall complexity of O(n) as proved in the following sections. The total running time of the algorithm is thus linear. Q.E.D.

## 4 - Calculating the unreachable region

Given a convex polygon P = $\{ p_1, p_2, \ldots , p_n \}$ and a line segment A = $[a_1, a_2]$ of length r, with the constraint that $a_1$ be *inside* or *on* the boundary of P, what is the unreachable region for $a_2$ ? Let CH be the convex hull of the n circles of radius r about each of the n vertices of P. The region of the plane outside CH is unreachable by $a_2$. There may also be such a region inside CH. If it exists, we prove that it is the intersection of the n circles. This intersection can be calculated in O(n) time using either Melville's algorithm [Mel] or the algorithm described in [Man] . The latter can be generalized to the case where each circle around a vertex has a different radius. Before sketching this algorithm, let us exhibit a few results. We show in [Man] that the unreachable region for $a_2$, inside CH($c_1$, $c_2$, ... , $c_n$), if it exists, is the strict interior of the intersection of all the $c_i$'s, $i$ = 1, 2, ... , n.

**Definition :** Let P = $\{ p_1, p_2, \ldots , p_n \}$ be a convex polygon, where the $p_i$'s are the vertices specified in terms of cartesian coordinates and given in counterclockwise order. Let $c_i$ be the circle about vertex $p_i$ of radius r. See figure 2. The *lune* of two circles $c_i$ and $c_j$, noted lune $(c_i,c_j)$, is $c_i \bigcap c_j$ .

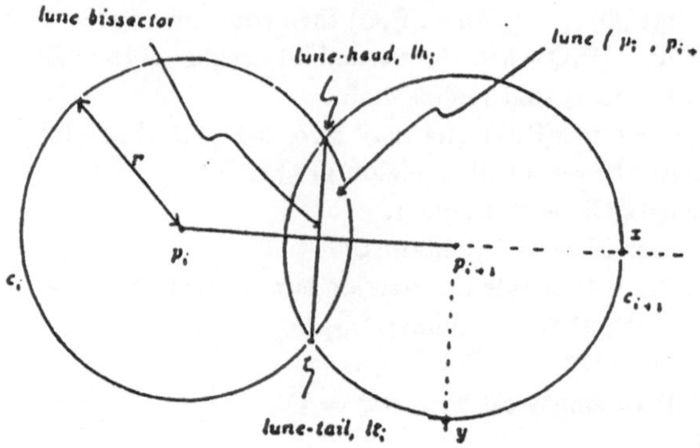


**Figure 2**

**Inner and outer arcs :**
We draw two types of arcs between lune heads according to their relative position. By convexity of P, $p_{i+2}$ must be in the region of the plane that is to the left of $(p_i, p_{i+1})$ and of $(p_1, p_2)$ and to the right of $(p_{i+1}, p_1)$. This region may be bounded or not. The angle $<p_{i+2}, p_i, p_{i+1}>$ is in the range $]\ 0\ ,\ \pi\ [$ and the lune bissector of $(p_{i+1}, p_{i+2}) = e_{i+1}$ makes with $(p_i, p_{i+1}) = e_i$, an angle in the range $]\ \pi/2\ ,\ 3 * \pi/2\ [$. We draw an *inner arc* on $c_{i+1}$, if $lh_{i+1}$ is on the arc $(lh_i, y)$ ( going counterclockwise). We draw a *outer arc* on $c_{i+1}$, if $lh_{i+1}$ is on the arc $(x, lh_i)$ ( going counterclockwise). A pattern is obtained by joining each lune head to the next one. We present a linear algorithm to compute the intersection of n circles $c_i$ , for $i = 1, 2, \ldots, n$, of radius r whose centers are the vertices of a convex polygon $P = \{ p_1, p_2, \ldots, p_n \}$.

**Notation :** I is the intersection of the n circles, diameter (P) is realized by $(p_k, p_l)$, *ca* is the total number of arcs in the pattern, *cg* is the number of inner arcs in the pattern.

*Algorithm Circle Intersection :*

**begin**

*Step 1* : If diameter(P) is
- $> 2 * r$ : stop, $I = \emptyset$ .
- $<= 2 * r$ : - for all circles $c_i$ do :
  - if $c_i$ contains entirely lune$(c_k, c_l)$, delete $c_i$.
  - if $c_i$ is outside lune$(c_k, c_l)$ : stop, I is empty.
  - Draw pattern and in the process :
    - * update *cg* and *ca*.
    - * for each outer arc on $c_i$, if lune$(c_{i-1}, c_{i+1})$ is outside $c_i$ : stop, $I = \emptyset$ .

*Step 2* : If $ca <> cg$ then for every outer arc on $c_i$ do :
- 1- if lune$(c_{i-1}, c_{i+1})$ is inside $c_i$ then
  - * delete $c_i$
  - * update pattern and counters

  else stop, $I = \emptyset$ .
- 2- if $cg < 2$ , stop , $I = \emptyset$ .

*Step 8* : If $ca = cg$ : the pattern is the intersection I.
**end.**

**Theorem 2 :** *Algorithm Circle Intersection* correctly computes the intersection of n circles of equal radius whose centers are the vertices of a convex polygon in linear time.
*Proof :* Refer to [Man].

## 5 - Calculating the reachable regions

Given two disjoint convex polygons $P=\{p_1, p_2, \ldots, p_n\}$ and $Q=\{q_1, q_2, \ldots, q_m\}$ with n and m vertices respectively and given a line segment $S = [a, b]$ of length r, we want to compute the regions in Q that $b$ can reach with the constraint that $a$ lies within the boundaries of P . Testing whether a reachability region *exists* for point $b$ in polygon Q, is a simple step of the complete algorithm. Therefore, in this section, we assume that such a region exists in Q, and describe the algorithm to calculate it. This is done in two main steps. The two problems we will solve are best visualized in figures 3-a and 3-b. Let CP denote the *convex hull* of n circles of radius r, with centers the n vertices of P respectively, and let UP denote their *intersection* . We will assume UP is not empty. Recall from the previous section that the reachable region for point $b$ when $a$ remains inside P is CP - UP, where the symbol - denotes the set difference. In other words the unbounded region of the plane outside CP is unreachable, as well as UP which is inside CP. Since we assumed that a reachability region exists then CP and Q must have a non-empty intersection Q'. Finding Q' will be the first problem. Also since UP exists, intersecting UP with Q' and obtaining Q" will be the second problem. The reachable regions for point $b$ will then be RQ = Q' - Q". Refer to figure 4 for a description of a *dome*, an *attic*, an *old* and a *new edge* and an *arc region.*

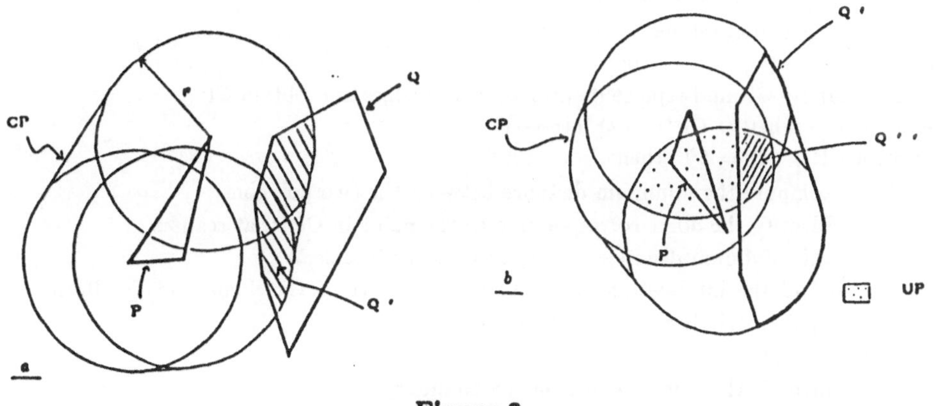


**Figure 3**

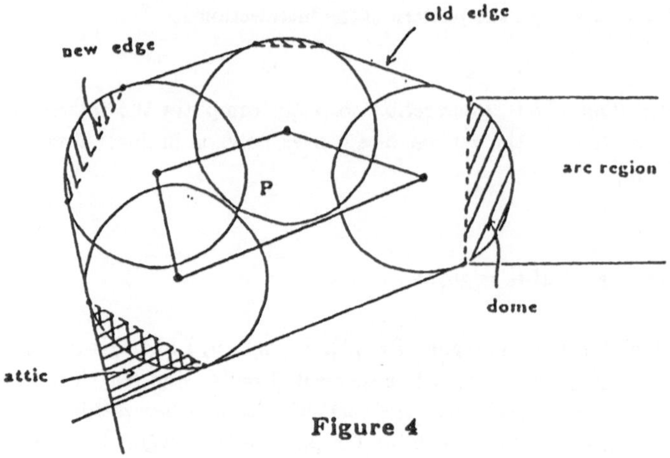


Figure 4

## 5.1 - The first problem

We want to compute the intersection region between CP and Q, denoted by Q′. Consider figure 5. With the assumptions made previously, only figures 5-a and 5-b are valid. The boundary of CP is an alternating sequence of arcs and edges. To have a linear running time algorithm, we first work with two convex polygons, then reinsert the arcs to update the intersection region. The two possible situations are that Q is entirely inside CP or that Q intersects CP and parts of Q are unreachable regions i.e. outside of CP.

*Algorithm Intersection-1:*
*Input* : two convex polygons P and Q, and a ladder S = [a, b] of length r.
*Output* : the intersection Q′of CP and Q.

*Assumption* : CP and Q intersect.
**begin**

*step 1* : compute CP and replace its arcs with new edges and obtain CP′.
*step 2* : detect whether CP′and Q intersect
*step 3* : if $CP' \cap Q = \emptyset$ then
- compute the minimum distance between the two polygons
- identify the dome corresponding to the point in CP′that realizes the minimum distance with Q ; this dome intersects Q
- find the intersection points between the arc of the dome and Q. If there aren't any, Q is strictly inside the dome.

*step 4* : if $CP' \cap Q <> \emptyset$ then
- identify the bridge points on the boundary
- for every outer subchain of Q, identify the intersection points with the arc in each arc region.

*step 5* : merge the inner chains of CP and Q, determined by the intersection points found in steps 3 and 4.

**end.**

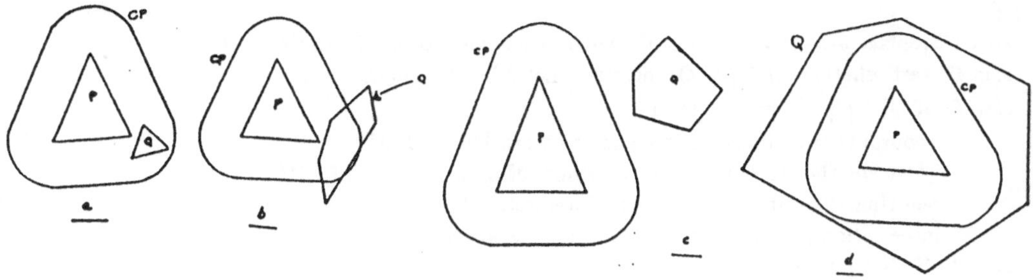


**Figure 5**

**Theorem 3 :** *Algorithm Intersection-1* correctly computes the intersection of CP and Q in linear time.

*Proof :* Refer to [Man] for the data structure and the correctness proof. Computing CP$'$ directly from P can be done in linear time. In step 2, detecting the existence of an intersection between two convex figures can be done in sublinear time using Chazelle and Dobkin's algorithm [CD]. In step 3, the minimum distance between two convex polygons can be obtained , together with the points realizing it in logarithmic time using Edelsbrunner's binary elimination technique [Ede]. Then once a dome is identified finding the intersection points with Q , if there are any , is done in at most O(n) time. For the general case in step 4, since we have the outer subchains of Q (they are ordered but this does not really matter) and the corresponding arc regions, updating the intersection region is a merge of two sorted lists each having a linear number of elements. Step 5 is also a linear merge of two sorted lists, if we leave pointers from the intersection points to the inner and outer chains in both directions. Therefore the total running time of the algorithm is bounded by O(n).

### 5.2 - The second problem

At this stage we have obtained the region Q$'$ in Q, in which the reachability regions are contained. The problem is to compute the intersection region IQ, between Q$'$, and UP. The reachability region in Q will be RQ = Q$'$- IQ. See figure 6 for an illustration.

*Algorithm Intersection-2 :*

*input :* - a convex figure UP that is the intersection of n circles of radius r , about vertices of a convex polygon P

- a convex figure Q$'$ whose boundary may contain arcs of CP, obtained from the algorithm in the previous section .

*output :* the reachability region for endpoint *b* of a ladder S = [*a*, *b*] ,

$Q' - (UP \cap Q')$

**begin**

*step 1* : replace arcs of UP and of Q′ with new edges, to obtain UP′and Q″

*step 2* : test whether UP′ ∩ Q″intersect. Let J be this intersection.

*step 3* : if UP′ ∩ Q″= ∅ then
- compute the minimum distance between UP′and Q″
- test whether the arc of UP, corresponding to a vertex of UP′ realizing the minimum distance, intersects Q″, if so compute the intersection points , otherwise UP and Q′are disjoint.

*step 4* : if UP′ ∩ Q″< > ∅ then
- compute CH ( UP′U Q″)
- replace new edges of Q″with its arcs
- build arc regions around UP′
- for every outer subchain of Q′, go from one arc region to the next and identify the intersection points

*step 5* : merge the two lists of UP and Q′and take only the regions determined by the intersection points, found in steps 3 and 4, and the outer chains of Q′

**end.**

**Theorem 4** : *Algorithm Intersection-2* correctly computes the intersection of UP and Q′ in linear time.

*Proof* : For the data structure and the correctness proof refer to [Man]. The complexity of steps 1, 2 and 3 is as in the previous algorithm. In step 4 , the CH can be computed using the rotating callipers of Toussaint in O(m + n) time [Tou3]. Replacing arcs and building arc regions can be done in O(n) time. We use a linear running time algorithm to compute the bridge points in sorted order . Since for every outer subchain of Q′we know which arc region we are in , the testing for intersection is merely a merge of these two lists , and since the number of edges and arcs in each list is bounded by n, this step of the algorithm is linear. Step 5 is also a mere traversal of two sorted lists and is carried out in O(n) time.

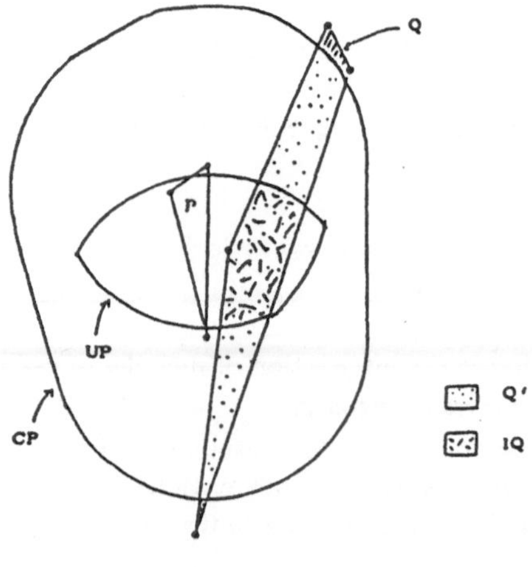


**Figure 6**

## 6 - Conclusion

We have solved the problem of finding all the reachability regions in two convex polygons, for the tips of a ladder, constrained to remain within the boundaries of the polygons, in time linear in the number of vertices. One interesting open question is how to compute, efficiently, the reachability region in n convex polygons for the vertices of a n-gon, free to move but with the restriction that each of its n vertices remain in a polygon. Another interesting problem that arises is the computation of these regions for a ladder in two simple polygons. The extension of these problems to the three dimensional space remains open , such as moving a ladder in two polyhedra, a triangle in three polyhedra , and a polyhedron in n polyhedra.

## References

[**Bro**] Brown Kevin Q., "*Geometric Transforms for Fast Geometric Algorithms*", PhD thesis, Carnegie-Mellon University, (Dec. 1979), 54-57.

[**BT1**] Bhattacharya B. and Toussaint G. T., "*A linear algorithm for determining the translation separability of two simple polygons*", Tech. Rep. No. SOCS 86.1, McGill University (Jan. 1986).

[**BT2**] Bhattacharya B. and Toussaint G. T.,"*Efficient algorithms for computing the maximum distance between two finite planar sets*", Journal of Algorithms,4, (1983), 121-136.

[**CD**] Chazelle B. and Dobkin D., "*Detection is easier than computation*", Proc. Twelfth Annual ACM Symposium on Theory of Computing, (April 1980) 146-153.

[**Cha1**] Chazelle B. and al., "*The complexity and decidability of separation*", Tech. Rep. CS-83-84, University of Waterloo, (Nov. 1983).

[**Cha2**] Chazelle B., "*Computational geometry and convexity*", Ph.D. thesis, Carnegie-Mellon University, (July 1980).

[**Daw1**] Dawson R. J., "*On the mobility of bodies in $R^n$*", manuscript,Cambridge University, England.

[**Daw2**] Dawson R. J., "*On removing a ball without disturbing the others*", Mathematics Magazine, vol. 57, no. 1, (Jan. 1984), 27-30.

[**Ede**] Edelsbrunner H., "*Computing the extreme distances between two convex polygons*", Journal of Algorithms 6,(1985) 213-224.

[**LP**] Lee D. T. and Preparata F. P., "*An optimal algorithm for finding the kernel of a polygon*", Journal of the ACM 26, (1979), 415-421.

[**Man**] Mansouri M., "*On the reachabillity region of a ladder in two convex polygons*", Master's Thesis, McGill University, Montreal (August 1986).

[**McT**] McKenna Michael and Toussaint Godfried T., "*Finding the minimum vertex distance between two disjoint convex polygons in linear time*", Comp. and Maths. with Applications, vol. 11, No. 12, (1985), 1227-1242.

[**Mel**] Melville R., "*An implementation study of two algorithms for the minimum spanning circle problem*", in Computational Geometry, G. T. Toussaint, Ed., North Holland, 1985.

[**O'R**] O'Rourke J., et al., "*A new linear algorithm for intersecting convex polygons*", Computer Graphics and Image Processing, Vol. 19,(1982), 384-391.

[**Sha**] Shamos M.I., "*Computational Geometry*", Ph.D. thesis, Yale University, (1978).

[**Tou1**] Toussaint G.T.,"*Movable Separability of sets*", in Computational Geometry, Ed., G. T. Toussaint, North Holland, (1985).

[**Tou2**] Toussaint G.T., "*The complexity of movement*", IEEE International Symposium on Information Theory, St. Jovite, Canada,(Sept. 1983).

[**Tou3**] Toussaint G.T., "*Solving geometric problems with the rotating calipers*", Proc. MELECON, Athens, Greece, (May 1983).

[**Whi**] Whitesides S. H.,"*Computational geometry and motion planning*", in Computational Geometry, Ed., G. T. Toussaint, North Holland, (1985).

# Computational Geometry with Restricted Orientations

Gregory J. E. Rawlins
Dept. of Computer Science
University of Indiana
Bloomington, Indiana, U.S.A.

Derick Wood
Dept. of Computer Science
University of Waterloo
Waterloo, Ont., Canada

**Abstract**

Given a set $\mathcal{O}$ of orientations (or angles) a line, ray, or line segment, in the plane, is said to be $\mathcal{O}$-oriented if the smallest angle it makes with a horizontal line is in $\mathcal{O}$. We are interested in planar objects that are formed by $\mathcal{O}$-oriented lines, rays, and line segments; we say that they are $\mathcal{O}$-oriented objects. Our interest in this area stems from the observation that orthogonal objects can, in general, be handled more efficiently than arbitrarily-oriented ones. As we demonstrate, as far as convexity is concerned $\mathcal{O}$-oriented geometry bridges the gap between orthogonal geometry and arbitrarily-oriented geometry.

## 1 Introduction

Convex sets are a comparatively recent but very fruitful concept in geometry having applications in optimization, statistics, geometric number theory, functional analysis and combinatorics ([7,10]) and this is one of the reasons for the inordinate interest in convex sets in computational geometry. But their study is also practically motivated since the convex hull of an object typically has much less complexity than the object itself and so it is much used in testing for intersections among objects ([10,18]). Convex polygons also occur in decomposition results since there are many good algorithms for convex polygons and so polygons are decomposed into convex subparts to answer various queries ([10,18]). Finally, the convex hull was one of the first concepts studied in computational geometry ([16]) and so deserves especial mention.

In this paper we investigate the more general concept of *restricted-orientation convexity* and apply it to *arbitrary* sets of points thereby generalizing all previous results (see [9,12]) and, also, verifying some otherwise unsupported observations in the literature (see also [11,13]).

Curiously, our investigation demonstrates that in the plane we may treat restricted-orientation convexity just as if it were orthogonal convexity. In other words, it is always possible to construct a case analysis which is only concerned with at most two orientations at a time. It seems natural to conjecture that this relationship also holds in three dimensions (that is, we only need results for three dihedral orientations) and so on to higher dimensions. This result is interesting on several levels, the most important of which is the purely practical one that, *in terms of convexity, there is no loss in going from orthogonal to arbitrarily many orientations.*

Because we refer to some of them in the body of the paper we list here some of the most salient properties of planar convex sets ([4]). In the following $\mathbf{P}$ is a planar convex set:

1. $\mathbf{P}$ is simply connected.

2. The intersection of $\mathbf{P}$ and any line is either empty or a connected set.

3. $\mathbf{P}$ is the intersection of all convex sets which contain it.

4. If $p \notin \mathbf{P}$ then there exists a line separating $p$ and $\mathbf{P}$.

5. **P** is the intersection of all halfplanes which contain it.

6. If $p, q \in \mathbf{P}$ then the line segment joining $p$ and $q$ is in **P**.

Except for property 1, all of these are defining characteristics of convex sets. In the concluding section of this paper we list the corresponding properties of our more general "convex" sets which include these as special cases.

This paper is subdivided into the following sections: Section 2 establishes the conventions that we adhere to in this paper. Section 3 contains the definition of these new "convex" sets and several of their more elementary properties. Section 4 contains two theorems; the "Separation Theorem" which gives exact conditions on when a point can belong to the "convex hull" of a set and the "Decomposition Theorem" which establishes a kind of incremental property of these new "convex" sets. Section 5 introduces the notion of a "stairline" which serves as a suitable analogue of a straight line for these sets. Section 6 contains two theorems; the "Characterization Theorem" which characterizes these new sets completely in terms of their boundary (using stairlines) in much the same way that planar convex sets have been so characterized and the "Visibility Theorem" which characterizes these sets in terms of a generalization of visibility. Finally, in section 7, we summarize the properties established in this paper and point the way to further work in this area.

# 2 Agreements

All of our results are described in $\Re^2$ since planar relationships admit of easy visualization, but the results are easily generalizeable to $\Re^n$ (and in fact to any finite-dimensional normed linear space) in the usual way. We assume the reader's familiarity with such elementary topological concepts as (path-) connectedness, closure, simplicity, separability, support, interior and boundary of planar figures.

We denote subsets of $\Re^2$ by bold face capital letters (e.g., **P** and **Q**) and elements of such sets by *lower case italic letters* (e.g., $p$ and $q$). We treat a subset of $\Re^2$ as a set of interior points together with its boundary. We use the symbol $O$, with or without subscripts, to refer to a set (possibly empty) of orientations. A collection of lines, segments and rays is said to be *O-oriented* if the set of orientations of the elements of the collection is a subset of $O$. Thus we speak of "$O$-lines", "$O$-segments" and "$O$-rays" to mean $O$-oriented lines, segments and rays. By extension, we call a polygon an "$O$-polygon" if its edges are $O$-segments.

Because we wish to preserve symmetry of direction in this paper we assume that the set $O$ is symmetric about the horizontal, that is, if it contains an orientation $\theta < 180°$ it also contains an orientation $\theta + 180°$ and similarly for $\theta > 180°$. The notion of $O$-orientation has been previously defined (for finite $O$) in [5,12,20,21] and, in a slightly related form, in [3]. As mentioned in the Introduction there is a vast literature concerning the special case of $O = \{0°, 90°\}$ (more exactly, $O = \{0°, 90°, 180°, 270°\}$) (see [10,22] for further references).

We assume that $O$ is representable as a list of disjoint *closed* ranges some of which may collapse to single orientations. For example, $O$ may be the set $\{\theta_1, [\theta_2, \theta_3], [\theta_4, \theta_5], \theta_6, \theta_7\}$ (all $\theta_i < 180°$). orientations greater than 180°.)

We call the open range $(\theta_1, \theta_2)$ *O-free* if there are no orientations in $O$ in the range $(\theta_1, \theta_2)$.

We use the notation $L(p, q)$ to mean the line passing through the points $p$ and $q$ and $LS(p, q)$ to mean the line segment with endpoints $p$ and $q$. We also use the notation $\Theta(L)$ (where $L$ is a line, segment or ray) to mean the orientation of $L$.

# 3 Restricted-Orientation Convexity

Property 2 of convex sets stated in the Introduction can be taken as a defining characteristic of convex sets (as are all the others except for property 1). A set is said to *convex* if its intersection with *any* line is empty or connected. Here however we are only interested in intersections with a particular class of lines, namely those whose orientations belong in some restricted set of orientations. As a result we speak of *restricted-orientation convexity*. The phrasing is somewhat unfortunate since it implies that it is a restriction of normal convexity when in fact the opposite is the case, restricted-orientation convexity includes (normal) convexity as a special case.

From this point on we assume that we have chosen some fixed set $O$ of orientations (none of our results depend on the particular set chosen).

**Definition 3.1** *We say that* **P** *is* $O$-convex *if the intersection of* **P** *and any* $O$*-line is either empty or connected.*

(Note that if **P** is a polygon then **P** is $\{\theta\}$-convex if and only if **P** is monotone in $\theta + 90°$.)

This is a natural generalization of the notion of orthogonal convexity — a set is orthogonally convex if its intersection with any horizontal or vertical line is either empty or connected. Orthogonal convexity has been defined not only in computational geometry ([22]) but also in digital picture processing ([14]) and for polyominoes ([1]).

Figure 1 contains some example figures which are $O$-convex for various $O$. Figure 1 (a) is not $O$-convex for any non-empty $O$, but is $O$-convex if $O = \emptyset$, as are all the other figures. Figures 1 (b) and (c) are convex with respect to any horizontal line, as are (d), (e) and (f), so they are all $\{0°\}$-convex besides being $\emptyset$-convex. Note that (b) and (c) are not convex in any other direction. Figures 1 (d), (e) and (f) are convex with respect to any vertical line as well and so they are also $\{0°, 90°\}$-convex. Note that (d) is not convex in any other direction. Figures 1 (e) and (f) are convex with respect to any line with orientation in the range $\{[90°, 180°]\}$ and so they are also $\{[90°, 180°]\}$-convex. Note that (e) is not convex in any other direction. Figures 1 (f) is $O$-convex for any $O$.

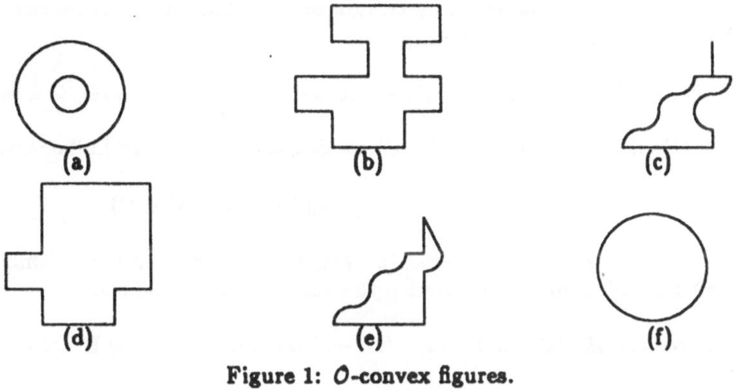

Figure 1: $O$-convex figures.

**Lemma 3.1** *All planar convex sets are* $O$*-convex.*

**[Note: We omit all proofs in this version; the reader is referred to [11] and [12] for the proofs of the stated results]**

Indeed, we can go further and completely characterize convex sets as a sub-class of $O$-convex sets:

**Observation 1** *A planar set is convex if and only if it is* $\{[0°, 180°)\}$*-convex.*

In fact it is easy to construct examples to show that this is true for no smaller set of orientations. For example if we delete just *one* orientation (say $\theta_1$) then any set consisting of just two distinct points on a $\{\theta_1\}$-line is $\{\theta\}$-convex for all $\theta \neq \theta_1$ but is, of course, not convex. In fact examples like these establish that the statement "for all **P**, **P** is connected if **P** is $O$-convex" holds if and only if $O = \{[0°, 180°)\}$.

Note that the following sets are convex, and hence $O$-convex for any $O$: the empty set, $\Re^2$, and, any point, line, segment, ray or halfplane in $\Re^2$.

**Lemma 3.2** *If $C$ is a non-empty collection of $O$-convex sets, then $\bigcap C$ is $O$-convex.*

We now introduce briefly the notion of abstract convexity spaces. Abstract convexity theory is concerned with collections of subsets of a set which obey two weak axioms. Such a collection of sets has been variously called a *convexity space* ([8]), *convexity structure* ([17]), *alignment* ([6]), or *algebraic closure system* ([2]). We use the term "convexity space" since we are more interested in the geometry, as opposed to the algebra, of such a collection. A convexity space, in the sense we use it here, is intended to be an abstraction of the more essential properties of convex sets in $\Re^n$.

**Definition 3.2** *Given a set, $S$, and a family, $\mathcal{C}$, of subsets of $S$ the structure $(S, \mathcal{C})$ is said to be a* convexity space *if*

1. $\emptyset, S \in \mathcal{C}$ *and*
2. $\forall C \subseteq \mathcal{C}$; $\bigcap C \in \mathcal{C}$, *where* $\bigcap C = \bigcap_{X \in C} X$.

$S$ is called the *groundset* of the convexity space and any element of the family $\mathcal{C}$ is said to be $\mathcal{C}$-*convex*. The interpretation of the family $\mathcal{C}$ is that it is the set of all convex sets over some space, where we have deferred the operational question of what we *mean* by convexity. The dominant characteristic of convex sets is taken to be closure under intersection. From our point of view, the important unifying result is the following:

**Theorem 3.3** *For all $O$, $(\Re, \mathcal{C}_O)$, where $\mathcal{C}_O$ is the set of all $O$-convex sets, is a convexity space.*

**Definition 3.3** *Given a convexity space $(S, \mathcal{C})$ we define the associated* hull operator, *$\mathcal{C}$-hull, as follows:*

$$\forall \mathbf{P} \subseteq S\,; \quad \mathcal{C}\text{-}hull(\mathbf{P}) = \bigcap \{\mathbf{Q} \mid \mathbf{P} \subseteq \mathbf{Q} \wedge \mathbf{Q} \in \mathcal{C}\}$$

It is straightforward to show that $\mathcal{C}$-*hull*(**P**) exists, is unique and is the 'smallest' $\mathcal{C}$-convex set which contains **P**. We now use this notion to define the $O$-*hull* of a set.

**Definition 3.4** *We call the intersection of all $O$-convex sets containing **P** the $O$-hull of **P**, and write $O$-hull(**P**).*

Observe that $\forall O, \mathbf{P}\,;\ \mathbf{P} \subseteq O\text{-}hull(\mathbf{P})$ even when $O = \emptyset$ or $\mathbf{P} = \emptyset$ (or both).

Because of Theorem 3.3 we can speak of *the* $O$-*hull* of any set and be assured of its existence and uniqueness. If $O = \emptyset$ then $O\text{-}hull(\mathbf{P}) = \mathbf{P}$, for all **P**, since **P** is the smallest set containing **P** which is not required to be convex in any direction. Similarly, if $\mathbf{P} = \emptyset$ then $O\text{-}hull(\mathbf{P}) = \mathbf{P}$, for all $O$, since the intersection of every $O$-line and **P** is empty. When $O = \{\theta\}$ and **P** is a polygon then the $O$-*hull* of **P** has been called the "$\theta$-visibility hull" of **P** ([15,19]).

Note that in Figure 1, (f) is the $O$-*hull* of (a) for any non-empty $O$ and (d) and (e) are the $\{90°\}$-*hulls* of (b) and (c) respectively.

**Lemma 3.4** $\forall O, \mathbf{P}\,;\quad$ **P** is $O$-convex if and only if $O\text{-}hull(\mathbf{P}) = \mathbf{P}$

Note that when $O$ is the set of all orientations this lemma reduces to property 3 stated in the Introduction.

**Lemma 3.5** *1.* $\forall O, \mathbf{P}$ ; $O\text{-}hull(O\text{-}hull(\mathbf{P})) = O\text{-}hull(\mathbf{P})$

*2.* $\forall O, \mathbf{P}, \mathbf{Q}$ ; $\mathbf{P} \subseteq \mathbf{Q} \Longrightarrow O\text{-}hull(\mathbf{P}) \subseteq O\text{-}hull(\mathbf{Q})$

**Lemma 3.6** *1. If $O$ is non-empty and* **P** *is connected, then $O$-hull(***P***) is simply connected.*

*2. If $O$ is non-empty and* **P** *is connected and $O$-convex, then* **P** *is simply connected.*

Compare this lemma with property 1 stated in the Introduction.

**Theorem 3.7** *A set is $O$-convex if and only if it consists of a set of disjoint connected components such that each component is $O$-convex and no $O$-line intersects any pair of components.*

Observe that if $O$ is the set of all orientations then for each pair of connected components there exists at least one $O$-line which intersects them. Hence, all $\{[0°, 180°)\}$-convex sets are connected.

# 4 Separation and Decomposition Theorems

Intuitively, we can think of forming the *O-hull* of a set **P** by sweeping a line of each orientation in $O$ across **P** and incrementing the hull formed so far so that it is convex in each direction in $O$. (Note that if $O$ is empty then we do not add anything to **P**.) Thinking of it in this way it seems reasonable that the hull we eventually produce is unchanged if we change the order of orientations in which we sweep. As we prove in Theorem 4.4 this is, in fact, the case but only for *connected* sets. For disconnected sets Lemma 4.3 is the strongest possible result.

The Decomposition Theorem (Theorem 4.4) can be established in the context of two or more convexity spaces over the same groundset (see [11]). However, in this paper we prefer to phrase our results in terms of $O$-convex sets.

**Lemma 4.1** *If* **P** *is connected and $p \in O$-hull(***P***), then each $O$-line through $p$ intersects* **P**.

**Theorem 4.2 (The Separation Theorem)** *Let* **P** *be connected and $p \notin$* **P**. *$p \in O$-hull(***P***) if and only if there exists a $\theta \in O$ such that the $\{\theta\}$-line through $p$ intersects* **P** *in, at least, two points on either side of $p$.*

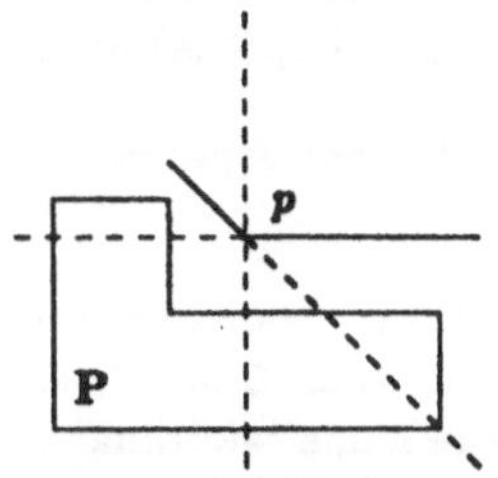


Figure 2: A halfplane containing **P** and not $p$.

Observe that if $O_1 \subseteq O_2$ then $O_1$-*hull*(**P**) $\subseteq$ $O_2$-*hull*(**P**) for all **P** since $O_2$-*hull*(**P**) contains **P** and is $O_1$-convex. In some sense as a set of orientations $O$ "grows" to include all possible orientations, the set $O$-*hull*(**P**) "grows" to the (normal) convex hull of **P**.

**Lemma 4.3** $\forall \mathcal{O}_1, \mathcal{O}_2, \mathbf{P}$ ; $\mathcal{O}_1\text{-}hull(\mathbf{P}) \cup \mathcal{O}_2\text{-}hull(\mathbf{P}) \subseteq \mathcal{O}_1\text{-}hull(\mathcal{O}_2\text{-}hull(\mathbf{P})) \subseteq (\mathcal{O}_1 \cup \mathcal{O}_2)\text{-}hull(\mathbf{P})$

This result also holds if we replace $\mathcal{O}_1\text{-}hull(\mathcal{O}_2\text{-}hull(\mathbf{P}))$ by $\mathcal{O}_2\text{-}hull(\ \mathcal{O}_1\text{-}hull(\mathbf{P}))$.

Simple counter-examples show that all these results are best possible, in that, there exists sets for which the respective converses are false. However, we can strengthen Lemma 4.3 considerably by restricting **P** to be connected.

**Theorem 4.4 (The Decomposition Theorem)** *If* **P** *is connected then* $\forall \mathcal{O}_1, \mathcal{O}_2$

$$\begin{aligned}(\mathcal{O}_1 \cup \mathcal{O}_2)\text{-}hull(\mathbf{P}) &= \mathcal{O}_1\text{-}hull(\mathcal{O}_2\text{-}hull(\mathbf{P})) \\ &= \mathcal{O}_2\text{-}hull(\mathcal{O}_1\text{-}hull(\mathbf{P})) \\ &= \mathcal{O}_1\text{-}hull(\mathbf{P}) \cup \mathcal{O}_2\text{-}hull(\mathbf{P})\end{aligned}$$

**Corollary 4.5** *If* **P** *is connected and* $\mathcal{O} = \bigcup \mathcal{O}_i$ *then* $(\bigcup \mathcal{O}_i)\text{-}hull(\mathbf{P}) = \bigcup(\mathcal{O}_i\text{-}hull(\mathbf{P}))$

This corollary verifies Toussaint and Sack's observation ([19]) that the (normal) convex hull is the union of the "visibility hulls" over all directions of visibility. Sack ([15]) showed, in the orthogonal case, that the horizontal hull of the vertical hull of an orthogonal polygon (or alternately the vertical hull of the horizontal hull) is equivalent to the union of both hulls. It was taken as self-evident that the union is the smallest horizontally and vertically convex polygon enclosing the orthogonal polygon. Corollary 4.5 validates that assumption.

This decomposition result immediately yields an algorithm to find the hull of any connected set given that we can find the hull in one direction. It turns out though that connected $\mathcal{O}$-convex sets have considerably more structure than this and we can exploit this structure to construct optimal algorithms to find the hull of any connected set (see [12] for the special case of finite $\mathcal{O}$, see [11] for the general case).

# 5 The Notion of a Stairline

To characterize $\mathcal{O}$-convex sets we need a new definition of "line" more appropriate to $\mathcal{O}$-convex sets. We call these generalized lines "stairlines" and we define and investigate them in this section. First though we need the concept of the *span* of a continuous curve in the plane.

**Definition 5.1** *We say that the continuous plane curve $S$ has* span $[\theta_1, \theta_2]$ *($\theta_1 \leq \theta_2$) if for any two distinct points $p, q \in S$ $\Theta(L(p,q)) \in [\theta_1, \theta_2]$.*

(Of course, $\theta_1 = \theta_2$ if and only if the curve is a line, segment or ray with orientation $\theta_1$.)

As an illustration: if $S$ is a continuous curve with span $[0°, 90°]$ and $(x_1, y_1), (x_2, y_2)$ are any two points on $S$ then either ($x_1 \leq x_2$ and $y_1 \leq y_2$) or ($x_1 \geq x_2$ and $y_1 \geq y_2$).

**Definition 5.2** *We say that a continuous curve in the plane with span $[\theta_1, \theta_2]$ is an* $\mathcal{O}$-stairline *if $(\theta_1, \theta_2)$ is $\mathcal{O}$-free.*

(Note that if $\theta_1 = \theta_2$ then $(\theta_1, \theta_2)$ is vacuously $\mathcal{O}$-free since there are *no* orientations in the range and so any line, segment or ray is an $\mathcal{O}$-stairline.)

We have chosen the portmanteau name "stairline" as a combination of (orthogonal) *staircase* ([22]) and (straight) *line*. By analogy with lines, segments and rays we also use the terms "$\mathcal{O}$-stairsegment" and "$\mathcal{O}$-stairray" with the obvious meanings. Note that a line, segment or ray of *any* orientation is an $\mathcal{O}$-stairline, $\mathcal{O}$-stairsegment or $\mathcal{O}$-stairray.

**Remark:** To avoid excessive terminology we assume for the rest of his paper that $\mathcal{O}$ is understood and we just refer to "stairlines" ("stairsegments" or "stairrays").

**Lemma 5.1** *1. If S is a stairline, then S is O-convex.*

*2. All stair-halfplanes are O-convex.*

*3. If* **P** *is connected and there exists a stair-halfplane which contains* **P** *and not the point p, then $p \notin$ O-hull(***P***).*

**Definition 5.3** *We say that a stairline composed of a sequence of connected line segments is a* polygonal stairline.

Polygonal stairlines have been previously defined for the special case of orthogonal objects (see [22] for references). See Figure 3 for examples of a stairsegment, a polygonal stairsegment, and an *O*-oriented polygonal stairsegment for *O* any subset of $\{[90°, 180°]\}$.

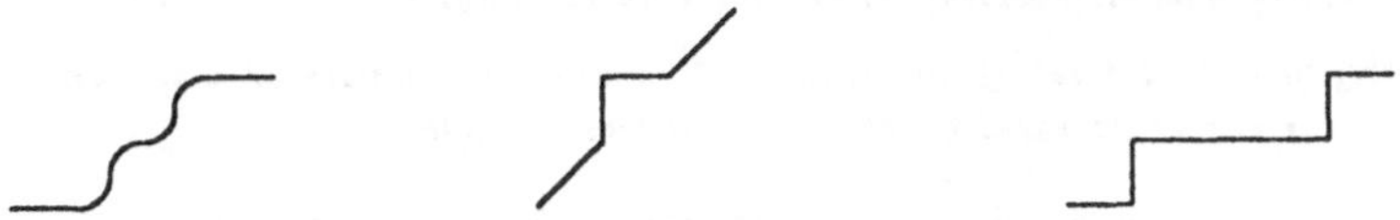

Figure 3: A variety of stairsegments.

**Definition 5.4** *We call the set of all stairsegments joining p and q the* O-region *of p and q and write O-region(p, q).*

Note that if $\Theta(LS(p,q)) \in O$ then *O-region*$(p,q) = LS(p,q)$. Of course, if $O$ consists of all orientations then, for all $p$ and $q$, *O-region*$(p,q) = LS(p,q)$. On the other hand if $O$ is empty then every range is $O$-free and so *any* continuous curve connecting $p$ and $q$ for any $p$ and $q$ is a "stairsegment", hence *∅-region*$(p,q) = \Re^2$.

With respect to $O$-convex sets stairlines are the most natural analogs of straight lines with respect to convex sets, in that: there exists a stairsegment which realises the shortest distance between any two points; an $O$-line meets a stairline at at most one point (unless collinear with some part of the stairline); and two stairlines with disjoint spans can only intersect at at most one point.

With stairlines standing for lines we can generalize convexity in other ways than the one we investigate in this paper. For example, we call a set **P** "strongly $O$-convex" if for every pair of points $p$ and $q$ in **P** *all* stairsegments with endpoints $p$ and $q$ lie in **P**. It is possible to prove that this definition of convexity always produces *convex* (in the normal sense) *O-oriented* sets. Indeed, when $O = \{0°, 90°\}$ then the strong $O$-convex hull of **P** is just the *bounding box* of **P**. We investigated the notion of strong $O$-convexity in a previous paper ([12]) and we show in [11] that both $O$-convexity and strong $O$-convexity along with many other natural definitions of convexity are essentially the same.

# 6 Other Characterizations of *O*-Convex Sets

In this section we characterize *connected* $O$-convex sets by deriving conditions on the form their boundary must take and proving a generalized version of property 6 (see the Introduction).

**Definition 6.1** *We say that p is an O-*extremal *of* **P** *if p is a point of support of* **P** *with respect to an O-line.*

We now show that the boundary of a closed connected $O$-convex set may be completely characterized in terms of stairsegments.

**Definition 6.2** *We say that a portion of a continuous curve in the plane is a* maximal stairsegment *in the curve if it is a stairsegment and it is not a proper subset of any other stairsegment in the curve.*

**Theorem 6.1 The Characterization Theorem** *A simply connected closed set is O-convex if and only if the portions of its boundary in between any two consecutive O-extremal points are maximal stairsegments.*

Observe that in the normal convex hull (that is, $O = \{[0°, 180°)\}$) *all* points are $O$-extremal and so the maximal stairsegments in the boundary shrink to points.

**Corollary 6.2** *1. A polygon is O-convex if and only if its boundary consists of a sequence of polygonal stairsegments meeting at convex interior angles.*

*2. An O-polygon is O-convex if and only if its boundary consists of a sequence of O-oriented polygonal stairsegments meeting at convex interior angles.*

For the special case of finite $O$ Corollary 6.2(1) has been stated without proof in [21] and it was proved in a different, more direct, way in [12].

Note that the characterization of the boundary of $O$-convex polygons as a sequence of polygonal stairsegments is a direct generalization of the case for orthogonal polygons ([22]).

In the theory of (normal) convex sets two points are said to be *visible* from each other in a set if the line segment joining them lies wholly in the set. Thinking of stairlines as the analogs of straight lines we are led to define a generalized form of visibility in which two points in a set are visible from each other if there exists *at least one stairsegment* joining them which lies wholly in the set. This leads to the next characterization of $O$-convex sets and againg it only applies to *connected* $O$-convex sets.

**Theorem 6.3 (The Visibility Theorem)** *If* **P** *is connected, then* **P** *is O-convex if and only if for all p and q in* **P** *there exists a stairsegment in* **P** *with endpoints p and q.*

Note that in normal convexity this theorem collapses to property 6 stated in the Introduction, since all (normal) convex sets are connected.

# 7 Conclusions

We have shown that $O$-convex sets contain both convex sets and orthogonally convex sets as sub-classes and that the properties of both can be explained as special cases of the properties of $O$-convex sets. The main characteristic of convex sets that we have lost in generalizing to $O$-convex sets is *connectivity*. A convex set is always connected.

Connected $O$-convex sets have all of the properties of convex sets listed at the beginning of the paper if we interpret a "line" as a stairline and generalize the betweenness relation to reflect the fact that the "line segment" joining two points is no longer necessarily unique and can be any stairsegment connecting them. And so, *any* point in $O$-*region*$(p, q)$ is "between" $p$ and $q$. In the following we assume that **P** is a *connected* $O$-convex set.

(1) If $O$ is non-empty then **P** is simply connected (Lemma 3.6(1)). Indeed, the connected components of any $O$-convex set are simply connected once $O$ is non-empty (Lemma 3.6).

(2) The intersection of **P** and any $O$-line is either empty or a connected set (by Definition). This is true even if **P** is allowed to be disconnected. One of the implications of this property, for convex sets, is that lines are themselves convex. We obtain the needed analogy by observing that

the intersection of any two $\mathcal{O}$-convex sets is again $\mathcal{O}$-convex (Lemma 3.2) (although observe that the intersection of two connected $\mathcal{O}$-convex sets may be disconnected).

(3) **P** is the intersection of all $\mathcal{O}$-convex sets which contain it (Lemma 3.4). This is true even if **P** is allowed to be disconnected.

(4) If $p \notin \mathbf{P}$ then there exists a stairline separating $p$ and **P** (Theorem 4.2 and Lemma 5.1(3)).

(5) **P** is the intersection of all stair-halfplanes which contain it (Lemma 5.1(3)).

(6) If $p, q \in \mathbf{P}$ then there exists a stairsegment in **P** connecting $p$ and $q$ (Theorem 6.3).

Restricted-orientation convexity is a generalization of orthogonal convexity and it is a useful vantage point from which to survey and unify many scattered results and observations in the literature of computational geometry. We have shown that restricted-orientation convexity is a reasonable generalization of convexity since properties analogous to those of normal convex sets hold for these more general "convex" sets.

## Acknowledgement

This work was supported under a Natural Sciences and Engineering Research Council of Canada Grant No. A-5692.

## References

[1] Bender, E. A.; "Convex $n$-ominoes", *Discrete Mathematics*, **8**, 219-226 (1974).

[2] Cohn, P.M.; *Universal Algebra*, Harper & Row, New York, 1965.

[3] Edelsbrunner, H.; *Intersection Problems in Computational Geometry*, Doctoral Dissertation, University of Graz, 1982.

[4] Grünbaum, B.; *Convex Polytopes*, Wiley-Interscience, New York, 1967.

[5] Güting, R. H.; *Conquering Contours: Efficient Algorithms for Computational Geometry*, Doctoral Dissertation, Universität Dortmund, 1983.

[6] Jamison-Waldner, R.E.; "A Perspective on Abstract Convexity: Classifying Alignments", *Lecture Notes in Pure and Applied Mathematics*, **76**, 113-150, (1982).

[7] Klee, V.; "What is a Convex Set?", *American Mathematical Monthly*, **78**, 616-631 (1971).

[8] Levi, F.W.; "On Helly's Theorem and the Axioms of Convexity", *Journal of the Indian Mathematical Society*, **15**, 65-76 (1951).

[9] Ottmann, Th., Soisalon-Soininen, E., and Wood, D.; "On the Definition and Computation of Rectilinear Convex Hulls", *Information Sciences*, **33**, 157-171 (1984).

[10] Preparata, F. P., Shamos, M. I.; *Computational Geometry*, Springer-Verlag, New York, 1985.

[11] Rawlins, G.J.E.; *Explorations in Restricted-Orientation Geometry*, Doctoral Dissertation, University of Waterloo, 1987.

[12] Rawlins, G.J.E. and Wood, D.; "Optimal Computation of Finitely-Oriented Convex Hulls", *Information and Computation*, **72**, 150-166 (1987).

[13] Rawlins, G.J.E. and Wood, D.; "Convexity Spaces: A Decomposition Theorem", unpublished manuscript, 1988.

[14] Rosenfeld, A. and Kak, A. C.; *Digital Picture Processing*, Academic Press, New York, 1976.

[15] Sack, J.-R.; *Rectilinear Computational Geometry*, Doctoral Dissertation, Carleton University, 1984.

[16] Shamos, M. I.; *Problems in Computational Geometry*, Doctoral Dissertation, Yale University, 1978.

[17] Sierksma, G.; "Extending a Convexity Space to an Aligned Space", *Indagationes Mathematicae*, 46, 429-435 (1984).

[18] Toussaint, G. T.; "Pattern Recognition and Geometrical Complexity", in *Proceedings of the International Conference on Pattern Recognition*, 2, 1324-1347 (1980).

[19] Toussaint, G. T. and Sack, J.-R.; "Some New Results on Moving Polygons in the Plane", in *Proceedings of the Robotic Intelligence and Productivity Conference*, Detroit, 158-164 (1983).

[20] Widmayer, P., Wu, Y. F., Schlag, M. D. F. and Wong, C. K.; "On Some Union and Intersection Problems for Polygons with Fixed Orientations", *Computing*, 36, 183-197 (1986).

[21] Widmayer, P., Wu, Y. F. and Wong, C. K.; "Distance Problems in Computational Geometry for Fixed Orientations", in *Proceedings of the ACM Symposium on Computational Geometry*, Baltimore, 186-195 (1985).

[22] Wood, D.; "An Isothetic View of Computational Geometry", in *Computational Geometry* (Toussaint, G., ed.). North Holland, Amsterdam, 429-459 (1985).

# A POPULATION-TYPE MODEL OF CAR MARKET DYNAMICS

Tomasz M. Romanowicz and Jan W. Owsiński
Polish Academy of Sciences
Systems Research Institute
Newelska 6
01 - 447 Warszawa, Poland

## 1. Introduction

Scientific thinking is based upon such primitive intellectual operations as distinguishing the different and associating the similar. These operations, applied to individual objects give rise to models: linguistic, intuitive, verbal or formalized. The same operations can now be performed on models. Association in the realm of models may lead to identification of analogy.

Validated analogies can function either on the linguistic level (e.g. "stream of vehicles", "life-cycle of a product"), or on a more formal, for instance, mathematical level.

Thus, a mathematical model describing initially certain physical processes can be taken and slightly modified to represent the dynamics of biological populations, von Foerster (1959). Since it turns out that this analogy works well, a number of studies going in the same direction follow, e.g. Aroesty et al. (1973), Rocklin and Oster (1976) or Rubinov and Lebowitz (1976). The present authors have broadened the framework of this model, including in it some aspects which result in nontrivial behaviour of the population and its cohorts, but at the same time made the model non-analytical and turned it into a simulation device, Romanowicz and Owsiński (1982).

This paper shows how the above analogy can be extended to encompass still another domain, namely that of cars count. The subject is assessment of dynamics of car numbers along their types and ages, according to various hypotheses concerning car market, car ageing, etc. First, the model is explained and presented, in its fundamental continuous and then computerized form, and afterwards its runs are shown related to a number of basic hypotheses.

## 2. The fundamental model

Imagine a population of any entities of the same kind (animals of one species, products serving the same purpose and the like). Total numbers of entities in the population, N, depend in general upon time. The entities can be classified into "types" or "varieties" and into age classes. Thus, N(t,T,v) shall denote all entities of age T and type v existing at the time moment t.

The von Foerster equation,

$$\frac{\partial N}{\partial t} + \frac{\partial N}{\partial T} = -Nm \tag{1}$$

refers exclusively to mortality- and age-related dynamics. Parameter m=m(T) denotes mortality rate function values. This equation is complemented with the following condition:

$$N(t,0) = s(t) \tag{2}$$

where s(t) denotes the "birth-rate" function. Having, additionally, the explicit form of m(T) one can start solving equation (1).

Dependence upon v may appear in s(t), reflecting various "birth--rates" of different types, as well as in m(T), where it reflects various "death-rates" for different types. Both these dependences are in reality often further differentiated according to T. In the case of m(T) it is insofar obvious as for a vast majority of entities death rate is very closely related to age. In the case of s(t), however, similar relation can be introduced only when a dependence of the birth--rate upon the numbers of existing entities is assumed. Such a dependence is quite natural for biological populations, where birth-rates depend upon the proportions of various age groups within a given population.

Having all these relations assumed and/or fitted to empirical data one can proceed to construction of a simulation model, as indicated in Romanowicz and Owsiński (1982).

## 3. Application to the automobile market

The automobile-oriented application presented here stems from the conviction, expressed e.g. by Marchetti (1983) that the car market dynamics is to a large extent governed by its internal death and ageing processes, over which only minor modifications are added by other forces.

The entities referred to in the previous descroption shall now be taken to be automobiles. Car "birth" moment is equivalent to the moment of its first sale as a brand-new product. Further resales are not taken into account. Car "death" is equivalent to its scrapping. Variable v may indicate car types, e.g. according to their broadly conceived size and/or durability. N(t,T,v) shall therefore denote the number of cars of age T and type v on the road at time t. Obviously, dynamics of this particular "population" cannot be described by the continuous version of equation (1), since at least along v there exists a natural discretization of the population. Moreover, car statistics are at least monthly or quarterly discretized as well, even though the model of the very production process could be approximated by continuous description.

Function $s(t)$ shall in this case reflect car sales in time t, eventually broken down into $s(t,v)$ according to types. It may be hypothesized that $s(t,v)$ depend in a way upon $N(t,T,v)$ on their appropriate sums, but other hypotheses may be equally legitimate. As to $m(t,T)$ - the death rates - they obviously depend upon the automobile age T and its type v.

Thus defined handy representation of car population dynamics in the form of a simulation model may be used to forecast certain aspects of market behaviour over a certain period ahead, this period being determined by rationality and/or validity of hypotheses made.

## 4. The computer model

The prerequisites outlined above have been implemented in a simple simulation model. This model has the following specifications:

time, t, is discretized into 15 one-year steps

age, T, is discretized also into one-year steps indexed i

type variable, v, is discretized into 5 distinct types indexed j.

Number of cars in denoted $N_{ij}$. Thus, a table is formed, containing elements $N_{ij}$, the table having dimensions $15 \times 5$, fully determining the state of a population at a given time moment $N_{ij}=N_{ij}(t)$, i.e. they vary with time due to the birth and death processes, i.e. sales and scrapping.

Mortality rates $m_{ij}$ are calculated on the basis of the assumption stipulating that ageing of cars is governed by the logistic type of curve,

$$f_j(i) = 1 - \frac{1}{1 + e^{-a_j i + b_j}} \qquad (3)$$

On the basis of values of $f_j(i)$ values of $m_{ij}$ are calculated:

$$m_{ij} = \frac{f_j(i+1) - f_j(i)}{f_j(i)} \qquad (4)$$

Having $m_{ij}$ one may define the cohort dynamics,

$$N^{t+1}_{i+1,j} = N^{t}_{ij}(1-m_{ij}) \tag{5}$$

where $N^t_{ij}$ denotes the number of cars in cohort (i,j) at time t. Formula (5) applies to i=2,...,15. For N(1,j), i.e. for the cars sold, a simple formula was chosen:

$$N^{t+1}_{ij} = s_j(1+\alpha)\sum_{i,j} N^{t}_{ij}m_{ij} \tag{6}$$

where $s_j$, j=1,...,5, is the vector of sale distribution among types, and $\alpha$ is the sales growth (or decrease) rate, and condition

$$\sum_j s_j = 1 \tag{7}$$

must hold.

This simple model was implemented as a conversational device on a personal computer. The conversational mode of work makes it possible to change virtually all the model parameters during a run. Hence, it becomes feasible to assess the influence exerted by hypothetical changes in various assumptions. These hypotheses may well concern car makes, market dependences and growth indices. In particular, $s_j$ may be redefined at every (annual) time step so as to reflect working of a hypothetical sales mechanism, relating sales distribution to various magnitudes appearing in and outside of the model.

## 5. Examples of model runs

Taking constant $s_j$ and $\alpha$=0 one obtains, ultimately, a quasi--static model, whose structure is reflected through a steady state to which the model converges after a limited number of steps, having started from an initial point of $N^0_{ij}$. Even such a "trivial" kind of run may bring in, however, interesting information regarding numbers of cars going in and out of the population during the convergence period. This may be of importance in cases when a stable market situation is perturbed by a sudden occurrence, e.g. dramatic oil price increase, a change in taxation schemes, announcement of new technological requirements. Certainly such changes do not cause overnight market shifts, but nevertheless they can appear in this way in the model, since time constants of market shifts may be comparable with the model time step and, additionally, there may, and usually is, a delay in recognition of the coming change.

In all the runs illustrated here the mortality curves (see (3)) are governed, for the five types distinguished, by the following parameters:

| $j$ | $a_j$ | $b_j$ | $s_j$ = const. |
|---|---|---|---|
| 1 | 0.7 | 5.0 | 10% |
| 2 | 0.6 | 4.0 | 20% |
| 3 | 0.5 | 3.5 | 40% |
| 4 | 0.5 | 3.5 | 20% |
| 5 | 0.4 | 3.0 | 10% |

The above table contains also the usually applied sales distribution, $s_j$, among types j, j=1,...,5.

As can easily be seen, types j are ordered from less to more durable cars, where durability results from both car produce and other, indirectly influencing factors, e.g. more durable cars can be understood to be more expensive and bigger, and therefore more care-worthy. Simultaneously, it was assumed for most of the runs that medium type is most sold, while the other types are sold less (e.g. j=1: small, inexpensive cars of limited durability and higher risk).

Thus, a market shift mentioned before would mean an immediate change of $s_j$. Results of such a change, starting from a stable situation characterized by the parameters quoted, and $\alpha=0$, are illustrated in Fig.1.

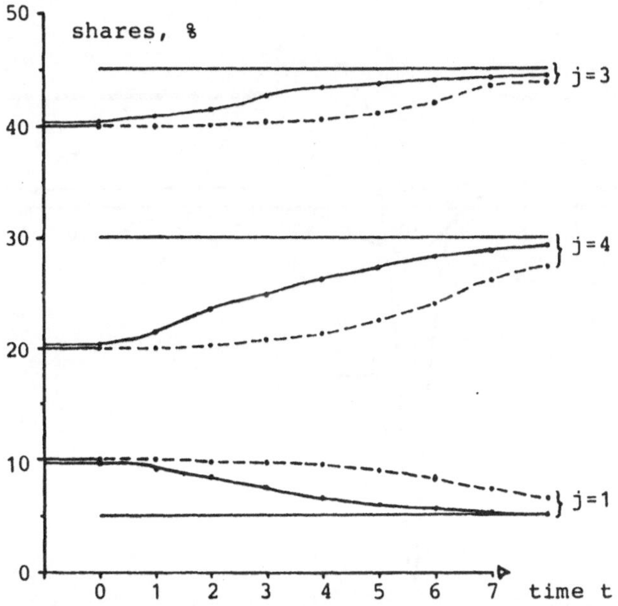


Fig. 1. Illustration of the sales distribution change; shares:
———, sales distribution
—•—•—, of cars on the road
-•-•-, of cars scrapped

The most important aspect of the results obtained is the difference of speed between the changes of numbers of cars on the road and the changes of numbers of cars scrapped. The very fact is obvious, but actual magnitudes are very important for planning of sales, spare parts, etc.

It might also be interesting to look at the consequences of another instantaneous change, related this time to the shifts in the $f_{ij}$ for the first two j's. Thus, in a stable population, $\alpha=0$, same as previously $a_j$ and $b_j$ are changed in the following way:

j=1 $a_j$ from 0.7 to 0.6 and $b_j$ from 5.0 to 4.0

j=2 $a_j$ from 0.6 to 0.55 and $b_j$ from 4.0 to 3.7

Illustrations of the results, as provided here, constitute, of course, just an aggregation of the model output, which specifies the dynamics of individual cohorts, i.e. $N^t_{ij}$.

It appears that decisive here are inter-type competition conditions, more than just the type-proper parameters. Changes are propagated over

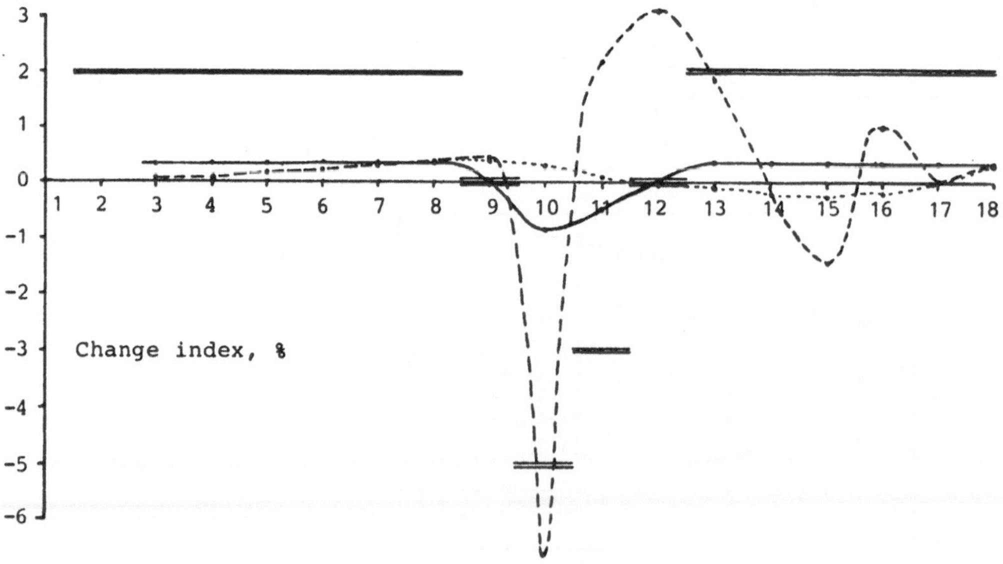


Fig.2. Year-to-year % changes of particular variables values
═══ car sales parameter
—•— total car numbers —·— —·— car sales numbers
···•···•· cars scrapped where differing from the sales numbers

all the types, but not to a high degree. Note that this particular run embodies an assumption as to a shift in characteristics of certain types of cars, e.g. in an effort to win more of sales, and also an assumption as to the mechanism generating consequences of such a shift. As can easily be seen, the consequences are by no means intuitively obvious. When fitting appropriate $f_j$ it would namely be of great importance to find out whether and where, if so - the curves cross each other.

Another type of market change and its consequences, as simulated by the model is shown in Fig.2. It consists primarily of the assumption as to the total sales dynamics, which proposes a certain temporal profile of total sales over a period of few years, in order to afterwards observe the consequences of this particular temporal sales profile. The case taken up and illustrated here concerns a period of steady, although feeble growth followed by a period of sales decline, expressed, however, not directly in terms of sales values for particular j, but by variability of $\alpha$.

One should note here two phenomena, which could be of interest to car dealers and producers: first, a relatively smooth nature of scrapping rates' curves when compared to other processes and, second, oscillations in absolute sales values. The latter property is especially important from the viewpoint of producers. Note that these oscillation do not result from "unfulfilled demand", which exists anyway, but from the internal dynamics of the analysed population, complemented with the assumption as to the simple mechanism of sales. When unfulfilled demand is added, it is possible to obtain even stronger oscillations.

The third, and final, illustration, concerns simulation of a dependence between distribution s(t) of sales and the current state of the automobile population. Results of appropriate runs are reported in Fig.3, 4.

In the first case, Fig.3, a simple model is applied in which $s_j^t$ depend directly upon the shares of cars j on the road in previous periods t-1. As could easily be anticipated, these types whose logistics were "worse" than their market shares in the steady state, started losing market, while those in the opposite situation - gained. In addition, those with bigger differences between the market share and the total number share moved at a quicker pace than those, for which these differences were small. That the process is important can be seen on the example of j=2, which over 10 periods (years) lost almost 20% of its market share, while j=5 increased its share in the same period by almost 30%.

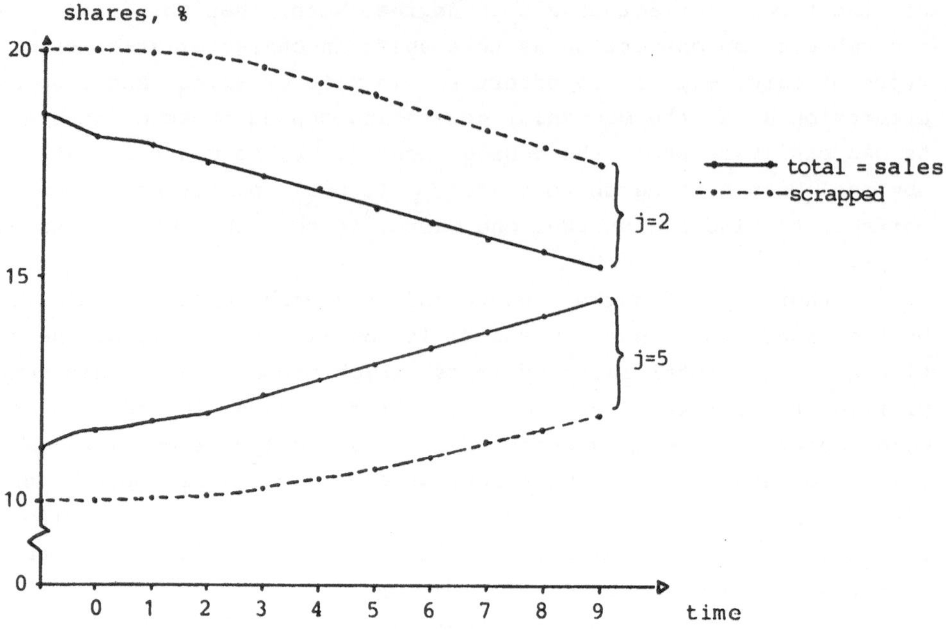


Fig.3. Dynamics of market shares and mortality rates for j=2 and j=5.

There is, however, one phenomenon which is not, at a first glance, obvious: that the processes, as far as market shares go, do not accelerate, at least not over the period of interest. This may be of course explained by an acceleration in the death rates, but nevertheless is not straightforward and constitutes an important feature of the population.

The second case of a sales "model" is the one in which $s_j^t$ are taken as averages of total number shares and scrapped cars shares. Since in the steady state market shares and scrapped cars shares have to be equal, this run was made against the background of shifting logistics. As before, there is an obvious inter-type competition run which ends up with a new steady state. Although, however, this new steady state is advantagous for j=2 in the pure case, this does not prevent from a drop in market shares of j=2 when the averaging model is applied (Fig.4). For a planning purpose it would be interesting to identify the logistic curve shift vs. sales model parameters relation, for which there is, over at least a period of few years, no change. Knowing actual sales model one would be capable of finding the economical car type change.

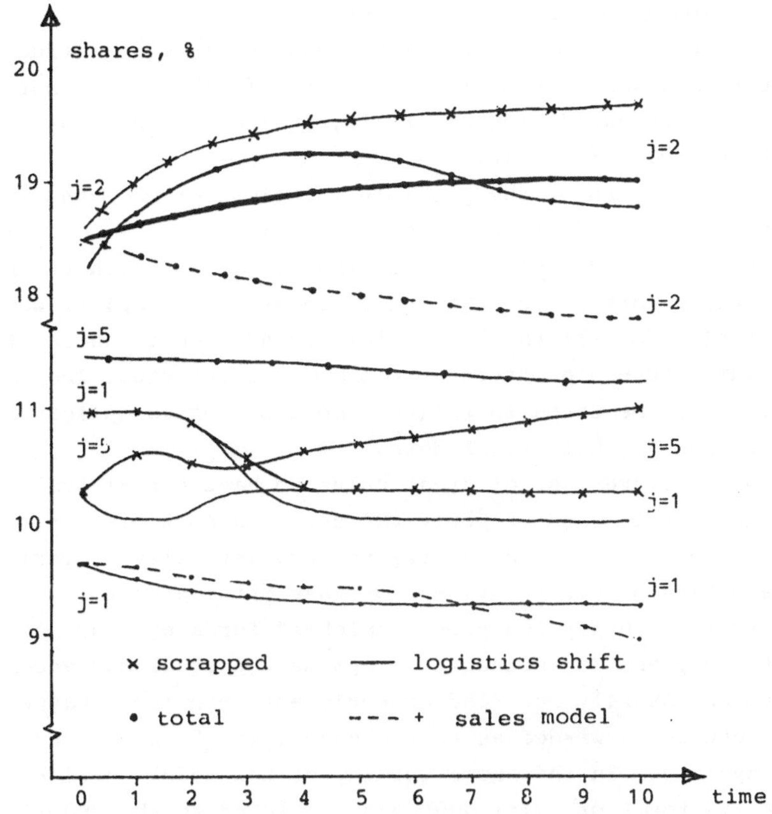


Fig.4. Effects of simultaneous logistics shift and averaging model, for j=1, 2 and 5.

6. Model application

This short paper intends to present just a calculation tool for assessing this aspect of an equipment market which is related to the state of population of the already sold equipment items. An ultimate hypothesis would consist in stating that as many items are being sold as there are the scrapped ones, provided the market is at least decently stable. Changes beyond this hypothetical sales level, up and down from it, would be due to "external", socio-economic variables.

The model readily represents the basic hypothesis and can be easily used to also represent "sales trend" submodel, relating some aspects of $s(t)$ and $\alpha$ in a more complex manner to $N_{ij}(t)$ and their functions, as well as to the external socio-economic indices.

Procedure of application of the model for a kind of equipment would consist of the following stages:

A. division into equipment sub-types,

B. fitting of the $f_j(i)$ function on the basis of existing data,

C. determination of the starting point (if the run is not meant for simple validation of $f_j(i)$, obtained in B., i.e. the values of $N_{ij}(0)$, $s(0)$, $\alpha(0)$,

D. identification of the possible relations of $s(t)$ and $\alpha(t)$ to endogeneous and/or exogeneous variables.

Obviously, the most difficult is the stage D., which at the same time, is very important for the model output. The simplest way is to take $s(t)=s(0)$ for all the $t$ simulated, and then to check other a priori assumed feasible alternative $s(t)$ trajectories. The $s(t)$ trajectory may also be taken to follow a more or less complicated trend line based upon the historical data.

It would, however, be of great value to have a real "sales trend" submodel, identified and properly validated on historical time series.

Having such a model one may try to forecast sales of various types of the same kind of equipment. By somewhat stretching the model one may as well try to identify the proper "niches" for a hypothetical new equipment subtypes. In another development the model may accomodate a market for an entirely new kind of equipment, provided reasonable assumptions can be forwarded as to the underlying "population", its "types" and dynamics. In this case working of the model shall heavily depend upon the forms of $s(t)$ and $\alpha(t)$, related to the underlying population.

References

[1] Aroesty, J., T. Lincoln, N. Shapiro and G. Boccia (1973): Mathematical Biosciences vol.17 p.243

[2] Foerster, von J. (1959): The Kinetics of Cell Proliferation. New York: Grune and Stratton.

[3] Marchetti, C. (1983): The automobile in a system context, the past 80 years and the next 20 years. Technological Forecasting and Social Change vol.23, p.3.

[4] Marchetti, C. (1984): Action curves and clockwork geniuses. Unpublished typescript.

[5] Rocklin, S., and G. Oster (1976): Journal of Mathematical Biology vol.3-4, p.225.

[6] Romanowicz, T. and J.W. Owsiński (1982): A conversational dynamic model of adaptive biological population. Int. J. Systems Sci. vol. 13, No.6, p.683.

[7] Rubinow, S.I. and J.L. Lebowitz (1976): Biophysical Journal, vol.8, p.891.

# MODEL FOR BLAST FURNACE ON-LINE SIMULATION

Saxén H., Uusi-Honko H., Kilpinen A.
Department of Chemical Engineering, Åbo Akademi
Biskopsgatan 8, 20500 Åbo, FINLAND
BITNET E-MAIL:VT_INST@FINABO

## ABSTRACT

A blast furnace simulation model developed primarily for on-line application is presented. The model, which describes the steady-state operation of the furnace in one spatial dimension, is adapted to data from the real process by adjusting a set of parameters. A thermodynamic process interface provides the boundary conditions. The model, which is shown to act as an intelligent measurement device providing information about variables that are impossible to measure reliably, can be used to predict the state of the furnace under new operational conditions.

## INTRODUCTION

Numerous mathematical models for the blast furnace (BF) process have been presented in the literature.[1] Most models have, however, been developed for off-line analysis only, and real process data is seldom utilized directly when the behavior of the furnace is studied by simulation. In an on-line model, special attention must be paid to the treatment of process data. Firstly, the model should use *standard measurements* that are available at the plant. Secondly, errors in these measurements should be taken into account. Thirdly, a comparatively simple treatment of the conditions in the furnace is preferred in order to reduce the modeling efforts and to avoid inclusion of several parameters the values of which are uncertain, *e.g.* porosities, particle diameters, *etc.* The model to be presented,[2,3] which is based on the above principles, describes a steady-state of the furnace in one spatial dimension. A set of adaptive parameters is included, which facilitates a tuning of the model to different operational conditions. The boundary conditions (BCs) used to determine the unknown parameters have been supplied by a thermodynamic interface program which "calibrates" some measurements in order to obtain a consistent set of BCs for a hypothetical steady-state of the process.

## MODEL STRUCTURE

The BF simulation model is based on the following concept:

- A steady-state of the process is assumed.
- The process characteristics are described by a set of consecutive chemical reactors (Fig. 1):
  1) In the uppermost part of the furnace the charged materials are heated followed by an evaporation of moisture.
  2) The rest of the furnace volume above the tuyere level is divided into a number of control volumes ("chemical reactors") of equal size. A chemical equilibrium problem and a set of energy balances are formulated for each reactor.
  3) In the lowermost of the control volumes mentioned above, some additional chemical reactions are considered describing the transfer of silicon to the metal.
  4) In the raceways of the furnace a combustion of fuels to CO, $H_2$, and SiO is considered.
  5) The final reduction reactions take place in the furnace hearth.
- An adaptive technique is used for tuning the model to process data.

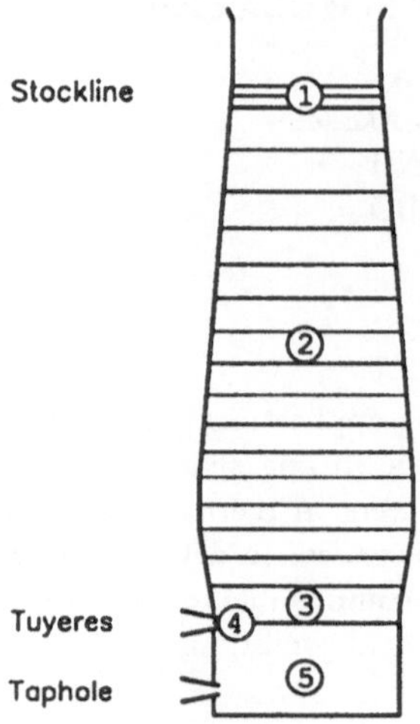


**Fig. 1** Division of the furnace volume.

Fig. 2 schematically represents a control volume. Solid or liquid materials, denoted by the molar flow vector $\underline{\dot{n}}_{s/l}$, descend as the result of the consumption of coke near the tuyeres and the melting and shrinking of iron and slag. (Underlined symbols are in the sequel used for vectors and matrices.) The descending phases meet an ascending reducing gas phase, $\underline{\dot{n}}_g$. The conditions at the upper and lower boundaries of the control volume have been denoted by "1" and "2", respectively.

Two control volumes in the model describe the uppermost part of the effective volume of the blast furnace. The charged materials are first assumed to be heated to 100°C followed by an evaporation of the accompanying moisture at this temperature. In this region no chemical reactions are considered, and, consequently, only two energy balance equations have to be solved. The heat exchange between the ascending and the descending phases is described by

$$\Delta \dot{Q} = \alpha_v \, \Delta V \, [\overline{T}_g - \overline{T}_{s/l}], \tag{1}$$

*i.e.*, it is modeled to be proportional to the mean temperature difference and the size, $\Delta V$, of the control volume using a volumetric heat transfer coefficient $\alpha_v$. The coefficient is considered to include the effect of both convective and radiative heat transfer and may be expressed as an empirical function of variables such as temperature, flow rate of gas, *etc.*

If $\alpha_v$ and the BCs at the stockline level are known, the equations can be simultaneously solved for the unknown temperatures of the gas entering the two regions and the volumes of the *heating* and *evaporation* zones.

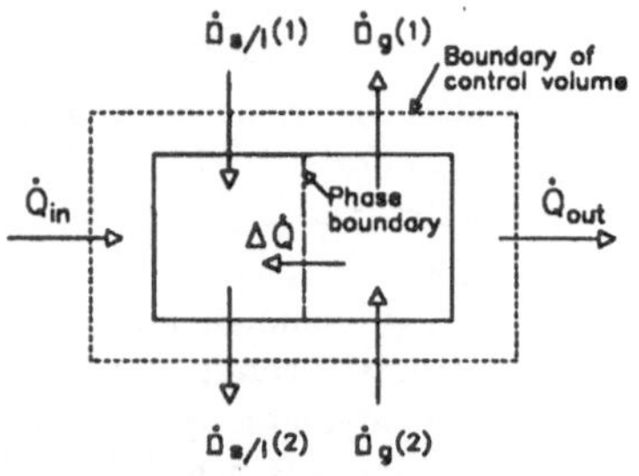


**Fig. 2** Schematic structure of a control volume.

The chemical conditions have been modeled by a set of $m$ (independent) chemical reactions. A chemical equilibrium problem is solved in each control volume, but only as an auxiliary step to give information about the limiting conditions. Assuming a reaction, $i$, to proceed towards equilibrium by a fraction, $\beta_i$, of the corresponding equilibrium extent, we obtain for the *actual* extent of reaction

$$e_i = \beta_i \, e_i^*, \tag{2}$$

where symbols denoted by an asterisk ($*$) refer to chemical equilibrium. Thus, $\beta_i$ can be considered a fractional extent of reaction. Naturally, its numerical value depends on numerous physical and chemical phenomena, *e.g.* diffusion, convection, chemical kinetics, *etc.*, and also upon the size of the control volumes. We therefore adopt the idea of estimating an overall reaction rate parameter, $\beta_{tot}$, based on **actual process data** instead of using reaction rates which are fixed *a priori*.
Using the notation introduced in Fig. 2, a material balance for the *equilibrium case* can be written

$$\underline{\dot{n}}^*_{out} = \begin{pmatrix} \underline{\dot{n}}^*_{s/l}(2) \\ \underline{\dot{n}}^*_g(1) \end{pmatrix} = \underline{\dot{n}}_{in} + \underline{N}\,\underline{e}^* = \begin{pmatrix} \underline{\dot{n}}_{s/l}(1) \\ \underline{\dot{n}}_g(2) \end{pmatrix} + \underline{N}\,\underline{e}^*, \tag{3}$$

where $\underline{N}$ is the stoichiometric matrix for the $m$ chemical reactions. Partitioning the matrix $\underline{N}^T = (\underline{N}_{s/l}^T\ \underline{N}_g^T)$ and introducing a matrix form of Eq. (2),

$$\underline{e} = \underline{\beta}\,\underline{e}^* = \mathrm{diag}(\beta_1, \beta_2, \ldots, \beta_m)\,\underline{e}^* = \beta_{tot}\,\underline{W}\,\underline{e}^*, \tag{4}$$

where $\underline{W}$ is a diagonal weighting matrix and $\beta_{tot}$ the overall rate parameter, the effluent flows can be written as functions of the conditions at "1" and of the equilibrium extents

$$\dot{\underline{n}}^*_{out} = \begin{pmatrix} \dot{\underline{n}}_{s/l}(1) \\ \dot{\underline{n}}_g(1) \end{pmatrix} + \begin{pmatrix} \underline{N}_{s/l} \\ \underline{N}_g(\underline{I} - \underline{\beta}) \end{pmatrix} \underline{e}^*. \tag{5}$$

This form is suitable for a solution procedure where the conditions at the upper surface of the control volumes are used as initial conditions for a sequential simulation "downwards" in the furnace. Correspondingly, the molar flows for the *non-equilibrium case* can be written

$$\dot{\underline{n}}_{out} = \begin{pmatrix} \dot{\underline{n}}_{s/l}(2) \\ \dot{\underline{n}}_g(1) \end{pmatrix} = \dot{\underline{n}}_{in} + \underline{N}\,\underline{e} = \begin{pmatrix} \dot{\underline{n}}_{s/l}(1) \\ \dot{\underline{n}}_g(1) \end{pmatrix} + \begin{pmatrix} \underline{N}_{s/l} \\ \underline{0} \end{pmatrix} \underline{e}. \tag{6}$$

We can now express as functions of the extent vector $\underline{e}^*$ the output molar flows of the condensed and gas phases for the equilibrium case, the input flows of gas and the non-equilibrium output flows of condensed components.
The chemical equilibrium problem is written as a system of nonlinear equations (NLEs)

$$\underline{f}_1 = \underline{N}^T \begin{pmatrix} \ln(\gamma_1 x_1) \\ \vdots \\ \ln(\gamma_j x_j) \end{pmatrix} - \begin{pmatrix} \ln K_1 \\ \vdots \\ \ln K_m \end{pmatrix} = \underline{0}, \tag{7}$$

where $j$ denotes the number of chemical components considered. The equilibrium constants, $\underline{K}$, can be calculated from thermodynamic data for the components, while the molar ratios, $\underline{x}$, are obtained by using the molar flows of Eq. (5). The equilibrium problem further includes the non-negativity constraints of $\dot{\underline{n}}$.
Next, energy balances for a control volume are formulated. Again referring to Fig. 2, there are four unknown temperatures; those of the effluent flows for the equilibrium case, $T^*_{s/l}(2)$ and $T^*_g(1)$, that of the effluent non-equilibrium mixture of the condensed phase, $T_{s/l}(2)$, and that of the entering gas phase, $T_g(2)$. Two energy balances for the gas phase and two for the condensed phases provide the four requisite equations, $\underline{f}_2 = \underline{0}$. The equilibrium problem is coupled with the unknown temperatures as the thermodynamic data (Gibbs energy) is calculated at the mean temperature of each control volume.

The silicon transfer in the blast furnace takes place both above the tuyere level and in the furnace hearth. SiO acts as an important intermediate product in the transfer reactions in the vicinity of the tuyeres. Some investigators[4,5] have reported astonishingly high contents of silicon in iron at the tuyere level — sometimes these values considerably exceed those corresponding to the tapped iron. Therefore, at least a local re-oxidation of silicon takes place somewhere between the tuyere and the taphole levels. The following two reactions are often considered

$$\mathrm{SiO}(g) + \underline{\mathrm{C}} \rightarrow \underline{\mathrm{Si}} + \mathrm{CO}(g) \qquad \mathrm{SiO_2} + \mathrm{C}(s) \rightleftharpoons \mathrm{SiO}(g) + \mathrm{CO}(g)\,, \tag{R1,2}$$

where underlined chemical symbols denote components in the iron phase. ($R2$) can act either as SiO sink or as SiO source; ($SiO_2$) may be reduced to SiO as slag descends in the high-temperature regions, or formed as ascending SiO (originating from the raceways) is cooled down and oxidized by CO. Taguchi *et al.*[6] claim the latter phenomenon takes place at the lower end of the cohesive zone.
In the model, the silicon transfer above the hearth is considered in the lowermost control volume

and in the raceways. All $SiO_2$ in coke ash is assumed to be gasified to SiO, which ascends and reacts with $\underline{C}$ or CO. In the former region, SiO can be produced by the forward reaction ($R2$).
In order to obtain an easy implementation of the above structure, a parameter, $f_{Si}$, denoting the fractional reduction of the slag $SiO_2$ at the tuyere level has been introduced. This fraction may be given negative values to imply a re-oxidation of SiO.

The raceway region of the furnace is modeled as a chemical reactor for combustion of coke and additional fuels to CO, $H_2$, and SiO. In the hearth model, the final chemical reactions and heat transfer are considered. The entering residuals of metal oxides, $M_vO_u$, are modeled to be reduced by an *excess* of silicon according to the reaction

$$\frac{u}{2}\,\underline{\mathrm{Si}} + \mathrm{M}_v\mathrm{O}_u \rightleftharpoons \frac{u}{2}\,(\mathrm{SiO_2}) + v\,\mathrm{M}\ , \qquad (R3)$$

where ($SiO_2$) denotes silica in the slag phase. The remaining silicon represents the tapped quantity. An energy balance for the furnace hearth is solved with respect to the mean temperature of pig iron. The above formulation of the equations for the furnace hearth provides a simple description of the conditions at the taphole used as a convergence criterion in the model.

The heat loss through the furnace wall is assumed to follow a distribution stated *a priori*,[10] where a major part of the energy is lost from the belly region. Constant values for the loss from the raceways and the furnace hearth have been used.

## FORMULATION OF THE MODEL PARAMETERS

The volumetric gas-burden heat transfer coefficient has been modeled to include a dependence of two state variables; the mass flow rate of gas and the temperature.

$$\alpha_v = \begin{cases} \alpha'_v & \text{if } T < T_{min}, \\ \alpha'_v\left(1.0 - \frac{T-T_{min}}{T_{max}-T_{min}}\Delta_T\right) & \text{if } T_{min} \le T \le T_{max}, \\ \alpha'_v(1.0 - \Delta_T) & \text{if } T > T_{max}; \end{cases} \qquad \alpha'_v = \alpha_{v,ref}\left(\frac{\dot{m}_g}{\dot{m}_{g,ref}}\right)^{0.8}. \qquad (8a,b)$$

The reference value, $\dot{m}_{g,ref}$, may be chosen as the normal mass flow rate of top gas for the BF in question. The temperature dependence, in turn, is similar to an expression proposed by Hatano *et al.*[7] The reason for not including a more specific description of the heat transfer coefficient is that the values of particle diameters, porosities, *etc.*, must then be estimated. Further, many papers on BF modeling[6-9] show that even though very detailed formulae are used to describe the heat transfer, correction factors still have to be introduced.
As the overall rate parameter $\beta_{tot}$ is to characterize the rates of the reduction reactions, it is modeled so as not to affect the rate of the "solution-loss" reaction (here numbered $m$). The weighting matrix, $\underline{W}$, which includes a linear temperature dependence, is expressed

$$\underline{W} = \mathrm{diag}(w_1, w_2, \ldots, \frac{w_m}{\beta_{tot}}); \qquad w_i = \begin{cases} 0.0 & \text{if } T < T_{r,i}, \\ \frac{T-T_{r,i}}{\Delta T_i} & \text{if } T_{r,i} \le T \le T_{r,i} + \Delta T_i, \\ 1.0 & \text{if } T > T_{r,i} + \Delta T_i\ . \end{cases} \qquad (9a,b)$$

## TREATMENT OF PROCESS DATA

A large number of process data is measured at the BF works, ranging from temperatures to complex chemical analyses. As the process data includes both systematic and stochastic measurement errors, an interface program for "calibration" of the measurement data was developed.[11] In the interface routine, the consistency of data is checked against balance equations for atoms (Fe, C, O, N, and H) and energy. The physical constraints of the process provide approximate limits for the accumulations. Therefore, as a simple approximation, the cumulative output values of atoms and energy can be set equal to the corresponding cumulative input values, if a sufficiently long integration period is used (see Appendix). The equations are solved by introducing a set of linear calibration coefficients operating on those of the measurements that are assumed to be inaccurate. If the number of measurements to be calibrated exceeds the number of available balance equations, an objective function, which includes weight factors for the different measurements, is minimized. If, conversely, the number of calibrated measurements is less than that of balance equations, accumulation terms are included and the steady-state assumption is relaxed.

The molar enthalpies for components in their pure states,[12–14] with the exception of pig iron, slag, coal, and oil, are used when calculating the thermodynamic properties.

The use of the interface is exemplified next. Firstly, data from a Finnish blast furnace, the BF2 of Rautaruukki Oy in Raahe, for the period January 1 – April 30, 1986, is studied. Six balance equations are formulated. The values of two central process variables — the flow rate of top gas and its humidity — are not measured and must therefore be calculated by using two balances. This leaves only four balance equations, thus four calibration coefficients are uniquely defined. The calibrated variables are the blast volume, the flow rates of coke and pig iron, and the top gas components ($CO$, $CO_2$, and $H_2$). In order to reduce the effect of disturbances caused by actual material accumulations, the cumulative balances were calculated over a period of 10 days, resulting in the calibration coefficients shown in Fig. 3.

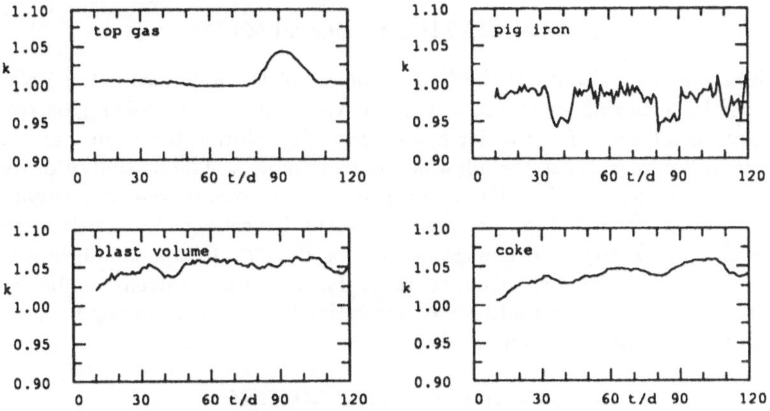


**Fig. 3** Calibration coefficients for top gas composition, blast volume, and flows of pig iron and coke (Rautaruukki Oy BF2, January 1 – April 30, 1986).

The calibrations are seen to be relatively stable, with the exception of a disturbance in the correction of the top gas analysis during the period 80 to 105 d. When the day reports at the works in question were scrutinized, it was confirmed that the top gas analyzer had indeed operated irregularly during this period of time. The three "valleys" in the calibration of the flow of pig iron starting at $t \approx 35$, 80 and 108 d and lasting for 10 days are caused by large single errors in the reported values of this variable. The fact that both the corrections of the blast volume and of the flow rate of coke assume values greater than unity, would indicate that some adjustments of the measuring instruments should be made.

Secondly, data from the BF2 of SSAB in Luleå, Sweden, is treated. Here the top gas $CO$ and $CO_2$ are calibrated separately, and an accumulation of materials in the process is also considered. The calculations are based on daily mean values of the measurements. Changes in the ore/coke-ratio of the charge are assumed to result in an accumulation of carbon and iron oxides in the furnace shaft.[15] It should be noted that the most extreme values of the calibration coefficients presented in Fig. 4 occur for days of irregular operation and shutdowns.

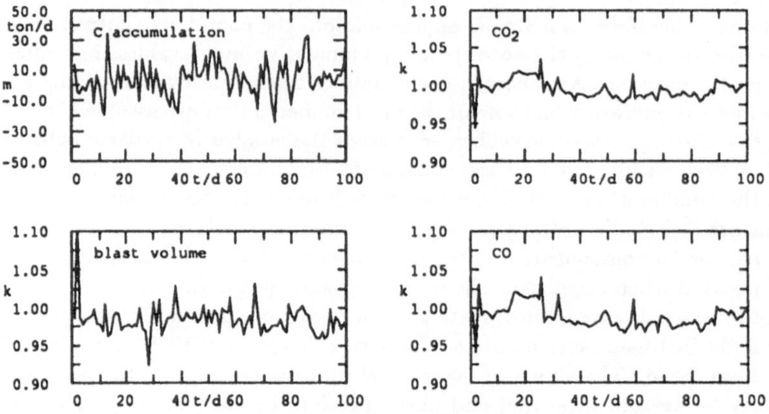


**Fig. 4** Accumulation of C, calibration coefficients for blast volume and top gas $CO_2$ and CO (SSAB Luleå BF2, September 1 – December 9, 1986).

Also, data from Dalsbruk Oy (formerly part of Ovako-Steel), Finland, has been analyzed. The results, not presented here, are very similar to those described above.

## MATHEMATICAL TREATMENT

In the numerical solution of the NLEs, guesses of the unknown extents of reaction are initially produced by a routine which places trial points within the feasible region constrained by the non-negativity conditions of the molar flows. Once the solution has been found, this may be used to produce initial guesses for the adjacent control volume. The feasibility of these starting points is provided by a step-length reduction until the molar flows become non-negative.
The NLEs, $\underline{f} = (\underline{f}_1^T\ \underline{f}_2^T)^T = \underline{0}$, have been solved by the Levenberg-Marquardt method,[16] using analytical gradients for $\partial\underline{f}/\partial\underline{e}^*$ and numerical ones for $\partial\underline{f}/\partial\underline{T}$. In the equilibrium problem, the activity coefficients, $\gamma$, have been set equal to unity. The resulting problem can be shown to have a unique solution.[17] The interface routine uses successive linear programming for the constrained minimization of the objective function.

## TUNING OF THE MODEL

In the modeling, two parameters, $\alpha_{v,ref}$ and $\beta_{tot}$, have been deliberately unspecified. However, once the BCs at the stockline level have been supplied by the interface routine, a sequential solution of the NLE systems yields the simulated analysis and temperature of the hot metal. Two functions are introduced

$$\underline{F} = \begin{pmatrix} \eta_T & 0 \\ 0 & \eta_{\dot{n}} \end{pmatrix} \begin{pmatrix} \Delta T_{tap} \\ \Delta \dot{n}_{tap} \end{pmatrix} = \underline{0}, \tag{10}$$

where scaling factors, $\eta$, are used for numerical reasons. $\Delta T_{tap}$ and $\Delta \dot{n}_{tap}$ denote the residuals between the simulated and the measured tapping temperatures and compositions, respectively. Eq. (10), which consists of a set of NLE systems as sub-problems, is solved with respect to the two unknowns using the Levenberg-Marquardt method.

In the examples to be presented we use daily mean values of the process data from the BF2 of Rautaruukki Oy for the time period January 1 – July 25, 1986. The interface routine, supplying the BCs consisting of molar flows, temperatures and pressures, has been applied to calibrate the four measurements discussed earlier (*cf.* Fig. 3). In the calculations, the temperature of the charged materials was assumed to follow a seasonal trend described by a sine function varying between 10°C and 50°C.

Twenty control volumes are used to describe the region above the tuyere level, where the iron oxides are reduced by CO according to $Fe_2O_3 \rightarrow Fe_3O_4 \rightarrow FeO \rightarrow Fe$, and the coke "solution-loss" reaction $C + CO_2 \rightleftharpoons 2\,CO$ is considered. In addition, the reactions for the transfer of silicon ($R1$–3) have been included. The parameters for the reaction rate and heat transfer expressions are given elsewhere.[10]

The results for a reference day, January 3, 1986, are shown in Fig. 5, where the main profiles are depicted. A thermal reserve zone and the step-wise reduction of the iron oxides are clearly seen. The rapid decrease in temperature of the ascending gas at $z \approx -27$ to $-23$ m is seen to be caused by the endothermic coke "solution-loss" reaction. In the furnace hearth, the energy released in the re-oxidation of silicon almost compensates for the heat loss. It should be pointed out that the molar flow denoted by $CO_2$ also includes the oxidized part of the hydrogen, which, for simplicity, has been considered inert in the modeling, and that the line denoted by C does not include the contribution of carbon burned in the raceways.

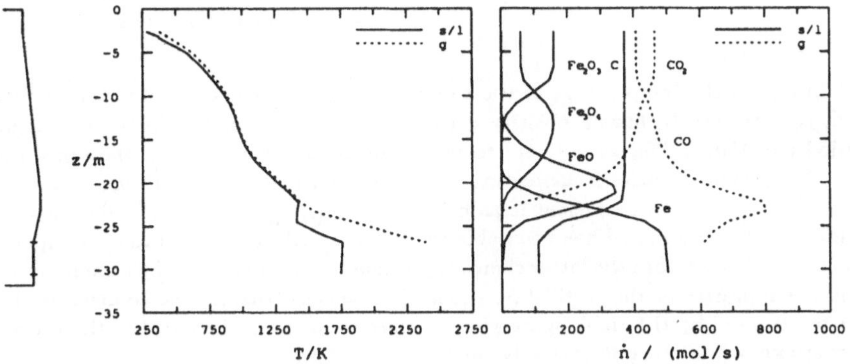


**Fig. 5** Temperatures and molar flows (Rautaruukki Oy BF2, January 3, 1986).

The results for the whole period are illustrated in Fig. 6. The parameters, which in general were solved in 5–8 main iterations, show a strong positive correlation and, moreover, vary considerably from day to day. The above observations suggest that some variable of the BCs is incorrectly treated and introduces noise into the results.

Two fixed quantities from which the noise could possibly stem are the heat loss and the temperature of the charged materials, $\vartheta_{s,top}$. A sensitivity analysis resulted in a rejection of the former hypothesis.[18] The second alternative was tested by fixing the heat transfer parameter at $\alpha_{v,ref} = 3.5\frac{\text{kW}}{\text{m}^3\text{K}}$, which closely corresponds to the mean of $\alpha_{v,ref}$ estimated above, and instead determining $\vartheta_{s,top}$ and $\beta_{tot}$ from the NLEs (10). Observe that the interface routine has to adjust the BCs during the iterative solution of the model parameters, since $\vartheta_{s,top}$ affects the energy balance equation. Fig. 7 shows that $\vartheta_{s,top}$ despite following the general seasonal trend (dashed line) used earlier, also gives evidence of considerable short-time variation. Such changes may still occur in practice, as the temperatures of sinter from the bins and coke from the yard sometimes vary substantially. Moreover, it should be remembered that the value of $\vartheta_{s,top}$ is affected by errors in the top gas temperature, the coke moisture, *etc.*

A noteworthy observation is that the variance of $\beta_{tot}$ has decreased by about 50%.

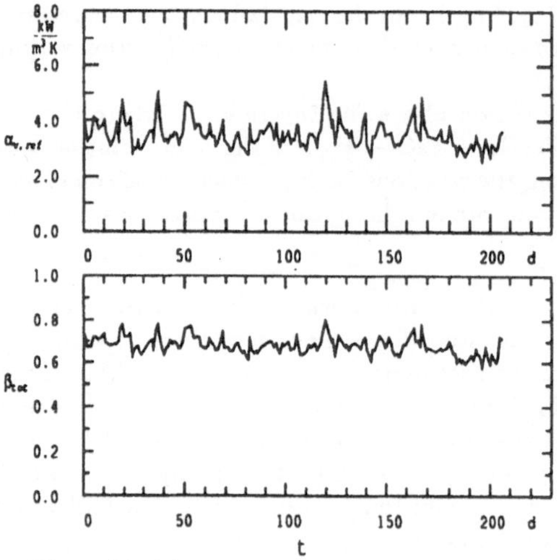


Fig 6 Model parameters $\alpha_{v,ref}$ and $\beta_{tot}$.

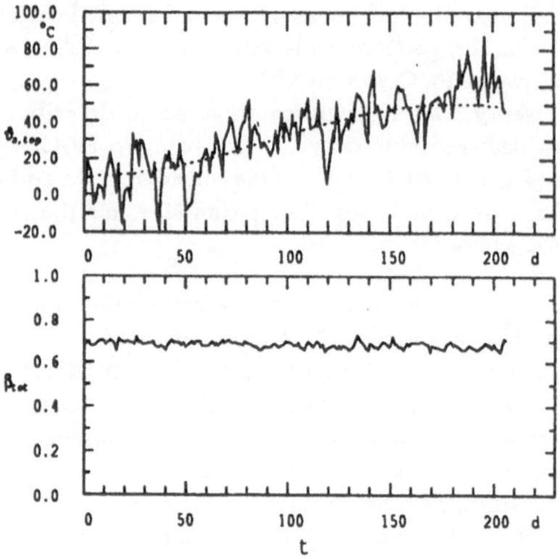


**Fig. 7** Model parameters $\vartheta_{s,top}$ and $\beta_{tot}$. The dashed line shows the earlier approximation used for $\vartheta_{s,top}$.

In Fig. 8, the temperature of the top gas, $\vartheta_{g,top}$, and the estimated temperature of the charge, $\vartheta_{s,top}$, are seen to have a positive correlation. The discrepancy in the trends during the time period $t = 100\ldots170$d were found to be the result of an increase in the moisture of coke. Moreover, $\vartheta_{s,top}$ shows a strong correlation with the ambient temperature $\vartheta_{amb}$, also plotted in the figure. The correlation coefficient is as high as $r_{\vartheta_{s,top}(t),\vartheta_{amb}(t-1)} \approx 0.74$ for the cold season, which indicates that the model has been able to reveal the effect of the outdoor temperature! Another clear indication that the latter choice of parameters is superior to the former one is supplied by Fig. 9, which illustrates the vertical positions of some isotherms for the condensed phases. The behavior represented by the right figure gives a more probable picture of the BF process, whose dynamics are known to be comparatively slow.

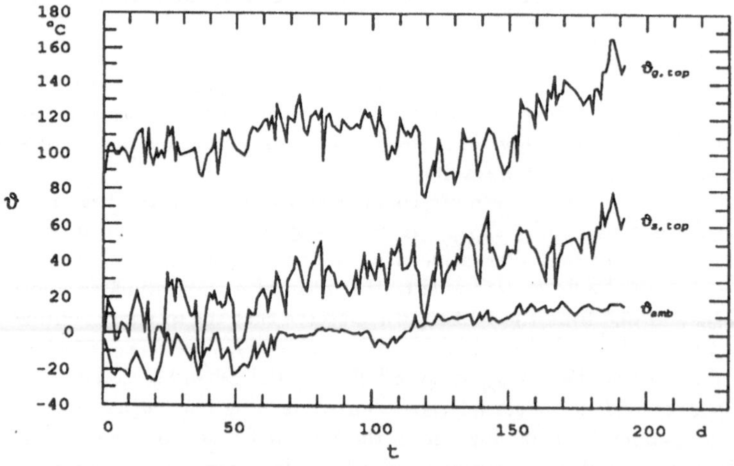


**Fig. 8** Top gas, charge, and ambient temperatures for January 1 – July 25, 1986.

The solution of the unknown temperature $\vartheta_{s,top}$ thus shows an important feature of the model,[19] that is, how it can be used as a device which gives information about variables that are hard, or impossible, to measure by standard techniques. Eq. (10) can, in fact, be solved with respect to several different variable pairs. For instance, $\alpha_{v,ref}$ and $\beta_{tot}$ may be fixed, and $\vartheta_{s,top}$ and $f_{Si}$ are determined. As a rule, $\vartheta_{s,top}$ should be included in the set of unknowns.[10]

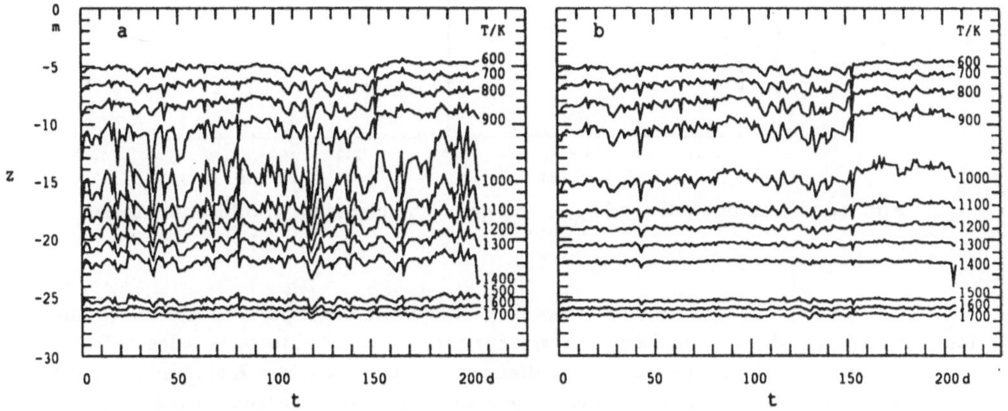


**Fig. 9** Vertical positions of isotherms for the condensed phases using a) $\alpha_{v,ref}$,$\beta_{tot}$ b) $\vartheta_{s,top}$,$\beta_{tot}$ as unknowns in Eq. (10).

## PREDICTION

The parameter values determined in the previous section can be used to predict the behavior of the furnace at different operational conditions. Only a slight modification of the treatment must be carried out. Firstly, the unknowns are chosen among the BCs, while the model parameters are given. Secondly, the interface program must now **determine** a set of unknown flows and temperatures. As the first step in the prediction, a base point and its corresponding model parameters are determined. Next, a change in some of the BCs is introduced and the simulation model is used to find the new state of the process. For the prediction alternatives to be treated here, the interface routine calculates the top gas temperature and analysis and the flow rate of pig iron using balances for atoms and energy. The functions of Eq. (10) are then solved with respect to the unknown BCs.

Only two special cases will be presented; a prediction of the productivity and the coke consumption when the tapping conditions are fixed **and** a prediction of the pig iron temperature and silicon content using fixed values for the input variables. For the sake of simplicity, the model parameters have been kept constant. The former case is illustrated by increasing the blast volume by 10% while keeping all other variables (including the oil injection rate) constant. The resulting increase in productivity was found to be approximately 9.5 %, while the specific coke consumption increased by 0.5%. The molar flow profiles for the reference case and the present case are shown in Fig. 10. The temperature profiles have not been plotted since they are almost unaffected by the change.

The problem of determining the output with given input is illustrated by an example where an increase of 20°C in the blast temperature has been introduced. The changes, which occur mainly in the lowermost parts of the furnace, result in an increase in the silicon content from 0.348 to 0.367% and an increase in the temperature of the hot metal of 21°C. It should be pointed out that some of the changes are due to fixing the inputs, *i.e.*, the productivity and the coke consumption, which over-emphasizes the effect on the tapping conditions.

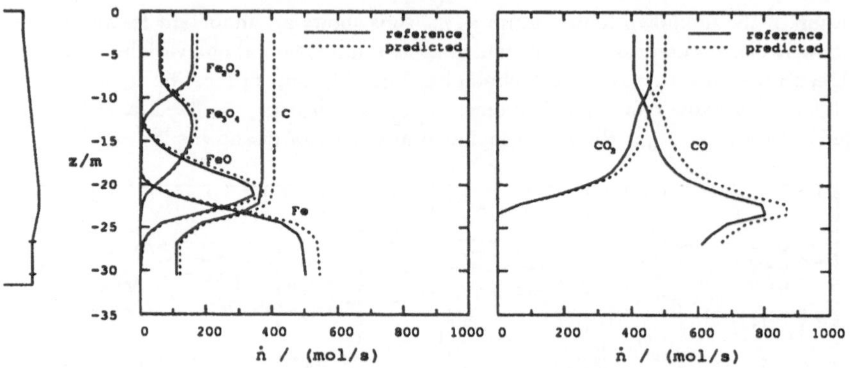


**Fig. 10** Molar flows of condensed and gas components for two blast volumes.

Different operational conditions can be predicted by the simulation model. Changes can be introduced in *e.g.* the blast moisture, oxygen enrichment, injected fuels, heat loss, *etc.* It should be observed that several changes can be simultaneously introduced, which is of interest in the BF process, where control by a single variable in general is impossible because of the coupling of the phenomena.

## DISCUSSION

The blast furnace simulation model outlined in the paper has been developed for on-line application by including a set of model parameters, which is adapted to the data from the process. These parameters, which should be considered specific for each furnace, and the corresponding description of the process will be scanned on-line to reveal trends in the BF operation. Constant parameter values were used in the predictions of the previous section, but it is straightforward to include a dependence of the model parameters upon the (boundary) conditions. Work is at present under way to refine the description of the reaction rates and the conditions in the furnace hearth. Moreover, an application of the model to process data for shorter time periods is studied. As a future outlook, the model parameters could be updated by time series analysis.

**Acknowledgements:** The authors wish to express their gratitude to the research personnel of the Nordic steel works for supplying the process data and valuable ideas. Financial support by Jernkontoret, Sweden, is gratefully acknowledged.

## APPENDIX

Consider a flow balance ($A1$) for a non-destroyable component, $X$, in which entering and effluent flows are calculated from measurement variables. Integrating ($A1$) gives ($A2$).

$$\dot{X}_{in} = \frac{dX_{acc}}{dt} + \dot{X}_{out}; \qquad \int_{t_0}^{t_1} (\dot{X}_{in} - \dot{X}_{out})\, dt = X_{acc}. \qquad (A1,2)$$

Calibration coefficients, $\underline{k}$, are introduced for a set of measurements, $\underline{y}$, yielding corrected measurements according to $\hat{y}_i(t) = k_i\, y_i(t)$. When solving these coefficients from the balance equations ($A2$), the accumulation terms, $\underline{X}_{acc}$, can be neglected if a sufficiently long period of time, $t_1 - t_0$, is used in the integration.

In order to allow for more calibration coefficients than balance equations, a quadratic minimization problem is formulated

$$\min_{\ln k_i \in R} \sum_{i=1}^{N} v_i \ln^2 k_i \qquad \text{s.t.} \quad \int_{t_0}^{t_1} \underline{f}(\underline{y}(t), \underline{k})\, dt = 0, \tag{A3}$$

where $v_i$ are weight factors for the $N$ measurements to be calibrated subject to the balance constraints (written in vector form). If the equalities are replaced by inequalities in the constraint equations, the minimization problem can be solved subject to user specified limits for the accumulation terms.

## REFERENCES


1) Saxén H., Report 86–1, Heat Eng. Lab., Åbo Akademi, Finland (1986).
2) Saxén *et al.*, Report 86–3, Heat Eng. Lab., Åbo Akademi, Finland (1986).
3) Saxén H., Report 86–6, Heat Eng. Lab., Åbo Akademi, Finland (1986).
4) Sugata *et al.*, "Some Discussions on Si Transfer in Blast Furnace Based on Raceway Measurement", Internal report (in Japanese), Nippon Steel Corp., Japan.
5) Taguchi *et al.*, "Discrimination Among Conditions at Lower Part of Blast Furnace by Distribution Ratios of Si, Mn and S Between Slag and Pig Iron", Internal report, Kawasaki Steel Corp., Japan.
6) Taguchi *et al.*, Proc. of 3$^{rd}$ Process Technology Conf., Pittsburg 1982, AIME, pp. 25–38.
7) Hatano *et al.*, *Trans. ISIJ* **22** (1982) pp. 448–456.
8) Togino *et al.*, *Trans. ISIJ* **20** (1980) pp. 639–645.
9) Kuwabara *et al.*, Proc. ISIJ-AUS IMM Joint Symp., Tokyo 1983, pp. 193–204.
10) Saxén H., Report 87–5, Heat Eng. Lab., Åbo Akademi, Finland (1987).
11) Uusi-Honko *et al.*, Proc. of XVIII Congr. of CEF, Naxos, Sicily, (1987) pp. 103–107.
12) Barin *et al.*, (1973) *Thermochemical Properties of Inorganic Substances*, Springer-Verlag, Berlin.
13) Barin *et al.*, (1977) *Thermochemical Properties of Inorganic Substances*, Supplement, Springer-Verlag, Berlin.
14) Kubaschewski *et al.*, (1967) *Metallurgical Thermochemistry*, Pergamon Press, London.
15) Uusi-Honko H., unpublished results.
16) Marquardt D.W., *J. SIAM* **11** (1963), pp. 431-444.
17) Westerlund *et al.*, *Chem. Eng. Sci* **42** (1987) pp. 188–190.
18) Kilpinen A., Report 87-82-A, Proc. Des. Lab., Åbo Akademi, Finland (1987) (in Swedish).
19) von Schalien *et al.*, Proc. of XVIII Congr. of CEF, Naxos, Sicily, (1987) pp. 565–570.

# A COVERAGE MONTE CARLO METHOD FOR RELIABILITY QUANTIFICATION

Tsunehiko Tanaka
Tokyo Research Laboratory
IBM Japan Ltd.
Tokyo 102, Japan

Hiromitsu Kumamoto
Faculty of Engineering
Kyoto University
Kyoto 606, Japan

and

Koichi Inoue
Faculty of Engineering
Kyoto University
Kyoto 606, Japan

## 1. Introduction

Network reliability and fault tree quantification are two major problems in the class of coherent structures. There has been a lot of work on the network reliability problem; some have given upper and lower bounds, and others have solved it approximately by Monte Carlo. There exists a linear time deterministic algorithm in size of the network (Satyanarayana and Wood [1]), which, however, is good only for a network of special structure. From the computational complexity, the problem is NP-hard (Provan and Ball [2]).

We cannot apply the method developed in [1] directly to the top event probability computation of a fault tree. Instead we can use a cut set inclusion-exclusion formula. Since enumerating all the min cut sets of a fault tree is as difficult as computing the top event probability, it is a common practice to eliminate some cut sets that make negligible contributions to the probability. Given $m$ significant min cut sets, deterministic algorithms must compute all the $2^m - 1$ terms in the inclusion-exclusion formula, which is impractical. The Monte Carlo approach is the powerful alternative.

Variance reduction techniques should be used especially for a rare-event problem to have a quick convergence to the true value with a reasonable number of trials. Van Slyke and Frank [3], and Kumamoto et al. [4] used information on bounds. Easton and Wong [5] devised a sampling plan which makes use of permutation properties. A new Monte Carlo method was introduced by Karp and Luby [6] . Their method requires only the computation of the sum of all the min cuts probabilities by deterministic algorithms and takes sample states that contribute to the top event occurrence. By restricting the sampling space further, we can obtain better estimate with smaller variances for the same number of trials. This idea leads to a general coverage Monte Carlo method. By general coverage Monte Carlo, we refer to order $k$ ($k = 1, \ldots, m$) coverage Monte Carlo. As we use higher order coverage Monte Carlo, preprocessing time drastically increases while variance of an estimator decreases. In other words, there is tradeoff between the preprocessing time and the variance of an estimator. By taking this fact into account and making a sampling plan more elaborate, we can get an even better estimate by the hybrid of various orders of general coverage Monte Carlo methods.

This paper is a continuation of our previous work (Kumamoto et al. [7], Tanaka et al. [8]) and contains most of the work done up to this point in the development of a coverage Monte Carlo method. In Section 2, a problem statement is given. In Section 3, a general coverage Monte Carlo method is presented. In Section 4, a hybrid coverage Monte Carlo method is developed and computation time and memory requirement are discussed.

## 2. Problem Statement

When a coherent fault tree is given, we would like to compute the top event probability from the significant min cut sets $C_j$ ($j = 1, \dots, m$) ($C_j$ represents the event that all the edges in the $j$th min cut set fail). Let

$$Q_l = \sum_{1 \le j_1 < \cdots < j_l \le m} Pr(C_{j_1} \cap \cdots \cap C_{j_l}) \tag{1}$$

i.e.,

$$Q_1 = \sum_{j_1 = 1}^{m} Pr(C_{j_1}) \tag{2}$$

$$Q_2 = \sum_{j_1 = 1}^{m} \sum_{j_2 > j_1}^{m} Pr(C_{j_1} \cap C_{j_2}) \tag{3}$$

etc.

then in terms of $Q_l$'s the top event occurrence probability $R$ can be written as:

$$R = Q_1 - Q_2 + \cdots + (-1)^{m-1} Q_m \, . \tag{4}$$

As is easily seen from (1), (2), (3) and (4), the computation of $R$ involves an exponential number of terms and becomes impractical. This paper presents the results of our efforts to circumvent this difficulty by computing only a partial sum in (4) and estimating the remaining quantity by Monte Carlo.

We assume the basic events to be statistically independent (occurrence probability of basic event $i = q_i$). Let $x = (x_1, \dots, x_n)$ denote a state vector of basic events where $x_i = 1$ if basic event $i$ is occurring and $= 0$ otherwise. Similarly let $\psi(x)$ denote a state of the top event where $\psi(x) = 1$ if the top event is occurring and $= 0$ otherwise. The occurrence probability $R$ of the top event can be expressed as:

$$R = Pr\{\psi(x) = 1\} \tag{5}$$

$$= \sum_{x} \psi(x) Pr(x) \, . \tag{6}$$

It follows immediately from (6) that direct Monte Carlo is given by:

$$\hat{R}_0 = N^{-1} \sum_{v=1}^{N} \psi(b_v), \quad b_v \sim Pr\{x = b_v\} \, . \tag{7}$$

The coefficient of variation of the estimator in (7) is equal to $[N^{-1}\{(1/R) - 1\}]^{1/2}$, which diverges to infinity as $R$ goes to zero. The direct Monte Carlo does not give an efficient estimator when the probability $R$ is very small. Therefore, the development of an efficient estimator is badly needed.

## 3. General Coverage Monte Carlo

The coverage $\alpha(x)$ ($\alpha$ for short) is defined as the number of min cuts supported by $x$. Also, let $g_j(x) = 1$ ( $g_j$ for short) if $x$ supports a min cut set $j$, and $= 0$ otherwise.

Now we are going to present an order $k$ coverage Monte Carlo method that takes sample states from $B_k$ where $B_k$ denotes the sampling space consisting of system states with coverage greater than or equal to $k$ ($k = 1, \ldots, m$).

We first present two lemmas which will be needed to prove *Theorem 1*.

*Lemma 1*

For any $b \in B_k$,

$$\sum_{1 \le j_1 < \cdots < j_l \le m} g_{j_1} \cdots g_{j_l} = \binom{\alpha}{l}, \quad k \le l \le \alpha . \tag{9}$$

The proof is given in Appendix A.

*Lemma 2*

For any k,

$$\sum_{l=k}^{\alpha} (-1)^{l-1} \binom{\alpha}{l} = (-1)^{k-1} \frac{k}{\alpha} \binom{\alpha}{k} . \tag{10}$$

The proof is given in Appendix B.

Now we are going to present an order k coverage Monte Carlo method that takes sample states from $B_k$ ($k = 1, \ldots, m$). *Theorem 1* gives a basic equation.

*Theorem 1*

$$R = W_{k-1} - D_{k-1} , \tag{11}$$

where

$$W_{k-1} = \sum_{l=1}^{k-1} (-1)^{l-1} Q_l, \quad W_0 = 0 , \tag{12}$$

$$D_{k-1} = (-1)^k \sum_{b \in B_k} \sum_{1 \le j_1 < \cdots < j_k \le m} \frac{k}{\alpha} g_{j_1} \cdots g_{j_k} Pr(b) , \tag{13}$$

$k = 1, \ldots, m$.

(proof)

$R$ can be expressed as a sum of $W_{k-1}$ and the remaining quantity, i.e.,

$$R = W_{k-1} + \sum_{l=k}^{m} (-1)^{l-1} Q_l \ . \tag{14}$$

On the other hand, using $g_j$'s $Q_l$ can be written by:

$$Q_l = \sum_{b \in B_l} \sum_{1 \leq j_1 < \cdots < j_l \leq m} g_{j_1} \cdots g_{j_l} Pr(b) \ . \tag{15}$$

Substituting (15) into (14),

$$R = W_{k-1} + \sum_{b \in B_k} \sum_{l=k}^{\alpha} (-1)^{l-1} \sum_{1 \leq j_1 < \cdots < j_l \leq m} g_{j_1} \cdots g_{j_l} Pr(b) \ . \tag{16}$$

Note that the summation of $b$ and $l$ should be done over $B_k$ and from k to $\alpha$ because of the product form of $g_j$'s. Applying ***Lemma 1*** and ***Lemma 2*** to (16):

$$R = W_{k-1} + \sum_{b \in B_k} \sum_{l=k}^{\alpha} (-1)^{l-1} \binom{\alpha}{l} Pr(b) \tag{17}$$

$$= W_{k-1} + \sum_{b \in B_k} (-1)^{k-1} \frac{k}{\alpha} \binom{\alpha}{k} Pr(b) \ . \tag{18}$$

Using ***Lemma 1*** with l=k again, (11), (12) and (13) can be obtained.

Q.E.D.

A joint probability of $b \in B_k$ and $k$ min cuts $C_{j_1}, \ldots, C_{j_k}$ ($j_1 < \cdots < j_k$) may be defined as:

$$P_k(b, j_1, \ldots, j_k) = g_{j_1} \cdots g_{j_k} Pr(x = b) / Q_k \ , \tag{19}$$

since $Q_k$ can be considered as a normalizing constant which can be observed from (15). Thus, (13) may be rewritten as:

$$D_{k-1} = (-1)^k Q_k \sum_{b \in B_k} \sum_{1 \leq j_1 < \cdots < j_k \leq m} \frac{k}{\alpha} P_k(b, j_1, \ldots, j_k) \tag{20}$$

$$= (-1)^k Q_k E\{\frac{k}{\alpha(b)}\} \ , \tag{21}$$

$$(b, j_1, \ldots, j_k) \sim P_k(b, j_1, \ldots, j_k) \ . \tag{22}$$

We have come to a point where we can define an order k coverage Monte Carlo estimator $\hat{R}_k$:

$$\hat{R}_k = W_{k-1} - (-1)^k Q_k N^{-1} \sum_{v=1}^{N} \frac{k}{\alpha(b_v)} \ , \tag{23}$$

$$(b_v, j_{1_v}, \ldots, j_{k_v}) \sim P_k(b, j_1, \ldots, j_k) \ . \tag{24}$$

$Q_l$ ($1 \leq l \leq k$) can be computed by:

$$Q_l = \sum_{1 \le j_1 < \cdots < j_l \le m} \left[ \prod_{i \in C_{j_1} \cup \cdots \cup C_{j_l}} q_i \right] \tag{25}$$

and the partial sum $W_{k-1}$ is obtained from (12). The variance of an order $k$ coverage estimator decreases as $k$ increases under reasonable assumptions, and its coefficient of variation approaches zero as system components become more reliable. (see Tanaka et al. [8] for more details.) The method for sampling $b, j_1, \ldots, j_k$ is given in Appendix C.

4. Hybrid Coverage Monte Carlo

There are three issues involved in comparing Monte Carlo methods; variance of estimator, preprocessing time, and memory requirements. Order $k$ coverage Monte Carlo required the computation of the quantities $Q_1, \ldots, Q_k$. In addition, a table used for sampling $x, j_1, \ldots, j_k$ needs to be stored. With general coverage Monte Carlo of reasonably small order $k$ that is of practical use, we have obtained variance reduction at the cost of preprocessing time and memory requirements.

Hybrid coverage Monte Carlo described in this section provides a method for saving computation time at the cost of variance. Order $k$ coverage Monte Carlo generates sample states that support at least $k$ min cuts. On the other hand, hybrid coverage Monte Carlo generates sample states that support various number of min cuts. In other words, a hybrid coverage Monte Carlo method uses various orders of coverage Monte Carlo to generate sample states.

Let $Z$ = { a set of all min cuts } and $Z_l$ = { a set of min cuts for which order $l$ coverage Monte Carlo is used }. We need to prove the following two lemmas for the development of hybrid coverage Monte Carlo.

***Lemma 3***

For any $b \in B_l$,

$$\{\alpha - (l-1)\}\, g_{j_1} \cdots g_{j_{l-1}} = g_{j_1} \cdots g_{j_{l-1}} \sum_{j_l = 1}^{m} g_{j_l} \qquad (l \ge 2) \tag{26}$$

where $j_l \ne j_r \ (j_r = 1, 2, \ldots, l-1)$.

The proof is given in Appendix D.

***Lemma 4***

For any $b \in B_l$,

$$\{\alpha - (l-1)\} \sum_{j_1 < \cdots < j_{l-1} \in Z_l} g_{j_1} \cdots g_{j_{l-1}} = \sum_{j_1 < \cdots < j_{l-1} \in Z_l} g_{j_1} \cdots g_{j_{l-1}} \sum_{j \in Z - Z_l} g_j$$

$$+ \, l \sum_{j_1 < \cdots < j_l \in Z_l} g_{j_1} \cdots g_{j_l} \,. \tag{27}$$

The proof is given in Appendix E.

In particular, when $l = 2$ in (27), we have:

$$(\alpha - 1)\sum_{j\in Z_2} g_j = \sum_{j\in Z_2} g_j \sum_{k\in Z_1} g_k + 2 \sum_{j<k\in Z_2} g_j g_k \,. \tag{28}$$

We consider hybrid coverage Monte Carlo that uses only order 1 and order 2 coverage Monte Carlo methods, i.e., $Z = Z_1 \cup Z_2$ . ***Theorem 2*** gives a basic equation. (Note $k = 0$ means that only order 1 coverage Monte Carlo is used.)

***Theorem 2***

$$R = Q_1 - \sum_{x\in B_1} \sum_j \sum_k f_h(b, j, k) g_j g_k Pr(b) \tag{29}$$

where

$$f_h(b, j, k) = \begin{cases} 1/\alpha & \text{if } j \in Z_2, k \in Z_1 \\ 2/\alpha & \text{if } j \in Z_2, k \in Z_2, j < k \\ (\alpha - 1)/\alpha & \text{if } j \in Z_1, k = 0 \,. \end{cases} \tag{30}$$

(proof)

When $k = 1$ in (11), we have:

$$R = -D_0 = Q_1 - D_h$$

where

$$\begin{aligned} D_h &= \sum_{b\in B_1} \sum_{j=1}^{m} \frac{\alpha - 1}{\alpha} g_j Pr(b) \\ &= \sum_{b\in B_1} \sum_{j\in Z_2} \frac{\alpha - 1}{\alpha} g_j Pr(b) + \sum_{b\in B_1} \sum_{j\in Z_1} \frac{\alpha - 1}{\alpha} g_j Pr(b) \,. \end{aligned}$$

Substituting (28) into $\sum_{j\in Z_2} (\alpha - 1) g_j$ and letting $g_0 \equiv 1$,

$$\begin{aligned} D_h &= \sum_{b\in B_1} \sum_{j\in Z_2, k\in Z_1} \frac{1}{\alpha} g_j g_k Pr(b) + \sum_{b\in B_1} \sum_{j<k\in Z_2} \frac{2}{\alpha} g_j g_k Pr(b) + \sum_{b\in B_1} \sum_{j\in Z_1} \frac{\alpha - 1}{\alpha} g_j g_0 Pr(b) \\ &= \sum_{b\in B_1} \sum_j \sum_k f_h(b,j,k) g_j g_k Pr(b) \,. \end{aligned}$$

Q.E.D.

Defining a joint probability of $b \in B_1$ and min cuts $C_j$ and $C_k$, and a normalizing constant $Q_h$ as:

$$P_h(b,j,k) = \frac{g_j g_k Pr(b)}{\sum_{b\in B_1} \sum_j \sum_k g_j g_k Pr(b)} \tag{31}$$

and

$$Q_h = \sum_{b \in B_1} \sum_j \sum_k g_j g_k Pr(b) \;, \tag{32}$$

we can obtain:

$$D_h = Q_h \sum_{b \in B_1} \sum_j \sum_k f_h(b,j,k) P_h(b,j,k) \;. \tag{33}$$

This leads to a new coverage Monte Carlo estimator which is a hybrid of order 1 and order 2 estimators:

$$\hat{R}_h = Q_1 - Q_h N^{-1} \sum_{v=1}^{N} f_h(b_v, j_v, k_v) \tag{34}$$

where

$f_h$ is given by (30) and $(b_v, j_v, k_v) \sim P_h(b,j,k)$ .

Below shown without proof are five properties of hybrid coverage Monte Carlo estimator.

In *Property 1*, we show that the variance of this hybrid estimator is less than that of an order 1 estimator and greater than that of an order 2 estimator under reasonable assumptions.

***Property 1***

$$Var(\hat{R}_2) < Var(\hat{R}_h) < Var(\hat{R}_1) \;. \tag{35}$$

The first inequality holds provided $Q_h > 2Q_2$ and the second inequality holds provided $2Q_h < Q_1$.

Hybrid Monte Carlo gives a estimator $\hat{R}_h$ with a smaller variance than $\hat{R}_1$ and requires less preprocessing time than $\hat{R}_2$.

*Property 2* and *Property 3* give upper bounds for the variance of the estimator; one with $R$ and the other without $R$.

***Property 2***

$$Var(\hat{R}_h) \le (Q_1 - R)(Q_h - Q_1 + R)/N \;. \tag{36}$$

***Property 3***

$$Var(\hat{R}_h) \le \frac{Q_h^2}{4N} \;. \tag{37}$$

*Property 4* and *Property 5* give upper bounds for the square of the coefficient of variation; one with $R$ and the other without $R$.

***Property 4***

$$Var\left(\frac{\hat{R}_h}{R}\right) \le \frac{\left(\frac{Q_1}{R} - 1\right)\left(\frac{Q_h}{R} - \frac{Q_1}{R} + 1\right)}{N} . \tag{38}$$

*Property 5*

$$Var\left(\frac{\hat{R}_h}{R}\right) \le \frac{Q_h^2}{4Q_1(Q_1 - Q_h)N} \tag{39}$$

provided $Q_1 > Q_h$.

Likewise, we can develop a more elaborate coverage Monte Carlo method.

## Appendix A Proof of Lemma 1

When l = 1, the above equality is:

$$\sum_{j_1=1}^{m} g_{j_1} = \alpha .$$

This is obviously true. Assume

$$\sum_{1 \le j_1 < \cdots < j_{l-1} \le m} g_{j_1} \cdots g_{j_{l-1}} = \binom{\alpha}{l-1}$$

holds, then

$$\text{the l.h.s. of (11)} = \frac{1}{l}\Big\{\sum_{1 \le j_1 < \cdots < j_{l-1} \le m} g_{j_1} \cdots g_{j_{l-1}} \sum_{j_l=1}^{m} g_{j_l}$$

$$- \sum_{1 \le j_1 < \cdots < j_{l-1} \le m} (g_{j_1}^2 g_{j_2} \cdots g_{j_{l-1}} + \cdots + g_{j_1} g_{j_2} \cdots g_{j_{l-1}}^2)\Big\}$$

$$= \frac{\sum_{1 \le j_1 < \cdots < j_{l-1} \le m} g_{j_1} \cdots g_{j_{l-1}} \{\sum_{j_l=1}^{m} g_{j_l} - (l-1)\}}{l}$$

$$= \frac{\binom{\alpha}{l-1}\{\alpha - (l-1)\}}{l} = \text{the r.h.s. of (9)}$$

Q.E.D.

## Appendix B Proof of Lemma 2

We know the identity:

$$(1 + a)^{\alpha} = \sum_{l=0}^{\alpha} \binom{\alpha}{l} a^{\alpha - l} .$$

Letting $a = -1$, we can get:

$$\sum_{l=1}^{\alpha}(-1)^{l-1}\binom{\alpha}{l} = 1 .$$

Thus, (10) holds when $k = 1$. Assume

$$\sum_{l=k-1}^{\alpha}(-1)^{l-1}\binom{\alpha}{l} = (-1)^{k-2}\frac{k-1}{\alpha}\binom{\alpha}{k-1}$$

holds, then

$$\begin{aligned}\text{the l.h.s. of (10)} &= (-1)^{k-2}\frac{k-1}{\alpha}\binom{\alpha}{k-1} - (-1)^{k-2}\binom{\alpha}{k-1} \\ &= (-1)^{k-1}\binom{\alpha}{k-1}\left(\frac{\alpha-k+1}{\alpha}\right) = \text{the r.h.s. of (10)} .\end{aligned}$$

Q.E.D.

Appendix C A Sampling Method for General Coverage Monte Carlo

The probability $P_k$ can be decomposed as:

$$P_k(b, j_1, \dots, j_k) = P_k(j_1)\, P_k(j_2|j_1) \dots P_k(j_k|j_1, \dots j_{k-1})\, P_k(b|j_1, \dots, j_k) .$$

Therefore, sample states can be obtained by first generating $j_1$ by:

$$P_k(j_1) = \frac{\sum_{j_1 < l_2 < \dots < l_k \le m} Pr(C_{j_1} \cap C_{l_2} \cap \dots \cap C_{l_k})}{\sum_{1 \le l_1 < \dots < l_k \le m} Pr(C_{l_1} \cap \dots \cap C_{l_k})}$$

and $j_l$ $(l = 2,\dots,k)$ by:

$$P_k(j_l|j_1, \dots, j_{l-1}) = \frac{\sum_{j_l < l_{l+1} < \dots < l_k \le m} Pr(C_{j_1} \cap \dots \cap C_{j_l} \cap C_{l_{l+1}} \cap \dots \cap C_{l_k})}{\sum_{j_{l-1} < l_l < \dots < l_k \le m} Pr(C_{j_1} \cap \dots \cap C_{j_{l-1}} \cap C_{l_l} \cap \dots \cap C_{l_k})} .$$

Finally, a sample state $b$ given k cuts is generated by:

$$P_k(b|j_1, \dots, j_k) = [\prod_{i \in C_{j_1} \cup \dots \cup C_{j_k}} b_i]\,[\prod_{i \notin C_{j_1} \cup \dots \cup C_{j_k}} q_i^{b_i}(1-q_i)^{1-b_i}] .$$

Appendix D Proof of Lemma 3

When $g_{j_1} \dots g_{j_{l-1}} = 0$, the equality clearly holds. When $g_{j_1} \dots g_{j_{l-1}} = 1$,

$$\begin{aligned}\text{the r.h.s. of (26)} &= (\sum_{j_{l-1}}^{m} g_{j_l}) - (\sum_{j=1}^{j_{l-1}} g_j) \\ &= \alpha - (l-1) = \text{the l.h.s. of (26).}\end{aligned}$$

Q.E.D.

Appendix E Proof of Lemma 4

From (26) for $j_1 < \cdots < j_{l-1} \in Z_l$,

$$\{\alpha - (l-1)\}\, g_{j_1} \cdots g_{j_{l-1}} = g_{j_1} \cdots g_{j_{l-1}} \sum_{j \in Z - Z_l} g_j + g_{j_1} \cdots g_{j_{l-1}} \sum_{j \in Z_l - \{j_1, \ldots, j_{l-1}\}} g_j \,.$$

By summing over $j_1 < \cdots < j_{l-1} \in Z_l$,

$$\begin{aligned}\{\alpha - (l-1)\} \sum_{j_1 < \cdots < j_{l-1} \in Z_l} g_{j_1} \cdots g_{j_{l-1}} &= \sum_{j_1 < \cdots < j_{l-1} \in Z_l} g_{j_1} \cdots g_{j_{l-1}} \sum_{j \in Z - Z_l} g_j \\ &+ \sum_{j_1 < \cdots < j_{l-1} \in Z_l} g_{j_1} \cdots g_{j_{l-1}} \sum_{j \in Z_l - \{j_1, \ldots, j_{l-1}\}} g_j \\ &= \sum_{j_1 < \cdots < j_{l-1} \in Z_l} g_{j_1} \cdots g_{j_{l-1}} \sum_{j \in Z - Z_l} g_j \\ &+ l \sum_{j_1 < \cdots < j_l \in Z_l} g_{j_1} \cdots g_{j_l} \,.\end{aligned}$$

Q.E.D.

## References


[1] A. Satyanarayana and R. K. Wood, Polygon-to-Chain Reductions and Network Reliability, ORC 82-4, Operations Research Center, University of California, Berkeley, 1982.

[2] J.S. Provan and M.O. Ball, The Complexity of Counting Cuts and of Computing the Probability That a Graph is Connected, SIAM J. *Comput.* **12** (4) (1983) 777-788.

[3] R.M. Van Slyke and H. Frank, Network Reliability Analysis: Part I, *Networks* **1** (1972) 279-290.

[4] H. Kumamoto, K. Tanaka and K. Inoue, Efficient Evaluation of System Reliability by Monte Carlo Method, IEEE *Trans. Reliability* **R-26** (5) (1977) 311-315.

[5] M.C. Easton and C.K. Wong, Sequential Destruction Method for Monte Carlo Evaluation of System Reliability, IEEE *Trans.Reliability* **R-29** (1) (1980) 27-32.

[6] R.M. Karp and M.G. Luby, A New Monte Carlo Method for Estimating the Failure Probability of an n-component System, Computer Science Division, University of California, Berkeley, 1983.

[7] H. Kumamoto, T. Tanaka and K. Inoue, A New Monte Carlo Method for Evaluating System-Failure Probability, IEEE *Trans.Reliability* to appear, 1987.

[8] T. Tanaka, H. Kumamoto and K. Inoue, System Reliability Analysis by a General Coverage Monte Carlo Method, *Reliability Theory and Applications* Proceedings of the China-Japan Reliability Symposium, S.Osaki and J.Cao (editors), pp.381-390, September, 1987, World Scientific.

# TRANSPORTATION INVESTMENT AND DYNAMIC EQUILIBRIUM IN A MULTIREGIONAL INPUT-OUTPUT SYSTEM

Domenico Campisi and Agostino La Bella
Istituto di Analisi dei Sistemi ed Informatica del CNR, Viale Manzoni, 30 - 00185 Roma

## 1. Introduction

The aim of this paper is that of investigating the relationships between investment in transportation and economic growth. The interest in this type of analysis has been stimulated by the magnitude of transportation expenditures as a quota of national income in both developed and developing countries. Actually, the bulk of the literature is concerned with two main issues: optimal investment planning, and the modeling of the socio-economic impact of large transportation projects. However, the techniques usually utilized are borrowed with little or none modifications from other fields of public and private decision making, without including in the analysis the particular characteristics that make transportation investments strongly connected with the development path of the economy. Among the conceptual schemes which have been used there are: the theory of spatial general equilibrium (Tulkens and Kioni Kabantu, 1983); the Input-Output model (Liew and Liew, 1984a and b,1985); the household production function approach (Sasaki, 1983); the econometric approach (Blum, 1982); the mathematical programming approach (Prastacos and Romanos, 1987); the welfare economics approach (Adler, 1971; Anand, 1975).

A formulation of the evaluation problem general enough to allow for a full economic appraisal of transportation plans has been proposed in La Bella (1986) and Campisi and La Bella (1987a and b). The proposed model is based on a multiregional multisectoral scheme and permits taking into account both elastic demand functions and the two main classes of instrument variables available to planning authorities, i.e. changes in the transport network and the pricing system. In Campisi and La Bella (1987a) it has been showed how the above approach can be applied in the short-run, to calculate diseconomies due to the mismatch between the spatial structures of production and consumption and to draw an "inefficiency map" of the current location and transportation patterns. The long run impact of large transportation investment is dealt with in Campisi and La Bella (1987b) where the conditions on the technical coefficients matrix A, the capital coefficients matrix B and the interregional trade matrix T considered together, under which the multiregional Leontief dynamic model has a balanced growth solution, have been estabilished. In mathematical terms, the existence of that kind of solution is related to the eigenvalues of the matrix $U=(I-TA)^{-1}TB$. It has been proved that, if U is irreducible, a balanced growth solution exists and the multiregional economy grows at a common rate, which is a function of the Frobenius eigenvalue of U. In this paper after a short discussion of necessary and sufficient conditions on A, B and T so that U is irreducible, we shall analyse the case of reducible U matrices, and establish conditions for inducing dynamic equilibrium through interventions on the transportation network. The conditions will be based on zero-nonzero structure of the matrices, and therefore will be derived considering the topological properties of their associated graphs allowing us to give particular evidence to the spatial structure of trade patterns as determined by the system of regional productions and demands. The corresponding problem of adding a minimum set of transportation links to the graph associated to the model so as to satisfy suitable connectivity conditions is then formulated in mathematical programming terms.

## 2. The Multiregional Model

In this section we introduce the structure of the multiregional Input-Output model which will be utilized in the sequel as the basis for our analysis. The model can be built according to the three different approaches (see Batten and Martellato, 1984):

- the first is the straightforward extension of the Leontief national model under the assumption that the sectoral and geographical origin of each delivery can be specified (Isard, 1951);
- the second is based on the theory of demand for products distinguished by place of production (Armington, 1969);
- the third is centered on the idea that the regional origin of goods is as irrelevant to users as their regional destination to producers, bringing about the concept of "trade pool" (Leontief and Strout, 1973).

The first approach implies the use of coefficients representing the pattern of interindustry shipments between production units located in different areas. Models with this structure have been defined "ideal" because the trade pattern is perfectly specified at both origin and destination. However, besides being virtually inoperable with currently available data (only recently the use of information theory and related techniques made the implementation possible) they need the very strong assumption of stability of both the technical and trade coefficient. In its turn, the trade pool (gravity) approach allows an explicit link between trade flows and transportation costs introducing spatial interaction coefficients, with a drastic increase in the number of equations. For the above reasons our model is based on the explicit representation of the share pertaining to each competing product, classified by origin within every regional market. Therefore, given the product mix in sector r, the corresponding market share in region j for the production in i is represented by $t_r^{ij}$. The resulting dynamic Input-Output model can therefore be written as:

$$x^i(k)=\Sigma_j T^{ij}\{A^j x^j(k)+B^j[x^j(k+1)-x^j(k)]\} \tag{1}$$

where $x^i$ is the vector of sectoral production; $A^i$ is the technological matrix related to region i, $B^i$ is the capital coefficients matrix in region i, $T^{ij}$ is the diagonal matrix of trade coefficients $t_r^{ij}$:

$$T^{ij} = \begin{bmatrix} t_1^{ij} & \cdots & 0 & \cdots & 0 \\ 0 & \cdots\; t_2^{ij} & \cdots & 0 & \cdots\; 0 \\ \cdots & \cdots & \cdots & \cdots & \cdots \\ 0 & \cdots & t_k^{ij} & \cdots & 0 \\ 0 & \cdots & 0 & \cdots & t_n^{ij} \end{bmatrix}$$

It follows from (1) that the following condition holds:

$$\Sigma_i t^{ij}_r = 1, \qquad \forall j, \forall r \tag{2}$$

The model in equation (1) is closed, i.e. non-investment final demand is assumed to consist only of personal consumption and households are treated like any other sector with consumption as its inputs requirements. In addition, it is assumed that no technical change takes place. It is certainly possible to conceive a more general formulation with time-varying coefficients; however, the study of long-run equilibrium is here intended more as a mean to explore the structural properties of system (1) than as a forecasting exercise. Moreover, all the coefficients are supposed to be exogeneously given, so that our problem could be reformulated as that of finding the optimum transportation pattern corresponding to a set of fully specified regional technology options. The regional coefficients can be either estimated using various survey and non-survey methods, or determined on the basis of future scenarios or planning goals whose consequences should be simulated. The estimation of the capital coefficient matrix B is more problematic; moreover, since it is usually singular (many sectors do not contribute capital goods) it poses serious questions on the possibilities of solving equation (1) forward in time. This problem will be discussed in detail in the next section. Another central problem in the above model is the estimation of matrix T. Rarely does information on regional trade flows allow a direct estimate of the coefficient matrix T. If sporadically little or no direct information is available, one usually resorts to a statistical approach such as the principle of minimum information. It is not the purpose of this paper to dwell on this type of problems; here, we focus on the conditions that T should satisfy in order to make system (1) able to substain a balanced growth process.

## 3. Dynamic equilibrium and topological properties

We are interested here in exploring the existence of positive solutions x(k) to equation (1). We assume that the technological matrix A and the capital coefficient matrix B are fixed over time. This implies that there is no changes neither in technology nor in the use of capital goods by sectors. Moreover, the trade coefficient matrix T is also assumed constant over time, implying that there is no change in trade and transportations patterns. Obviously, we are however interested in analyzing different solutions corresponding to different T matrices. It will be stated that, if the diagonal elements at A are less than unity and the column sums are at most one, with a least one of them less than one, the inverse of (I-TA) exists and is non-negative. Moreover, the Hawkins-Simon condition holds (all the principal minors of A are positive; Nikaido, 1968). It can be realistically assumed that the above conditions are verified in our case, after a transformation of matrix A, if necessary, to an appropriate set of units. Therefore, since T is stochastic, we can conclude that there are no restrictions in assuming the existence of an inverse matrix for (I-TA) as well. The extension to the multiregional case, therefore, creates no problem under this respect. Then, equation (1) can be rewritten as:

$$(I - TA + TB)\, x(k) = TB\, x(k+1) \qquad (3)$$

which can easily be solved backward in time (see for instance Luenberger and Arbel, 1977). The possibility of solving it forward in time (Duchin and Szyld, 1986) is often hindered by the singularity of matrices B corresponding to empirical observations. Nevertheless even if B is singular, i.e. contains entire rows of zeros corresponding to regional sectors which do not produce intermediate goods (agriculture and services, for example), non-negative solutions for (1) might exist and the study is related to the eigenvalues of matrix $U=(I-TA)^{-1}TB$ (Szyld, 1985). It is hower interesting to investigate the balanced growth properties of the linear dynamic system (1). Since both $(I-TA)^{-1}$and TB are non-negative, so is their product U. If in addition U is irreducible, it satisfies the hypothesis of the Perron-Frobenius theorem (Horn and Johnson, 1986) stating that there exists a simple dominant eigenvalue $\lambda^*$ of U, with a corresponding eigenvector $v^*$, which determines the balanced growth solution of (1). Here we want to explore some properties arising from the peculiar structure of matrix T related to the topological properties of U. We note that a weighted directed graph G(L) can be associated to every matrix L. Each entry $l_{pq}$ is represented by an oriented arc (or edge) going from vertex $V_p$ to $V_q$ whose weight is $l_{pq}$ itself. Entries of zero are associated with absent arcs, while the diagonal entry $l_{pp}$ requires a loop in its associated vertex. The matrix $M(L)=\{\delta_{pq}\}$, where $\delta_{pp} = 1$ if $l_{pq}\neq 0$ and $\delta_{pq}=0$ if $l_{pq}=0$, is called the incidence matrix of L. If L is irreducible, this means that in G(L) every vertex can be reached by a directed walk from every other vertex; such a directed graph is called "strongly connected".

It is interesting to note that in the case of Leontief systems (where only one regional technology is assumed) the concepts of irreducibility, indecomposability, and connectedness, reduce to the same indecomposability concept from the Frobenious theory on nonnegative square matrices. This problem has been tackled in the literature with reference to the traditional dynamic Leontief model. For instance, to make U irreducible, Leontief (1977) assumed that both A and B were irreducible. Meyer (1978) relaxed the conditions on B, assuming its singularity and reducibility under the hypothesis of irreducibility of matrix A. Szyld (1985) firstly relaxed the conditions on the single matrices A and B, investigating the irreducibility of $(I-A)^{-1}B$ on the basis of the topological properties of their associated graphs, whereas the multiregional extension was investigated by Campisi and La Bella (1987b) who proved the following theorem :

THEOREM 1: *Let T, A and B be respectively the trade coefficient matrix, the block diagonal matrix of regional technologies and the block diagonal matrix of regional capital coefficients. Assume also that each component matrix has at least a nonzero entry for each column (this condition is verified for T by definition), then* $(I-TA)^{-1}TB$ *is irreducible if and only if the graph* $G(T) \cup G(A) \cup G(B)$ *is strongly connected.*

This theorem means that even if A, B and T are reducible, $(I-TA)^{-1}TB$ may not be, if some suitable conditions on the nonzero structure of the three matrices are satisfied. What we require, basically, is that: 1) every sector in each region has at least one input and one output from and to other sectors in the same region; and 2) every sector in each region has a capital input from at least another sector in the same region. These assumptions seem to be realistic. In the case of the technological matrix A, condition 1) automatically holds if the model is closed with respect to consumption. As far as condition 2) is concerned, it may seem very restrictive to assume that capital inputs must come from the same region. It must be noted, however (Campisi and La Bella 1987a), that such condition has been posed only to guarantee that TB has at least one nonzero entry for each column. For our purpose it would be sufficient to assume directly that the latter holds. This is quite reasonable, since it is equivalent to assume the existence of at least one capital input for each activity, no matter of its provenence. Besides being realistic, this condition is also crucial because otherwise U would have an entire column of zeros.

## 4. Conditions for balanced growth

In the previous section we have discussed some necessary and sufficient conditions for the irreducibility of matrix U. As observed before, if U is irreducible, being also by construction non-negative, the Perron-Frobenius theorem holds and there exists a simple dominant eigenvalue $\lambda^*$ of U, with a corresponding positive eigenvector $v^*$. We shall explore here the relations between the properties of U and the balanced growth solution of system (1). We start observing the equivalence between the following two eigen-problems:

$$Uv = \lambda v \tag{4}$$

$$Hu = \mu TBu \tag{5}$$

where H = I-TA +TB and the eigenvectors of (5) are solutions at (1). It is not difficult to see that problems (4) and (5) share their eigenvectors and that

$$\mu_i = (\lambda_i + 1)\ /\lambda_i, \quad \forall i \tag{6}$$

It follows that in the long run all sectors in all regions grow up at the same rate $\mu^*=(\lambda^*+1)/\lambda^*$ and propotions among them are established according to the components of eigenvector $v^*$. Note that the irreducibility of U is only a sufficient condition for the existence of a balanced growth path for system (1). We shall discuss now the case of reducible U matrices, and show that conditions can be given for the existence of balanced growth. If U is reducible, it is well known (Gantmacher, 1959) that, after a suitable permutation P, it can be written in its normal form:

$$PUP^T = \begin{bmatrix} U_{11} & 0 & . & 0 & . & 0 \\ 0 & U_{22} & \cdot & 0 & \cdot & 0 \\ \cdot & \cdot & \cdot & \cdot & \cdot & \cdot \\ \cdot & \cdot & \cdot & U_{gg} & \cdot & 0 \\ U_{g+1\,1} & U_{g+1\,2} & \cdot & U_{g+1\,g} & \cdot & 0 \\ \cdot & \cdot & & \cdot & & \cdot \\ U_{z1} & U_{z2} & \cdot & U_{zg} & \cdot & U_{zz} \end{bmatrix} \qquad (7)$$

where the diagonal blocks are irreducible matrices, and in each row $U_{f1}$, $U_{f2}$,....... $U_{f\,g-1}$, with f = g+1, ..., z, at least one matrix is different from zero. It can be shown also that the normal form of a matrix U is uniquely determined to within a permutation of the blocks and permutations within the diagonal blocks (the same for rows and columns). Making use of the normal form, it can be proved (Gantmacher, 1959) that a balanced growth solution exists also in the case of reducible U if and only if each of the square matrices $U_{11}$, $U_{22}$, .... $U_{gg}$ has $\lambda^*$ as the dominant eigenvalue and none of the matrices $U_{g+1\,g+1}$, .... $U_{zz}$ has this property. It follows from the proof of the above proposition that under the same conditions we get $\lambda^*_k < \lambda^*$ for k=g+1, ...., z (Gantmacher, 1959). In economic terms this amount to say that balanced growth is possible if the various groups of isolated regional sectors all have a common balanced growth rate, whereas the groups that deliver goods to regional sectors in other groups have, as isolated groups, a different growth rate. The latter would be higher since $(1+\lambda^*_k)/\lambda^*_k>(1+\lambda^*)/\lambda^*$.

## 5. Augmentation problems

As we have previously discussed, in our frame a balanced growth can be achieved if either U is irreducible or same rather complicated conditions on its structure are satisfied. We limit our analysis here to the first case, i.e. we shall explore how to add an optimal set of links to the transportation network so as to make U irreducible. This is equivalent to add new edges to G(T) in order to make

G(T)UG(A)UG(B) strongly connected. Several feasible solutions, in terms of different set of new arcs may be able to produce strong connectivity when added to G(T). Two optimization problems can then be formulated, one for seeking the set with the minimum number of edges, and the other for seeking the minimum cost set. Obviously, the latter requires the definition of a weight for each edge. Since the edges to be added must belong to G (T), they will have the form [(r,i),(r,j]. The associated weight f[(r,i),(r,j)] is chosen to represent the investiment cost for the corresponding infrastructure, i.e. for the supply of transportation that, according to a separate model for estimating multicommodity flows, will generate a trade coefficient $t_T^{ij} > 0$. Let therefore V be the set of vertices (r,i) and $\varepsilon_0$ a finite set of edges such that all elements are ordered pairs of distinct elements of V. The associated directed graph is $G_0=(V,\varepsilon_0)$ which in our case coincides with G(T)UG(A)UG(B). We say that $G_0(V,\varepsilon_0)$ is a spanning subgraph of $G_\alpha=(V,\varepsilon_\alpha)$ if $\varepsilon_0 \subseteq \varepsilon_\alpha$, the sets of nodes coinciding.

*The strong augmentation problem is in general the problem of finding a minimum cost set edges* $\varepsilon_1$ *such that* $G_1 = (V, \varepsilon_0 \cup \varepsilon_1)$ *is strongly connected.* The above problem is well posed since it clearly admits feasible solutions, the only difficulty being related to the order of the computational algorithms. Eswaran and Tarjan (1976) proved that the directed Hamiltonian cycle problem is reducible to the strong connectivity augmentation problem, implying strong computational difficulties in its solution (the directed Hamiltonian cycle problem is NP-Complete). Easier computational solutions can, however, be found if the given graph has some particular structures. For instance, if the graph we wish to strongly connect has the property that some vertex $\ell$ can be reached from every vertex, then the problems reduces to that of finding a minimum weight spanning arborescence with root $\ell$, which is efficiently solvable (Edmonds, 1967). Other indications on the minimum number of edges to be added can be found in Kajitani and Ueno (1986).

Consider now the unweighted version of the strong connectivity augmentation problem, where we want to minimize the number of edges to be added. We can reduce this problem to a simpler one converting $G_0$ into a directed graph $G'_0$ which contains one vertex for each strongly connected component (subgraph) of $G_0$. $G'_0$ is acyclic and is called a condensation with respect to strong components of $G_0$. It is obvious that, given a set edges which strongly connects $G'_0$, $G_0$ is strongly connected by the corresponding set of edges. Therefore, we can restrict our attention to the acyclic graph $G'_0$. The following theorem gives a lower bound on the number of edges to be added to make $G'_0$ strongly connected.

THEOREM 2: *Let* $G'_0$ *be an acyclic directed graph with s sources, t sinks and q isolated vertices, where* $s+t+q > 1$. *Then at least* $\max(s,t) + q$ *edges are needed to make* $G'_0$ *strongly connected.*

The proof although straightforward can be found in Eswaran and Tarjan (1976).

We are now ready to come back to our dynamic I-O model, noting that adding a set of arcs to the given graph to make it strongly connected is equivalent to transform the original reducible matrix U into an irreducible one, and that there is a complete correspondence between each added arc and the transformation of some original zero elements of U into non-zero ones. Moreover, the latter result concerning the condensation problem of the unweighted augmentation problem is equivalent to partition U in its normal form as in section 4, and compact the aggregated pools of regional productions into a simplier aggregated model. The given lower bound on the number of edges to be added in theorem 2 is therefore comparable with the number of the zero element in the obtained matrix to be transformed into positive entries. Note, in this case, that sinks, sources and isolated elements of the corresponding graph are of obvious interpretation in terms of zero, non-zero structure of the corresponding matrix.

*Letting E be the set of all possible edges having the form [(r,i),(r,j)], the strong augmentation problem associated to our multiregional system (MULTRAP) can now be defined as that of finding a minimum cost set of edges $\varepsilon_2 \in E$ such that $G_2 = (V, \varepsilon_0 U \varepsilon_2)$ is strongly connected.* The problem can be equivalently formulated assuming that $f(\ell)$ assumes finite (positive) values for all the edges $\ell \in G(E)$ and infinite otherwise. The given constraints on the edges of the MULTRAP problem will not guarantee the existence of feasible solutions as in the general case. It is easy to see that the particular structures of our matrices greatly reduces the computational burden. For instance, given m regions and n sectors, we would have mn nodes in G(E), and in general, a maximum of $(mn)^2$ possible links; that figure reduces to m(m-1)xn once we take into account that E has the form previously discussed. It may be of some relevance a mathematical programming formulation of MULTRAP in the weighted case. We shall indicate with $w^{ij}_{rs}$ the weights associated to each link [(r,i),(s,j)], and assume $w^{ij}_{rs}=0$, $\forall$[(r,i)(s,j)]$\in$G(E), and $w^{ij}_{rs}>0$, $\forall$[(r,i), (s,j)]$\notin$G(E). MULTRAP can now be formulated as follows:

$$\min_{\{x^{ij}_{rs}\}} \ \Sigma_i \Sigma_j \Sigma_r \Sigma_s \ x^{ij}_{rs} w^{ij}_{rs} \tag{8}$$

$$x^{ij}_{rs} = 0, 1 \qquad \forall i,j,r,s \tag{9}$$

$$x^{ij}_{rs} = 0 \qquad \forall r \neq s \tag{10}$$

$$x^{ij}_{rs} = 1 \qquad \forall i,j,r,s : r=s, \ i=j \tag{11}$$

$$[I + M \sum_{k=0}^{\infty} (xA)^k xB)]^{mn-1} > 0 \tag{12}$$

The optimal solution $x^*$ of the above problem is equivalent to the incidence matrix $M(T^*)$ of the augmented interregional trade matrix. Therefore its corresponding graph represents the optimally augmented "transportation" network. It should be noted that constraint (12) ensures that the resulting $T^*$ makes the network $G(T^*)UG(A)UG(B)$ strongly connected. That condition in matrix terms (see theorem 1, before) is equivalent to say that $(I-TA)^{-1}TB$ is irreducible. It can also be proved (see for instance Horn and Jonhson, 1985) that an nxn matrix is irreducible if and only if $[I + M(D)]^{n-1} > 0$. Since $(I-TA)^{-1} = TB \sum_{k=0}^{\infty}(T A)^k T B$, the equivalence between constraint (12) and the strong connectivity condition on the resulting graph can be easily be verified. Since the above mathemathical formulation is not easily solvable for large scale problems, we note that a reasonable alternative (in terms of complexity) is to design efficient algorithms that yield near-optimal, or approximate soltutions. Approximation algorithms for augmentation problems and upper bounds for the performance of the algorithms in terms of worst-ratio of the near optimal augmentation solutions to the cost of the optimal solutions, can be found, for example, in Frederickson and Ja'ja (1984).

## 6. Conclusions

In this paper we have analyzed the relationships between economic growth and transportation supply in a multiregional multisectoral system. We have provided tools for identifying possible bottlenecks in transportation infrastructure which hamper development possibilities for some regions, and proposed a general method for tackling the problem of augmenting the transportation network in order to achieve balanced growth with either the minimum cost or the minimum number of links added. Our work has been based on the analysis, in terms of dynamic equilibrium properties of the system, of the trade patterns resulting from the matching of regional demands and the provision of transportation infrastructure. This does not limit the generality of the results since we have mainly dealt with the topological structure of the problem. However, the dimensions of trade coefficentes and physical flows determine the actual solution of the eigenvalue problem: although of great relevance in practice, this subject has not been tackled here, leaving room for further work.

## References


H. ADLER (1971): *Economic Appraisal of Transport Projects: A manual with Cases Studies*. Indiana University Press, Bloomington, Indiana.

S. ANAND (1975): Appraisal of Highway project in Malaysia: Use of the Little-Mirles Procedure. Working Paper 213. International Bank for Reconstruction and Development, Washington, D.C.

P.S. ARMINGTON (1963): A Theory for Products Distinguished by Place of Production. *International Monetary Fund Staff Papers*, 16, 159-176.

D. BATTEN, D. MARTELLATO (1984): Classical Versus Modern Approaches to Interregional Input-Output Analysis. Paper presented at the *$24^{th}$ European Congress of RSA*. Milan, 28-31 August.

U. BLUM (1982): Effects of Transportation Investments on Regional Growth: a theoretical and Empirical Investigation. *Papers of The Regional Science Association*, 49, 169-174.

D. CAMPISI, A. LA BELLA (1987a): An Input-Output Based Approach to the Short-Term Evaluation of Transportation Plans. *Applied Mathematical Modelling*, 11, 2, 127-132.

D. CAMPISI, A. LA BELLA (1987b): Transportation supply and economic growth in a multiregional system. IASI Research Report R 181. In press: *Environment and Planning A*.

F. DUCHIN, D.B. SZYLD (1986): A Dynamic Input-Output Model with Assumed Positive Output, *Metroeconomica*, 269-282.

J. EDMONS (1967): Optimum Branching, *J. Res. Nat. Bur. Standards*, Sect B, 71, 233-260

K.P.E. ESWARAN, E. TARJAN (1976): Augmentation Problems - *Siam J. Comp.*, 5, 653-665

G.N. FREDERICKSON and J. JA'JA (1984): Approximate Algorithms for Several Graph Augmentation Problems. *Siam J. Comput.*, Vol. 10, No 2, 270-283.

F.R. GANTMACHER (1959): *Applications of the Theory of Matrices*. N.Y. Interscience.

R.A. HORN, C.A. JOHNSON (1985): *Matrix Analysis*, Cambridge University Press.

Y. KAJITANI, S. UENO (1986): The Minimum Augmentation of a Directed Tree to a K-Edge Connected Directed Graph. *Networks*, 16, 181-197.

Y. KANEMOTO and K. MERA (1983): *General Equilibrium Analysis of the Benefits of large Transportation Improvements*, Discussion Paper, Institute of Socio-Economic Planning, University of Tsukuba.

W. ISARD (1951): Interregional and Regional Input-Output Analysis: A Model of a Space-Economy, *Review of Economics and Statistics* 33, 318-328.

A. LA BELLA (1986): Integrated Transportation Planning: Physical Expansion Versus Economic Constraints, in (P. Nijkamp, S. Reichmann eds.): *Transportation Planning in a Changing World*. Gower, London, 243-253.

W. LEONTIEF (1977): The Dynamic Inverse, in *Contribution to Input-Output Analysis* (A.P. Carter and A. Brody eds.) Amsterdam: North-Holland Publishing Co. 1970, pp.17-46, reprinted in Wassily Leontief, *Essays in Economics, Volume 2*. White Plans, N.Y.: M.E. Sharpe 1977.

W. LEONTIEF and A. STROUT (1963): Multiregional Input-Output Analysis, in (T. Barna, ed.), *Structural Interdipendence and Economic Development*, McMillan, London.

C.K. LIEW and C.J. LIEW (1984 a): Measuring the Development Impact of a Proposed Transportation System, *Regional Science and Urban Economics*, 14, 175-198.

C.K. LIEW and C.J. LIEW (1984 b): Multi-Modal, Multi-Output, Multiregional Variable Input-Output Model, *Regional Science and Urban Economics*, 14, 265-281.

C.K. LIEW and C.J. LIEW (1985): Measuring the Development Impact of a Transportation System: A Simplified Approach, *Journal of Regional Science*, 25, 241-258.

D.G. LUENBERGER and AMI ARBEL (1977): Singular Dynamic Leontief Systems, *Econometrica*, 45, 991-995.

U. MEYER (1982): Why Singularity of Dynamic Leontief System Doesn't Matter, in *Input-Output Techniques, Proceedings of the third Hungarian Conference on Input-Output Techniques*, 3-5 November 1981. Budapest: Statistical Publishing House, 181-189.

H. NIKAIDO (1968): *Convex Structures and Economic Theory* - Academic Press.

P. PRASTACOS and M. ROMANOS (1987): A Multiregional Model for Allocating Transportation Investments. *Transportation Reasearch - B*, 2, 133-148.

K. SASAKI(1983): A Household Production Approach to the Evaluation Of Transportation System Change. *Regional Science and Urban Economics*, 13, 241-249.

D. B. SZYLD (1985): Conditions for the Existence of a Balanced Growth Solution for the Leontief Dynamic Input-Output Model, *Econometrica* , 53, 6, 1411-1419.

H. TULKENS and T. KIONI KABANTU (1983): A Planning Process for the Efficient Allocation of Resources to Transportation Infrastructure, in J.F. Thisse and H.G. Zoller (Eds.), *Locational Analysis of Public Facilities*, North-Holland, 127-152.

# MODELLING AND OPTIMIZATION OF PUBLIC TRAFFIC LARGE SCALE SYSTEMS

Zhang Qiren
Graduate School of Chinese Academy of Social Science, Beijing, PR China

Xiong Guilin
Dept. Of Electronic Engng., Changsha Railway Institute, Changsha, Hunan Province, PR China

ABSTRACT The modelling and optimization of public traffic large scale systems (PTLSS) are discussed in this paper. Some results from practice while the first auther directing and the second author participating the R & D Item of the Public Traffic Systems Engineering for Changsha City in PR China are also given.

## I. INTRODUCTION

Urban public traffic system what a complex, stochastic, of multi-objects and multi-functions and dynamic large scale system is is a wellknown thing at all. The theories and methodologies of large scale system, therefore, should be used to study synthetically such a system in several levels and directions. For reason given above, we have briefly discussed the modelling and optimization of this system in theory and practice and penetratively expressed some experiential results from the R & D Item of the Public Traffic Systems Engineering for Changsha City in our country which is directed by the first author and investigated by a few researchers including another author. For the sake of concision, the contents and structure of the paper are shown in Fig. 1.

## II. THE AHP DECISION MODEL OF PUBLIC TRAFFIC LARGE SCALE SYSTEM

AHP (Analytical Hierarchy Process)[1] is a new decision-making method which combinning qualitative and quantitative analysis. By using AHP, the subject of optimizing public traffic large scale system of Changsha City is divided into 5 levels as general object, strategy object, tactics object, evaluation criteria and technical devices, and then the AHP decision model of the system is built up. This model contains 28 determination matrices which are constructed by the Delphi consultation. With the help of AHP algorithm and the identity test formula, we have accomplished the single level and general level arrangement computation on a PC-1500 computer in terms of BASIC language, and obtained 4 general level arrangement tables.

## III. THE OD MODEL OF THE INHABITANTS' TAKING BUS

In order to know the state of all the inhabitants' taking bus, we have accomplished the origin and destination investigation of the inhabitants taking, i.e. the OD

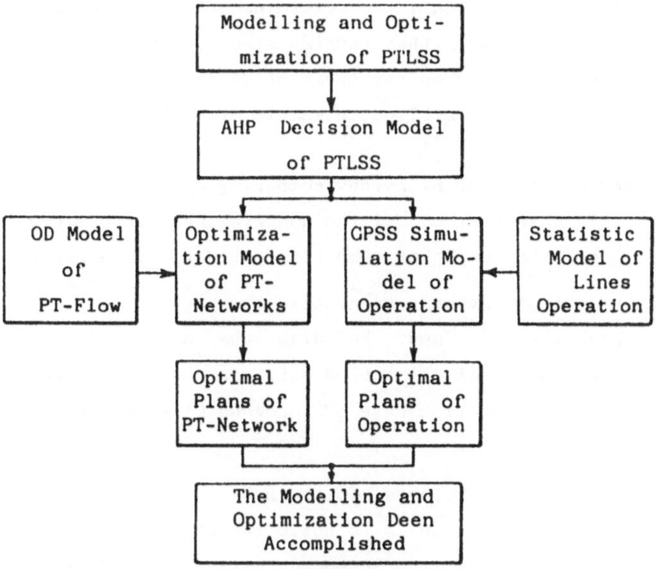


Notes: PTLSS——Public Traffic Large Scale System;

PT- ——Public Traffic.

Fig 1 Modelling and Optimization Diagram Of PTLSS

investigation. Thus the necessary information for the programming of the public traffic network is obtained. The OD investigation and information processing are done by a IBM 4331 computer. The output result from computer are some OD matrices, i.e. the OD model of the inhabitants' taking. Because of enormous amount of information data, we use the sampling investigation method.

1. OD Sampling Investigation

(1) The investigation range must be overall the field that the bus can reach. Generally, the following aspects should be considered:

a. Statistic field must be as congruent with other social statistic field as possible (e.g. vital statistics).

b. Size of statistic field must be appropriate.

c. There must be bus stops which can be used as core in the field.

d. The statistic field can not cover more than two main traffic lines, but can contain the cross of the main traffic lines.

e. Statistic fields divided should be convenient to synthesize every kind of information.

According to principles shown above and the actual conditions of Changsha City, we divide the whole field into 41 smaller statistic fields for OD investigation.

(2) Sample Ratio

The first problem to be solved is to determine the sample size of OD sample investigation. As well known, the sample size depends upon accuracy needed. Traditionally, it is determined on the basis of demographic data, and if the number of population is more than one million, a sample of 4% have to be taken .

(3) Sample Method

We used the sample method of hierarchy comparing estimations in the OD investigation of Changsha City. In this method, all the undertakings are divided into 3 levels by its stuff number, then the sample of comparing estimations is proceeded in every level.

2. Processing of the OD Investigation Data

This work contains the data input, building the disk file, classification and arrangement, the computation of all the OD matrices and its store into disk. The processing of the OD investigation data of Changsha City is accomplished on a IBM 4331 computer.

3 The OD Model of the Inhabitants' Taking Bus

The purpose of OD investigation is to build the OD model of the inhabitants' taking, i.e. the OD matrices output by the computer using the OD investigation statistic information. It is generally shown by OD distribution table:

| O \ D | 1 | 2 | ... | n | Σ |
|---|---|---|---|---|---|
| 1 | $r_{11}$ | $r_{12}$ | ... | $r_{1n}$ | $U_1$ |
| 2 | $r_{21}$ | $r_{22}$ | ... | $r_{2n}$ | $U_2$ |
| ⋮ | ⋮ | ⋮ | | ⋮ | ⋮ |
| n | $r_{n1}$ | $r_{n2}$ | ... | $r_{nn}$ | $U_n$ |
| Σ | $V_1$ | $V_2$ | ... | $V_n$ | R |

where:

$$\sum_{j=1}^{n} r_{ij} = U_i \quad (i=1,2,\cdots,n)$$

$$\sum_{i=1}^{n} r_{ij} = V_j \quad (j=1,2,\cdots,n)$$

$$\sum_{i=1}^{n} U_i = \sum_{j=1}^{n} V_j = R$$

## IV. THE NETWORK OPTIMIZATLON MODEL OF PUBLIC TRAFFIC LARGE SCALE SYSTEM

The performance of the public traffic system depends mainly on the goodness of the distribution of the public traffic large scale system network. It should not reach anyway its optimal operation state if the distribution of the network was unreasonable[2,3].

The network optimization model of public traffic large scale system contains the passenger-flow assignment and optimal selection of the lines, its structure is shown in Fig.2

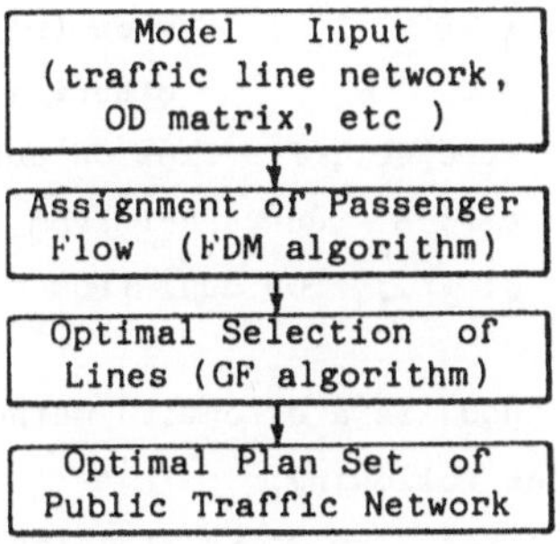


Fig.2 Optimization Model Diagram of PT-Network

1. Model Input

Model input concerns with traffic networks, OD matrices, traffic time of every section, length of road section, oil spending in road section and the capacity for passengers.

2. The Optimal Assignment of Traffic Flow

(1) Problem Statement

As shown in the first Wardrop principle, every passenger will select the shortest path. But the traffic time is related with the crowding, i.e. the traffic time $TA_i$ of every curve i in traffic network should be expressed as the increasing monotonuosly function of the passenger flow $f_i$ on this curve i:

$$TA_i = TA_{0i} + \beta \frac{f_i}{c_i - f_i} \quad (f_i < c_i)$$

Where: $TA_i$ is the fundamental traffic time; $\beta$ is the time ratio; $c_i$ is the largest capacity on the curve i.

So the total traffic time sum $T_i$ of passengers passing curve i is:

$$T_i = f_i \cdot TA_i = f_i \cdot (TA_{0i} + \beta f_i / (c_i - f_i))$$

The averaged traffic time of all passengers on the whole public traffic network is:

$$T = \frac{1}{R} \sum_{i \in A} T_i \qquad \forall i \in A$$

Where R is the total traffic taking between all cores; A is the curve set. The problem of passenger flow assignment is to select $f_i$ such that T to be minimal, i.e.

$$\min T = \frac{1}{R} \sum_{i \in A} f_i \cdot (TA_{0i} + \beta f_i / (c_i - f_i))$$

$$\text{s.t. a.} \sum_{i \in A_j} f_i^r - \sum_{i \in B_j} f_i^r = R_j^r, \ \forall j \in V, \forall r \in C$$

$$\text{b.} \ f_i - \sum_{i \in C} f_j^r = 0, \ \forall i \in A$$

c. $f_i < c_i$, $\forall i \in A$

d. $f_i \geq 0$, $f_i^r \geq 0$, $\forall i \in A$, $\forall r \in C$

where:

$$H_j^r = \begin{cases} \sum_{K \in O_r} D^k & \text{when } j=r \\ -D^{k^r} & \text{when } (r,j) \in K \\ 0 & \text{otherwise} \end{cases}$$

A—curve set; V—node set; C—core set; $f_i$—flow on curve i; $f_i^r$—flow on curve i from core r; $O_r$—OD dual set at core r; $A_j$—curve set from node j; $D^k$—traffic taking of OD dual K; $B_j$—curve set to node j; (r,j)—OD dual from r to j.

(2) FDM ALgorithm

FDM (the Flow Deviation Method) is a effective method for the optimal assignment of traffic flow. Its steps are as following:

a. Finding initial flow $\underline{f}^o$: set $f_i$=o, $\forall i \in A$, then calculate $\frac{\partial T}{\partial f_i}\Big|_{f_i=0}$ and use it as weight to find shortest path flow $\underline{f}^*$, thus $f^* \to f^o$ and use $\underline{f}^o$ as initial to calculate $T^o$, o→n.

b. Finding $\underline{f}^*$, $\lambda$, use $\frac{\partial T}{\partial f_i}\Big|_{f_i=f_i^n}$ as weight to find shortest path flow $\underline{f}^*$, then to find $\lambda$.

c. Using $f_i^{n+1} = f_i^n + \lambda(f_i^* - f_i^n)$ iteratively compute new $\underline{f}^{n+1}$ and $T^{n+1}$.

d. If $|(T^{n+1} - T^n)/T^{n+1}| < \varepsilon$ and $f_i \leq ac_i$, $f^{n+1}$ is the optimal solution, when $f_i > ac_i$ we may increase "a" and repeat the computation. If $|(T^{n+1}-T^n)/T^{n+1}| \not< \varepsilon$, set n+1→n and go to step b. ($0 < a < 1$).

(3) Program and Example

Using FDM algorithm, we have programmed the computer program with FORTRAN language for the optimal traffic assignment. It has 3 subprograms: FDM for every coordination computation; FLITYD for the computation of $\underline{f}^*$; GSM for the optimization of $\lambda$. These computations are accomplished on a IBM4331 computer.

3. Optimal Assignment of Traffic Lines

(1) Problem Statement

The optimal assignment of traffic lines is expressed as:

$$\max_{L} J = \sum_{l \in L} RD_l$$

s.t. $m \leq M_L$ (for the number of lines)

$RD_l \geq R_L$ $\forall l \in L$ (for passenger flow)

$FZ_j \geq F_L$ $j \subset l$, $\forall l \in L$ (for the flow in section plane)

$Q_\lambda \subseteq Q$ (for the park place)

$M_\lambda^i \leq M_q^i$, $\forall i \in Q$ (for the capacity of parks)

$l \in TRL$, $\forall l \in L$ (for the lines to be selected)

$L_s < L_l < L_B$, $\forall l \in L$ (for the length of lines)

$RC_j < C_L$, $j \subset l$, $\forall l \in L$ (for the repeated number of lines)

where: L—selected lines set; $DR_L$—through traffic passengers' number on line L; $M_L$—permitted most number of lines; $FZ_j$—passenger flow in section plane j of selected lines; $Q_\lambda$—terminal set of selected lines; $M_q^i$—capacity of park at node i; TRL—line set to be selected; $L_s$—permitted shortest length of lines; $RC_j$—repeated number of lines in section j; l—selected line; m—number of selected lines; $R_L$—least number of through traffic passengers; $F_L$—least flow of the line's section plane; Q—node set permitted to used as park; $M_A^i$—number of lines its terminal is at node i; $L_l$—length of selected line l; $L_B$—permitted largest length of lines; $C_L$—permitted largest number of repeated lines.

(2) GF Algorithm

We have proposed the GF (finding a Good, Feasible set of lines) algorithm in order to build the good and feasible line set:

GF1: Produce line set to be selected: it contains time, distance and oil spending shortest lines and some experienced lines.

GF2: optimal selection of lines: calculate the through traffic passengers number of every line to be selected and select one which has most through traffic passengers and stands to all limitation. Then subtract respective number from the passenger flow of the OD dual and road section covered by the selected line, and add 1 to the capacity of respective terminals, add 1 to the repeated number of lines covered by the selected line.

GF3: logical transfer: if all roads which have nonzero flow are covered, go to GF5; if the number of selected lines equals the least number of predicted lines, go to GF5; if the number of selected lines equals the least number of predicted lines, go to GF4; otherwise go to GF2.

GF4: Add uncovered road sections to lines:

a. If the terminal number of uncovered road sections is larger than 2 and the respective terminal capacity is not full, such line is selected as has this kind of terminals and contains most uncovered road sections. Subtract respective number from the passenger flow of OD dual and road sections covered by this line, and subtract 1 from the terminal capacity. Add 1 to the line repeated number of covered road sections. This procedure is repeated until the terminal number of uncovered road sections is equal to or larger than 2 or all terminal capacity is full. If all nonzero-flow-road sections are covered, go to GF5; otherwise go to b.

b. Select such new line from line set to be selected as contains most uncovered road sections and stands to the capacity limitation of terminal. This procedure is repeated until all nonzero-flow-road sections are covered or the procedure is limited by terminal capacity.

GF5: Input results: if all nonzero-flow-road sections are covered, the obtained line set is the results to find. Otherwise, some new terminals or terminal capacity should be increased and the GF algorithm is repeated.

(3) Example and Program:

The optimal assignment program of traffic lines with GF algorithm is a significant part of the whole traffic network optimization program. It contains 4 subprograms: GF accomplishes the maincomputation and coordinates the whole algorithm; OTA accomplishes the input of distance and oil-spending matrix; SP produces shortest line set to be selected of line, distance and oil spending; GFT selects such a line that has most through traffic passengers and stands to all limitation.

## V. THE OPERATION STATISTIC MODEL OF THE PUBLIC TRAFFIC LINES

### 1. Passenger Flow Statistic Model

Investigation statistic of passenger flow is an essential step in the optmization of public traffic system. It influences the reasonable organization and use of transportation ability, as well as the operation and economic effect, and the service quality. Through it, we can know the distribution of passenger flow in time and space.

(1) Number of Passengers Getting On or Off

In order to know and analyse the passenger flow in whole a day, the statistic of number of passengers getting on or off at every stop should be proceeded by different time section. The time section is taken to be 15 minutes.

(2) Passengers' Arriving Law

The number of arriving passengers at any time is independent on any previous number, i.e., it satisfies the independent condition of Poisson flow. And the generality and limit are also satisfied. The stationary condition is not satisfied, but in a time section it is satisfied. Thus we can use Poisson flow to describe the passengers' arriving law at stop i and in time section j:

$$\lambda_{ij} = \frac{PU_{ij}}{T_j}$$

Where: $\lambda_{ij}$—intensity of arriving flow at stop i and in time section j; $PU_{ij}$—total number of arriving passengers at stop i and in time section j; $T_j$—the length of time section j.

(3) Distribution Model of Passenger Flow

It is expressed as a distribution matrix:

$$\begin{bmatrix} P_{11} & P_{12} & \cdots & P_{1n} \\ P_{21} & P_{22} & \cdots & P_{2n} \\ \vdots & \vdots & & \vdots \\ P_{n1} & P_{n2} & & P_{nn} \end{bmatrix}$$

Where n is the stop number of a line, $P_{ij}$ is the number of passengers who get on at stop i and off at stop j. So we have $P_{ij}=0$ $(j\leq i)$ and usually $P_{ij}\neq 0$ $(j>i)$.

(4) Possibility Distribution of Passengers' Getting Off

It depends upon the passenger flow distribution. The possibility distribution function $PF_{ij}$ of passengers who get on at stop i and off at stop j is:

$$\begin{cases} PF_{ij} = PF_{i,j-1} + \dfrac{P_{ij}}{PU_i} & \left(\begin{matrix} i=1,2,\ldots,n-1 \\ j=2,3,\ldots,n \end{matrix}\right) \\ PF_{ij} = 0 & (j\leq i) \end{cases}$$

### 2. Statistic Model of Bus Operation

(1) One-Way Traffic Time

If the dispatch time of bus i is $TK_i$ and its end time is $TD_i$, the one-way traffic time $TR_i$ of this bus is

$$TR_i = TD_i - TK_i$$

For the statistic of m buses (m>30, in large sample case), the averaged one-way traffic time is

$$TM = \frac{1}{m} \sum_{i=1}^{m} TR_i$$

The variance of one-way traffic time is

$$TS = \sqrt{\frac{1}{m} \sum_{i=1}^{m} (TR_i - TM)^2}$$

The change coefficient is

$$CV = TS/TM$$

When CV is less than 0.33, it can be described by normal distribution.

(2) Section Driving Time

Let $TA_{ij}$ be the leaving time at the last stop of section i and $TB_{ij}$ be the arriving time at the next stop, the section driving time $TI_{ij}$ of bus j in section i is

$$TI_{ij} = TB_{ij} - TA_{ij}$$

For the statistic of m buses ($m > 30$), the averaged section driving time in section i is

$$TM_i = \frac{1}{m} \sum_{j=1}^{m} TI_{ij}$$

The variance of it is

$$TS_i = \sqrt{\frac{1}{m} \sum_{j=1}^{m} (TI_{ij} - TM_i)^2}$$

The change coefficient is

$$CV_i = TS_i/TM_i$$

(3) Stop Time

The number of passengers getting on or off influences the stop time. The stop time $TP_{ij}$ of bus j at stop i is

$$TP_{ij} = T_o + (PU_{ij} + PD_{ij}) \cdot T_a$$

Where: $T_o$—open-close time and preparing time for start; $T_a$—averaged time for every passenger to get on or off; $PU_{ij}$—number of passengers getting on bus j at stop i; $PD_{ij}$—number of passengers getting off bus j at stop i.

## VI. GPSS SIMULATION MODEL OF PUBLIC TRAFFIC LARGE SCALE SYSTEM

Mathematically, urban publc traffic system is a typic stochastic service system. Its main properties yield only statistic law, and the relation among variables can not be expressed by analytical formula. The only way to describe and analyse is by the aid of computer simulation.[4,5,6]

1. System Boundary

Using the decentralised control theory of large scale system, we divide the public traffic large scale system into several subsystems by lines. The interaction of lines is shown in the arriving law of passengers at every stop. But the arriving law is processed by time section with the real investigation data of passenger flow. Thus the coupled relations of lines have been considered.

Every line has two directions. But the dispatch in every direction is controlled by two terminal stops respectively, and the arriving laws of two directions are independent. So we only need to study an unidirectional line system which contains starting stop, some middle stops and end stop.

2. System Variables

(1) Uncontrollable Variables

a. The number of stops n;

b. The number of sections (n-1);

c. Arriving rate of passengers at every stop;

d. Probability distribution of passengers' getting off.

(2) Decision Variables

a. Bus dispatch interval;

b. Number and type of equipped buses;

c. Number and dispatch time of additional buses;

d. Number, dispatch time and stops of main-stop buses;

e. Number, dispatch time and interval length of interval buses.

3. Objective Function

The objective is selected to minimize the sum of passengers' loss and operational cost, i.e.

$$\min Z = \min (T_p \cdot C_t + L_s \cdot C_v)$$
$$\text{s. t.} \quad m \leqslant M$$

Where: Z—the total expense (yuan); $T_p$—the total waiting time and taking time of all passengers (hour); $C_t$—the averaged productivity of the urban workers (yuan/hour); $L_s$—the total distance run by all buses (km); $C_v$—the unit operation cost (yuan/km); m—the number of real equipped buses; M—the largest possible number of equipped buses.

4. The Diagram of Model

The logical relation of system model is based on the time sequence. The system operation is just the motion forward of three kinds of transactions flow of buses, passengers and control signals. So the simulation models are built up independently with these three kinds of transactions in time section, and then connected each other by the same names of logic switch and user chain such that a complete dynamic system is constructed.

The model of whole system is composed of 9 parts and every part has independent actions and functions. The partition of the model and their logical relations are shown in Fig.3.

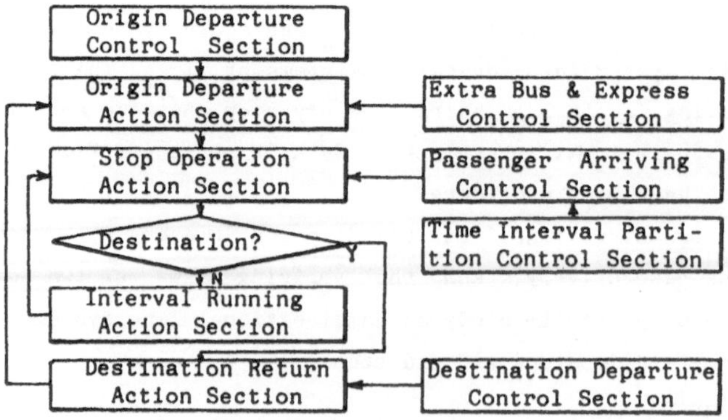


Fig.3. Logic Diagram of GPSS Simulation Model of Unidirectional Line System

5. Program and Examples

Based on the real statistic data of operation lines in Changsha City, the model program is accomplished by GPSS/1100 language. And the simulation computation is carried out on an UNIVAC 1100 computer.

## VII. CONCLUDING REMARKS

Urban public traffic system is a complex socio-economic large scale system. It has widely bright feature for using the theory of large scale system to study the optimization of this system. In this paper, we have only made some initiative work, and much work have to be done in no distant future[7].

## REFERENCES

1. Saaty, T.L., The Analytical Hierarchy Process, McGraw-Hill Inc., 1980.
2. Mandl, C., Applied Network Optimization, Academic Press, 1979.
3. Kennington, J.K., Algorithms for Network Programming, John Wiley & Sons, Inc., 1980.
4. Gordon, G., The Application of GPSS V to Discrete System Simulation, Prentice-Hall, 1975.
5. Schriber, T.I., Simulation Using GPSS, Wiley-Interscience, 1974.
6. Bly P.H. & R.I. Jacksen, Evaluation of Bus Control Strategies by Simulation, TRRL Report, 1974.
7. Zhang Qiren & Xiong Guilin, Introduction to Public Traffic Systems Engineering, Publ. of Hunan Univ., 1988 (to be printed, in Chinese).

# INCENTIVE STACKELBERG STRATEGIES IN LINEAR QUADRATIC DIFFERENTIAL GAMES WITH TWO NONCOOPERATIVE FOLLOWERS

Koichi MIZUKAMI and Hansheng WU

Department of Information and Behavioral Sciences
Faculty of Integrated Arts and Sciences
Hiroshima University, Hiroshima 730, Japan

ABSTRACT

This paper is mainly concerned with the derivation of the sufficient conditions for the incentive Stackelberg strategy in the two-level hierarchical differential games with two noncooperative followers, characterized by a classof linear state dynamics and quadratic cost functionals. In the paper, we first give some concepts in the two-level hierarchical games with two noncooperative followers, and by a simple numerical example, show a general method of solving such a two-level incentive static game problem. Then, we construct a new form of the incentive Stackelberg strategy $\bar{\gamma}_0=\{\bar{\gamma}_{01},\bar{\gamma}_{02}\}$ of the leader P0, and also derive the sufficient conditions which are satisfied by this strategy $\bar{\gamma}_0=\{\bar{\gamma}_{01},\bar{\gamma}_{02}\}$.

## 1. Introduction

An appropriate equilibrium solution concept for hierarchical multi-criteria decision-making problems is the Stackelberg solution concept which was first introduced in economics within the context of static competition [1]. Its dynamic version later entered the control literature through the works of Chen and Cruz [2] and Simaan and Cruz [3,4], and found applications in nonzero-sum differential games where one player has enough ability or power to enforce his strategy on the other player. Within the context of two-player differential games, the more powerful player is called a leader, and the other one is called a follower. Generally, such two-player differential games in which one player is leader and the other one follower are called Stackelberg games, and an equilibrium solution in Stackelberg games is called a Stackelberg strategy. However, an incentive Stackelberg strategy is the announced strategy by which the leader can achieve his team-optimal solution in the Stackelberg games.

In the recent years, there were many papers dealing with incentive Stackelberg strategies, e.g., many aspects of incentive problems were investigated in [5], and the approach to obtaining the closed-loop Stackelberg solution of linear quadratic dynamic games was developed in [6]. In [7], another method to construct the Stackelberg strategy was presented with rather relaxed restriction. In [8], by using a geometric approach, the existence and derivation of affine incentive Stackelberg strategy were addressed. In [9], the authors gave an extensive discussion and derivation of closed-loop Stackelberg strategies and incentive policies in dynamic decision-making problems of some types. In [10], the sufficient conditions for the incentive Stackelberg strategies in linear quadratic differential games were derived. In [11]-[13], the concept of "inducible region" was introduced and some applications of this concept to the multi-stage Stackelberg games were given. The incentive Stackelberg games (i.e., Stackelberg games where the leader seek his team-optimal solution) with two players can be extended to the ones with three players in two-level or three-level hierarchy [5].

In this paper, we will consider in detail another form of Stackelberg games which has not been considered in detail by the authors mentioned in [1]-[13], so far. That is, we will only discuss the two-level hierarchical games with two noncooperative followers and one leader. The hierarchical structure of such games is depicted in Fig.1, where there are three players P0, P1 and P2; P0 is a leader, and P1 and P2 two followers, under the stipulation that two followers act noncooperatively. The hierarchical structure depicted in Fig.1 means that P0 announces his strategy, ahead of time to the other two players P1 and P2; then P1 and P2, in view of the strategy of P0, decide their optimal strategies respectively.

In this paper, for the two-level hierarchical games with two noncooperative followers, we will mainly discuss such differential games characterized by a class of linear state dynamics and quadratic cost functionals. Being different from the methods which have been used in the other papers, we construct new form of incentive Stackelberg strategy $\bar{\gamma}_0$ for P0 to achieve his team-optimal solution in such differential games, and also obtain the sufficient conditions which are satisfied by this incentive Stackelberg strategy $\bar{\gamma}_0$.

In Section 2, we give some concepts, and show a general method of solving such a two-level incentive game problem by a simple numerical example. In Section 3, we state such a two-level incentive differential game problem characterized by a class of linear state dynamics and quadratic cost functionals, and define a new form of the incentive Stackelberg strategy $\gamma_0$. In Section 4, we give the sufficient conditions for $\bar{\gamma}_0$ to be the incentive Stackelberg strategy, and its proof. The paper will be concluded in Section 5 with a very brief discussion on above results and some possible further research.

## 2. Two-level Incentive Games with Two Noncooperative Followers

### 2.1. Two-level incentive games with two noncooperative followers.

Consider the hierarchical structure depicted in Fig.1. We assume that P0 is a leader, P1 and P2 are two followers. Here, P1 and P2 are under the noncooperative mode of action. Specifically, we stipulate that P1 and P2 choose their decisions under Nash equilibrium solution concept. Furthermore, we assume that the cost functions of Pi, i=0,1, 2, are respectively $L_i(u_{01},u_{02}; u_1,u_2)$ where $u_1 \in R^{m_1}$ and $u_2 \in R^{m_2}$ are respectively the decisions of P1 and P2, and $u_0=(u_{01},u_{02}) \in R^{m_0}$ (here, $m_0=m_{01}+m_{02}$) is P0's decision. We also assume that

i) P0 announces his strategy $\gamma_0=\{\gamma_{01},\gamma_{02}\}$, where $\gamma_{01} : R^{m_1} \to R^{m_{01}}$ and $\gamma_{02} : R^{m_2} \to R^{m_{02}}$, ahead of time to the other players P1 and P2;

ii) then P1 and P2 decide their optimal decisions after knowing the announced strategies of P0.

We will then call the games ordered in such a way the two-level hierarchical games with two noncooperative followers. Here, if P0 can obtain the team-optimal solution by his strategy $\gamma_0=\{\gamma_{01},\gamma_{02}\}$ announced, we then call $\gamma_0$ incentive Stackelberg strategy of P0 in such game. Furthermore, such two-level hierarchical games to find the incentive Stackelberg strategy $\gamma_0$ are called the two-level incentive games with two noncooperative followers.

If we assume that $u_0^*=\{u_{01}^*,u_{02}^*\} \in R^{m_0}$, $u_1^* \in R^{m_1}$ and $u_2^* \in R^{m_2}$ are the team-optimal solutions of the leader P0, such two-level incentive game problem can then be simply stated as:

Find $\gamma_0=\{\gamma_{01},\gamma_{02}\} \in \Gamma_0$, where $\gamma_{01} : R^{m_1} \to R^{m_{01}}$, and $\gamma_{02} : R^{m_2} \to R^{m_{02}}$, such that

$$\arg\min_{u_1} L_1(\gamma_{01}(u_1),\gamma_{02}(u_2^*); u_1,u_2^*) = (u_1^*, u_2^*) \ , \tag{1a}$$

$$\arg\min_{u_2} L_2(\gamma_{01}(u_1^*),\gamma_{02}(u_2); u_1^*,u_2) = (u_1^*, u_2^*) \ , \tag{1b}$$

and

$$\gamma_{01}(u_1^*) = u_{01}^* \quad ; \quad \gamma_{02}(u_2^*) = u_{02}^* \ , \tag{1c}$$

where $\Gamma_0$ is the class of admissible incentive Stackelberg strategies of P0.

Actually, the equation (1) is a set of sufficient conditions for $\gamma_0=\{\gamma_{01},\gamma_{02}\}$ is incentive Stackelberg strategy in such a game problem.

## 2.2. An illustrative example.

In order to show a general method of solving such game problems, we consider the two-level incentive static games with two noncooperative followers described by the following form (see [5], p.171).

$$\begin{aligned} L_0 &= u_{01}^2+u_{02}^2+u_1^2+u_2^2 \quad , \\ L_1 &= u_{01}-3u_{02}+(u_1-1)^2+(u_2-1)^2 \quad , \\ L_2 &= u_{01}+u_{02}+(u_1+1)^2+(u_2+1)^2 \quad . \end{aligned} \tag{2}$$

According to (1), the main steps to find the incentive Stackelberg strategy $\gamma_0=\{\gamma_{01},\gamma_{02}\}$ is as follows.

(i) Finding the team-optimal solution of P0, we obtain that

$$u_{01}^* = u_{02}^* = 0 \ ; \quad u_1^* = 0 \ ; \quad u_2^* = 0 \ . \tag{3}$$

(ii) Here, let us consider the incentive Stackelberg strategy $\gamma_0 = \{\gamma_{01},\gamma_{02}\}$ of the form

$$u_{01} = \gamma_{01}(u_1) = u_{01}^*+k_1\cdot(u_1-u_1^*) \ ; \quad u_{02} = \gamma_{02}(u_2) = u_{02}^*+k_2\cdot(u_2-u_2^*), \tag{4}$$

where $k_1$ and $k_2$ are two undetermined coefficients.

Substituting for $u_{01}^*$, $u_{02}^*$, $u_1^*$ and $u_2^*$ from (3) in (4), further, we can have that

$$u_{01} = \gamma_{01}(u_1) = k_1\cdot u_1 \ ; \quad u_{02} = \gamma_{02}(u_2) = k_2\cdot u_2 \ . \tag{5}$$

(iii) Solve the following problem (i.e., find Nash solution):

For i=1,2,

$$\begin{aligned} &\min_{u_i} L_i(u_{01},u_{02};\ u_1,u_2) \\ &\text{s.t.} \quad u_{01}=\gamma_{01}(u_1)=k_1\cdot u_1 \ ; \quad u_{02}=\gamma_{02}(u_2)=k_2\cdot u_2 \quad . \end{aligned} \tag{6}$$

Since $L_i$, i=1,2, are strictly convex in $u_i$, the necessary and sufficient conditions for the problem above are given as follows.

$$(\partial L_1/\partial u_1)=k_1+2(u_1-1)=0 \ ; \quad (\partial L_2/\partial u_2)=k_2+2(u_2+1)=0 \ . \tag{7}$$

(iv) When we choose that $k_1$=2 and $k_2$=-2, then both $u_1^*$=0 and $u_2^*$=0 must be the solution of the problem given in (iii). Thus, we can obtain the incentive Stackelberg strategies of P0 as follows.

$$\gamma_{01}(u_1) = 2u_1 \ ; \quad \gamma_{02}(u_2) = -2u_2 \ . \tag{8}$$

Factly, by the strategies (8) announced ahead of time, the cost functions of P1 and P2 will become that

$$\begin{aligned} L_1(\gamma_{01}(u_1),\gamma_{02}(u_2);\ u_1,u_2) &= u_1^2+(u_2+2)^2-2 \quad , \\ L_2(\gamma_{01}(u_1),\gamma_{02}(u_2);\ u_1,u_2) &= (u_1+1)^2+u_2^2-2 \quad . \end{aligned} \tag{9}$$

By (9), the Nash equilibrium solution of P1 and P2 must be that $u_1^*$=0 and $u_2^*$=0. Thus, the strategy $\gamma_0=\{\gamma_{01}(u_1),\gamma_{02}(u_2)\}$ given by (8) satisfy the condition (1), they are then indeed the incentive Stackelberg strategies announced by P0 in such games.

In the subsequent parts of this paper, using the idea and method stated in this section, we will mainly discuss the two-level incentive differential games with two noncooperative followers, characterized by a class of linear state dynamics and quadratic cost functionals.

## 3. Incentive Stackelberg Strategies in LQ Differential Games with Two Noncooperative Followers

Here, we consider the two-level hierarchical games with two noncooperative followers, characterized by a class of linear state dynamics and quadratic cost functionals. The evolution of such games is described by the linear differential equation of the form

$$(dx(t)/dt) = A(t)x(t)+\sum_{i=1}^{2}\{B_{0i}(t)u_{0i}(t)+B_i(t)u_i(t)\}\ , \qquad x(t_0) = x^\circ\ , \quad t\in[t_0,t_1]\ , \tag{10}$$

where $x(t)\in R^n$ is the state vector, $u_0(t)=\{u_{01}(t),\ u_{02}(t)\}\in R^{m_{01}+m_{02}}$ is P0's decision variable, and $u_i(t)\in R^{m_i}$, i=1,2, are Pi's decision variables; $A(t)$, $B_i(t)$, i=1,2, and $B_0(t)=\{B_{01}(t),B_{02}(t)\}$ are matrices of appropriate demensions and with components piecewise continuous on the time interval $[t_0,t_1]$.

Here, we assume that

i) the player P0 announces his strategy of the following form ahead of time to the other players P1 and P2,

$$u_{01}(t) = \gamma_{01}(x(t),u_1(t),t) = \eta_{01}(t)x(t)+\eta_{11}(t)u_1(t)\ , \tag{11}$$

$$u_{02}(t) = \gamma_{02}(x(t),u_2(t),t) = \eta_{02}(t)x(t)+\eta_{22}(t)u_2(t)\ , \tag{12}$$

where $\eta_{0i}(t)\in R^{m_{0i}\times n}$, i=1,2, and $\eta_{ii}(t)\in R^{m_{0i}\times m_i}$, i=1,2, are strategy parameter matrices and their components are piecewise continuous functions of time on the interval $[t_0,t_1]$;

ii) then, P1 and P2 decide their optimal strategies, after knowing the announced strategy of P0;

iii) P1 and P2 at same level act noncooperatively, and choose their decisions under the Nash equilibrium solution concept.

In this differential game problem, the cost functionals of Pi, i=1, 2, are given by $J_i$, i=1,2, respectively, where

$$J_i = (1/2)\cdot x^T(t_1)F_i(t_1)x(t_1)+(1/2)\cdot\int_{t_0}^{t_1}\{x^T(t)Q_i(t)x(t) + \sum_{j=1}^{2}[u_{0j}^T(t)R_{i0j}(t)u_{0j}(t)+u_j^T(t)R_{ij}(t)u_j(t)]\}dt\ , \tag{13}$$

where the matrices $Q_i=Q_i^T\geq 0$, $F_i=F_i^T\geq 0$, $R_{ij}=R_{ij}^T\geq 0$, $R_{i0j}=R_{i0j}^T\geq 0$, $R_{00j}>0$, $R_{0j}>0$, $R_{jj}>0$, i=0,1,2, j=1,2, are piecewise continuous function of time on $[t_0,t_1]$, and with appropriate dimensions, respectively.

We also assume that each player has access to the perfect state information and can utilize it in the choice of his decision. Further, we denote the space of the admissible strategies for P0 by $\Gamma_0$; and let the strategy spaces $\Gamma_1$ and $\Gamma_2$ of P1 and P2 be $R^{m_1}$ and $R^{m_2}$, respectively. For each pair $(\gamma_0, u_1, u_2) \in \Gamma_0 \times \Gamma_1 \times \Gamma_2$, the equation (10) has an unique solution on $[t_0, t_1]$, for all $x^\circ \in R^n$, and the values of $J_i$, i=0,1, 2, are well defined.

In such a game problem, if $\eta_{0i}$ and $\eta_{ii}$, i=1,2, satisfy some conditions such that P0 can obtain a team-optimal solution by $\bar{\gamma}_0=\{\bar{\gamma}_{01},\bar{\gamma}_{02}\} \in \Gamma_0$, defined in (11) and (12), we call $\bar{\gamma}_0$ the incentive Stackelberg strategy of P0. Further, we call this two-level hierarchical differential game to find the strategy $\bar{\gamma}_0$ the two-level incentive differential games with two noncooperative followers, in this paper later.

Now, the question is to find the conditions which are satisfied by $\eta_{0i}(t)$ and $\eta_{ii}(t)$, i=1,2, such that $\bar{\gamma}_0$ is the incentive Stackelberg strategy of P0. In the next section, we will give such conditions, and prove them.

## 4. Sufficient Conditions for the Incentive Stackelberg Strategies of the Leader P0

In this section, we first directly give the sufficient conditions (Theorem 4.1) for the incentive Stackelberg strategies of P0. Then, we state the main steps by which these conditions can be derived.

### 4.1. The sufficient conditions.

Theorem 4.1. The strategy $\bar{\gamma}_0=\{\bar{\gamma}_{01},\bar{\gamma}_{02}\}$, where

$$\bar{\gamma}_{01}(x(t),u_1(t),t) = \eta_{01}(t)x(t)+\eta_{11}(t)u_1(t) \quad , \tag{14}$$
$$\bar{\gamma}_{02}(x(t),u_2(t),t) = \eta_{02}(t)x(t)+\eta_{22}(t)u_2(t) \quad , \tag{15}$$

is the incentive Stackelberg strategy in the two-level incentive differential games, described in Section 3, if $\eta_{0i}(t)$ and $\eta_{ii}(t)$, i=1,2, satisfy

$$\eta_{0i} = -R_{00i}^{-1}B_{0i}^TK+\eta_{ii}R_{0i}^{-1}B_i^TK \quad , \tag{16}$$
$$\eta_{ii}^T\{B_{0i}^TP_i-R_{i0i}R_{00i}^{-1}B_{0i}^TK\} = R_{ii}R_{0i}^{-1}B_i^TK-B_i^TP_i \quad , \qquad i=1,2, \tag{17}$$

where $K(t)$ and $P_i$, i=1,2, are the solution of the following differential equations:

$$(dK(t)/dt) = -KA-A^TK-Q_0+KZ_0K \quad ; \qquad K(t_1) = F_0 \quad , \tag{18}$$
$$(dP_1(t)/dt) = -P_1\Lambda-\Lambda^TP_1-Q_1-Z_1-KB_2R_{02}^{-1}B_2^TP_1$$
$$-KB_2R_{02}^{-1}\eta_{22}^T\{B_{02}^TP_1-R_{102}R_{002}^{-1}B_{02}^TK\} \quad ; \quad P_1(t_1)=F_1 \quad , \tag{19}$$
$$(dP_2(t)/dt) = -P_2\Lambda-\Lambda^TP_2-Q_2-Z_2-KB_1R_{01}^{-1}B_1^TP_2$$
$$-KB_1R_{01}^{-1}\eta_{11}^T\{B_{01}^TP_2-R_{201}R_{001}^{-1}B_{01}^TK\} \quad ; \quad P_2(t_1)=F_2 \quad , \tag{20}$$

where

$$\Lambda(t) = A-\sum_{j=1}^{2}\{B_{0j}R_{0j}^{-1}B_{0j}^T+B_jR_{0j}^{-1}B_j^T\}K \quad , \tag{21}$$

$$Z_0 = A+\sum_{j=1}^{2}\{B_{0j}R_{00j}^{-1}B_{0j}^T+B_jR_{0j}^{-1}B_j^T\} \quad , \tag{22}$$

$$Z_i = K\{B_iR_{0i}^{-1}R_{ii}R_{0i}^{-1}B_i^T+\sum_{j=1}^{2}B_{0j}R_{00j}^{-1}R_{i0j}R_{00j}^{-1}B_{0j}^T\}K \quad . \tag{23}$$

4.2. The derivation of the sufficient conditions.

According to the general method shown in Section 2.2, we state the main steps of the derivation of the sufficient conditions given by Theorem 4.1.

(i) The team-optimal solution of P0.

In this step, we find the team-optimal solution of P0, called Problem (A). Therefore, Problem (A) will be as follows.

$$\min_u J_0(u)$$

$$\text{s.t. } (dx(t)/dt) = A(t)x(t)+\sum_{j=1}^{2}\{B_{0j}(t)u_{0j}(t)+B_j(t)u_j(t)\} \quad ,$$

$$x(t_0) = x^\circ \quad ; \quad t \in [t_0,t_1] \quad ,$$

$$u_{0i}(t) \in R^{m_{0i}} \quad ; \quad u_i(t) \in R^{m_i} \quad , \quad i=1,2,$$

where $u(t)=\{u_{01}(t),u_{02}(t);\ u_1(t),u_2(t)\}$.

Assume that $\{u_{01}^*,u_{02}^*;\ u_1^*,u_2^*\}=u^*$ is a team-optimal solution of Problem (A) for P0. Using the results of the linear quadratic problem in optimal control theory, we can then obtain that

$$u_{0i}^*(t) = -R_{00i}^{-1}B_{0i}^TK(t)x^*(t) \quad ; \quad u_i^*(t) = -R_{0i}^{-1}B_i^TK(t)x^*(t), \ i=1,2, \tag{24}$$

where $x^*(t)$ is the optimal state trajectory corresponding to $u^*(t)$, and $K(t)$ is the solution of the matrix Riccati differential equation of the form

$$(dK(t)/dt) = -KA-A^TK-Q_0+K\{A+\sum_{j=1}^{2}[B_{0j}R_{00j}^{-1}B_{0j}^T+B_jR_{0j}^{-1}B_j^T]\} \quad ,$$
$$K(t_1) = F_0 \quad , \quad t \in [t_0,t_1], \tag{25}$$

whereby, we can obtain (18) given in Theorem 4.1.

Let $\phi(t,t_0)$ be the solution of the following differential equation

$$(\partial\phi(t,t_0)/\partial t) = \Lambda(t)\phi(t,t_0) \quad ; \quad \phi(t_0,t_0) = I \ , \ t \in [t_0,t_1] \ . \tag{26}$$

The optimal state trajectory $x^*(t)$ and values of $u_{0i}^*(t)$ and $u_i^*(t)$, i= 1,2, are then given by

$$x^*(t;\ t_0,x^\circ) = \phi(t,t_0)x^\circ \ ; \tag{27}$$

$$u_{0i}^*(t) = -R_{00i}^{-1}B_{0i}^TK\phi(t,\ t_0)x^\circ \ , \tag{28}$$

$$i=1,2.$$

$$u_i^*(t) = -R_{0i}^{-1}B_i^TK\phi(t,\ t_0)x^\circ \ , \tag{29}$$

If we require that

$$\eta_{0i}(t) = -R_{00i}^{-1}B_{0i}^TK+\eta_{ii}R_{0i}^{-1}B_i^TK \quad ; \quad i=1,2, \tag{30}$$

which is (16) in Theorem 4.1, we then have that

$$u_{0i}^*(t) = \bar{\gamma}_{0i}(x^*(t),\ u_i^*(t),\ t) \quad , \quad i=1,2. \tag{31}$$

Therefore, we can have that

$$u_{0i}(t) = \bar{\gamma}_{0i}(x(t),\ u_i(t),\ t)$$
$$= \eta_{0i}(t)x(t)+\eta_{ii}(t)u_i(t) \quad ; \quad i=1,2, \tag{32}$$

where $\eta_{0i}(t)$ and $\eta_{ii}(t)$, i=1,2, satisfy (30).

(ii) Solve the following problem, called Problem (B), i.e., find the Nash equilibrium solution of P1 and P2.

For the player Pi, i=1,2,

$$\min_{u_i} J_i(u_{01},u_{02};\ u_1,u_2)$$

$$\text{s.t. } (dx(t)/dt)=A(t)x(t)+\sum_{j=1}^{2}\{B_{0j}(t)u_{0j}(t)+B_j(t)u_j(t)\},$$
$$x(t_0) = x^{\circ},\quad t\in[t_0,t_1],$$
$$u_{0i}(t) = \bar{\gamma}_{0i}(x(t),\ u_i(t),\ t)\ ;\quad u_i(t)\in R^{m_i}.$$

Since $J_i$, i=1,2, are strictly convex in $u_i(t)$, by making use of the Pontryagin's minimum principle, the sufficient condition and necessary condition for Problem (B) are given by (10) and the equations of the form

$$(\partial H_i/\partial u_i)=\eta_{ii}^T\{R_{i0i}u_{0i}+B_{0i}^T\lambda_i\}+R_{ii}u_i+B_i^T\lambda_i = 0, \tag{33}$$

$$(d\lambda_i(t)/dt)=-Q_ix-A^T\lambda_i-\sum_{j=1}^{2}\eta_{0j}^T\{R_{i0j}u_{0j}+B_{0j}^T\lambda_i\}, \tag{34a}$$

$$\lambda_i(t_1) = F_ix(t)\big|_{t_1},\qquad i=1,2, \tag{34b}$$

where $\lambda_i(t)\in R^n$, i=1,2, are the adjoint vector functions, and $H_i$ is the Hamiltonians as follows.

$$H_i = (1/2)\cdot\{x^TQ_ix+\sum_{j=1}^{2}[u_{0j}^TR_{i0j}u_{0j}+u_j^TR_{ij}u_j]\}$$
$$+\lambda_i^T\{Ax+\sum_{j=1}^{2}[B_{0j}u_{0j}+B_ju_j]\} \quad . \tag{35}$$

(iii) The conditions for the incentive Stackelberg strategy of P0.

Now, we will ensure that $u_1^*(t)$, $u_2^*(t)$ and $x^*(t)$ are the solution of Problem (B). For this, we need, together with (30), that they hold

$$\eta_{ii}^T\{R_{i0i}u_{0i}^*+B_{0i}^T\lambda_i\}+R_{ii}u_i^*+B_i^T\lambda_i = 0\ ;\quad i=1,2, \tag{36}$$

for all $t\in[t_0,t_1]$, where $\lambda_i(t)$, i=1,2, are respectively the solutions of the following differential equations

$$(d\lambda_i(t)/dt)=-Q_ix^*-A^T\lambda_i-\sum_{j=1}^{2}\eta_{0j}^T\{R_{i0j}u_{0j}^*+B_{0j}\lambda_i\}, \tag{37a}$$

$$\lambda_i(t_1) = F_ix^*(t)\big|_{t_1},\qquad i=1,2. \tag{37b}$$

Let

$$\lambda_i(t) = P_i(t)x^*(t),\quad i=1,2. \tag{38}$$

Substituting respectively for $\lambda_i(t)$, i=1,2, $x^*(t)$, $u_i^*(t)$ and $u_{0i}^*(t)$, i=1,2, form (38), (27)-(29) in (37), we can obtain that Pi, i=1,2,will be the solutions of the following coupled matrix linear differential equations

$$(dP_i(t)/dt)=-P_i\Lambda-A^T P_i-Q_i-\sum_{j=1}^{2}\eta_{0j}^T\{B_{0j}P_i-R_{i0j}R_{00j}^{-1}B_{0j}^T K\} \quad , \qquad (39a)$$

$$P_i(t_1) = F_i \ , \quad i=1,2. \qquad (39b)$$

Similarly, substituting respectively for $\lambda_i(t)$, i=1,2, $x^*(t)$, $u_i^*(t)$ and $u^*_i(t)$, i=1,2, from (38), (27)-(29) in (36), we can then obtain (17) in Theorem 4.1.

For the purpose of simplicity, here we further discuss (39). By making use of (30) and (17), we can then obtain directly (19) and (20).

Thus, we complete the derivation of Theorem 4.1.

Remark:

(i) The incentive Stackelberg strategy $\bar{\gamma}_0=\{\bar{\gamma}_{01},\bar{\gamma}_{02}\}\in\Gamma_0$ is linear for $x(t)$, $u_1(t)$ and $u_2(t)$ in Theorem 4.1.

(ii) In the light of the procedures by which Theorem 4.1 was derived, we can know that this theorem is only the sufficient conditions for $\bar{\gamma}_0$ being the incentive Stackelberg strategy in such a game problem.

(iii) By the procedures stated above, $\bar{\gamma}_0$ can be directly determined and this calculation is straightforward and feasible.

## 5. Conclusion

This paper mainly discusses the two-level hierarchical differential games with two noncooperative followers, characterized by a class of linear state dynamics and quadratic cost functionals. In the paper, we first give some concepts in such games. Then, for such a two-level incentive differential game problem, we construct new form of the incentive Stackelberg strategy $\bar{\gamma}_0=\{\bar{\gamma}_{01},\bar{\gamma}_{02}\}$ of P0, and also derive the sufficient conditions which are satisfied by $\bar{\gamma}_0=\{\bar{\gamma}_{01},\bar{\gamma}_{02}\}$.

As stated in the paper, $\bar{\gamma}_0=\{\bar{\gamma}_{01},\bar{\gamma}_{02}\}$ is linear for $x(t)$, $u_i(t)$, i=1,2, in such a strategy. In the light of the conditions given in Theorem 4.1, $\bar{\gamma}_0$ can be easily determined, and the calculation is straightforward and feasible.

With the aid of the approach and concepts given in this paper, we may also obtain some results for the two-level incentive differential games with noncooperative N-follower ($N > 2$), characterized by linear state dynamics and quadratic cost functionals, which will be further studied in our research.

## References

[1] Von Stackelberg, H.: Marktform und Gleichgewicht, Springer, Berlin 1934.

[2] Chen, C.I. and J.B. Cruz Jr.: Stackelberg Solution for Two-person Games with biased Information Patterns, IEEE Trans. Automatic Con-

trol, Vol.AC-17, pp.791-798, 1972.

[3] Simaan, M. and J.B. Cruz Jr.: On the Stackelberg Strategy in Nonzero-sum Games, Journal of Optimization Theory and Applications, Vol.11, pp.535-555, 1973.

[4] Simaan, M. and J.B. Cruz Jr.: Additional Aspects of the Stackelberg Strategy in Nonzero-sum Games, Journal of Optimization Theory and Applications, Vol.11, pp.613-620, 1973.

[5] Ho, Y.C., P.B. Luh, and G.J. Olsder: A Control-theoretic View on Incentives, Automatica, Vol.18, pp.167-179, 1982.

[6] Basar, T. and H. Seibuz: Closed-loop Stackelberg Strategies with Applications in the Optimal Control of Multilevel Systems, IEEE Trans. Automatic Control, Vol.AC-24, pp.166-178, 1981.

[7] Tolwinski, T.: Closed-loop Stackelberg Solution to Multi-stage Linear Quadratic Game, Journal of Optimization Theory and Applications, Vol.34, pp.485-501, 1981.

[8] Zheng, Y.P. and T. Basar: Existence and Derivations of Optimal Affine Incentive Schemes for Stackelberg Games with Partial Information: A Geometric Approach, International Journal of Control, Vol.35, pp.997-1011, 1982.

[9] Zheng, Y.P., T. Basar, and J.B. Cruz Jr.: Stackelberg Strategies and Incentives in Multiperson Deterministic Decision Problems, IEEE Trans. Systems ,Man and Cybernetics, Vol.SMC-14, pp.10-24, 1984.

[10] Basar, T. and G.J. Olsder: Team-optimal Closed-loop Stackelberg Strategies in Hierarchical Control Problems, Automatica, Vol.16, pp.409-414, 1980.

[11] Luh, P.B., S.C. Chang, and T.S. Chang: Solutions and Properties of Multi-stage Stackelberg Games, Automatica, Vol.20, pp.251-256, 1984.

[12] Chang, T.S. and P.B. Luh: Derivation of Necessary and Sufficient Conditions for Single-stage Stackelberg Games via the Inducible Region Concept, IEEE Trans. Vol.AC-29, pp.63-66, 1984.

[13] Luh, P.B., T.S. Chang and T. Ning: Three-level Stackelberg Decision Problems, IEEE Trans., Vol.AC-29, pp.280-282, 1984.

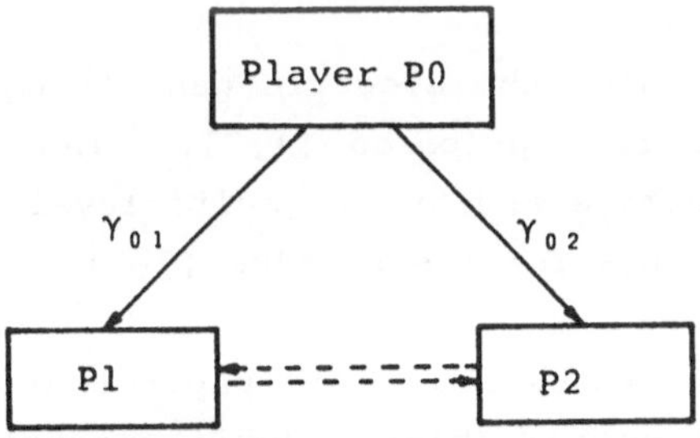


Fig.1. Hierarchical decision-making structure with two noncooperative followers.

# STACKELBERG GAME MODEL INCORPORATING PUBLIC PARTICIPATION FOR ELECTRIC GENERATING SYSTEM DEVELOPMENT

Katsuya OGINO
Dept. of Applied Mathematics & Physics, Faculty of Engineering, Kyoto University, Kyoto 606 JAPAN

## ABSTRACT

The present paper represents a two-level multi-follower Stackelberg game linear quadratic model for the electric generating system development. For stable electric supply, the model incorporates the public participation from the point of view of social siting concern. The model analyzes the effects of social siting concern on the generating system development. The interperiod generating capacity balance and several evaluation factors such as the social siting concern play fundamental roles in the model. Its characteristics are investigated by simulation analyses.

## 1. Introduction

The growing concern in siting electric facility forms the contentious problem and the important effect on meeting electric demand. The siting is thus the one of the urgent problems to be integrated in the electric generating system development. With special emphases on the social siting concern, the present paper represents a two-level multi-follower Stackelberg game model for the electric generating system development. The salient feature of the model is that it incorporates the public participation in the system development.

Regarding the electric generating system development, various models have been proposed [1,2]. The author has also adopted the multi-objective and/or the multi-level multi-follower Stackelberg game approaches for the problem [3-6].

It is assumed in this paper that there exist, in a region under investigation, plural public groups with different social siting concern. An interperiod balance equation of electric generating capacity is firstly investigated. The factors fundamental for evaluating the social siting concern, the electric supply and the cost are secondly introduced, where the social negative preference provides a comparative standard in selecting alternative power plant type mix from the point of view of social siting concern. The problem of electric generating system development is then formulated as a two-level multi-follower Stackelberg game in a linear quadratic form.

In the model, the generating system development decision unit (leader) in the first level explores effective development strategy of electric generating system, permitting the public participation as the second level decision units. Under the leader's development strategy, the public group decision units (followers) in the second level pursue their respective development strategy to reflect their respective social siting concern in the development decision making. The social siting concern is thus actively treated through the second level decision units. The model evaluates interactions between the electric supply and the social siting concern. Finally, simulation analyses are illustrated to show the characteristics and the effectiveness of the model.

## 2. Generating Capacity Balance

Consider an overall planning period [0, T], broken down into n periods of equal time interval dt as

$$T = n\ dt. \tag{1}$$

Then, the electric generating capacity in a period k is equal to the effective amount of the capacity available from the previous period plus the expanded amount of capacity in the period k. Thus, the interperiod generating capacity balance can be represented by a linear difference equation

$$c_k = (I - A_k)c_{k-1} + e_k, \qquad (k=1,2,\ldots,n), \qquad (2)_1$$

$$c_0 \text{ : given,} \qquad I \text{ : mxm-unit matrix,} \qquad (2)_2$$

where

m : number of power plant types under investigation such as nuclear type, fossil type and so on,

$c_k$ : electric generating capacity by each power plant at the end of the period k; col.$(c_{1k},\dots,c_{mk})$,

$e_k$ : amount of the generating capacity by each power plant type, expanded in the period k; col.$(e_{1k},\dots,e_{mk})$,

$A_k$ : mxm-diagonal matrix with the diagonal element $a_{ik}$ of the attrition rate of each power plant type i in the period k.

## 3. Evaluation Factors

The siting problem is one of the fundamental problems in recent electric generating system development. The electric supply shortage might be caused by the social siting concern. The present section represents main evaluation factors in the electric generating system development.

The social siting concern mainly about the environmental problems around electric power plant site can be investigated by the social siting concern evaluation factor in the most general form

$$F_{sk} = F_{sk}(np_k, c_k, e_k), \tag{3}$$

where

$np_k$ : comparative measure of the social negative preference from the point of view of siting problem to each power plant type in the period k; col.$(np_{1k},\dots,np_{mk})$.

The social negative preference, measured by the social poll and/or the questionnaire, gives an order in the social siting concern to alternative power plant types and provides basis for a comparative standard in selecting alternative power plant type mix in the system development.

One of the concrete forms of the social siting concern evaluation is given as

$$F_{sk} = np_k'e_k, \qquad (4)_1$$

$$F_{sk} = np_k'CF_kc_kdt, \qquad (4)_2$$

where

$CF_k$ : mxm-diagonal matrix with the diagonal element $cf_{ik}$, representing the capacity factor of each power plant type i in the period k.

$(4)_1$ and $(4)_2$ represent the siting concern respectively about the capacity expansion and the electric generation. The " ' " denotes the transposition of vectors.

The social negative preference is based mainly on various types of pollution discharge from each power plant type, which can be evaluated in an input-output form as

$$P_k = \begin{bmatrix} p1_{1k} & \cdots & p1_{mk} \\ \cdots & \cdots & \cdots \\ \cdots & \cdots & \cdots \\ pq_{1k} & \cdots & pq_{mk} \end{bmatrix}, \qquad (5)$$

where

q : number of pollution types,

$pj_{ik}$ : discharge amount of pollution j per unit power output from power plant type i in the period k.

The electric supply-demand deviation can be evaluated by

$$F_{ek} = (1,\ldots,1)c_k - pd_k, \qquad (6)_1$$

$$F_{ek} = cf_k'c_kdt - ed_k, \qquad (6)_2$$

where

$pd_k$ : peak demand in the period k,

$ed_k$ : energy demand in the period k,

$cf_k$ : capacity factor of each power plant type in the period k; col.$(cf_{1k},\ldots,cf_{mk})$.

The electric cost can be evaluated by

$$F_{ck} = cp_k{}'e_k, \tag{7$_1$}$$

$$F_{ck} = om_k{}'CF_kc_kdt, \tag{7$_2$}$$

where

$cp_k$ : present worth of the capital cost per unit capacity of each power plant type in the period k; col.$(cp_{1k},\ldots,cp_{mk})$,

$om_k$ : present worth of the O & M cost per unit power output of each power plant type in the period k; col.$(om_{1k},\ldots,om_{mk})$,

## 4. Stackelberg Game Linear Quadratic Model

Based on the previous investigations, the present section integrates the social siting concern as public participation and represents a Stackelberg game model for the electric generating system development.

In order to incorporate the social siting concern into the electric generating system development, the model, in this paper, assumes that there exist several players, a leader and plural followers (plural public groups with different social siting concern), and that $e_k$ in (2) consists of the leader's expansion ($e1_k$) and the follower's expansion ($e2_{ik}$, i=1,...,f) as

$$e_k = e1_k + e2_{1k} + \ldots + e2_{fk}, \tag{8}$$

f : number of public groups.

By selecting $e2_{ik}$, the followers intend to reflect respective social siting concern on the capacity expansion pattern.

The electric generating system development problem can then be described as a Stackelberg Linear Quadratic Game as follows :

" For the interperiod generating capacity balance equation (2) with (8), the leader and the followers seek optimal expansion strategies $e1_k$ and $e2_{ik}$ which minimize respective objective functions;

$$J_i = 1/2 \sum_{k=0}^{n-1} [(y_{ik+1}-\bar{y}_{ik+1})'Q_{ik}(y_{ik+1}-\bar{y}_{ik+1})$$

$$+ z_{ik}'R_{ik}z_{ik}], \quad (i=1,1,2,...,f), \tag{9}$$

$$y_{ik} = D_{ik}c_k, \qquad z_{ik} = E_{ik}e_k, \tag{10}$$

$\bar{y}_k$ : given desired state such as demand. "

$D_{ik}$ and $E_{ik}$ in (10) are matrices of proper dimension, and $y_{ik}$ and $z_{ik}$ consist of various evaluation factors of player i in the period k. $Q_{ik}$ and $R_{ik}$ in (9) are weighting matrices, reflecting player i's intention in the system development.

In the model, the leader permits the public participation as the followers in the system development, while the followers intend to reflect their siting concern in a Leader-Multifollower setting.

## 5. Simulation Analyses

To investigate the characteristics of the present model, simulation analyses are illustrated in Fig. 1 - Fig. 5 in this section. The main data adopted are T=20.0, n=4, m=3 and f=2. The third power plant type III is assumed to be new and available from the third period. The characteristic of cost adopted in the simulation is such that $cp_{1k} < cp_{2k}$, $om_{1k} > om_{2k}$, and the third type power plant III is assumed to be quite costly but socially preferable, compared with the other power plant types I and II. Fig. 1 and Fig. 2 represent the social negative preference, assumed in the present simulations. Fig. 2 reads that the power plant type I is more preferable to type II, and the power plant type I becomes socially less preferable in the periods 3 and 4 to that in the first two periods. Fig. 3 represents the optimal installed capacity for the case of one player (leader) optimization without any follower, where the player mainly takes the cost minimum strategy. Fig. 4 represents the optimal installed capacity for the case, where one follower with the negative preference in Fig. 1 is assumed in the second level. Fig. 1 indicates that the Type II or III is preferred to Type I throughout the overall development period. In comparison with Fig. 3, the

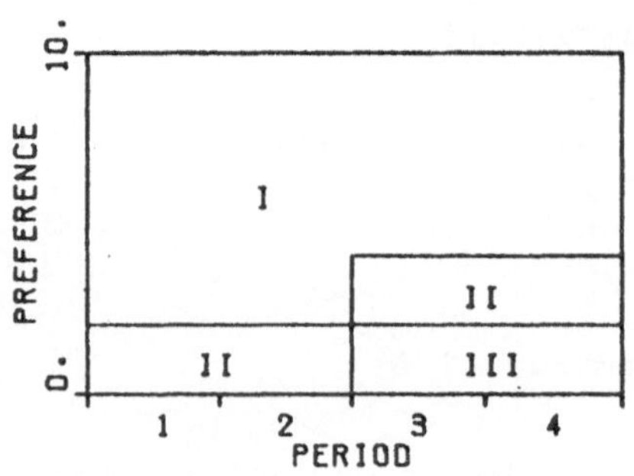


Fig. 1 Negative Preference

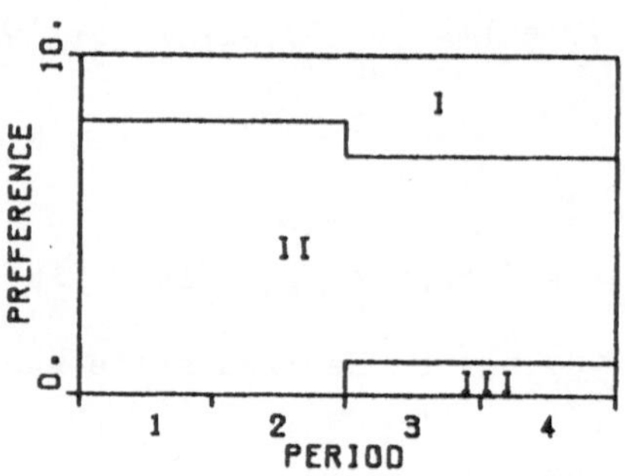


Fig. 2 Negative Preference

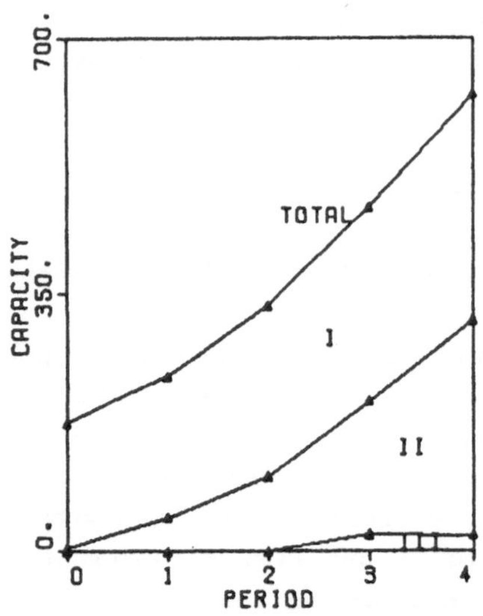


Fig. 3 Installed Capacity

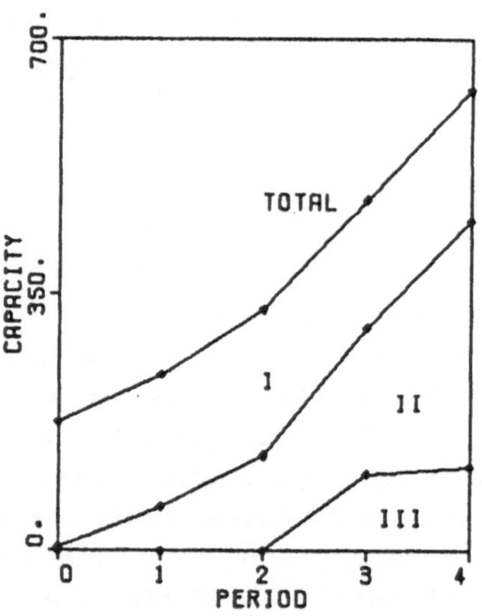


Fig. 4 Installed Capacity

decrease in installed capacity of Type I and also the increase of Type II and III are indicated in Fig. 4. The effect of the follower's negative preference is thus reflected in the capacity expansion pattern. Fig. 5 represents the optimal installed capacity for the case when two followers are assumed in the second level (the same follower of Fig. 4 case with Fig. 1, and another follower with negative preference in Fig. 2.) The second follower prefers Type I to Type II unlike the first follower.

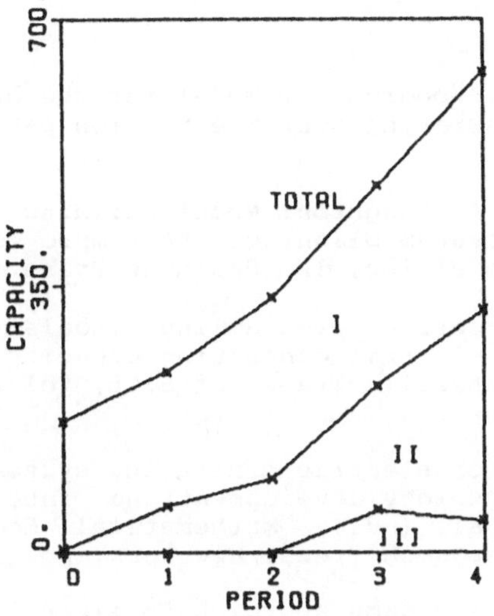


Fig. 5 Installed Capacity

In comparison with Fig. 4, Fig. 5 represents the increase in expanding Type I and also the decrease in expanding Type II especially at periods 2, 3 and 4. This difference between Fig. 4 and Fig. 5 shows the effect of the second follower's preference on the capacity expansion pattern. Simulation analyses thus clearly indicate the effects of public participation (followers) on the generating system development.

## 6. Conclusions

Permitting the public participation, the present paper develops a two-level multi-follower Stackelberg game model for the electric generating system development. Following the interperiod capacity balance, several evaluation factors are investigated. Simulation analyses indicate the effectiveness of the model. Thus, the present linear quadratic or linear programming Stackelberg game approaches enable various interaction analyses between the electric supply, the social siting concern, the cost and so on in the electric generating system development.

## References

[1] D.L. Farrar and F. Woodruff: A model for the determination of optimal electric generating system expansion pattern, NTIS, Springfield (1974).

[2] F. Begrali and M. A. Laughton: Model building with particular reference to power system planning ; the improved Z-substitutes method, in Energy Modelling, IPC Business Press (1974).

[3] K. Ogino: Multiobjective programming models introducing social siting concern to electric generating capacity expansion policy analysis, (in Japanese), Trans. of SICE, Vol.15, pp.839-844 (1979).

[4] K. Ogino: Models for electric generating system development - acceleration of technology development and interaction analysis -, in X. J. Avula et al. (ed.), Mathematical Modelling in Science and Technology, Pergamon Press, New York, pp.254- 259 (1984).

[5] K. Ogino: Stackelberg game approach to electric generating system development, Proc. of the 1986 IFAC Workshop on Modelling Decision and Game with Application to Social Phenomena, pp.384-392 (1986).

[6] K. Ogino: Stackelberg game linear programming model for electric generating system development, Proc. of the 26th SICE Annual Conference, Vol. II, pp. 1341-1344 (1987).

[7] T. Basar and G. J. Olsder: Dynamic noncooperative game theory, Academic Press, New York (1982).

# COOPERATIVE FUZZY GAME WITH SIDE-PAYMENTS

**Atsushi Yamada**

Department of Administration Engineering
Chuo University
Kasuga, Bunkyo-ku, Tokyo 112
Japan

## 1. Introduction

This fuzzy game theory starts rather from coalitions of players than individual players. The set $\mathcal{M}$ of fuzzy coalitions is that of bounded measurable functions m's with $0 \leqq m \leqq 1$ defined on a $\sigma$-finite measure space $(X, \mathfrak{F}, \mu)$ . A new interpretation of an element of X in this paper is a constituent element of personality. Then a player as well as a coalition is represented by $m \in \mathcal{M}$ . The set of individual players is specified as a certain subset $\mathcal{M}^1$ of $\mathcal{M}$ according to the situation. From the measurability of sums of measurable functions, of course, addition of individual players yields a fuzzy coalition unless the sum of all $\mathcal{M}^1$-functions fails to belong to $\mathcal{M}$ . However a great difference from the classical theory is that an individual player is infinitely decomposable to fuzzy coalitions as the microscopic units which may be considered as a schizothymic phenomenon. This idea is called a holonic method in the game theory.

This game theory is constructed on the fundamental skelton consisting of the two concepts, namely characteristic functional v and allocation functional a . Characteristic functional v called simply game v characterizes the game itself. Allocation functional a called simply allocation a represents shares for each coalition. By the terminology of economics, v describes the products of the coalitions, while an allocation a the incoms of the coalitions.

A general game is a general functional v on $\mathcal{M}$ with $v(0) = 0$ . Let $\mathcal{G}$ [ $\mathcal{S}$, $\mathcal{V}$, $\mathcal{C}$ and $\mathcal{A}$ ] denote the set of all general [ general superadditive, superadditive, convex and additive resp.] games. Then $\mathcal{A} \subset \mathcal{V} \cap \mathcal{C} \subset \mathcal{V} \cup \mathcal{C} \subset \mathcal{S} \subset \mathcal{G}$ . The restriction of v to $\mathcal{M}^1$ corresponds to the characteristic function of the game in the classical theory.

Allocation a is not proper to the considered game v , but it is merely an additive functional on $\mathcal{M}$. Properly speaking, products and incomes are different things and are not always mutually dependent. Allthough an allocation a represents shares for each coalition, mathematically a is an unessential game. Hence the set of all allocations is considered as $\mathcal{A}$ which is not proper to the considered game v . However allocations are imposed some restrictions in reality. These restrictions are imposed usually depending on the considered game v . We consider two restrictions, civil minimum and Pareto frontier which gives a bound from below and above respectively. The set of all such allocations as v-relative unessential games is denoted by $\mathcal{A}(v)$.

## 2. Games

1. Let $(X, \mathcal{F}, \mu)$ be a $\sigma$-finite measure space. A bounded measurable function m is called a coalition if $0 \leq m \leq 1$ $\mu$-a.e.

Let $\mathcal{M}$ denote the set of all coalitions.

We consider that m and n are equivalent if $m = n$ $\mu$-a.e. Precisely the members of $\mathcal{M}$ are equivalence classes of functions differing from each other on a set of $\mu$-measure zero. No distinction will be made, however, between these equivalence classes and their representative functions. In the sequel "$\mu$-a.e." will be omitted in the formulas with respect to $\mathcal{M}$-functions.

2. A real valued functional v on $\mathcal{M}$ is called a general game if it satisfies (G1):

(G1) $v(m) = 0$ whenever $m \in \mathcal{M}$ and $m = 0$ .

Let $\mathcal{G}$ be the set of all general games.

3. A real valued functional v on $\mathcal{M}$ is called a general superadditive game if it satisfies (S1) — (S4):

(S1) [Separative superadditivity] $v(m + n) \geq V(m) + v(n)$ whenever $m, n \in \mathcal{M}$ and $m \wedge n = 0$ ,

(S2) [Upper semicontinuity] $v(m) \geq \overline{\lim}_k v(m_k)$ , whenever $m, m_k \in \mathcal{M}$ $(k=1,2,\cdots)$ and $\lim_k m_k = m$ ,

(S3) [Total boundedness] $\sup_{\{X_i\} \in \Pi_X} \sum_i | v(1_{X_i}) | < \infty$ ,

where $\Pi_A$ denotes the set of all partitions of a $\mathcal{F}$-set to finite or countably infinite $\mathcal{F}$-sets and $1_A$ denotes the indicator function of A .

(S4) [Positive subhomogeneity] $v(\alpha 1_A) \geqq \alpha v(1_A)$ where $\alpha$ is a constant with $0 < \alpha < 1$ and $A \in \mathcal{F}$ .

Let $\mathcal{S}$ be the set of all general superadditive games.

4. For any $v \in \mathcal{S}$ it is easily seen that

(S5) $=$ (G1) ,

(S6) $\quad \sup_{\{A_i\} \in \Pi_A} \sum_i |v(1_{A_i})| < \infty$ ,

(S7) $\quad v(1_A) \geqq \sum_i v(1_{A_i})$ for any $\{A_i\} \in \Pi_A$ .

In fact, (S5), (S6) and (S7) follow from (S1)+(S2)+(S4) , from (S3) and from (S1)+(S2)+(S6) , respectively.

5. A real valued functional v defined on $\mathcal{M}$ is called an additive game if it satisfies (S2), (S3) and the following (A0) — (A2):

(A0) [Continuity] $\lim_k v(m_k) = v(m)$ whenever $m, m_k \in \mathcal{M}$ $(k=1,2,\cdots)$ and $\lim_k = m$ ,

(A1) [Separative additivity] $v(m + n) = v(m) + v(n)$ whenever $m. n \in \mathcal{M}$ and $m \wedge n = 0$ ,

(A2) [Positive homogeneity] $v(\alpha 1_A) = \alpha v(1_A)$ whenever $\alpha$ is a constant with $0 < \alpha < 1$ and $A \in \mathcal{F}$ .

Let $\mathcal{A}$ be the set of all additive games.

6. Proposition 1. $(\forall A \in \mathcal{F}, 0 < \forall\alpha < 1)$ $v(1_A) \geqq v(\alpha 1_A) + v((1 - \alpha)1_A)$ and (S4) implies (A2) .

Proof. Let $A \in \mathcal{F}$ and $0 < \alpha < 1$ . Put $\beta = 1 - \alpha$ . From (S4), $v(\alpha 1_A) - \alpha v(1_A) \geqq 0$ , $v(\beta 1_A) - \beta v(1_A) \geqq 0$ . $0 = v(1_A) - v(1_A) \geqq v(\alpha 1_A) + v(\beta 1_A) - v(1_A) = v(\alpha 1_A) - \alpha v(1_A) + v(\beta 1_A) - \beta v(1_A) \geqq 0$ . Hence $v(\alpha 1_A) = \alpha v(1_A)$ .

7. A real valued functional v defined on $\mathcal{M}$ is called an unessential game if there exists a $\mu$-absolutely continuous signed measure $\lambda_v$ on $\mathcal{F}$ such that

$$v(m) = \int_X m \, d\lambda_v$$

for all $m \in \mathcal{M}$ .

8. Theorem 1. (i) For any $v \in \mathcal{S}$ , there exists a $\mu$-absolutely continuous signed measure $\lambda_v$ on $\mathcal{F}$ such that

$$v(m) \geq \int_X m \, d\lambda_v$$

for all $m \in \mathcal{M}$ and $\lambda_v$ is maximal in the sense that if $\lambda$ is a $\mu$-absolutely continuous signed measure on $\mathcal{F}$ and if $v(1_A) \geq \lambda(A)$ for all $A \in \mathcal{F}$ , then $\lambda_v \geq \lambda$ on $\mathcal{F}$ .

(ii) If $v \geq 0$ on $\mathcal{M}$ , then $\lambda_v \geq 0$ on $\mathcal{F}$ .

(iii) For $v \in \mathcal{S}$ , $v \in \mathcal{A}$ iff $v$ is unessential.

Proof. Similuarly to the proof of Theorem 2.1 in [5] .

Signed measure $\lambda_v$ the existence of which is guaranteed in theorem 1 is called a signed measure associated with $v$ .

9. Proposition 2. For any $v \in \mathcal{S}$ , $v \in \mathcal{A}$ iff $v$ satisfies (A0) and (A3):

(A3) [Additivity] $v(m + n) = v(m) + v(n)$ whenever $m, n$ and $m + n \in \mathcal{M}$ .

Proof. From (iii) of theorem 1 and from proposition 1.

10. A real valued functional $v$ on $\mathcal{M}$ is called superadditive game if it satisfies (S2) — (S4) and (V1):

(V1) [Superadditivity] $v(m + n) \geq v(m) + v(n)$ whenever $m, n$ and $m + n \in \mathcal{M}$ .

Let $\mathcal{V}$ be the set of all superadditive games.

11. Proposition 3. If $v \in \mathcal{V}$, $v$ satisfies (A2) .

Proof. From proposition 1.

12. A real valued functional $v$ on $\mathcal{M}$ is called a convex game

if it satisfies (G1), (S2) — (S4) and (C1):

(C1) [Convexity] $v(m \vee n) + v(m \wedge n) \geqq v(m) + v(n)$

whenever $m, n \in \mathcal{M}$.

Let $\mathcal{C}$ be the set of all convex games.

Proposition 4. $\mathcal{A} \subset \mathcal{V} \cap \mathcal{C} \subset \mathcal{V} \cup \mathcal{C} \subset \mathcal{S} \subset \mathcal{G}$.

Proof. Directly from the definitions.

## 3. Allocations

1. The set of all allocations is considered as $\mathcal{A}$ which is not proper to the considered game $v$. However allocations are imposed some restrictions in reality. These restrictions depend on the considered game $v$.

The first restriction on a general allocation $a \in \mathcal{A}$ is

(i) [Civil minimum restriction] $(\forall m \in \mathcal{M})\ a(m) \geqq \int_X m \, d\lambda_v$.

This gives a bound of $a$ from below.

The next restriction which gives a bound from above is

(ii) [Pareto restriction] $a(1_X) = \lambda_v(X) + M(v)$, where

$$M(v) = \sup_{m \in \mathcal{M}} \left( v(m) - \int_X m \, d\lambda_v \right).$$

For $v \in \mathcal{S}$ the set of allocations depending on $v$ is defined by

$$\mathcal{A}(v) = \{ a \mid a \in \mathcal{A},\ a \text{ satisfies (i) and (ii)} \}.$$

Remark: In the above definition we assume $M(v) < \infty$.

2. Proposition 5. $M(v) \geqq 0$ for $v \in \mathcal{G}$. If $v \in \mathcal{S}$ satisfies $v(1_X) \geqq v(m) + v(1_X - m)$ for any $m \in \mathcal{M}$, then $M(v) < \infty$. For any $v \in \mathcal{A}$, $M(v) = 0$.

Proof. Considering $m = 0$, $M(v) \geqq 0$. Let $v \in \mathcal{S}$ and $m \in \mathcal{M}$, $v(1_X) \geqq v(m) + v(1_X - m)$. Using (i) of theorm 1 for $1_X - m$,

$$v(1_X) \geqq v(m) + \int_X (1_X - m)\, d\lambda_v .$$

Then for any $m \in \mathcal{M}$ ,

$$v(1_X) - \lambda_v(X) \geqq v(m) - \int_X m\, d\lambda_v .$$

Hence $M(v) = v(1_X) - \lambda_v(X) < \infty$ . Finally let $v \in \mathcal{A}$ . From (iii) of theorem 1 , $M(v) = 0$ .

3. Proposition 6. If $v \in \mathcal{A}$, $\mathcal{A}(v) = \{v\}$ . For any $v \in \mathcal{S}$ with $M(v) < \infty$ , $\mathcal{A}(v)$ is not empty.

Proof. If $v \in \mathcal{A}$ , then

$$\mathcal{A}(v) = \{ a \mid a \in \mathcal{A},\ a \geqq v \text{ on } \mathcal{M},\ a(1_X) = v(1_X) \} .$$

Then $v \in \mathcal{A}(v)$ , or $\{v\} \subset \mathcal{A}(v)$ . Let $a \in \mathcal{A}(v)$ . From $a(1_X) = v(1_X)$ , $\lambda_a(X) = \lambda_v(X)$ . Then for any $A \in \mathcal{F}$ ,

$$(\lambda_a(A) - \lambda_v(A)) + (\lambda_a(A^c) - \lambda_v(A^c)) = \lambda_a(X) - \lambda_v(X) = 0 .$$

From $a \geqq v$ on $\mathcal{M}$, $\lambda_a \geqq \lambda_v$ . Then $\lambda_a = \lambda_v$ . Hence $a = v$ . Then for $v \in \mathcal{A}$, $\mathcal{A}(v) = \{v\}$ .

Let $v \in \mathcal{S}$ with $M(v) < \infty$ . Take a $\sigma$-absolutely continuous probability measure $p$ on $\mathcal{F}$ and put

$$(\forall m \in \mathcal{M}) \quad a(m) \underset{\text{def}}{=} \int_X m\, d\lambda_v + M(v) \cdot \int_X m\, dp .$$

Then $a \in \mathcal{A}$ . Let $m = 1_X$ , $a(1_X) = \lambda_v(X) + M(v)$ . Since $M(v) \geqq 0$ from proposition 5 ,

$$(\forall m \in \mathcal{M}) \quad a(m) \geqq \int_X m\, d\lambda_v .$$

Hence $a \in \mathcal{A}(v)$ , i.e. $\mathcal{A}(v)$ is nonempty.

4. For $v \in \mathcal{S}$ the following subset of $\mathcal{A}(v)$ is called the core of $v$ :

$$\mathcal{C}(v) = \{ a \mid a \in \mathcal{A}(v) ,\ a \geqq v \text{ on } \mathcal{M} \} .$$

If $v \in \mathcal{A}$ , $v \in \mathcal{C}(v)$ . Then $\mathcal{A}(v) = \{v\} \subset \mathcal{C}(v) \subset \mathcal{A}(v)$. Then $\mathcal{C}(v) = \{v\} = \mathcal{A}(v)$ .

5. A general game v is called a balanced game if it satisfies the following property:

(B1) For any natural number $N \neq 0$, for any $(r_i)_{i=1}^{N}$ with $r_i \in R^+$ and for any $(m_i)_{i=1}^{N}$ with $m_i \in \mathcal{M}$,

$$\sum_{i=1}^{N} r_i m_i = 1_X \longrightarrow \sum_{i=1}^{N} r_i v(m_i) \leqq v(1_X) .$$

Note that $r_i m_i \in \mathcal{M}$ . i.e. $0 \leqq r_i m_i \leqq 1_X$ since $r_i m_i \geqq 0$ .

6. Theorem 2. If $v \in \mathcal{S}$ is a balanced game, then v has a nonempty core $\mathcal{C}(v)$ .

Proof. We can prove using Hahn-Banach theorem. [6]

REFERENCES

[1] J.-P. Aubin, Coeur et valeur des jeux flous a paiements lateraux, C. R. Acad. Sci. Paris, Serie A, 279 (1974), 891 894.
[2] R. J. Aumann, Markets with a continuum of traders, Econometrica 32 (1964), 39 50.
[3] R. J. Aumann, Existence of competitive equilibria in markets with a continuum of traders, Econometrica 34 (1966), 1 17.
[4] R. J. Aumann and L. S. Shapley, Values of non-atomic games, Princeton Univ. Press, 1974.
[5] A. Yamada and S. Tsurumi, Mathematical foundation of general cooperative fuzzy games, Tohoku Math. Journ. 35 (1783), 53 63.
[6] A. yamada, On fuzzy games, (in Japanese) The 100th Aniversary Bull. of Chuo Univ. (1985), 657 684.

# DYNAMIC SIMULATION AND OPTIMAL CONTROL OF URBAN SOCIO-ECONOMIC AND ECO-ENVIRONMENT SYSTEM

Wang Yuji  Xiang Yuanwang
Hunan Academy of Social Sciences.ChangSha City.
Hunan Province.CHINA

## ABSTRACT

Multitudinous factors such as population, economy, science, technique and culture are gathered in cities. This is a complex socio-economic and ecologic large scale system, whose structure is complicated and function is comprehensive. The problems of this system are intricate and complex, covering many fields. Based on long range planning of Zhuzhou, a middle city of Hunan Province, this article aims at analysis in the structure, function and social behaviour of urban socio-economic and eco-environment system, standing in systematic macro-control, mainly from points of ecologic economics. What's more, by developing USES-SDM(Urban Socio-Economic and Eco-Environmrnt System-System Dynamics Model), using for investment decision control, the authors discuss simulation and control of coordinated development of urban sociaty, eonomy,science and technique and ecology. They also carry out an analogue and policy analysis of Zhuzou Long Range Planning as an example. Practice has made us known, USES-SDM using for urban development strategy research has accuracy, stability, feasibility and practicality, and got to the expected objective and results.

## 1.INTRODUCTION

The theory of benefits from investment in capital construction is one of the important components in the Marxist-Leninist theory of reproduction. The process of socialist reproduction depends to a great extent on the correct strategy and effective usage of investment in capital construction, and also on the situation of its implementation. Consequently, to control the direction, the structure and the rhythm of investment is one of the major measures in realizing the strategy for urban development. based on this idea and starting from the eco-economic view, we have tried to research the ruban socia-economy-eco-environment system approximatively like an overall macro-control system of information feedback, in which the investment strategy is indicated by control Variable U, and the synthetical benefits of economy, society and environment by controlled Variable Y(as showed in Figure 1).

Figure 1:

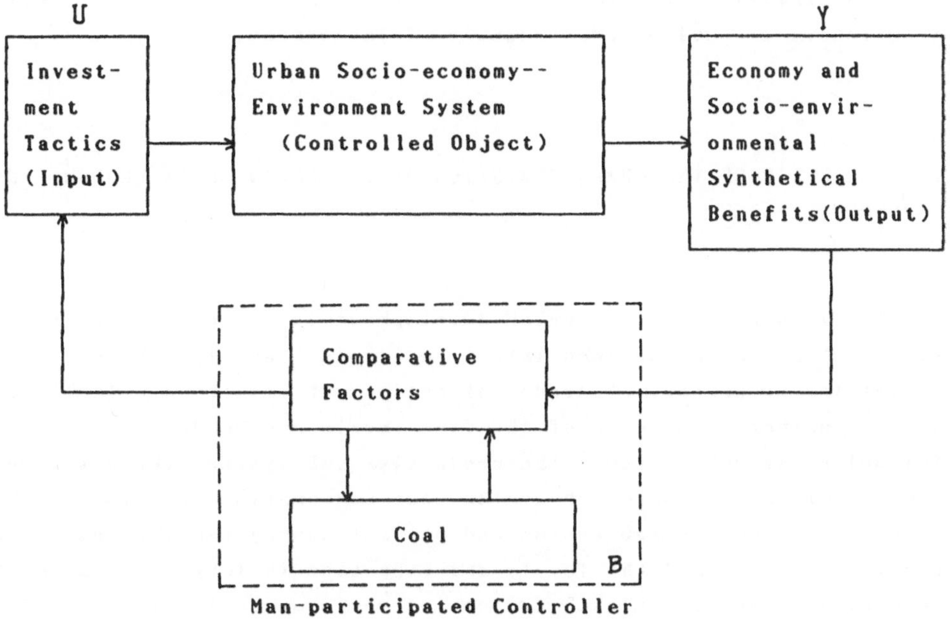


We also tried to carry out macroscopic adjustment and control through the settlement of the following problems so as to achieve the coordination of the equilibratory relation between economy and ecology, and to transfer the opened urban system continuously towards an ordered status far from the equilibratory point, and towards the most suitable direction satisfying the three major benefits,that is, the benefits of economy, society and environment,and thus attain the ideal circulation gradually.For this purpose, we decompose the problem into the four aspects that follow:

(1).The evaluation and analysis of the synthetical benefits of economy, society and environment of the ruban eco-economic system under the present status of structural function.

(2).The simulation of the future and the exploitation of the optimal control locus for reaching the targets established for the three major benefits of economy, society and environment.

(3).The analysis of the structural function and contributing factors in the relation of coordinate development between the urban socio-economic system and the eco-economy, and of its mechanism.

(4). The analysis of strategies for investment ( for example, politics, effeciency, direction and structure of investment, etc. ).

## 2.AN ANALYSIS OF THE CHARACTERISTICS OF THE URBAN SOCIA-ECONOMY-ECO-ENVIRONMENT SYSTEM

(1).The leading factor of urban development is human socio-economic activities,which can be devided into five levels, that is, the socio-economic activities of person, of group, of region, of country,and lastly of the world. Further-more, each of the five levels can be decomposed into the following six sub-systems: the productive sub-system, the distributive sub-system,the exchange sub-system, the consumption sub-system, the non-material productive sub-system and concept sub-system.The intricate connections of them and the levels and structure in length and breadth are harmoniously unified within the entire socio-economic system, and the regional socio-economic activities among them is of great significance because they are situated in the middle of the system and play the part of forming a connecting link between the higher levels and the lower ones.The scope involved in the problem of urban developing strategy on which we are now discussing, for this reason, has been continuously extending from the category of production relations and productive forces limited by traditional economics into other social domains and the domains of technology, ecology and the crust, to follow the enlargement of the system's borders.

(2).The urban socio-economy-eco-environment system is a complicated large-cale one with human activities as its pricipal part, involving a great number of factors in various domains such as society, economic, science and technology, and eco-environment. For this reason it has some specific characteristics of its own, other than the common ones, namely large scale, complicated structure, compound function and multi-factors, by which all large systems are characterized.

(1).Human being is the main body of socio-economic activities and a coercive factor influencing the structural function and social behaviour of the system. The subjective consciousness varies with different persons, and, is of great randomness and difficult to control concentratedly.

(2).The system state appears only once at every unit of time in the course of urban development and cannot be repeated because of its unreversibility.

(3).The control tactics put into the system belong to the kind of variables that change slowly, and the corresponding effect gained through them needs a long period of time, the time-blocking effect being great.

(4).The connection between urban socio-economy and eco-environment is loose, and the boundaries of the two are indistinct. The mechanism of the motive law is unclear. Tfe change is slow, the counter-audio-visual character is strong, and the measurability and comparability are comparatively poor. And in a word, it is considerably difficult to identufy this system.

## 3.THE STRUCTURE AND MECHANISN OF THE USES-SDM MODEL

On the basis of the above-mentioned idea for modelling and analysis of the characteristics of the urban eco-economy system, we have constructed a model for investment decision, which is called USES-SDM (Urban Socio-Economic And Eco-Environment System-System Dynamics Model ). The model consists of ten groups of sub-models, whose mutual relations are showed in Figure 2.

All the sub-models are connected in an interweaved way through various variables. The netted feedback structure of the model is formed in accordance with the causality and matual actions of the variables.

In addition, it ought to be pointed out that attention has been paid to the following problems in the course of setting up the model in consideration of the fact that urban eco-environment is linked closely with nearly all human socio-economic activities.

Figure 2:

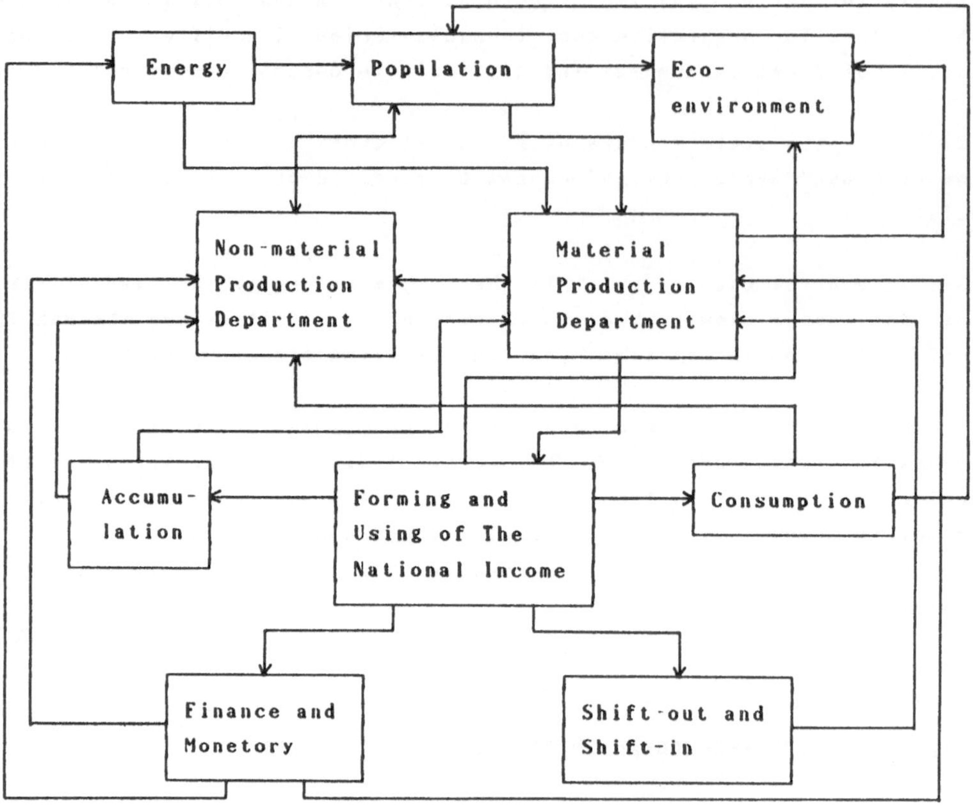


(1).This model is built on and thus supported by the SD simulation model of the large system of overall urban socio-economy, and so the entireness and precision for simulation of the USES-SDM model system is guaranteed.

(2).Human being is the main body in the urban eco-economic system and also the decision-maker who remoulds, controls,adjusts and makes use of the system for acertain economic purpose. For this reason, by adding the man-participated controller B to the model it is possible to utilize some technical methods such as decision-making for multi-targets, Delphi, fuzzy,judgement, expert system, in order to increase the quantity of information feedback between eco-environment and socio-economy, and to give play to the ability of human being of recognition of complicated systems.

## 4. THE RESULTS OF SIMULATION AND THE POLICY ANALYSIS

The medium-sized city of zhuzhou, the largest hub of communications in South China, is one of the heavy-industrial bases of Hunan Province, occupying an important place both in the province and in our country's overall developing strategy of step-by-step progression from the east to the west.

The following is some of the simulated results and the policy analysis cited in combination with some examples for research of the lomg-term planning of the Zhuzhou City. The model was programmed with Continuous System Simulation language (Micro-Dynamo ), then operated and debugged on an IBM-PC/XT micro-computer. The simulation time ranges from 1980 to 2000, and for the period from 1985 to 1995 an analysis of simulation precision has been made.

### (1). Selection Among The Patterns Of Urban Development Strategy

In connection with the overall target for 2000 of the Zhuzhou City to stive to complets the task by increasing the total output value of industry and agriculture by four-fold 2.5-3 years ahead of time, or in other words, by increasing it to 11400 million yuan in 1998, or earlier in 1997, and to satisfy the requirment realizing a personal average of national income over 2100 yuan and apersonal average of consumption lvel over 800 yuan, and through both a model comparason and a simulation analysis, the former was the three kinds of models for development strategy, that is, the speed planning type for preserving the economic structure now available, the type of structure for giving priority to light industry, and the type of compound coordination, and the latter was on the varied schems about investment structure, one of the optimum schemes for the developing model of the compound corrdinate type has been obtained.

### (2). Adjustment Of Investment Tactcs And Simulation Of Variations In Eco -Environmental Quality.

In the light of the characteristics of eco-environmental polution in the Zhuzhou City we have made a causility link within the system of polution by industrial economic effluent, and simulated the process of dynamic changes in eco-environmental quality caused by adjustment of the structural function of the system on some different tactics of investment in accordance with the target for 2000 of 0.50 thousand tons per ten thousand yuan( a load index of polution by industrial effluent) forecast by the exploitation model for the development of environmental protection in the Zhuzhou City. The quantitative variations in the main technical index when the optimum scheme for the exploitation model of the compound coordinate type was adopted, displayed fairly good results of simulation forecast. For instance, the yearly average of the relative error between the forecast value and the real value is below 3.8%.

(3). An Analysis Of Investment Tactics

Investment effect exercises a powerful influence on the development of the national economy, and on the improvement of the people's living standard as well . The simulation results make it clear that, if the delivery ratio of investment in fixed assets for heavy industry and light industry is increassed by 12.5% so as to bring the funds in play as early as possible, the total output value of industry will increase by 17.5 %, and the personal average of the national income will also increase by 14.8% As the proportion between accumulation and consumption influences immediately the production and the people's living standard, some suitable accmulation ratio should be chosen to guarantee the balanced development of both production and life and the comprehensive achievement of the three benefits of society,economy and environment. Again, the simulation results state clearly that an accumulation kept to 45% or so (Accmulation ratio of a heavy industrial city is generally higher than the country's average standard. ) may basically lead to simultaneous improvement of both production and living standard urder the present circumstance of the economic structure in the Zhuzhou City.

## 5. CONCLUDING REMARKS

The research for the long-term urban development strategy is one of extremely complicated systems engineering.To resolve these problems by taking advantage of the system dynamic, comparing with other methods to set up a parameter model and then to analyze the problems of the system by seeking the optimum value through the solution of a group of linear equations or nonlinear ones, possesses the following distinguishing features;

(1) .It seeks none of the optimum solutions and does not make statistical forecast, only constructs the behaviour of the system according to the structure of the microscopic causality link, thus the difficulty to set up a parameter model of nonlinear complicated system of higher degree such as socio-economy and to seek its solutions has been avoided.

(2). Facing the problems, and starting with the causality link, it adopts a structure of combinating type to set up the model. Some equations can be added, deleted and inlaid at will according to the different characters and significances of the problems, and it is convenent and flexible to adjust the scale of the model and to amend or combinate the overall model.

Figure 3.

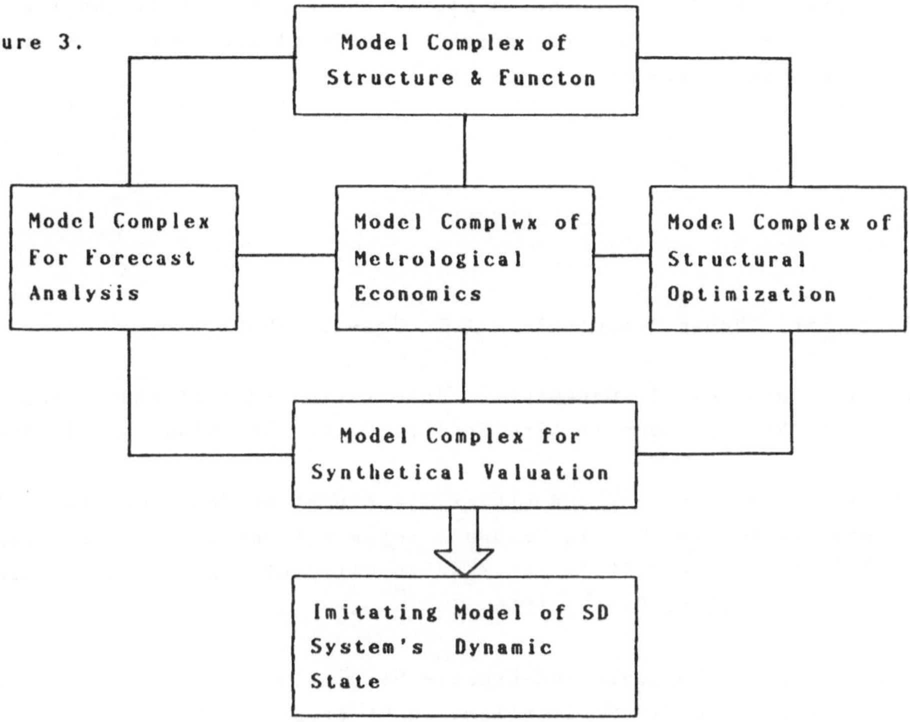


(3). The structure of the system decides its function and social behaviour. As the system dynamic sets up the model in accordance with the netted causality structure within a more complicated system, it is able to describe the system behaviour in a comoaratively authentic way, to obtain various kinds of index quantities at the same time, and to give a fairly good overall function to the model.

The aboved mentioned cistinguishing features have methodologically guaranteed the scientific and advanced character of research of the strategy for the long-term urban development by the system dynamics, while the limitations of the method itself have been compensated by combinating the model through adoption of many other methods(input output tech nique, economics, mathematical programming economic cybernetics and so on). The practics of the Research of the Long-Term Planning for Develop ment in the Zhuzhou City has indicated that the USES-SDM model for investment decision control supported on this kind of combinated model group(See Figure 3 ) possesses quite good accuracy, stability, feasibility and practicability, and has achieved the expected goal and the desired results when utilized to research the development strategy of the urban socio-economy-eco-environment system.

## REFERENCES

J.W.Forrester, "Urban Dynamics", M.I.T. Press, 1969.

Hu Yukui, "Systemical Dynamics", Forecast & Exploitation Company, Science & Technology Consultative Service Center of China, April, 1984.

Wang Yuji, Xiang Yuanwang and others, "A Report On The Model Technology Of Planning Of Development In Economy, Science,Technology And Society In The Period Of 1986 To 2000 In The Zhuzhou City",Office of the municipal Government of Zhuzhou, March, 1986.

Xiang Yuanwang, Xu Hongfu, "The Dynamic Simulation Of The Eco-Economic System In TheZhuzhou City", A Collection Of Treatises From The National Academic Discussion On The Mathematical Eco-Theory And Its Application, The Publishing House Of Wuhan University, Maech, 1987.

Wang Yuji, Xiang Yuanwang and others, "Systems Engineering For Regional Planning",Hunan University Publishing House, April,1986.

# APPLICATION OF MULTICRITERIA OPTIMIZATION TO STRUCTURAL SYSTEMS

Manfred Weck
Frank Förtsch
Laboratorium für Werkzeugmaschinen und Betriebslehre der RWTH Aachen
Lehrstuhl für Werkzeugmaschinen
Steinbachstr. 53/54, D - 5100 Aachen

## 1. Introduction

Systems for Structural Optimization have been around since the 60-ties and with the development of more effective methods of mathematical programming and optimality criteria their application has grown since then. Nearly all of these approaches consider the optimization problem as a problem of minimizing a single scalar function of the design parameters subject to several constraints /1, 2, 3, 4, 5/. The constraints as well as the objectiv function may include mechanical quantities like deformations, stresses, eigenfrequencies etc. or the material volume whereas the optimization parameters in general are quantities of geometry. As a matter of fact the process of finding an optimal design in general is not a scalar optimization problem but a vetor optimization problem because there are several objectives which are required to be optimal. This leads to the concept of multicriteria optimization, a subject which has received attention only recently /6, 7, 8, 9/.

In the following some basics of multicriteria optimization are briefly reviewed and two software-systems which are based on these foundations are described. Efficiency of the presented methods is discussed by means of the application of the multicriteria approach to structural examples.

## 2. Some basics of multicriterion optimization

The general multicriteria optimization problem may be formulated in the following way

$$\begin{aligned} &\text{Min} \quad \underline{f}(\underline{x}) \\ &\text{s. t.} \quad c_i(\underline{x}) = 0 \qquad && i = 1, 2, \dots, m' \\ &c_i(\underline{x}) \geq 0 \qquad && i = m'+1, m'+2, \dots m \end{aligned}$$

where $\underline{f}(\underline{x})$ is the vector-objective function containing the q criteria which have to be considered. The constraints may be formulated either as equality or inequality constraints or both.

The condition for optimality is defined by using the feasible set

$$\mathbb{D} = \left\{ \underline{x} \in R^n \;\middle|\; \begin{array}{l} c_i(\underline{x}) = 0,\ i = 1, 2, \ldots, m' \\ c_i(\underline{x}) \geq 0,\ i = m'+1, m'+2, \ldots, m \end{array} \right\}$$

Definition

A vector $\underline{x}^* \in \mathbb{D}$ is pareto-optimal for the vector-optimization problem if there is no vector $\underline{x}$ for which

$$\underline{f}(\underline{x}) \leq \underline{f}(\underline{x}^*)$$

holds with at least one component j with

$$f_j(\underline{x}) < f_j(x^*)$$

Obviously the last definition states that there is no unique solution $\underline{x}^*$ for which all the criteria are optimal simultaneously but there are a large number of pareto-optimal solutions in general. On the other hand if $\underline{x}^*$ is pareto-optimal it is not possible to decrease a criteria without increasing the value of at least one other. The multicriteria optimization problem to be considered is now to find the set of pareto optimal points and to decide which one of them to choose as the optimal design for the actual design task.

3. Methods for solving multicriteria optimization problems

One way to solve the vector optimization problem is to scalarize the vector function and solve the scalar optimization problem obtained through the use of mathematical programming. Out of the large number of methods only two shall be mentioned here.

The norm methods measure the distance between two points with a so called distance function of order p which is based on the lp-norm

$$g\ (\underline{f}(\underline{x})) = \left( \sum_{i=1}^{q} \left| \frac{f_i\ (\underline{x}) - f_i^*}{f_i^*} \right|^p \right)^{1/p}$$

with $1 < p < \infty$ . $f_i{}^*$ being arbitrary but commonly choosen as the single minimum of the function $f_i(\underline{x}^*)$. Omitting the denominator leads to an absolute measure of the distance whereas the given distance function must be considered as a relative one. A useful feature of this approach is the representation of the generally non-commensurable criteria in a non-dimensional way. Of course other denominators for this normalization are possible.

The weighting method may be viewed as a special form of the utility functions. Weighting coefficients are introduced in order to define the relative importance of the criteria.

$$g\ (\underline{f}(\underline{x}) = \sum_{i=1}^{q} w_i\ n_i\ f_i\ (\underline{x})$$

Usually they are normalized in the following way

$$\sum_{i=1}^{q} w_i = 1$$

The normalization factors $n_i$ again are arbitrary but often choosen in the way that

$$n_i\ f_i(\underline{x}^\circ) = 1$$

where $\underline{x}^\circ$ is the design vector of the starting design.

The graphic representation of both methods in picture 1 shows some basic features. Assuming the set

$$\mathbb{W} = \{\ \underline{z}\ |\ \underline{z} = \underline{f}(\underline{x});\ x \in \mathbb{D}\ \}$$

is convex. Then it is possible to generate all pareto-optimal points using the weighting method just by varying the weighting coefficients. In the nonconvex case the weighting method may not generate the complete pareto optimal set. On the other hand using norm-methods it is possible to compute pareto optimal solutions even in the nonconvex case. For different values of $f_i{}^*$ different pareto optimal points are generated but care must be taken because it may happen that for a bad choice of $f_i{}^*$ solutions are obtained which are not pareto optimal. Another point to be mentioned is that for a fixed point

$\underline{f}^* = \{ f_1^*, f_2^*, \ldots, f_q^* \}^T$ a variation of parameter p does not generate the complete pareto optimal set in general.

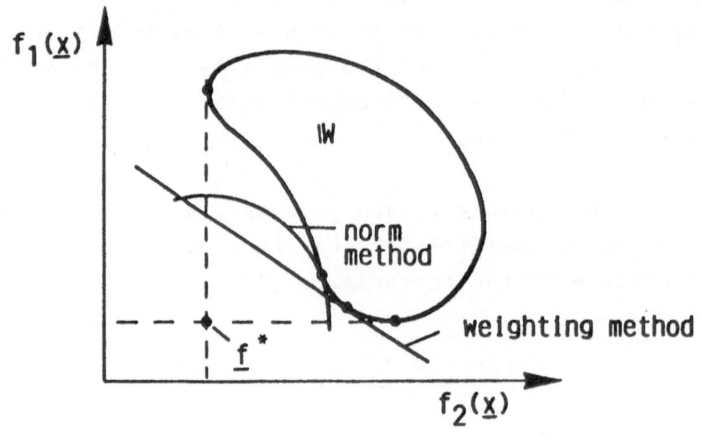


Fig. 1: Attainable set for a problem with two criteria

Eschenhauer and his co-workers /7, 8, 10/ examined in a comparative study combinations of methods for the transformation of the vector optimization problem into a scalar formulation and mathematical programming methods. It was shown that there is no optimal algorithm for any type of structural multicriteria optimization problem. In this approach after some test the weighting method was applied. It can be used very easily and allows an easy interpretation of the obtained results which is of great importance for its practical use.

The so obtained scalar optimization problem is solved by an SQP-Algorithm developed by Powell /11, 12/. This methods solves a quadratic subproblem with linearized constraints. The Hessian of the Lagrange function of the problem is recursively approximated by a modified BFGS-formula using the first derivatives of the Lagrange function.

## 4. Calculation of mechanical quantities and their derivatives

For the calculation of the mechanical quantities like deformations, eigenfrequencies etc. the finite element method (FEM) is used. For static loads the deformations will be the solution of the equation system

$$\underline{K}\ \underline{u} = \underline{F}$$

If the material law is linear elastic and if no other nonlinearities are included this is a linear equation system. The derivatives of the deformations with respect to the optimization parameters are derived by differentiation of the given formula

$$\underline{K} \frac{\partial \underline{u}}{\partial \underline{x}} = - \frac{\partial \underline{K}}{\partial \underline{x}} \underline{u} + \frac{\partial \underline{F}}{\partial \underline{x}}$$

where the right handside of the equation can be viewed as a speudo load.

Calculation of the eigenfrequencies leads to the eigenproblem

$$( - \omega^2 \underline{M} + \underline{K}) \underline{y} = 0$$

and the derivatives with respect to the optimization parameters will be obtained by

$$2\omega \frac{\partial \omega}{\partial \underline{x}} = \underline{y}^T (- \omega^2 \frac{\partial \underline{M}}{\partial \underline{x}} + \frac{\partial \underline{K}}{\partial \underline{x}}) \underline{y}$$

using the well known normalization

$$\underline{y}^T \underline{M} \underline{y} = 1$$

Especially in dynamic analysis substructuring and static condensation to decrease the computational effort are often used. Having master and slave degrees of freedom stiffness and mass matrices are partitioned and with the transformation matrix of static condensation

$$\underline{L} = \underline{K}_{ii}^{-1} \underline{K}_{ig}$$

where i is the index of slave degrees of freedom and g is the index of master degrees of freedom the reduced matrices are obtained in the following way

$$\underline{K}_k = (\underline{K}_{gg} - \underline{K}_{gi} \underline{L})$$

and $$\underline{M}_k = \underline{L}^T \underline{M}_{ii} \underline{L} - \underline{M}_{gi} \underline{L} - \underline{L}^T \underline{M}_{ig} + \underline{M}_{gg}$$

The gradients of these matrices are obtained by

$$\frac{\partial \underline{K}_k}{\partial \underline{x}} = \frac{\partial \underline{K}_{gg}}{\partial \underline{x}} - \frac{\partial \underline{K}_{gi}}{\partial \underline{x}} \underline{L} - \underline{K}_{gi} \frac{\partial \underline{L}}{\partial \underline{x}}$$

and

$$\frac{\partial \underline{M}_k}{\partial \underline{x}} = \frac{\partial \underline{L}^T}{\partial \underline{x}} (\underline{M}_{ii} \underline{L} - \underline{M}_{ig}) + (\underline{L}^T \underline{M}_{ii} - \underline{M}_{gi}) \frac{\partial \underline{L}}{\partial \underline{x}} +$$

$$- \underline{L}^T \frac{\partial \underline{M}_{ig}}{\partial \underline{x}} + \underline{L}^T \frac{\partial \underline{M}_{ii}}{\partial \underline{x}} \underline{L} - \frac{\partial \underline{M}_{gi}}{\partial \underline{x}} \underline{L} + \frac{\partial \underline{M}_{gg}}{\partial \underline{x}}$$

## 5. Software system for multicriteria structural optimization

Two software systems have been developed. One serves for the multicriteria optimization of spindle-bearing systems /13, 14/ whereas the other system is based on a general dynamic analysis finite element program meaning problems of a more general type can be handled /14/.

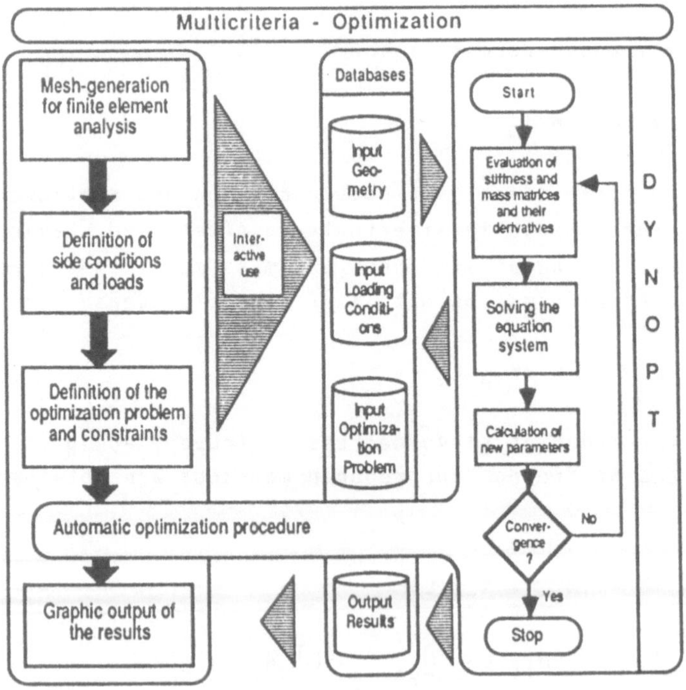


Fig. 2: Software system for multicriteria structural optimization

Picture 2 shows the optimization procedure for the latter case but the system for multicriteria optimization of spindle bearing systems is similar.

When optimizing a structure the first step is to build up the finite element mesh of the part which is to be optimized together with side conditions and loads. In the second step the optimization problem is formulated. This means the choice of criteria to be optimized, their weighting coefficients and the parameters of the geometry which serve as optimization parameters. After having defined the optimization problem and after preparation of the finite element mesh which is done interactivly the automatic optimization procedure is started. It runs with no further interaction by the user and stops automatically if the optimal solution is achieved. The procedure contains the finite element analysis with evaluation of the gradients of the mechanical quantities and the calculation of new optimization parameters by the optimization algorithm together with a check for convergence imbeded in an iterative circle. After the optimization procedure has stopped the user may wish to have a graphic output of the results.

## 6. Application of multicriteria optimization to practical problems

### 6.1 Optimization of a spindle bearing system

In the design of a machine tools spindle bearing systems play an important role. Minimizing the weight of the spindle helps to decrease the drive power and the costs connected with that. A good static and dynamic behaviour on the other hand must be realized in order to achieve sufficient machining quality. With these objectives the task to find an optimal design of a spindle is a multicriteria optimization problem. Picture 3 shows the principal drawing of the spindle bearing system which is to be optimized.

The optimization parameters defined include inner and outer diameters and the length of the spindle sections lying between the two bearings. Because of constructive reasons no geometric modifications are allowed at the spindle head or the spindle end. In order to achieve a reasonable solution the inner diameter is bounded by 0 and of course must not be bigger than the outer diameter. As the bearings are outside the geometry which may be varied, no constraints have to be established which allow for assembly of the system. However, it is possible to do so. The outer diameters are restricted to 200 mm.

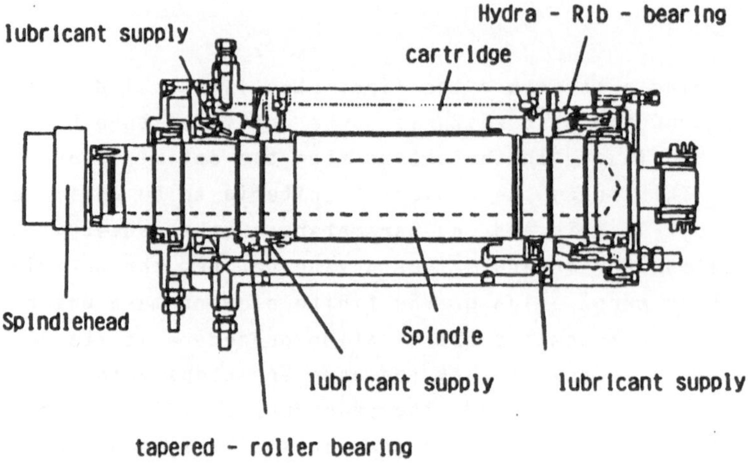


Fig. 3: Principal drawing of a spindle bearing system

In the first step an optimization was carried out with only one criteria: the static response at the spindle head. As an additional constraint the material volume was restricted to that of the starting solution. Picture 4 shows the optimized geometry in the upper half which has a stairlike look. The static deformation could be reduced by 15.8 %.

Optimization in order to maximize the first eigenfrequency of the system together with upper bounds on the material volume and the static deformation leads to an optimized spindle as shown in the middle of picture 4. In this case the distance between the bearings is longer than that in the first optimization. The first eigenfrequency could be increased by about 5.5 %. The reason for the small increase in the first eigenfrequency is that the mode shape of this frequency is a bending of the spindle head. However, for constructive reasons the geometry of the spindle head was fixed and therefor its' mass could not be variied.

The optimization with respect to minimum material volume together with bounds on the static deformations leads to the spindle shown in the lower part of picture 4. The material volume could be reduced by 20.1 %.

I MINIMAL STATIC RESPONSE AT POINT A
( ADDITIONAL CONSTRAINT : MATERIALVOLUME )

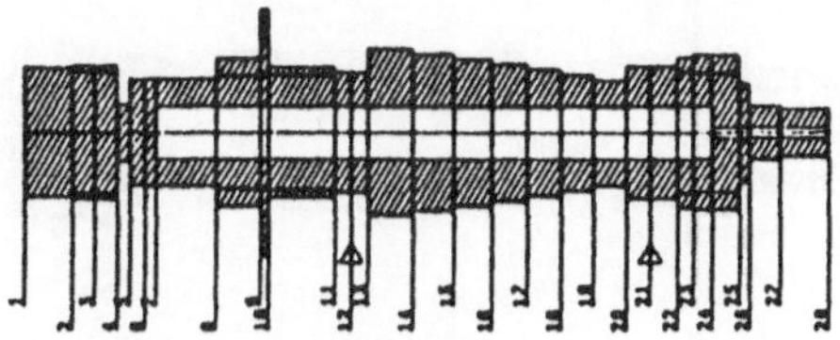

II MAXIMUM FIRST EIGENFREQUENCY
( ADDITIONAL CONSTRAINTS : STATIC DEFORMATION )

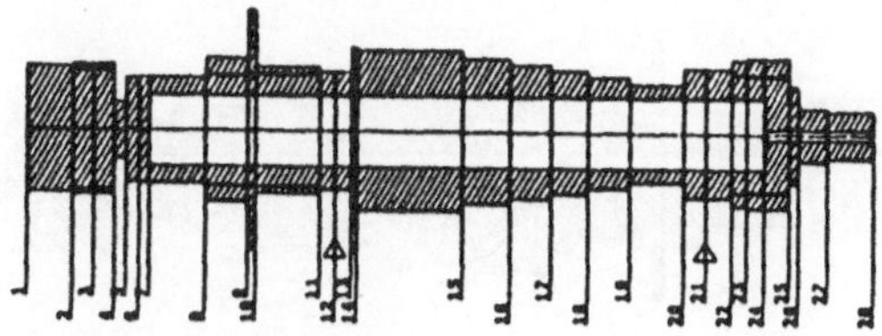

III MINIMUM MATERIALVOLUME
( ADDITIONAL CONSTRAINTS : STATIC DEFORMATION )

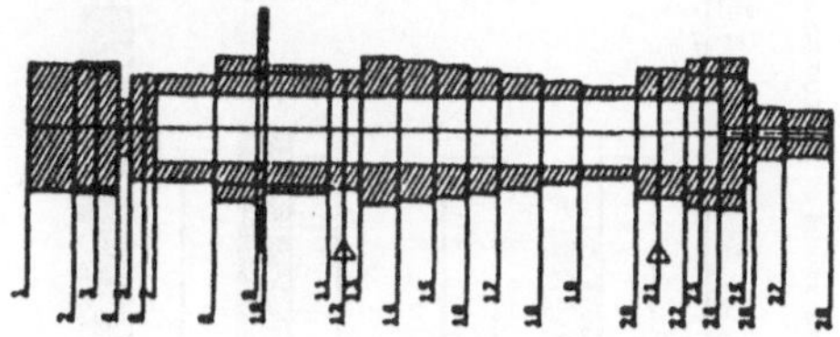

Fig. 4: Results of the optimization with a single criteria

Optimizing for all criteria simultaneously using the weighting method leads to different results depending on which weighting coefficients are choosen. Picture 5 shows the results for an optimization calculation with most relative importance given to the static response at the spindle head (40 %) and at the middle of the spindle (40 %) whereas the material volume and the first eigenfrequency was given only 10 % relative importance each. Corresponding to that the static response could be reduced by 27,4 % at the spindle head and 99,94 % at the middle of the spindle. The latter is in fact of no importance because this deformation is very small compared with that at the spindle head even for the starting geometry. To obtain this result the material volume was increased by 57.2 %. The first eigenfrequency could be increased by 10.9 %.

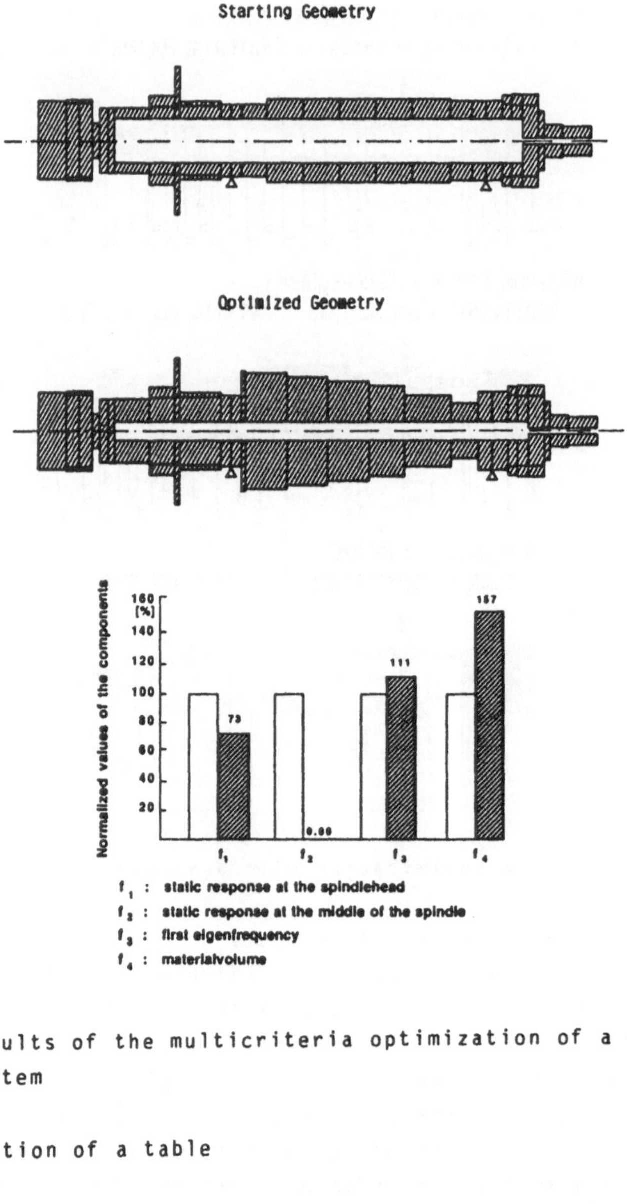


Fig. 5: Results of the multicriteria optimization of a spindle bearing system

6.2 Optimization of a table

The secound example - a table - was calculated with the system DYNOPT. Picture 6 shows the finite-element model of the table consisting of 4-noded plane shell elements and beam elements with circular cross sectional area. The static load is acting at the center of the table. For the opitmization only a quarter of the table has been calculated using the symmetry of the problem. The nodes marked by circulars have been completely condensed using static condensation whereas the nodes marked by black points remain completly in the model. For this optimization the 4th eigenfrequency has been considered. The mode shape is a plate bending mode of the table. As the static behaviour as well as the mode shape is symmetric with respect to the two symmetry planes the optimization can be carried out using the model, shown in the lower part of picture 6.

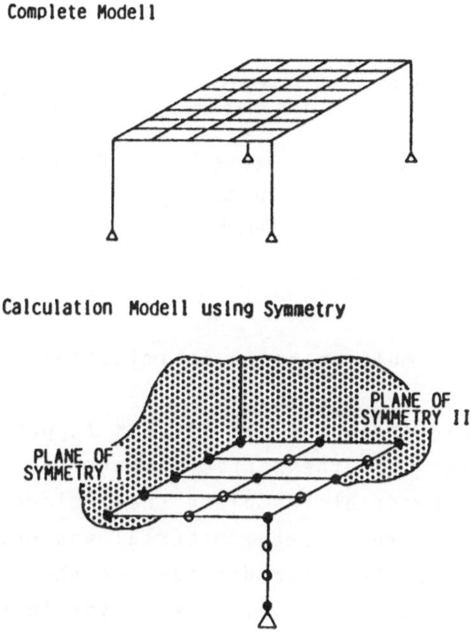


Fig. 6: Finite element modell of a table

The thickness of the shell elements are defined as optimization parameters as well as the inner and outer diameters of the beam elements giving a total of 14 optimization parameters. The static deformation at the center of the table and the material volume are to be minimized whereas the eigenfrequency was maximized. Lower bounds have been defined for the thickness as well as for the inner diamters. The difference between outer and inner diameter was bounded too. The calculation was carried out with different weighting coefficients /14/. Picture 7 shows the results for two selected weightings.

Giving relative importance to the static response at the center of the table (80 %) leads to a decrease in that omponent of 91,4 % and an

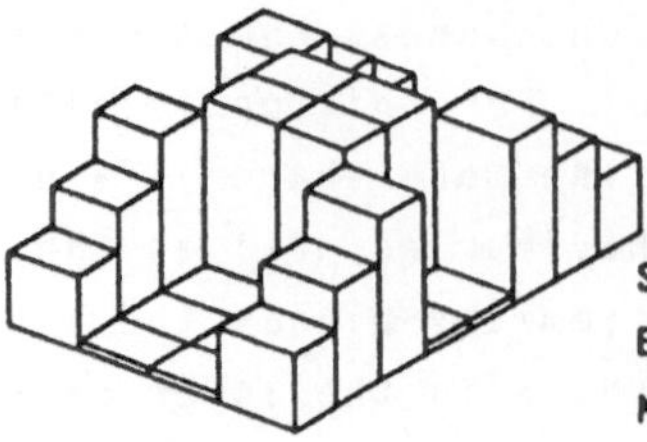

| Criteria | Weight | Relative Performance |
|---|---|---|
| Static Deformation | ( 0.8 ) | 8.6% |
| Eigenfrequency | ( 0.1 ) | 224.5% |
| Materialvolume | ( 0.1 ) | 190.6% |

| Criteria | Weight | Relative Performance |
|---|---|---|
| Static Deformation | ( 0.1 ) | 145.5% |
| Eigenfrequency | ( 0.1 ) | 101.6% |
| Materialvolume | ( 0.8 ) | 61.3% |

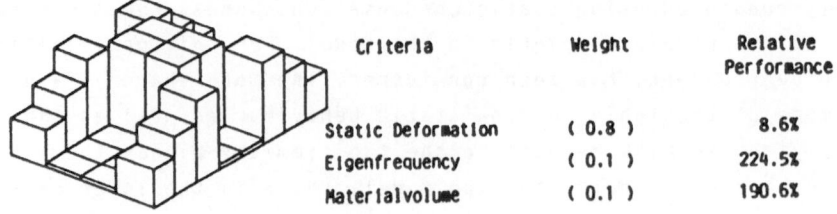

Fig. 7: Results of the multicriteria optimization of a table

increase in the 4th eigenfrequency of 124.5 % (upper part of picture 7). This is due to the fact that the static response and the mode shape look similar. The thickness of the plane shell elements are plotted normal to the table. As can be seen, material was put on the side and the center of the table. The inner diamters of the beam elements were set to zero whereas the outer diameters were set to the upper bound. Thus the material volume was increased from 100 % of the starting geometry to 190.6 %. A decrease of 38.7 % in that component was obtained

References


/1/ Ward, P., Patel, D., Wakeling, A., Weeks, R.: Application of Structural Optimization using Finite Elements. Computer Aided Optimal Design Structural and Mechanical Systems, ed. C.A. Mota Soares, Springer Verlag, Berlin, 1987.

/2/ Steinke, P.: Verfahren zur Spannungs- und Gewichtsoptimierung von Maschinenbauteilen. Fortschritt-Berichte VDI-Z, Reihe 1, Nr. 107, VDI-Verlag, Düsseldorf, 1983.

/3/ Hörnlein, H.R.E.M.: Take-off in Optimum Structural Design Computer Aided Optimal Design: Structural and Mechanical Systems. ed. A. Mota Soares, Springer-Verlag, Berlin, 1987.

/4/ Berke, L., Khot, W. S.: Structural Optimization using Optimality Criteria. Computer Aided Optimal Design: Structural and Mechanical Systems, ed. A. Mota Soares, Springer-Verlag, Berlin, 1987.

/5/ Weck, M.,Förtsch, F.: Spannungsoptimierung offener Ausrundungen in Maschinenbauteilen. Konstruktion 38, Heft 6, 1986, S. 213 - 219.

/6/ Stadler, W.: Multicriteria Optimization in Mechanics (A Survey). Applied Mechanics Reviews Vol. 37, No. 3, March 1984.

/7/ Eschenhauer, H. W.: Multicriteria Optimization Procedures in Application on Structural Mechanics Systems. Vortrag anläßlich der International Conference on Vector Optimization in Darmstadt, Aug. 1986.

/8/ Eschenhauer, H.: Rechnerische und experimentelle Untersuchungen zur Strukturoptimierung von Bauweisen. Forschungsbericht des Instituts für Mechanik und Regelungstechnik der Universität-Gesamthochschule Siegen, 1985.

/9/ Baier, H.: Mathematische Programmierung zur Optimierung von Tragwerken insbesondere bei mehrfachen Zielen. Dissertation, Universität Darmstadt, 1978.

/10/ Sattler, H. J.: Ersatzprobleme für Vektoroptimierungsaufgaben und ihre Anwendung in der Strukturmechanik. VDI-Fortschritt-Berichte Reihe 1, Nr. 88, VDI-Verlag, Düsseldorf, 1982.

/11/ Powell, M. J. D.: A Fast Algorithm for Nonlinearly Constrained Optimization Calculations. in Lecture Notes in Mathematics No. 630, ed. G. A. Watson, Springer-Verlag, Berlin, 1978.

/12/ Powell, M. J. D.: Extensions to Subroutine VF02AD. Proceedings of the 10th IFIP Conference on System Modelling and Optimization. Lecture Notes in Control and Information Sciences, Springer-Verlag, Berlin, 1982.

/13/ Weck, M., Förtsch, F.: Multikriterienoptimierung von Spindel-Lager-Systemen. VDI-Z, Bd. 129, Heft 10, 1987, S. 113 - 116.

/14/ Förtsch, F.: Entwicklung und Anwendung von Methoden zur Optimierung des mechanischen Verhaltens von Bauteilen. Manuscript for dissertation, Aachen, 1987. to be published

# APPLICATION OF OPTIMIZATION METHODS IN STRUCTURAL SYSTEMS RELIABILITY THEORY

P. Thoft-Christensen
University of Aalborg
Aalborg, Denmark

## 1. INTRODUCTION

Structural systems reliability theory and structural optimization theory are two areas where impressive progress has taken place in the last two decades. This progress is first of all based on the development of fast and relatively inexpensive computers, but the incitement to do research in this area is also due to the need for design of structures which are cheaper and safer. With the recent advances a rational design of a large set of structures, e.g. offshore steel structures, is possible.

In this paper the application of optimization methods in structural systems reliability theory is discussed. The emphasis is put on applications to optimal design and to optimal maintenance strategies. Some previously published examples will be briefly mentioned and some new results on shape optimization will be discussed in detail.

In section 2 a brief presentation of some of the most important formulas in structural systems reliability theory are presented and in section 3 non-linear optimization algorithms are briefly discussed. Optimal design with special emphasis on shape optimization is treated in section 4. The possibility of using expert knowledge is also mentioned in section 4. In section 5 derivation of optimal inspection and repair strategies is discussed.

## 2. RELIABILITY OF STRUCTURAL SYSTEMS

In the books by Thoft-Christensen & Baker [1] and Thoft-Christensen & Murotsu [2] a detailed description of the reliability of structural elements and structural systems is given. In this section only some of the most important concepts and formulas are presented.

Let the basic variables (e.g. material properties, load parameters and geometrical quantities) be $\bar{X} = (X_1, \ldots, X_n)$, where $X_i$, $i = 1, \ldots, n$ are random variables.

Assume that $\bar{X}$ is transformed into a set of independent standardized normally distributed variables $\bar{Z} = (Z_1, \ldots, Z_n)$. For a single element the failure with regard to a certain failure mode is determined by the failure function $f(\bar{z})$ which divides the $\bar{z}$-space into a failure region ($f(\bar{z}) \leq 0$) and a safe region ($f(\bar{z}) > 0$). The reliability index $\beta$ is then defined by

$$\beta = \min_{f(\bar{z}) = 0} (\bar{z}^T \bar{z})^{\frac{1}{2}} \tag{1}$$

This non-linear optimization problem with only one constraint can easily be solved by the well-known optimization methods (see e.g. Thoft-Christensen [3]). The probability of failure $P_f$ for the element in question with regard to the failure mode considered is equal to the probability that outcomes of $\bar{z}$ are in the failure region. It can be shown that

$$P_f \approx \Phi(-\beta) \tag{2}$$

where $\Phi$ is the standard normal distribution function. $M = f(\bar{Z})$ is called the safety margin.

For structural systems a number of different failure definitions has been suggested (see e.g. Thoft-Christensen [4]). Here, only estimation of the structural systems reliability at level 1 will be mentioned. Let a structural system consist of m single failure elements (failure modes). Then at level 1 the structural system is considered to be in a state of failure when a single failure element fails. A good estimate of the corresponding probability of failure $P_f^s$ is (see Thoft-Christensen [5])

$$P_f^s = 1 - \Phi_m(\bar{\beta}; \bar{\bar{\rho}}) \tag{3}$$

where $\Phi_m$ is the m-dimensional standard normal distribution function, $\bar{\beta} = (\beta_1, \ldots, \beta_m)$ and $\bar{\bar{\rho}}$ is the correlation matrix for the safety margins. $\beta_i$, $i = 1, \ldots, n$ is the reliability index of failure element i. By using (2) a formal systems reliability index $\beta^s$ can be defined.

## 3. OPTIMIZATION METHODS

In the applications described in sections 4 and 5 a general non-linear programming algorithm NLPQL implemented by Schnittkowski [6] (see also Sørensen [7]) is used. This algorithm is based on an optimization method by Han, Powell & Wilson (see [6]) which has proved to be very effective for optimal design and optimal strategy problems (see [7]). It is an iterative method where each iteration step consists of two parts. First, a search direction is determined, and then a line search is performed. The search direction is determined by solving a quadratic optimization problem formed by a quadratic approximation of the Lagrangian function of the non-linear problem and a linearization of the constraints of the current design point. The line search is performed using an augmented Lagrangian merit function.

In principle any non-linear optimization method can be used but often convergence problems occur. The most important ones are described in detail by Vanderplaats [8].

## 4. OPTIMAL DESIGN OF STRUCTURES

Reliability-based optimal design of structures has been treated by several authors, see e.g. Murotsu [9], [10], Frangopol [11], [12] Thoft-Christensen & Sørensen [13], [14] and Thoft-Christensen & Murotsu [2]. In the last-mentioned book an extensive number of references to reliability-based optimal design can be found.

Initially in this section the optimal design problem is briefly stated and illustrated with an example taken from [13]. Next some new results on reliability-based shape optimization are presented. These results are based on an M.Sc. thesis by Frisk & Poulsen [15].

In reliability-based optimization the objective function is often chosen as the weight F of the structure. The constraints can either be related to the reliability of the single elements or to the reliability of the structural system. In the last-mentioned case the optimization problem for a structure with h elements is

$$\min F(\bar{y}) = \sum_{i=1}^{h} \varphi_i l_i A_i(\bar{y})$$

$$\text{s.t.} \quad \beta^s(\bar{y}) \geq \beta_o^s \tag{4}$$

$$y_i^l \leq y_i \leq y_i^u \qquad i = 1, 2, \ldots n$$

where $A_i$, $l_i$ and $\varphi_i$ are the cross-sectional area, the length, and the density of element no.i. $\bar{y} = (y_1, \ldots, y_n)$ are the design variables. $\beta_o^s$ is the target systems reliability index. $y_i^l$ and $y_i^u$ are lower and upper bounds for the design variable $y_i$, $i = 1, \ldots, n$.

Example 1:

Consider the three-dimensional truss model of a steel-jacket offshore structure shown in figure 1. The load and the geometry are described in detail in [16] (see also [13]). The load is modelled by two random variables and the yield capacities of the 48 truss elements are modelled as random variables with expected values $270 \times 10^6$ $\mathrm{Nm}^{-2}$ and coefficients of variation 0.15. The correlation structure of the normally distributed variables is described in [16]. In this example the design variables $y_i$, $i = 1, \ldots, 7$ are the cross-sectional areas ($\mathrm{m}^2$) of the seven

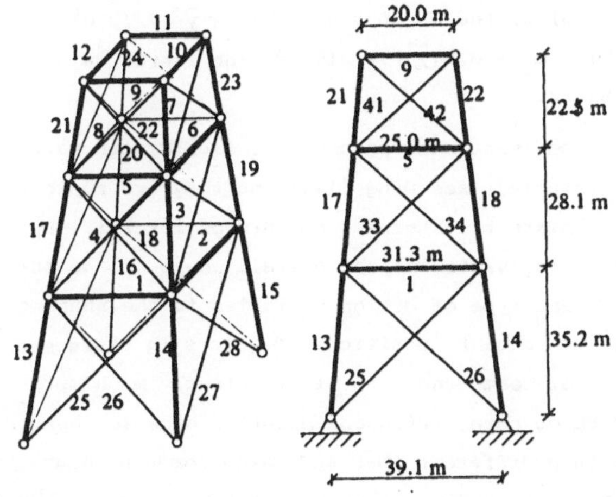


| Area | Elements |
|---|---|
| $y_1$ | 1, ..., 4 |
| $y_2$ | 5, ..., 8 |
| $y_3$ | 9, ..., 12 |
| $y_4$ | 13, ..., 24 |
| $y_5$ | 25, ..., 32 |
| $y_6$ | 33, ..., 40 |
| $y_7$ | 41, ..., 48 |

Figure 1. Space truss tower with design variables.

structural elements (see figure 1). The optimization problem is

$$\begin{aligned} &\text{min. } F(\bar{y}) = 125y_1 + 100y_2 + 80y_3 + 384y_4 + 399y_5 + 319y_6 + 255y_7 \ (m^3) \\ &\text{s.t.} \qquad \beta^s(\bar{y}) \geqq \beta_o^s = 3.00 \\ &\qquad 0 = y_i^\ell \leqq y_i \leqq y_i^u = 1 \end{aligned} \tag{5}$$

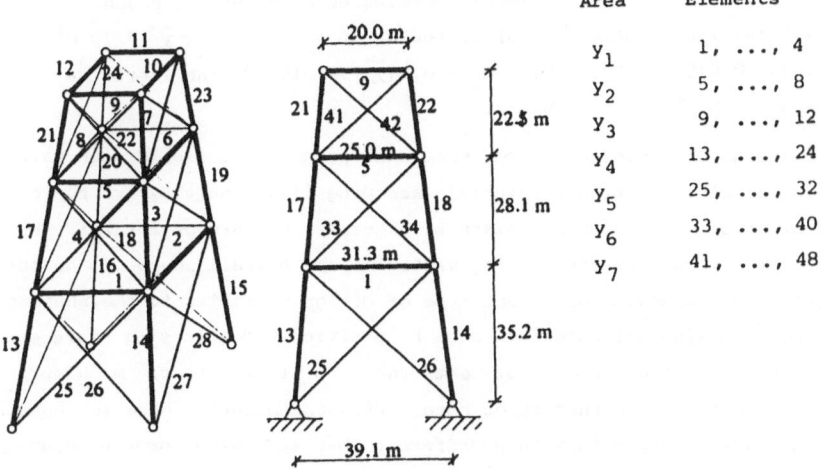

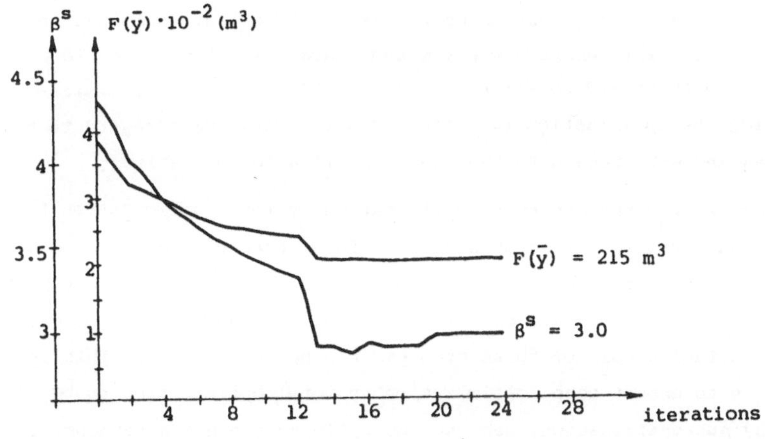


Figure 2. Iteration history

$\beta^S$ is the systems reliability index at level 1 evaluated by using the Hohenbichler approximation. The solution of (5) by the NLPQL algorithm is $\bar{y}^* = (0.01, 0.001, 0.073, 0.575, 0.010, 0.009, 0.011)$ m$^2$ and $F(\bar{y}^*) = 215$ m$^3$. The iteration history for $F(\bar{y})$ and $\beta^S$ is shown in figure 2.

In example 1 the design variables are cross-sectional parameters. Such design variables are often called sizing design varaiables (see Ding [17]) and they are typically cross-sectional areas of beams or bars, plate thicknesses, moments of intertia. During optimization of a structure with sizing variables the overall shape of the structural system remains unchanged. A different type of design variables is the shape design variables. Shape design variables can e.g. be positions of joints in truss structures (coordinate variables) or lengths of beams and bars. It is clearly much more complex to do shape optimization than to do pure sizing optimization because changing the shape of the structure may result in a different load and, even for non-redundant structures, a redistribution of internal forces. Structural optimization is therefore usually performed by choosing a given shape of the structure and then do sizing optimization. Even in classical deterministic structural optimization it is not all that simple to perform shape optimization.

To the author's knowledge the work by Frisk & Poulsen [15] is the first reliability-based shape optimization investigation. In the same work it is emphasized that it is useful to combine shape optimization with expert knowledge techniques.

It is useful to divide the different methods used to solve shape optimization problems in at least two different approaches, namely

- shape optimization by total optimization
- shape optimization by sub-optimization

In shape optimization by total optimization the sizing design variables $\bar{y} = (y_1, \ldots, y_e)$ and the shape design variables $\bar{z} = (z_1, \ldots, z_m)$ are combined in a design vector $\bar{x} = (x_1, \ldots, x_n)$, where $n = e + m$ and where the optimization is performed simultaneously with regard to all n design variables. In shape optimization by sub-optimization the optimization is performed alternately with regard to the sizing design variables and with regard to the shape optimization variables.

It is important to bear in mind that in shape optimization by total optimization two different types of design variables are combined. Therefore, this can result in convergence problems when the magnitude of the corresponding gradients differs very much or if the design variables themselves are of different order of magnitude. However, by scaling the design variables some of these problems can be reduced. When this is done it seems reasonable to expect shape optimization by total optimization to be more attractive than by sub-optimization, because sub-optimization has a tendency to find non-global optimum points. In total optimization it will only depend on the optimization algorithm whether or not a global optimum is obtained.

In example 2 (taken from [15]) an optimal design problem with 6 shape design variables and 4 sizing design variables is solved. The optimization method used is based on the NLPQL algorithm and total optimization is used.

Example 2:

Consider the simple frame structure in figure 3 made of tubular steel members. The loading is two uncorrelated loads $P_1$ and $P_2$. $P_1$ and $P_2$ are normally distributed with $\mu_{P_1} = 25$ kN/m$^2$, $\sigma_{P_1} = 5$ kN/m$^2$ and $\mu_{P_2} = 10000$ kN, $\sigma_{P_2} = 500$ kN. The horizontal load is concentrated in the joints and is assumed to be proportional to the diameter of the relevant tubular member (see [15]). The yield stress Y is assumed to be normally distributed with $\mu_Y = 3400 \cdot 10^2$ kN/m$^2$ and $\sigma_Y = 340 \cdot 10^2$ kN/m$^2$ and the correlation coefficient between the members is $\rho = 0.3$. The modulus of elasticity for the steel material is deterministic and equal to $2 \cdot 10^8$ kN/m$^2$.

10 design variables are considered, namely 6 shape variables $x_1, \ldots, x_6$ and 4 sizing variables $x_7, \ldots, x_{10}$. The upper horizontal beam has a constant length and cross-sectional area. The structure has 19 tubular members and each of them 3 failure elements (failure modes), namely a yield failure element at the end of the beams and a stability failure element. The effective length of the beams is 0.6 times the total length.

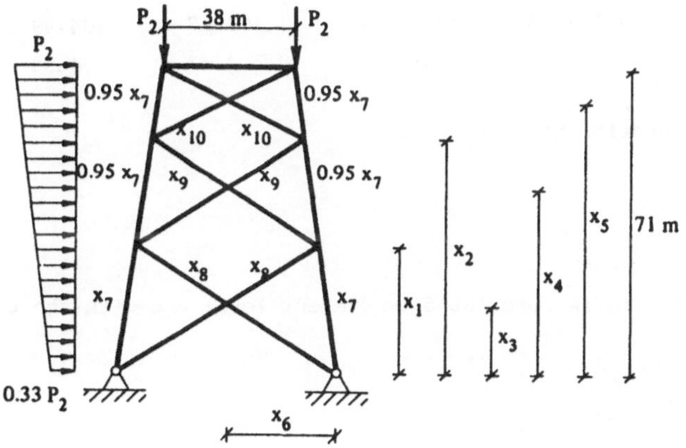


Figure 3. Frame structure with 6 shape design variables $x_1, \ldots, x_6$ and 4 sizing design variables $x_7, \ldots, x_{10}$.

| k | $x_k^\ell$ | $x_k^u$ | $x_k$-start | $x_k$-end | Dimension |
|---|---|---|---|---|---|
| 1 | 0.001 | 35 | 20 | 21.34 | m |
| 2 | 36 | 70.99 | 45 | 48.84 | m |
| 3 | 0.001 | 35 | 15 | 14.15 | m |
| 4 | 20 | 60 | 35 | 34.14 | m |
| 5 | 45 | 70.99 | 55 | 57.45 | m |
| 6 | 16 | 40 | 20 | 18.29 | m |
| 7 | 0.05 | 0.09 | 0.09 | 0.0694 | $m^2$ |
| 8 | 0.0001 | 0.04 | 0.03 | 0.0209 | $m^2$ |
| 9 | 0.0001 | 0.03 | 0.03 | 0.0164 | $m^2$ |
| 10 | 0.0001 | 0.03 | 0.015 | 0.0074 | $m^2$ |

Table 1.

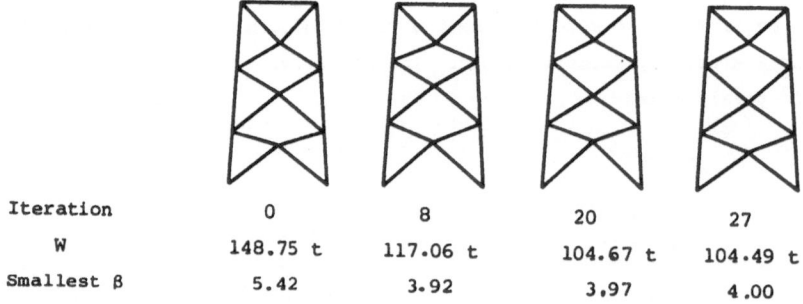

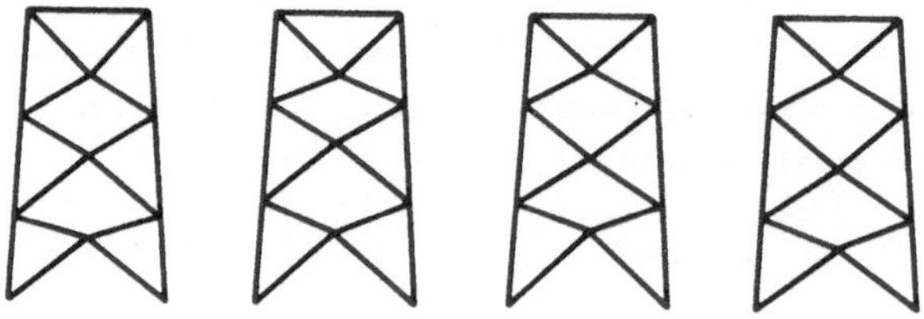

| Iteration | 0 | 8 | 20 | 27 |
|---|---|---|---|---|
| W | 148.75 t | 117.06 t | 104.67 t | 104.49 t |
| Smallest β | 5.42 | 3.92 | 3.97 | 4.00 |

Figure 4. Iteration history.

The optimization problem is formulated on element level and with the weight as objective function

$$\begin{aligned} \text{min. } & W(\bar{x}) = 7.85 \sum_{i=1}^{19} l_i(\bar{x})\, a_i(\bar{x}) \\ \text{s.t. } & \beta_j(\bar{x}) \geq 4.00,\ j = 1, \ldots, 38 \quad \text{yielding} \\ & \beta_j(\bar{x}) \geq 4.00,\ j = 39, \ldots, 57 \quad \text{stability} \\ & x_k^\ell \leq x_k \leq x_k^u,\ k = 1, \ldots, 10 \end{aligned} \qquad (6)$$

where $\bar{x} = (x_1, \ldots, x_{10})$, $l_i$ is the length of beam i, $a_i$ is the cross-sectional area of beam i, and $\beta_j$ is the reliability index of failure element j. Upper and lower bounds for $x_k$, k = 1, ..., 10 are shown in table 1. In the same table initial values for $\bar{x}$ and the result of the solution of (6) are also shown. The start value of W is 148.75 tons and the smallest reliability index for any failure element is $\beta$ = 5.42. After 27 iterations with the NLPQL algorithm the optimal values of $\bar{x}$ are as shown in table 1 ($x_k$-end). The minimum weight is W = 104.49 tons and the smallest $\beta$-value is $\beta$ = 4.00. This lowest acceptable reliability index $\beta$ = 4.00 is obtained for 7 stability failure elements. The shape of the structure in the initial state (iteration 0), after 8 iterations, after 20 iterations and the optimal shape (after 27 iterations) is shown in figure 4. The computer time was 108 CPU sec. on a VAX 8700 computer.

The same optimization problem has been solved using 6 different sets of start values for the design variables. The initial weight of the structure and the smallest $\beta$-value for these different computer runs are in the intervals [75.22 tons, 148.75 tons] and [-9.70, 5.42]. All 6 computer runs gave the same optimum solution and the computer times were in the interval [79 sec., 261 sec.]

It is obvious from figure 4 that the optimal solution from an economic point of view is unfavourable. It is expensive to produce the 3 tubular joints in the symmetry line. This result is typical for shape optimization of structures where the weight is used as an objective function. It is, however, not expedient to reformulate the optimization problem so that the production costs of e.g. tubular joints are included. The objective function will namely in such a case be very complicated. It seems to be much more natural to use expert knowledge in the way described below. As a simple example of expert knowledge consider the brace in figure 5, where, depending on the position of the joint, a K-brace or an X-brace is considered most economic. For the optimal shape in figure 4 application of this simplified expert knowledge will result in a structure, where the upper two braces become X-braces and the lowest brace becomes a K-brace. Shape optimization with application of expert knowledge is illustrated in example 3, taken from [15].

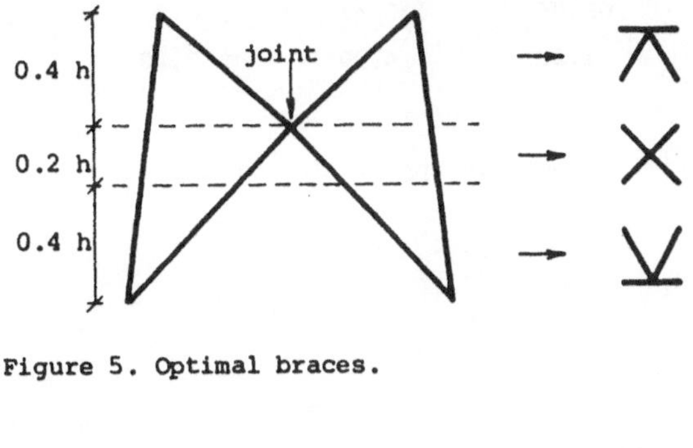


Figure 5. Optimal braces.

<u>Example 3</u>:

The same problem as in example 2 is considered again, but now the expert knowledge illustrated in figure 5 is included. The result of the optimization is shown in figure 6. The states 1 and 2 are identical with the initial state and the optimal state in figure 4. In state 3 the middle brace is fixed as an X-brace. By continued iteration state 4 is then obtained. Next the lowest brace is fixed as a K-brace (state 5) and by renewed optimization state 6 is obtained. Finally the upper brace is fixed as an X-brace (state 7) and by renewed optimization state 8 is obtained.

In figures 7 and 8 the variation of the smallest β-index and the weight W during the iteration are shown. In the optimal design (state 8) only 4 failure elements have β = 4.00.

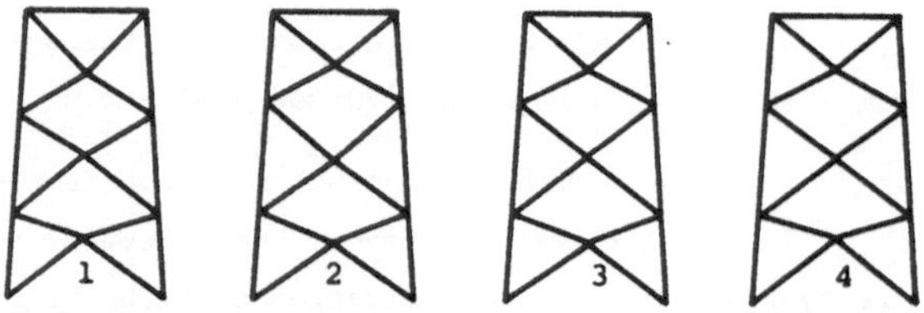

| | | | | |
|---|---|---|---|---|
| State | 1 | 2 | 3 | 4 |
| Weight (tons) | 148,25 | 104.49 | 104.47 | 104,94 |
| Smallest β | 5.42 | 4.00 | 3.43 | 4.00 |

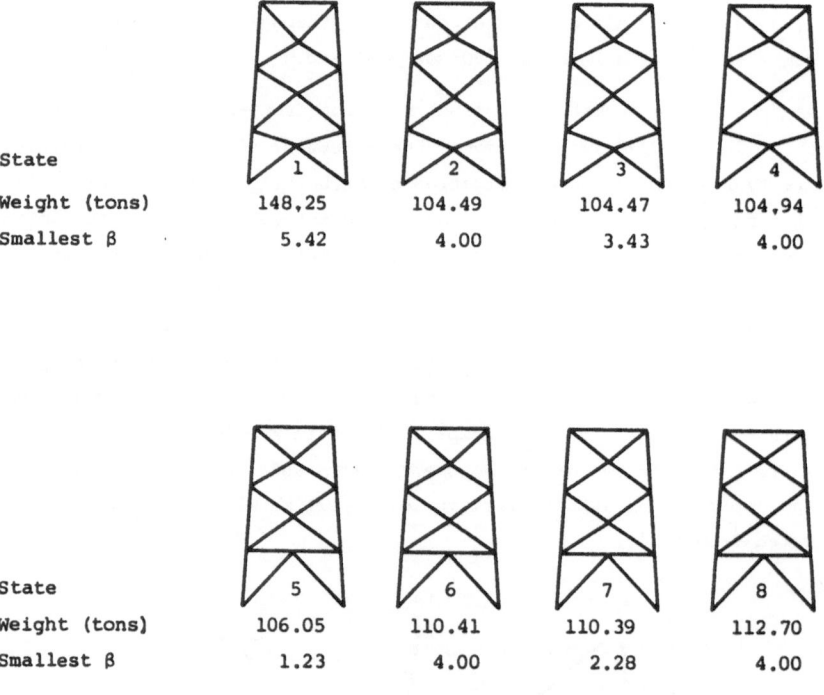

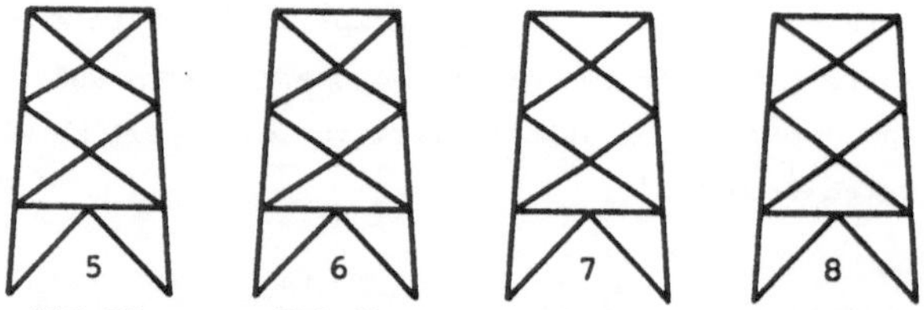

| | | | | |
|---|---|---|---|---|
| State | 5 | 6 | 7 | 8 |
| Weight (tons) | 106.05 | 110.41 | 110.39 | 112.70 |
| Smallest β | 1.23 | 4.00 | 2.28 | 4.00 |

Figure 6. Shape optimization with application of expert knowledge.

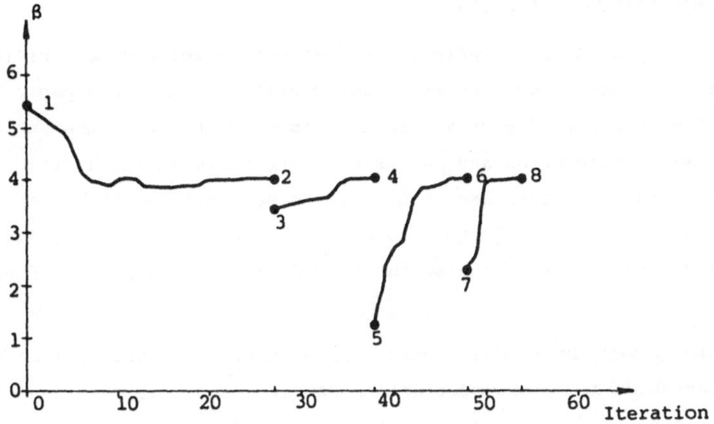


Figure 7. Variation of the smallest β-index.

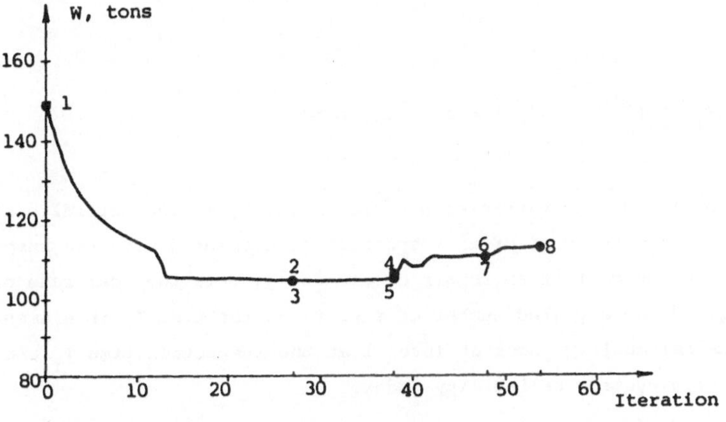


Figure 8. Variation of the structural weight.

5. OPTIMAL MAINTENANCE STRATEGIES

An interesting application of optimization methods is related to deriving optimal strategies for inspection and repair of structural systems. In a paper by Thoft-Christensen & Sørensen [18] such a strategy was suggested with the intention of minimizing the cost of inspection and repair of a structure in its lifetime T under the constraint that the structure has an acceptable reliability. In a later paper by Sørensen & Thoft-Christensen [19] this work was extended by including not only inspection costs and repair costs but also the production (initial) cost of the structure in the objective function.

In this section a very brief description based on [19] of this application is given. The design variables are cross-sectional parameters $z_1, \ldots, z_m$, inspection qualities $q_1, \ldots, q_N$ and $t_1, \ldots, t_N$ where m is the number of cross-sections to be designed and N the number of inspections (and repairs). The optimization for a structural system modelled by s failure elements can then be formulated in the following way

$$\min_{\substack{q_1 \ldots q_N \\ t_1 \ldots t_N}} \; C(\bar{z}, \bar{q}, \bar{t}) = C_I(\bar{z}) + \sum_{i=1}^{N} \sum_{j=1}^{s} C_{IN,j}(q_i) e^{-rT_i} + \sum_{i=1}^{N} \sum_{j=1}^{s} C_{R,j}(\bar{z}) E[R_{ij}(\bar{z}, \bar{q}, \bar{t})] e^{-rT_i} \tag{7}$$

$$\text{s.t.} \quad \beta^s(T_i) \geq \beta^{min} \qquad i = 1, 2, \ldots, N, N+1$$

where bounds on the design variables not are shown. $C_I$ is the initial cost of the structure, $C_{IN,j}(q_i)$ the cost of an inspection of element j with the inspection quality $q_i$, $C_{R,j}$ is the cost of an repair of element j. r is the real rate of interest and $E[R_{ij}(\bar{z}, \bar{q}, \bar{t})]$ the expected number of repairs at the time $T_i$ in element j. $\beta^s(T_i)$ is the systems reliability index at level 1 at the inspection time $T_i$ and $\beta^{min}$ the lowest acceptable systems reliability index.

In example 4 the solution of the optimization problem (7) for a plane model of an offshore platform is shown. This example is taken from [19], where all details are described.

Example 4:

Consider the plane model of a steel jacket platform shown in figure 9. Due to symmetry only the 8 fatigue failure elements indicated by x in figure 9 are considered. Design variables are the tubular thicknesses of the 6 groups of elements indicated by 0 in figure 9. Using $\beta^{min} = 3.00$, T = 10 years, r= 0 and N = 6 the following optimal solution is determined.

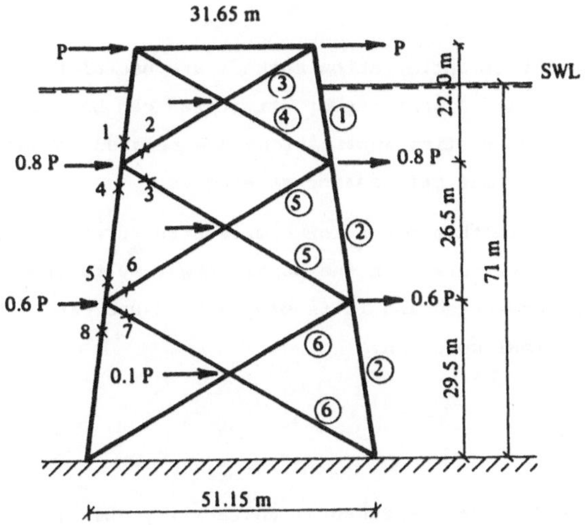


Figure 9. Plan model of a steel jacket platform.

$\bar{t}$ = (2, 2, 2, 1.19, 0.994, 0.925) years

$\bar{q}$ = (0.1, 0.133, 0.261, 0.332, 0.385, 0.423)

$\bar{z}$ = (68.9, 62.7, 30.0, 30.0, 50.4, 32.0) mm

The minimum cost is C = 29.7 · $10^6$ DKK. The corresponding variation of the systems reliability index $\beta^S$ with the time t is shown in figure 10.

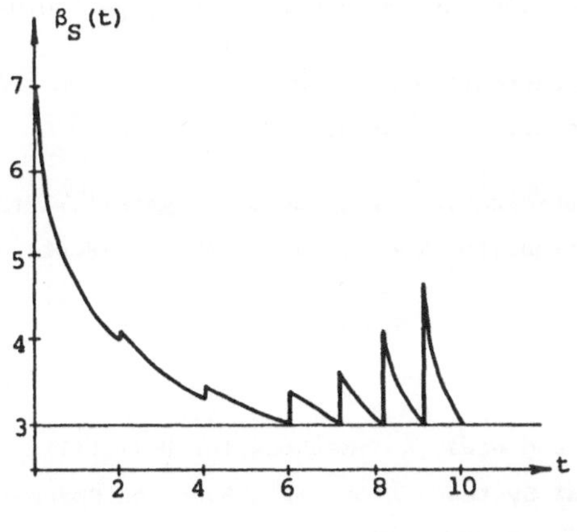


Figure 10. $\beta_S$ as a function of the time t for optimal design and inspection variables and for N = 6.

## 6. CONCLUSION

In this paper it is demonstrated that optimization methods are useful tools for structural engineering. Estimation of the reliability of structures can be formulated as optimization problems, but more interesting applications are related to optimal design of structures and derivation of optimal maintenance strategies.

Such applications are illustrated in this paper for simple structures, but the same techniques can be used for real structures. In the paper new results from reliability-based shape optimization are presented and it is proposed to use expert knowledge techniques for improving the optimal solution.

## REFERENCES

[1] Thoft-Christensen, P. & M.J. Baker. Structural Reliability Theory and Its Applications. Springer-Verlag, 1982.

[2] Thoft-Christensen, P. & Y. Murotsu. Application of Structural Systems Reliability Theory. Springer-Verlag, 1986.

[3] Thoft-Christensen, P. Some Experience from Application of Optimization Technique in Structural Reliability. Bygningsstatiske Meddelelser, Vol. 48, 1977, pp 31-44.

[4] Thoft-Christensen, P. Recent Advances in Structural Systems Reliability Theory. IABSE Symp. Safety and Quality Assur. Civil Engng. Struct., Tokyo, 1986. IABSE Report, 51, 101-108.

[5] Thoft-Christensen P. Reliability Modelling of Structural Systems. Proc. IFIP Conf. on "System Modelling and Optimization" (editors A. Prékopa, J. Szelezsan & B. Strazicky), Budapest, Hungary, 1985. Springer-Verlag 1986, pp 970-981.

[6] Schittkowski, K.: NLPQL: A FORTRAN Subroutine Solving Constrained Non-Linear Programming Problems. Annals of Operation Research, 1986.

[7] Sørensen, J.D. Reliability-Based Optimization of Structural Systems. 13th IFIP Conf. on System Modelling and Optimization, Tokyo, Japan, Aug. 31-Sept. 4, 1987.

[8] Vaanderplaats, G.N. Numerical Optimization Techniques for Engineering Design. McGraw-Hill, 1984.

[9] Murotsu, Y., Yonezawa, M., Oba, F. and Niwa, K.: Methods for Reliability Analysis and Optimum Design of Structural Systems. Proc. 12th Symp. on Space Technology and Science, Tokyo, Japan, 1977, pp 1047-1054.

[10] Murotsu, Y., Kishi, M., Okada, H., Yonezawa, M. and Taguchi, K. "Probabilistic Optimum Design of Frame Structure." Proc. IFIP Conf. on System Modelling and Optimization (editor P. Thoft-Christensen), Copenhagen, Denmark, 1983. Springer-Verlag 1984, pp 545-554.

[11] Frangopol, D.M. Sensitivity Studies in Reliability-Based Analysis of Redundant Structures. Structural Safety, Vol. 3, 1985, pp 13-22.

[12] Frangopol, D.M.: Sensitivity of Reliability-Based Optimum Design. J. Structural Engineering, Vol. 111, 1985, pp 1703-1721.

[13] Thoft-Christensen, P. & Sørensen, J.D. Recent Advances in Optimal Design of Structures from a Reliability Point of View. Quality & Rcliability Management, Vol. 4 1987, pp 19-31.

[14] Thoft-Christensen, P. & Sørensen, J.D. Reliability Analysis of Tubular Joints in Offshore Structures. To be published in Reliability Engineering, 1987.

[15] Frisk, L. & Poulsen, P.: Formoptimering med pålidelighedssidebetingelser. M.Sc.-thesis, University of Aalborg, Denmark, June 1987 (in Danish). Advisors were P. Thoft-Christensen and J.D. Sørensen.

[16] Sørensen, J.D., Thoft-Christensen, P. and Sigurdsson G. Development of Applicable Methods for Evaluating the Safety of Offshore Structures, Part 2. Institute of Building Technology and Structural Engineering. The University of Aalborg, Paper No. 11, 1985.

[17] Ding, Y. Shape Optimization of Structures: A Literature Survey. Computers & Structures, Vol. 24, 1986, pp 985-1004.

[18] Thoft-Christensen, P. & Sørensen, J.D. Optimal Strategy for Inspection and Repair of Structural Systems. Civ. Engng. Syst., Vol. 4, 1987, pp 94-100.

[19] Sørensen, J.D. Integrated Reliability-Based Optimal Design of Structures. Proc. IFIP Conf. on "Reliability and Optimization of Structural Systems", Aalborg, Denmark, May 1987 (editor P. Thoft-Christensen). Springer-Verlag, 1988, pp 385-398.

# SAFETY ASSESSMENT OF BRIDGES DURING CONSTRUCTION

Hitoshi Furuta and Naruhito Shiraishi
Department of Civil Engineering
Kyoto University, Kyoto 606, Japan

## ABSTRACT

In the safety assessment of bridges during construction, various uncertainties due to the properties peculiar to individual structures should be considered. To establish an appropriate safeguard system, it is important to utilize qualitative information based on engineers' intuition and experience, as well as quantitative information such as the measured stress and displacement. Using the concept of fuzzy reasoning, an attempt is made in this paper to develop an evaluating system which is implemented on the basis of modus ponens. The ultimate objective of this prototype system is to monitor the change of structural safety during construction. To illustrate this system, a numerical example is presented herein.

KEYWORDS: Bridge, Construction, Fuzzy Reasoning, Monitoring System, Safety Assessment

## INTRODUCTION

From the viewpoint of applying the theory of structural reliability[1)], it is important to evaluate the safety of individual structures. To discuss the safety of a structure in consideration, it is inevitable to assess its safety during construction, because structures, especially bridge structures often failed during construction.

In general, safety assessment of bridge structures during construction have the following difficulties[2)]:

1)various construction methods are possibly chosen according to the size and type of bridge and the site condition.

2)special machines and temporary facilities are used.

3)the structural characteristics generally change as completion approaches.

4)the structure is likely to suffer from several natural environmental and loading conditions as typhoons, earthquakes, tsunami and floods.

Thus, various uncertainties due to the properties peculiar to individual structures should be considered in the safety assessment during construction. This implies that the evaluation cannot be done in a unified manner similar to those in the process of designing or fabrication. In order to account for this peculiar characteristic, it is useful to utilize the engineering judgment or intuition of experienced engineers[3]. To establish an appropriate safeguard system, it is, thus, important to use qualitative information based on engineers' intuition and experience, as well as quantitative information such as the measured stress and displacement.

Using the concept of fuzzy reasoning[4], an attempt is made to develop an evaluating system which is implemented on the basis of modus ponens. Fuzzy reasoning is an extension of the ordinal reasoning with binary logic. Using th concept of fuzzy set[5], it is possible to consider the vagueness or ambiguity involved in inference process and rules to be used. So far several calculating methods have been proposed for implementing the fuzzy reasoning[6]. In this paper, those methods are compared and discussed from the standpoint of the applicability and easiness of calculation. Based on the result of comparison, a safety assessment system is developed herein. The ultimate objective of this prototype system is to monitor the change of structural safety during construction. To illustrate this system, a numerical example is presented herein.

## FUNDAMENTALS OF FUZZY REASONING

In general, process of reasoning may be divided into two types; modus ponens and modus tollens. Modus ponens is expressed by an antecedent P and consequent Q:

| | | |
|---|---|---|
| Proposition | P | Q |
| Observation | P | |
| Conclusion | | Q |

(1)

where this reasoning is executed if and only if the antecedent of a rule (P→Q) and a fact supplied by observation or testing are completely equivalent. On the other hand, the deduction of modus tollens has the following form:

| | | |
|---|---|---|
| Proposition | P | Q |
| Observation | | Q* |
| Conclusion | P* | |

(2)

in which the superscript * denotes the negation.

These syllogism can be extended to a case with fuzzy antecedent/consequent[7]. In this paper, modus ponens is used for the fuzzy reasoning. Assuming that the antecedent and consequent are specified by fuzzy sets, Eq. 1 is expressed as follows:

$$\begin{array}{l} \text{IF X is F THEN Y is G} \\ \underline{\quad \text{X is F'} \qquad\qquad\qquad} \\ \qquad\qquad \text{Y is G'} \end{array} \tag{3}$$

in which P and Q are replaced by (X is F) and (Y is G). Here, F and G denote fuzzy sets, respectively. It should be noted that F and F' are similar but not equivalent. Namely, the fuzzy reasoning can be implemented even if the antecedent (X is F) and a fact (X is F') are not equivalent. This can be done by using the concept of partial matching.

According to Zadeh's work[4], the fuzzy modus ponens is implemented using the possibility distribution $\Pi(\;)$.

$$\begin{array}{ll} \Pi(X/Y) & = (F \times G) \cup (F^* \times V) \\ \Pi(X) & = F' \\ \hline \Pi(Y) & = F' \circ ((F \times G) \cup (F^* \times V)) \end{array} \tag{4}$$

where $\Pi(X/Y)$ means the conditional rule and the symbols x and U denote the cartesian product and union, respectively. The operation defined by the symbol $\circ$ follows the compositional rule.

$$A \circ B = \int_{U \times W} \min_{v \in V} (\mu_a(u,v),\ \mu_b(v,w))/(u,w) \tag{5}$$

in which $\mu_a$ and $\mu_b$ are the membership functions of fuzzy sets A and B which are defined on UxV and VxW, respectively. Other than the above reasoning method, several methods are proposed to construct a fuzzy relation[6].

$$R_c = F \times G = \int_{U \times V} \min(\ \mu_F(u),\ \mu_G(v))/(u,v) \tag{6}$$

$$R_s = F \times V \underset{S}{\rightarrow} V \times G = \int_{U \times V} (\ \mu_F(u) \underset{S}{\rightarrow} \mu_G(v))/(u,v) \tag{7}$$

where

$$\mu(u) \underset{S}{\rightarrow} \mu(v) = \{1:\ \mu_F(u) \le \mu_G(v),\ 0:\ \mu_F(u) > \mu_G(v)\} \tag{8}$$

$$R_a = (F^* \times V) \oplus (U \times G) = \int_{U \times V} \min(1,\ (1-\mu_F(u)+\mu_G(v))/(u,v) \tag{9}$$

$$R_m = (F \times G) \cup (F^* \times V) = \int_{U \times V} (\min(\mu_F(u),\ \mu_G(v)),\ (1-\mu_F(u)))/(u,v) \tag{10}$$

in which $\oplus$ is a bounded sum.

**SAFETY ASSESSMENT BASED ON FUZZY REASONING**

To establish a safety assessment method, rules relating input data and evaluation are firstly collected, which are stored as a data-base. The data-base is built as follows:

1)prescribe all possible factors ($X_i$, $Y_j$, $Z_k$, ••••••) affecting structural safety.

2)enumerate possible consequences which are derived from a factor $X_j$, i.e., $X_i \rightarrow Y_j$, $X_j \rightarrow Y_j$, ••••••••.

3)give several kinds of evaluation to the factors $X_j$, $Y_j$, $Z_j$,•••••. These evaluations are determined by means of fuzzy sets F, G,•representing verbal information.

4)from the above data, useful rules are constructed and stored in the memory as a data-base.

Based on the fuzzy modus ponens, each fuzzy reasoning is executed as shown in Eq.3. Using the fuzzy relation R, the output result is obtained in the form of membership function .

$$\mu_{g'}(v) = \mu_{F'}(u) \circ R \quad (11)$$

This reasoning procedure is continued until the final evaluation is obtained. In this process, a number of rules are used in combination with each other. Fig. 1 depicts an example of reasoning, where the input information is transmitted from X through S via X, Z. If a node Y has more than two different input sources, i.e. $X_i$ and $X_j$, their contribution to Y is evaluated by the concept of intersection.

$$\mu_G(u) = \min(\mu_{Gi}(u), \mu_{Gj}(u)) \quad (12)$$

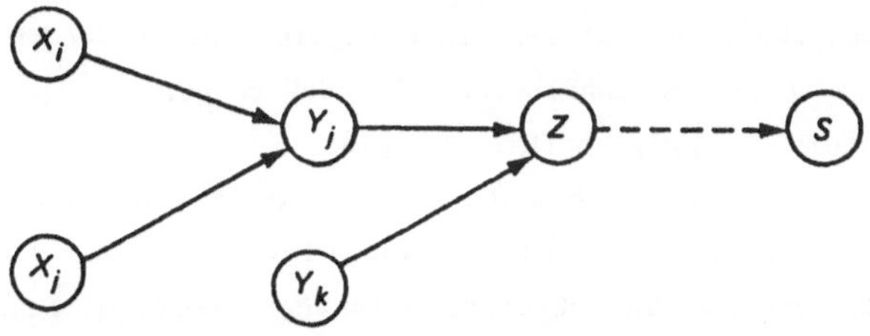


Fig. 1 Inference Related Diagram

**NUMERICAL EXAMPLE**

Consider the safety assessment of a bridge under construction. The type and geometry of the bridge are shown in Fig. 2[8]. It is assumed that this bridge is assembled by using a cable erection method. the outline of the erection is shown in Fig. 3. For the sake of simplicity, we only consider the construction process from the beginning to the erection of the fifth block of the lower chords. To construct available rules, 126 items are taken into account. They are classified into several groups such as supporting tower, staying wire, bent, crane, shoe, camber, bolt, weld, etc. As an example, items considered in the supporting tower are shown in Table 1.

The rules which are constructed and accumulated in the data-base are used to deduce the final evaluation. Fig. 4 presents a representative reasoning procedure which is used for assessing the change of structural safety until the first block of the lower chords is erected. In this figure, circle nodes denote input nodes which receive necessary information. Since in this method the information is represented by fuzzy sets, the final evaluation is obtained in the form of membership functions. In this example input data are expressed in terms of seven linguistic variables such as 7-Very Large, 6-Large, 5-More or Less Large, 4-Medium, 3-More or Less Small, 2-Small and 1-Very Small. For instance, 7-Very Large means that the degree of safety is very large. Fig. 5 shows the membership functions used for those seven linguistic variables. Using the inference diagram shown in Fig. 4, the calculating methods $R_c$, $R_s$, $R_a$ and $R_m$ are compared and examined. Fuzzy relations are shown in Table 2, where the figures of 5, 6, and 7 correspond to the above linguistic variables, respectively. Numerical results are summarized in Table 3, in which IN=7 means that all input data are assumed to be 7-Very Large. From Table 3 it is seen that $R_s$, $R_a$ and $R_m$ provide the same result in the case of IN=4. All the membership grades are obtained as one, whereas $R_c$ provides the result that all the membership grades are zero. In the cases of IN=5, IN=6 and IN=7, $R_c$ provides the result that all the membership grades are 6, and RPvsPv shows that the membership grades are 5, 6 and 7 corresponding the values of IN. This implies that $R_c$ and $R_s$ are reasonable, especially $R_s$ can deduce exact conclusions for all cases. On the contrary, $R_a$ and $R_m$ are not satisfactory to apply to this example.

Using the inference process based on the calculating methods $R_c$ and $R_s$, the safety assessment is done here. Fig. 6 and Fig. 7 present the final results obtained by using $R_c$ and $R_s$. The abscissa u is the support of fuzzy sets such that u=0 indicates the collapse and u=1 indicates entire safety. The construction process of the lower chords are divided into 22 steps. Until Step 7, $R_c$ and $R_s$ present the same result whose membership grades are high near u=1. This indicates

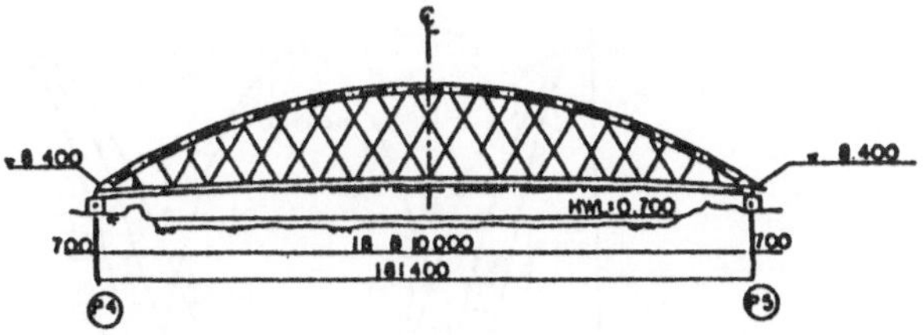


Fig. 2 Bridge Example

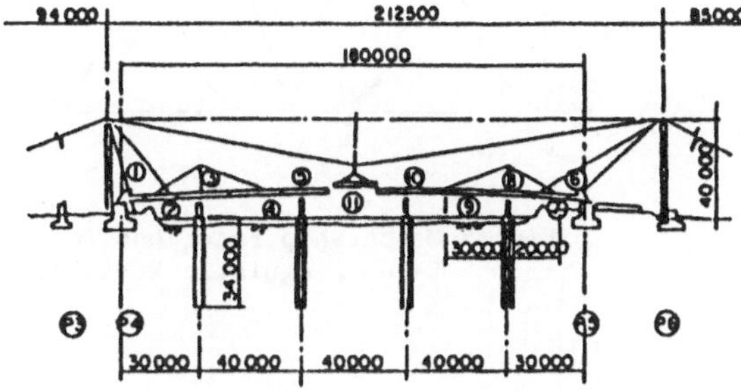


Fig. 3 Constructing Procedure for Lower Chord

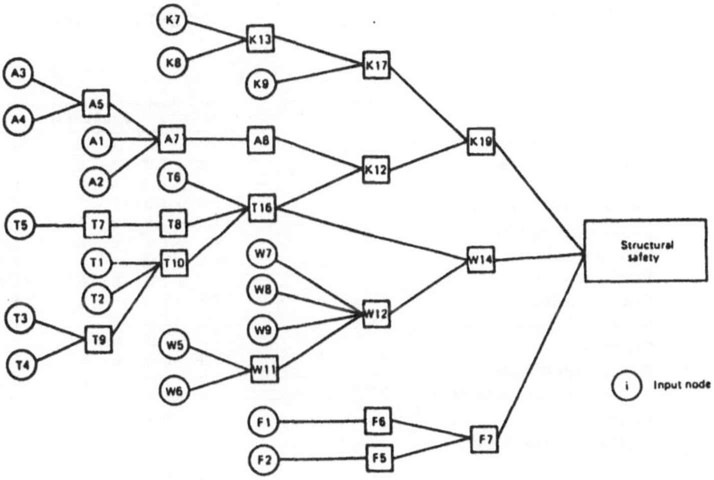


T1 to T10, T16: see table 1

A1: Angle of main cable
A2: Connection
A3: Ground condition
A4: Slide or overturn
A5: Anchor
A7: Adjustment of back stay
A8: Anchor system
K7: Damage
K8: Lateral stability
K9: Clearances
K12: Stability for wind load
K13: Back staying cable
K17: Adjustment of cable
K19: Cable system
W5: Damage
W6: Lateral stability
W7: Clearances
W8: Connections
W9: Adjusting facilities
W11: Material property
W12: Adjustment of wires
W14: Wire system
F1: Elongation
F2: Horizontal stability
F5: Capacity
F6: Influence of applied load
F7: Shoe

Fig. 4 Inference Process

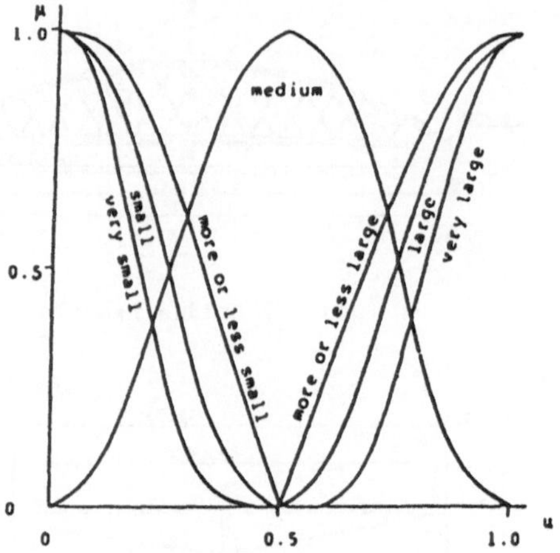


Fig. 5 Membership Functions for Seven Linguistic Variables

Table 1 Safety Factor regarding Supporting Tower

| Number | Factor |
|---|---|
| 1 | Angle of main cable |
| 2 | Supporting wire |
| 3 | Connection of staying wire |
| 4 | Clip of trucking wire |
| 5 | Sink of foundation |
| 6 | Bracing member |
| 7 | Ground condition |
| 8 | Stability of substructure |
| 9 | Connections |
| 10 | Adjustments of wires |
| 11 | Stability for moving loads |
| 12 | Hinges |
| 13 | Stability for wind loads |
| 14 | Saddle |
| 15 | Condition of cable system |
| 16 | Stability of tower |

Table 2 Example of Rules

| Rule | Evaluation | | Rule | Evaluation | |
|---|---|---|---|---|---|
| A → B : | A | B | A → B : | A | B |
| T 1 → T 10 | 5 | 5 | K 7 → K 13 | 6 | 5 |
| T 2 → T 10 | 5 | 5 | K 8 → K 13 | 6 | 5 |
| T 3 → T 9 | 7 | 7 | K 9 → K 17 | 6 | 5 |
| T 4 → T 9 | 7 | 7 | K 12 → K 19 | 6 | 5 |
| T 5 → T 7 | 6 | 6 | K 13 → K 17 | 6 | 6 |
| T 6 → T 16 | 6 | 6 | K 17 → K 19 | 6 | 5 |
| T 7 → T 8 | 6 | 6 | K 19 → S | 7 | 7 |
| T 8 → T 16 | 6 | 6 | F 1 → F 6 | 7 | 6 |
| T 9 → T 10 | 6 | 5 | F 2 → F 5 | 6 | 5 |
| T 10 → T 16 | 6 | 5 | F 5 → F 7 | 6 | 6 |
| T 16 → K 12 | 6 | 5 | F 6 → F 7 | 6 | 6 |
| T 16 → W 14 | 6 | 6 | F 7 → S | 7 | 7 |
| A 1 → A 7 | 7 | 6 | W 5 → W 11 | 6 | 5 |
| A 2 → A 7 | 6 | 5 | W 6 → W 11 | 6 | 5 |
| A 3 → A 5 | 6 | 6 | W 7 → W 12 | 6 | 5 |
| A 4 → A 5 | 7 | 7 | W 8 → W 12 | 6 | 6 |
| A 5 → A 7 | 6 | 5 | W 9 → W 12 | 7 | 7 |
| A 7 → A 8 | 6 | 5 | W 11 → W 12 | 6 | 6 |
| A 8 → K 12 | 6 | 6 | W 12 → W 14 | 6 | 5 |
| | | | W 14 → S | 7 | 7 |

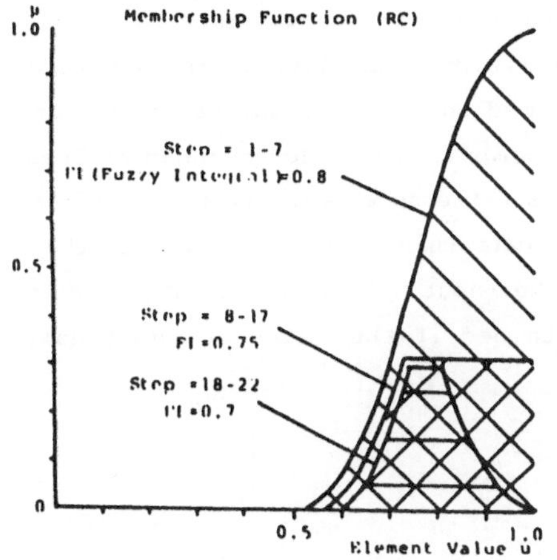


Fig. 6 Numerical Result by $R_c$

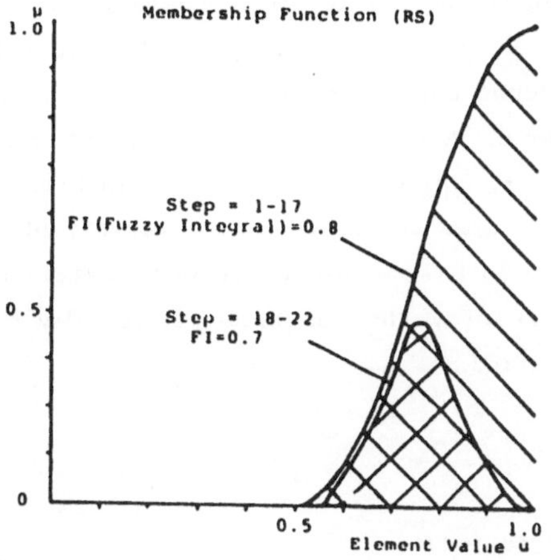


Fig. 7 Numerical Result by $R_s$

Table 3 Numerical Results

| Input Data | R | Element Value 0.1 | 0.2 | 0.3 | 0.4 | 0.5 | 0.6 | 0.7 | 0.8 | 0.9 | 1.0 |
|---|---|---|---|---|---|---|---|---|---|---|---|
| IN=1~3 | Rc | 0.0 | 0.0 | 0.0 | 0.0 | 0.0 | 0.0 | 0.0 | 0.0 | 0.0 | 0.0 |
| | Rs | 1.0 | 1.0 | 1.0 | 1.0 | 1.0 | 1.0 | 1.0 | 1.0 | 1.0 | 1.0 |
| | Ra | 1.0 | 1.0 | 1.0 | 1.0 | 1.0 | 1.0 | 1.0 | 1.0 | 1.0 | 1.0 |
| | Rm | 1.0 | 1.0 | 1.0 | 1.0 | 1.0 | 1.0 | 1.0 | 1.0 | 1.0 | 1.0 |
| IN=4 | Rc | 0.0 | 0.0 | 0.0 | 0.0 | 0.0 | 0.08 | 0.32 | 0.32 | 0.32 | 0.32 |
| | Rs | 1.0 | 1.0 | 1.0 | 1.0 | 1.0 | 1.0 | 1.0 | 1.0 | 1.0 | 1.0 |
| | Ra | 1.0 | 1.0 | 1.0 | 1.0 | 1.0 | ɨ.0 | 1.0 | 1.0 | 1.0 | 1.0 |
| | Rm | 1.0 | 1.0 | 1.0 | 1.0 | 1.0 | 1.0 | 1.0 | 1.0 | 1.0 | 1.0 |
| IN=5 | Rc | 0.0 | 0.0 | 0.0 | 0.0 | 0.0 | 0.08 | 0.32 | 0.68 | 0.92 | 1.0 |
| | Rs | 0.0 | 0.0 | 0.0 | 0.0 | 0.0 | 0.28 | 0.57 | 0.82 | 0.96 | 1.0 |
| | Ra | 0.68 | 0.68 | 0.68 | 0.68 | 0.68 | 0.71 | 0.82 | 0.86 | 0.96 | 1.0 |
| | Rm | 0.57 | 0.57 | 0.57 | 0.57 | 0.57 | 0.57 | 0.57 | 0.82 | 0.92 | 1.0 |
| IN=6 | Rc | 0.0 | 0.0 | 0.0 | 0.0 | 0.0 | 0.08 | 0.32 | 0.68 | 0.92 | 1.0 |
| | Rs | 0.0 | 0.0 | 0.0 | 0.0 | 0.0 | 0.08 | 0.32 | 0.68 | 0.92 | 1.0 |
| | Ra | 0.64 | 0.64 | 0.64 | 0.64 | 0.64 | 0.64 | 0.68 | 0.86 | 0.94 | 1.0 |
| | Rm | 0.50 | 0.50 | 0.50 | 0.50 | 0.50 | 0.50 | 0.50 | 0.68 | 0.92 | 1.0 |
| IN=7 | Rc | 0.0 | 0.0 | 0.0 | 0.0 | 0.0 | 0.08 | 0.32 | 0.68 | 0.92 | 1.0 |
| | Rs | 0.0 | 0.0 | 0.0 | 0.0 | 0.0 | 0.01 | 0.10 | 0.46 | 0.85 | 1.0 |
| | Ra | 0.50 | 0.50 | 0.50 | 0.50 | Q 50 | 0.58 | 0.67 | 0.85 | 0.94 | 1.0 |
| | Rm | 0.46 | 0.46 | 0.46 | 0.46 | 0.46 | 0.46 | 0.50 | 0.68 | 0.92 | 1.0 |

that the safety is high. Namely, the erection of the bridge was carried out without any error or accident before Step 8, but afterwards, some trouble in later erection reduced the safety level. Here, it is assumed that the strong storm occurred at Step 8. According to this assumption, the membership grades changed at Step 8 in Fig. 6, while no change arose in Fig. 7. This difference means that in this example $R_S$ is more sensitive to the variety of input data than $R_C$ is. This is why the rules used in this example are well matched to the input data in the calculation of $R_C$. Therefore, the converse result may be obtained if the rules or input data are modified.

## CONCLUSIONS

In this paper, an attempt was made to develop a safety assessment method of bridges during construction, using the concept of fuzzy reasoning. From the numerical example, the following conclusion were derived:

1) Using the fuzzy reasoning, it becomes possible to make use of such qualitative information as intuition or engineering judgment of engineers in the safety assessment of structures during construction, as well as such quantitative information as the measured stress or displacement. Then, it is important to accumulate and store the useful knowledge acquired from experienced engineers.
2) The comparison of several methods constructing the fuzzy relations leads to the conclusion that $R_C$ using the cartesian product is the most appropriate at least for the problem treated here, because of the validity and easiness of calculation.
3) Although this method is not yet applicable for practical use, its concept is useful to monitor the change of structural safety during construction. It is necessary to improve the inference diagram by using the brain storming and the structural modeling.

## REFERENCES

1) P. Thoft-Christensen and M. J. Baker, Structural Reliability Theory and Its Applications, Springer-Verlag, 1982.
2) Special Issue on Bridge Construction, Bridge and Foundation Vol. 16, No. 8, pp.47-55, 1980. (in Japanese)
3) J. T. P. Yao, Safety and Reliability of Existing Structures, Pitman Advanced Publishing Program, 1985.
4) L. A. Zadeh, Fuzzy Logic and Approximate Reasoning, Synthesis, Vol. 30, pp.407-428, 1975.

5) L. A. Zadeh, Fuzzy sets, Information and Control, Vol. 8, pp.338-353, 1965.
6) M. Mizumoto, S. Fukami and K. Tanaka, Some methods of Fuzzy Reasoning, M. M. Gupta (Ed.) Advances in Fuzzy Set Theory and Applications, North-Holland, p.117-136, 1979.
7) D. I. Blockley, The Nature of Structural Design and Safety, Ellis Horwood, 1980.

# CONVENTIONAL & RELIABILITY COMPARISON STUDY OF VARIOUS DESIGNS OF A TYPICAL 2-WAY R.C. FLAT SLAB INTERIOR PANEL

Hak-Fong Ma* and Shou-Pin Chen**

*Impell Corporation, Lincolnshire, Illinois, 60015, U.S.A.
**Lienho Junior College of Technology, Taiwan, Republic of China

## 1. INTRODUCTION

The 2-way flat slab is one of the most efficient structural system for economy. Its main advantages lie in the ease of construction with minimum field labor because of simple framework and reinforcing steel layout. Flat slab also results in minimum storey height for required clear headroom and provide for flexibility in the layout of columns, partitions and small openings. Since the 2-way slab system is statically highly indeterminate, simplifying methods are developed to expedite practical designs. The purpose of this study is to investigate into the economy and reliability of a 2-way reinforced concrete flat slab system designed based on 5 commonly used methods of design.

The scope of this study is limited to a typical office building interior panel reinforced concrete slab (23 feet square) with solid uniform depth (7 inches) reinforced in two directions to transfer loads to the supporting columns (2 feet square), with drop panels of 8 feet square and 11 inches thick. The concrete nominal strength, $f_c$, is 3000 psi and the yield strength of the steel, $f_y$, is 40 ksi. The story height is 12 feet and the design live load is 60 psf. The thickness of slab and drop panel adopted in this study satisfied the provisions required by the ACI Code [3].

## 2. TRADITIONAL METHODS OF ANALYSIS AND DESIGN

Among the many acceptable methods of analysis and design available to the designer, 5 commonly used methods are considered in this study. They are briefly described in the following sections.

2.1 ACI's Direct Design Method [3] - This method is applicable provided certain limiting requirements are satisfied. In essence, the total static moment $M_0$ in a strip bounded by the center line of panel on each side of the center line of supports (usually columns) is given by the equation:

$$M_0 = W_u L_2 L_n^2 / 8 \tag{1}$$

where: $W_u$ is the factored load per unit area, $L_2$ is the length of span transverse to the direction in which moments are being determined, $L_n$ is the clear span in the direction that moments are being determined.

Empirical rules are then followed in which certain percentages of $M_0$ are then considered as the critical negative factored moment and positive factored moment. Furthermore, the moments are then distributed to column strip and middle strip according to ACI Code and based on which positive and negative steel in each strip can be designed.

2.2 ACI's Equivalent Frame Method [3] - This method is applicable in almost all situations. The structure is considered to be made up of equivalent frame on column lines taken longitudinally and transversely through the slab system. Each frame consists of a row of equivalent columns and slab-beam strips bounded by the center line of panel on each side. Rules in ACI Code are followed to model the stiffnesses of the equivalent columns and slab-beams. The simplified two dimensional equivalent frame is then analyzed to determine the positive and negative moments in critical locations. Percentage distribution of these critical positive and negative moments to column strip and middle strip is the same as the Direct Design Method. It should be noted that the two design methods suggested by the ACI Code are based on ultimate design method with due consideration in elastic behavior of the system under service load conditions.

2.3 Finite Element Method Using Flat Plat Elements - With the rapid advance in computer technology, modeling of 2-way slab system using flat plate finite elements is no longer impractical. Based on elastic theory of analysis, the accuracy of this analysis can be improved with the number of elements used. For this study, 121 elements with 144 nodal points are used [8]. Simplifying symmetric boundary conditions of the interior panel are assumed. Critical positive and negative moments are determined using SAP4 [2] Computer program. Based on these critical moments, the required reinforcements are then designed using the ultimate design method described in the ACI Code.

2.4 Grid Beam Elements Including Torsional Effects - Instead of the more sophisticated flat plat elements adopted in the above section, simple grid beam elements (assumed to be easily available in all engineering firms) are used. The slab is divided into many strips in both directions and each strip of slab with width b is modeled by a beam element with gross moment of inertia $I_g = b h^3/12$, where h is the slab thickness, and torsional rigidity $J = X^3Y/3$; where X and Y are the smaller and larger dimensions of the grid beam respectively. Once the critical positive and negative moments are determined (e.g. using SAP4 [2]), reinforcing steel in the slab can be designed per ACI Code. This method represents a conservative but simplified method for slab design when the more sophisticated elements are not available.

2.5 Grid Beam Elements Without Torsional Effects - This method is similar to the method described in Section 2.4 except that the torsional capability of the slab is (conservatively) ignored and hence J = 0 is used in the model.

Based on the 5 design methods briefly described above, the total weight of steel required for each design may be summarized in Fig. 1 It can be seen that the methods described in Sections 2.5 and 2.3 have the most and least amount of steel used respectively. Their ratio of total steel is (2178/1685) = 1.293, indicating that up to 30% difference may be used based on 2 different acceptable design methods. The maximum steel ratio for column strip in this study is (1507/1014) = 1.486. In general, it can be observed that the more sophisticated analysis method may lead to an economy of material used.

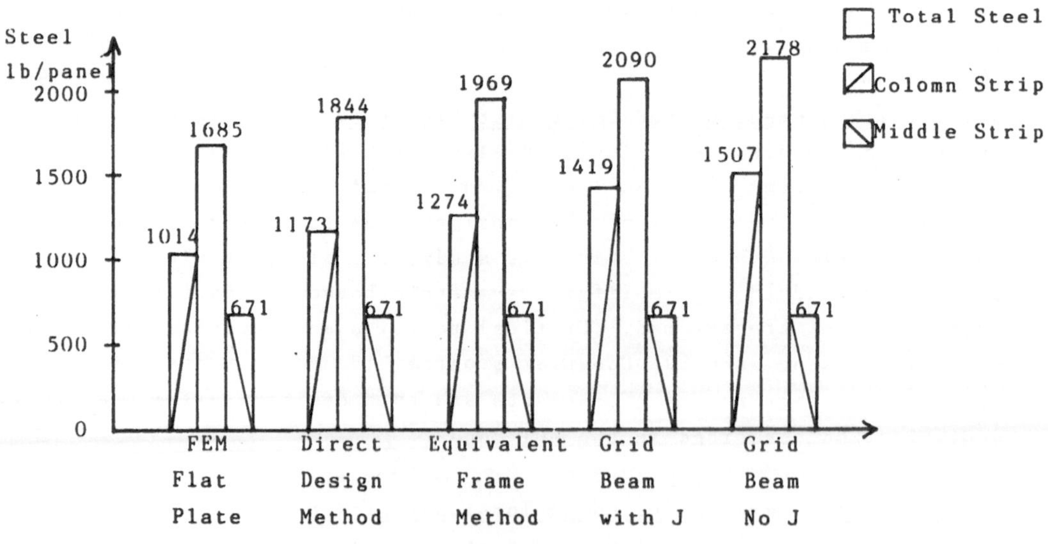


Figure 1. Reinforcements Based On 5 Design Methods

## 3. RELIABILITY ANALYSIS OF 2-WAY SLAB SYSTEM

In contrast to the traditional deterministic approach in structural engineering, reliability analysis is based on a probabilistic point of view which takes into account the presence of uncertainties arising from design parameters, imperfect modeling or lack of experience. There are many methods involving different levels of sophistication [1,4] to incorporate these uncertainties. This study adopted the simpliest level of methods to highlight the probabilistic view of 2-way slab design. The bases of the method are briefly described in this section.

The main concern in this study is the safety of the slab system. Therefore, the ultimate limit state is defined as the structural collapse of the panel. Failure mode approach is adopted to estimate the system reliability and 3 steps are required in the failure mode analysis of a structure.

1) Identification of all significant modes,
2) Reliability analysis of the individual failure mode,
3) Synthesis of individual failure modes to obtain the system collapse probability.

Step 2 will first be discussed in the section below, followed by step 1 and step 3.

3.1 Reliability Analysis of Individual Failure Mode - For each individual failure, we may define a performance function, or limit state function $Z = g(X_1, \ldots, X_n)$ such that $Z \leq 0$ represents failure and $X_1, \ldots, X_n$ are the basic random variables of the structural system such as loads, concrete strengths, effective depths etc. in this study.

The "First Order Second Moment" approach [1,4] would expand the performance function in a Taylor series at a point $(x_1^*, \ldots, x_n^*)$ on the failure surface such that $g(x_1^*, \ldots, x_n^*) = 0$.

Define the set of reduced variate as follow:

$$X_i' = \frac{X_i - \mu X_i}{\sigma X_i} \quad ; \ i = 1, 2, \ldots, n \tag{2}$$

where $\mu X_i$ and $\sigma X_i$ are the mean and standard deviations of $X_i$ respectively. The reliability index $\beta$ may be obtained by iteration such that:

$$x_i'^* = -\alpha_i^* \beta \tag{3}$$

in which $\alpha_i^*$ are the direction cosines as follows:

$$\alpha_i^* = (\partial g/\partial X_i')_* / \sqrt{\Sigma(\partial g/\partial X_i')_*^2} \tag{4}$$

where the derivatives are evaluated at $(x_1'^*, \ldots, x_n'^*)$. The reliability index may be interpreted as the distance from the tangent plane of the failure surface at $(x_1'^*, \ldots, x_n'^*)$ to the origin of the reduced variates [1].

For the special case when the performance function Z is a linear function of the normal (Gaussian) basic random variables $X_i$, i = 1,..., n, then Z is also of normal (Gaussian) distribution and the modal structural failure probability is given by:

$$P_f = 1 - \Phi(\beta) \tag{5}$$

where $\Phi$ is the cumulative distribution of the Standard Gaussian variate.

For the more general case where the performance function Z is not a linear function of the basic variables $X_i$ or that not all variables $X_i$, i = 1, ..., n, are of normal distribution, equation (5) is not exact. Although other methods [6,7] may be used to improve the estimation of $P_f$, this study limits itself to the use of equation (5) as a very rough estimate of $P_f$. Additional Monte Carlo simulation results are used to demonstrate the correctness of $P_f$. In any circumstances, reliability index $\beta$ is a good indication of $P_f$.

3.2 Identification of Significant Failure Mode - To identify all the significant modes of a structural system, the analyst must have strong background, good understanding of the behavior of the system under various loading conditions and good insights. In this study, the 2-way slab system is limited to gravity loads. Failure is defined as partial or complete collapse of the system (panel).

3.2.1 Bending Failure - One possible collapse mechanism of 2-way slab is due to the formation of yield lines. Four yield patterns are considered significant and they are shown in Figures 2 to 5. Based on the Principal of Virtual Work, the performance function for such mechanisms can be written as:

$$Z = \text{Internal Virtual Work} - \text{External Virtual Work} \tag{6}$$

Following ACI Code equation, the ultimate moment capacity M of a cross section may be written as:

$$M = A_s f_y \left( d - \frac{A_s f_y}{1.7 f_c b} \right) \tag{7}$$

where $A_s$ = area of tension steel ($in^2$); $f_y$ = yield stress of steel (psi); d = effect depth (in); $f_c$ = compressive strength of concrete (psi); b = width of slab (in).

By combining equations (6) and (7), performance functions of mechanism 1 (folding type yield line), 2 (yielding inside drop panel), 3 (yielding outside drop panel) and 4 (diagonal type yield line) may be simplified in the following equations [8].

$$Z_1 = \frac{4}{276} [ A_{s1} f_y (d_1 - \frac{A_{s1} f_y}{1.7 f_c b_1}) + A_{s2} f_y (d_2 - \frac{A_{s2} f_y}{1.7 f_c b_2}) +$$

$$A_{s3} f_y ( d_3 - \frac{A_{s3} f_y}{1.7 f_c b_3})] - 264.5 (D+L) \quad (8)$$

$$Z_2 = \frac{2\pi}{12} [ A_{s4} f_y (d_1 - \frac{A_{s4} f_y}{1.7 f_c b_4})] - 526.6 (D+L) \quad (9)$$

$$Z_3 = \frac{2\pi}{12} [ A_{s5} f_y (d_1 - \frac{A_{s5} f_y}{1.7 f_c b_5}) + A_{s6} f_y (d_3 - \frac{A_{s6} f_y}{1.7 f_c b_6}) ]$$

$$- 495.5 (D+L) \quad (10)$$

$$Z_4 = \frac{0.348}{12} [ A_{s7} f_y (d_1 - \frac{A_{s7} f_y}{1.7 f_c b_1}) + A_{s8} f_y (d_2 - \frac{A_{s8} f_y}{1.7 f_c b_8}) +$$

$$A_{s9} f_y (d_3 - \frac{A_{s9} f_y}{1.7 f_c b_9})] - 176.33 (D+L) \quad (11)$$

where variables $A_s$, $f_y$, $f_c$, b and d have the same meanings as in equation (7); D and L are dead and live load in psi respectively.

3.2.2 Shear Failure - a) Punching shear - Test and actual slab failure results indicated that failure usually occurred at column area (that is why drop panel is used to reinforce the column area). Nevertheless, the existing equations predicting punching shear failure suggested by various researchers show significant variabilities [5] against experimental results. For the sake of simplicity, only 3 equations predicting punching shear failure are used in this study; namely the ACI equation ($Z_5$), equation suggested by Moe [9] ($Z_6$) and by Yitzhaki [10] ($Z_7$). They are summarized as follows:

$$Z_5 = 16 \sqrt{f_c} (cd_1 + d_1^2 ) - 529 (D+L) \quad (12)$$

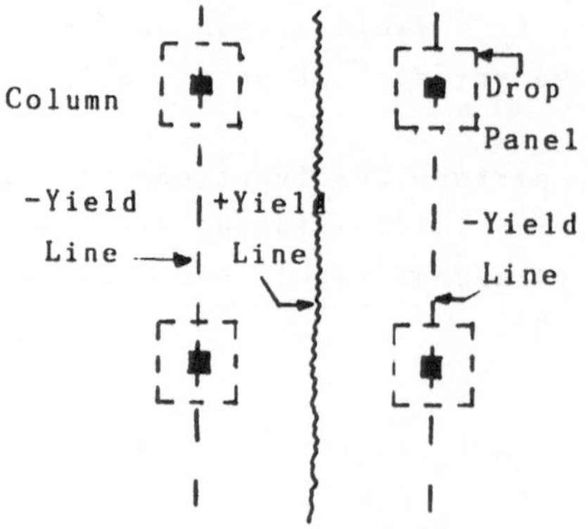


Figure 2. Mechanism 1

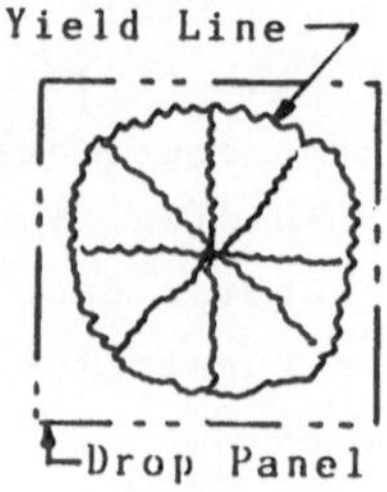


Figure 3. Mechanism 2

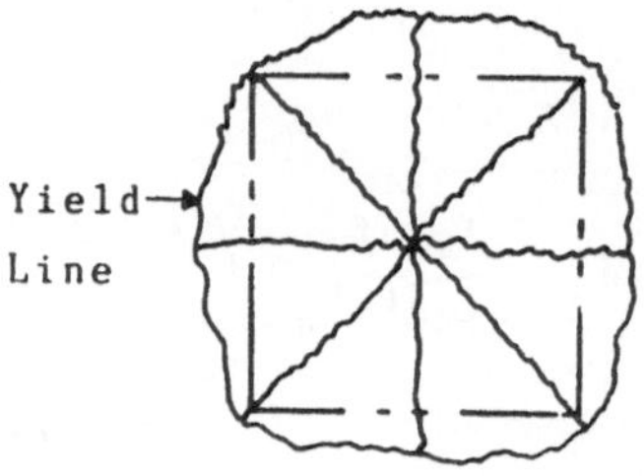


Figure 4. Mechanism 3

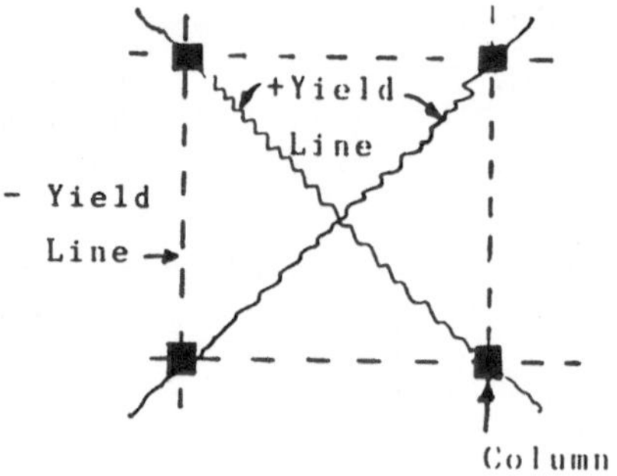


Figure 5. Mechanism 4

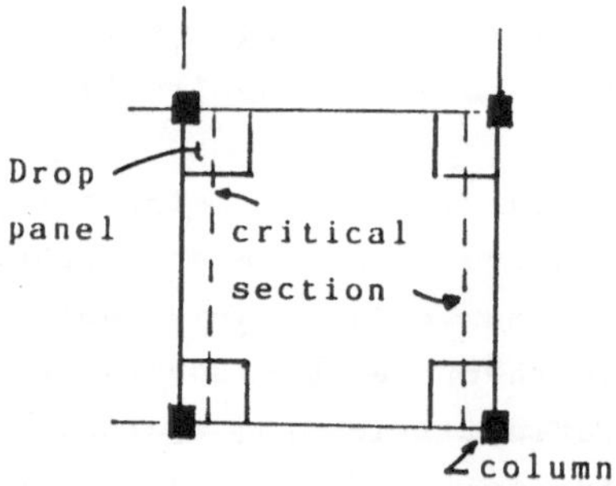


Figure 6. Critical Section 1

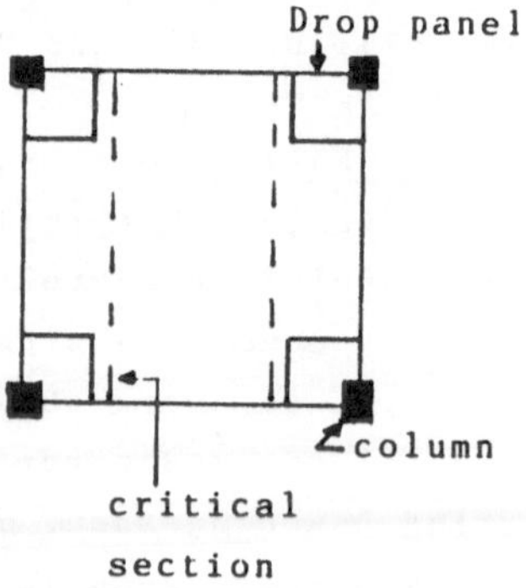


Figure 7. Critical Sect. 2

$$Z_6 = (9.23 - 1.12c/d_1)\sqrt{f_c}\ (4cd_1) - 529\ (D+L) \qquad (13)$$

$$Z_7 = 8\left(1 - \frac{A_{s4}\ f_y}{2f_c b_4 d_1}\right)\left(149.3 + 0.164\ \frac{A_{s4}\ f_y}{b_4 d_1}\right)\left(1 + \frac{c}{2d_1}\right)d_1^2 - 529\ (D+L) \qquad (14)$$

variable c represents column size in inches and other variables have the same definitions as in equation (11).

b) One-way shear - Based on ACI Code formulation, the 2 assumed critical locations are shown in Figures 6 and 7. Their respective performance functions can be written as:

$$Z_8 = 2\sqrt{f_c}(b_1 d_1 + b_2 d_2) - 230\ (D+L) \qquad (15)$$

$$Z_9 = 2\sqrt{f_c}\, b_3 d_3 - 162.92(D+L) \qquad (16)$$

The mean values, standard deviations ofthe variables in performance functions $Z_1$ to $Z_9$ may be found in [8]. Table 1 summarizes the results of analysis.

3.3 System Reliability - To determine the reliability (or alternatively probability of failure) of a system, the correlations among the major mechanisms are required. Let $E_i$ be the event that the system failed in mode i, then for a system having n major failure modes, the system failure probability may be written as:

$$P_F = P\ (E_1 \cup E_2 \cup \ldots \cup E_n) \qquad (17)$$

Many approximateins [6,7] exist to account for the correlations between major failure modes. Since the purpose of this study is to compare practical design safety margins from both traditional and probabilistic point of view, the simplist method is used here. By grossly assuming events $E_i$ are statistically independent or perfectly correlated, system failure probability $P_F$ can be written as:

$$\mathrm{Max}\ P_{fi} \leq P_F \leq 1 - \prod_1^n (1 - P_{fi}) \approx \sum_1^n P_{fi} \qquad (18)$$

## 4. DISCUSSION AND CONCLUSION

It can be seen from this study that significant differences in moment may exist between 2 acceptable design methods. While traditional design methods do not offer a consistent measure of safety margins of various designs, reliability analysis offers a unified approach which express the safety margins in terms of failure probability (or reliability index).

In the context of optimization, both approaches may again differ in perspective. If one is to formulate an optimization problem in the design of a 2-way flat slab with the objective function being the minimum amount of steel and the restraints being ACI Code requirements, then finite element method using flat plate elements gives the optimal design in traditional approach. However, if one changes the objective function to minimizing cost which includes not only material cost but also the designer's cost, then probably ACI Code's Direct Design method may become the optimal design method. In general, it can be observed that the reliability of a well designed system increases with the increase in amount of steel used. In this study, the reliability of the panel increases 100 times with a 30% increase in steel. Thus the optimal design may depend on the tradeoff between safety and cost. Regardless of approach, this study demonstrates the importance of the formulation of an optimization problem on the optimal solution later determined.

## 5. ACKNOWLEDGMENT

This study reported herein is the results of a research program supported by the National Science Council grant number NSC74-0410-E011-03. Support for this work is gratefully acknowledged.

## 6. REFERENCES

1. Ang, A.H-S, and Tang, W. H., Probability Concepts in Engineering Planning and Design, Vol. 2: Decision, Risk, and Reliability, New York , John Wiley & Sons (1984)

2. Bathe, K.J., Wilson E. L., and Peterson, F. E., "SAP4- A Structural Analysis Program for Static and Dynamic Response of Linear Systems" College of Engineering, U . of California, Berkely, (1973).

3. "Building Code Requirements for Reinforced Concrete [ACT 318-83]," American Concrete Institute, Detroit (1983).

4. Christensen, P. T. and Baker, M. J., Structural Reliability Theory and its Applications, Springer-Verlag (1982).

5. Criswell, M. E., and Hawkins, N.W., "Shear Strength of Slabs: Basic Principals and Their Relation to Current Methods of Analysis," Shear in Reinforced Concrete, Vol.2 (SP-42), ACI, P.641-676 (1974).

6. Ditlevsen, O., "Narrow Reliability Bounds for Structural Systems," Jour. of Struct. Mech., Vol.1, No.4 p.453-472 (1979).

7. Ma, H-F, "Reliability Analysis of Redundant Ductile Structural Systems ", Phd Dissertation, Dept. of Civil Eng., U. of Illinois (1981).

8. Ma, H-F and Chen S-P, " Investigation on Design and Analysis Methods for Reinforced Concrete 2-way Flat Slab System", a report to National Science Council of Republic of China, (1985) (in Chinese).

9. Moe, J., "Shearing Strength of Reinforced Concrete Slabs and Footings Under Concentrated Loads," Development Department Bulletin D47, Portlnad Cement Association, p. 130 (1961).

10. Yitzhaki, D., "Punching Strength of Reincorced Concrete Slabs," Proc. ACI Journal, Vol. 63, p 527-542 (1966).

Table 1 Reliability Analysis Results

| | | Flat Plate Element | Direct Design Method | Equivalent Frame Method | grid Beam with J | Grid Beam without J |
|---|---|---|---|---|---|---|
| $Z_1$ | $\beta_1$ | 3.13 | 3.35 | 4.01 | 4.40 | 4.62 |
| | $P_{f1}$ | $8.7x10^{-4}$ | $4.1x10^{-4}$ | $3.1x10^{-5}$ | $5.5x10^{-6}$ | $2.0x10^{-6}$ |
| $Z_2$ | $\beta_2$ | 6.92 | 6.92 | 5.79 | 6.84 | 7.07 |
| | $P_{f2}$ | - | - | - | - | - |
| $Z_3$ | $\beta_3$ | 4.30 | 4.67 | 4.82 | 5.26 | 5.58 |
| | $P_{f3}$ | $8.6x10^{-6}$ | $1.5x10^{-6}$ | $7.2x10^{-7}$ | - | - |
| $Z_4$ | $\beta_4$ | 6.88 | 6.23 | 6.25 | 6.62 | 6.74 |
| | $P_{f4}$ | - | - | - | - | - |
| $Z_5$ | $\beta_5$ | 5.72 | 5.72 | 5.72 | 5.72 | 5.72 |
| | $P_{f5}$ | - | - | - | - | - |
| $Z_6$ | $\beta_6$ | 8.95 | 8.95 | 8.95 | 8.95 | 8.95 |
| | $P_{f6}$ | - | - | - | - | - |
| $Z_7$ | $\beta_7$ | 9.11 | 9.11 | 9.39 | 9.67 | 9.79 |
| | $P_{f7}$ | - | - | - | - | - |
| $Z_8$ | $\beta_8$ | 11.06 | 11.06 | 11.06 | 11.06 | 11.06 |
| | $P_{f8}$ | - | - | - | - | - |
| $Z_9$ | $\beta_9$ | 11.54 | 11.54 | 11.54 | 11.54 | 11.54 |
| | $P_{f9}$ | - | - | - | - | - |
| Monte Carlo Results | | $10^{-3}$ | $6x10^{-4}$ | $3x10^{-5}$ | $1x10^{-5}$ | $5x10^{-6}$ |

Note: $P_f$ is based on the assumption that Z is normal variate

# ENERGY OPTIMIZATION IN INDUSTRIAL MODELS

R. Kümmel, H.-M. Groscurth

Physikalisches Institut der Universität Würzburg
D-8700 Würzburg, West Germany

W. van Gool

Energy Science Project, Department of Inorganic Chemistry
State University Utrecht, Utrecht, Netherlands

**Abstract:**

An optimization formalism is established for energy conservation in industrial systems. The objective function of optimization is the amount of primary energy saved by maximizing the use of secondary energy within the limits established by the laws of thermodynamics. Secondary energy results from the inputs of primary energy and would be directly discharged into the environment without energy conservation measures.

The general optimization problem is nonlinear, involving integral equations. It can be made accessible to linear programming by forgoing the use of some secondary energy, mainly in exhausts. In a special case study we examine the impact of heat exchangers, heat pumps and cogeneration of heat and electricity on the conservation of primary energy in the Netherlands, West Germany and Japan. The Simplex Algorithm leads to the following results: Heat exchangers are the most important instrument for energy conservation. The possibility of conservation depends heavily on the shape of the energy demand profile. This profile shows the amount of energy (in enthalpy units) demanded at the various quality (≃ temperature) levels. The Japanese profile, which shows steadily decreasing demand from higher to lower qualities, results in the biggest savings, up to 60%. West Germany with its peak demand of high quality heat has only a potential ov saving 50% of the primary energy input. The Netherlands with a high demand of low temperature heat may save as much as Japan.

The above results were obtained under idealizing assumptions and give therefore only the extreme limits of energy conservation. The method can be applied to more realistic models. It may help to design efficient strategies for the abatement of environmental pollution. If one excludes cost related to security of supply and varying currency exchange rates, cost optimization can also be performed with the same mathematics.

## 1. Introduction

An industrial system is characterized by its energy demand profile. Energy demand and energy supply are reported by their heat value H (enthalpy) in official statistics. Optimization is only meaningful if the exergy content of enthalpy H, i.e. the potential to perform mechnical or electrical work, is taken into account. Thus, it is useful to introduce the concept of quality Q of enthalpy H [1] and define

quality Q := amount of exergy / amount of enthalpy

If one bases the definiton of Q on the optimum process for the conversion of heat at temperature T into work, it is simply given by the Carnot factor

$$Q = 1 - \frac{T_0}{T}, \tag{1.1}$$

where $T_0$ is the temperature of the reservoir to which the waste heat is rejected; usually one takes $T_0$=298.15K as the temperature of the environment.

Van Gool et al. introduced the HSQD-diagram which shows the enthalpy (H) supplied (S) to industrial processes versus the quality (Q) of energy demanded (D) by these processes [1]. Examples for the economies of the Netherlands, West Germany [1] and Japan [2] are given in Fig.1 . Since no energy is lost in any process (first law of thermodynamics), it can be reused. However, its quality never increases but mostly decreases in a process (second law of thermodynamics); thus, reuse can only occur at lower quality levels (heat cascading via heat exchangers). Delivery to higher quality levels is possible, though, if heat pumps are employed which require extra exergy input. Additionally it may be energy saving to produce electricity and heat together (cogeneration). Within these limits drawn by the laws of physics, we want to explore strategies of energy optimization in the sense of servicing a given energy demand profile eihter by a minimum amount of primary energy (of Q=1) or at minimum cost.

It is the purpose of this paper to present the mathematical framework for the optimization procedure including heat exchangers, heat pumps and cogeneration. The results will be applied to simplified models of the Dutch, West German and Japanese industrial economies.

## 2. The General Framework for Optimization

### 2.1 Basic Data Required

It is convenient to define an amplified quality scale by

$$q := 10 \cdot Q = 10 \cdot \left[1 - \frac{T_0}{T}\right]. \tag{2.1}$$

This means that energy like electricity which can be transformed into work completely has quality q=10, while useless energy in form of heat at temperature $T_0$ of the environment has quality q=0. As primary energy carrier (PEC) we have one sort of fuel F

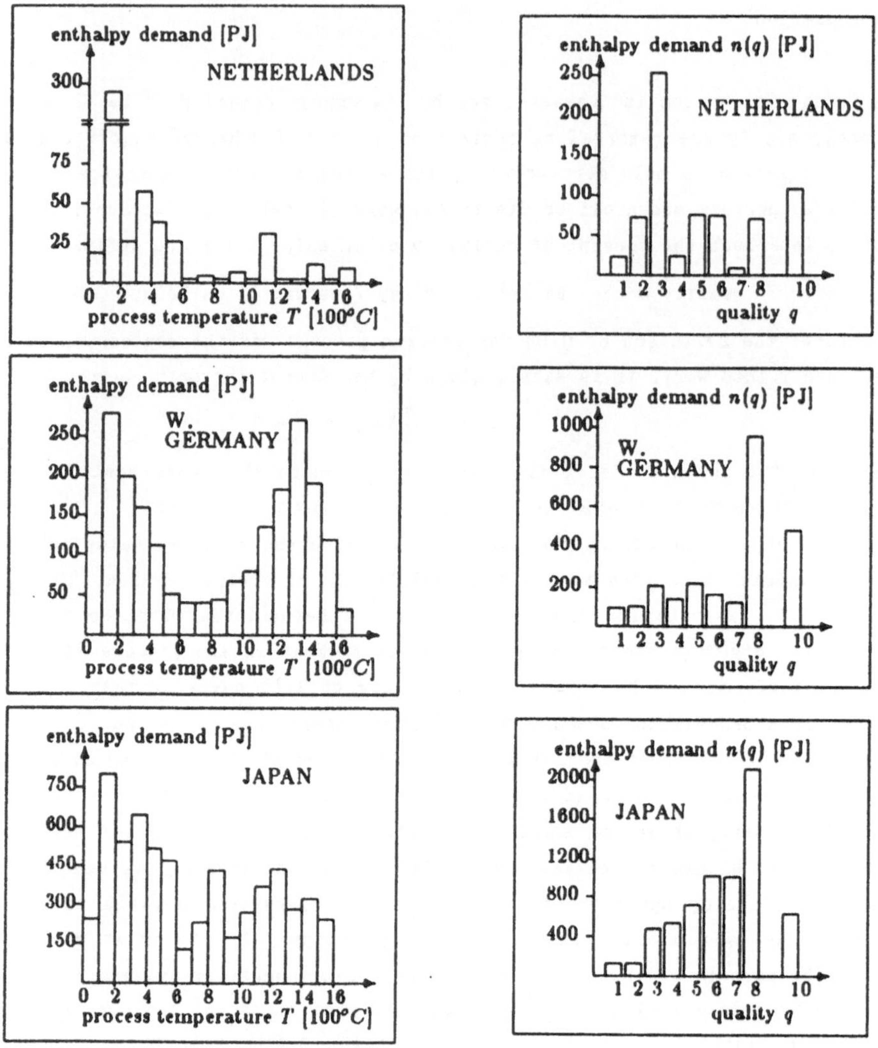


Figure 1: Enthalpy demanded at the different temperature and quality levels by the Dutch, West German and Japanese industry. The conversion from temperature to quality intervals is done according to $H(Q)\Delta Q = H(T)\Delta T$, $\Delta Q = (T_0/T^2)\Delta T$ and $q:=10\cdot Q$. Quality Q=1 (q=10) refers to electricity.

which has quality q=10. Let the energy demand profile of an industrial system be given by the demand vector $|n(q)\rangle = |n(q_1), n(q_2), \ldots, n(10)\rangle$, with

$n(q)$ = number of annually demanded enthalpy units of quality q.

In the following if we talk about an enthalpy q, we mean an amount of enthalpy (H) of quality q. In order to be able to do any calculations we have to assume knowledge of some technical and economical quantities:

1) Conversion efficiencies

$\eta(q,F)$ = fraction of one unit of demand enthalpy q generated from one unit of fuel F (primary conversion efficiency)

$\eta'(q,q')$ = fraction of one unit of demand enthalpy q generated from one unit of secondary energy q' (secondary conversion efficiency)

By definition secondary energy is the energy which, without energy conservation measures, would be waste energy discharged to the environment in the considered conversion and production processes. It comprises process heat dissipated by cooling as well as energy losses by friction, exhausts and transportation.

2) Secondary energy generation ratios

$\nu(q',q,F)$ = fraction of one unit of secondary energy q' generated during the process of *supplying* one unit of fuel F to demand level q

$\nu'(q',q,q'')$ = fraction of one unit of secondary energy q' generated during the process of *supplying* one unit of secondary energy q" to demand level q

$\mu(q',q)n(q)$ = amount of secondary energy q' recovered from level q after its demand n(q) has been satisfied by fuel F and/or secondary energy

For example in a factory melting glass the amount N(q,F) of fuel F is used to produce the necessary heat n(q). The fraction $\eta(q,F)N(q,F)$ of the enthalpy contained in the fuel melts the glass and increases its internal energy. The fraction $\nu(q',q,F)N(q,F)$ will be released to the environment as hot fumes (or may be used again as secondary energy elsewhere). In case secondary energy N'(q,q") from other processes q" is used to melt the glass the corresponding fraction would be $\nu'(q',q,q'')N'(q,q'')$. The internal energy of the glass will be decreased when it becomes solid again. The fraction $\mu(q',q)n(q)$ gives the released energy that can be used again. Obviously the magnitude of this fraction depends only on the internal energy of the glass which is a state function. If q' is equal to zero in one of the conversion efficiencies this means that reuse of this part is not possible and the enthalpy has to be discharged to the environment. The restrictions on the respective sets $(\eta,\nu)$, $(\eta',\nu')$ and $\mu$ are a consequence of the first law of thermodynamics.

## 2.2 Primary and Secondary Energy Supply

In order to derive the equations for the total amount of fuel used in the process and the total amount of available secondary energy, we need some more definitons.

N(F) = total number of enthalpy units of fuel F (q=10)

S(q') = number of enthalpy units of secondary energy of quality q' generated by supplying fuel F and secondary energies S(q") to all demand levels q according to the optimization procedure

Every demand level receives enthalpy units either from direct heating or in the form of secondary energy from other levels.

N(q,F) = number of units of PEC F supplied directly to demand sector q

N'(q,q') = number of secondary energy units q' supplied to q

The demand n(q) is satisfied by primary and/or secondary energy:

$$n(q) = \eta(q,F)N(q,F) + \sum_{q''} \eta'(q,q'')N'(q,q'') \tag{2.2}$$

We can now write the generated secondary energy as

$$S(q') = \sum_{q} \left\{ \mu(q',q)n(q) + \nu(q',q,F)N(q,F) + \sum_{q''} \nu'(q',q,q'')N'(q,q'') \right\}. \tag{2.3}$$

The secondary energy is available for servicing different demand levels. Using

x(q,q')S(q') = fraction of total amount of available secondary energy q' supplied to demand sector q

the number of units of demand enthaly q to be provided directly from F is

$$\delta n(q) = n(q) - \sum_{q'} \eta'(q,q')\ x(q,q')\ S(q')\ . \tag{2.4}$$

Thus the total amount of PEC F required by the industrial system is

$$N(F) = \sum_{q} N(q,F)\ , \tag{2.5}$$

where

$$N(q,F) = \frac{\delta n(q)}{\eta(q,F)} = \frac{1}{\eta(q,F)} \left\{ n(q) - \sum_{q'} \eta'(q,q')\ x(q,q')\ S(q') \right\}. \tag{2.6}$$

From the definitons it follows that

$$N'(q,q') = x(q,q')\ S(q')\ . \tag{2.7}$$

Inserting this and Eq.2.6 into Eq.2.3, thereby changing $q' \to q''$, one arrives at the integral equation

$$S(q') = \sum_{q} \left\{ \left[ \mu(q',q) + \frac{\nu(q',q,F)}{\eta(q,F)} \right] n(q) - \sum_{q''} \left[ \frac{\nu(q',q,F)\eta'(q,q'')}{\eta(q,F)} - \nu'(q',q,q'') \right] x(q,q'')\ S(q'') \right\}. \tag{2.8}$$

The equations 2.5, 2.6 and 2.8 represent the *fundamental supply equations*. Once the x(q,q') have been determined by minimizing N(F), all interesting quantities are known. Clearly, since one cannot distribute more secondary energy than is produced, the equations are subject to the restrictions

$$\sum_{q} x(q,q')\ S(q') \le S(q') \qquad \text{for all } q'. \tag{2.9}$$

These equations are too complicated to be solved exactly. One way of treating the system of equations is to make some special technical assumptions in order to simplify the mathematics.

## 3. A Special Case Study: Heat Cascading, Heat Pumps and Cogeneration

### 3.1 Some Simplifying Assumptions

We are now leaving the general optimization scheme in order to build a simplified model of the energy flows. The purpose of this model is to explore the outmost thermodynamic limits to energy conservation. We will introduce special technical devices that transport energy from one level to another. There will be ten quality levels for heat, running from zero to nine, to which all demand has been aggregated, see Fig.1 .

As a first step we assume that we may neglect all secondary energy generated when fuel F and secondary energy are supplied to the demand levels:

$$\nu(q',q,F) = \nu'(q',q,q'') = 0 \qquad (3.1)$$

For a technically reasonable value of $\eta(q,F)=0.85$ [1] the losses by friction, exhausts etc. are less than 15%. Thus, if primary or secondary energy is delivered to a demand level q, the fractions $[1-\eta(q,F)]N(q,F)$ and $[1-\eta'(q,q')]N'(q,q')$ are considered to be unrecoverably lost. What we gain mathematically is that now secondary energy $S(q')$ is a function of known quantities and not subject to optimization. Eq.2.3 turns into

$$S(q') = \sum_{q} \mu(q',q)\ n(q) \ . \qquad (3.2)$$

The next step is a rather crude assumption: all *in*-enthalpy of quality q is also available as *out*-enthalpy of the same quality. This means that *all* enthalpy delivered to a demand level of quality q is available as secondary energy of the *same* quality. But we will not allow the available secondary energy to be used on the same level. In this way we simulate quality losses in the process as they occur according to the second law of thermodynamics. Using the Kronecker symbol $\delta$ we can write

$$\mu(q',q) \rightarrow \delta_{qq'} \ ; \quad \delta_{qq'}=1 \text{ if } q=q' \text{ and 0 otherwise,} \qquad (3.3)$$

$$\eta'(q,q'=q) = 0 \ . \qquad (3.4)$$

These simplifications make the calculations a lot easier, and since we try to explore the outmost limits to energy saving, we should be allowed to act like this. A future task will be to use a specific $\mu \leq 1$ for each pair q,q'. Inserting Eq.3.3 into Eq.3.2 results in

$$S(q') = n(q') \ . \qquad (3.5)$$

In general the conversion efficiencies $\eta'(q,q')$ are functions describing the devices that convert secondary energy, such as heat exchangers and heat pumps, as well as the mechanisms of transportation from the location where the energy q' is avai-lable to the one where q is needed. As a further simplification we assume that all processes take place in the *same location* at the *same time*; thus, there are no losses of energy due to transportation.

## 3.2 Electricity in the Process

Before specifying the conversion devices we have to introduce electricity into our model.

$n(10)$ = fixed demand of enthalpy units of electricity ($q=10$)

$N(E,F)$ = amount of fuel used to produce electricity in all-electric power plants (i.e. without cogeneration)

For reasons given in the following sections we have to make a distinction between $n(10)$ and

$S(10)$ = electricity produced in all-electric power plants; (The concept of secondary energy is formàlly extended to include electricity which results from burning fuel F.)

The total amount of fuel required is now

$$N(F) = \sum_q N(q,F) + N(E,F) \ , \quad \text{with} \quad N(E,F) = \frac{S(10)}{\eta(10,F)} \ . \tag{3.6}$$

According to Ref.[1] the efficiencies to produce electricity and heat from fuel may be taken as $\eta(10,F)=0.4$ and $\eta(q,F)=0.85$ respectively.

## 3.3 Conversion Devices

In the following we will use indices for the examined conversion devices to indicate that a quantity refers to a scenario where a special device is being employed:

ex : heat exchangers ; p : heat pumps ; c : cogeneration

### a) Heat Cascading

Heat cascading means that heat from a certain level q' is used again on a *lower* level q. The conversion is done by heat exchangers which use a medium like hot steam or hot water to transport heat. In Ref.[1] their enthalpy efficiency was estimated to be $\eta'_{ex}(q,q')=0.90$. It can be shown that this will not lead to a decrease of quality q' by more than one unit. Since delivery to the same quality level is already excluded, no further measures have to be taken to deal with quality (or temperature) losses.

### b) Heat Pump

Heat pumps transport energy from a level q' to a *higher* level q. For doing so they need an extra exergy input. We assume that exergy is provided in the form of electricity, though it would be feasible to use fuel instead. $S(10)$ now services not only $n(10)$ but also the heat pumps:

$x_{q'}(q,10)\cdot S(10)$ = fraction of electricity S(10) used to pump heat from q'<q to q

$x(10,10)\cdot S(10)$ = amount of electricity S(10) servicing fixed demand n(10)

Obviously the sum of all fractions $x_{q'}(q,10)$ and $x(10,10)$ must be equal to unity.

We treat heat pumps as Carnot engines operating between the temperature levels T and T'<T. The quality levels q,q' corresponding to the temperatures T,T' are obtained from Eq.2.1. In this case the Carnot factor, the inverse of which gives the heat pump efficiency [3], is

$$\eta_{Carnot} = \frac{\text{electrical energy}}{\text{heat delivered to q from q'}} = \frac{T - T'}{T} \tag{3.7}$$

$$= \frac{x_{q'}(q,10)S(10)}{\eta'_p(q,q')x_p(q,q')n(q')} = \frac{q - q'}{10 - q'} . \tag{3.8}$$

Here, $x_p(q,q')n(q')$ is the fraction of secondary energy $S(q')=n(q')$ pumped to level q. One should not mix up the heat pump efficiency as defined above and the heat conversion efficiency $\eta'_p(q,q')$ which is a measure of the ratio of the amount of heat delivered to q to the amount of heat pulled out of q'. This ratio is greater than 1 since exergy in the form of electricity is added. Straightforward calculations using the first and second law of thermodynamics show that the heat conversion efficiency $\eta'_p(q,q')$ of the heat pump is given by

$$\eta'_p(q,q') = \frac{T}{T'} = \frac{10 - q'}{10 - q} , \tag{3.9}$$

so that Eq.3.8 turns into

$$x_p(q,q')\ n(q') = x_{q'}(q,10)\ S(10)\ \frac{10 - q}{10 - q'} . \tag{3.10}$$

It is clear that employing a heat pump only makes sense if one saves more enthalpy units of PEC for heat production than are needed to produce electricity to drive the pumps.

c) Cogeneration

Producing electricity creates a lot of waste heat that is normally thrown away. It is, however, possible to use this heat as secondary energy. One looses some of the efficiency in both electricity and process heat production, but the overall enthalpy efficiency is higher than before. In cogeneration plants the conversion efficiencies for the production of electricity and heat q from fuel have typically values of $\eta_c(10,F)=0.25$ and $\eta_c(q,F)=0.55$ respectively, thus leaving only a waste heat fraction of $\eta_c(0,F)=0.20$ [1]. We define

$x_c(q)n_c(10)$ = amount of electricity cogenerated with heat of quality q

and assume that total amount of electricity produced in all cogeneration processes

$$n_c(10) = \sum_q x_c(q)\ n_c(10) , \tag{3.11}$$

is exclusively used to satisfy fixed demand n(10) and not for heat pump demand. This leads to a reduction of the amount of electricity generated in all-electric power plants:

$$x(10,10)\ S(10) = n(10) - n_c(10) \tag{3.12}$$

Although $n_c(10)$ is variable it is completely determined by Eq.3.11, allowing us to treat $x_c(q)n_c(10)$ as a single variable that is subject to optimization, thus keeping the whole probler linear.

Cogeneration is only used up to a certain heat level $q_0=6$ in order not to loose too much electricity efficiency. We express this in Eq.3.13 by using the Heavyside function $\theta|7-q|$ which is equal to one if $q<7$ and zero otherwise.

## 3.4 The Energy Optimization Problem

a) Required Primary Fuel

Adding up the effects of the three discussed conservation devices the total amount of fuel N(F) needed to fulfill all demand of electricity and process heat is given by

$$N(F) =$$

$$\sum_{q=1}^{9} \frac{1}{\eta(q,F)} \left\{ n(q) - \sum_{q=q+1}^{10} \eta'_{ex}(q,q')\ x_{ex}(q,q')\ n(q') \right.$$

$$\left. - \theta|7-q|\ \frac{\eta_c(q,F)}{\eta_c(10,F)}\ x_c(q)\ n_c(10) - \sum_{q=0}^{q-1} \theta|q,q'|\ \frac{10-q'}{10-q}\ x_p(q,q')\ n(q') \right\}$$

$$+ \frac{n_c(10)}{\eta_c(10,F)} \tag{3.13}$$

$$+ \left[ \frac{n(10) - n_c(10)}{\eta(10,F)} + \frac{1}{\eta(10,F)} \sum_{q=1}^{9} \sum_{q'=0}^{q-1} \theta|q,q'|\ \frac{q-q'}{10-q}\ x_p(q,q')\ n(q') \right]$$

The sum of the terms in curly braces gives the fuel required for direct heating, where the savings by heat exchangers, cogeneration and heat pumps are subtracted from the demand of heat n(q). The ratio $n(q)/\eta(q,F)$ represents the amount of fuel needed when no conservation measures are taken. The term in the next row gives the fuel fed into cogeneration plants and the last sum in brackets is the fuel for all-electric power plants, the latter consisting of one part serving the remaining electricty demand, not satisfied by cogeneration, and the other serving the heat pumps. The func tion $\theta|q,q'|$ is zero, if the heat pump does not deliver more enthalpy than it uses, and one otherwise.

The parameters to be determinded by optimization are the 45 $x_{ex}(q,q')$, the 6 $x_c(q)n_c(10)$ and up to 45 $x_p(q,q')$.

b) Restrictions

Minimization of N(F) in Eq.3.13 is subject to a number of restrictions which are due to the first law of thermodynamics (conservation of energy); as a matter of fact, all parameters have to be non-negative:

- The amount of all secondary energy taken from level q' in order to be exchanged or pumped must not exceed n(q'), i.e. the sum of all fractions $x_{ex}(q,q')n(q')$ and $x_p(q,q')n(q')$ has to be less than or equal to n(q').
- All conservation devices together are not allowed to deliver more enthalpy units to a level q than are demanded by n(q).
- Electricity production in cogeneration is not allowed to exceed the fixed demand n(10).

c) The Simplex Algorithm

The energy optimization problem is one of linear programming and fits exactly into the *SIMPLEX* algorithm as described by Foulds [4]. The *objective* function is N(F) from Eq.3.13. Inequality constraints are changed into equalities by adding *slack* variables. We can identify these slack variables with the unrecoverable waste heat, the necessary direct heat and the fraction of the fixed electricity demand that is served from all-electric power plants, i.e. x(10,10)S(10).

## 3.5 Results for The Netherlands, West Germany and Japan

In an earlier paper [1] we examined demand profiles of the Netherlands and West Germany, which are shown in Fig.1. They are some years old and one may doubt the validity of the numerical values, but the important thing that determines the possibilities of energy conservation is the shape of the profile. This at least should be roughly correct, though passive conservation measures like insulation may have led to changes in the low quality levels in recent years. On the other hand our goal is to demonstrate the capabilities of the model, so we treat these profiles just as examples.

In [1] we used technical and physical criteria to reduce the number of options. For example heat pumps were only allowed to pump over one level. In Tab.1 we show these results and the results from our new program.

The numbers in the last row of Tab.1, for example, indicate that employing heat exchangers, heat pumps and cogeneration (ex,p,c) and allowing the use of secondary energy from electricity (+n(10)) in the Netherlands the PEC input is reduced from 976 PJ/a to 351 PJ/a, that is to 36% of the former PEC input; in West Germany, under the same conditions, 50% of the old input still has to be provided, while in Japan the number is only 33%. The difference in the possible savings of West Germany and Japan results from the demand profile. From Fig.1 one sees, that the ratio of the highest peak to the next peak is much bigger in the case of West Germany. Since the largest peak is at q=8, nearly all of this has to be provided by primary energy and this is more than can be reused on all lower levels.

Comparing the results of Ref.[1] with the present ones for the "+n(10)" case, we find that there is no difference if one uses only exchangers or exchangers together with cogeneration. In the use of the heat pump there are differences of 20% in case of the Netherlands and 6% in case of West Germany, which are due to the fact that in the present program the maximum amount of heat available to the heat pump on a certain level is just the demand on that level, whereas in Ref.[1] all excess heat from higher levels was assumed to be available, too. The advantage of the new model lies in its flexibility against change of parameters and in the possibility to expand it to more realistic scenarios.

Heat exchangers turn out to be the main source of energy conservation under our idealizing assumptions, especially that of space and time unity, whereas heat pumps and cogeneration contribute only small amounts. However, in more realistic scenarios these devices will play a much bigger role without changing the savings potentials dramatically. This will be reported elsewhere.

We are aware that we did not include the energy demand that will result from the production of the technical devices that save energy. This is also a task for the future. Cost optimization, which excludes the cost of redundancies and variations of currency exchange rates, is possible with the same mathematics. Essentially one has to multiply S(q') in Eq.2.6 by the cost difference between direct heating and the use of secondary energy q' and maximize the total amount of saved cost represented by the second term on the right hand side of Eq.2.6 summed over all demand levels q.

| | NETHERLANDS | | | WEST GERMANY | | | JAPAN | |
|---|---|---|---|---|---|---|---|---|
| | +n(10) | -n(10) | Ref.[1] | +n(10) | -n(10) | Ref. [1] | +n(10) | -n(10) |
| no saving | 976<br>100% | | | 3544<br>100% | | | 8765<br>100% | |
| ex | 483<br>50% | 597<br>61% | 483<br>50% | 1816<br>51% | 2325<br>66% | 1816<br>51% | 3381<br>39% | 4049<br>46% |
| ex+p | 353<br>36% | 458<br>47% | 421<br>43% | 1754<br>50% | 2263<br>64% | 1637<br>46% | 2926<br>33% | 3595<br>41% |
| ex+c | 393<br>40% | 494<br>51% | 393<br>40% | 1816<br>51% | 2325<br>66% | 1816<br>51% | 3381<br>39% | 4049<br>46% |
| ex+c+p | 351<br>36% | 452<br>46% | 393<br>40% | 1755<br>50% | 2263<br>64% | 1637<br>46% | 2926<br>33% | 3595<br>41% |

Table 1: Optimization results for three countries. The absolute numbers show the PJ per year of primary energy required to service the demand of Fig.1, if optimization is performed with combinations of heatexchangers (ex), heat pumps (p) and cogeneration (c); the percentages are taken with respect to the 'no savings' scenario; +n(10) indicates inclusion, -n(10) exclusion of waste heat from electricity. The results of Ref.[1] are shown for comparison.

The next step will be to make the model more realistic by including heat transportation losses and losses of process heat during use (e.g. in chemical plants). This will lead to regionalization of the model. Also one has to use seperate profiles for the demand of energy and the available secondary energy.

**Acknowledgement**

We would like to thank Kokichi Itoh for providing the data for the Japanese energy demand profile.

**References**

[1] W. van Gool, R. Kümmel, in: Energy Decisions for the Future, Vol.I, (Miyata, Matsui Eds.), Tokyo, 1986, pp. 90-106

[2] Study Committee for Long-Term Energy Strategies, Electric Power Demands and Electrification, Tokyo, 1980

[3] J. Fricke, W.L. Borst, Energie, Oldenbourg, München, 1984, pp. 77-78

[4] L.R. Foulds, Optimization Techniques, Springer, New York, 1981,

# DISCRETIZATION AND MARKOV MODELING OF A STATE VARIABLE IN DYNAMIC PROGRAMMING

---

## IMPACTS ON THE RESULTS OF OPTIMAL OPERATION AT ELECTRICITE DE FRANCE OF THE FRENCH PRESSURIZED WATER REACTORS (MODEL GARP)


Marie MOATTI

Electricité de France - Etudes Economiques Générales

2, rue Louis Murat - 75384 PARIS CEDEX 08



**ABSTRACT**

This paper presents some of the conclusions drawn from the study of GARP, a dynamic programming model that is used to optimize the energy production of French nuclear generating units within a given year.

The aim of this paper is to study the impact of different factors on the optimal values given by GARP.

After a brief description of GARP, we'll determine the effect of energy and time discretization.

Then, we'll measure the consequences of modeling the state of each reactor (either running or going through a forced outage) with a two-state Markov process within the framework of the precited discretization.

The conclusions of this paper can be more generally applied to a dynamic programming model using state variables discretization and using a two-state Markov process for another state variable.


## I. DESCRIPTION OF THE PROBLEM

### 1. Description of GARP

GARP optimizes the operation of French nuclear generating units within a given year.

In this model, each reactor can be considered as a reserve of energy that has to be optimally released before the next outage for maintenance and refueling ; the maintenance and refueling schedule has been previously determined by another model (RELAX).

The method used in GARP to solve this problem is dynamic programming.

Because of the high dimension of this problem (there are about 50 PWR in France now), a method of relaxation is used ; GARP can then be considered as a succession of local problems, each one pertaining to the determination of optimal energy production for a given reactor.

In this paper we'll study discretization and Markov modeling within the frame of one of these local problems only. Thus, we'll be interested in the optimization of the energy generated by a single reactor, all other control variables being held constant.

## 2. Description of the local problem

The year is made of 52 weeks and for each week t, we want to determine the optimal energy generation $U_t$. Thus, the control variables are the $U_t$'s.
These $U_t$'s are determined so as to maximize a concave function that is the valorization of the energy generated by the reactor within the year.
This criterion can be written the following way :

$$\max_{(U_t)t=1,52} \sum_{t=1}^{52} f_t(U_t) + \text{boundary conditions}$$

where $f_t(U_t)$ is the valorization of $U_t$ for each t.
The constraints are the following :

$$\forall\, t, \quad 0 \leqslant U_t \leqslant U_{max} \qquad (1)$$

$$\sum_{t=1}^{52} U_t \leqslant L \qquad (2)$$

(1) derives from a power constraint and expresses the fact that the energy released during a given week is limited by the maximum power of the reactor.
(2) is an energy constraint and expresses the fact that the reactor is a reserve of energy, i.e., the total amount of energy that can be generated during the year is upper bounded by a limit L that can be much smaller than 52 x $U_{max}$!

In this problem, the state variables are :

1) the time t
2) the amount of energy $l_t$ left in the reactor on week t
3) the state $e_t$ of the reactor on week t, either running ($e_t$=R) or going through a forced outage ($e_t$=F) ; a two-state Markov process is used to model this last state-variable. The transition probability matrix defining this Markov process is

$$\begin{bmatrix} P_{FF} & P_{RF} \\ P_{FR} & P_{RR} \end{bmatrix}$$

where : $P_{FF}$ is the transition probability from state F to state F
$P_{RR}$ is the transition probability from state R to state R
$P_{FR} = 1 - P_{FF}$
$P_{RF} = 1 - P_{RR}$.

Outages that last for less than a week are also taken into account in the model but in a deterministic way : we simply derate the unit capacity with an average availability factor $d_c$ standing for short outages. Thus, when releasing energy, the reactor power is actually upper bounded by $d_c P$, where P is the unit installed capacity.

Of course, state variables 1 and 2 are continuous variables. But in order to solve the problem here, discrete state variables are used. Actually, the time-period chosen in GARP is the week, and the step chosen for the reserve of energy is the maximum amount of energy that can be released during a week.

Let S be the optimal value function. $S(e_t, l_t, t)$ represents the best energy valorization that can be obtained between week t and the end of the year, if the amount of energy left in the reactor on week t is $l_t$, and if the state of the reactor on week t is $e_t$. Thus, the criterion written previously is equal to $S(R, l_1, 1)$ if we suppose that the reactor is running on week 1.
This criterion and the optimal operation of the reactor during the year are found using the recurrence relation that characterizes dynamic programming :

$$S(F, l_t, t) = P_{FF}\, S(F, l_t, t+1) + P_{FR}\, S(R, l_t, t+1).$$

$$S(R, l_t, t) = \underset{U_t}{\text{Max}}\ f_t(U_t) + \underset{e_{t+1}/e_t=R}{E} \left[ S(e_{t+1}, l_t - U_t, t+1) \right]$$

$$= \underset{U_t}{\text{Max}}\ f_t(U_t) + P_{RR}\, S(R, l_t - U_t, t+1) + P_{RF}\, S(F, l_t - U_t, t+1)$$

## II. IMPACT OF TIME AND ENERGY DISCRETIZATION ON OPTIMAL VALUES

### 1. The deterministic model

We saw that the amount of energy left in the reactor is a discretized state variable, and that the step chosen for this variable is the maximum amount of energy that the reactor can release during one week.
In GARP, we assume that, whatever t is, the optimal value function is linear between two successive discretization points of the amount of energy left in the reactor.
Thus, if we need the optimal value for state $(e_t, l-\alpha, t)$ with $\alpha \in [0,1]$, we'll assume that we have :

$$S_l\, (e_t, l-\alpha, t) = (1-\alpha)\, S(e_t, l, t) + \alpha\, S(e_t, l-1, t).$$

GARP may need to use this optimal value by starting for instance at level l on week t-1 and releasing the energy α during week t-1.
But we know that for a given week t, the curve showing the optimal values as a continuous function of the level of energy in the reserve is a concave curve (see Figure 1).

Figure 1 :

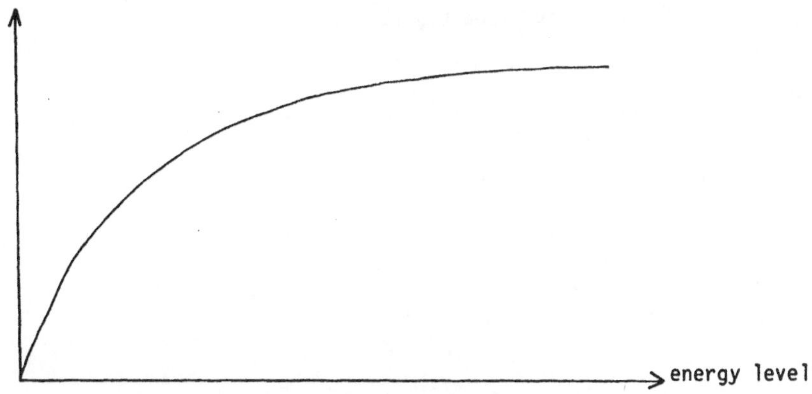


The true optimal value function for a given week t.

Thus, because of the optimal value function concavity, $S_1(e_t,1-\alpha,t)$ is a low estimate of the true optimal value $S_2(e_t,1-\alpha,t)$ (see Figure 2).

Figure 2 :

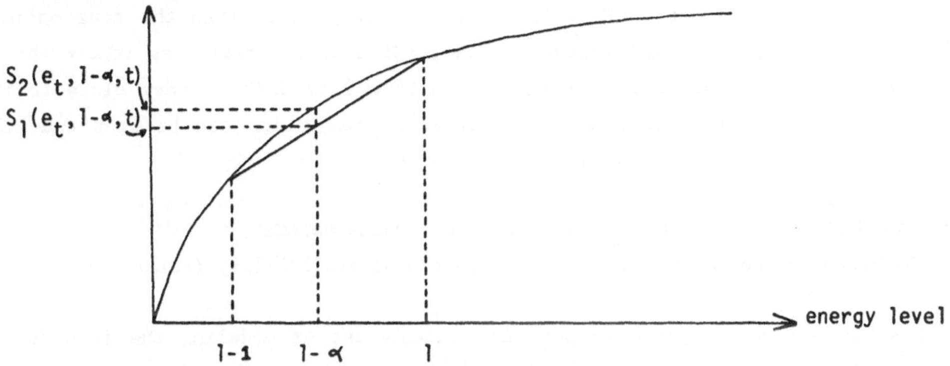


The linearity assumption.

Therefore, since the recurrence relation inevitably leads GARP into computing optimal values for intermediary energy levels, the optimal values computed by GARP will be smaller than the true optimal values that we would obtain without using the linearity assumption.

An easy way to have an estimation of the error induced by the linearity assumption is to suppress the Markov process that models the long forced outages : instead, we include these outages in the average availability factor used to derate the reactor capacity.

The problem becomes fully deterministic. Its resolution is straightforward in this case and we can find the true optimal values by hand. Then, we can compare them to the ones resulting from GARP (see Figure 3).

Figure 3 :

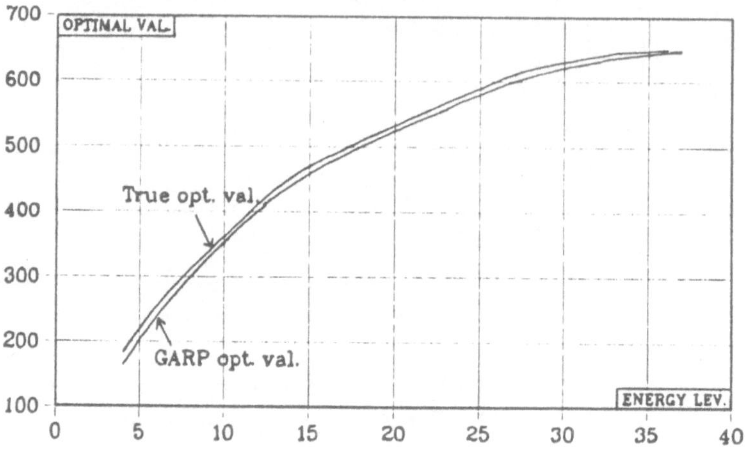


GARP optimal values and true optimal values
for different energy levels on week 1.

As we perceived it before, GARP optimal values are smaller than the true optimal values. However, GARP optimal values are very close to the real ones (the error is about 0,5 %). Thus, the local criterion resulting from GARP is very close to the real optimal criterion. Therefore, the linearity assumption can be kept and GARP enables to operate the reactor on a satisfying basis.

## III. INTRODUCING A MARKOV PROCESS FOR THE LONG FORCED OUTAGES

### 1. Different Markov processes for the same global availability factor

In this section, we're going to see the consequences of modeling the long forced outages with a Markov process, within the framework of the discretization and linearization method used in GARP.

The Markov process introduces a random factor in the model, but the weight of this random factor can vary according to the transition probabilities chosen to model the forced outages.

Whatever Markov process is used, the model is devised so as to keep the global average availability factor constant. We're going to compare the optimal values obtained for different Markov processes.

First let's define the "weight" of the random factor introduced by the Markov process : it is the average availability factor $d_1$ standing for long outages, i.e., the unconditioned stationary probability of being in state R.

Thus we have

$$\begin{bmatrix} 1 - d_1 \\ d_1 \end{bmatrix} = \begin{bmatrix} P_{FF} & P_{RF} \\ P_{FR} & P_{RR} \end{bmatrix} \begin{bmatrix} 1 - d_1 \\ d_1 \end{bmatrix}$$

and therefore :

$$d_1 = \frac{1 - P_{FF}}{2 - P_{FF} - P_{RR}}$$

when $P_{FF}+P_{RR} \neq 2$, which will always be the case for the different Markov processes we'll consider here.

We saw that $d_c$ is the availability factor standing for short outages. Let d be the global average availability factor resulting from both the long forced outages and the short outages. Whe have :

$$d = d_1 \, d_c$$

where d is a constant. Thus, if we suppress the Markov process by setting $P_{FF}=0$ and $P_{RR}=1$, the whole weight of the forced outages will be transferred to the short outages (we'll have $d_1=1$ and $d_c=d$). In this case, when releasing energy, the reactor power will be upper bounded by dP.

When there is a Markov process, i.e., when $d_1 \neq 1$, the expectation of the optimal value for state $(e_t, l_t, t)$ is given by :

$$\underset{e_t}{E}\left[S(e_t, l_t, t)\right] = d_1 \; S(R, l_t, t) + (1 - d_1) \; S(F, l_t, t)$$

When there isn't any Markov process, let $S(l_t, t)$ be the optimal value computed for state $(l_t, t)$.

We're going to see how the optimal values vary when we modify the respective weights of long and short outages by changing the transition probabilities $P_{FF}$ and $P_{RR}$. Thus, we'll compare the expectations of optimal values resulting from stochastic models defined for different Markov processes to the optimal values resulting from the deterministic model defined for $P_{FF} = 0$ and $P_{RR} = 1$.

## 2. Comparing $E\left[S(e_t, l_t, t)\right]$ and $S(l_t, t)$ for different Markov processes

Using model GARP, we observe the following results, whatever l and t are :

- if $P_{FF} + P_{RR} = 1$ $\quad E\left[S(e_t, l_t, t)\right] = S(l_t, t)$
- if $P_{FF} + P_{RR} > 1$ $\quad E\left[S(e_t, l_t, t)\right] \leqslant S(l_t, t)$
- if $P_{FF} + P_{RR} < 1$ $\quad E\left[S(e_t, l_t, t)\right] \geqslant S(l_t, t)$

Two factors may change the optimal values when introducing a Markov process for the long outages. First, the operation of the reactor becomes more difficult because we are now in a stochastic environment, and that should have a decreasing effect on the expectations of the optimal values. Second, when the reactor is in state R, it can generate power $P_{max} = d_c P$ (where P is the reactor capacity), whereas it can only generate power $P_{max} = d_1 d_c P$ in the deterministic model ; in other words, <u>when running</u>, the reactor generates more power in the stochastic model than in the deterministic one ; thus the reactor needs less generation decisions to empty itself, and therefore the error stemming from the linearity assumption should be smaller in the stochastic case than in the deterministic one : that should have an increasing effect on the expectations of the optimal values.

Apparently, the first factor prevails on the second one when $P_{FF} + P_{RR} > 1$, and it's the contrary when $P_{FF} + P_{RR} < 1$.

When $P_{FF} + P_{RR} = 1$, the effects of these two factors seem to annihilate, and the expectations of the optimal values are equal to the optimal values obtained in the deterministic case.

## 2.a What happens when $P_{FF} + P_{RR} = 1$

In this case, we have :

$$P_{RR} = d_1,\ P_{FR} = 1 - P_{FF} = P_{RR},\ P_{RF} = P_{FF}$$

and the recurrence relation previously established becomes :

$$\begin{aligned} S(F,l_t,t) &= P_{FF}\ S(F,l_t,t) + P_{RR}\ S(R,l_t,t+1) \\ &= (1 - d_1)\ S(F,l_t,t+1) + d_1\ S(R,l_t,t+1) \\ &= E\left[S(e_{t+1},l_t,t+1)\right] \end{aligned}$$

$$\begin{aligned} S(R,l_t,t) &= \underset{U_t}{Max}\ f_t(U_t) + P_{RR}\ S(R,l_t-U_t,t+1) + P_{RF}\ S(F,l_t-U_t,t+1) \\ &= \underset{U_t}{Max}\ f_t(U_t) + E\ \left[S(e_{t+1},l_t-U_t,t+1)\right] \end{aligned}$$

We notice here that the transition probabilities defining the Markov process don't intervene at all in the recurrence relation. Thus, knowing that we are in state R or in state F at week t doesn't help here. A two dimension state vector (energy and time) would be enough to solve the problem, since $E\left[S(e_t,l_t,t)\right]$ can be computed just by knowing the unconditioned expectation $E\left[S(e_{t+1},l_t,t)\right]$.

Actually, it can be demonstrated that we have :

$$\forall\ (l,t)\ E\left[S(e_t,l,t)\right] = S(l_t,t)$$

where $S(l_t,t)$ is the optimal value computed for state $(l_t,t)$ in the deterministic version.

## 2.b What happens when $P_{FF} + P_{RR} > 1$ or $P_{FF} + P_{RR} < 1$

To understand the property pertaining here to Markov processes with $P_{FF} + P_{RR} > 1$ or $P_{FF} + P_{RR} < 1$, let's look at Figure 4.

Figure 4 :

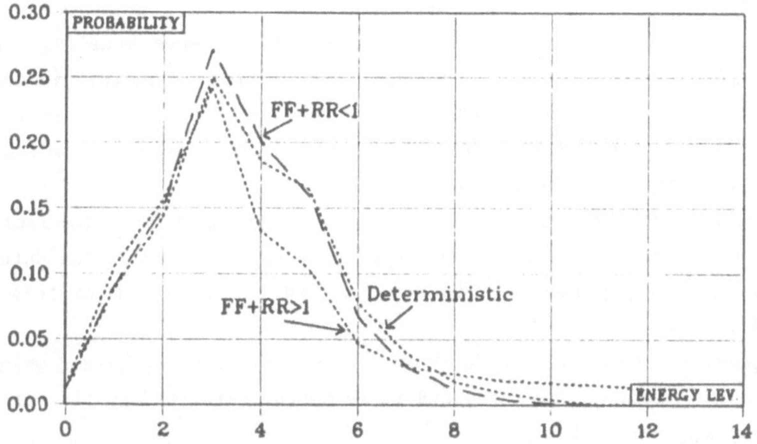


Three density functions of the level of the reserve on week 41 for different Markov processes. They give the probability of being on each level, whether running or not.

Figure 4 displays three density functions of the level of the reserve in the reactor for a given week (week 41). These density functions result from the simulation and give the probability of having an amount of energy l left in the reactor for each week and for each l. One of the density functions was obtained for $P_{FF} = 0.9$ and $P_{RR} = 0.99$ (Markov process fulfilling the condition $P_{FF} + P_{RR} > 1$), the second density function was obtained for $P_{FF} = 0.11$ and $P_{RR} = 0.8$ (which fulfills the condition $P_{FF} + P_{RR} < 1$), and the third density function was obtained in the deterministic case.

Of course, one would rather expect to get a deterministic trajectory from the simulation in the deterministic version (i.e., we would expect to observe only one level of energy each week with probability 1 instead of a spread density function), but the density function is due here to the modeling of continuous state variables into discrete ones. It is obtained with a Fokker-Planck simulation : for example, if we start at week 1 with a reserve of 30.6, we'll say that the probability of being on level 30 at week 1 is 0.4, and the probability of being on level 31 is 0.6 ; this approach is perfectly consistent with the linearization method used to compute the optimal value function.

Of course, in the deterministic case, the "probabilities" obtained each week are not affiliated to any real random factor. They just provide a means to compute the trajectory of the level of energy left in the reserve along the weeks, and this trajectory is obtained by merely computing the expectation of the level of energy each week.

Nevertheless, an analogy can be developped here between modeling continuous state variables into discrete ones and introducing a "random factor" in the model, since the density functions obtained in any case have to be seen as a result of the combination of the random factor introduced by the Markov process and the "random factor" stemming from the discretization.

Apparently, from what we can see on Figure 4, the global random factor issued from this combination is more important when $P_{FF} + P_{RR} > 1$ than in the deterministic case (since the density function is more spread out), and it is less important when $P_{FF} + P_{RR} < 1$.

Consequently, the nuclear reactor generation is better optimized when $P_{FF} + P_{RR} < 1$ than in the deterministic case, and it is better optimized in the deterministic case than when $P_{FF} + P_{RR} > 1$.

## IV. CONCLUSION

Modeling continuous state variables into discrete state variables and using a linearization method to compute optimal values has a slight but undeniable effect on the results of GARP, such as the optimal values.

Actually, the discretization-linearization method acts like a random factor since it is partly responsible for the diffusion of the density function obtained at the simulation.

The weight of the total random factor involved in the model (Markov process plus discretization) can be measured through the diffusion of the density function.

This diffusion is less important when $P_{FF} + P_{RR} < 1$ than when there is no Markov process. This is why the reactor generation is better optimized in this case than in a seemingly deterministic situation. On the other hand, the diffusion is more important when $P_{FF} + P_{RR} > 1$ than when there is no Markov process, and the operation of the reactor is more difficult here.

Whenever $P_{FF} + P_{RR} = 1$, the Markov process is useless because there is no correlation between the states of the reactor in two successive time-steps. This property holds independantly of the discretization-linearization technique used here. But this discretization-linearization technique is at the root of the equality of the results of GARP observed when $P_{FF} + P_{RR} = 1$ and when there isn't any Markov process.

## POWER GENERATOR CONTROL BY VARIABLE STRUCTURE CONTROL THEORY

Sigeru Omatu*, Kunio Matsushita** and Katsuo Isaka***

* Department of Information Science and Systems Engineering, Faculty of Engineering, University of Tokushima, Tokushima, 770, Japan.

** Power Transmission Section, Technical Research Center, Shikoku Electric Power Corporation, Yashima,Takamatsu, 761-01, Japan.

*** Department of Electrical Engineering, Faculty of Engineering, University of Tokushima, Tokushima, 770, Japan

ABSTRACT

In order to enhance the dynamic performance of power generators we propose a new control algorithm based on the variable structure control theory. The algorithm considered here is to control both governor's angle and excitation voltage of the magnetic field coil of a synchronous generator by switching the control parameters according to the position of the system in the phase plane. After deriving the stable switching parameters, we show some numerical simulation results for the real power generator control system.

### 1. INTRODUCTION

We sometimes encounter the electrical short circuit phenomena caused by lightning on the power transmission lines. In this case the electrical power output from the generator becomes zero, and then, the relative load angle is increased by a constant mechanical power supply according to Newton's motion law. Such an unpredictable change in electrical power output affects the quality of generated power via offsetting the desired frequency value. Thus, the control of electrical power generation attracts the attention of the workers on power systems operation.

Several attempts have been done to enhance the performance of the integral controller by supplementing control action[1] or by reformulating the control problem according to a dynamic concept[2]. With the development of optimal control theory, Forsha and Elgerd[3] initiated the attempt of optimizing the frequency control operation. Several works have followed[4-6] but their feasibility for implementation is still questionable. Bengiamin and Chan [7] proposed a new control algorithm based on the variable structure system theory for the linearised system. This algorithm has a simple structure to control the governor but it can be applied to the case where only small load change occurs. Furthermore, in the real plant the nonlinearity plays an important role in the behavior of the generator system, and both governor's position and voltage of a magnetic field circuit are control variables.

In this paper, we consider the control problem of the electric power generator by controlling both the governor's position and the voltage of magnetic field coil based on the variable structure control system theory. The variable strucutre system switches from one structure to another according to a certain switching logic. This facilitates combining the useful new properties of each of the structures and posessing new properties not present in any of the structures used. Furthermore, operating the system in what is known by the sliding mode makes it insensitive to parameter variation and nonlinearity of the system.

## 2.THE VARIABLE STRUCTURE CONTROL

To illustrate the basic idea of the variable structure control, we consider the simple system described by

$$\ddot{x}(t)-\xi\dot{x}(t)+\Psi x(t)=0,\ \xi>0 \qquad (1)$$

where $\Psi$ is given by the following switching law

$$\Psi = \begin{cases} \alpha & \text{if } xs>0. \\ -\alpha & \text{if } xs<0 \end{cases} \qquad (2)$$

Here,s is given by

$$s=\tilde{c}x+\dot{x}$$
$$\tilde{c} = -\lambda \ =-\xi/2+\sqrt{\xi^2/4+\alpha}$$

Note that $\lambda$ denotes the eigenvalue of the system which is unstable, that is ,has the positive real part of the eigenvalue. If the value of $\Psi$ is positive $\alpha$ greater than $\xi/2$, then the solution x(t) of (1) has the trajectories as shown in Fig.1 in the phase plane $(x,\dot{x})$. But if the value of $\Psi$ is negative, then the solution x(t) of (1) has the trajectories as shown in Fig.2. According to the switching rule of (2) we have the trajectories of Fig.3. These trajectories show that the combination of the unstable modes such as in Figs.1 and 2 may be possible to stabilize the original unstable system. This is an important property of variable structure system theory. Another property is robustness of the control design based on the variable structure system theory. In order to show the robustness, let us consider the case where the line s is given by

$$s=cx+\dot{x},\ c > \tilde{c}$$

Then the trajectories are given in Fig.3 where all the trajectories are stable and go to the origin (0,0).

Therefore, the design procedure is to determine the values of $\alpha$ and c such that the representative point could be brought from any initial position in the phase plane to the switching line and then slides on that line towards the origin where the system settles down. Thus, the switching line s=0 determines the mode of operation during the transient period. That is what is known by the sliding mode. Whenever the representative point leaves the sliding line the controller changes its structure to force that point back. Operating the system according to the above description requires that the following three specifications are satisfied: (1)A desired sliding mode is formed, (2)the existence of a sliding mode at every point of the switching line is guaranteed, and (3)the system states could be steered towards the sliding line. Selecting the appropriate value of c could meet the first specification which originates from physical constraints on system operation. The second and third conditions require that the following condition should be imposed in the neighborhood of s=0:

$\dot{s}s < 0.$

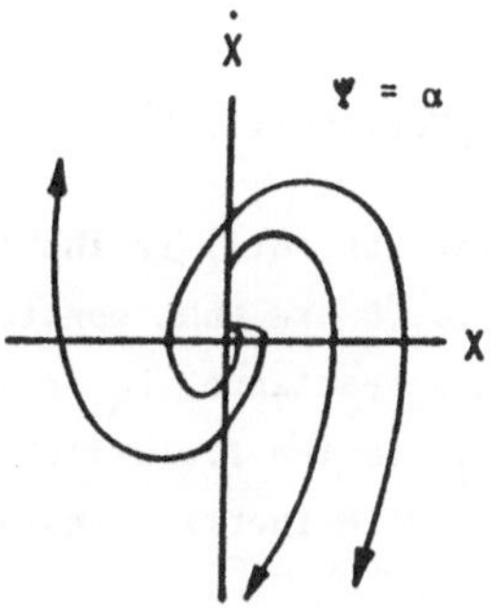


Fig.1. Unstable mode I.

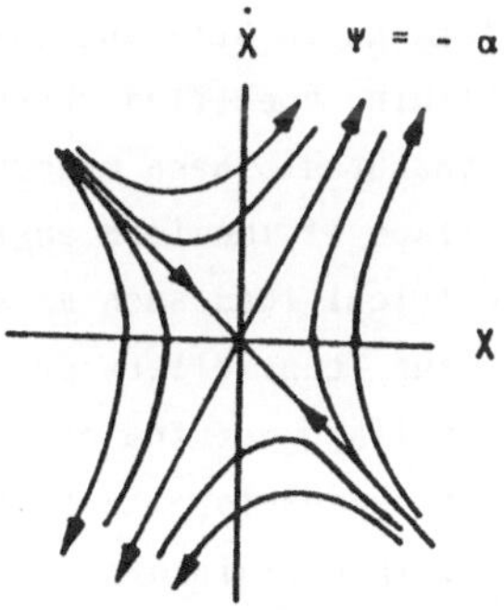


Fig.2. Unstable mode II.

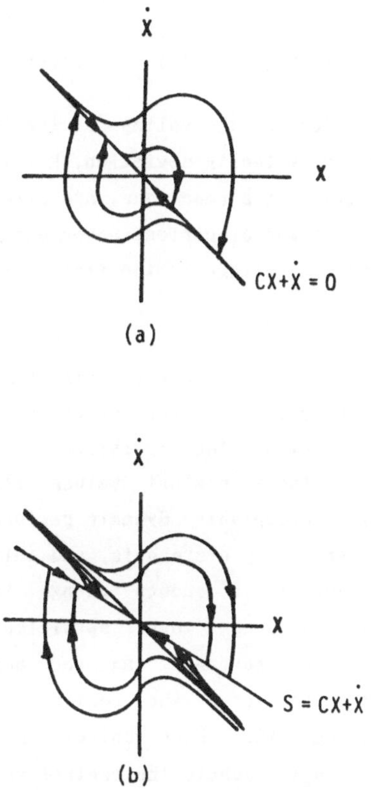


Fig. 3. Stabilization of the unstable system where (a) is stabilization and (b) is a sliding mode.

## 3. POWER GENERATOR SYSTEM MODEL

The system is modeled by its power balance equation which relates the mismatch in power generation and consumption to the change in accelerating power and frequency as shown in Fig.4. In this paper, we consider the case of fault on the transmission line such as lightning.Hence,the load change corresponds to that of electrical output. The equation describing the system dynamics shown in Fig.4 can be given by

$$\frac{d\delta(t)}{dt} = \Delta\omega(t)$$

$$\frac{d\Delta\omega(t)}{dt} = \frac{1}{M}(P_M - P_E)$$

$$\frac{d\psi_{fd}(t)}{dt} = \frac{1}{T_{do}}(e_{fd}+K_1\psi_{fd}(t) + k_2V_b\cos\delta(t))$$

$$P_E = K_3\psi_{fd}V_b\sin\delta(t) + K_2V_b^2\sin2\delta(t)$$

$$v_t = (K_5\psi_{fd}^2+K_6\psi_{fd}V_b\cos\delta(t) + K_7V_b^2\cos^2\delta(t) +K_8V_b^2\sin^2\delta(t))^{1/2}$$

where $V_b$ denotes the voltage of the infinite bus, $\delta(t)$ is the load angle, $\Delta\omega(t)$ is the angular velocity deviation, $K_i$, $i=1,2,\ldots,8$ are some constants which depend on $X_e$, quadrature axis reactance, and direct axis reactance, $P_M$ is the mechanical power input, $P_E$ is the electrical power output, $\psi_{fd}$ is the field flux in the exciter, $e_{fd}$ is the applied voltage to the field coil, and M is inertia constant of the rotor.

## 4. PROBLEM STATEMENT

The electrical power generator control is to retain the power balance in the system such that the deviations of the load angle $\delta(t)$ and the angular velocity $\Delta\omega(t)$ from their nominal values range within specified bounds according to practically acceptable dynamic performance measures. These measures are determined by the rise times, overshoots, and settling times of the load angle and the angular velocity due to a sudden change in electrical load such as a short circuit by lightning. Based on these specifications the controllers act on the speed gear change of the governor and the applied voltage of the excitation coil of the generator. Therefore, the control problem is to develop the control siganls $u_1(t)$ and $u_2(t)$ such that these control signals lead to certain dynamic performance where $u_1(t)$ and $u_2(t)$ denote the control variables of mechanical output from the turbine and the applied voltage of the excitation coil.

## 5. LINEARIZATION OF THE SYSTEM

Let us denote $\delta(t)$, $\Delta\omega(t)$, and $\psi_{fd}(t)$ by $x_1(t)$, $x_2(t)$, and $x_3(t)$, respectively. Then we get

$$\frac{dx_1(t)}{dt} = x_2(t)$$

$$\frac{dx_2(t)}{dt} = \frac{1}{M}(P_M - P_E)$$

$$\frac{dx_3(t)}{dt} = \frac{1}{T_{do}}(e_{fd}+K_3x_3 + K_4V_b\cos x_1(t))$$

Let us define $u_1(t)$ and $u_2(t)$ by

$$P_M(t)=\overset{0}{P_M}(t)+u_1(t)$$

$$e_{fd}(t)=\overset{0}{e_{fd}}(t)+u_2(t)$$

where

$$\overset{0}{P_M}(t) = \begin{cases} 1.0, & t\leq T+0.27,\ t>T+7.87 \\ P_E^0(t), & T+0.27<t<T+7.87 \end{cases}$$

$$u_1(t)=0, \quad t\leq T+0.87$$

$$P_E^0(t)=K_3x_3^*(t)V_b\sin x_1^*(t)+2K_2V_b^2\sin 2x_1^*(t)$$

$$e_{fd}^0(t) = \begin{cases} 2.164, & t\leq T \\ \left(\frac{2.298-2.164}{7.87}\right)(t-T)+2.164, & T<t\leq T+7.87 \\ 2.298, & T+7.87<t \end{cases}$$

$$u_2(t)=0, \quad t\leq T$$

Here, T denotes the fault initiation time, $x_1^*(t)$ and $X_2^*(t)$ are given in Fig.5, and $x_2^*(t)=0$. Thus, after the fault initiation time T we can cut off the short circuit between the transmission lines and the earth during 0.07 seconds and from 0.27 seconds later we can begin to control the mechanical output $P_M(t)$ of the turbine. In this case, we can decrease the mechanical output $P_M(t)$ down to the values of 0.4 power unit.

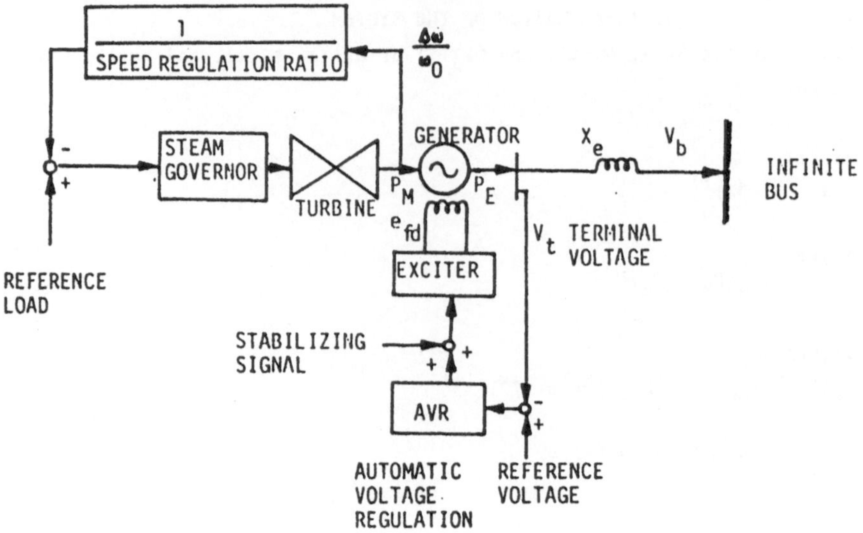


Fig. 4. Power generator control system.

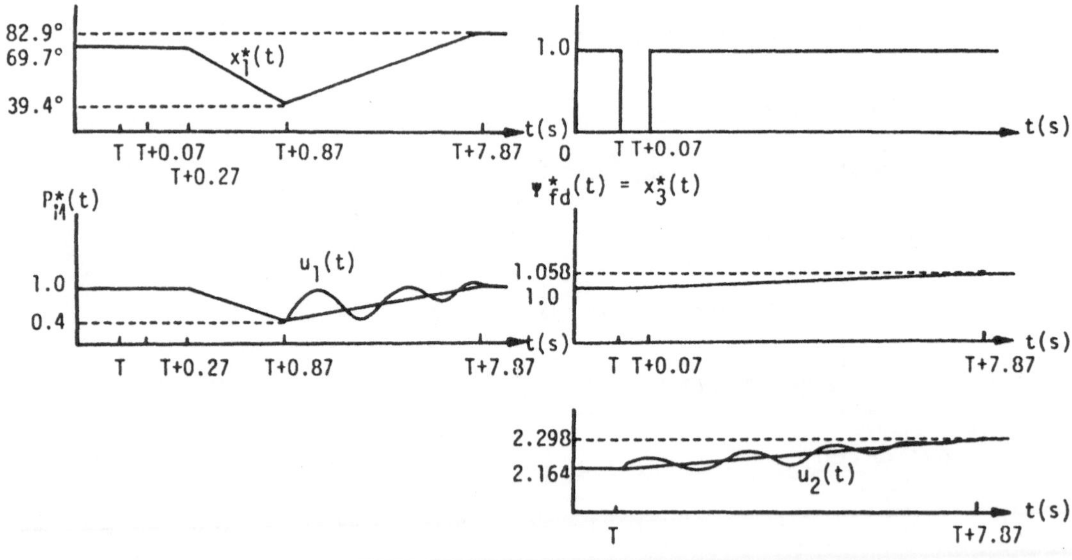


Fig.5. Time chart of the control variables.

Let us denote the error signals by

$$e_1(t)=x_1(t)-x_1^*(t)$$

$$e_2(t)=x_2(t)-x_2^*(t)$$

$$e_3(t)=x_3(t)-x_3^*(t).$$

Thus, the problem is to control $u_1(t)$ and $u_2(t)$ such that $e_i(t), i=1,2,3$ go to zero. Using the Taylor expansion method, we obtain the first approximation model of $P_E(t)$ as

$$P_E(t)=K_3x_3(t)V_b\sin(x_1^*(t)+e_1(t))+K_4V_b^2\sin2(x_1^*(t)+e_1(t))$$

$$\cong P_E{}^0(t)+F_1(t)e_1(t)+F_2(t)e_3(t)$$

where

$$F_1(t)=K_3x_3^*(t)V_b\cos x_1^*(t)+2K_4V_b{}^2\cos2x_1^*(t)$$

$$F_2(t)=K_3V_b\sin x_1^*(t).$$

Letting

$$a_1(t)=dx_1^*(t)/dt,\ a_3(t)=dx_3^*(t)/dt,$$

we obtain

$$\frac{de_1(t)}{dt}=e_2(t)-a_1(t)$$

$$\frac{de_2(t)}{dt}=\frac{1}{M}(u_1(t)-F_1(t)e_1(t)-F_2(t)e_3(t))$$

$$\frac{de_3(t)}{dt}=\frac{1}{T_{do}}(u_3(t)+K_1e_3(t)-K_2V_b\sin x_1^*(t)e_1(t))-a_3(t)$$

## 6. VARIABLE STRUCTURE CONTROL SCHEME

In order to solve the above control problem, we introduce a variable structure control scheme by introducing the switching surfaces,

$$s_1=c_1e_1(t)+e_2(t),\ c_1>0$$

$s_2=c_2e_1(t)+e_3(t),\ c_2>0$

where $c_i$ are constants whose values will be determined to satisfy the specifications 1-3 of the variable structure control in Section 2. Then the control scheme can be given by

$$u_1(t)=\psi_1e_1(t)+\psi_2e_2(t)+\delta_0(t)$$

$$u_2(t)=\phi_1e_1(t)+\phi_2e_3(t)$$

where

$$\psi_i=\begin{cases}\alpha_i & \text{if } e_is_1>0\\ -\alpha_i & \text{if } e_is_1<0\end{cases}$$

$$\phi_i=\begin{cases}\beta_i & \text{if } e_is_2>0\\ -\beta_i & \text{if } e_is_2>0\end{cases}$$

The parameters $\alpha_i$ and $\beta_i$ are determined such that $s_is_i<0$, that is, $s_i=0$ are stable switching lines.

## 7.SIMULATION STUDY

In this section, we show the numerical simulation results for the following case:

$$K_1=-(x_d+x_e)/(x_d'+x_e),\ K_2=(x_d-x_d')/(x_d'+x_e))$$

$$k_3=1/(x_d'+x_e),\ K_4=-(x_q-x_d')/(2(x_q+x_e)(x_d'+x_e))$$

$$K_5=x_e^2/(x_d'+x_e)^2, K_6=2x_ex_d'/(x_d'+x_e)^2,$$

$$K_7=x_d'^2/(x_d'+x_e)^2,\ K_8=x_q^2/(x_q+x_e)^2$$

where

$$x_d=x_q=1.6,\ x_d'=0.3,\ M=9.6/377,$$

$$x_e=0.43(t<T),\ 0.68(t>T),\ T_{do}=6.2,$$

$v_t=1.04$, $V_b=0$ $(T<t<T+0.07)$, 1.0(otherwise),

$p_M=1.0$, $\delta=69.7$ $(t<T)$, 82.9 $(t>T)$

$\psi_{fd}=1.00$ $(t<T)$, 1.058 $(t>T)$

$e_{fd}=2.164$ $(t<T)$, 2.298 $(t>T)$

Fig.6 illustrates the simulation results by the variable structure control when $c_1=0.2$, $\alpha_1=0.02$, $\alpha_2=-0.015$, $C_2=0.05$, $\beta_1=1.0$, $\beta_2=2.0$. As a comparison, we show the simulation result by using the linear quadratic control method in Fig.7 for two kinds of weighting functions:Q=0.0001 and R=1.0;Q=0.0001 and R=2.0.

Computational time for the linear quadratic control method is required very much compared with that for the variable structure control theory and the stability performance realized by the variable structure control method is better than that of the linear quadratic control method. Furthermore, the sensitivity of the control parameters used for the variable structure control method is more insensitive than that for the latter method.

## 8.CONCLUSIONS

The power generator control problem is formulated and a new control scheme is developed based on the variable control theory. Operating the system in a sliding mode is advantageous in reducing its sensitivity to variations in system parameters and uncertainty in load variations. Appropriate selection of the switching surface parameters introduces a direct means to steer the system towards a practically desirable mode of operation. The switching logic of the developed controller is simple and seems amenable to implementation. The future problem is to consider the control problem of the governor angle and stabilizing voltage of the automatic voltage controller.

## REFERENCES

[1] L.H.Fink,Advances in instrumentation, 26, 501(1971).
[2] O.I.Elgerd, "Electric energy systems theory:An introduction", McGraw-Hill, (1971), 315-389.
[3] C.E.Forsha and O.I.Elgerd, IEEE Trans.,PAS-89, 4,(1970), 563-577.
[4] M.Calovic, IEEE Trans. PAS-91,6 (1972), 2271-2285.
[5] S.M.Miniesy and E.V.Bohn, IEEE Trans.,PAS-91,5,(1972), 1910-1915
[6] M.Saeki,A.Hara, and M.Araki, Trans. IEE of Japan, 105-B,1,(1985), 31-38.
[7] N.N.Bengiamin and W.C.Chan, IEEE Trans. PAS-101,2,(1982), 376-380.

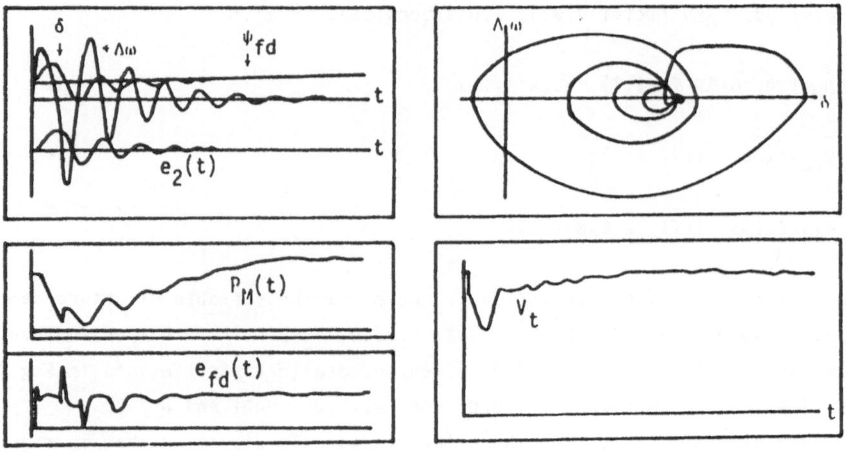


Fig.6. The control results by the variable structure control theory.

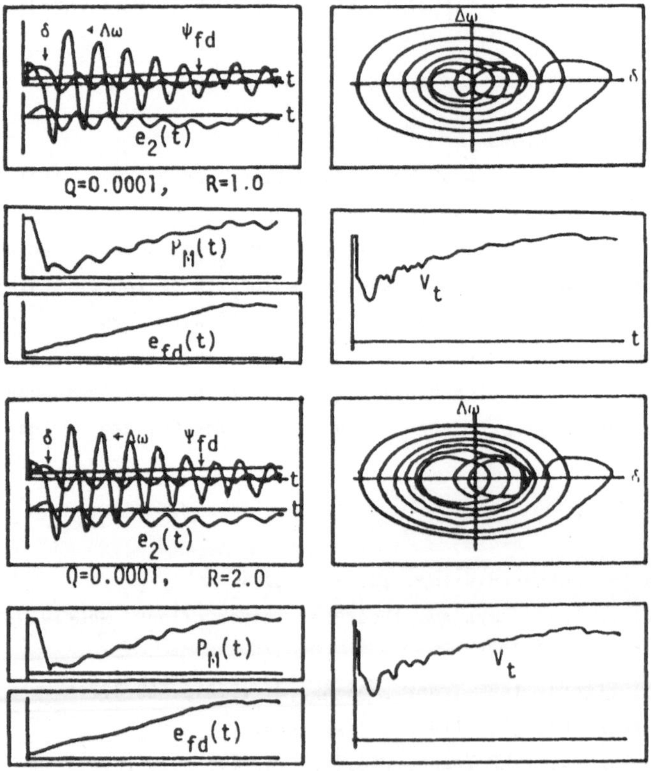


Fig. 7. The optimal control results by the linear quadratic control theory.

# AUTOMATIC GENERATION CONTROL IN HYDRO-THERMAL ELECTRIC POWER SYSTEM

Dido M. Dervišević

Dept. of Mathematical Methods Modelling Development

ENERGOINVEST IRIS-SISTEMI, Tvornička 3, Sarajevo, Yugoslavia

This paper deals with some aspects of automatic generation control of a two-area mixed hydro-thermal electric power system. The linear discrete time state space model with a thermal unit and a hydro unit per area, is used. The speed governor deadband and the generator rate constraints are being taken into consideration. Automatic Generation Control (AGC) of hydro units equipped with classical generation controllers is used for load change regulations from moment to moment. Discrete mode optimisation of such controllers using Integral Squared Error (ISE) concepts have been attempted. While AGC of thermal units equipped with self-tuning predictors is for regulations of sustained load changes.

## INTRODUCTION

In the conventional load-frequency control, regulating plants are controlled by PI controllers with both the frequency deviation signal and the tie-line load signal. According to the operation data, hydro-plants are able to follow the load fluctuation with response time shorter than 1 minute, whereas the response time of the thermal plant depends upon the type of a unit and a size of load disturbances. Hence, it is difficult to maintain the power system frequency within an allowable range in the dry season when hydro plants are scarcely available. To overcome this difficulty, new thermal units in the Yugoslav power system are introduced into supplementary control, because old thermal units are unsuitable for supplementary control (turbine-following mode, for instance).

One of difficultes of the power control is to follow the ever changing system load in a satisfactory way. In order to facilitate this task, numerous attemps have been made to forecast the unknown future demand. A large number of varios algorithms have been designed for general prediction problem. Especially well suited to the problem of predicting power system loads whit a short time horizont is a form of the Kalman filter [1]. However still simpler methods can be chosen as well. Since time series of Area Control Error (ACE) is a discrete time stochastic processes, it is assumed that ACE can be derived from a white noise process by the rational spectrum filter. For the prediction of the increment ACE was using Wittenmark's self-tuning predictor.

Two different controllers are mutualy compared, a proportional plus integral controller and a combination of a proportional plus integral controller and a predictor. Four cases are simulated: the hydro units in supplementary control with and without thermal units in supplementary control at deterministic and stochastic load disturbance.

## SYSTEM STUDIES

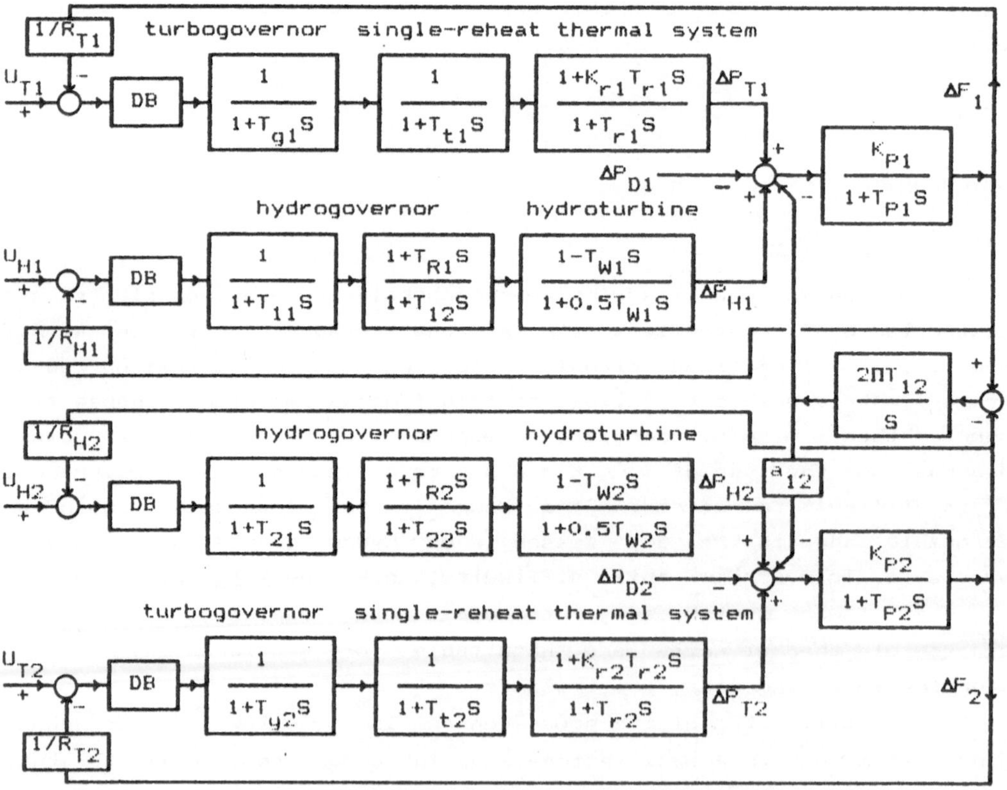


Figure 1: Transver-function model of a two area hydro-thermal system

A two-area interconnected power system has been represented as shown in Figure 1 (to study the performance of the proposed controllers). The system representation includes governor deadband (see [2]) and generation rate constraints (see [3]). The nominal parameters are given in Appendix.

## OPTIMISATION OF HYDRO UNITS PI CONTROLLER

The PI algorithm is represented by the following control law:

$$U = K_P \; ACE + K_I \int ACE \; dt \qquad (1)$$

where $K_P$ is the proportional gain, $K_I$ is the integral gain settings of the hydro units. In order to optimise the $K_I$ setting, the discrete time version of the ISE techniques is used. Cost functions such as

$$J = \sum_{K=0}^{\infty} \left( \Delta P^2_{tie1} + \Delta F^2_1 \right) \qquad (2)$$

is tried out.

A practical sampling period of T=2 secs is chosen.

## DETERMINISTIC LOAD DISTURBANCE

Figure 2. (thick line) shows the plot of J against $K_I$ for several values of $K_I$, considering a step load perturbation is in the first

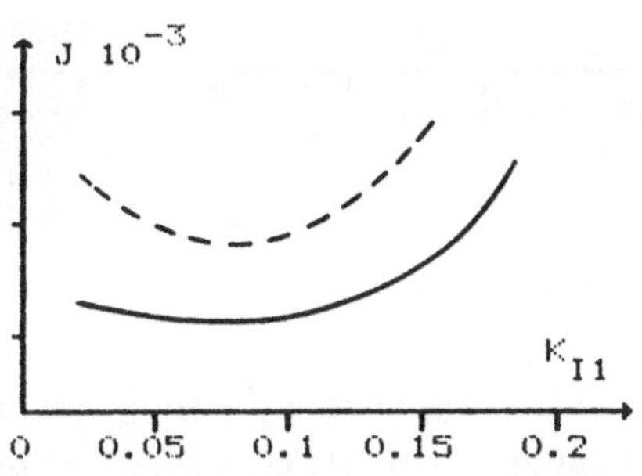


Figure 2. Cost function $J=f(K_{I1})$
—— 1% step load perturbation
- - stochastic load perturbation

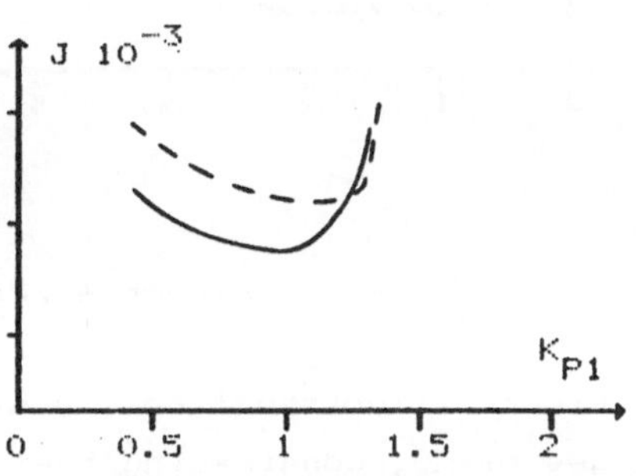


Figure 3. Cost function $J=f(K_{P1})$
—— 1% step load perturbation
- - stochastic load perturbation

area. The optimum integral gain setting is found to be $K_I$=0.07 .

Having optimized the integral gain setting of the supplementary controllers, attempt is made to optimize the proportional gain $K_P$, keeping the integral gain setting at its optimum value of $K_I$=0.07 . Evaluating the cost function J for several values of $K_P$ (see Figure 3 thick line) considering a step load perturbation is in the first area, it is found that for $K_P$=1.1 , the cost function attains the minimum value. Thus a proportional plus integral supplementary controller with $K_P$=1.1 and $K_I$=0.07 may be considerred as an optimum PI controller for a step load perturbation.

STOCHASTIC LOAD DISTURBANCE

The random load variations for each AGC area is modelled similarly as in [4] standard deviation of about 0.03 % of the initial area load. Load disturbance of the first area was synthetized by adding a ramp load change and a step load change to a random load variation, while second area load disturbance was consisted of random load variation. The disturbance first area considered is shown in Figure 4 .

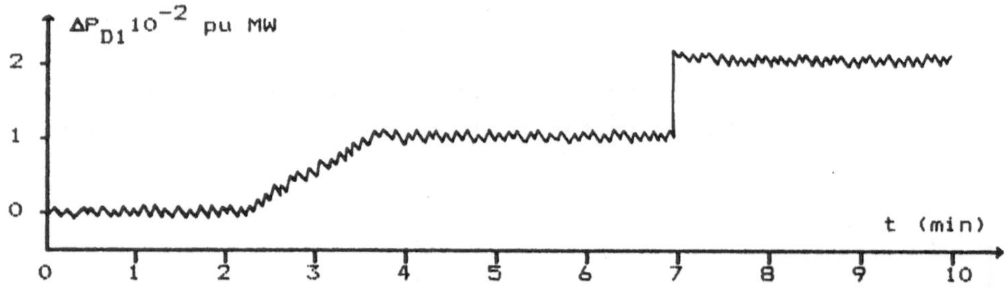


Figure 4 $\Delta P_{D1}$ used in the study

Figure 2 (dotted lines) shows the plot of J against $K_I$ for several values of $K_I$, considering there are stochastic load disturbance in both areas. The optimum integral gain setting is found to be $K_I$=0.08 . Figure 3 (dotted lines) shows a plot of J against $K_P$, keeping the integral gain setting at its optimum value of $K_I$=0.08 , for several values of $K_P$, considering there are stochastic load disturbance in both area. The optimum integral gain setting is found to be $K_I$=0.07 .

Having optimized the integral gain setting of the supplementary controllers, attempt is made to optimize the proportional gain $K_P$, keeping the integral gain setting at its optimum value of $K_I$=0.07 . Evaluating the cost function J for several values of $K_P$ (see Figure 3 thick line) considering a step load perturbation is in the first area, it is found that for $K_P$=1.1 , the cost function attains the minimum value. Thus a proportional plus integral supplementary controller with $K_P$=1.1 and $K_I$=0.07 may be considerred as an optimum PI controller for a step load perturbation.

STOCHASTIC LOAD DISTURBANCE

The random load variations for each AGC area is modelled similarly as in [4] standard deviation of about 0.03 % of the initial area load. Load disturbance of the first area was synthetized by adding a ramp load change and a step load change to a random load variation, while second area load disturbance was consisted of random load variation. The disturbance first area considered is shown in Figure 4 .

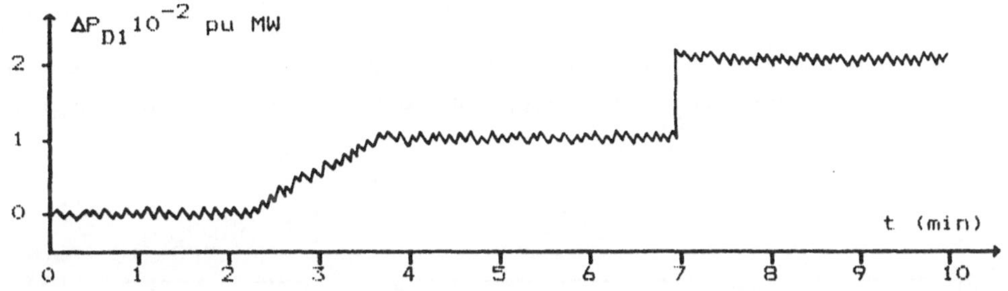


Figure 4 $\Delta P_{D1}$ used in the study

Figure 2 (dotted lines) shows the plot of J against $K_I$ for several values of $K_I$, considering there are stochastic load disturbance in both areas. The optimum integral gain setting is found to be $K_I$=0.08 . Figure 3 (dotted lines) shows a plot of J against $K_P$, keeping the integral gain setting at its optimum value of $K_I$=0.08 , for several values of $K_P$, considering there are stochastic load disturbance in both in the first area is used.

Figure 5 shows the plot of I against n (K=1) for several values of n considering stochastic load disturbance is in the first area.

Figure 5 shows that representation of an area by a fifth order predictor is adequate. Number of predictor parameters is selected to be n=5 .

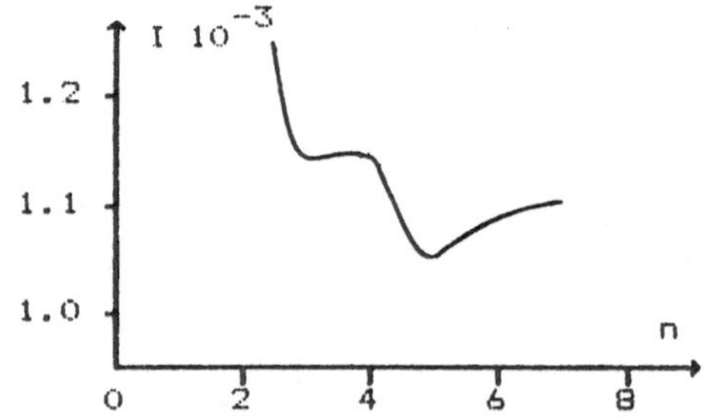


Figure 5 I=f(n)

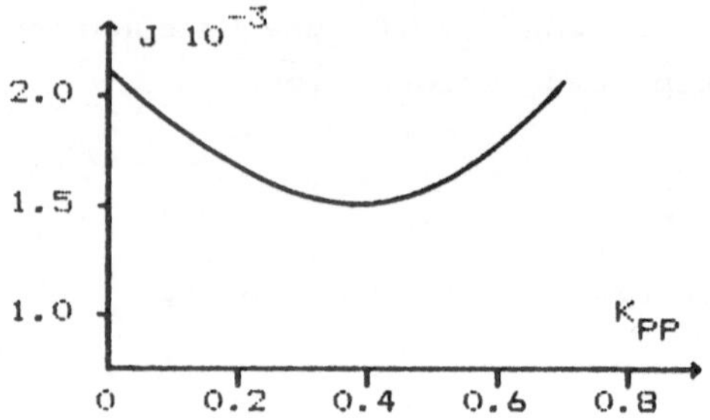


Figure 7 Cost function $J=f(K_{PP})$

THERMAL UNITS CONTROLLER

Figure 6 shows the way in which are introduced thermal units into supplementary control in this study. Where $K_{PP}$ is the proportional gain.

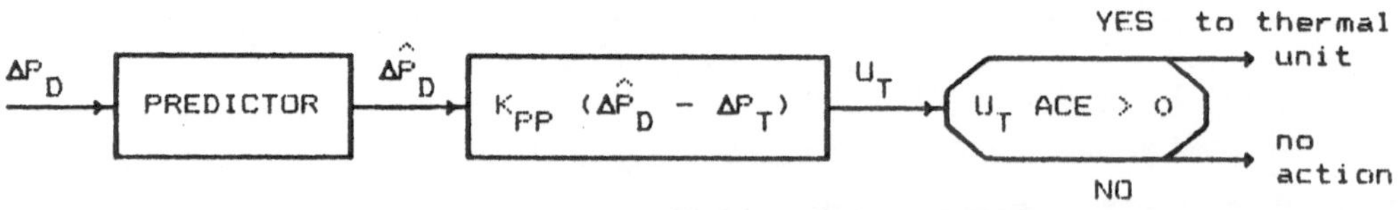


Figure 6 Supplementary control of a thermal unit

In order to optimise the $K_{PP}$ setting cost function J is tried out. Figure 7 shows the plot of J against $K_{PP}$ for several values of $K_{PP}$, considering stochastic load disturbances are in both areas. The optimum value of $K_{PP}$ is 0.4 .

SIMULATION TEST

Figure 8 shows generation change of the hydro unit in the first area with and without thermal units in supplementary control at stochastic load disturbance .

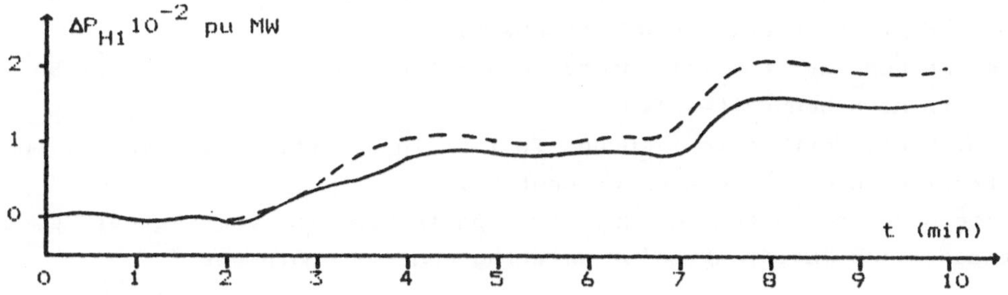


Figure 8 Generation change of hydro unit first area at stochastic load disturbances
——— hydro units and thermal units in supplementary control
– – – hydro units in supplementary control

Figure 9 shows generation change of the hydro unit in the first area with and without thermal units in supplementary control at deterministic load disturbance.

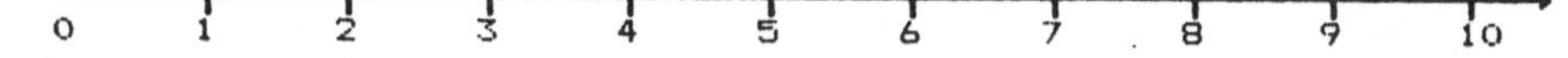

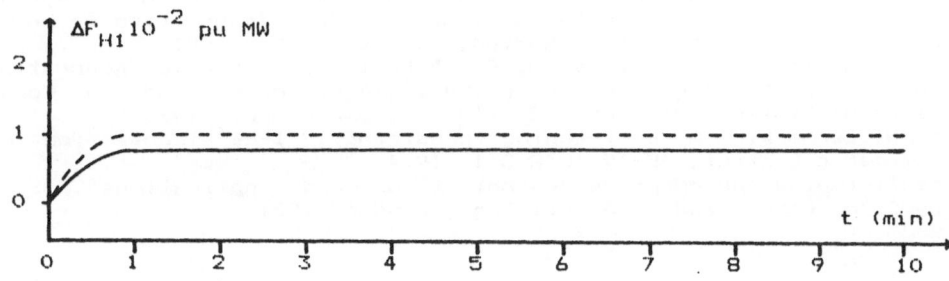

Figure 9 Generation change of hydro unit first area at deterministic load disturbances
——— hydro units and thermal units in supplementary control
– – – hydro units in supplementary control

## CONCLUSION

A state-space model for a typical two-area mixed hydro-thermal system including two different controllers for both kind units has been used.

Analysis reveals that the optimum proportional gain and integral gain achieved at deterministic load disturbance is similar to one achieved at stochastic load disturbance.

A method of thermal units controllers optimization using ISE concepts have been attempted.

The hydro unit control burden is less in a system with thermal unit controller than in a system without it.

The proposed control system is applicable to the actual system without major modification of existing control equipment.

## REFERENCES

[1] T. Bohlin, "Four cases of identification of changing systems", Report 7503, The Royal Institute of Technology Stockholm, 1975

[2] S. C. Tripathy, T. S. Bhatti, C. S. Jha, O. P. Malik, C. S. Hope, "Sampled data automatic generation control analysis with reheat steam turbines and governor dead-band effects", IEEE Transaction on Power Apparatus and Systems Vol. PAS-103, 1045-1051, May 1984

[3] M. L. Kothari, P. S. Satsangi, J. Nanda, "Sampled data automatic generation control of interconnected reheat thermal system considering generation rate constraints", IEEE Transaction on Power Aparatus and Systems Vol. PAS-100, 2334-2342, May 1981

[4] F. P. deMello, R. J. Mills, W. F. B'Rells, "Automatic generation control, part. I-process modeling", IEEE Transaction on Power Aparatus Systems, Vol. PAS-92, 710-715, March/April 1973

[5] B. Wittenmark, "A self-tuning predictor", IEEE Transaction on Automatic Control, AC-19, 848-851, 1974

[6] "Self-tuning and adaptive control: theory and applications", Edited by C. J. Harris and S. A. Billings, London, 1981

## APPENDIX

Nominal system parameters :

$f = 50$ Hz $\quad T_{12} = 0.545 \quad P_{r1} = P_{r2} = 4000$ MW $\quad H_1 = H_2 = 10$ sec.

$D_1 = D_2 = 0.02$ pu MW / Hz $\quad R_{T1} = R_{T2} = R_{H1} = R_{H2} = 2.5$ Hz / pu MW

$T_{g1} = T_{g2} = 0.08$ sec. $\quad T_{t1} = T_{t2} = 0.3$ sec. $\quad K_{r1} = K_{r2} = 0.5$

$T_{r1} = T_{r2} = 10$ sec.

$T_{11} = T_{21} = 34$ sec. $T_{12} = T_{22} = 0.588$ sec. $T_{R1} = T_{R2} = 5$ sec.

$T_{W1} = T_{W2} = 1$ sec.

Generation rate constraints (GRC) of thermal unit is 30 % / min

GRC of hydro unit for raising the generation is 270 % / min

GRC of hydro unit for lowering the generation is 360 % / min

Linearized deadband (DB) is $0.8 - 0.063662\ d/dt$

# STOCHASTIC MODEL FOR OPTIMAL GAS EXPLOITATION

Pranas ALEKSIUNAS, Antanas BURKAUSKAS,
Algis GARLIAUSKAS

Institute of Mathematics and Cybernetics
Lithuanian SSR Academy of Sciences
Akademijos 4, 232600 Vilnius, USSR

## ABSTRACT

The probabilistic constrained stochastic programming model to the optimal developing of gas extracting area is considered. The distribution function of the resource increments is known. Gamma, lognormal,stable, chi-square, exponential and normal distributions were used. The interfluence among the total cost, reliability, possible demand and distribution was analysed.

## 1. INTRODUCTION

This report is an attempt to apply a probabilistic constrained stochastic programming model to the optimal developing of gas extracting area. The approach is based on the following considerations : the physical situation of gas extracting contains quite natural randomness , moreover, the investigations in natural resource exploration allow to assume that the distribution function of the resource increments is known.From the mathematical point of view it was interesting to proceed the numerical evaluation of this type of mathematical programming problems in the case of various distributions of random parameters.

## 2. THE MODEL

Assume that we have N of gas production regions and the

amount of the explored resource in the region i is Q , i=1,...,N (fig.1).

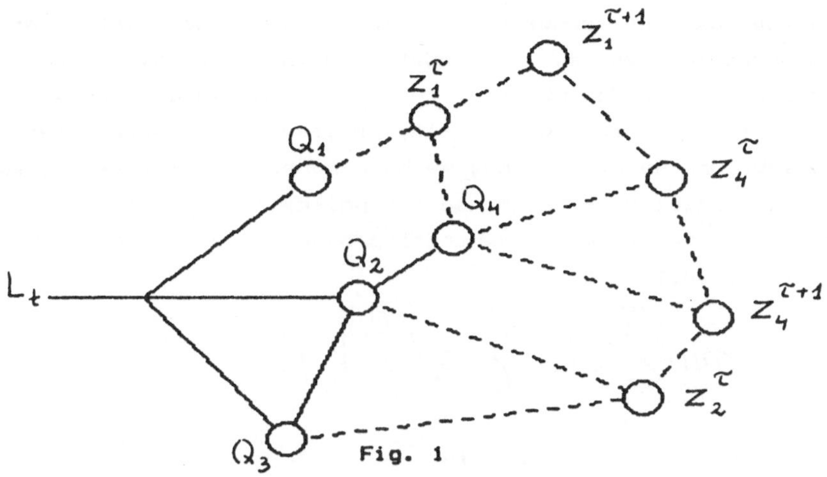


Fig. 1

In the time period t we extract the amount $X_i^t$ to meet the total demand $L_t$, t=1, ... ,T. Starting from the period $\tau$ , $1 \leq \tau \leq$ T, the investments for prospecting $I_i^t$ give random increment of the resources $Z_i^t(I_i^t, \omega)$, $t \geq \tau$. We assume, that the random function $Z_i^t(I_i^t, \omega)$ is known (either analytically or by tabulation ). The dependence Z= Z(I) is concave (fig.2), moreover, in the reasonable interval of I varying it provides a linear approximation. It is quite natural that the capital investments have certain range limitations ( upper bound $I_u$ and lower bound $I_l$ respectively ), so the linearity of Z(I) provides the calculation possibility of the probability functions.

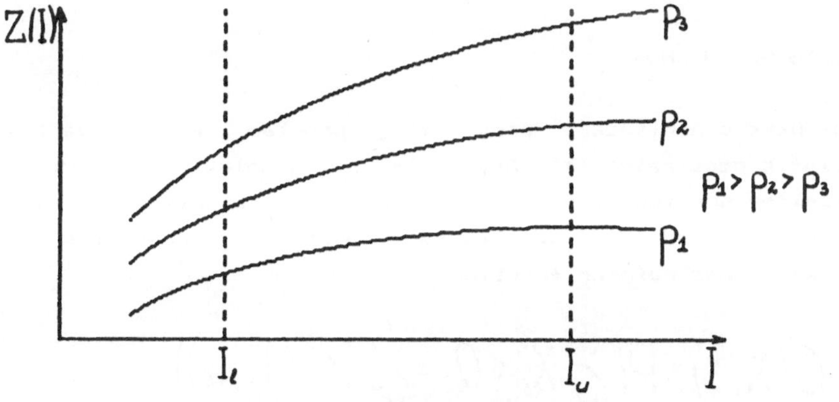


Fig. 2

The problem is formulated in the following way. Since the investments give additional resources we may force the present extraction towards future prospecting and satisfy the total regional balance constraints. The future increments of the resources are random, so we must take into account the reliability considerations. We require the random balance constraints to be satisfied with the probability no lower than certain reliability level p. On the other hand, we have monetary limitations, too. The optimal decision is to choose the policy $(X_i^t, I_i^t)$, $i = 1, .. N$, $t = 1, .. T$ to minimize the total production costs and discounted capital investments

$$\min \sum_{i,t=1}^{N,T} C_i^t X_i^t + \sum_{t=\tau}^{T} a_t \left( \sum_{i=1}^{N} I_i^t \right)$$

subject to demand constraints

$$\sum_{i=1}^{N} g_i^t X_i^t \geq L_t , \quad t = 1, \dots, T$$

and resource constraints

$$\sum_{s=1}^{t} X_i^s \leq Q_i , \quad t = 1, \dots, \tau - 1$$

$$P\left( \sum_{s=1}^{t} X_i^s \leq Q_i + \sum_{s=\tau}^{t} Z_i^s ( I_i^s , \omega ) \right) \geq p, \quad t = \tau, \dots, T \qquad (*)$$

## 3. ANALYSIS OF THE MODEL

We have a nonlinear programming problem with special probabilistic constraint (*). Any well-running solution algorithm may be applied provided the two main problems are solved: 1.exact expressiability and 2. acceptable concavity properties of probabilistic constraining function

$$G(X, I) = P\left( \sum_{s=1}^{t} X_i^s \leq Q_i + \sum_{s=\tau}^{t} Z_i^s ( I_i^s , \omega ) \right)$$

Essentially it depends on the selecting of probability

distribution of the random resource increments. The investigations in natural resource exploration [1,2,3,4]give clare indications to use lognormal, Gamma, stable, chi - square, exponential and normal distributions. Assuming linearization of the dependence $Z = Z(I)$, the first problem reduces to the knowledge of distribution of the linear combination of random variables. In the case of normal and stable distributions we have no problems since they are closed under linear operations. In other cases ( except lognormal one) assuming stochastic independence of increments we may use characteristic function technique and inverce formula. In the case of exponential distribution we have exact finite expression, in the cases of chi-square and Gamma distributions we must use the row expressions of characteristic function and integration by terms. For the lognormal case the approximation technique was proposed recently [5, 6].

The concavity question of constraining function is much more complicated, since the shortage of exact expressions in the most cases makes the analysis very difficult. Various generalized concavity properties were obtained for the normal [ 7, 8 ], stable [ 8] and lognormal [ 5, 7 ] cases. Nevertheless, the computer runnings were proceeded for the cases of other distributions too with empirical guarantee that the obtained local solution is a global one, too. The numerical and graphical analysis of constraining function has shown, that the typical behaviour of the probability function $f(x,y)=P(\xi x+\eta y\leq c)$ is bell-shaped upon certain (sometimes high enough) level p. The typical level sets are shown in fig.3.

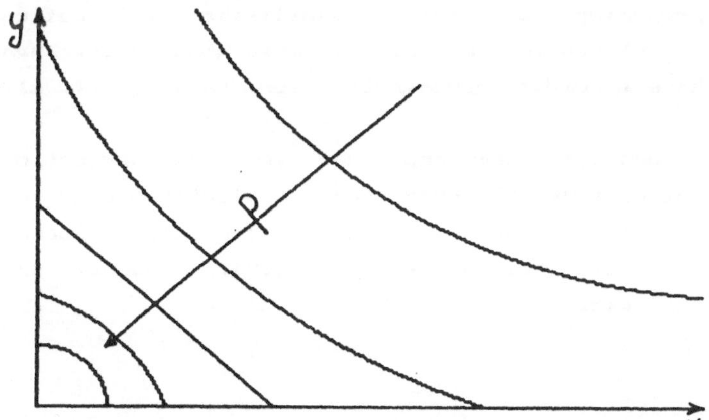


Fig. 3

Since the reliability level p under logical considerations is usually close to unity we may expect the convexity of the model without analytical proof.

It must be mentioned that the presence of probabilistic function in nonlinear programming problem imposes certain requirements for the choosing of method, initial solution, accuracy. Since the range of the constraining function is relatively small one must be especially precise to avoid the falling into domains of trivial probabilities zero and unity, where the iterational process is stopped. Whereas possible, the analytical evaluation of gradients is adviced instead of numerical differenciation.Special attention must be paid to the selection of initial solution to quarantee a good descent.

## 4.NUMERICAL EVALUATION

The problem was solved for the above mentioned distributions of random resource increments and varying all initial parameters (fig.4). It may be noticed that we have not only principal possibility of the solution of this type problems, but a tool for preparing flexible recomendations for the decision maker in stochastic situation as well :at various directively posed indices (objective level, reliability, demand) we can show how the winning in one acts on the loosing in another . For example , certain numerical characteristics ( as variance ) may be treated as preciseness of preparing of the prbabilistic information (measurements, samples), so they can be estimated monetarily. Then in some sense we have a common denominator for varying of all initial data.

The computer runnings were proceeded for 3 gas production regions and 5 time periods, the shapes of distributions used for the resource increment are shown in fig.5,6 . The interinfluence among the total cost, reliability, possible demand and distribution was analysed.

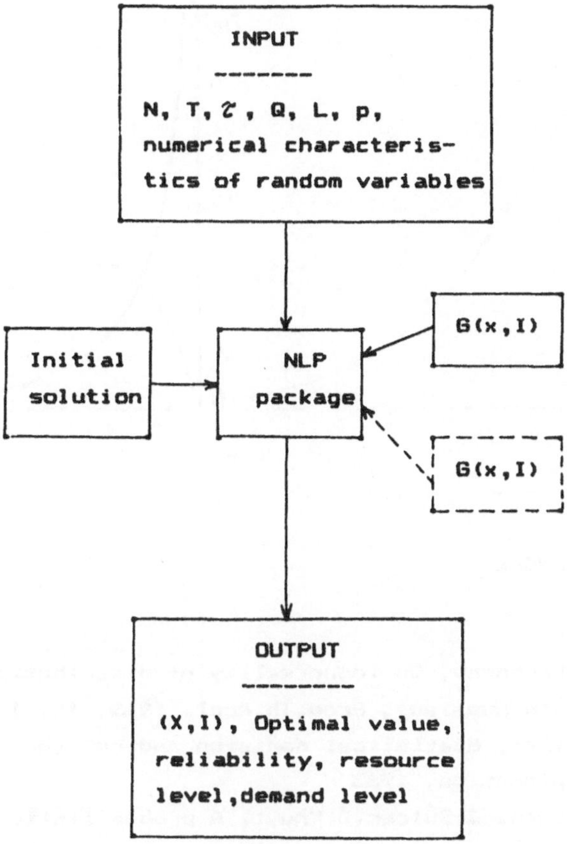
INPUT
--------
N, T, $\tau$, Q, L, p,
numerical characteristics of random variables
Initial solution
NLP package
G(x,I)
G(x,I)
OUTPUT
---------
(X,I), Optimal value, reliability, resource level, demand level

Fig. 4

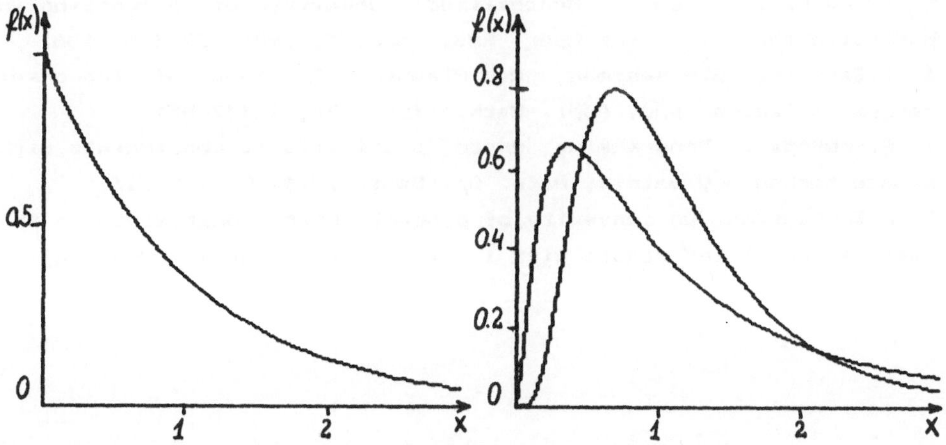
f(x)
0.5
0
1
2
X
f(x)
0.8
0.6
0.4
0.2
0
1
2
X

Fig. 5

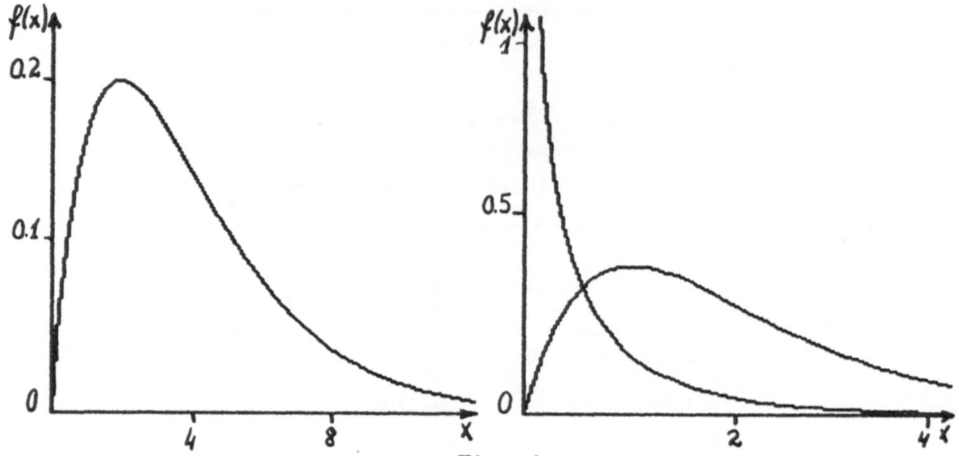


Fig. 6

## REFERENCES

1. Yu. V. Prohorov, On lognormality of distribution in geochemical problems, (in Russian), Prob.Th.Appl. 1965, 10, 1 (184-187)

2. G.M.Kaufman, Statistical decision and related technigue in oil and gas exploration, 1963

3. G.M.Kaufman, J.Bulcer,D.Kruyt, A probabilistic model of oil and gas discovery, in Studies in Geology, 1975 vol.1

4. B.Mandelbrot, The Pareto - Levy random functions and the multiplicative variation of income, IBM Res. Centr. Rept. 1960

5. S.Shaible, W.Ziemba, Generalized concavity of a function in partfolio theory, Z. fur Oper. Res., ser. A, 1985, 29 (161-186)

6. E.Barouch, G.M.Kaufman, M.L.Glasser, On sums of lognormal random variables, Sud. Appl. Math., 1986, 75, 1 (37-55)

7. A.Prekopa , Programming under probabilistic constraints with random technology matrix, Math. Op. Stat. 1974, 5 (109-116)

8. A.Burkauskas, On convexity of probabilistic constraints in the case of normal and stable with 1 2 distributions (in Russian)

# A RESOURCE PLANNING AND MANAGEMENT SYSTEM FOR PINE PLANTATIONS IN FIJI

**A G D WHYTE**
**School of Forestry**
**University of Canterbury**
**CHRISTCHURCH NEW ZEALAND**

## ABSTRACT

*A traditional O.R. systems approach to evolving ways of planning and managing plantations of Caribbean pine in Fiji is described. The overall system was first devised, then made acceptable to resource managers before various information-gathering, data-processing, modelling, analytical and report-writing routines were developed in a carefully structured sequence. The utility of the information gathering procedures is explained and then each technique is briefly outlined. An interesting evolution of data-processing capabilities is described next to illustrate benefits of introducing sophisticated technology in a way that is acceptable to people in a developing country.*

*The paper deals mainly with descriptions of various modelling systems which are used in assisting with the planning process. Examples of modelling include:*

*(1) simulation of crop growth and assorted yields for administratively convenient and biologically homogeneous crop aggregations;*

*(2) bucking felled trees to maximise financial returns subject to matching the resource characteristics with forecasted market demands through dynamic and linear programming;*

*(3) scheduling annual harvests of individual stands for up to between 5 and 10 years ahead with linear programming;*

*(4) providing forecasts of long-term assorted wood supplies and costs of production using simulation;*

*Reference is also made to a further stage in the coordinated modelling, namely the characreisation of the whole forest sector in Fiji.*

*The importance of integrating the information, forecasting, modelling and decision-making components of the system and with complementary research studies is emphasised. Ways of encouraging managers and researchers to cooperate ,and ways of training and educating staff to implement the systems are explained briefly. The paper concludes with a review of the benefits gained from adopting this approach.*

## Introduction

Planning and management systems in forestry need to focus simultaneously on short, medium and long-term horizons. Plantation forestry, particularly that which occurs in the Southern Hemisphere, is concerned with fewer years ahead than the long time spans which must be contemplated for natural forests, but even so, 30 to 50 years hence is commonplace. Day to day decisions, consequently, can have major impacts on what may be feasible many years from now, and *vice-versa*. If such considerations are not taken into acount, there are great dangers that major adverse repercussions could eventuate, to which Man's profligate use of prized forest resources over the years already bears ample testimony.

Adoption of a traditional, comprehensive systems approach through formally characterising the environment in which decision-making about resources and industrial use of the raw material they can provide has to be carried out, can confer considerable benefits. A case history for a relatively small forestry firm in a developing country is set out here to illustrate one instance in which such an approach has reaped considerable benefits.

As is often the case nowadays, analysis of complex systems such as the one described here has been carried out by several individuals, each making his or her own specialist contribution. The main points of interest in this paper are not likely to be any one of these individual efforts, but the manner in which they have all been brought together and co-ordinated for the good of the organisation concerned as a whole.

## Evolution of the Fiji Pine Commission

Plantations of Caribbean pine, *Pinus ceribaea* Mor. var. *hondurensis* Barret Golf., and slash pine, *Pinus elliottii* Engel. were established in the 1950's and 60's by the Fiji Forestry Department to reduce erosion on highly degraded, fire-induced grasslands in low rainfall areas of Viti Levu. These early plantings showed considerable promise as potentially productive crops as well as for soil stabilisation. Consequently, planting increased in the late 1960's and, by 1972, a separate Pine Scheme, funded through loans from the Commonwealth Development Corporation and aid from New Zealand, among other countries, had been set up and made directly responsible to the Minsiter of Agricutlture, Fisheries and Forests.

In 1976, an independent statutory body called the Fiji Pine Commission was created to facilitate and develop an industry based on the growing, harvesting, processing and marketing of pine products. The Commission entered into a partnership with the Fijian owners of land to increase the potential area for plantations and to encourage participation of the owners in the industry. The owners receive rentals on the leased land and royalties for felled timber together with a share of the profits from downstream processing. A fuller account is provided in Whyte (1983).

By the end of 1986, about 45 000 ha out of a total target area of 55 000 ha had been established almost entirely of Carribbean pine, despite several setbacks with severe cyclones in 1972, 1983 and 1985 (see Fiji Pine Commission, 1987). A large sawmill/chipmill complex capable of processing up to 215 000 $m^3$ per annum has recently been completed, about half the sawn out-turn of around 50 000$m^3$ from which will be sold domestically, and the other half as exports along with nearly 140 000$m^3$ of chips. Longer term, the sustained yield potential could well exceed 1.1 $Mm^3$ of log supply per year.

In 1975, I was invited to Fiji as part of New Zealand's aid programme to advise the Commission on refining their management modelling and forward planning capability, but ended up devising first of all data-gathering, data processing and wood supply forecasting systems as a platform on which to build a comprehensive management package. The approach taken as is described in the next section, was to identify what information was really needed by managers to allow them to manage the resources successfully; that is, the traditional systems analysis way of tackling problems.

## Design of the Planning and Management System

Managers of plantation resources need up to date information on: the quality of establishment (site preparation, planting, weeding, survival and blanking); the growth and development of crops over time with or without tending (thinning, pruning and fertilising); the impact of disasters (fire and cyclone damage); the sustainable quantity and quality of wood production; where, when and how to harvest crops; where and in what amounts and forms to supply raw material to ulilisation outlets; the quantities and kinds of raw material actually delivered and left in the forest; and the costs of all operations. With this information, planners and managers can forecast the short, medium and long-term supplies of each kind of log by length and diameter class to keep these plants supplied with the desired amounts and kinds of raw material as efficiently (in the economic sense) as possible, and to prescribe and control logging and raw material delivery operations.

The planning and management system depicted in Fig. 1 was designed to provide all this information in an appropriate form and with due timeliness. For example, on each day when seedlings are planted, the supervisor of operations formally assesses the amount of work done and its adequacy in terms of spacing and quality of seedlings, firmness and depth of planting and an index of weediness. Such information is used immediately by the supervisor to improve the effectiveness of this supervision and ensure that any remedial action needed is taken immediately; the information is passed to the local forest manager weekly to provide him with regular progress reports and with advance warning on where weed problems might arise in the future; aggregated information is also sent to the Commission headquarters for storage in the management information system records (see Fickes & Dunn 1980) and for later retrieval when stratifying crops for other assessment purposes. After 6 months or so, formal counts are made of the numbers of surviving seedlings, with this information being passed on as before so that any weeding and blanking that needs to be done can be prescribed, and so that management records can be updated. At age two and a half years, stocking per hectare is again assessed, as, by then, the pine crop should be above any weed competition and thus can provide a reliable indication of future yielding capability for a given locality. If stocking is greater than 800 per hectare, the total yield will likely reach its full potential for the site, while stockings below that figure will result in proportionately lower yields. The stocking estimates are mapped and stored on the management information system records, then retrieved later for stratifying single age-class populations as part of routine inventory (or stock-taking) operations.

Routine inventories are conducted at five yearly intervals throughout the life of each crop, or whenever a change in stocking occurs either by design (i.e. intermediate thinning) or by accident (e.g. wind and fire damage). Prior information is, as stated earlier, used to stratify any single age class/locality combination into yield capability classes. Information is collected from sampling units, some of which are matched and others unmatched on successive occasions; that is, the technique of sampling with partial replacement (SPR) as described in standard texts, e.g. Cochran (1977).

The total quantum of inventory data can then be used to forecast present and future wood supply capabilities (up to 10 years ahead) in detail, as will be described shortly, while data from matched plots are utilised to monitor the reliability of the growth functions built into the growth and yield model (see next section). Information on stem straightness is also collected, as this characteristic has a considerable bearing on the quality of final yield.

A year or so before crops are harvested, a detailed inventory of stem frequency by size and defect class is undertaken. This information is required in the short term to prescribe and co-ordinate logging operations, estimate realisable volumes by log type, and to choose the best sequence and kind of log-making strategies. The mensurational information is also useful in the long term for checking a wide range of functions that feature in the growth and yield model, namely tree height, volume and taper. Fuller accounts of the measuring procedures are given by Whyte (1978) and Patel (1985).

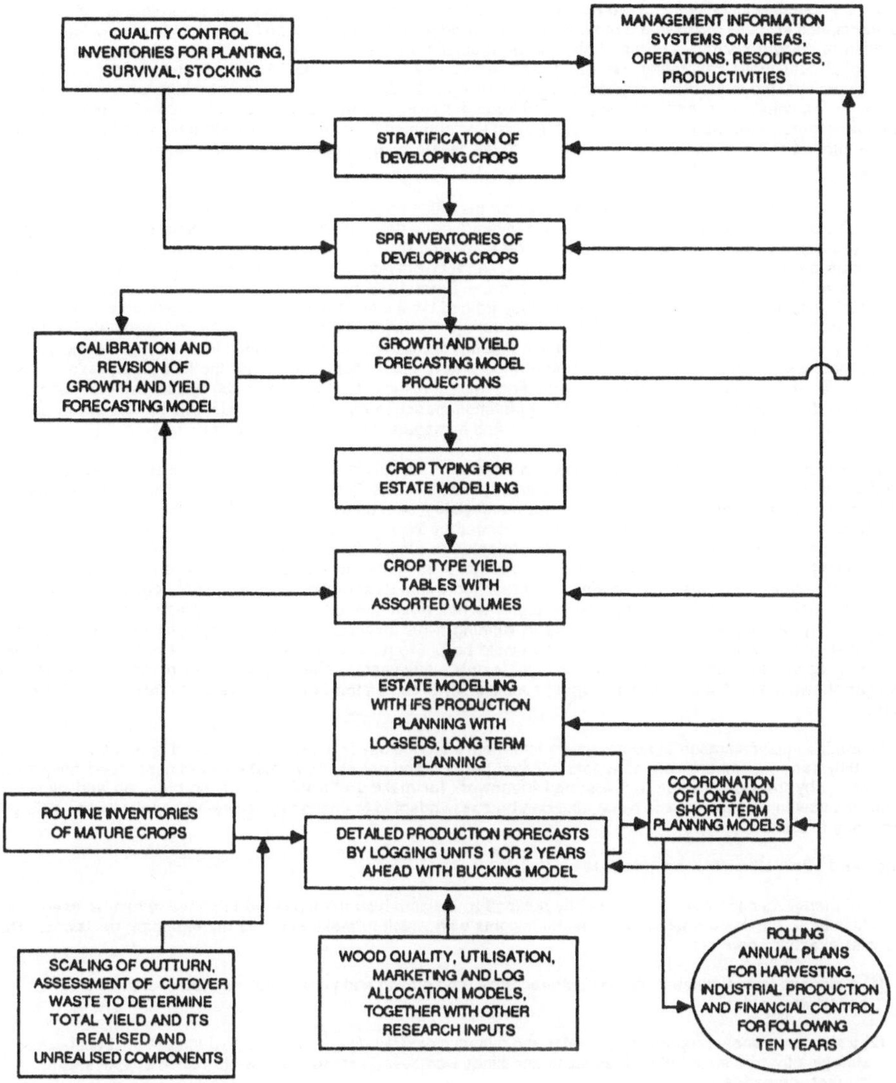


**FIG. I. Inventory, Growth and Yield, Information and Planning Systems for Fiji Pine Commissi**

The two last components just described are combined so that long and short term planning is co-ordinated. Another part of this co-ordination involves assessment of actual outturn by volume piece scaling and weighbridge monitoring and the determination of cutover waste. Long and short term forecasts, therefore, are calibrated against actual production. This process of continually calibrating, reviewing and adjusting knowledge about the present and future plantation resources is crucial to the succsssful adoption of the various operational research procedures which form the main body of this paper. The consistency and stability achieved through having retained a single set of information gathering and processing procedures over the years and through having implemented them routinely prior to a compelling need for the results, has proved to be of invaluable assistance to all levels of management and to processing plant design consultants.

It may be of considerable interest to operational research practitioners to digress a little at this point in order to explain how local staff were trained and supervised in routine inventory and data-processing tasks. As part of the initial report, it was suggested that undergraduate students could be employed over their long vacation in training local staff in the inventory, data-processing and analytical techniques and in providing ongoing research support for modelling developments and other relevant areas of investigation. From 1975 to 1986, at least three students each year undertook such asignments over 3 month periods: they trained local staff in new techniques or refinements to existing ones, collected research data, analysed these data during the following academic year and presented a formal dissertation, thesis or report on their findings. These findings were then integrated within the existing inventory, data-processing, forecasting and planning systems and formed part of the training projects for the following year's students to undertake. The continuity of effort and a genuine findings. If nothing else, this approach has shown one successful way of implementing research knowledge in routine plantation management, a not inconsiderable benefit for an applied discipline where the gap worldwide between researchers and managers is becoming ever wider.

Acceptance of modern data processing has also been effected in an individual way. In the first instance, software was developed on mainframe computers in New Zealand and then adapted bit by bit for the programmable calculators that the staff in Fiji had initially acquired. One run of the growth and yield model on a programmable calculator took over an hour to complete but it did give locals the taste of what computers could do, and they soon became addicted to their capabilities. Subsequently, the software was further adapted to run on the government mainframe computer in Suva, about four hours journey away from the Commission headquarters. Later, when the service from this machine could be provided only at irregular intervals, an agreement was reached with a local organisation with spare computer capacity to rent time on its machine. By the time that spare capacity had been exhausted, the Pine Commission had built up a considerable body of software, much experience in running it on a range of machines with its own staff, and the knowledge that any machines of its own it acquired would be put to good use. The number of micro-computers the Commission owns and runs has risen from two to eight in only a few years. This sequence has resulted in nearly all of the packages shown in Fig. 1 having been adapted successfully to run on the Commission's own machines by their own local staff.

Successful routine implementation of appropriately informative technology is the essence of good operational research. Agreement between researchers and managers is never perfect, and not all the former's recommendations have been put into practice by the Commission, but a sound framework for making effective use of the planning and decision-making models now to be outlined is in place, and can be easily adapted to change the system's capabilities as different needs continue to arise.

### Planning and Decision-Making Models

Four forms of modelling capability are now briefly outlined to indicate how the amassed information can be used in an integrated fashion to provide managers with useful insights with which to make sensible management decisions. The four examples examined here are:

(1) a medium-term simulation model to characterise crop growth and yield trends and forecast outturns of assorted yields;

(2) a short-term linear programming master and dynamic programming sub-model to generate the best bucking stategies for tree stems of various sizes and kinds fromseveral stands in order to match supplies with market demands;

(3)a medium-term linear programming model for scheduling , to best economic advantage, annual harvests from crops, yield forecasts for which were derived through the procedure explained in (1) , but which are further subject to logging and port-handling constraints;

(4)long-term simulation models to forecast assorted wood supplies and costs of production for one or two crop rotations.

**Growth and Yield Modelling**

Discussions with decision-makers and allied staff indicated that separate estimates would be needed for sawlogs, chip logs, post and pole roundwood, chip residues and bark + sawdust. Although no intermediate thinnings were originally envisaged, the growth and yield models were designed to cater flexibly for a range of changed silvicultural practices. Specimen output tables were drawn up to ensure that all managers' and planners' needs were adequately met. This issue having been resolved, it was clear that an implicit stand level model with derived diameter at breast height (dbh) distribution to provide size class information and compatible tree taper and volume equations for estimating assortments would be an appropriate form. Crop rotations envisaged could range from 12 to 25 years, and certainly anything older would be most unlikely.

The state variables for estimating current yield at a given age are:
minimum dbh/of crops, $d_{min}$; mean dbh of crops, $d_{mean}$; variance of the crop dbh's, $d_{var}$ and stems per hectare, N. The growth model contains equations for predicting each of the future crop diameter state variables, viz:

$$d_{min} / dT = f_1 ( d_{min}, d_{mean}, d_{var}, N, T )$$

$$d_{mean} / dT = f_2 ( d_{mean}, N, T )$$

$$d_{var} / dT = f_3 ( d_{mean}, d_{min}, N, T )$$

As stems per hectare do not decrease by more than 2 or 3 unless catastrophic cyclones occur, regular mortality is asumed to be nil over a forecasting period, and so N does not alter. The state variables can be manipulated at any age, T years, to derive the shape, scale and location parameters (a, b and c respectively) of the Weibull probability density function. By starting with the minimum dbh class, numbers of stems are derived for each successive class from the Weibull cumulative distribution,

$$F(x) = 1 - e^{-[(x-a)^c / b]}$$

as explained in Whyte 1986.

Stand tables (numbers of stems per hectare by dbh class) are converted into stock tables (average assorted volumes per tree and per hectare by dbh class) in two steps:

(1) average height, $h_{ij}$ of each class, $d_{ij}$, is predicted from the equation,
$h_{ij} = \exp ( b_0 + b_1 d_{ij} + b_2 d_{ij}^{-1} + b_3 T )$;

(2) total stem volume inside and outside bark, together with potential volume assortments are derived from,
$v_{ij} = c_0 + c_1 ( d_{ij}^2 h_{ij} )$

$$v_{xk} = v_{ij} / h_{ij} \int [ a_1 ( l / h_{ij} ) + a_2 ( l / h_{ij} )^2 + \ldots\ldots + a_5 ( l / h_{ij} )^5 ] \, . \, dl$$

where $v_{xk}$ is the volume of log $x_k$ of length $l_2 - l_1$, and $a_i$, $b_i$ and $c_i$ are weighted least-squares regression coefficients appropriate for the species and locality.

Outputs from modelling growth and yield include, for each of the 10 years following inventory inputs, forecasts of volumes (outside and inside bark) by dbh class and log assortment. As mentioned previously, the diameter prediction equations are monitored regularly through remeasurement of sampling units in routine inventories, while the height, volume and taper functions are checked at time of pre-harvesting inventory. Each year the current state variables are updated on a part of the whole resource, while each single crop will be updated within any five years. This intensity of calibration and review is essential because so little of the resource has reached maturity, on account of severe cyclone damage and early clearfelling. Early versions of the model were prepared by Geiser (1977) and Broad (1978), with major revisions of the equations being provided by Reid (1986).

The growth and yield model forecasts production as though all stems were free of any defects. Such is not the case in reality, however, particularly as the Pine Commission's resources have ben hard hit by cyclone damage. An ancillary model called LOGSEDS (see Leitch, 1984), based on analyses of large quantities of stem defect measurements and derivation of eight distinct defect patterns, allows realistic deductions to be made for unmerchantable portions of the stem. Revised estimates can then be obtained of the number and volume of logs per hectare by small end diameter and length classes for up to ten years ahead for the whole resource. Such information was extremely valuable in designing the new sawmill and choosing appropriate equipment.

**Bucking Model**

Bucking (or cross-cutting) is the process of cutting up the full merchantable length of a stem into shorter lengths so that the logs can be conveniently transported and demands for various kinds of logs can be met in financially rewarding ways. Single stem optimisation, however, may result in a serious mismatch of logs supplied with what the markets desire. Both supply and demand considerations must be made, and short term prescriptions should be fully co-ordinated with medium-term estimates from LOGSEDS, as explained in the previous section. The nature of the standing resource is characterised in the modelling process from data collected in pre-harvesting inventions carried out one or at most two years prior to clear-felling of the crops, while the market demands are specified in terms of the most informed mill production targets for the following year.

The general approach adopted follows the basic procedure developed by Gilmore and Gomory (eg in 1966) in which a DP algorithm is explicitly interfaced with a LP model through price-directed Dantzig-Wolfe decompostion. The DP sub-problem generates activities for the LP so that bucking strategies can reflect properly the opportunity costs arising from critical constraints on demands and reosurces. In the formulation, developed by Eng (1982) and reported subsequently by Eng, Daellenbach and Whyte (1986), the plantation resources which are candidates for being logged in the coming year, are classified into J classes which depend on the size and quality of the stems. If $x_{ij}$ denotes the number of stems of class j bucked by pattern i, $r_{ij}$ the corresponding financial return, $a_{ijk}$ the volume of log type k produced through adopting pattern i on class j, $s_j$ the number of stems of class j available in the resources and $b_k$ the estimated demand for log type k, then the aim is to determine optimal values of $x_{ij}$ so as to

(1) maximise $\sum_i \sum_j r_{ij} x_{ij}$

subject to (2) $\sum_i \sum_j a_{ijk} x_{ij} = b_i$

(3) $\sum_j x_{ij} < s_j$

with (4) $x_{ij} > 0$ and integer, all i,j.

The co-efficients $r_{ij}$ and $a_{ijk}$ are obtained through dynamic programming with the following recursive relation

(5) $f(L) = \text{maximum}\,( r(y_k, L) + f(L - y_k) )$

where $y_k$ denotes the length of a short log of type k cut at a distance $L-y_k$ from the base of the whole stem, $r(y_k, L)$ represents its value, $f(0) = 0$
$Y(L)$ is the set of feasible short logs at L for all K end-use categories
and $0 < L < LMAXj$ (the total usable length of stem).

The value of a short log depends on its location, length and freedom from defect. Fig 2 illustrates the computational scheme.

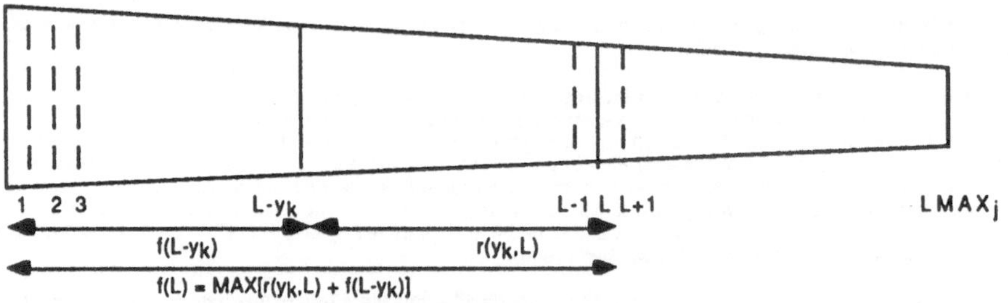


Fig 2. **Discretization of stem and DP representation**

The output of the DP bucking problems for each class j comprises the vector of log type volumes { $a_{cj1}, \ldots, a_{cjk}$ } and the associated returns per stem, $r_{ij} = F(LMAXj)$. The outputs from the DP recursion then become the inputs to the LP problem, the global optimum for which is obtained iteratively through Dantzig-Wolfe decompostion when

(6) $F(LMAXj) > \hat{P}_j$

where $\hat{P}_j$ is the shadow price of supplying class j

Recursion (5) is employed to find new bucking patterns that satisfy condition (6); the LP is then solved again iteratively until no class can generate a pattern that satisfies (6), at which point the optimal solution has been found.

Typically, about 100 classes for each of several logging blocks would be considered each year. From the analyses, bucking rules can be prescribed for each block and sensitivity analysis conducted to examine the departures from optimality if alternative rules that are managerially more convenient were to be adopted. The modelling capability is not now being used in this way, however, as FPC managers have chosen to employ simpler heuristics. The technique was, however, used to help in identifying medium term supply characteristics that greatly assisted in choosing appropriate machinery fork and layout of, a major sawmill that has just been built. Because this use for the methodology is not altogether suitable, an alternative but related modelling approach, as was alluded to in the previous section, was devised to help refine the design of the milll. Neverthless, the decomposition approach could be gainfully employed muchmore widely than it is, but only if managers are prepared either to allow local staff to be adequately trained in using the technique or else to pay for a qualified consultant to help run the model and interpret the outputs. This is one area in the whole Fiji Pine management system, where it has simply not been possible to simplify matters sufficiently to allow procedures to be implemented routinely.

## Harvest Scheduling

Yield forecasts for substantial areas of maturing resources allow harvests to be scheduled to test economic advantage. One such study involved maximisation of not discounted profits from export log sales over a period of years subject to constraints on minimum contractual supplies, utilisation of port facilites, available manpower, logging technique and transport capabilibies, and on administrative convenience.

$$\text{Maximise } \sum p_j \left( \sum_{I}\sum_{k}\sum_{m} g_{ij}.X_{ijkm} - \sum_{i}\sum_{j}\sum_{k} n_{ijk}.N_{ijk} - \sum_{i}\sum_{j}\sum_{k} c_{ijm} . \sum_{k} g_{ij}.X_{ijkm} \right)$$

where $p_j$ is discounted price / $m^3$ in year j

$g_{ij}$ is yield for area i, year j in $m^3$

$X_{ijkm}$ is fraction of area i cut in year j by logging method k and shipped through port m

$N_{ijk}$ is fraction of a year that method k is used in area i during year j $n_{ijk}$ is cost of logging method k for afull year in area i during year j

$c_{ijm}$ is transport cost / $m^3$ from area i to port m during year j

Subject to:

(1) $\sum_{j}\sum_{k}\sum_{m} X_{ijkm} < 1$; $\quad i = 1,2,..........., I$

(2) $\sum_{m} r_i . X_{ijkm} < f_{ij} . N_{ijk}$; $\quad$ all i, j, k

(3) $\sum g_{ij}.\sum X_{ijk1} - Y_{j1} = d_{j1}$; $\quad$ all j

$\sum g_{ij}. \sum X_{ijk2} + Y_{j2} = d_{j1}$; $\quad$ all j

$Y_{j1} - Y_{j2} > D_j - d_{j1} - d_{j2}$; $\quad$ all j

where $Y_{j1}$ represents an annual surplus at port 1, $Y_{j2}$ a shortfall at port 2, $D_j$ the combined annual requirement for both ports and $d_{jm}$ the individual port requiremtns in year m.

(4) $r_i . \sum_{j}\sum_{m} X_{ij2m} < F_{i2}$; $\quad$ all i

$r_i . \sum_{j}\sum_{m} X_{ij4m} < F_{i4}$; $\quad$ all i

whereF$_{ik}$ represents maximum or minimum number of hectares to be cleared by machine types 2 and 4 respectively.

(5) $\sum N_{ij2} < 1$; $\quad$ all j.

The main use of this kind of modelling has been to indicate possible solutions that could be worth examining in the more detailed simulation methodology outlined in the next section. Results of running the LP have shown that a large number of alternative "near-optimal" solutions can be generated and that minor smoothing of the use of logging machinery and re-allocation of the harvesting sequence among adjacent areas and/or years can be made with little financial sacrifice. Experience has baeen that managers are happier specifying their own harvesting schedules, but, if it can be demonstrated that the loss through implementing them is relatively minor compared with ;the theoritical optimum, the expense of modelling in the manner just shown is likely to be more than justified, as all concerned may rest assured that there is little likelihood of finding greatly improved sequences for cutting. Also, useful insights can be gained through sensitivity analysis of infra-structural constraints representing port-handling, roading and logging equipment capabilities. A fuller account of this modelling is covered in de Kluyver, Whyte, Baird *et al.* 1979

### Forest Estate Models

The term "forest estate model" is used in New Zealand to describe the process whereby whole forest capacities and yielding abilities are described so that simulated management strategies for the total resource over a considerable period of time can be compared. The management plans involve specification of intermediate and final fellings, replanting or selling off felled areas, and planting new areas. The estate model employed by the Fiji Pine Commission is that called IFS - Interactive Forest Simulator, developed by Garcia (1981), and modified in-house to incorporate costs and revenues from implementing the prescribed strategies. The value of including such a facility is that the forest- wide consequences of implementing the short and medium-term term plans arising from the previous three modelling components can be examined in reasonable detail over a considerable period of time (say two to two and a half rotations), thus assisting in co-ordinating the overall planning and management. The disadvantage is that the unit of planning, the crop type, is a substantial aggregation of crops of many ages which purport to conform to the same growth trends, silvicultural treatments, kinds of harvesting and costs of production. A certain amount of individual crop sensitivity is lost, therefore, during this stage of modelling.

### Extension of the System

Fiji has had a long history of harvesting its natural forests which contain many fine timbers as well as various all-purpose ones . This newly created pine plantation resource will obviously be in direct competition with the existing industry based on raw material from the natural forest. The Fiji Pine Commission has long been aware of this possibility and has been pressing for fully integrated sector-wide studies. A framework has been constructed and is soon to emerge in the form of a post-graduate thesis that has adapted a New Zealand approach to resolving sector issues (see Whyte, 1987 and Baird and Whyte, 1987). Progress with this Fiji-wide modelling has been extremely slow, because there has been no equivalent commitment to evolving integrated systems of inventory, forecasting planning and decision-making in ownerships other than the Pine Commission's.

Co-ordinated strategic studies into examining the interactions among the independent plans of individual forestry and forest industry organisations need to show ways of allowing effective co-operation among them so as to avoid wasteful use of scarce national and regional resources and needless competition, so as to serve the good of the nation as a whole. Such an ideal could be achieved through extending the inventory, planning and management system for a single organisation (the Fiji Pine Commission) into one that covers forestry and the forest industry throughout the whole country.

### Conclusions

Considerable benefits have accrued to the Fiji Pine Commission over the years through having adopted a traditional systems approach for constructing an appropriate framework to integrate information-gathering, data-processing, forecasting, planning and decision-making components of managing a young pine plantation resource. While a few small elements here and there have been dropped over the years and other new ones added to the system, the main structure remains intact, and that has thus provided a stability which has greatly assisted the management process.

Important features that have contributed materially to this success include:

- overall coherency of the integrated parts;
- recognition of evolving managerial needs;
- acknowledgement of local available skill, behaviour patterns,customs and culture during the design stage of systems;
- awareness of the prevailing and likely future environment of most relevance to management;
- choice of modelling forms which; match the management needs specifically so that management and not the methods assume dominance;
- active involvement of local staff in implementing and monitoring the system components;
- regular contributions of young undergraduate students to calibrating performance of local staff, and to training regularly in ongoing programmes.

Nevertheless, success of this kind can only be achieved with the help of well-motivated managers, who are committed to the practising of management. The full value from having an integrated planning and management system can never be realised without such a commitment. That has a strong bearing on how OR practitioners should approach problem-solving.

**References**


Baird, F.T. and Whyte, A.G.D., 1987: *"Flexibility in plantation harvesting in the 1990's"* NZ Forestry Council Working Paper No. 10 27 pp

Broad, L.R., 1978: *"A revised yield forecasting model for Pinus caribaea in Fiji"* School of Forestry, University of Canterbury, M for Sc. Report 56 pp.

Cochran, W.G., 1977: *"Sampling Techniques"* , Wiley Press, New York, 3rd edition, 344 -55.

de Kluyver, C.A., Whyte, A.G.D., Baird, F.T. *et al*. 1980: *"Forest harvest scheduling in Fiji: a comparative study of lnear programming, heuristics and decomposition techniques"* NZ Oper. Res. 8 (i) 33-72

Eng, G., 1982: *"Amethodology for forest outturn assessment and optimal tree bucking"* University of Canterbury, M.Sc. Thesis 108 pp.

Eng, G., Dallenbach, M.G. and Whyte, A.G.D. 1986: *"Bucking tree-length stems optimally"* Can. Jour. For. Res. 16 1030-5

Fickes, J. and Dunn, J., 1980: *"FRRAP (Forest Reporting, Recording and Planning"* FPC Report 121 pp

Fiji Pine Commission, 1987: *"Annual Report for 1986"*

Garcia, O., 1981: *"IFS, an interactive forest simulator for long-term planning"* N.Z.Jour. For. Sci. 11(1) 8-22

Geiser, M., 1977: *"A yield forecast model for Pinus caribaea in Fiji."* School of Forestry;, University of Canterbury M. For. Sc. Report 66 pp

Gilmore, P.C. and Gomory, R.E., 1966: *"The theory and computation of knapsack functions"* Oper. Res. 14 1045-74

Leitch, J., 1984: *"Resource Analysis for Sawmill Design and Development"* FPOC Report 19 pp + Appendices (unpublished)

Patel, N., 1985: *"Forest Inventory System Manual"* Fiji Pine Commission Lautoka, Fiji 108 pp

Reid, S.B., 1986: *"Revisions of equations to forecast yields of Caribbean pine in Lololo Forest Fiji"* School of Forestry, University of Canterbury B. For. Sc. Dissertation 72 pp.

Whyte, A.G.D., 1978: *"Inventory and yield forecasting systems for Caribbean pine in Fiji"*. N.Z. Forest Service FRI. Symposium #20 (ed. D.A. Elliott) 41 - 7.

Whyte, A.G.D., 1983: *"Development of the Fiji Pine Commission: an account of the emergence of a major new industry in a develping country"* N.Z. Jour. For. 28 (3) 412-22

Whyte, A.G.D., 1986: *"Modelos de crescimento e de producao""* EMBRAPA - CNPF, Curitiba March 1986 61 pp

Whyte, A.G.D., 1987: *"An appropriate framework for modelling the New Zealand forest sector"* In Roebe, M.M., and R. Fodder (eds) proceedings 56th ANZAAS Congress, Palmerston North. Jan. 1987 44–61

# A STOCHASTIC GROWTH MODEL FOR EVEN-AGED SINGLE-SPECIES FOREST STANDS

Tatsuo Sweda
Department of Forestry, Nagoya University
Chikusa, Nagoya, 464 Japan

**Summary**

Applying the theories of Markov process a pure theoretical model describing growth of the single-species even-aged forest stand as an ever-changing process of frequency distribution of diameters of constituent trees was constructed. The model consists of two mathematical expressions named the diameter distribution function and diameter transition probability function. Assuming that all the trees are of diameter zero at the establishment of the single-species even-aged stand, the former expression gives the frequency distribution of tree diameters at any future time. According to this model, the diameter distribution develops from Dirac's delta function at the beginning of stand growth into an inversely J-shaped distribution at early stages of stand growth, then into skew bell-shaped one at intermediate stages, and finally approaches a normal distribution as the stand matures. Given an actual frequency distribution of tree diameters at any time of stand growth, the latter function gives projection of future diameter distributions. An application of the diameter distribution function to an observed series of diameter distributions revealed satisfactory agreement between the two; thus indicating the validity of the proposed model and the majority of the assumptions underlying it.

## Assumptions underlying the Model

In this paper, the diameter distribution function $a_n(t)$ is defined as the probability of finding trees of diameter n in the even-aged stand of age t, while the diameter transition probability function $p_{j,k}(r,s)$ as the probability of diameter growth from size j at stand age r to size k at age s. With this definition these two probability functions were derived from the following assumptions:

Major Assumption;

Diameter growth in an even-aged single-species stand is a

discrete-state continuous-time parameter Markov process, where the terms state and time refer to diameter of trees and stand age respectively.

Minor Assumptions;

1. Trees either can attain only one unit of diameter increase or maintain the same diameter during an infinitesimal time interval.
2. The probability to gain any given amount of diameter increase during any given time interval is a function of time only and the same for all the constituent trees of the stand regardless of their current diameters.
3. The average diameter growth rate of any single-species even-aged stand is proportional to the reach from the current average diameter to the average asymptotic diameter which the stand will attain eventually.
4. The number of trees dying out of the stand in the course of stand growth depends only on the stand age and is free from the size of the trees.

Of these five assumptions in all, the first, third and fourth appear explicitly in the following derivation. The others, including the major assumption remain rather implicit, but constitute important bases of the model.

## Diameter Distribution Function

Let $b_o(\Delta t)$ be the probability that no diameter increase occurs in an even-aged single-species stand during a minute time interval $\Delta t$; $1-b_o(\Delta t)$ then represents the probability of diameter increase during the same interval. Supposing that probability $b_o(\Delta t)$ is continuous with respect to time, instantaneous probability of diameter increase can be defined as

$$c = \lim_{t \to 0} \frac{1-b_o(\Delta t)}{\Delta t} \quad . \tag{1}$$

According to the first minor assumption, the trees of the stand have only two possibilities during this minute interval of $\Delta t$, either to grow by one unit or maintain the same diameter. Using equation (1), the probability of the first contingency is approximated by

$c\Delta t+o(\Delta t)$ ,

and the probability of the second by

$1-c\Delta t+o(\Delta t)$ .

The quantity $o(\Delta t)$ in the above expressions represents the error associated with the linear approximation. This error term is a quantity of smaller order of magnitude than $\Delta t$, and thus, even when divided by $\Delta t$ approaches zero as $\Delta t$ tends to zero, i.e.

$$\lim_{t\to 0} \frac{o(\Delta t)}{\Delta t} \longrightarrow 0 . \quad (2)$$

Now consider the process in which a tree attains diameter of size n by time $t+\Delta t$. Again according the first minor assumption, the process can be realized in one of the following two mutually exclusive ways:

1. diameter being of size n by time t, and no increment added during the succeeding interval $\Delta t$,

2. diameter being of size n-1 by time t, and one unit of increment added during $\Delta t$.

The probability of the first contingency is given by the product of the probabilities $a_n(t)$ of finding the diameter to be n at time t and $1-c\Delta t+o(\Delta t)$ of no diameter increase, resulting thus in the joint probability

$$a_n(t)\{1-c\Delta t+o(\Delta t)\} . \quad (3)$$

Similarly, the probability of the second contingency is given by the product

$$a_{n-1}(t)\{c\Delta t+o(\Delta t)\} , \quad (4)$$

where the first element represents the probability of finding the diameter to be n-1 at time t, and the second element the probability of unit diameter increase. Finally the unconditional probability $a_n(t+\Delta t)$ of finding diameter to be n at time $t+\Delta t$ is given by the sum of probabilities (3) and (4). Thus

$$a_n(t+\Delta t) = a_n(t)\{1-c\Delta t+o(\Delta t)\}+ a_{n-1}(t)\{c\Delta t+o(\Delta t)\}, \quad (5)$$

results. By sorting out similar terms, diviving both sides by $\Delta t$ and passing to the limit of $\Delta t \longrightarrow 0$, we get

$$\frac{da_n(t)}{dt} = -ca_n(t) + ca_{n-1}(t) . \quad (6)$$

Generally, any stochastic process described by equation (6) is called the Poisson process. For the ordinary Poisson process coefficient c is a constant. However, for the diameter growth process under consideration it should vary with time, and for equation (6) to be solved c has to be expressed as a function of time in such a manner that it would reflect the growth of trees.

The third minor assumption serves this purpose. It is formulated as

$$\frac{dA(t)}{dt} = K\{M-A(t)\} \quad , \tag{7}$$

where A(t) stands for the current average diameter, M the asymptotic average diameter which the stand would ultimately attain, and K the intrinsic rate of growth. Solving (7) we get

$$A(t) = M\{1-\exp(-Kt)\} \quad , \tag{8}$$

which upon differentiation with respect to time yields the average diameter growth rate of the stand

$$\frac{dA(t)}{dt} = MK\exp(-Kt) \quad . \tag{9}$$

Thus for trees of diameter n at time t, their most probable diameters at time $t+\Delta t$ is estimated by

$$n+MK\exp(-Kt)\Delta t+o(\Delta t) \quad , \tag{10}$$

where $o(\Delta t)$ represents the error associated with linear approximation as defined by (2).

On the other hand, it follows from the first minor assumption that trees of diameter n at time t may either be of the same diameter or of diameter n+1 at time $t+\Delta t$. The respective probabilities $1-c\Delta t+o(\Delta t)$ of no diameter change and $c\Delta t+o(\Delta t)$ of unit increase play weighing factor for each of the contingencies. Thus another expression for the average diameter at time $t+\Delta t$ is given by

$$(n+1)\{c\Delta t+o(\Delta t)\} + n\{1-c\Delta t+o(\Delta t)\} \quad . \tag{11}$$

Equating expressions (10) and (11), rearrenging the resultant equation and passing it to the limit of $\Delta t \rightarrow 0$, we get

$$c = MK\exp(-Kt) \quad . \tag{12}$$

Substituting this in (6), the final equation characterizing

the stochastic diameter growth is obtained:

$$\frac{da_n(t)}{dt} = -MKe^{-Kt}a_n(t) + MKe^{-Kt}a_{n-1}(t) \ . \tag{13}$$

This difference-differential equation can be solved with the initial condition

$$a_n(0)=1 \qquad \text{if} \quad n=0 \ ,$$
$$a_n(0)=0 \qquad \text{if} \quad n=0 \ . \tag{14}$$

Since the mathematical manipulation leading to the final solution is rather lengthy, it is spared here and only the result is given below:

$$a_n(t) = \frac{M^n(1 - e^{-Kt})^n}{n\,!} \exp\{M(e^{-Kt} - 1)\} \ . \tag{15}$$

This is the diameter distribution function which gives, for any fixed time t, the diameter distribution, i.e. the probability of finding trees of diameter n in a single-species even-aged stand. As the name Poisson process implies, the diameter distribution function (15) gives Poisson distribution which changes its shape as

the mean diameter

$$A(t) = M\{1 - \exp(-Kt)\} \ , \tag{16}$$

and the standard deviation

$$S(t) = \sqrt{M\{1 - \exp(-Kt)\}} \ , \tag{17}$$

change with the passage of time t. A graphical example of the diameter distribution function is shown in Fig.1, in which changes in distribution from inverse-J through skewed bell-shape to normality is well depicted.

**Diameter Distribution with Simple Death Process Incorporated**

Incorporating the age-dependent simple death process of individual trees depicted by the forth minor assumption, equation (15) can be further modified to

$$N_n(t) = b\exp(-pt)\frac{M^n(1 - e^{-Kt})^n}{n!} \exp\{M(e^{-Kt} - 1)\} \ , \tag{18}$$

which instead of the probability gives the number of trees of diameter n at time t in a even-aged stand in which the initially established b trees decrease with the death rate p.

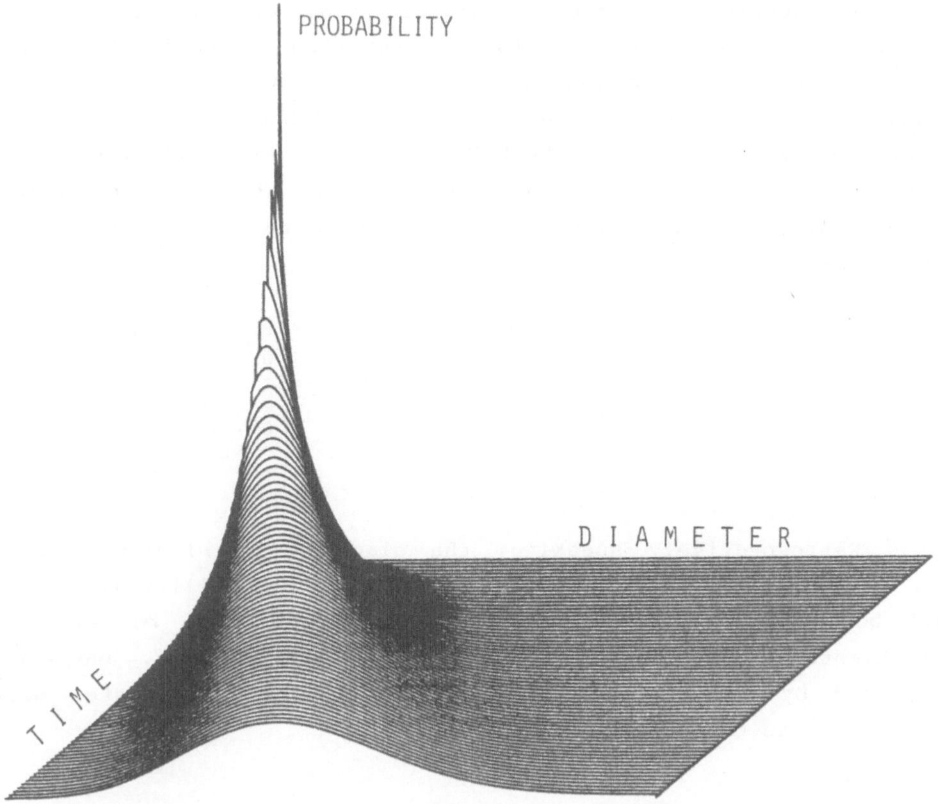


Figure 1: The diameter distribution function, which gives the probability of finding trees of any desired diameter n at any desired time t.

**Diameter Transition Probability Function**

The same logic applied to the derivation of the diameter distribution function applies to the diameter transition probability function $p_{j,k}(r,s)$ which denotes the probability that the trees of diameter j at time r grow to attain another diameter k by any later time s. Mathematical formulation of the assumptions excluding the forth minor and the subsequent manipulation yield

$$p_{j,k}(r,s) = \frac{\{Me^{-Kr}(1 - e^{-K(s-r)})\}^{k-j}}{(k - j)!} \exp\{Me^{-Kr}(e^{-K(r-s)}-1)\} \quad \text{for } k \geq j, \qquad (19)$$

$$p_{j,k}(r,s) = 0 \quad \text{otherwise.}$$

By substituting successive diameter class values 0, 1, 2, 3, . . . for the state parameters j and k, the transition probability $p_{j,k}(r,s)$ can be developed into a matrix whose element at the $j^{th}$ row and $k^{th}$ column represents the diameter transition probability from the $j^{th}$ diameter class at time r to the $k^{th}$ at time s. Multiplication by this transition matrix of a diameter distribution at any given time r, developed into a vector gives a projected diameter distribution at any future time s as follows;

$$[a_0, a_1, a_2, \ldots]_s = [a_0, a_1, a_2, \ldots]_r \begin{bmatrix} p_{00} & p_{01} & p_{02} & \cdots \\ 0 & p_{11} & p_{12} & \cdots \\ 0 & 0 & p_{22} & \cdots \\ 0 & 0 & 0 & \cdots \\ \cdot & \cdot & \cdot & \cdots \end{bmatrix} \tag{20}$$

Retrospectively speaking, the diameter distribution function is a special case of the diameter transition probability function in which both the initial time r and diameter j are set to zero. The coefficients M and K are common to both the equations, and can be determined by least-squares fitting of the average diameter growth equation (4) to the observed counterpart.

## Validity of the Model

Validity of the model was tested against the observed development process of diameter distribution restored from increment cores extracted from an even-aged single-species stand.

Within an even-aged pure larch (*Larix lariciana* Du Roi K. Koch) stand regenerated after forest fire in the Province of New Brunswick, Canada, a square sample plot was established in which growing conditions were deemed fairly uniform. Increment cores were bored at breast height (1.3 m above ground) from all the existing individuals within the plot, resulting in a total of 265 stems. Using an 'Addo-x' increment measuring machine, the radius of every consecutive annual ring on each of the boring was measured to an accuracy of 0.01 cm. Then the average diameter at each consecutive age of the stand was calculated, and a series of diameter distribution reconstructed by counting the number of readings in each diameter class.

Since the coefficients M and K of the model are common to all the model equations (15) through (19), they can be determined, at least from theoretical point of view, through any one of these equations. However, from practical point of view, they are most easily determined through the simplest equation, i.e. (16). The lesat squares fitting of this average diameter growth equation to the

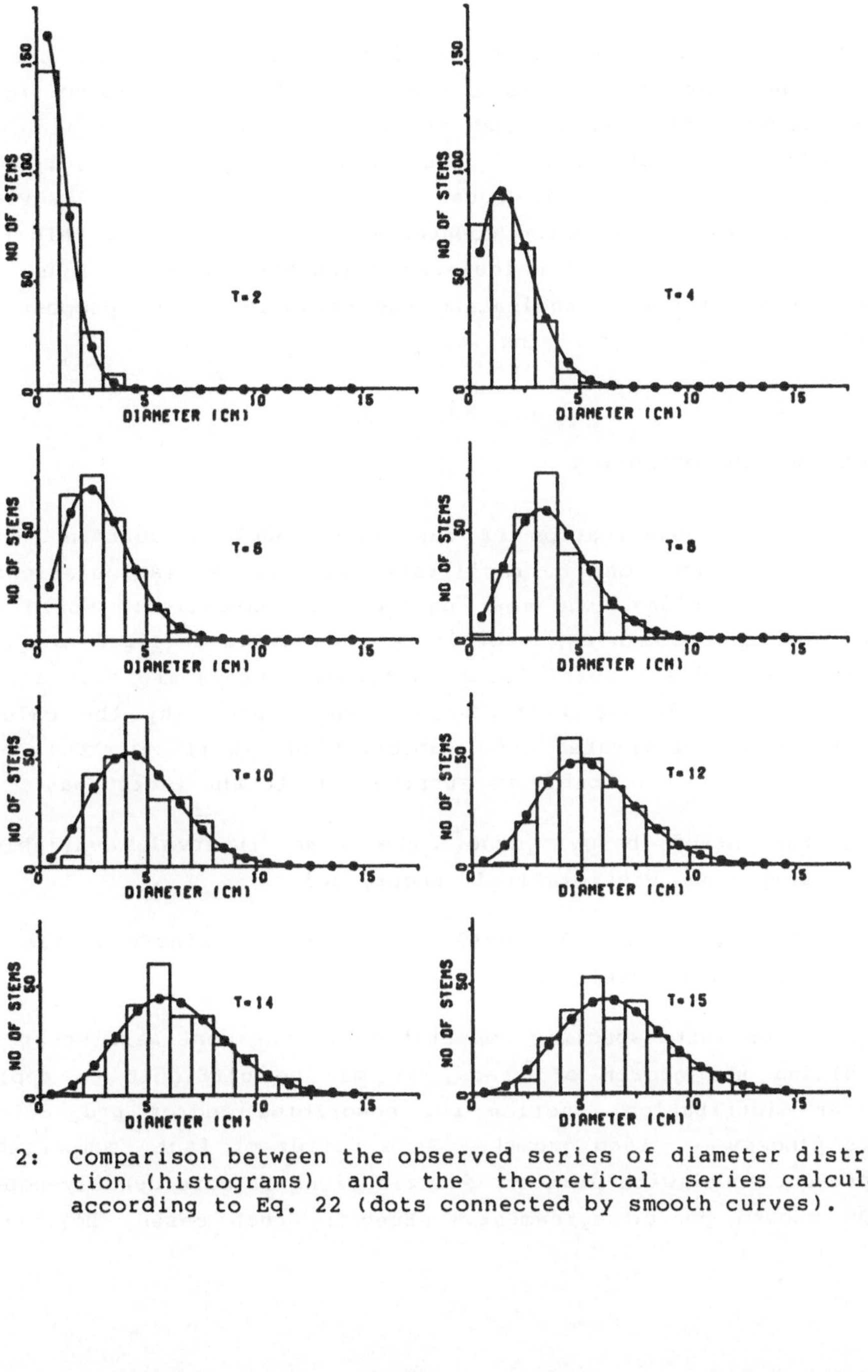


Figure 2: Comparison between the observed series of diameter distribution (histograms) and the theoretical series calculated according to Eq. 22 (dots connected by smooth curves).

observed counterpart resulted in

$$M=27.722\ \text{cm} \quad \text{and} \quad K=0.0182\ \text{(dimensionless)} \tag{21}$$

Substituting the above result in equation (15), the diameter distribution function for this particular larch stand is given by

$$a_n(t)=\frac{\{27.72(1-e^{-0.0182t})-0.5\}^n}{n!}\ \exp\{0.5+27.72(e^{-0.0182t}-1)\}. \tag{22}$$

The numerical term amounting to 0.5 which appears in the above equation but not in (15) is a correction for disagreement in class median between the theory and practice. The theoretical diameter distributions were obtained by substituting n=0, 1, 2, 3, . . . cm for t=1, 2, 3, . . . years in equation (22) and are shown in Fig.2 in comparison with the observed counterparts. As seen from this figure, the observed and the calculated diameter distributions showed satisfactory agreement, indicating the validity of the proposed model and the assumptions underlying it.

## Discussion and Conclusion

In the construction of the above model, a certain sacrifice was made of assumptions to facilitate mathematical manipulations. Of the five assumptions, the most unlikely or unrealistic would be the second and fourth minors, especially when intra-specific competition is tight, and as a result, lesser individual trees are more likely to be supressed by larger individuals. The reason why the calculated distributions represented the observations well in the present comparison may most probably be attributable to the facts that

1. the stand being young, the time interval available for comparison was relatively short, and

2. the stand was relatively sparse with seemingly little intra-specific competition.

Since intra-specific competition is the most persistent factor controlling the growth of trees, it may be difficult to apply the diameter distribution function for describing and/or projecting the entire process of stand growth. As a matter of fact, comparisons of the present model with diameter distributions of more densely-populated stands showed poorer agreement. Even in such cases, however, the agreement improved as lesser individuals were deleted from the observed distributions, indicating that the principles governing the present model holds at least for the subset of dominant individuals which are not in direct competition with each other.

It can be concluded that for such a relatively short period of time as the one between two successive thinnings during which competition is not too serious, all the present assumptions will reasonably comply with the actual growth process, and thus the diameter transition probability function would be successfully applied as an effective management tool for describing and projecting the growth of even-aged stands.

# MODELLING FOREST MOSAICS

**David G. Green**
Research School of Pacific Studies, Australian National University,
GPO Box 4, Canberra 2601, Australia

## Abstract

A useful way of modelling forests is to represent them as cellular automata, that is as mosaics in which each cell represents an area of the land surface. Theoretical studies using such models have revealed aspects of fire behaviour and forest dynamics that are not apparent in other models. They also mirror both traditional quadrat sampling and pixel-based satellite imagery, making them ideal tools for forest management.

## Introduction

Simulation is essential to the understanding and management of forest systems. Individual trees are so long-lived that even a simple experiment could take several human generations to complete and the impact of a single fire may be felt for up to 1000 years afterwards. An important limitation of most forest simulation models is that they ignore the spatial distribution of trees. In this account I describe ways of introducing spatial pattern into forest models. Spatial simulations make it possible to explain patterns and behaviour that cannot be examined otherwise (Green 1983a; Green et al., 1983, 1985) and are potentially important tools in forest management.

## Model structure

Spatial simulations of forests link trees or tree stands with specific locations (Green et al., 1985). The most accurate way to achieve this linkage is to maintain an inventory of all trees, recording the location of each one. However this representation is impractical, even for moderately large populations, because it handles spatial processes clumsily.

A more practical approach is to represent the land surface as a cellular automaton. **A cellular automaton** is an array of "cells", each of which functions as an automaton (Wolfram, 1984). In such models, each cell represents a discrete area of the land surface. Cells are classified by "states" indicating vegetation type, fuel abundance, etc. The scale of the grid varies according to the model: individual cells may represent single trees or whole forest stands. This approach has distinct practical advantages. For example, it represents spatially extensive processes, such as fire and seed dispersal, both naturally and easily. It also mirrors both traditional quadrat sampling methods and pixel-based satellite imagery. Thus if cells in the model are equated with pixels in an image, then satellite data can be fed directly into models of environmental resources (Minnich, 1983, Green et al, 1985). Different topologies for the cell population also form the basis for useful models. In particular, modelling the growth of individual trees call for automata in which the "cell" population grows and in which neighbouring cells respond to one another (Fig. 1). Suitably calibrated growth models of this kind can provide accurate estimates of the average size of trees of any given age and can thus be useful adjuncts to an overall community model.

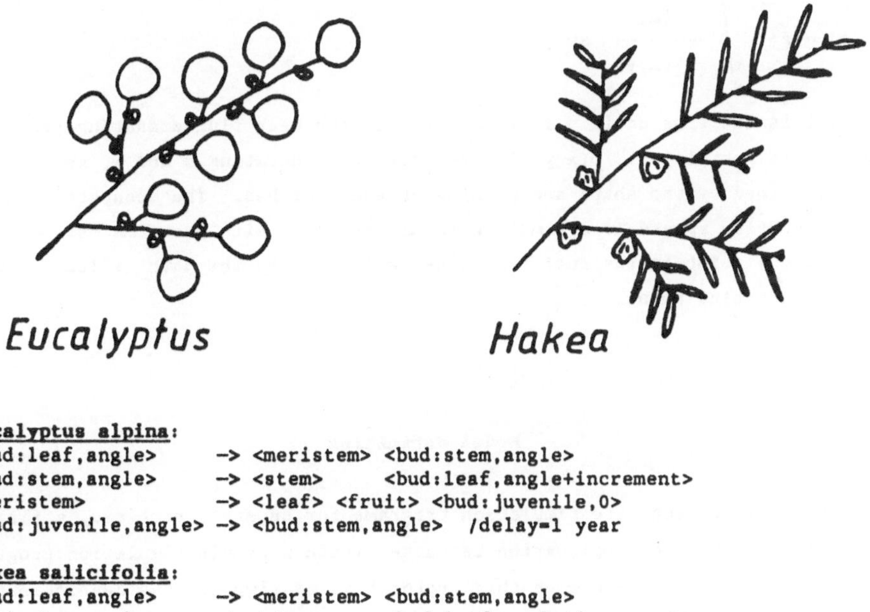


**Eucalyptus alpina**:

```
<bud:leaf,angle>       -> <meristem> <bud:stem,angle>
<bud:stem,angle>       -> <stem>     <bud:leaf,angle+increment>
<meristem>             -> <leaf> <fruit> <bud:juvenile,0>
<bud:juvenile,angle> -> <bud:stem,angle>  /delay=1 year
```

**Hakea salicifolia**:

```
<bud:leaf,angle>       -> <meristem> <bud:stem,angle>
<bud:stem,angle>       -> <stem>     <bud:leaf,angle+increment>
<meristem>             -> <leaf>     <bud:juvenile,0>
<bud:juvenile,angle> -> <fruit>    <bud:stem,angle>  /delay=1 year
```

**Figure 1.** Growth patterns for two tree species. Syntactic models of these patterns are given below. Names in angle brackets denote growth structures and arrows denote growth processes. The states of particular structures are indicated by colons. The models differ syntactically in the timing of fruit production and semantically in the shapes of leaves and fruit.

The most accurate method of representing continuous time flow is to record a list of events, ordered according to the precise times at which they occur but, as above, this method is awkward when dealing with spatial relationships. A discrete approach is again more practical. Discrete time steps are represented by scanning through the model grid and updating the following information for each cell: vegetation type and age, time since the last fire, and values of environmental parameters. Time steps may vary from (say) one minute, for a fire spread model, to whole years for long-term management problems, but must be sufficiently small (in relation to the phenomenon concerned) to prevent round-off errors from accumulating.

Fire spread is simulated by computing the effect that ignited cells have on their neighbours. The most practical approach is to use an "ignition template", which is laid over burning cells when they ignite and specifies the time delay before each surrounding cell ignites (Frandsen & Andrews, 1979; Green, 1983a). Many different mechanisms of fire spread may occur (Green et al., 1983). The "heat accumulation" model assumes that fuel at point **y** absorbs heat from fuel already ignited (either by radiation or convection), until it becomes hot enough to ignite. Thus

$$T(\mathbf{y}) = f\left( \int \frac{H(\mathbf{x})}{D^r(\mathbf{x},\mathbf{y})} \, d\mathbf{x} \right) ,$$

where T(**y**) is the time until **y** ignites, H(**x**) is the heat flux emanating from point **x** (zero if **x** is not alight), D(**x**,**y**) is the distance separating **x** and **y**, and r is a constant determined by the shape and packing of the fuel bed. The "contact" model assumes that fire spread is a simple epidemic process, with the time delay until ignition for any point in the fuel bed being determined by the path of least time from the fire's starting point.

## Model definition

Rather than develop separate simulation programs for separate problems, a flexible approach is to embody an application language within a single simulation program (Green et al., 1987). The aim of this approach is to simplify the development of simulations by making the language and concepts used to model forest processes the same as those used by foresters to describe them. For example, the program **MOSAIC** embodies an "**ECO**system Simulation **LANG**uage" ("ECOSLANG"), which deals with concepts such as species characteristics, including longevity, fecundity, and age at maturity; environmental gradients, that is, changes in either time or space of rainfall, soil moisture, salinity, altitude etc.; reproduction and dispersal strategies; and fire regimes. Particular models are defined by sets of commands in the language (Fig. 2). Features not specifically referred to receive default (usually null) settings. Fires, for instance, do not occur unless a fire regime is explicitly defined. Other syntactic approaches have also been developed too. For example, the "vital attributes" scheme concentrates on plant growth and species replacement patterns (Noble & Slatyer, 1980).

For use in management, the task of specifying a model can be made "user friendly" by having an expert system ask for relevant details. Because application languages define conceptual frameworks, basing programme design around them also simplifies the task of producing such expert systems.

```
MODEL
     TAXA       grass   euc     rf          (Names for the species modelled)
     SYM        g       e       r           (Symbols for each species)
     AGE        5       9999    9999        (Life expectancy of each species)
     ADULT      0       10      20          (Age when reproduction begins)
     SEED   local 10                        (Select dispersal mechanism)
     DISP       5       5       5           (Maximum dispersal distances)
     FEC        1       1       1           (Relative fecundities)
     FTOL       0       0       2           (Fire tolerance codes)
     FIRE   10    0   0.1   0     0.5       (Values of fire regime parameters)
     FTAX   grass   0    0     5    10      (Post-fire species responses)
     FTAX   euc     5    5    50    50
     FTAX   rf     50   50  1000  1000
     EGRAD  space 1 99                      (Define environmental gradient)
     ETAX   grass   0    0   100   100      (Responses to gradient)
     ETAX   euc     0    0    60    80
     ETAX   rf     20   40   100   100
     COVER   total   euc   0    0           (Initial plant cover)
PROB                                        (Begin defining run options)
     MAP freq fire end                      (Map printing options)
     TRACE all                              (Details to record during run)
     TIME 500 (years)                       (Time period to model)
RUN                                         (Run the model)
END                                         (End this job)
```

**Figure 2.** A typical model definition file for the spatial simulation program MOSAIC. This model simulates two dominant vegetation types ("EUC" and "RF") competing on an environmental gradient (results in Fig. 6). Modelling commands are given in uppercase bold letters; options offered by the program are shown in lowercase bold letters. Comments are given in brackets.

Fire behaviour

Cellular automaton models of fire spread through discrete fuel beds (i.e. where trees, bushes, or grass tussocks do not form a continuous fuel bed) confirm many basic observations (Gill et al., 1981), but have also revealed some surprising properties of fires burning in patchy fuels (Fig. 3). For example, sensitivity analyses (Frandsen & Andrews, 1979; Green 1983a; Green et al., 1983) show that fire shapes vary greatly in response to fuel type, fuel density, and wind speed. Hot fires are seen to be more regular and predictable in shape and size than mild fires. Certain fire patterns are a consequence of fuel patchiness alone. For instance, it had been assumed that the "tear-drop" shape (Fig. 3), common in heath fires, was caused by shifting wind direction. However, simulation (Green, 1983a) shows that it is a simple consequence of extreme fuel patchiness.

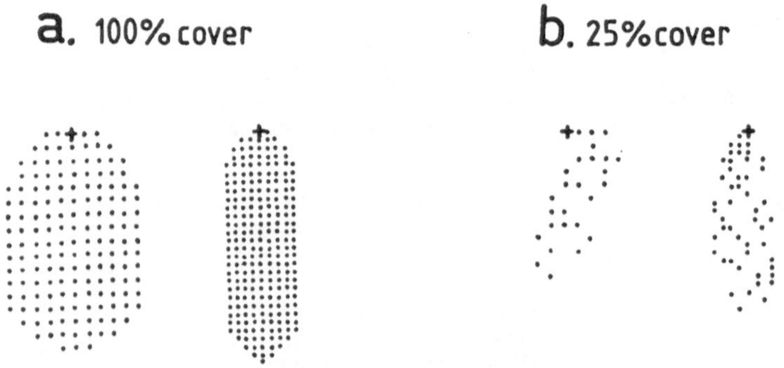


**Figure 3.** Patterns of fire spread in (a) uniform fuel bed, (b) patchy fuel bed. After Green (1983a).

Another finding is that different mechanisms of fire spread apply in different types of fires (Green et al., 1983). In mild fires heat accumulation seems to be the dominant mechanism, whereas hotter fires tend to spread by direct contact. In extremely hot fires, burning fuel elements interact by heating each other (i.e. H(**x**) varies with time), thus increasing the overall fire intensity. This same heating effect tends to make fires hotter as they grow, so the dominant mechanism of spread changes with time.

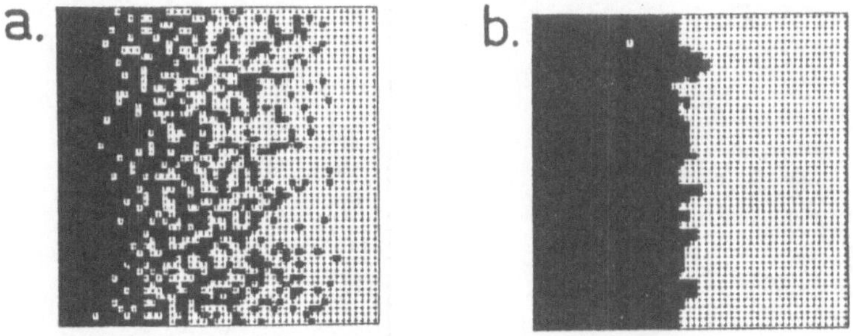


**Figure 6.** Vegetation zones on an environmental gradient (e.g. altitude). The gradient runs from left to right on both maps, with the two vegetation types favoured at opposite ends of the gradient (cf. Fig. 2). The fire regime is identical in both cases, but the fire "strategies" differs: (a) both species are fire-tolerant and can seed onto any vacant site; (b) one species is fire-sensitive and seeds must originate from a nearby source.

## Forest dynamics

Long-term forest behaviour is modelled by making each time step correspond to a whole year. Such models are ideal for testing hypotheses about forest dynamics and have yielded results with implications for forest management:

(a) They confirm that a small number of environmental processes, notably fire, dispersal, and environmental change, could account for much of the large-scale pattern seen in some forest communities. In particular, these processes combine to create clear zonation patterns in simulated communities (Fig. 6).

(b) Spatial simulations emphasize the inherent unpredictability of short-term population changes. Even if fires are made to occur at fixed intervals, they still ignite at random locations, so the effect of each fire on individual populations cannot be pre-determined. Fires can also make plant communities highly unstable, especially in a changing environment (Fig. 7).

## Applications

Application of spatial models to forest management requires both field validation and the input of mapped information about fuel abundance and moisture, topography, etc. (Fig. 8). The main obstacle is the sheer volume of field data required by the time and spatial scales.

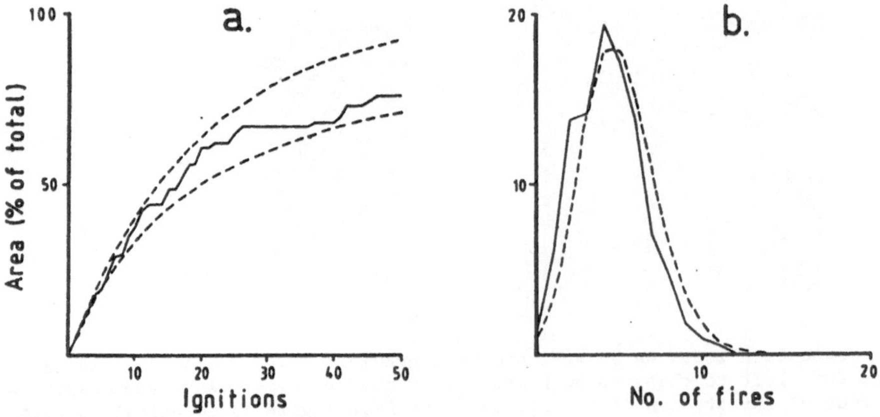


**Figure 4.** Frequency/area relationships for randomly ignited fires in a region of homogeneous fuel. (a) Total area burnt as a function of the number of ignitions. (b) Area as a function of the number of times that the sites involved have been burnt, assuming 1 fire per year for 100 years (with fuel replacment); the resulting vegetation pattern is shown in Fig. 5.

<u>Fire occurrence</u>

Simulation reveals the total areas that may be expected to be burnt, and the frequency of burning, under different fire regimes (Fig. 4). If fires ignite at random times and locations, then even in a uniform environment different sites experience an astonishing variety of fire regimes (Fig. 5). Starting from random vegetation patterns, this variety causes homogeneous communities to break up into patches dominated by fire-tolerant (i.e. short recovery time) and by fire-sensitive (i.e. long recovery time) vegetation. Environmental gradients in space (e.g. changes in soil moisture on a hillside) tend to fix such vegetation patches, giving rise to vegetation zones (Fig. 6).

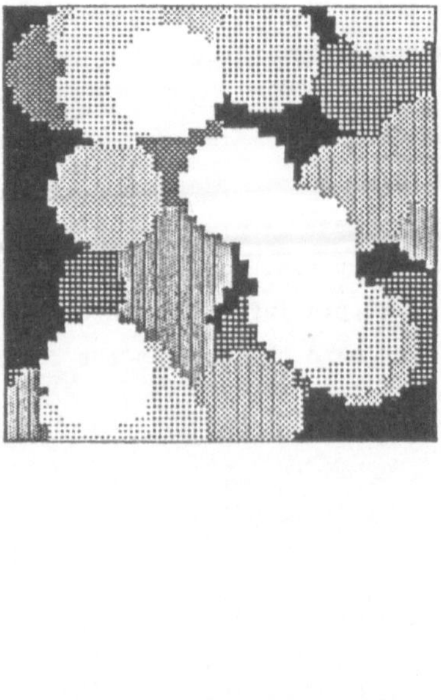

**Figure 5.** Vegetation map for the area modelled in Fig. 4b. The intensity of shading indicates the time since sites were last burnt, ranging from <10 yrs (white) to >90 yrs (black).

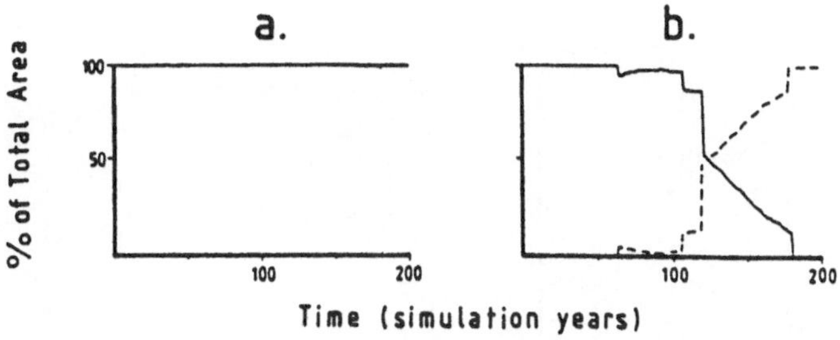


**Figure 7.** Community dynamics in a changing environment: (a) in the absence of fire, the prevailing vegetation resists invasion; (b) fires create large openings in which the invader can compete on equal terms with the established vegetation, leading to sudden changes in community composition.

## Validation

Spatial data on fire and forest patterns can be obtained in abundance by remote sensing (Minnich, 1983). To exploit satellite data, cells in a forest model must correspond with pixels in the satellite data. Validation therefore requires extensive ground-truthing and image rectification. Historical data on fires and forest change in the recent past can now be obtained by fine-resolution palynology (Green et al., 1987; Green & Dolman, 1988). Under ideal circumstances, it is now possible to obtain forest histories, with single-year sampling resolution, from microscopic pollen, charcoal and other components preserved in lake, dam or swamp sediments.

## Fire and forest management

Because simulations can be set up and run with any set of assumptions, they are ideal for performing "what-if" exercises, both in environmental management and in education. As we have already seen, theoretical studies with MOSAIC have produced results with potential management implications. Directly practical studies require only that MOSAIC's purely artificial environments be replaced with maps of real plant communities and landscapes within a region of interest.

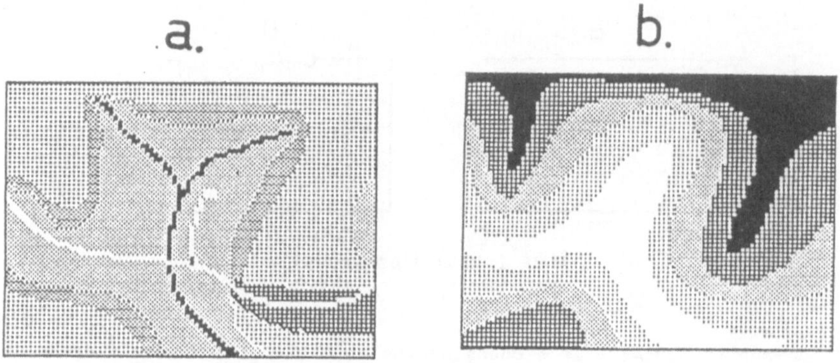


**Figure 8.** Vegetation and topography maps used by the model TINDER. The vegetation map also shows a road and a stream.

Cellular automaton models of vegetation are already being used in fire and forest management (Kessell & Good 1985). Uses of such models include logistic planning, assessment of fire risk, and testing the likely effects of various management strategies. For instance, a simple modification of MOSAIC for use in planning and education is the program TINDER. The program reads maps of vegetation, topography, and soil moisture (Fig. 8) and offers several possible fire control policies to choose from. TINDER then runs for several years, simulating plant growth, day-by-day changes in weather, etc., and models fires as they occur.

## Conclusion

Cellular automaton models are ideal for testing hypotheses about forest dynamics and yield results with implications for forest management. As we have seen, discrete models of fire spread reveal surprising behaviour for fires in patchy fuels and successional models confirm that a small number of environmental factors and processes can account for much of the large-scale pattern seen in forest communities. Personal computers nowadays are powerful enough to run sophisticated models, making it feasible for managers to develop models of their local region and for models to be used for educational purposes.

**References**

FRANDSEN, W.H. and ANDREWS, P.L. (1979). Fire behaviour in nonuniform fuels. USDA Forestry Service Research Paper INT-232, Washington, DC.

GILL, A.M., NOBLE, I.R. and GROVES, R. (Eds.) (1981). Fire in the Australian Biota. Australian Academy of Science, Canberra.

GREEN, D.G. (1983a). Shapes of simulated fires in discrete fuels. Ecol. Model. 20, 21-32.

GREEN, D.G. (1983b). The ecological interpretation of fine resolution pollen records. New Phytol. **94**, 459-477.

GREEN, D.G., BRADBURY, R.H. & BAINBRIDGE, S.J. (1987). Embodiment of formal languages. Mathematics and Computers in Simulation, to appear.

GREEN, D.G. & DOLMAN, G.S. (1988). Fine resolution pollen analysis. Journal of Biogeography, to appear.

GREEN,D.G., GILL, A.M. & NOBLE, I.R. (1983). Fire shapes and the adequacy of fire-spread models. Ecological Modelling **20**, 33-45.

GREEN, D.G., HOUSE, A.P.N. & HOUSE, S.M. (1985). Simulating spatial patterns in forest ecosystems. Mathematics and Computers in Simulation **27**, 191-198.

KESSELL, S.R. and GOOD, R.B. (1985). Technological advances in bushfire management and planning. In Natural Disasters in Australia, Australian Academy of Technological Sciences, Parkville, Victoria.

MINNICH, R.A. (1983). Fire mosaics in Southern California and Northern Baja California. Science **219**, 1287-1294.

NOBLE, I.R. & SLATYER, R.O. (1980). The use of vital attributes to predict successional changes in plant communities subject to recurrent disturbance, Vegetatio **43**, 5-21.

WOLFRAM, S. (1984). Cellular automata as models of complexity. Nature **311**, 419-424.

# FORESTRY APPLICATIONS OF OPERATIONS RESEARCH

L.R. Foulds, Department of Management,
University of Waikato, New Zealand

## 1. INTRODUCTION

For the last thirty years there have been many applications of operations research (OR) in the forestry industries. In this paper we analyze the current state of the art in the applications of OR in this important area, and review major trends among these applications.

The rapid expansion in the use of OR methodology in forestry over the last three decades has been brought about by three factors: the increasing efficiency of the digital computer, the significant improvement in the technical tools of OR, and the increasing demand for forest products and the land upon which they are harvested.

There have been two recent surveys of this topic, with different objectives. The first was by Bare, et al. (1984) which concentrated on reviewing the current state of OR applications in forest land management and in the production of forest products. It traced the growing sophistication in the use of OR in forestry since the inception of OR. The paper also explored reasons for the barriers which hinder wide acceptance of management science in both the public and private sectors of forestry. It then went on to explore ways to reduce these barriers. A second survey by Harrison and de Kluyver (1984) did not attempt to duplicate the extensive review of work mentioned in the previously mentioned survey. Rather it attempted to discuss the future trends and major issues that are emerging in the application of OR in forestry. The purpose of the present paper is not to repeat the content of those two papers, but rather to bring to light work that has been done since they were published and also to highlight future directions. We concern ourselves with two areas: the management of forest stands, and the manufacture of forest products.

## 2. FOREST STAND MANAGEMENT AND FOREST PRODUCTS MANUFACTURE

Realistic models of forestry-related activities are, of necessity, extremely complex. Workers in the field have found that it is very difficult to manipulate these models to produce useful solutions. So, as in many other applications of OR, a number of simplifying assumptions must be made in order to render the models capable of producing worthwhile results. Common tactics are to aggregate various parts of the system, and to make approximations in constructing the model. This means the models that are successfully used today do not cover all of the forestry-related operations of a particular system. They usually fall into one of two categories: forest stand management (including transportation) or forest products manufacture.

The models concerned with forest stand management usually cover one or more of the following five topics: nursery operations, administration of a growing stand, growth and yield, global administration concerns, or protection of the forest. Once a stand is harvested attention turns to the conversion of the raw felled trees into finished forest products. Topics which find OR applications in this area include: the trimming of paper (one of the earliest applications of OR), cross cutting into logs, the manufacture of timber, the production of veneer, and finally other further production of a secondary nature.

Many OR techniques have been used successfully in both of the two general areas mentioned above. Simulation has been used extensively in many aspects of forest stand management. More specifically, stand growth has been modelled to predict average diameter stocking levels, stand age, and volume per unit area. It is also been used in most areas of forest protection.

Another successful area of OR to be applied in forestry is that of mathematical programming. Linear programming (LP) models and techniques have been especially useful in global management issues. Often a number of specific sub-models are used to model various parts of the overall operation, with the output of one model being used as input to the next. Also decomposition techniques are being to an increasing extent to separate large-scale models into smaller problems with tractable structure. The techniques of integer and nonlinear programming are also being used to an increasing extent due to the rapid expansion of computing power available to forest planners. Finally recursive optimization techniques have been found especially useful in forest stand management. Dynamic programming (DP) is the most commonly used of these techniques and is especially suitable for small problems where there is a set of time periods and hence some dynamic nature to the management problems. This completes our general review of the most promising areas of forestry applications of OR. We now review recent work on forest stand management.

## 3. FOREST STAND MANAGEMENT

We begin this section with a review of some general management models. As mentioned previously, one of the most important OR techniques in this endeavour is simulation.

There have been a number of papers which use simulation to model the phenomena of a growing forest with view to efficient management. Leech (1985) has introduced a simulation model which is an essential part of the management planning tool kit of the forestry department in South Australia. The forestry system that uses this model manages a large area of radiata pine plantation and provides logs to a number of plants, sawmills, a pulpmill, a plymill, and a preservation plant. The model is used to predict future log outturn with a seventy year horizon. It simulates growth and harvesting patterns according to a variety of different silvicultural management strategies. The ultimate objective is to identify the optimal cutting level and to evaluate the effects of possible changes in silvicultural management.

There has been further work on developing simulation models among other models, such as linear programming and dynamic programming, by Whyte (1987). He reports on a resource planning and management system model for the pine plantations in Fiji. Part of this is a simulation model which predicts growth and forecasts assorted yields for administratively convenient and biologically homogeneous crop aggregations. Moving to another part of the world, Kobayashi (1984) reports on a different simulation model which was developed to explore the future resources of a training forest in Japan. It was assumed in the model that the yield had to be sustained over a long term and the results of the model were used to calculate allowable yield volume. A very different use of simulation has been reported by Green et al. (1985) and Green (1987). This use of the technique has yielded a series of simple spatial simulation models which attempt to recognise that the spatial distribution of trees in the forest is important. Attempts have been made to simulate forest spread, bush fires, breeding systems in a tropical rain forest, and aspects of forest behaviour that are not apparent in other models.

We turn now to reports on various other optimization problems in forest management. The first two, both by Hellman (1986) are concerned with growth models. They comprise a system of differential equations which assume that growth is slowed by contact between the crowns of trees, and also by shading. The resulting optimal control problems have been analysed in practice to produce useful results. The purpose is to output data which can be used as the input for an economic model for timber production. The next two reports are concerned first with even-aged forest stands and then secondly with uneven-aged forest stands. Sweda (1987) has used the theory of Markov processes to create a model describing the growth of a single species stand to project future diameter distribution. Hotvedt (1987) has modelled uneven-aged stands under dynamic return coefficients. Dynamic programming is used in the model to determine optimal conversion strategies and steady states.

We turn now to linear models: both continuous and integer. Giese and Jones (1984) have recently presented an economic model of short rotation forestry which uses Lemke's linear complementarity algorithm to compute optimal sustainable harvesting strategies. The method is applied in an illustrative numerical example using existing data to find optimal harvesting strategies for a two-species problem incorporating a land depletion constraint. Rumpf et al. (1985) have presented a mathematical programming model which determines the lowest cost set of flight plans for a spraying program to limit the destruction of forests by pests. Their method is illustrated in an application to the Maine Forest Service area of operation in the northeast of the United States. The spruce budworm is a destructive defoliater of spruce fir in this area and this model has been successful in determining optimal strategies in terms of cost. It also provides 'what if' analyses for operating conditions to improve management decisions. Weintraub et al. (1986), (1987) have examined different approaches to LP aggregation for large scale forest planning models. These are to be part of a more global forest management model. They discuss implementation characteristics, CPU time savings, and also the quality of the solutions obtained. A further paper on the same theme has been presented by two of the authors in the papers just mentioned. Navon and Veiga (1984) have looked at the effects of aggregation in LP models. They have become concerned with the loss of optimality and have investigated the effects of reducing the level of detail in the LP models of forest planning. They discuss the trade-off between reducing computational cost and losing specificity. One of the most important large scale LP forest planning models to date is called FORPLAN. It has been used extensively by US national forest planners. Kent and Bare (1987) have reported on a detailed evaluation of its utility. This is especially useful as Kent was one of the original developers of this model and Bare has previously developed an OR evaluation of FORPLAN. Another LP based planning system is termed FOLPI. It has been developed by Garcia (1984) and has been used widely by the New Zealand Forest Service. It has a great advantage in that it is not necessary for the manager using it to possess a detailed knowledge of LP. It thus allows the user to state a particular problem in forestry terms. This original statement is then automatically translated by the system into an LP format. The output to the LP problem is then interpreted in terms of the original statement and reports can be generated interactively. It is also useful for engaging in sensitivity analysis. Choi and Nauumo (1984) attempt to integrate the long term and medium term planning of forest management. The authors have developed a LP model for long term planning and a zero-one linear model for medium term planning. The result is an integration which has been developed into a computerized planning system which includes labour constraints and cutting plans. We end this report on the use of mathematical programming techniques by mentioning a paper by Rose (1984), which critically reviews mathematical programming in forest resource planning. This paper contains valid criticism of the technological limitations of the computing hardware which are currently available to solve the LP models. As has been mentioned earlier in this paper, this has forced planners to aggregate data in order to obtain resolutions. The value of these derived solutions is questioned. The paper goes on to discuss a new procedure which solves extremely large, disaggregated models using simulation at the University of Minnesota. It is suggested that this approach is far more useful to forest planners than that based on LP.

We turn now to discuss briefly the use of decision support systems (DSS) in forest operations. Hendricks et al. (1986) have developed a microcomputer-based DSS for use in multi-criteria forest management planning. An extremely useful feature of their system is the ability to view solutions graphically. A decision map is available to allow the user to trace through the decision making process. Another such system has been developed by Toda and Kamio (1984). Its aim is to assist in the decision making process for future forest planning, improvement, and management. It has been used successfully in Gifu prefecture in Japan. The increasing trend to use the DSS approach in forest stand management has been reviewed and discussed by Robak (1987).

We end this section with two studies on planning methods for forest road networks. Sakia (1984) has investigated a method for optimal forest road planning in a mountainous region where the terrain presents serious planning difficulties. The road is planned using DP with evaluated equations for road length and average skidding distance incorporated. Finally Kitagawa (1984) has presented a different method for the same problem. It is based on a newly devised index for evaluating the necessity for the road from the view point of long term forest management. Digital cardigraphic data has been used to apply the system to the forest road network for an actual small township in an mountainous area. The system appeared to serve the long term planning need of the forest road network and it produced a variety of alternatives quickly and cheaply for the planners.

It is apparent that OR has made a significant contribution in forest stand management and will continue to do so. It is heartening to see that there is an increasing tendency to merge the sub-models in an effective way in order to produce useful solutions to global problems of stand management. The increasing tendency towards the integrated models has come about mainly through the availability of increasingly more powerful computing equipment and more efficient management information systems.

## 4. FOREST PRODUCTS MANUFACTURE

Turning to OR applications in forest products manufacture, we begin by noting work on log bucking. (The operations of cutting a felled tree into shorter logs is termed bucking the tree). Faaland and Briggs (1984) have shown how to use DP in computation for log bucking and lumber manufacturing strategies. This paper considers the activities of bucking and sawing the bucked logs into lumber in a single production system. The DP model allows for variations in tree shape and quality which can be recorded by modern electronic scanners. A further application to bucking has been devised by Eng and Daellenbach (1985) who use Dantzig-Wolfe decomposition and DP column generation. They show how Dantzig-Wolfe decomposition with delayed column generation, by means of dynamic programming, can be used to determine the optimal set of bucking policies for a stand. If the bucking policies are determined by Dantzig-Wolfe decomposition they are consistent with constraints on end-use quality and numerical quantities required, along with supply and demand constraints.

The following two papers were presented at the proceedings of the Symposium on Forest Management Planning and Managerial Economics at the University of Tokyo in 1984. We begin with that by von Gadow (1984) which was concerned about the application of LP in forestry operations scheduling. The model attempts to maximise the utility in allocating forestry operations to different months and different work teams. The other paper by, Hosokawa (1984), is a report on a control system for forest production management in Brazil.

We end this section with three papers by Wyk and Eng (1986), Wyk and Ward (1986), and Ward (1986). They were presented at a general meeting of the Conversion Planning Project team, held in April (1986), at the Forest Research Institute in Rotorua, New Zealand. These papers describe a suite of sawmill analysis programmes which were specifically developed to aid planners who have to deal with changes in technology, consumer demand, and log supply. Specifically, a sawmill simulation model and also a similar model for a plymill have been developed.

This ends our report on recent work in the area of forest products manufacture. There is a wide variety of OR optimization techniques that are being used in this area, including goal programming and other multi-objective optimization methods. The utility of OR in the manufacturing side of forestry has been aided considerably by the availability of portable, personal computers. This coupled with the trend towards user- friendly programs that require no advanced training, has meant that OR has a bright future in the forestry industry.

## 5. SUMMARY AND CONCLUSIONS

It should be clear that the use of OR has an increasingly important part to play in the wise administration of forestry resources. In the last few years there have not been major technical breakthroughs in the use of OR in either forest stand management nor in forest products manufacture. There has been rather, an aggregation of submodels and also an increasing tendency to employ portable personal computers with interactive programs. These two tendencies are somewhat at variance in the sense that the integrated and more realistic models need very large computing power in order to be implemented. However, at the same time, the use of personal computers has led to some modularisation of the modelling approach. Thus we see two trends in parallel that are somewhat opposite. Large scale, realistic, integrated models implemented on large computers; and also submodels implemented in personal computers. There is not necessarily a conflict in this phenomenon as both tendencies are of great utility in forest planning management. However it is clear that the output of the latter models should be used as guides in formulating the general, aggregated models. It is also vital that OR analysts should remain in close contact with the actual managers of forest stands in order to ensure that OR analysts work on problems of genuine importance in the industry.

Another tendency is the increasing number of papers appearing concerned with the economic considerations of forest management. One such example has been provided recently by Zimmermann et al. (1985) in which an attempt was made to forecast the Swiss timber market.

It is also clear that not just the use of computers to implement cleverly designed OR models is going to have a major impact in the future on forest management. The continuing evolution of management information systems, in the form of decision support systems and expert systems, is also going to be of major import. An investigation into useful data bases, imaginative types of models to be used in conjunction with these bases, and the interplay between models of various time horizons must be investigated.

Further problems worthy of research include the increasing tendency towards deforestation and the increasing demand of the public in many countries to use the public forests for recreational purposes. Only if issues such as those outlined in this section are addressed, will OR and related disciplines be able to continue to make a significant contribution to the problems that forest managers will face in the next century.

## REFERENCES


1. Bare, B.B., Briggs, D.G., Roise, J.P. and G.F. Schreuder, "A Survey of Systems Analysis Models in Forestry and the Forest Products Industries", European Journal of Operational Research, 18 (1984) 1-18.

2. Choi, J.C., and H. Nauumo, "A Computerized Planning System for a Private Forest", Proceedings IUFRO Symposium on Forest Management Planning and Managerial Economics, Oct. 15-19 (1984) University of Tokyo, 332-342.

3. Eng, G., and H.G. Daellenbach, "Forest Outturn Optimization by Dantzig-Wolfe Decomposition and Dynamic Programming Column-Generation", Operations Research, 33 (1985) 459-465.

4. Faaland, B., and D. Briggs, "Log Bucking and Lumber Manufacture Using Dynamic Programming", Management Science, 30 (1984) 245-257.

5. Garcia, O., "FOLPI, A Forestry-Orientated Linear Programming Interpreter", Proceedings IUFRO Symposium on Forest Management Planning and Managerial Economics, University of Tokyo, Oct. 15-19 (1984) 293-305.

6. Giese, R.F., and P.C. Jones, "An Economic Model of Short Rotation Forestry", Mathematical Programming, 28 (1984) 206-217.

7. Green, D.G., "Modelling Forest Mosaics", 13th IFIP Conference on System Modelling and Optimization, Tokyo, Japan, Aug. 31-Sept. 4 (1987).

8. Green, D.G., House, A.P.N., and S.M. House, "Simulating Spatial Patterns in Forest Ecosystems", Mathematics and Computers in Simulation, 27 (1985) 191-198.

9. Harrison, T.P., and C. de Kluyver, "MS/OR and the Forest Products Industry: New Directions" Interfaces, 14 (1984) 1-7.

10. Hellman, O., "An Optimization Problem in Forest Management", European Journal of Operational Research, 23 (1986) 37-47.

11. Hellman, O., "On the Optimal Control of a Forest which Consists of Several Species of Trees", TIMS Studies in the Management Sciences, 21 (1986) 429-437.

12. Hendricks, G.L., and T.P Harrison, "Discussion of Graphic Aids for Multi-Criteria Forest Management", ORSA/TIMS Joint National Meeting, Miami Beach, USA. Oct. 27-29 (1986).

13. Hosokawa, R.T., "Control System for the Forest Production Management in Brazil", Proceedings IUFRO Symposium on Forest Managemet Planning and Managerial Economics, University of Tokyo, Oct. 15-19 (1984) 467-482.

14. Hotvedt, J.E., "Optimal Steady States for Uneven-aged Forest Stands under Dynamic Return Function Coefficients", Presented at TIMS/ORSA, Joint National Meeting, New Orleans, USA, May 4-6 (1987).

15. Kent, B., and B.B. Bare, "Evaluation of a Large Scale LP Forest Plannig Model: Forplan", Presented at TIMS/ORSA, Joint National Meeting, New Orleans, USA, May 4-6 (1987).

16. Kitagawa, K., "A New Logical System for Forest Road Network Planning", Proceedings IUFRO Symposium on Forest Management Planning and Managerial Economics, University of Tokyo, Oct. 15-19 (1984) 353-362.

17. Kobayashi, S., "Computer Simulation for Yield Prediction in the Training Forest of Niigata University", Proceedings IUFRO Symposium on Forest Management Planning and Managerial Economics, University of Tokyo, Oct. 15-19 (1984) 444-450.

18. Leech, J.W., "A Management Planning Model of a Large Plantation Forest:, Mathematics and Computers in Simulation, 27 (1985) 199-206.

19. Navon, D., and G. Veiga, "Estimating Optimality Loss in Linear Programming Aggregation: Application to Forest Planning Models", Proceedings IUFRO Symposium on Forest Management Planning and Managerial Economics, University of Tokyo, Oct. 15-19 (1984) 431-443.

20. Robak, E.W., "Increasing O.R. Applications in Forest Operations: The DSS Approach" 13th IFIP Conference on System Modelling and Optimization, Tokyo, Japan, Aug. 31-Sept 4 (1987).

21. Rose, D.W., "A Critical Review of Mathematical Programming in Forest Resource Planning", Proceedings IUFRO Symposium on Forest Management Planning and Managerial Economics, University of Tokyo, Oct. 15-19 (1984) 306-321.

22. Rumpf, D.L., Melachrinoudis, E., and T. Rumpf, "Improving Efficiency in a Forest Pest Control Spray Program", Interfaces, 15 (1985) 1-11.

23. Sakai, T., "Studies on Planning Method of Forest Roads Network" Proceeding IUFRO Symposium on Forest Management Planning and Managerial Economics, University of Tokyo, Oct. 15-19 363-370.

24. Sweda, T., "A Stochastic Growth Model for Even-aged Single-Species Forest Stands", 13th IFIP Conference on System Modelling and Optimization, Tokyo, Japan, Aug. 31-Sept 4 (1987).

25. Toda, K., and K. Kamio., "Forest Planning and Computerized Forest Resource Information System In Gifu Prefecture", Proceedings IUFRO Symposium on Forest Management Planning and Managerial Economics, University of Tokyo, Oct. 15-19 (1984) 399-407.

26. Von Gadow, K., "Applications of Linear Programming in Foresty Operations Scheduling", Proceedings IUFRO Symposium on Forest Management Planning and Managerial Economics, University of Tokyo, Oct. 15-19 (1984) 270-284.

27. Van Wyk, J.L., and G. Eng. "Computer Aided Sawmill Design and Evaluation", Conversion Planning Project Team General Meeting, April 8-11 (1986).

28. Van Wyk, J.L., and N. Ward, "General Introduction to Process Simulation Models", Conversion Planning Project Team General Meeting, April 8-11 (1986).

29. Ward, N.H., "Plywood Mill Simulation Model", Conversion Planning Project Team General Meeting, April 8-11 (1986).

30. Weintraub, A., Guitart, S., and V. Kohn, "Strategic Planning in Forest Industries", Operations Research, 24 (1986) 152-162.

31. Weintraub, A., Navon, D., Hrubes, R., and G. Viega, "Aggregation in Large Scale Forest Planning Models", Presented at TIMS/ORSA Joint National Meeting, New Orleans, USA, May 4-6 (1987).

32. Whyte, A.G.D., "A Resource Planning and Management System for Pine Plantations in Fiji", 13th IFIP Conference on System Modelling and Optimization, Tokyo, Japan, Aug.31-Sept. 4 (1987).

33. Whyte, A.G.D., "Management Modelling of the Forest Sector in New Zealand" Proceedings IUFRO Symposium on Forest Management Planning and Managerial Economics, University of Tokyo, Oct. 15-19 (1984) 25-34.

34. Zimmermann, A.J., Zweifel, P., and E. Kofler, "Application of the Linear Partial Information Model to Forecasting the Swiss Timber Market", Journal of Forecasting, 4 (1985) 387-398.

# AGGREGATION OF ECOLOGICAL NETWORKS BASED ON INFORMATION THEORY

Hironori Hirata and Yasuo Sugai

Department of Electronics, Chiba University

1-33 Yayoi-cho, Chiba-shi 260 JAPAN

**Abstract**

Regarding an ecological network as a communication channel, the information contained in the structure of an ecological network may be theoretically defined by the concept of mutual information. Optimal aggregation minimizes the difference of mutual information between object system and aggregated model. The information cannot increase (it generally losses) during the process of network aggregation.

First, this paper studies what kind of patterns of structure of ecological networks result in no aggregation loss. Parallel structures will be shown as special patterns of structure which result in no loss information. Parallel structures are generally tend to minimize aggregation loss.

Second, it is shown how grouping compartments so as to minimize the loss of mutual information creats collections of species, i.e., how aggregation identifies the macro-structure (e.g. hierarchical structure) of ecological networks. Approximate aggregation is applied to several real ecological networks.

## 1. Introduction

Aggregation is one of increasingly interesting issues in ecological field (Hirata, 1987, etc.). Aggregation theory plays an important role in constructing reasonable macroscopic models. There are two types of aggregation: perfect aggregation and approximate aggregation. Perfect aggregation means that there is no loss between object system and aggregated model. Generally there occurs some loss during the process of aggregation. This case is referred as the approximate aggregation.

Although the aggregation is practically very useful, it generally has the difficulty of long computation time due to combinatorial problem. Analytical study of aggregation before using computer may be helpful to complete aggregation of large-scale systems within reasonable time. First, this paper studies what kind of patterns of structure of ecological networks result in no aggregation loss. Parallel structures will be shown as special patterns of such structure. Parallel structures are generally tend to minimize aggregation loss. Second, it is shown how aggregation identifies the macro-structure (e.g. hierarchical structure) of ecological networks, i.e., how grouping compartments so as to minimize the loss of mutual information creats collections of species. Approximate aggregation is applied to several real ecological networks.

Regarding an ecological network as a communication channel, the information contained in the structure of an ecological network may be theoretically defined by the concept of mutual information. Optimal aggregation minimizes the difference of mutual information between object system and aggregated model. The information cannot increase (it generally losses) during the process of network aggregation.

## 2. Preliminary

### 2.1. Ecological Networks

Consider an ecological network consisting of n compartments each of which is characterized by its throughflow of some medium, such as carbon, nitrogen, mass, energy etc. The detailed structure of the k-th compartment is characterized as shown in Figure 1 for which the symbols are defined as follows:

$T_{kj}$: the flow leaving the k-th compartment and directly contributing to the j-th compartment; $T_{kj} \geq 0$.

$D_k$ : the dissipated flow leaving the k-th compartment; $D_k \geq 0$.

$E_k$ : the useful export leaving the k-th compartment; $E_k \geq 0$.

$I_k$ : the input flow to the k-th compartment from the outside world: $I_k \geq 0$.

The throughflow of the k-th compartment, $T_k$, is defined as

$$T_k = \sum_{j=1}^{n} T_{kj} + D_k + E_k \qquad (1)$$

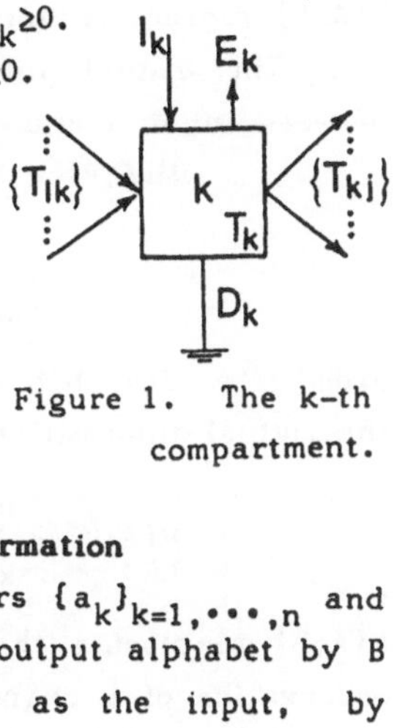


Figure 1. The k-th compartment.

### 2.2. Communication Channel and Mutual Information

Denote the input alphabet by A containing the letters $\{a_k\}_{k=1,\cdots,n}$ and associated probabilities defined by $P(a_k)$. Denote the output alphabet by B and its letters, which need not the same in number as the input, by $\{b_j\}_{j=1,\cdots,r}$ with probabilities $P(b_j)$. A memoryless channel is now completely specified by giving $P(b_j/a_k)$, i.e., the probability of output $b_j$ when the input is $a_k$, for $j=1,\cdots,r$ and $k=1,\cdots,n$. For each input there will always be an output letter so that

$$\sum_{j=1}^{r} P(b_j/a_k) = 1 \qquad (2)$$

The probabilities $P(b_j/a_k)$ which indicate how the transition from input to output takes place will be named transition probabilities. $H=[P(b_j/a_k)]$ is the communication matrix of this channel.

The probability of the output letter being $b_j$ is the sum of the probabilities of the events favorable to this occurrence and so is given by

$$P(b_j) = \sum_{k=1}^{n} P(b_j/a_k)P(a_k) \tag{3}$$

The entropy of A is defined by

$$H(A) = - \sum_{k=1}^{n} P(a_k)\log P(a_k) \tag{4}$$

H(A) represents uncertainty about A before we know B.

The entropy H(X) generally can be thought of as a measure of the folowing things about X: (a) The information provided by an observation X. (b) The uncertainty about X. (c) The randomness of X.

Meanwhile the conditional entropy of A, given B is defined by

$$H(A/B) = - \sum_{j=1}^{r} P(b_j) \sum_{k=1}^{n} P(a_k/b_j)\log P(a_k/b_j) \tag{5}$$

H(A/B) represents uncertainty about A after we know B.

The mutual information of the channel, i.e., the mutual information between input A and output B, is defined by

$$M(A;B) = H(A) - H(A/B) = H(B) - H(B/A) \tag{6}$$

$$= \sum_{k=1}^{n} \sum_{j=1}^{r} P(b_j/a_k)P(a_k)\log[P(b_j/a_k)/P(b_j)] \tag{7}$$

From $$P(a_k,b_j) = P(b_j/a_k)P(a_k) = P(a_k/b_j)P(b_j) \tag{8}$$
the mutual information can be rewritten as

$$M(A;B) = \sum_{k=1}^{n} \sum_{j=1}^{r} P(a_k,b_j)\log[P(a_k,b_j)/P(a_k)P(b_j)] \tag{9}$$

M(A;B) represents the amount of information about the input provided by the observation of a channel output.

### 2.3. Information of Ecological Networks

#### 2.3.1. Theoretical Form

In ecological channel, 0-th compartment is input; (n+1)-st compartment, export; (n+2)-nd compartment, dissipation; from 1st to n-th compartments; regular elements like species. Although one could consolidate input, export and dissipation into one global compartment, they are treated separately to facilitate extension of the theory.

Now time is introduced. The time interval for flow from one compartment to another is taken to be $\theta$.

Let $a_k$ be the event that a single given medium (such as carbon, nitrogen, mass, energy etc.) passes through the k-th compartment at time $t_1$, and $b_k$ be the event that the medium passes through the k-th compartment at time $t_1+\theta$.

$P(a_k)$ : the probability that the medium passes through the k-th compartment at time $t_1$.

$P(b_j)$ : the probability that the medium passes through the j-th compartment at time $t_1+\theta$.

$P(b_j/a_k)$ : the probability that the medium which passes through the k-th compartment will be taken up by the j-th compartment.

$P(a_k/b_j)$ : the probability that the medium passed from the k-th compartment to the j-th compartment, given that the medium has been taken up by the j-th compartment.

From equation (9) the mutual information of this ecological channel can be expressed as

$$M(N) = M(A;B) \tag{10}$$

$$= \sum_{k=0}^{n+2} \sum_{j=0}^{n+2} P(a_k,b_j)\log[P(a_k,b_j)/P(a_k)P(b_j)] \tag{11}$$

M(N) shows the information contained in the structure of the ecological network.

### 2.3.2. Applicable Form

We shall now apply the theoretical form (11) to a real ecological network. Although we may theoretically think of several ways to experimentally measure and define the probabilities $P(a_k)$, $P(b_k)$ and $P(b_j/a_k)$ from the meanings given in the previous section , if measurable data is limited one shown in Figure 1, the percentage of flow distribution should be used as a surrogate for the true probabilities.

$Q_k$ : the percentage of the total flow which passes through the k-th compartment at time $t_1$; $Q_k \geq 0$ $(k=0,\cdots,n)$, $Q_{n+1}=Q_{n+2}=0$.

$P_j$ : the percentage of the total flow which passes through the j-th compartment at time $t_1+\theta$; $P_0=0, P_j \geq 0$ $(j=1,\cdots,n+2)$.

$f_{kj}$ : the percentage of total flow through the k-th compartment at time $t_1$ that flows into the j-th compartment at time $t_1+\theta$; $f_{kj} \geq 0$.

$e_k$ : the percentage of the flow through the k-th compartment which is expected as useful flow; $e_k \geq 0$.

$r_k$ : the percentage of the flow through the k-th compartment which is dissipated; $r_k \geq 0$.

The relations between these variables are provided by the equations

$$P_j = \sum_{k=0}^{n} f_{kj}Q_k \quad (j=1,\cdots,n), \quad P_{n+1} = \sum_{k=1}^{n} e_k Q_k, \quad P_{n+2} = \sum_{k=1}^{n} r_k Q_k \tag{12}$$

where

$$\sum_{j=1}^{n} f_{kj} + r_k + e_k = 1 \quad (k=1,\cdots,n), \qquad \sum_{j=1}^{n} f_{0j} = 1 \tag{13}$$

In ecological network N, one uses $Q_k$, $P_j$ and $f_{kj}$ as $P(a_k)$, $P(b_k)$, and $P(b_j/a_k)$ as

$$P(a_k) = Q_k, \qquad P(b_j) = P_j, \qquad P(b_j/a_k) = f_{kj} \tag{14}$$

and one may identify variables $Q_k$, $f_{kj}$, $r_k$ and $e_k$ as follows

$$Q_k = T_k/T \quad (k=1,\cdots,n), \qquad Q_0 = I/T \tag{15}$$

$$f_{kj} = T_{kj}/T_k \ (k,j=1,\cdots,n), \ f_{0j} = I_j/I \ (j=1,\cdots,n), \ f_{k0} = 0 \ (k=0,\cdots,n) \tag{16}$$

$$r_k = D_k/T_k \qquad e_k = E_k/T_k \tag{17}$$

where

$$T = T^* + I \tag{18}$$

and

$$T^* = \sum_{k=1}^{n} T_k, \qquad I = \sum_{k=1}^{n} I_k \tag{19}$$

From equations (11) and (14)-(19), the mutual information of the ecological channel, i.e., the information, M(N), contained in the structure of ecological network, can be expressed in terms of flows as follows:

$$M(N) = (1/T) \sum_{k=0}^{n} \sum_{j=1}^{n+2} T_{kj} \log (TT_{kj}/T_k T_j) \tag{20}$$

where $T_{kn+1}=D_k$, $T_{kn+2}=E_k$, $T_{0k}=I_k$.

## 3. Basic Concepts of Aggregation

The following discussion outlines some basic principles.

**Definition 1**

Let N and $\overline{N}$ be the sets of elements

$$N = \{\alpha_k\}_{k=1,\cdots,n} \tag{21}$$

$$\overline{N} = \{\beta_i\}_{i=1,\cdots,m} \qquad (m \leq n) \tag{22}$$

such that a homomorphic mapping $\phi$ is made from N to $\overline{N}$ as

$$\phi : N \rightarrow \overline{N} \tag{23}$$

then N is called the original network; $\overline{N}$, the aggregated network of N; and $\phi$, the aggregation mapping.

$\phi$ may be conveniently represented by a matrix as follows:

**Definition 2**

Define an mxn ($m \leq n$) aggregation matrix S as

$$S = [s_{ik}]_{i=1,\cdots,m,k=1,\cdots,n} = \begin{bmatrix} s_{11} & s_{12} & \cdots & s_{1n} \\ s21 & & & \cdot \\ \cdot & & & \cdot \\ \cdot & & & \cdot \\ s_{m1} & & \cdots & s_{mn} \end{bmatrix} \qquad (24)$$

where

$$0 \leq s_{ik} \leq 1 \quad \text{and} \quad \sum_{i=1}^{m} s_{ik} = 1 \qquad (25)$$

If all $s_{ik}$ are either 0 or 1, the mapping is referred to below as a "discrete aggregation" (Hirata, 1978). Otherwise, it is refered as a "weighted aggregation". Those positions of $s_{ik}$ which are not zero signify which elements should be aggregated into the same group, i. Because the aggregated network ($\overline{N}$) depends on the aggregation matrix (S), it may be expressed as a function of S, $\overline{N}(S)$.

During the process of network aggregation (both weighted and discrete) the following relation is maintained (Hirata and Ulanowicz, 1985)

$$M(N) \geq M(\overline{N}) \qquad (26)$$

Equation (32) shows that information cannot increase (it is generally lost) during the process of aggregation, i.e., the difference $M(N)-M(\overline{N})$ is never negative. This loss of information may be regarded as the cost, J, of the aggregation:

$$J = M(N) - M[\overline{N}(S)] \qquad (27)$$

and J is minimized by choosing the optimal S* such that

$$J^* = \min_{\{S\}} J = M(N) - M[\overline{N}(S^*)] \qquad (28)$$

## 4. Patterns for Perfect Aggregation

In this section we shall only consider discrete perfect aggregation (the aggregation without any loss, i.e., J=0). For simplicity, we shall first discuss the case of the aggregation of two elements.

**Lemma 1**

The quantity $J_{KL}$ (the cost of aggregating two elements K and L ) equals 0, if and only if the following three conditions hold for $i=0,1,\cdots,n$ ($i \neq K,L$) and $j=1,\cdots,n+2$ ($j \neq K,L$).

$$T_{iK}/T_K = T_{iL}/T_L \tag{29}$$
$$T_{Kj}/T_K = T_{Lj}/T_L \tag{30}$$
$$T_{KK}/T_K^2 = T_{KL}/(T_K T_L) = T_{LK}/(T_L T_K) = T_{LL}/T_L^2 \tag{31}$$

From Lemma 1 several special patterns for perfect aggregation are derived as follows:

**Proposition 1**

There are the following two cases of perfect aggregation. In both cases (a) and (b), the situation of elements K and L to be aggregated may be called a parallel position.

(Case a) There is no flow between elements K and L (i.e., $T_{KL}=T_{LK}=0$) and no self-loops of elements K and L (i.e., $T_{KK}=T_{LL}=0$) as in Figure 2(a), and conditions (29) and (30) in Lemma 1 should be satisfied for i=0,1,•••,n (i≠K,L) and j=1,•••,n+2 (j≠K,L).

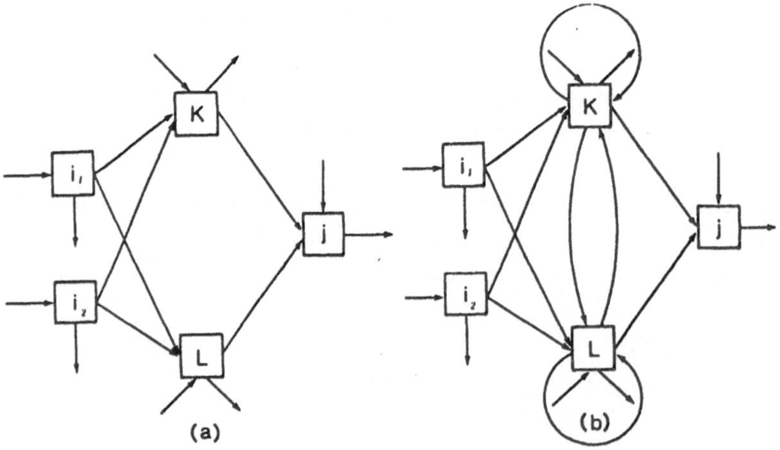


Figure 2. Parallel position: Two types of perfect aggregation in Proposition 1.

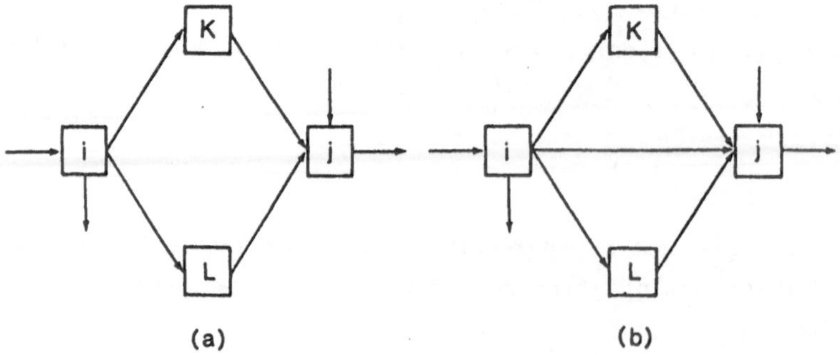


Figure 3. Perfect parallel position.

(Case b) There are complete flow connections between K and L (i.e., $T_{KL}\neq 0$ $T_{LK}\neq 0$) and self-loops of elements K and L (i.e., $T_{KK}\neq 0, T_{LL}\neq 0$) as in Figure 2(b), and conditions (29)-(31) in Lemma 1 should be satisfied for $i=0,1,\cdots,n$ $(i\neq K,L)$ and $j=1,\cdots,n+2$ $(j\neq K,L)$.

The following is a special, however, importance case of Proposition 1.

**Proposition 2**

There is no aggregation loss, i.e., $J_{KL}=0$, when elements K and L have only one inflow from a common element and one outflow to a common element, i.e., they hold a perfect parallel position as in Figure 3(a) and (b).

Proposition 2 can be derived from Proposition 1. That elements K and L have only one inflow and one outflow respectively means that $T_{iK}=T_{Kj}=T_K$ and $T_{iL}=T_{Lj}=T_L$. Therefore conditions (29) and (31) are satisfied. Since there is no flow between elements K and L and no self-loop of elements K and L, thecondition (31) is also satisfied. Although Propositions 1 and 2 can be extended to the case of N aggregating elements, they are omitted due to the limit of pages. These results show that parallel structures is important for perfect aggregation. Although Propositions 1 and 2 are only for perfect aggregation, it is easy to demonstrate that parallel structures generally tend to minimize aggregation loss.

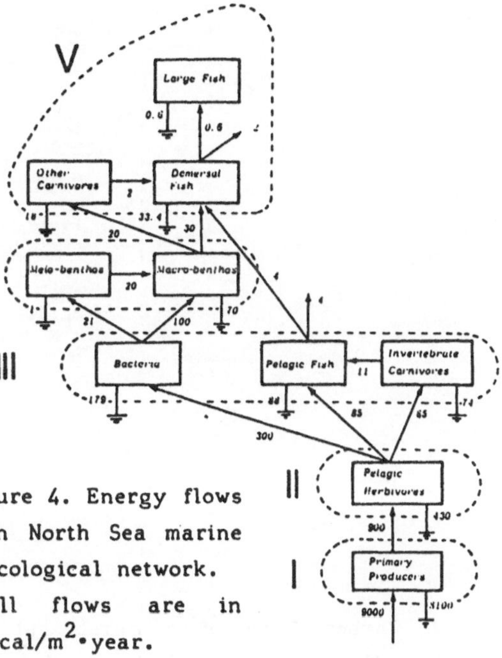


Figure 4. Energy flows in North Sea marine ecological network. All flows are in kcal/$m^2$•year.

## 5. Examples of Aggregation

As shown in Figure 4 a food web based on the main groups of organisms in the North Sea has been quantitatively estimated by Steel (1974). It consists of ten compartments. The resulting ecological network of 10 elements was aggregated into 5 groups as indicated by the dotted perimeters in Figure 4

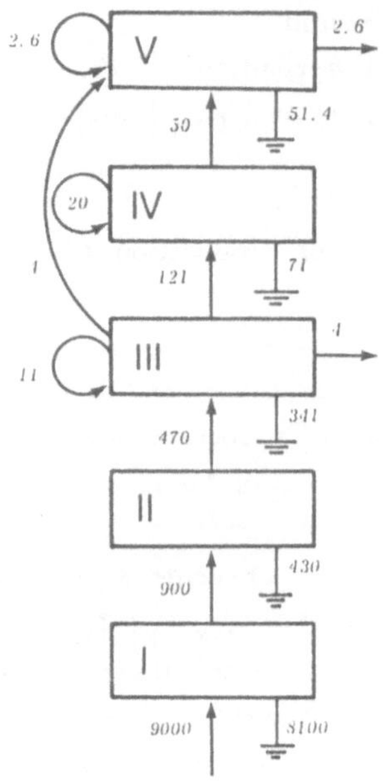


Figure 5. Macro-network after aggregation into five groups.

producing the final network macroscopic structure illustrated in Figure 5.

Flows among 17 compartments of a tidal marsh stream ecosystem (in miligrams carbon/m$^2$-d) were measured by Homer and Kemp (unpublished ms) and are shown schematically in Figure 6. The ecological network of 17 elements was aggregated to 7 groups indicated by the dotted perimeters in Figure 6. The organization imposed by the aggregation may be characerized as trophic hierarchy. If one distinguishes pathways of active feeding (heavy lines) from passive detrital flows (fine lines) it becomes possible to decompose the network into two acyclic subgraphs as shown in Figures 7(a) and 7(b). The trophic identities of the aggregated compartments in Figure 7(a) are apparent: the microphytes (I) and detritus (III) provide a food base for the pelagic herbivores (IV) and bentic herbivores (V); these in turn are fed upon by the carnivores (VI), and the top carnivores-omnivores (VII). It is seen from the

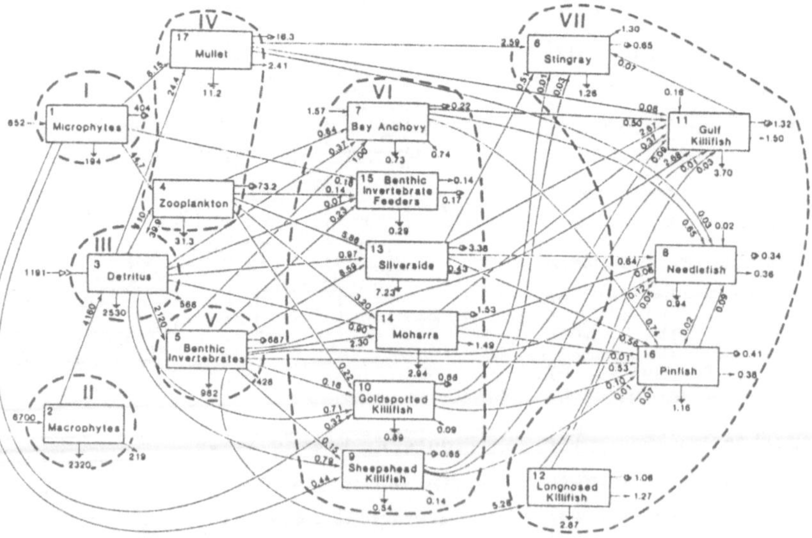


Figure 6. Carbon flows of a marsh gut ecological network, Crystal River, Florida. All flows are in mg carbon/m$^2$•day.

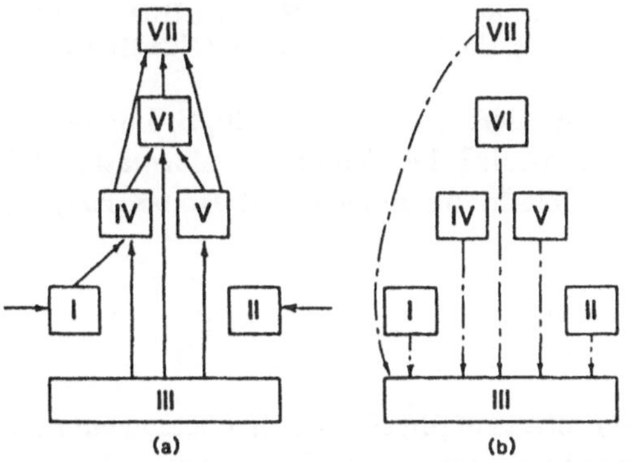


Figure 7. A decomposition of the macro-ecological network in Figure 6 into acyclic subgraphs. (a) the structure of the grazing chain, (b) the detrital returns.

discussion of section 4 why minimizing the decrease information of ecological network should lump compartments in a fashion similar to trophic levels.

## 6. Conclusion

Aggregation based on information theory is useful to identify macro-structure of large-scale networks like ecological systems. Theoretical discussion of aggregation derived special patterns of structure for perfect aggregation: parallel structures. The derived rules of thumb not only are theoretically important but also may be useful to solve the approximate aggregation problem of ecological systems.

## References

Hirata, H., 1978. Aggregation method for linear large-scale systems with random coefficients and inputs. Int. J. Systems Science, **9**: 515-529.

Hirata, H. and Ulanowicz, R.E., 1985. Information theoretical analysis of the aggregation and hierarchical structure of ecological networks. J. Theor. Biol., **116**: 321-341.

Hirata, H., 1987. Modelling and analysis of ecological systems: Large-scale system point of view. Int. J. Systems Science, (in press).

Steel, J.H., 1974. The structure of marine ecosystems, Harvard University Press, Cambridge, Mass.

# CONSTRAINED CROSS-VALIDATION APPLIED TO ESTIMATION OF KINETIC PARAMETERS OF CELL POPULATIONS IN PERTURBED GROWTH

A. Bertuzzi, A. Gandolfi, S. Lucidi
Istituto di Analisi dei Sistemi ed Informatica del CNR
Viale Manzoni 30, 00185 Roma, Italy

## 1. INTRODUCTION

In order to study the effects of physical or chemical agents over the proliferation of a cell population, both in culture and in experimental model systems, the flow cytometry measurements of the cell DNA content are widely used [1]. In this technique the cells of a sample drawn from the population are stained with a DNA-specific fluorochrome and forced to flow through a laser beam; during transit each cell emits a flash of fluorescence light, whose intensity is proportional to the amount of fluorochrome bound to the DNA. Thus a histogram is obtained which, for each discrete value of the intensity measured, gives the corresponding number of cells in the sample. The fluorescence histogram (FCM histogram) is a rough picture of the underlying DNA distribution, that is masked by staining and instrumental dispersive effects. Suitable algorithms allow to find from the histogram an estimate of the DNA distribution by a deconvolution procedure [2,3,4].

A time sequence of flow cytometric histograms from a proliferating population shows the time evolution of the DNA distribution and, in presence of perturbing agents (drugs, X-ray irradiation, hyperthermia), indicates how the different phases of the cell cycle are affected by the perturbation. A better understanding of the relationship between the perturbing action and the observed sequence can be obtained by using mathematical models of the cell population, in order to estimate the kinetic parameters of the cellular cycle from the available data [5,6,7].

In previous papers we developed a mathematical model of a cell population suitable to the analysis of sequential FCM histograms [6]. We developed also a procedure for estimating some cell kinetic parameters when a sequence of histograms and the growth curve of the cell population are available [8, 9]. This procedure utilized an unconstrained regularization method for filtering the growth curve data: the regularization parameter was chosen according to Mallows [10], and required the knowledge of the variance of the noise affecting the data of population size.

In the present paper we propose an improved version of our procedure for the estimation of the kinetic parameters. This version: (i) takes into account the positivity constraints over the estimated kinetic quantities, and (ii) determines the regularization parameter by the method of the constrained cross-validation [11], that does not require the variance of the noise on the data. The results of the application of this new procedure to the analysis of simulated data are presented.

## 2. MATHEMATICAL MODEL

The cell cycle is an ordered sequence of biochemical events leading to cell division (mitosis). According to these events, the cycle is divided into four phases, called G1, S, G2, and M. In G1, proteins and RNA are synthesized in preparation for DNA duplication, that occurs in the S phase. Then the cell progresses through G2, and finally enters the mitotic phase (M). Each of the two daughter cells enters the G1 phase of a new cycle. In a proliferating population at a given time, cells are distributed along all the cycle phases, and a distribution of the cells with respect to the DNA content can be defined.

The cell population model, in view of the analysis of sequences of DNA distributions, was structured in terms of DNA content [6]. It is assumed that there is no intrapopulation variability of the rate of DNA synthesis, and that there is no cell loss from the growing population. The model equations are continuity equations for the subpopulations of cells with DNA content equal (conventionally) to 1 (cells in G1 or in a quiescent state called G0), within 1 and 2 (cells in S), and equal to 2 (cells in G2+M). The model, that is linear but in general not stationary, has the form:

$$\frac{d}{dt}N_1 = -\lambda_1(t)N_1(t) + 2\lambda_2(t)N_2(t) \tag{1}$$

$$\frac{\partial}{\partial t}n(x,t) + \frac{\partial}{\partial x}[v(x,t)n(x,t)] = 0, \qquad v(1,t)n(1,t) = \lambda_1(t)N_1(t) \tag{2}$$

$$\frac{d}{dt}N_2 = v(2,t)n(2,t) - \lambda_2(t)N_2(t), \tag{3}$$

where $N_1(t)$ and $N_2(t)$ denote the number of cells with DNA content at $t$ equal to 1 and 2, respectively; $n(x,t)$ is the cell density in S with respect to the DNA content $x$, $1 < x < 2$; $v(x,t)$ is the DNA synthesis rate at $t$ of a cell with DNA content $x$; $\lambda_1(t)$ and $\lambda_2(t)$ denote, respectively, the exit rates from G1 and from G2+M at time $t$.

We are mainly interested in determining the cellular flux into the S phase $f(t) = \lambda_1(t)N_1(t)$, the rate of DNA synthesis $v(x,t)$, and the mitotic rate $g(t) = \lambda_2(t)N_2(t)$. The transition of cells from G1 to S is governed by complex biochemical events, in part unknown, and is a main regulation point in determining the length of cell cycle. The influx $f$ characterizes well the effects of such events on the proliferative activity of the population. The progression of cells through S, G2 and M phases occurs in a more "deterministic" fashion; the rate $v(x,t)$ describes the rate of DNA duplication at the various stages of the S phase, and its determination is allowed by the information contained in the FCM histogram. The following expressions for $f(t)$ and $v(x,t)$ can be obtained from the model equations:

$$f(t) = \frac{d}{dt}\left[N(t)\left(2-\theta_1(t)\right)\right] \tag{4}$$

$$v(x,t) = \frac{1}{N(t)\tilde{n}(x,t)}\frac{d}{dt}\left[N(t)\left(2-\theta_1(t)-\int_1^x \tilde{n}(z,t)dz\right)\right] \tag{5}$$

where $N(t) = N_1(t) + \int_1^2 n(x,t)dx + N_2(t)$ is the total population size, $\theta_1(t) = N_1(t)/N(t)$ is the G1 phase fraction, and $\tilde{n}(x,t) = n(x,t)/N(t)$ is the DNA distribution in S. The description of the cycle is completed by the mitotic rate $g$ that, because of the absence of cellular loss, is equal to $dN/dt$.

For the distributed parameter dynamic system (1)-(3), measurements are available at discrete times (usually 5-10 data points) for the total population size and for the DNA distribution ($\theta_1$ and $\tilde{n}(x)$, estimated from the corresponding FCM histogram). For each of the data points, a distinct cell culture is prepared with a given initial cell concentration in the culture medium, and it is grown in identical experimental conditions. At the measurement time, one aliquot of the cells is used for the flow cytometry measurement, and the other aliquot for the determination of the total population size. Such a complex experimental procedure involves a number of error sources: (i) differences in initial concentration and growth rate among cultures, (ii) error in cell counting, (iii) variability in the staining procedure, (iv) estimation error in the deconvolution process of the fluorescence histogram.

## 3. ESTIMATION PROCEDURE

First we note that, if the cell population is in balanced exponential growth with known growth rate $d(lnN)/dt = \alpha$, the rate of DNA synthesis is a time-independent $v(x)$, and can be estimated from a single FCM histogram [see equation (5)]. The balanced exponential growth can occur, in experimental model systems, over limited time intervals of unperturbed growth preceding the approach to the saturation.

In an arbitrary growth condition, if $v(x,t)$ can be approximated as a time-invariant $v_i(x)$ in the interval $[t_i, t_{i+1}]$ between two successive histograms of the FCM sequence (measurement times $t_i$, $i = 1, 2, \ldots, n$), then equation (5) through an integration with respect to time gives:

$$v_i(x) = \ln\left(\frac{N(t_{i+1})}{N(t_i)} \frac{2 - \theta_1(t_{i+1}) - \int_1^x \tilde{n}(z, t_{i+1})dz}{2 - \theta_1(t_i) - \int_1^x \tilde{n}(z, t_i)dz}\right) \Big/ \int_{t_i}^{t_{i+1}} \frac{\tilde{n}(x,t)dt}{2 - \theta_1(t) - \int_1^x \tilde{n}(z,t)dz} \tag{6}$$

Equation (6) can be further transformed as follows. Let $y_i = lnN(t_i)$; let also $\Delta_i = y_i - y_{i-1}$, $i = 2, \ldots, n$, and $\Delta_1 = y_1$. Then, by applying the trapezoidal rule to the integral with respect to time in (6), we obtain

$$v_i(x) = a_i(x)\Delta_{i+1} + b_i(x), \quad i = 1, \ldots, n-1, \tag{7}$$

where

$$a_i(x) = \frac{2}{t_{i+1} - t_i}\left[\frac{\tilde{n}(x,t_i)}{2 - \theta_1(t_i) - \int_1^x \tilde{n}(z,t_i)dz} + \frac{\tilde{n}(x,t_{i+1})}{2 - \theta_1(t_{i+1}) - \int_1^x \tilde{n}(z,t_{i+1})dz}\right]^{-1}$$

$$b_i(x) = a_i(x) ln \frac{2 - \theta_1(t_{i+1}) - \int_1^x \tilde{n}(z,t_{i+1})dz}{2 - \theta_1(t_i) - \int_1^x \tilde{n}(z,t_i)dz}.$$

Further, we change the function $v_i(x)$ to its piecewise constant approximation given by

$$v_{ij} = a_{ij}\Delta_{i+1} + b_{ij}, \quad j = 1, \ldots, L \tag{8}$$

where L is the number of S-phase subcompartments, and $a_{ij}$, $b_{ij}$ are respectively the averages of $a_i(x)$, $b_i(x)$ over such subcompartments.

The average S-phase influx in $[t_i, t_{i+1}]$ is given by

$$f_i = \frac{N(t_{i+1})\,[2-\theta_1(t_{i+1})] - N(t_i)\,[2-\theta_1(t_i)]}{t_{i+1}-t_i}. \tag{9}$$

The direct determination of $v_{ij}$ and $f_i$ by (8)-(9) from successive histograms and the corresponding cell counts is not possible because of the high sensitivity to the experimental errors on the population size $N$, as expected since (8) and (9) contain a finite-difference approximation of a derivative. On the contrary, the estimation errors on $\theta_1$ and $\tilde{n}(x)$ were found to have not a severe influence [8]. Thus we will consider in the following the estimated DNA distributions as error-free, and will determine filtered values of $N(t_i)$ by means of the regularization technique [12].

It was found in previous papers, both on simulated and experimental data [8,9], that a good stabilizing term can be constructed on the basis of the rates $v_{ij}$. Let $z_i$ be the logarithm of the measured population size at $t_i$; we assume $z_i = y_i + \varepsilon_i$, where $\varepsilon_i$ is an uncorrelated zero-mean noise with variance $\sigma$. Our regularized estimates of $N(t_i)$ are obtained by the minimization with respect to $\Delta_1, \ldots, \Delta_n$ of the smoothing functional

$$J = \sum_{i=1}^{n}(z_i - \sum_{k=1}^{i}\Delta_k)^2 + \lambda\sum_{i=1}^{n-2}\sum_{j=1}^{L}(v_{i+1,j} - v_{ij})^2, \tag{10}$$

where $\lambda > 0$ is the regularization parameter.

The minimization of index (10) is subjected to the following constraints. The assumption of absence of cellular loss implies that $\Delta_i \geq 0$, $i = 2, \ldots, n$. Moreover, the rates $v_{ij}$ and the average fluxes $f_i$ are nonnegative quantities, due to the unidirectionality of the progression of cells along the cycle. In order to have $v_{ij} \geq 0$, it must be

$$\Delta_{i+1} \geq \max_{1\leq j\leq L}\left\{-\frac{b_{ij}}{a_{ij}}\right\}, \quad i = 1, \ldots, n-1.$$

The condition $f_i \geq 0$ is equivalent to

$$\frac{f_i}{N(t_i)} = \frac{1}{t_{i+1}-t_i}\left[\frac{N(t_{i+1})}{N(t_i)}(2-\theta_1(t_{i+1})) - (2-\theta_1(t_i))\right] \geq 0,$$

implied by

$$\Delta_{i+1} \geq ln\frac{2-\theta_1(t_i)}{2-\theta_1(t_{i+1})} = \gamma_i, \quad i = 1, \ldots, n-1.$$

All the above constraints are taken into account by the conditions

$$\Delta_{i+1} \geq \max_{1\leq j\leq L}\left\{-\frac{b_{ij}}{a_{ij}}, 0, \gamma_i\right\} = \rho_{i+1}, \quad i = 1, \ldots, n-1. \tag{11}$$

In view of the linear dependence of the quantities $v_{ij}$ on $\Delta_{i+1}$, see equation (8), we rewrite the index (10) as follows. Defining the vectors $z = [z_1 \cdots z_n]^T$, $\Delta = [\Delta_1 \cdots \Delta_n]^T$, $y = [y_1 \ldots y_n]^T$, and the $n \times n$ lower triagular matrix $K$ with unit entries, the problem becomes: find the solution $\hat{\Delta}_\lambda$ to

$$\min_{\Delta} J(\Delta; \lambda) = \|z - K\Delta\|^2 + \lambda(\Delta^T A \Delta + b^T \Delta + c) \tag{12}$$

$$\text{s.t.} \quad \Delta_i \geq \rho_i, \quad i = 2, \ldots, n,$$

where the matrix $A$, the vector $b$, and the scalar $c$ can be computed from the quantities $a_{ij}$ and $b_{ij}$ defined above, and depend only on the estimated DNA distributions.

## 4. CHOICE OF $\lambda$ AND OPTIMIZATION TECHNIQUE

The estimate of $\Delta$ obtained as solution of problem (12) depends critically on $\lambda$. Very small values of this parameter cause the estimated $N(t_i)$ to reproduce the measured values, with possible large deviations on the reconstructed patterns of the synthesis rate as given by (8). For a very large $\lambda$, estimates are obtained that could cause a compression of the actual changes of the DNA synthesis rate along the sequence.

In order to determine an optimal value of $\lambda$, we use the method of the cross-validation [11,13,14] that, with respect to other methods, for instance [10], has the advantage of not requiring the knowledge of the variance of the noise affecting the data on the population size. We applied this methods in the two versions known as ordinary cross-validation (OCV) and generalized cross-validation (GCV).

According to the OCV, the optimal $\lambda$ is the minimizer of

$$V_o(\lambda) = \sum_{i=1}^{n} (z_i - y_{\lambda,i}^{[i]})^2, \tag{13}$$

where $y_{\lambda,i}^{[i]}$ is the i-th component of the estimate of $y = K\Delta$ obtained when the i-th point of the data is deleted. This estimate is determined, for each given $\lambda$, from the constrained minimization of the index $J^{[i]}(\Delta; \lambda)$ obtained from (12) by simply deleting the term $(z_i - (K\Delta)_i)^2$ in $\|z - K\Delta\|^2$. The minimizer $\lambda_o$ of (13) is found by global search over a suitable set of discrete values of $\lambda$, and the estimate of $\Delta$, $\hat{\Delta}_{\lambda_o}$, is then determined as the constrained minimum of $J(\Delta; \lambda_o)$. This estimate is finally used to compute the estimates of the DNA synthesis rate and of the S phase influx.

In order to apply the GCV method, we follow the procedure indicated in [11,15], finding the minimizer $\lambda_g$ of

$$V_g(\lambda) = \frac{\|z - \hat{y}_\lambda\|^2}{(1 - \frac{1}{n} Tr Q_\lambda)^2} \tag{14}$$

where $\hat{y}_\lambda = K\hat{\Delta}_\lambda$ has the analytical expression given by:

$$\hat{y}_\lambda = Q_\lambda z + p_\lambda \tag{15}$$

with $Q_\lambda = KE(E^TK^TKE + \lambda E^TAE)^{-1}E^TK^T$, $p_\lambda = -\frac{\lambda}{2}KE(E^TK^TKE + \lambda E^TAE)^{-1}E^Tb$; the $n \times m$ matrix E ($m$ being the number of non-active constraints) identifies the set of active constraints in the minimization problem (12). As expected, the functional (14) presented discontinuities due to the changes with $\lambda$ of the active constraint set. In order to reduce the incidence of such discontinuities, we substituted $n$ in (14) with the number $m$ of non-active constraints. The effect of this modification is shown in Fig. 1, where the behaviour of the CGV functional is reported for one of the simulated data (to be described in section 5).

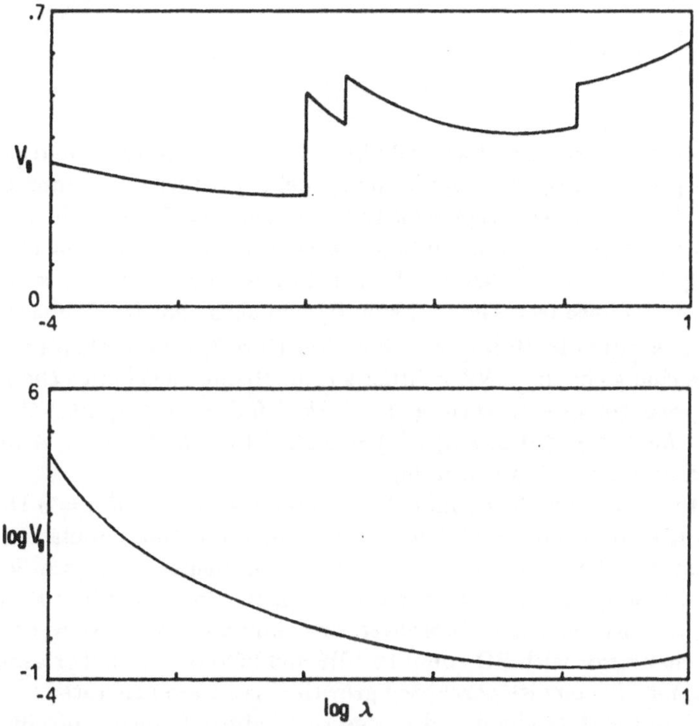


Fig. 1. Plots of $V_g(\lambda)$ according to equation (14) (top), and after substitution of $n$ with the number $m$ of non-active constraints (bottom), for a realization of test 2 at 20% noise level.

As seen above, the OCV requires that, for each $\lambda$ value, $n$ constrained minimizations of the indexes $J^{[i]}(\Delta;\lambda)$ are performed, whereas the GCV requires for each $\lambda$ one constrained minimization of $J(\Delta;\lambda)$ for determining the active constraint set. Therefore, in order to find the regularized estimate of the vector $\Delta$ we must solve a large number of constrained minimization problems. Each of these problems consists of minimizing a quadratic convex objective function subject to linear constraints. For solving these particular optimization problems, usually called quadratic programming problems, many efficient algorithms have been proposed in the literature, see [16,17,18]. Here we use a very simple approach based on a transformation of variables.

First of all we transform the variables $\Delta_i$ of the problem in the new variables $\xi_i$ according to $\Delta_1 = \xi_1$, $\Delta_i = \xi_i^2 + \rho_i$, $i = 2, \ldots, n$. Thus we transform the original constrained problem into an unconstrained one; it was proved in [19] that every stationary point of this unconstrained problem for which the Hessian matrix is positive semidefinite is also a solution of the original quadratic programming problem.

To the unconstrained problem so obtained, we apply a recently developed Newton-type algorithm [20] based on nonmonotone line searches. Under broad assumptions, this algorithm produces a sequence of points converging to a stationary point (which cannot be a maximum point) of the unconstrained problem. Moreover, from the computational point of view, it can allow a considerable saving both in the number of line searches and in the number of function evaluations [20].

## 5. RESULTS

The estimation procedure described above was tested on two simulated data sets. Both sets were generated by the population model (1)-(3) with constant $\lambda_1 = 0.25h^{-1}$ and $\lambda_2 = 1.0h^{-1}$. The two sets represent two different transients of the cell population growth. The first set (test 1) represents a desynchronization transient of a population residing initially in G1. The DNA synthesis rate $v$ is time invariant, with a parabolic pattern along the S phase (see Fig. 2, panel B) and such that the S phase transit time $T_S = \int_1^2 dx/v(x)$ is equal to 10 h. The second set (test 2) represents a transient due to a stepwise time change of the DNA synthesis rate. Before the change the population is in balanced exponential growth with $v(x) = 0.2h^{-1}$ for $1 \le x \le 2$; after the change it is $v(x) = 0.05h^{-1}$ for $1 \le x \le 1.5$, and $v(x) = 0.2h^{-1}$ for $1.5 < x \le 2$. A more detailed account of data generation is given in [8].

For both test 1 and test 2 the model generated a sequence of $n = 6$ DNA distributions, at times $t_i$ equally spaced of 4 h, and the corresponding population sizes $N(t_i)$. In order to simulate the errors present on the experimental data (see Section 2), we proceeded by obtaining for each DNA distribution the corresponding FCM histogram through a Monte Carlo method. Moreover, the simulated $N(t_i)$ were corrupted by a zero-mean gaussian error with SD equal to 10% and 20% of $N(t_i)$. For each noise level, twenty realizations of the noise corrupted growth curve were generated.

Starting from the data simulated as described above (which reproduces the information usually available when an experimental sequence is analyzed), the estimation procedure consists of the following steps:

1. From the histograms of the sequence, the DNA distributions are estimated by deconvolution [3] dividing the S phase into $L = 8$ subcompartments. The number of subcompartments is limited by the need of avoiding ill conditioning problems. From the reconstructed DNA distributions the elements of the matrix $A$ and of the vector $b$ in the index (12) are computed.
2. For a set of discrete values of $\lambda$, chosen in a suitable interval with a step equal to 0.1 on $log_{10}\lambda$, the OCV functional (13) or the GCV functional (14) is computed, finding the minimum $\lambda_o$ or $\lambda_g$.
3. The constrained minimum of $J(\Delta; \lambda_o)$, or of $J(\Delta; \lambda_g)$, is used to compute the estimates of the DNA synthesis rate $v_{ij}$, $i = 1, \ldots, n-1$, $j = 1, \ldots, L$, and the estimates of the S-phase influx $f_i$, $i = 1, \ldots, n-1$, according to equations (8) and (9).

Figure 2 shows, for one of the cases of test 1 with noise level of 10% and when the GCV was applied, the results of the estimation procedure.

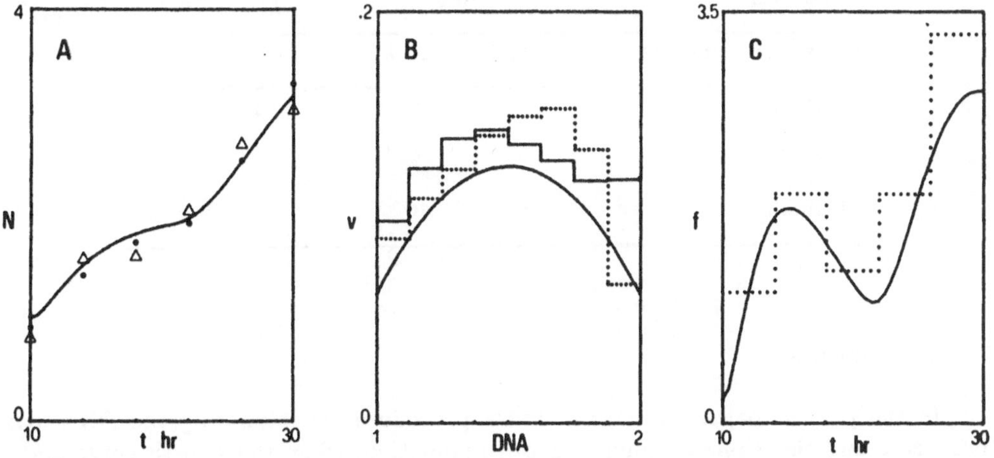


**Fig. 2.** Test 1: estimated kinetic parameters from a realization at 10% noise level. Panel A: estimated population sizes (circles) together with the simulated population size measurements (triangles). Simulated noise-free growth curve (continuous). $N$ is expressed in $10^5$ cells. Panel B: estimated DNA synthesis rate in two different time intervals (dotted, continuous) and true DNA rate pattern (continuous). $v$ is expressed in 1/hr. Panel C: estimated S-phase influx (dotted) and true S-influx versus time (continuous). $f$ is expressed in $10^4$ cells/hr.

For all the estimates of the S-phase influx, the DNA synthesis rate, and the S-phase transit time, the relative errors averaged over the sequence were computed. Table I (for test 1) and Table II (for test 2) give a general view of the performance of the estimation procedure. The tables report, for both OCV and GCV, the number of cases (over the 20 realizations analyzed for each noise level) in which all the averaged errors do not exceed 20% and 30%. The results obtained by the unconstrained method of Mallows [8,10] for the choice of $\lambda$ are also reported for comparison.

Table I. Number of cases in which $E_f$, $E_v$, and $E_{T_S}$ do not exceed 20 % and 30 % in test 1. 20 cases examined for each noise level.

| Noise level | 10% | | 20% | |
|---|---|---|---|---|
| Estimation error (<) | 20% | 30% | 20% | 30% |
| OCV | 14 | 14 | 10 | 14 |
| GCV | 14 | 15 | 9 | 14 |
| Mallows | 14 | 15 | 12 | 15 |

Table II. Number of cases in which $E_f$, $E_v$, and $E_{T_S}$ do not exceed 20 % and 30 % in test 2. 20 cases examined for each noise level.

| Noise level | 10% | | 20% | |
|---|---|---|---|---|
| Estimation error (<) | 20% | 30% | 20% | 30% |
| OCV | 4 | 9 | 1 | 6 |
| GCV | 4 | 9 | 1 | 7 |
| Mallows | 6 | 10 | 1 | 6 |

## 6. CONCLUSIONS

In the present paper we have tested a procedure for estimating the DNA synthesis rate and the S-phase influx in a cell population, when the growth curve and a sequence of FCM histograms are available. This procedure gives nonparametric regularized estimates of the above kinetic quantities, taking directly into account the positivity constraints imposed by their biological meaning.

We used for the choice of the regularization parameter the cross-validation method, that does not require the knowledge of the variance of the noise affecting the population size data. Two versions of this method, the ordinary and the generalized cross-validation have been implemented and applied to two simulated data sets. The performances of OCV and GCV, evaluated by means of suitable error indexes, appear to be substantially equivalent, with a very slight improvement when passing from OCV to GCV as could be expected according to theoretical considerations [13,14]. We note, however, that the OCV does not require any particular structure for the functional whose minimization gives the estimate $\hat{\Delta}_\lambda$, and does not require the knowledge of the influence matrix $Q_\lambda$. Thus the OCV could be used when the functional that produces $\hat{\Delta}_\lambda$ is not a quadratic functional as it occurs, for instance, in the case of a population model that takes into account the cellular loss.

The results obtained by OCV and GCV are comparable with those obtained by means of the unconstrained method of Mallows [10], that requires as an additional information the variance of the noise on population size data. We also note that the results presented in this paper, with sequences of 6 data points spaced at 4 h, are generally not worse than those reported in [8], where 11 data points spaced at 2 h were used. Although the performance of the procedure in the test 2 appears to be poor, the analysis of actual experimental data is expected to give reliable results, since the error on the population size data is usually of 5-15%.

*Acknowledgements.* We thank Dr. G. Gatti for help in writing some of the computer programs and performing a part of the numerical tests. This research was supported by Consiglio Nazionale delle Ricerche, Progetto Finalizzato "Oncologia", Grant No. 102312-104348.

## REFERENCES


1. S. Zietz and C. Nicolini, Flow microfluorometry and cell kinetics: a review. In *Biomathematics and Cell Kinetics* (A.J. Valleron and P.D.M. Macdonald, eds.), pp. 357-394, Elsevier/North-Holland Biomedical Press, Amsterdam, 1978.
2. H. Baisch, H.P. Beck, I.J. Christensen, N.R. Hartmann, J. Fried, P.N. Dean, J.W. Gray, H.J. Jett, D.A. Johnston, R.A White, C. Nicolini, S. Zietz and J.V. Watson, A comparison of mathematical methods for the analysis of DNA histograms obtained by flow cytometry. *Cell Tissue Kinet.* 15: 235-249, 1982.
3. A. Bertuzzi, A. Gandolfi, A. Germani, M. Spanó, G. Starace and R. Vitelli, Analysis of DNA synthesis rate of cultured cells from flow cytometric data. *Cytometry* 5: 619-628, 1984.
4. C. Bruni, L. Capurso, G. Koch, M. Koch, F. Lampariello, S. Lucidi and L. Teodori, Automatic analysis of flow cytometrically determined DNA distributions in the presence of abnormal stemlines *Math. Modelling* 7: 1325-1338, 1986.
5. J.W. Gray, Cell cycle analysis of perturbed cell populations: computer simulation of sequential DNA distributions. *Cell Tissue Kinet.* 9: 499-516, 1976.
6. A. Bertuzzi, A. Gandolfi, A. Germani and R. Vitelli, A general expression for sequential DNA fluorescence histograms. *J. theor. Biol.* 102: 55-67, 1983.
7. M.K. Sundareshan and R.A. Fundakowski, Stability and control of a class of compartmental systems with application to cell proliferatin and cancer therapy. *IEEE Trans. Automat. Contr.* AC-31: 1022-1032, 1986.
8. A. Bertuzzi, A. Gandolfi and R. Vitelli, A regularization procedure for estimating cell kinetic parameters from flow-cytometry data. *Math. Biosci.* 82: 63-85, 1986.
9. G. Starace, G. Badaracco, A. Bertuzzi, A. Gandolfi, C. Greco, M.D. Totaro, R. Vitelli and G. Zupi, Kinetic and survival response of the M14 cell line to lonidamine associated with adriamycin or hyperthermia. *J. Cancer Res. Clin. Oncol.* 113: 451-458, 1987.
10. C.L. Mallows, Some comments on $C_P$. *Technometrics* 15: 661-675, 1973.
11. G. Wahba, Constrained regularization for ill posed linear operator equations, with applications in meteorology and medicine. In *Statistical Decision Theory and Related Topics III, Vol. 2* (S.S. Gupta and J.O. Berger, eds.), pp.383-418, Academic Press, New York, 1982.
12. A.N. Tikhonov and V.Y. Arsenin. *Solutions of Ill-Posed Problems.* Wiley/Winston, London, 1977. 15: 661-675, 1973.
13. G.H. Golub, M. Heath and G. Wahba, Generalized cross-validation as a method for choosing a good ridge parameter. *Technometrics* 21: 215-223, 1979.
14. P. Craven and G. Wahba, Smoothing noisy data with spline functions. *Numer. Math.* 31: 377-403, 1979.
15. G. Wahba, Ill posed problems: numerical and statistical methods for mildly, moderately and severely ill posed problems with noisy data. *Tech. Rep. No. 595, Department of Statistic University of Wiscosin, Madison*, 1980.
16. P.E. Gill, W. Murray and M.H Write, *Practical Optimization.* Academic Press, New York, 1981.
17. D.P. Bertsekas, Projected Newton methods for optimization problems with simple constraints. *SIAM J. Contr. Optim.* 20: 221-246, 1982.
18. P.H. Calamai and J.J. Moré, Projected gradient methods for linearly constrained problems. *Tech. Rep. No. 73, Mathematics and Computer Science Division, Argonne National Laboratory*, 1986.
19. F.S. Sisser, Elimination of bounds in optimization problems by transforming variables. *Math. Prog.* 20: 110-121, 1981.
20. L. Grippo, F. Lampariello and S. Lucidi, A nonmonotone line search technique for Newton's method. *SIAM J. Numer. Anal.* 23: 707-716, 1986.

# ON THE CHOICE OF PARAMETERS IN MATHEMATICAL MODELS OF IMMUNOLOGICAL TOLERANCE

J. Doležal

Institute of Information Theory and Automation
182 08 Prague, Czechoslovakia

T. Hraba

Institute of Molecular Genetics
142 20 Prague, Czechoslovakia

## 1. Introduction

Our mathematical model of immunological tolerance [1] was developed on the basis of experimental findings in tolerance induced to human serum albumin (HSA) in hatched chickens. In this experimental model, suppressor cells were not found to play any significant role, and the observed inhibition of anti-HSA antibody production in tolerant chickens seemed to be caused by elimination or irreversible functional inactivation of lymphocytes reacting specically to the tolerance inducing antigen. Under these conditions, the escape from tolerance would be effected by spontaneous maturation from stem cells of new lymphocytes reacting to HSA.

The available experimental data suggested further that elimination or inactivation of B lymphocytes was the major mechanism responsible for this state of tolerance. However, the computed rate of escape from tolerance was much faster than observed experimental values; on the other hand, it was in good agreement with experimental data concerning the duration of the B cell tolerance to HSA obtained in tolerant chickens immunized with a cross-reacting antigen. This fact was interpreted as a proof of an additional mechanism superimposed on B cell tolerance. This mechanism was probably T cell tolerance and we incorporated it in the original mathematical model [2].

The mathematical model was compared also with experimental data of other authors on tolerance induced in adult mice [2, 3]. The calculated values agreed reasonably well with the experimental ones on B cell tolerance, but the scarcity of experimental data on T cell lifespan and duration of T cell tolerance did not allow a more definite choice among the proposed modifications of the model either in mice or in chickens.

As the data on population dynamics of B cells are most extensive in mice, we decided to compare the data calculated according to our model with experimental results in idiotype or isotype suppression induced in mice by monoclonal antibodies /mAb/ binding the respective determinants of immunoglobulins. Neonatal injections of such antibodies induce often chronic suppression mediated by suppressor cells. However, in other instances, the suppression of idiotypes [4] or isotypes [5] is of short duration. In these cases it is probably due to a direct elimination of B cells carrying these markers. The recovery from this suppressinn is effected by differentiation from precursors of new B cells posessing the respective markers after disappearance of the injected mAb from the organism. The mechanism of this suppression is thus analogous to that of B cell tolerance. Both these phenomena are induced by binding of immunoglobulin receptors of B cells: in the case of tolerance, it is the binding of the specific antigen by the immunoglobulin cell receptors, and in the case of idiotype or isotype suppression, the binding of the injected antibody to the immunoglobulin receptors of B cells.

## 3. Mathematical model

Two developmental compartments of B lymphocytes reactive to the tolerated antigen are considered:

(i) the immature cell compartment - I cells;

(ii) the mature cell compartment - X cells.

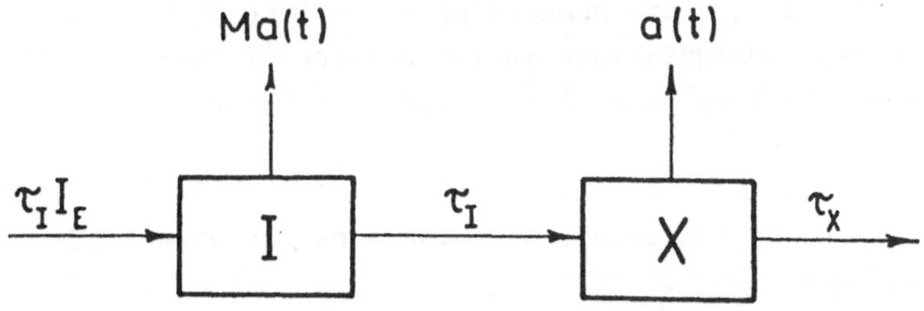


Fig. 1. Compartments of I and X cells with transition rates

The I cells arise by an antigen-independent differentiation process from their precursors and mature, independnetly of antigen, too, into X cells (Fig. 1). The parameter $\tau_I$ is the rate (all rates are scaled in day$^{-1}$) of maturation of I into X cells, $\tau_X$ the rate of natural death of X cells, $\tau_I I_E$ the rate of the differentiation of I cells from their precursors ($I_E$ being the number of the respective cells at the steady state). The quantity a(t) is the rate of irreversible inactivation of X cells by the tolerizing dose of antigen (suppressive dose of mAb) and analogously Ma(t) that of I cells, with M>1 reflecting the higher sensitivity of I cells to tolerance (suppression) induction compared to X cells. Further

$$a(t) = \begin{cases} 0, & 0 \leq t < t_1, \\ a_0, & t_1 \leq t \leq t_2, \\ a_0 \exp(-\beta(t-t_2)), & t \geq t_2 \end{cases} \quad (1)$$

where $a_0$ depends on the amount of antigen injected, $\beta$ is the rate of its non-immune elimination, $t_1$ is the day of antigen (mAb) administration, and $t_2$ is the day, when the antigen (mAb) concentration starts to decrease below the tolerizing (suppressive) level $a_0$. In fact, time course of a(t) for $t_1 \leq t \leq t_2$ does not influence the recovery from tolerance (suppression), as far as $a(t) \geq a_0$. It serves only to simulate the retarded recovery from tolerance, or idiotype and isotype suppression, and for the sake of simplicity a constant value $a(t) = a_0$, $t_1 \leq t \leq t_2$, was chosen for simulation runs.

Sizes of I and X cell compartments are described by the following differentiation equations with the given initial values:

$$dI(t)/dt = \tau_I [I_E - I(t)] - Ma(t)I(t), \qquad I(0) = I_0 \quad (2)$$

$$dX(t)/dt = \tau_I I(t) - \tau_X X(t) - a(t)X(t), \qquad X(0) = X_0 \quad (3)$$

where I(t) and X(t) are the numbers of I and X cells at time t. From the steady-state considerations in the absence of antigen (mAb), i.e. $a_0=0$, it simply follows that $\tau_I I_E = \tau_X X_E$ with index E denoting the steady-state values.

Denote $X_c(t)$ the number of X cells in the controls at time t, which is obtained as the solution of the model equations (2) - (3) with $a_0 = 0$. Then

$$r(t) = 100\ [X(t) / X_c(t)] \quad (4)$$

is the percent measure of X cell recovery from tolerance (idiotype or isotype suppression).

## 3. Comparison with experimental results

In our previous papers, we tested also lifespans of B cells which seem to be too long according to the accumulated experimental evidence. Therefore, in this paper, we use the lifespan value of 5 days both for the immature ($\tau_I$=0.2) and mature B cells ($\tau_X$=0.2) which is the maximal acceptable one on the basis of experimental evidence [6]. Besides these, the other parameters used in the simulation runs were as follows: $I_0=I_E=100$, $X_0=100$ (tolerant), $X_0=0$(suppression), $X_E=100$, M=5, $a_0=2.4$, $t_1=1$. The elimination rates of antigen (mAb) are given in each case according to experimentally established values. Therefore, only the parameter $t_2$ was to be selected for satisfactory fit.

First we compared the calculated and experimental values on B cell tolerance to human gamma-globulin (HGG) induced in adult mice [7]. The elimination rate of HGG in mice is characterized by $\beta$ = 0.1. A reasonably good fit of the simulated curve with experimental data was obtained with $t_2$=12 (Fig. 2; crosses indicate experimental values here and in the vollowing figures).

Next we modelled the recovery from the idiotype suppression induced by neonatal injection of mAb Ac38, directed against the Ac38+ idiotype of anti-(4-hydroxy-3-nitrophenyl)acetyl antibody B1-8[4]. At the time of the recovery from suppression, the elimination rate of Ac38 is $\beta$ =0.2. The best fit with this value was obtained for $t_2$=70 (Fig. 3 - full line). A better fit would be obtained for $\beta$ = 0.1 and $t_2$=47 (Fig. 3 - dashed line). However, the elimination rate of Ac38 was

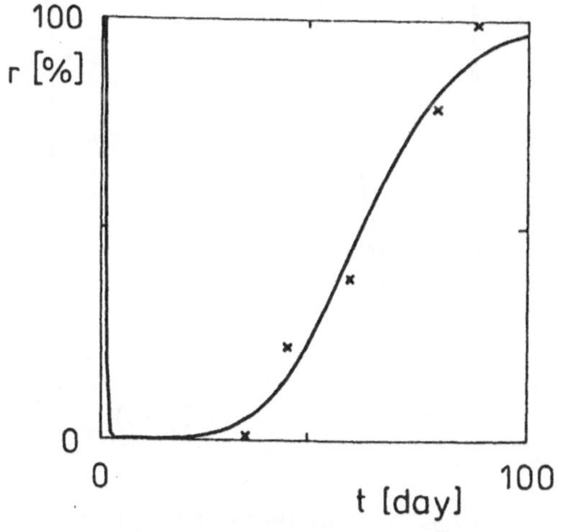


Fig. 2. Recovery from HGG tolerance

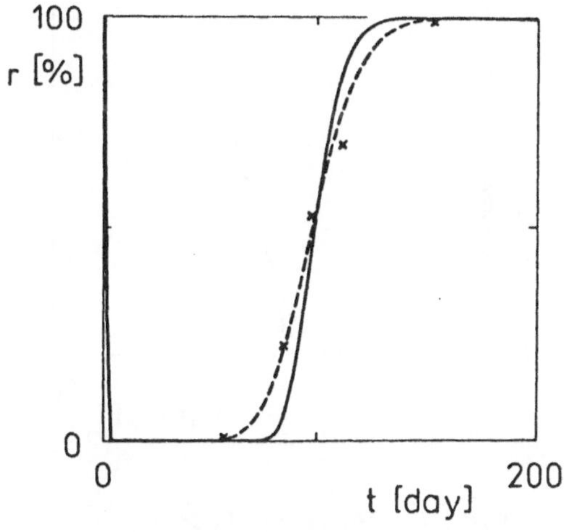


Fig. 3. Recovery from $Ac38^+$ suppression

throughly studied and established to be $\beta$=0.2 at the age of escape from suppression [4].

The $\lambda_1$ - isotype suppression was induced by neonatal injection of anti-$\lambda_1$ mAb Ls136 [5]. We are not aware of data on the elimination rate of Ls136, but as it is of the same isotype ($IgG_1$) as Ac38, we accepted it to be $\beta$=0.2. The best fit for this value was obtained with $t_2$=31 (Fig. 4 - full line). A better fit could be obtained with $\beta$=0.3, $t_2$=38 (Fig. 4 - dashed line).

The inhibitation of IgM antibody formation was induced by neonatal injection of anti-IgM mAb AF6 [5]. We also do not know the elimination rate of this antibody, but because it belongs to the $IgG_1$ class, too, we used for the simulation runs the previous value $\beta$=0.2. The best fit was obtained with $t_2$=1 (Fig. 5 - full line). We obtained a better fit for the experimental values with $\beta$=0.5 and $t_2$=13 (Fig. 5 - dashed line).

The last experimental data compared with calculated values were those on the inhibition of $\delta^+$ B cells in spleens of mice injected with anti-$\delta$ mAb 41407 after birth [5]. We do not know either the elimination rate or the isotype of 41407, so at first by analogy we used the value $\beta$=0.2 and the best fit was obtained with $t_2$=7 (Fig. 6 - full line). However, a better fit was obtained sith $\beta$=0.5 and $t_2$=20 (Fig. 6 - dashed line).

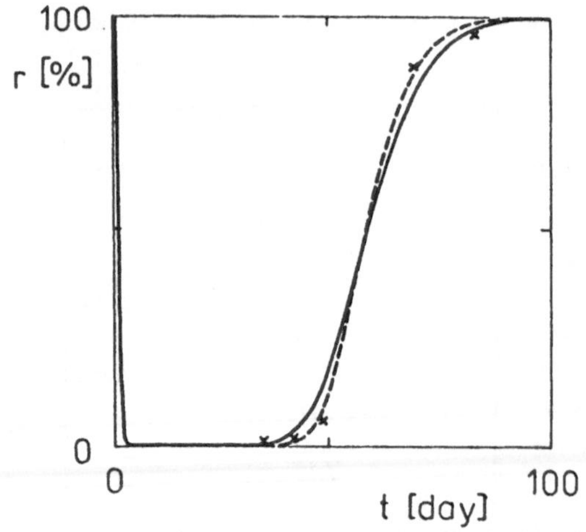


Fig. 4. Recovery from $\lambda_1$ suppression

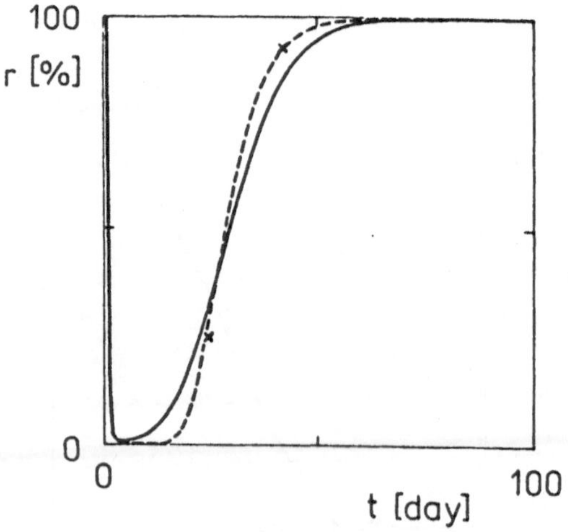


Fig. 5. Recovery from IgM suppression

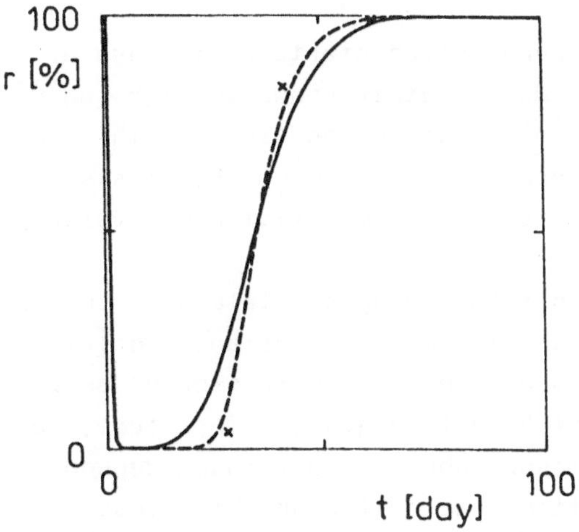


Fig. 6. Recovery from $\delta^+$ suppression

4. Discussion

Although our mathematical model agrees reasonably well with experimental observations in B cell tolerance [1,3], we did not obtain a satisfactory agreement between the experimental and simulated data in any case of the compared states of immunoglobulin suppression induced by the respective antibodies. In the simulation of $\lambda_1$ and IgM suppression a better fit was obtained with higher elimination rates of the injected mAb than that observed in mAb of the same isotype. The difference was more pronounced in the case of IgM suppression. It is of interest that a similar recovery rate was observed in $\delta^+$ B cell suppression. The recovery from IgM suppression induced by AF6 occurs at an age, when the elimination rate of Ac38 was observed to be substantially slower than $\beta = 0.2$[4]. As AF6 possesses the same isotype as Ac38, the same elimination rate can be expected, and therefore, the escape rate from the suppression should be even slower than that simulated with $\beta = 0.2$.

One possible explanation of this discrepancy could be that the number of receptors to which the anti-isotype mAb binds is much higher than that in the case of the anti-idiotype mAb. This factor could accelerate the elimination of the anti-isotype mAb, when their concentration drops and the relative proportion of the mAb molecules to those of the corresponding immunoglobulin receptors of B cells is low. However, it is impossible to use ths explanation for the kinetics of the idiotype suppression induced by mAb Ac38. Although this elimination rate was found to be $\beta = 0.2$, better fit of calculated values was obtained with a slower one.

The recovery from Ac38 idiotype suppression starts at the age of 7-10 weeks, i.e. in young adult mice. On the other hand, the IgM and $\delta^+$ cells isotype suppressions start to vanish during the first month after birth. The recovery from the $\lambda_1$ suppression is intermediate, it starts at the beinning of the second month of age; the best fitting $\beta$ value ($\beta$ =0.3) is also intermediate.

There are several observations that immunological tolerance to the same antigen disappears faster i n younger than in older animals [8]. It seems that the mechanism of the immunoglobulin suppression of short duration is not different from that of B cell tolerance. Therefore, we wonder, if the rate of escape from immunoglobulin suppression depends primarily on age and not the elimination rate of the antibody inducing the suppression.

In contrast to the on the first glance obvious assumption that excape from tolerance (immunoglobulin suppression) is governed by the elimination rate of antigen (mAb), some other, age dependent, mechanism dominates. Our explanation is that tolerogenic (suppressive) activity of antigen (mAb) is lost, when it drops below a certain level, although the kinetics of the recovery behaves as being mediated by the "virtual" rate of elimination $\beta$ . Such v alues characterize the mentioned age dependent mechanism of recovery and need not be related to the actual elimination rate of antigen (mAb).

Thus the obtained results suggest approaches which are amenable to experimental analysis, and at the same time, are a stimulus for designing alternative modifications of the mathematical model assuming other mechanisms governing kinetics of immunological tolerance and immunoglobulin suppression.

## 5. References


[1] Klein, P., Hraba, T. and Doležal, J.: The use of immunological tolerance to investigate B lymphocyte replacement kinetics in chickens. J. Math. Biology 16, 131-140, (1983).

[2] Hraba, T. and Doležal, J.: On mathematical model of immunological tolerance. In: System Modelling and Optimization (A. Prékopa et al., Eds.), Springer-Verlag, Berlin 1986, pp. 340-349.

[3] Klein, P., Hraba, T. and Doležal, J.: Mathematical model of B lymphocyte replacement kinetics: its application to the recovery from tolerance in adult mice. Math. Biosci. 73, 227-238, (1985).

[4] Takemori, T. and Rajewsky, K.: Specificity, duration and mechanism of idiotype suppression induced by neonatal injection of monoclonal anti-idiotype antibodies into mice. Eur. J. Immunol. 14, 656-667, (1984).

[5] Saito, T., Tokuhisa, T. and Rajewsky, K.: Induction of chronic idiotype suppression by ligands binding to the variable (not the constant) region of the idotypic target. Eur. J. Immunol. 16, 1419-1425, (1986).

[6] Freitas, A.A., Rocha, B and Coutinho, A.A.: Lymphocyte population kinetics in the mouse. Immunol. Rev. 91, 5-37, (1986) .

[7] Parks, E. E. and Weigle, W.O.: Maintenance of immunological unresponsiveness to human $\gamma$-globulin: evidence for irreversible inactivation in B lymphocytes. J. Immunol. 124, 1230-1236, (1980).

[8] Hraba, T.: Mechanism and Role of Immunological Tolerance. S. Karger, Basel 1968.

# A MATHEMATICAL MODEL OF TROPHIC INTERACTION OF BIOPOPULATIONS OF BIOLOGICAL SYSTEMS

Algis GARLIAUSKAS, Sofija KOVARSKAJA*, Valdas LIMANAUSKAS

Istitute of Mathematics and Cybernetics
Institute of Botanics*
Lithuanian SSR Academy of Sciences
Akademijos 4, 232600 Vilnius, USSR

ABSTRACT

The mathematical models of the ecosystem of the lake-cooler of the nuclear power station are considered. The first model is based on the principle of homology and relation. The other one is founded on the method of matter conservation, it involves coefficients which depend on light intensity and temperature change. The comparison of the observation data and digital solutions is presented.

## 1. INTRODUCTION

In order to study all the aspects of ecological monitoring of a nuclear power station one must solve the problems of control, regulation and management, to predict the behavioristic effects of the bio-communities. The analysis of the state of the lake-cooler has shown that very complicated trophic interactions are taking place. In spite of the fact that many mathematical models of ecosystems have been suggested or even partially realised a lot of problems still do exist, the majority of which concern the questions of stability, adaptivity and correspondence of the models to the actual behavioristic characteristics [1-5].

Taking into account an exceptional importance of examining the whole complex of the monitoring system and particularly approaching the problem systematically, we shall deal with a concrete mathematical model of the lake-cooler ecosystem, taking into consideration the structural analysis and thermal influence on the bio-communities of the natural lake.

## 2. STRUCTURAL ANALYSIS OF ECOSYSTEM

The structural analysis for ecosystems is important from two points of view. At first, it is necessary to present precisely the mechanism of biopopulations interaction not separately, but in groups, classes. Here it is important to know the structural interactions (counteractions) on the level of the relation matrix analysis. Secondly, it is very important to catch the main of ecosystems, i.e. the structural changes, forms. It is necessary to realise the modelling of structural changes and to evaluate the structural stability. The differencial equations are modelling the exactly fixed structures of species collections of an ecosystem.

Taking as basis the theory of a simplicial complex [6], homological structural analysis [7,8] and an additional reasoning of a combinatorial topology, complex ecosystems can be presented on the basis of the following methodology.

For the concrete ecological situation either the matrix of interactions on the principle "predator-victim": $\|\Lambda\|$ or the one "victim-predator": $\|\Lambda^*\|$ - the transpositional matrix of relations is formed. To this there is a set $\Lambda \subset X \times Y$ . Further on the simplexes $\sigma_p$ from a complex $K_Y(X;\Lambda)$ or $K_X(Y;\Lambda^*)$ are formed.

The chains of $q$ -connectivity with common face of the dimension $\beta$ according to the requirement

$$q = \min\{i, \beta_1, \beta_2, \dots, \beta_{n-1,j}\} \tag{1}$$

for $i$ and $j = 1, 2, \dots, n-1$ are defined in the complexes $K_X$, $K_Y$. After that for all of $q = 0, 1, \dots, \dim K$ the classes of equivalence $Q_q = \mathrm{card}\, K/q$ and the first structure vector $Q = (Q_{\dim K}, \dots, Q_1, Q_0)$ are defined. It is important that with the help of vector $Q$ the disjoint blocks of matrix $\|\Lambda\|$ are found according to which one can evaluate the information flow between the species classes of an ecosystem.

By vector $\hat{Q}$ the abstract vector is defined

$$\hat{Q} = Q - U \quad \text{, where} \quad U = (1, 1, \ldots, 1) \qquad (2)$$

and when $Q_0 = K\ (>1)$ complex is presented as

$$K = \bigcup_{\ell}^{k} K_\ell \qquad (3)$$

and

$$\hat{Q} = \sum_{\ell}^{k} \hat{Q}_\ell \qquad (4)$$

For the estimation of the importance degree of structural relations of complexes $K_X$ and $K_Y$ the first structure coefficient $h$ of relations $\Lambda$ is used:

$$h = \frac{(Q, Q')}{\|Q\| \, \|Q'\|} \qquad (5)$$

where $Q$ and $Q'$ are the structure vector of $K_X$ and $K_Y$ respectively. These vectors can be presented in Euclidian $n+1$ space $R^{n+1}$ in such a way:

$$\left.\begin{aligned} Q &= \sum_{i=0}^{n} Q_i e_i \\ (Q, Q') &= \sum_{i=0}^{n} Q_i \cdot Q_i' \; ; \; \|Q\| = \sqrt{\sum_{i=0}^{n} Q_i^2} \; ; \; \|Q'\| = \sqrt{\sum_{i=0}^{n} Q_i'^2} \end{aligned}\right\} \qquad (6)$$

Further the definition of eccentricitet

$$ecc(\sigma) = \frac{\hat{q} - \check{q}}{\check{q} + 1} \qquad (7)$$

is introduced, where $\hat{q}$ - the dimension of $\sigma$ , $\check{q}$ - the biggest of $q$ , when $\sigma$ is connected with at least one of other simplexes from $K$ . In accordance to the paper [8] the eccetricitet is offered to be used for estimation of degree of the simplicial integration. And the bigger $ecc(\sigma)$ the more isolated is $\sigma$ from all the others.

Further on the basis of the theory of homology either an isomorphism $H_0 \simeq J$ or 1-dimensional hole $H_1 \simeq J$ (at $H_k = \sum_{i=1}^{k} \oplus J_i$ some $k$ -dimesional holes will be) are defined, or the complex will have torsion according to

$$H_p = G_p^0 \oplus Tor\, H_p$$

where $G_p$ - a free group, $Tor\, H_p$ - a subgroup of a torsion group $H_p$ .

The dynamics of the ecosystem structure changes can be presented by changes in time of a ranging image

$$P = P_0 \oplus P_1 \oplus \ldots \oplus P_N \qquad (9)$$

where $P_i\, \sigma^{(i)} \to k$ . Here $\sigma^{(i)}$ – i-th simplex in $K_Y$ . $N = \dim K_Y$ – the biggest dimension of simplexes in $K_Y$ .

For this the changes of the whole image $P$ in time are found, or more exactly, the remarking of simplicial complex $K_Y^{(t)}$ to $K_Y^{(t+1)}$ is realised. Every such remarking expresses the image $|w| : |K_Y^{(t)}| \to |K_Y^{(t+1)}|$ which corresponds to the simplicial image of polyhedra $|K_Y^{(t)}|$ to $|K_Y^{(t+1)}|$.

## 3. STRUCTURAL ANALYSIS OF CONCRETE ECOSYSTEM

The concrete ecosystem of the natural lake-cooler of the nuclear power station is regarded. The matrix of relations was formed for the ecosystem with 29 species. According to (1) all classes of equivalence $Q_q$ were formed in the simplicial complex $K$ . The results of $q$ -analysis are presented in table:

| $\sigma_q^i$, i | q | $Q_q$ | Disjunctional blocks and $\sigma_q^i$ simplexes |
|---|---|---|---|
| $i=1$ | 16 | 1 | $\{x_{12}\}$ |
| $i=\overline{1,3}$ | 15 | 3 | $\{x_{12}\}\{x_{18}\}\{x_{19}\}$ |
| $i=\overline{1,3}$ | 14-10 | 2 | $\{x_{18}, x_{19}\}\{x_{12}\}$ |
| $i=\overline{1,3}$ | 9 | 1 | $\{x_{18}, x_{19}, x_{20}\}$ |
| $i=\overline{1,5}$ | 8 | 2 | $\{x_{18,19}, x_{11,12}\}\{x_8\}$ |
| $i=\overline{1,7}$ | 7 | 3 | $\{x_{18,19}, x_{11,12}, x_8\}\{x_{20}\}\{x_{13}\}$ |
| $i=\overline{1,12}$ | 6 | 5 | $\{x_{18,19}, x_{7,8}, x_{10-13}\}\{x_{14}\}\{x_{16}\}\{x_{25}\}\{x_{20}\}$ |
| $i=\overline{1,16}$ | 5 | 5 | $\{x_{18,19}, x_{10-13}, x_{7,8}, x_{16}\}\{x_{14}\}\{x_{15}\}\{x_{20}\}\{x_{25,27-29}\}$ |
| $i=\overline{1,24}$ | 4 | 3 | $\{x_{7-13}, x_{16}, x_{18,19}, x_{27-29}, x_{24,25}\}\{x_{14,15}, x_{1-6}\}\{x_{20}\}$ |
| $i=\overline{1,24}$ | 3 | 1 | $\{x_{1-16}, x_{18-20}, x_{24,25}, x_{27-29}\}$ |
| $i=\overline{1,25}$ | 2 | 1 | $\{x_{1-20}, x_{24,25}, x_{27-29}\}$ |
| $i=\overline{1,26}$ | 1 | 1 | $\{x_{1-20}, x_{24-29}\}$ |
| $i=\overline{1,29}$ | 0 | 1 | $\{x_{1-29}\}$ |

By formula (2) the components of the abstract vector were defined and are shown in fig.1.

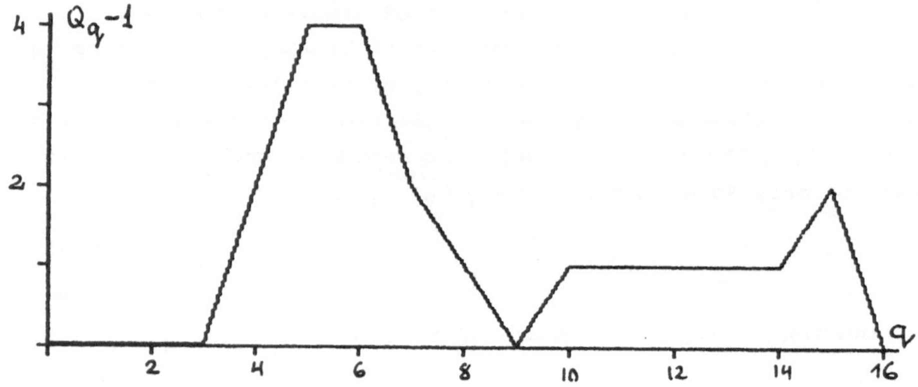


Fig.1.

Fig.1 shows that the worst connectivity and accordingly the information exchange take place on rows 5,6 of matrix $\|\Lambda\|$ . In the ecological sense the bacterioplancton, bacteriobenthos, filters, sedimentary organic material and detritophages have been isolated from other species which happen to be in the other block.

The calculated value of the eccentricitet according to (7) confirmed that it was the index of species flexibility according to food, i.e. the index of their capacity to survive.

Taking into consideration the conditions of isomorphism for this example, we defined that the torsion according to (8) does not exist, but $H_1 = J$, and so the predators complex has the 1-dimension hole restricted by 1-dimension simplexes.

## 4. A MATHEMATICAL MODEL OF TROPHIC INTERACTION OF ECOSYSTEM

After structure analysis of the whole relation matrix and according to the practical observations the possibility has been found to reduce the matrix considerably and to use the mathematical model which is not complex but adequate enough to real biocenosis processes of the ecosystem of the natural lake-cooler of the nuclear power station.

The mathematical model is described in the following system of differential equations:

$$\dot{x}_i = \frac{1}{S}\Big(a_{in}x_i x_n - x_i \sum_{j\in K_i}(1+e_{ij})a_{ij}x_j - g_i x_i^2 - d_i x_i\Big),\ i=\overline{1,m};$$

$$\dot{x}_i = \frac{1}{S}\Big(x_i \sum_{j\in P_i}{}' a_{ij}x_j - x_i \sum_{j\in K_i}(1+e_{ij})a_{ij}x_j - r_i x_i - g_i x_i^2 - d_i x_i\Big),\ i=\overline{m+1,n-2};$$

$$\dot{x}_{n-1} = \frac{1}{S}\Big(\sum_{j=1}^{n-2}(g_j x_j + d_j x_j) - x_{n-1}\sum_{j\in K_{n-1}} a_{n-1,j}x_j - x_{n-1} + \sum_{i,j} e_{ij}a_{ij}x_i x_j\Big);$$

$$\dot{x}_n = \frac{1}{S}\Big(\sum_{j=m+1}^{n-2} r_j x_j + x_{n-1} - x_n\sum_{i=1}^{m} a_{ni}x_i\Big);$$

where $X_i$ – components which unite populations according to the functional groups:

$X_1,\dots,X_m$ – producers (macrophits, phytoplancton),
$X_{m+1},\dots,X_{n-2}$ – consumers (zooplancton, bentos, fish),
$X_{n-1}$ – detritus and bacteria,
$X_n$ – biogene (the most limiting substance).

The given system (10) involves the following assumptions: most limiting is one substance, the behaviour of components is determined by food resources, not by the inner state (age, rate of sexes, etc.); the intensity of the interaction is proportional to the production of the quantity of the interacting components and does not depend on the total quantity $S'$, i.e. the most limiting substance in the lake.

All $X_i$ are estimated by the units of the limiting substance and $\sum x_i = S'$ .

$a_{ij}$ - coefficients of assimilation-utilization,
$e_{ij}$ - the part of food transformed to excrements,
$z_i$ - coefficients of respiration,
$g_i$ - coefficient of interspacial competition,
$d_i$ - coefficient of natural mortality,
$P_i$ - set of indices of the components which are used by the i-th,
$K_i$ - set of indices of the components using the i-th,
$\alpha$ - coefficient of the decomposition of detritus.

Matrix $\{a_{ij}\}$ is symmetrical, in matrix $\{e_{ij}\}$ : $e_{ij}=0$ , if $e_{ji} \neq 0$ . All the coefficients of the system are periodical in time as they depend on temperature and light intensity.

The main difference between the energetic model [9] and given system is that the latter allows to find out the steady perennial oscillations of numerical solutions within a wide range of meanings of parameters that are independent of the initial data.

For the numerical realization n=13 is taken, the matrix of interaction depends on light intensity and temperature which changed harmoniously within a one-year period.

The coefficient of mortality rate was found using the formula

$$d_i = \frac{c_i}{(T - T_{i\,min})(T_{i\,max} - T)}$$

where T is the temperature at a given moment, $T_{i\,max}$ and $T_{i\,min}$ are temperatures of the complete mortality of the i-th due to over-heat and over-freezing.The variables are interpreted in the following way:

$x_1$ - aerial , $x_2$ - floating, $x_3$ - bottom dwelling macrophyts;
$x_4$ -diatomic, $x_5$ - blue-greenish algae; $x_6$ - protozooplancton,
$x_7$ - placid zooplancton, $x_8$ - predatory zooplancton;
$x_9$ - fish; $x_{10}$ - bacterioplancton; $x_{11}$ - bentos;
$x_{12}$ - detritus; $x_{13}$ - biogen.

Some results of the numerical integration are presented in fig.2, where the behaviour of the system during a seven year period is shown. Since the end of the second year till the beggining of the sixth one the temperature had been rising by 5 C per year. Light intensity and the amplitude of annual thermal oscillations did not change. The rise of the temperature has led to the extinction of the "predator-zooplancton" .

The comparison of the solution results with the observations

Fig. 2

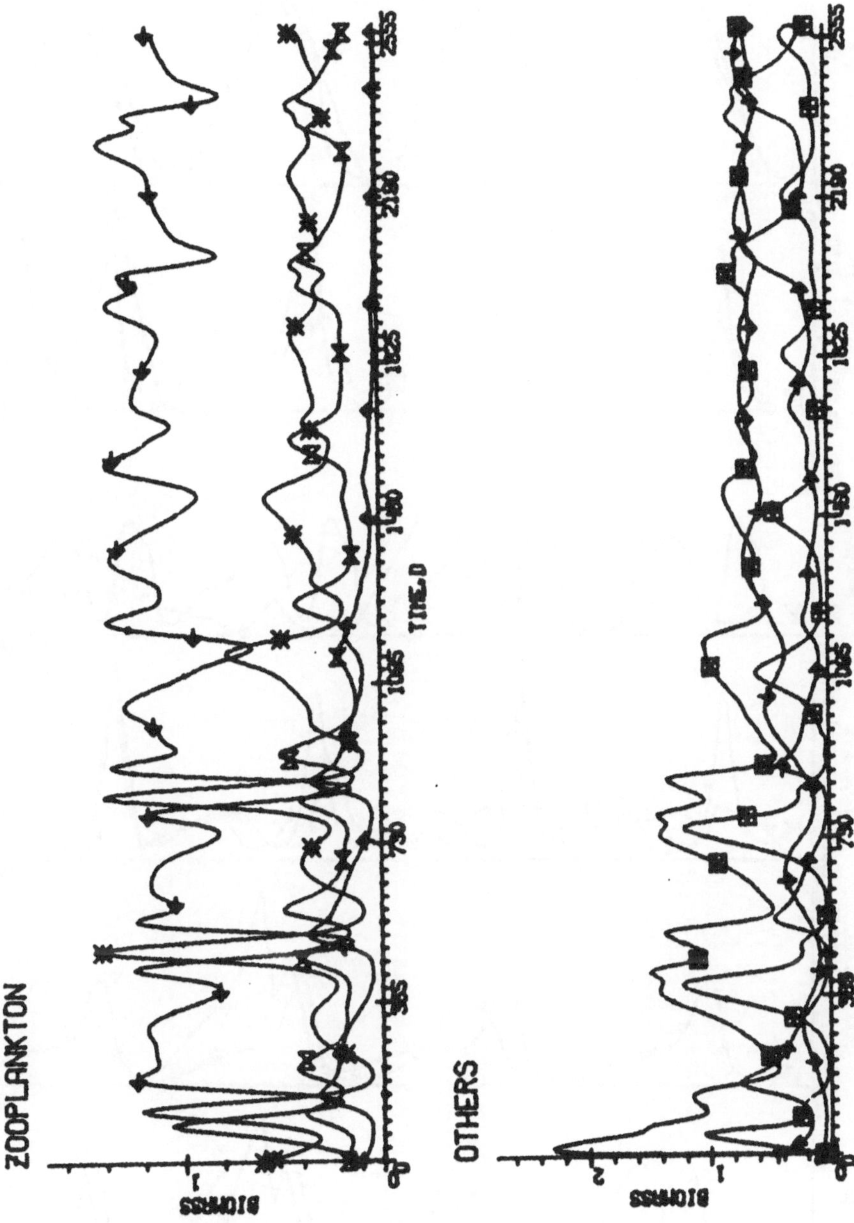

Fig. 3

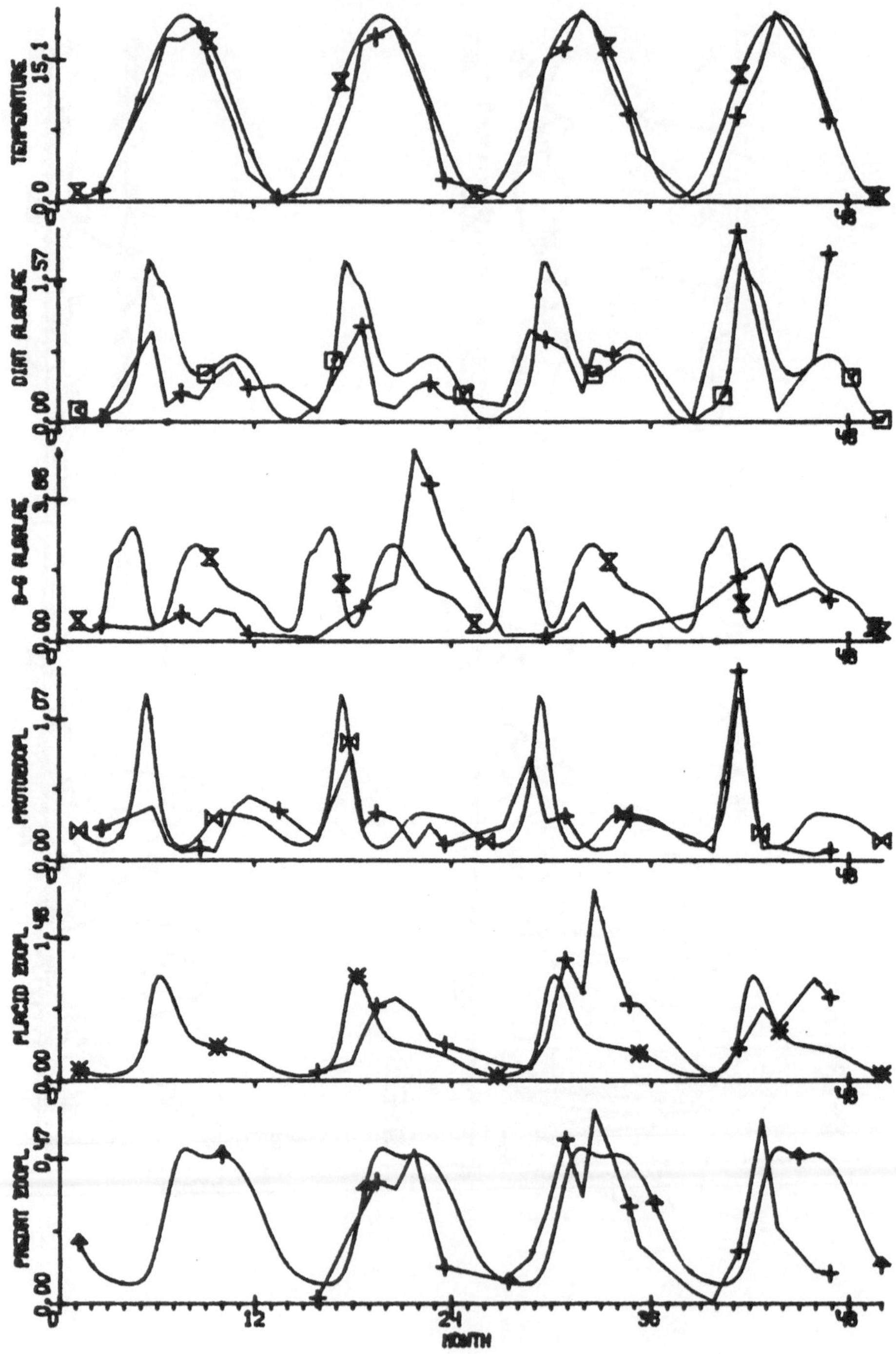
TEMPERATURE
15.1
0.0
DIRT ALGALAE
1.57
0.00
B-G ALGALAE
3.66
0.00
PROTOZOOPL
1.07
0.00
PLACID ZOOPL
1.46
0.00
PRESENT ZOOPL
0.47
0.00
0
12
24
36
48
MONTH

has been realized in the system (10) consisting 14 equations: was presented by two types of detritus: $X_{12}$ - "approach" and $X_{14}$ - "not approach" detritus.

The results of solutions and their comparison with the observation results since 1979 to 1982 are presented in fig.3. In the figure the numerical solution of the system (10) is shown with a smooth curve expressing te changes of biomass (miligramm per litre) in time, while the results of the observations are shown with a broken curve.

With the chosen parameters the periodicity of the numerical solutions has an one-year period and they have good correspondence to the periodicity of the observation data. The essential difference between the solution of the model and the real data remained only for the component "other algaes", where the observation data have not clearly expressed periodicity.

## CONCLUSIONS

1. The structural analysis based on simplicial complexes let to explain the general structure interactions of complex ecosystems.

2. The given mathematical model based on matter balance and first realised for the ecosystem of a lake-cooler shows the identity of the behaviour of solutions to the actual processes of biocenosis.

3. The given model differs from the used energetical models in that it allows to get steady perennial oscillations of the numerical solutions within a wide range of changing parameters, which are independent of the initial data.

4. The results of simulation show the possibility to use the given model for the purposes of predicting the behaviouristic characteristics of an ecosystem under thermal change and other external factors.

5. The identification of the parameters of the model was realised on the basis of the many-year processed observations data and let to obtain good coincidence.

## REFERENCES


1. Svirezev Y.M.,Logofet D.O. Stability of biological populations. - Moscow: Science, 1978 (in Russian).
2. Odum Y. The bases of ecology. - Moscow: MIR, 1979 (in Russian).
3. Garliauskas A. Mathematical modelling of biomedical systems. "Mathematical models in biology and medicine". - Vilnius, 1985, No.1, p.5-29 (in Russian).
4. Park R.A. et al. Simulation, 1974, vol.3, No.2, p. 33-50.
5. Garliauskas A., Garliauskiene A. Simulation of biomedical systems. Proceedings of the 12th IFIP Conference. "System Modelling and Optimization" Springer - Verlag, Berlin, Heidelberg 1986, p. 264-281.
6. Aleksandrov P.S. Combinatorial Topology. - Moscow-Leningrad, 1974 (in Russian).
7. Atkin R.H. Mathematical Structure in Human Afairs. Heineman, London, 1973, p. 211.
8. Casti J. Connectivity, Complexity and Catastrophe in Large-scale Systems. Chichester, N.York, Brisbane. Toronto, 1979, p.216.
9. Krishev I.I. Mathematical models of dynamics of ecosystem of reservoir - cooler. - Experimental and field research of biological grounds of productivity of lakes. -Leningrad, 1979, p.131-146 (in Russian).

# OPTIMAL PLANNING FOR OPERATING A PRODUCTION PROCESS WITH A BUFFER AND AN ELECTRIC POWER PLANT


Nobuo Sannomiya*, Koichiro Watanuki* and Takashi Tsuda**

* Department of Electronics, Kyoto Institute of Technology, Matsugasaki, Sakyo-ku, Kyoto 606, Japan

** FUJI FACOM Co., Fuji-machi, Hino-shi, Tokyo 191, Japan


## 1. Introduction

In the last few years, computer control systems have been widely used in various kinds of production processes. In these systems, optimization techniques play an important role for making a long range production planning with due regard to energy saving and production cost reduction. For a process with a buffer or a storage device, the planning problem is large scale, because the constraint of the problem has an interconnection between two adjacent time periods. Therefore, solving it exactly is difficult in the case of executing the real time control and planning with use of a process computer. Several decomposition procedures have been proposed for solving mixed-integer linear programs with special structure[1,2], and have been applied to short range planning problems[3].

This paper deals with an optimal planning for operating a production process which produces some products by consuming electric power. The products are stored in a buffer. For energy saving, an independent electric power plant is operated. The electric power demand condition in the factory is satisfied by operating the power plant and by purchasing electric power from an electric power company. The purchase price of electric power changes with appropriate time intervals during a whole day. Therefore, we have to make a planning for operating the power plant and for using the buffer so as to minimize the total operating cost. Since the minimum load of the power plant is not zero but takes a positive value, the feasible region for the decision variables of the problem becomes disconnected. Consequently, the problem needs to be formulated as a mixed-integer program. The planning problem without electric power plant is simply formulated as a linear program, and has been considered previously[4]. The problem treated here is its extension.

In this paper, a decomposition algorithm is proposed for solving the problem easily by using a process computer. The original problem is decomposed into several subproblems based upon a heuristic idea. The master program is solved by applying the dynamic programming technique. The subproblems are mixed-integer programs of smaller sizes than the original problem. They are solved by the greedy method modified suitably for the present problems. As a case study, an optimal planning for operating a production mill at cement works is considered. By examining an illustrative example, it is observed that the proposed algorithm is efficient, requiring less computation time than the branch and bound method.

## 2. Problem Statement

As shown in Fig. 1, a product(called the product A) is produced and stored in a buffer in order to be shipped on a prearranged date and to produce another product(called the product B). At each time $i \in (1, 2, \ldots, T)$, the quantity of the product A, $y_{pi}$[kg/h], and its shipment $y_{si}$[kg/h] are known. Let w[kg] be the total quantity of the product A which is necessary to produce the product B in the whole period. The value of w is given. For producing the product B, we consume electric power which is supplied by operating an independent electric power plant and/or by purchasing from an electric power company. The operating cost d[¥/MWh] of the power plant is constant, but the purchase price $c_i$ [¥/MWh] of electric power varies with the time i.

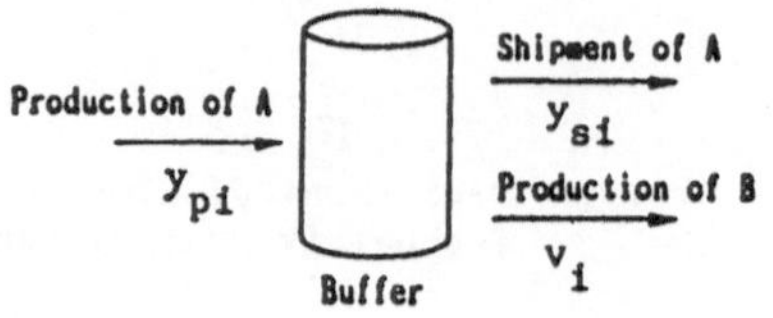


Fig. 1 Production process with buffer.

We define the following variables for the time $i \in (1, 2, \ldots, T)$:

$v_i$; quantity[kg/h] of the product A which is necessary to produce the product B at the time i.

$u_i$; electric power[MWh/h] supplied at the time i by the electric power plant.

$x_i$; binary variable representing the operating state of the electric power plant ($x_i$ = 1 when the electric power plant is operated at the time i, and $x_i$ = 0 otherwise).

$H_i$; silo level[kg] in the final stage of the time interval i.

The silo level $H_i$ is given by

$$H_i = H_0 + \Sigma_{j=1}^{i} (y_{pj} - y_{sj} - v_j) \Delta t_j \qquad i = 1, 2, \ldots, T \tag{1}$$

where $H_0$ is the initial silo level[kg] and $\Delta t_j$ is the j-th time interval[h]. Due to the limited silo capacity, the following constraint must be satisfied:

$$H_{min} \leqq H_i \leqq H_{max} \qquad i = 1, 2, \ldots, T \tag{2}$$

where $H_{min}$ and $H_{max}$ are the lower and upper bounds[kg] of the silo capacity, respectively. By defining

$$\left.\begin{aligned} a_i &\triangleq H_0 + \Sigma_{j=1}^{i} (y_{pj} - y_{sj}) \Delta t_j - H_{min} \\ b_i &\triangleq H_0 + \Sigma_{j=1}^{i} (y_{pj} - y_{sj}) \Delta t_j - H_{max} \end{aligned}\right\} \tag{3}$$

(2) is rewritten as

$$b_i \leqq \Sigma_{j=1}^{i} v_j \Delta t_j \leqq a_i \qquad i = 1, 2, \ldots, T-1 \tag{4}$$

Further, we have at i=T

$$\Sigma_{j=1}^{T} v_j \Delta t_j = w \tag{5}$$

The relationship between supply and demand of electric power is given by

$$p_i \Delta t_i + u_i \Delta t_i = \alpha v_i \Delta t_i + f_i \Delta t_i \tag{6}$$

where

$p_i$; electric power[MWh/h] purchased at the time i.

$\alpha$; unit electric power[MWh/kg] necessary for producing the product B.

$f_i$; the quantity[MWh/h] of power demand at the time i for other purposes in the factory.

Due to the non-negativeness of $p_i$ and the constraints for the performance of the equipments, we have at each time

$$\left.\begin{array}{l} 0 \leqq v_i \leqq v_{max} \\ x_i u_{min} \leqq u_i \leqq x_i u_{max} \\ x_i = 0 \text{ or } 1 \\ u_i - \alpha v_i \leqq f_i \end{array}\right\} \quad i = 1, 2, \ldots, T \tag{7}$$

where

$v_{max}$; the upper bound[kg/h] of $v_i$.

$u_{min}$, $u_{max}$; the lower and upper bounds[MWh/h] of $u_i$, respectively. Note that $u_{min} > 0$.

The objective function to be minimized is the total cost of electric power, which is given by

$$z = \Sigma_{i=1}^{T} (c_i p_i + du_i) \Delta t_i \tag{8}$$

By eliminating $p_i$ with use of (6), we have the following mixed-integer program:

$$\text{P1:} \quad \min z = \Sigma_{i=1}^{T} [\alpha c_i v_i + (d - c_i) u_i] \Delta t_i$$

subject to (4), (5) and (7)

### 3. Algorithm

When we consider the production planning for one month, the number of the constraint (4) is more than two hundreds. However, the active components are very few. Then we try to decompose P1 into several subproblems so as to be able to neglect (4). For this purpose, the set of time {1, 2, ..., T} is decomposed into L blocks. The subset of time belonging to the $\ell$-th block is

$$\tau(\ell) = \{T_{\ell-1}+1, T_{\ell-1}+2, \ldots, T_\ell\} \qquad \ell = 1, 2, \ldots, L \tag{9}$$

where $T_0 = 0$ and $T_L = T$. Corresponding to this decomposition, the total production quantity $w$ is distributed to each block. The production quantity for the $\ell$-th block is

$$w_\ell = \Sigma_{i \in \tau(\ell)} v_i \Delta t_i \tag{10}$$

From (4) and (5), the following relations must be satisfied.

$$\Sigma_{\ell=1}^{L} w_\ell = w \tag{11}$$

$$b_{T_\ell} \leqq \Sigma_{j=1}^{\ell} w_j \leqq a_{T_\ell} \qquad \ell = 1, 2, \ldots, L-1 \tag{12}$$

When we assign appropriate values to L, $\tau(\ell)$ and $w_\ell$ ; $\ell = 1, 2 \ldots, L$, the following L subproblems are obtained.

$$\text{P2;} \qquad \min z_\ell (w_\ell) = \Sigma_{i \in \tau(\ell)} [\alpha c_i v_i + (d - c_i) u_i] \Delta t_i$$

subject to (7) for $i \in \tau(\ell)$ and (10)

$\ell = 1, 2, \ldots, L$

Further, when we assume the value of L and the sets $\tau(\ell)$; $\ell = 1, 2, \ldots, L$, the optimal values of $w_\ell$ ; $\ell = 1, 2, \ldots, L$ are determined by solving the following master program:

$$\text{P3;} \qquad \min z = \Sigma_{\ell=1}^{L} z_\ell (w_\ell)$$

subject to (11) and (12)

For a given value $w_\ell$, each subproblem P2 is solved to obtain $z_\ell (w_\ell)$. By using the values $z_\ell (w_\ell)$; $\ell = 1, 2, \ldots, L$ thus obtained, the master program P3 is solved. The subset of time for each block, $\tau(\ell)$, is determined iteratively on the basis of a heuristic idea. Thus, we have the following algorithm:

Step 1. Set L = 1. Solve P2 by putting $w_1 = w$.
Step 2. If (4) holds for all $i \in \{1, 2, \ldots, T-1\}$, stop. If not, the set of time is decomposed at the point in which (4) is violated to the greatest degree. Set $L \rightarrow L + 1$.
Step 3. Solve P3 with the subproblem P2. Return to Step 2.

Note that $z_\ell (w_\ell)$ is a nonlinear function of the scalar variable $w_\ell$. Then, the solution of P3 is obtained by applying the dynamic programming technique. For this purpose, the feasible region for $w_\ell$ is discretized so as to have the following K points:

$$w_\ell \in W \triangleq \{w^j = (j-1) \frac{w}{K-1} ; j = 1, 2, \ldots, K\}$$
$$\ell = 1, 2, \ldots, L \tag{13}$$

For convenience, we set

$$b_{T_\ell} = \frac{w}{K-1} K_\ell^- , \qquad a_{T_\ell} = \frac{w}{K-1} K_\ell^+ \tag{14}$$

where $K_\ell^-$ and $K_\ell^+$ are positive integers.

In order to derive the functional recurrence formula for the optimal objective function, we introduce

$$F_\ell (k) \triangleq \min \{\Sigma_{i=1}^{\ell} z_i(w_i) \mid w_i \in W , \Sigma_{i=1}^{\ell} w_i = \frac{w}{K-1}(k-1)\}$$
$$K_\ell^- + 1 \leqq k \leqq k_\ell^+ + 1 \quad ; \quad \ell = 1, 2, \ldots, L \tag{15}$$

Accordingly, $F_L(K)$ is the optimal objective value to be determined. From the principle of optimality, the following relations are obtained:

$$F_1(k) = z(w_1^k) \qquad \text{for } K_1^- + 1 \leq k \leq K_1^+ + 1 \tag{16}$$

$$F_\ell(k) = \min_{i \in I_{\ell-1}} [F_{\ell-1}(k-i) + z_\ell(w_\ell^i)] \tag{17}$$

$$\text{for } K_\ell^- + 1 \leq k \leq K_\ell^+ + 1 \quad \text{and} \quad \ell = 2, 3, \ldots, L$$

where

$$I_{\ell-1} \triangleq \{i \mid i \in \{1, 2, \ldots, K\},$$
$$k-i \in \{K_{\ell-1}^- + 1, K_{\ell-1}^- + 2, \ldots, K_{\ell-1}^+ + 1\}\} \tag{18}$$

In order to calculate (16) and (17), we need to seek the value of $z_\ell(w_\ell)$ for various values of $w_\ell$. That is to say, we have to solve P2 many times.
A greedy method modified suitably for the present problem is proposed to obtain a suboptimal solution quickly. The procedure is as follows:

The problem P2 has no constraints representing an interrelation among $u_i$ for $i \in \tau(\ell)$. Then we assign the value of $u_i$ independently so as to minimize the objective function. Consequently,

$$u_i = \begin{cases} 0 & \text{for } c_i < d \text{ or for } c_i > d \text{ and } \alpha v_i + f_i < u_{min} \\ \min\{u_{max}, \alpha v_i + f_i\} & \text{otherwise} \end{cases} \tag{19}$$

Under the assumption that $u_i$ is given by (19), the following ratio is calculated for each $i \in \tau(\ell)$:

$$\gamma_i \triangleq \frac{\text{effect of } v_i \text{ to objective function}}{\text{effect of } v_i \text{ to the constraint (10)}} \tag{20}$$

which is obtained, in the case where $u_{max} - u_{min} > \alpha v_{max}$, as

i) for $c_i < d$

$$\gamma_i = \alpha c_i v_{max} \tag{21}$$

ii) for $c_i > d$ and $f_i > u_{max} - \alpha v_{max}$

$$\gamma_i = \alpha c_i v_{max} - (c_i - d) u_{max} \tag{22}$$

iii) for $c_i > d$ and $u_{min} - \alpha v_{max} < f_i < u_{max} - \alpha v_{max}$

$$\gamma_i = \alpha d v_{max} - (c_i - d) f_i \tag{23}$$

iv) for $c_i > d$ and $f_i < u_{min} - \alpha v_{max}$

$$\gamma_i = \alpha c_i v_{max} \tag{24}$$

Then, the number i is rearranged in the ascending order of $\gamma_i$. Corresponding to the new order of i, we assign

$$v_i = \begin{cases} v_{max} & \text{for } i = 1, 2, \ldots, i^*-1 \\ (w_\ell - \Sigma_{i=1}^{i^*-1} \Delta t_i v_{max}) / \Delta t_{i^*} & \text{for } i = i^* \\ 0 & \text{for } i = i^*+1, \ldots, T_\ell - T_{\ell-1} \end{cases} \quad (25)$$

where $i^*$ is the number satisfying the following relation:

$$\Sigma_{i=1}^{i^*-1} v_{max} \Delta t_i \leqq w_\ell \quad \text{and} \quad \Sigma_{i=1}^{i^*} v_{max} \Delta t_i > w_\ell$$

Consequently, a suboptimal solution of P2 is obtained as (19) and (25).

## 4. Numerical Result

As a case study, we consider an optimal operation planning of a production mill at cement works. A production management system has been constructed in a cement works[4]. The purpose of the system is to make an optimal operation planning for the production process so as to minimize electric energy consumption taking into consideration of the constraints for the equipments and the preassigned production plan. Since the production process in cement works includes large silos, an effective utilization of silos is important for long range planning. Consequently, the production management problem becomes large scale. The process is now working under an operation planning whose algorithm is based on a heuristic method. By applying the proposed algorithm to this problem, we aim at showing an improvement of the existing algorithm.

In the present case study, the product A corresponds to clinker, and the product B corresponds to cement. The purchase price of electric power varies three times on a weekday. It varies once on Sunday. Then, the total number T of time intervals is more than one hundred when we consider the production planning for one month.

As an example, we set the following values of the problem parameters:

$$v_{max} = 612\text{x}10^3 \text{ kg/h}, \quad u_{min} = 24 \text{ MWh/h}, \quad u_{max} = 56 \text{ MWh/h}$$

$$\alpha = 37\text{x}10^{-6} \text{ MWh/kg}, \quad d = 4.7\text{x}10^3 \text{ ¥/MWh}$$

The other parameters are changed corresponding to several cases. First of all, we decide an appropriate value for K, i.e. the number of the discrete values for w. Table 1 shows the influence of K on the objective value and the computation time obtained for the case of applying the proposed algorithm to a typical example. The result obtained by the branch and bound method is also shown in the table. The program code of the branch and bound method utilized is MPS/X which is provided by Fujitsu Ltd. at the Data Processing Center of Kyoto University. The optimal solution is obtained by the branch and bound method, but it requires much computation time. On the contrary, the proposed algorithm gives a suboptimal solution quickly. According to Table 1, the large value of K yields an exact solution but needs much computation time. Therefor, the value of K should be decided in consideration of both accuracy and computation time. From the result of Table 1, an appropriate value is set to be K = 201.

Let us compare the results obtained by the proposed algorithm with those by the branch and bound method. Table 2 shows the computational results obtained for eight numerical problems. It is observed that the merit of the proposed algorithm is about 80% reduction of computation time and the demerit is about 1% degradation of accuracy, as compared with the branch and bound method.

Figure 2 shows the variation of the optimal silo level $H_i$ with the time i during the period of one month in the case of Problem 1, where $w = 2.28625 \times 10^8$ kg and $f_j$ = 24 MWh/h for all j. In this case, the silo level $H_i$ attains the upper or the lower bound four times. Then the problem P1 is decomposed into five subproblems as the result of convergence of the algorithm. Figure 3 shows the solution of $u_i$ and $v_i$ for Problem 1.

Table 1 Variation of the computational results with K

| | CPU time (sec) | Objective value ($10^7$¥) |
|---|---|---|
| Proposed algorithm | | |
| K = 31 | 0.07 | 2.9793 |
| K = 51 | 0.10 | 2.9470 |
| K = 101 | 0.32 | 2.9292 |
| K = 201 | 1.15 | 2.9263 |
| K = 301 | 2.52 | 2.9253 |
| K = 401 | 4.40 | 2.9249 |
| K = 601 | 6.77 | 2.9243 |
| Branch and bound method | 7.20 | 2.9123 |

Table 2 Comparison of the computational results (K = 201)

| Problem No. | CPU time | Objective value |
|---|---|---|
| 1 (T=112) | 16 | 0.48 |
| 2 (T=112) | 16 | 0.49 |
| 3 (T=112) | 16 | 0.75 |
| 4 (T=112) | 18 | 1.35 |
| 5 (T=112) | 25 | 1.16 |
| 6 (T=110) | 13 | 0.45 |
| 7 (T=116) | 4 | 0.03 |
| 8 (T= 78) | 41 | 0 |

Note: CPU time = $t_{DP}/t_{BBM} \times 100$ [%]

$t_{DP}$ ; CPU time obtained by the proposed algorithm.

$t_{BBM}$; CPU time obtained by the branch and bound method.

Objective value = $(z^* - z_{opt})/z_{opt} \times 100$ [%]

$z^*$ ; Objective value obtained by the proposed algorithm.

$z_{opt}$; Optimal objective value.

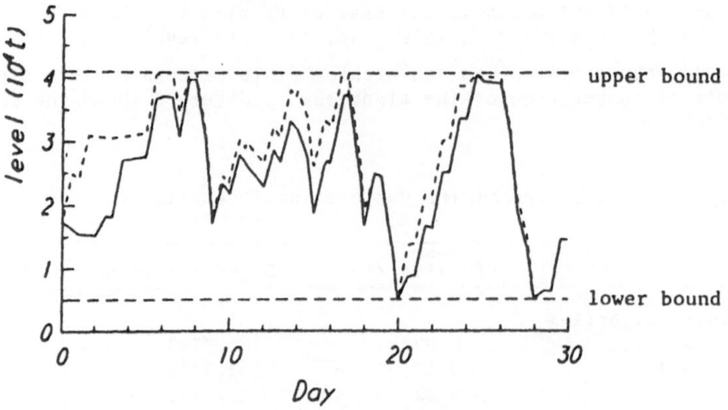


Fig. 2 The optimal silo level (Problem 1).
Solid line ; by the proposed algorithm,
Dotted line ; by the branch and bound method.

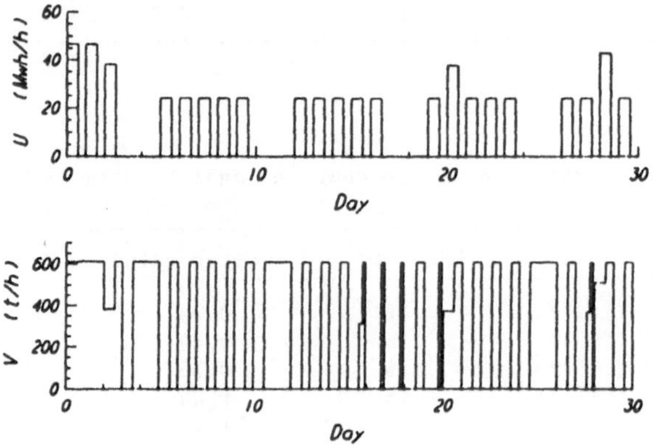


(a) by the proposed algorithm

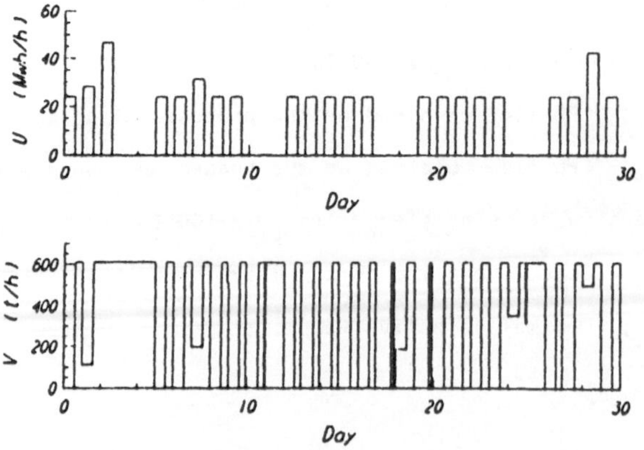


(b) by the branch and bound method

Fig. 3 The solution of Problem 1.

## 5. Conclusion

An optimal operation planning of a production process with a buffer and an electric power plant has been considered. The problem is formulated as a mixed-integer linear program with special structure. A decomposition algorithm has been proposed for solving the problem easily by using a process computer. According to the numerical results, the proposed algorithm is expected to be applicable to real situation such as real time control and planning with use of a process computer.

This work is partly supported by the Grant-in-Aid for Scientific Researches of the Ministry of Education, Science and Culture of Japan under Grant: Energy Research (1)61040031. All the numerical computations were made by FACOM M-382 at the Data Processing Center of Kyoto University.

## References

[1] N. Sannomiya and M. Tsukabe : A method of decomposing mixed-integer linear programming problems with angular structure, Int. J. Systems Science, Vol. 12, pp. 1031-1043 (1981).

[2] N. Sannomiya and K. Okamoto : A method for decomposing mixed-integer linear programs with staircase structure, Int. J. Systems Science, Vol. 16, pp 99-111 (1985).

[3] N. Sannomiya, Y. Nishikawa, H. Akagi, Y. Takeyama, T. Tsuda and T. Yashima : Optimal planning for operating an oxygen plant, Lecture Notes in Control and Information Sciences 67, Real Time Control of Large Scale Systems, Springer-Verlag, pp. 559-566 (1985).

[4] N. Sannomiya, T. Tsuda, M. Watanabe and K. Ohya : Schedule optimization package for production planning, Proc. Int. Workshop on Industrial Automation Systems, pp. 225-230 (1987).

# COMPUTER AIDED SAWMILL DESIGN AND EVALUATION

G. ENG
Applied Mathematics Division
Department of Scientific and Industrial Research
P O Box 1335, Wellington
NEW ZEALAND

**Abstract**

**This paper describes a suite of sawmill analysis models which have been developed to assist sawmillers who have to plan for changes in log supply, processing technologies, and market needs. Linear programming, probability laws, queueing and simulation techniques have been combined with sawlog geometry and machine demand calculations to form an integrated sawmill design and evaluation model. The individual models have been used independently and collectively to improve productivity in existing sawmills, help with the planning of modifications required to cope with changes in the log resource, and as an aid in the design of new sawmills.**

## 1 Introduction

### 1.1 Background

The forestry sector in New Zealand (NZ) is soon to begin an expansionary phase which will see it double its current harvest of around 10 million $m^3$ a year to 20 million $m^3$ a year by the turn of the century. Roughly half this volume will be available for primary conversion into solidwood products, particularly lumber. Since domestic demand for the predominantly (around 95%) Pinus Radiata resource will remain essentially static, virtually all of the increased output will be available for export in one form or another.

Since New Zealand will be a perfect competitor, i.e. price-taker, on the international scene, we will need to supply our exports as efficiently as possible.

Toward this end, the Forest Research Institute, Ministry of Forestry set up a Conversion Planning Project Team [1], which between 1983 and 1986 has developed a modelling system to assist in the management and utilization of the 'new crop' resource. The Sawmill Design and Evaluation suite of models is a component of the modelling system.

### 1.2 Why Our Own Models?

The main reason for developing our own models is that others are not (ideally) suited to New Zealand's unique conditions. Many are not comprehensive enough. Others are simply not available. Most overseas models cannot handle:

i. The range of timber grades obtained from radiata pine.
ii. The complex sawmill layouts required to process the range of log diameters and log qualities available from the new crop radiata pine.

## 2 Purpose

Sawmill analysis models are required because:

1. Mill Studies
   - are limited to locally available technologies.
   - are labour intensive, time-consuming and hence, expensive.
   - may interfere with normal production.
   - may not be futuristic.
2. A large number of interacting variables are difficult to investigate under operating conditions.
3. Changes in the log resource, processing technologies and markets together require an integrated investigative approach.
4. A modelling approach provides insight through repeated experiments based on numerous assumptions without the need for extensive 'live' observation and analysis.

[1] The author was on secondment to FRI from February 1985 until April 1986

# 3 The Model Framework

There are three, essentially separate, components in the sawmilling analysis framework,

- the determination of sawpatterns,
- the selection of sawpatterns, and
- the processing of selected sawpatterns (logs) through a mill.

Figure 1 summarizes the overall model's components and shows how they are linked to form the sawmill design and evaluation system.

## 3.1 Sawing Simulation

The determination of sawpatterns is analysed using a sawing (or sawlog) simulation model. The physical process of converting sawlogs into sawn timber (lumber) is influenced by sawmill type, log geometry, sawkerfs and log positioning.

A wide variety of approaches to this problem exist and are in use. The best-opening-face approach [7] is probably the best known and widely used in North America. Other sawlog simulation programs available include SAWSIM [6] and [1]. Because these are not entirely suitable for NZ conditions, their most useful features have been adapted and included in a model that meets local needs.

The model, SAWING, is an enhanced version of an earlier sawlog and sawpattern simulation program, SIMSAW [13], [14] which in turn was based on the glass-log model [11]. SAWING takes

i. sawmill machines data, e.g. number of saws, sawkerfs,
ii. log data, e.g. size distribution (stratified into log batches, log geometry, log positioning,
iii. grading rules, green sawn allowances and sawn timber prices, and
iv. 'parent' sawpatterns,

and simulates the yield from sawing the logs. The sawing simulation operates at the most disaggregated level of the sawing process, giving volumetric and grade yields on board-by-board, log-by-log, batch to batch (esp. log diameter classes) and possibly population, bases. The 'parent'

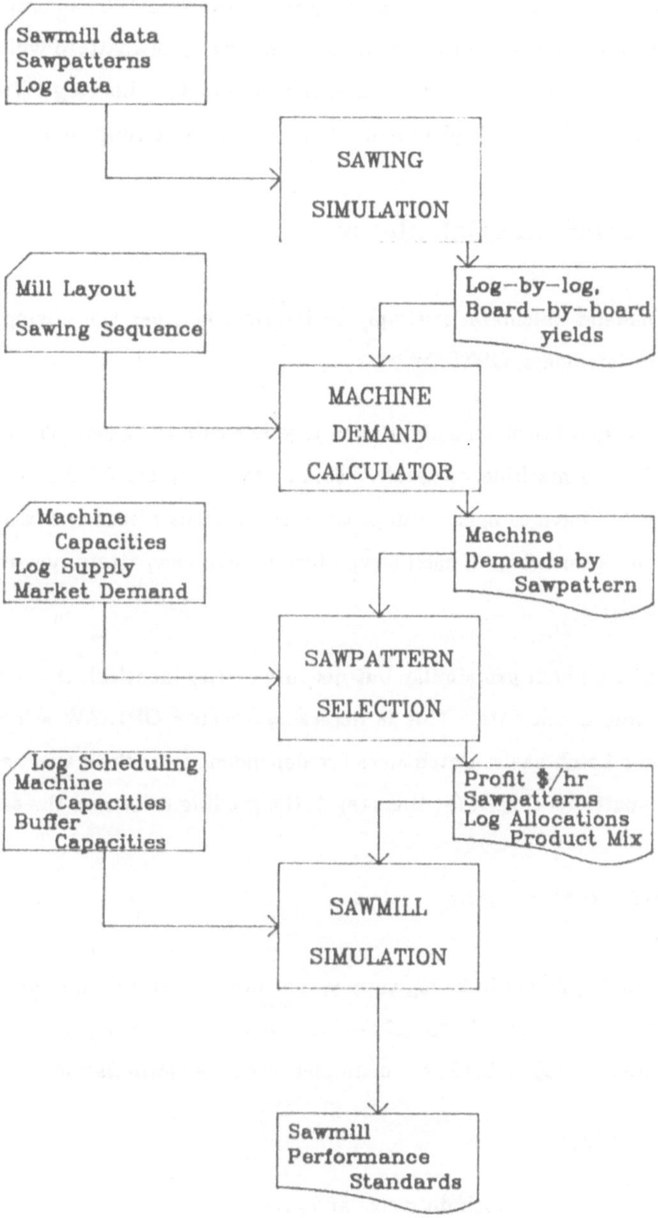


**Figure 1: Structure of the Model System**

sawpatterns are a set of feasible patterns supplied to the model. The simulation calculates the results of sawing a log to each of these sawpatterns. As any sawpattern will not be applicable to the whole range of sawlogs, the inferior ones are eliminated at this stage. The retained patterns will satisfy criteria such as high yield and robustness across a range of log sizes.

## 3.2 Machine Demand Calculator

The Machine Demand Calculator is simply an interface between the sawing simulation and the sawpattern selection model, OPTSAW.

Log-by-log output from the sawing simulation together with the sawmill's machine configuration are then used by the machine demand calculator to determine (i) the sequence in which the specified mill would have to operate to produce the various pieces of sawn timber and (ii) the machine demands (passes/log) of each sawpattern (where sawpattern now refers to a particular log size).

Given that logs in a batch are similar but not necessarily identical, the program also converts log-by-log data into batch data. This is necessary because OPTSAW selects sawpatterns and allocates logs on a batch basis. Batch sizes are dependent on such factors as the diversity of the log mix, the sawpatterns chosen (at this stage), the grading rules and the sawing system.

## 3.3 Sawpattern Selection

A linear programming model links log supply, machine capacities, and demand for sawn timber by determining the most valuable mix of products to produce from a given log supply under given market conditions. [8], [9] and [12] are examples of similar formulations.

The formulation includes:

- Log supply constraints including size and grade distribution.
- Sawn timber product specifications and constraints.
- Machines processing rates (converted to passes).
- Product recovery standards (most of the coefficients).

It has the following form:

Max (contribution by sawpatterns)

s.t.

| | | | |
|---|---|---|---|
| Logs used | $<$ | Supply | (for each log class) |
| Sawn Timber | $<, =, >$ | Demand | (for each grade and/or size) |
| Machine utilization | $<$ | Capacity | (for each machine center) |

The 'menu' of sawpatterns retained from the sawing simulation are the activities of the LP. The yield and machine usage coefficients are determined by the machine demand calculator from the SAWING results. Generally, the machines data are expressed in passes and the log supply, timber yields and timber demands in m$^3$. The most common snapshot used for sawpattern selection is an hour so that we have machine availability, etc, defined on an hourly basis.

The objective function can be defined in a number of different ways depending on the objectives of the user. If the model is to be used to investigate an existing sawmill with a long-term log supply contract, virtually all mill costs can be regarded as fixed and the appropriate objective will be to maximize revenue. If the system is to be used to evaluate a new sawmill to process graded logs, all costs may be regarded as variable. The appropriate objective will then be to maximize profit. If, on the other hand, a log sale specifies that all logs of sawlog specification are to be sold to, and sawn by one sawmiller, the cost of the logs can be viewed as a fixed cost and the objective will be to maximize contributions to fixed costs and profit.

The assumptions can have a profound impact on the sawpatterns selected and logs allocated. It should also be apparent that a judicious choice of alternative sawpatterns is important in order to provide meaningful results or to avoid having to repeat the 'optimization' exercise.

## 3.4 Sawmill Simulation

'Optimization' is a misnomer in the previous section in that the LP technique does not address aspects of sawmill operations such as the (uneven) flow of pieces between machines, in-process buffer utilization, bottlenecks between machines, random down-time occurrences and the order of log inputs, eg. random, sorted and scheduled. In fact, the sawpatterns selected by OPTSAW are only realistic if the mill has full control over log scheduling and if the in-process buffer capacities between machine centres are large enough to prevent interference. In reality, this is invariably not the case and simulation is required to reduce the optimistic production standards implied by the linear programming to more realistic levels.

For the analysis of the dynamic aspects of mill operations, simulation and related techniques are more appropriate. Models developed by Martin [10], Aune [2], [3] and Carino and Bowyer [4], [5], are examples of sawmill simulation models but which are unsuitable for our purposes.

Sawmill simulation is the last computer-based step in the full design and evaluation procedure. Simulations are run by taking a set of logs with associated sawpatterns, organizing the logs in some way (possibly, randomly) and simulating their processing sequences through the model. Note that the patterns need not be chosen by OPTSAW. In some cases experienced sawmill managers prefer to simulate their own sets of sawpatterns. In other instances the simulation of arbitrary or hypothetical sawpatterns is of interest.

The simulation model can range from one which is essentially deterministic to one which incorporates a number of stochastic elements. Currently, because of data and experience limitations, only the primary inputs' (logs) sequences can be non-deterministic. Piece flow routes are generally known as are the number and type of pieces (e.g. log, board) emanating from a process action. Machine process rates are regarded as fixed and no downtime is incorporated in the model. In other words, a sawlog has a pre-determined sawing 'recipe' as it enters the mill.

It is easy to include variable machine outputs, stochastic process (service) rates and certain types of downtime. Such enhancements would increase the robustness of simulations but would also need to be verifiable in practice. Incorporating in-mill intelligence to more accurately simulate reality or to seek improvements to prevailing real time practices is part of the ongoing process of extending the model's range of possible applications. For example, the model's management of piece flow adequately approximates reality and variations on this theme effectively seek to improve upon the existing methods of mill-floor operators or to look at alternative designs.

At this stage, the simulation model is primarily concerned with providing insight into the physical operations of a sawmill. It does not directly include any product or financial analysis. These are largely incorporated in the other modules of the overall model.

## 4 Illustration of a Sawmill

Figure 2 illustrates a sawmill according to the diagrammatic conventions used. Note that the sorters need not represent decision points as, often, different piece types exiting a machine centre have no choice where to go next. Sorters and mergers have no capacity and may only be conceptual in nature but are quite helpful in clarifying sawmill layout for modelling purposes.

The (capacitated) directed links represent conveyors or buffer areas.

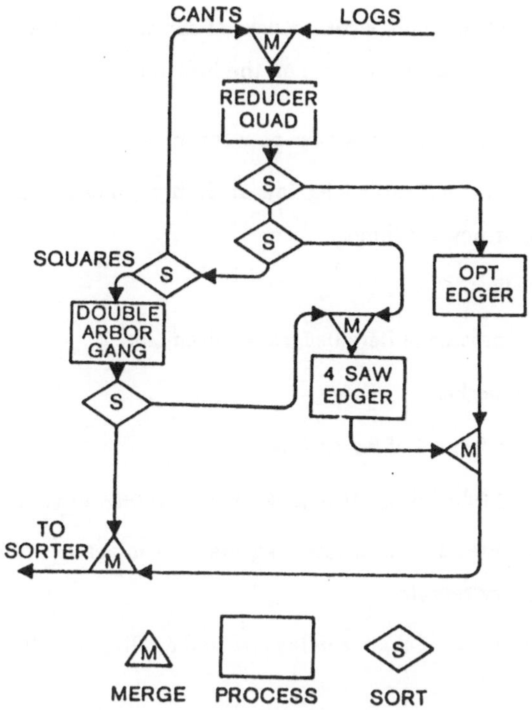


Figure 2: Reducer Quad Sawmill Flow Diagram

The reducer quad band sawmill shown in Figure 2 is a type of sawmill capable of processing logs of virtually any size. 'Unsquared' cants return to the quad reducer to be squared on their second pass. At each of the two passes, chippers flatten off opposite sides of the log and 4 boards are produced which are finished off at the optimizing edger. Logs and cants are 'minibatched' in that the reducer will alternately cut five (say) of each to fixed saw and reducer head settings for each batch. The double arbor gang saw receives 'squares' (rectangles, strictly) from the reducer and produces a number of boards which may be further sawn into 2-5 narrower boards at the 4 saw edger.

Thus, in terms of a general production process, a sawmill may be thought of as a reverse car assembly plant, i.e. a disassembly process.

# 5 Applications

The following are some of the more common uses of the sawmill models. It should be apparent that each use has varying levels of reliance on the individual models.

(a) Evaluate alternative proposals for new sawmills.

(b) Investigate mill modifications eg. layout changes, machine performance modifications, replacement or new machines.

(c) Analyse mill flow.

(d) Determine in-process buffers usages/requirements.

(e) Identify bottlenecks.

(f) Investigate the impact of log sorting.

(g) Investigate log scheduling strategies, eg. (mini) batching, scheduling, random.

(h) Examine the effects of alternative sawpattern mixes on mill performance, including volume recovery targets.

(i) Examine the effects of change in log size and quality on mill performance and products produced.

(j) Simulation reduces LP results to more realistic levels.

(k) Investigate sawpattern selection.

(l) Investigate trade-offs between value, constraints, recovery and throughput.

(m) Investigate the costs of restricting choice of log types, sawpatterns, etc.

(n) Test foreign/new technologies with local resource and product specifications.

# References

[1] Airth, J.M. and Calvert, W.W. (1973)*Computer Simulation of Log Sawing*, Dept of the Environment, Canadian Forestry Service, Ottawa, Canada.

[2] Aune, J.E. (1974) *System Simulation - A Technique for Sawmill Productivity Analyses and Designs*, The Forestry Chronicle, Vol 50, No 4, pp66-69.

[3] — (1977) *Computerized Sawmill Design – Model Versus Reality*, Forintek Canada Corp., Vancouver, Canada.

[4] Carino, H.F. and Bowyer, J.L. (1981) *Sawmill Analysis Using Queueing Theory Combined with a Direct Search Optimizing Algorithm*, Forest Products J., Vol 31, No 6, pp31-40.

[5] — (1982) *DSIM (Direct Search Minimization): A Queueing Based Interactive Computer Model for Wood Products Mill Design and Productivity Analysis*, University of Minnesota, Agricultural Experiment Station, Minnesota, USA, Technical Bulletin 334, 52pp.

[6] Carroll-Hatch Systems Ltd (1985) *Sawsim Sawmill Simulation Program*, North Vancouver, British Columbia, Canada.

[7] Hallock, H. and Lewis, D.W. (1971) *Best Opening Face Programme*, Australian Forest Industries J., November 1974, pp21-31.

[8] Jackson, N.D. and Smith, D.W. (1961) *Linear Programming in Lumber Production*, Forest Products J., Vol 11, pp272-274.

[9] McKillop W. and Hoyer-Nielson S. (1968) *Planning Sawmill Production and Inventories Using Linear Programming*, Forest Products J., Vol 18, No 5, pp83-88.

[10] Martin J. (1971) *General Purpose Sawmill Simulator for Hardwood Forests in the Northeast*, M.Sc Thesis, Sch. Forest Res. Penn. State Univ., University Park, PA, USA.

[11] Reynolds, H.W. (1970) *Sawmill Simulation*, USDA Forest Service Research Paper NE-152, Northeastern Forest Experiment Station, Upper Darby, PA., USA.

[12] Sampson and Fasick (1970) *Operations Research Application in Lumber Production*, Forest Products J., Vol 20, No 5, (April) pp12-16.

[13] Singmin, M. (1978) *A Simulation Program to Evaluate the Effect of Sawing Patterns on Log Recovery*, National Timber Research Institute, CSIR, Pretoria, South Africa, Project no TP/43289, 30pp.

[14] van Wyk, J.L. and Koekemoer, P.H. (1976) *An Investigation into the Possible Applications of SIMSAW - A Computer Programme for Sawlog Simulation and Sawpattern Evaluation*, NTRI, CSIR, Pretoria, South Africa, 51pp.

## Acknowledgement

The collaboration of Dr J Louw van Wyk, Forest Research Institute, Rotorua, New Zealand, in the development of the material of this paper is gratefully acknowledged.

# ONLINE APPLICATION OF LARGE SCALE LINEAR PROGRAMMING TO PULP AND PAPER MILL BY MINI-COMPUTER

H.Hara and K.Yamashita
Software Systems Department
Fuchu works,Toshiba Co.
Fuchu,Tokyo,Japan

T.Watanabe
Heavy Apparatus Laboratory
Fuchu works,Toshiba Co.
Fuchu,Tokyo,Japan

## 1.Introduction

This paper presents online and real-time application results of large scale linear programming (LP) to scheduling problems in actual pulp and paper mill by using a mini-computer. There exist several difficulties in applying large scale LP in real-time, such as limited computing time, memory size and power of computer available. In addition to these, a close link with the production process has to be maintained, since the problem to be solved changes at every execution in accordance with the variation in the process and external conditions. We developed a system that optimizes production schedule as well as monitoring and supervising the production process. The system has been in operation since March, 1986 and reducing the production cost by more than 1% on the average. In the followings, the outline of the system and scheduling problems are first described, then the problem formulation is given. The main part of the paper discusses difficulties we confronted and measures we took in applying large scale LP on real-time.

## 2.System configuration

The system described in this paper controls and supervises the pulping and blending process in the pulp and paper mill. The following functions are provided and realized by hundreds of real-time tasks, which are executed periodically or at request, by a mini-computer.

(1) Process monitoring and supervision
process data gathering, filing, alarm processing and user friendly man-machine communication by color graphics and printers.

(2) Pulp requirement calculation
Based on paper making schedule provided by plant management computer, the required quantity for each pulp is calculated for the peried of 3~5 days.

(3) Pulp production and blending schedule function
Optimal pulp production schedule and pulp blending schedule for 24 hours are provided every 8 hours to meet with paper making demand.

(4) Pulp bleaching schedule function
Pulp bleaching schedule is provided to minimize bleaching cost by using nonlinear programming algorithm.

(5) Quality management
Paper quality monitoring, control, and forecasting functions are provided.

(6) Production monitoring and control
Various production monitoring functions are provided, including production efficiency monitoring and production progress monotoring. Simulation function is also provided to make production adjustment easier.

## 3. Scheduling problems

In this section, the outline of the plant is first explained. Then, the summary of the scheduling problems and the problem formulation will be discussed.

### 3.1. Outline of the plant

Fig.1 shows a conceptual diagram of the pulp and paper mill. The major processes in the pulp mill are pulping, bleaching and blending. Several types of materials are used, including chips, pulp woods, used papers, and pre-processed pulps. These materials go through different type of pulping processes such as chemical, mechanical and thermomechanical. The pulp characteristics depends on the materials and type of processing. Some of the pulps are bleached to obtain required whiteness. In the blending process, several types of pulps are mixed to produce the pulp stock required by paper making machines. Pulps are blended to meet with each paper quality, including whiteness and color tone.

The buffer tank, which is called the cushion chest in Fig.1, is originally for stable operation, but plays an important role in optimizing schedule. The best use of cushion chest capacity is important to reduce production cost, since the power contract is made on time-of-use basis. In this plant, 17 types of pulps are produced and there exist 10 paper making machines. This complexity makes the mathematical programming approach worthwhile.

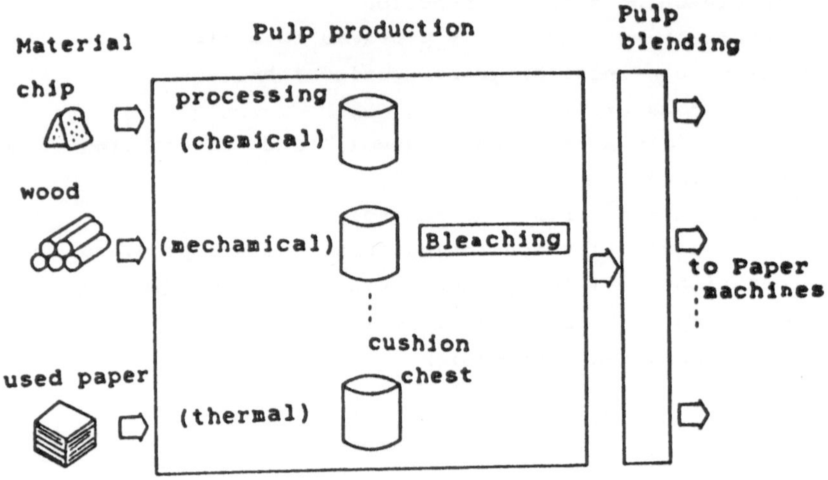


Fig 1. Paper making process

3.2.Summary of the scheduling problems

Pulp production and blending schedule can be formulated as one scheduling problem. However, it is divided into two problems: pulp production schedule problem and pulp blending schedule problem. The reason is that the interference between two schedules is found to be weak and formulation into one problem makes the problem size too big to solve on real-time. Fig.2 is a simplified diagram for pulp production and blending processes, where dividing line between production schedule and blending schedule is indicated.

(1) Pulp blending schedule problem
The problem here is to find cost-minimum pulp blending schedule of 24 hours to meet with the pulp stock requred by 10 different paper making machines. Once the paper making schedule is given, the pulp blending ratio can be calculated for each paper product. In this function,blending ratios are varied within the

prespecified limited ranges to reduce pulp cost. Mathematical programming approach is needed since the blending ratio has to be decided for every paper products which spread over the time horizen, and various restrictions, such as paper quality, pulp supply limit and pulp cost, have to be taken into consideration.

(2) Pulp production schedule problem

The cost minimum pulp production schedule of 24 hours must be obtained when the prescribed pulp blending schedule is given. The major factor in pulp production cost is the electric power cost, which varies depending on the time-of-use. Due to power supply limit and cushion chest capacity limit, the pulp production in each pulping process cannot always be maximized at cheapest power cost time zone. Hence, mathematical programming approach is employed to provide minimum cost schedule.

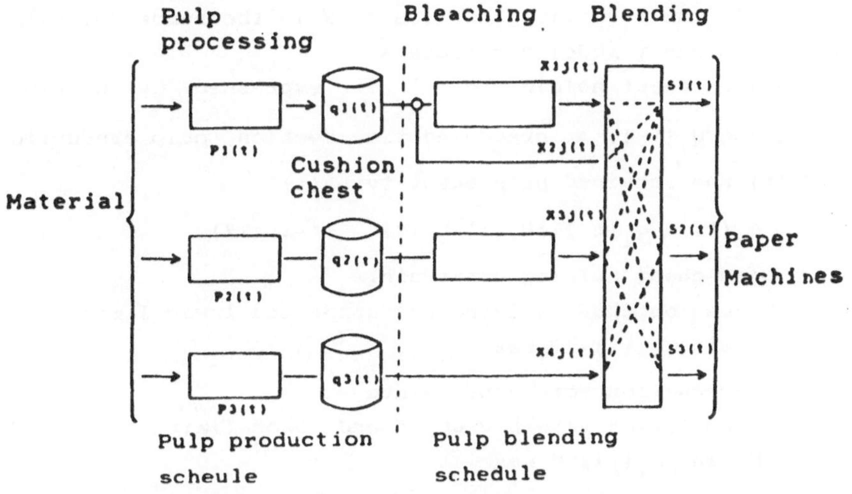


Fig 2. Simplified flow diagram of pulp mill

3.3.Problem formulation

Pulp blending and production schedule problems are formalized in linear programming framework. In the following, description will be given mainly on pulp production schedule problem, although pulp blending scheduling will also be discuseed in parallel. The pulp production schedule problem is formulated as follows.

Variables

Variables are pulp production rate and cushion chest holding at each time section which is chosen to be two hours, since pulp production rate should not be changed every one hour. For the mill in Fig.3, the elements of x are $P_i(t)(i=1, ,3;t=1, ,12)$ and $q_i(t)(i=1, ,3;t=1, ,12)$.

Objective function

The objective function is the pulp production cost in the production process of Fig 1. The coefficient vector is the pulp production cost.

Constraints

There are 4 kinds of constraints.

(1) Electric power constraints

These are inequality constraints on electric power comsumption in each time section.

$$\sum_{i=1}^{N} E_i(t) \times P_i(t) \quad Emax \quad (t= 1,\cdots,12)$$

where $E_i$ is electric power needed to produce unit ton of pulp and Emax is the electric power limit. N is the number of pulp lines.

(2) Cushion chest model constraints

Cushion chest holding $q_i(t)$ is expressed by cushion chest holding $q_i(t-1)$ at preceding time section, pulp production rate $P_i(t)$ and required pulp quantity $L_i(t)$.

$$q_i(t) = q_i(t-1)+P_i(t)-L_i(t) \quad (t=1,\cdots,12)$$

(3) Cushion chest holding constraints

Each cushion chest holding has upper and lower limit,

$$Q_i min \leqq q_i(t) \leq Q_i max$$

(4) Pulp production rate constraints

Pulp production rate has upper and lower limit,

$$P_i min \leq P_i(t) \leqq P_i max$$

In the actual pulp production scheduling problem,there are 492 variables and 624 constraints in scheduling period of 24 hours. It is noted that constraints (3) and (4) can be removed from the coefficient matrix by using bounded variable algorithm.

Both scheduling problems have a nice coefficient matrix structure as in Fig 3. By using the notation in Fig 3, the problem is expressed as follows,

$$\text{minimize} \quad Z = \sum_{i=1}^{p} c_i^T x_i$$

$$\text{subject to} \quad \sum_{i=1}^{p} A_i x_i = b_0$$

$$B_i\, x_i = b_i,\ x_i \geqq 0,\ i=1,\cdots,p$$

where p is the number of diagnal blocks in Fig.3.
Table.1 shows the problem size of two scheduling problems.

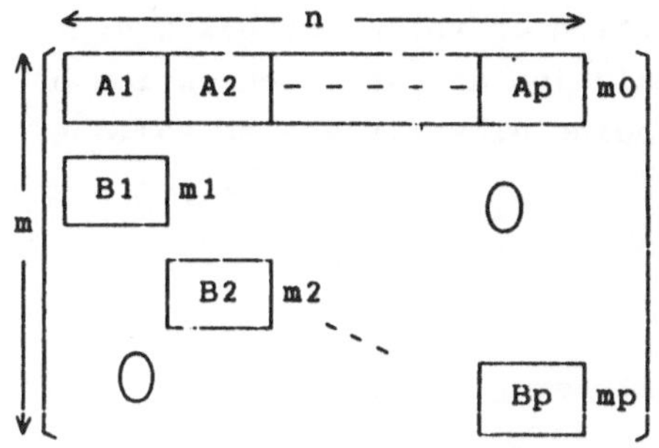


Fig 3. Structure of A matrix

Table 1 Program size

| | Pulp product. | Pulp blending |
|---|---|---|
| Variables n | 492 | 2400 |
| Constraints m | 624 | 6002 |
| Diagnal blocks p | 8 | 24 |

## 4. Real-time optimal scheduling

In this section, we discuss the difficulties we faced and the counter measures taken in applying large scale linear programming to real-time optimization of production schedule. The effect of the countermeasures is also discussed.

### 4.1. Real-time requirement

In order to realize real-time optimization of production schedule, the follwings are required.

(1) Limited computing time
Although the period of scheduling calculation is 8 hours in this system, the computing time is required to be within an hour to incorporate the latest process and external conditions into the problem formulation.

(2) Limited memory size
The minicomputer that carries out schedule optimization also

runs various other functions including process monitoring, production monitoring and quality control. These functions consist of more than 400 tasks. Consequently, efficient use of memory is inevitable, especially that of main memory. It was necessary to confine the program size well within 1 MB. It is noted that reduction of program size also contributes to the reduction of computing time.

(3) Close link with production process

The conditions, which affect production schedule, change on real-time. The up-to-date information of these changes have to be incorporated into problem formulation at every execution of optimization function. Major changing elements are as follows.

(a) paper making schedule

(b) machine maintainance schedule

(c) power supply limit

(d) material supply limit

Note that some of the above elements even affect the structure of the problem.

## 4.2.Conutermeasures

To cope with the requirements and restrictions described in the preceding section, we introduced the following measures.

(1) Dantzig-Wolfe's decomposition principle [1)]

The coefficient matrix A in the two linear programming problems has a nice block-angular structure shown in Fig 3. We applied well known decomposition principle, by which the number of rows of the A matrix can be reduced to $m_0+p$ from $m_0+\sum_{i=1}^{p} m_i$. In our problems, the reduction is 624 to 164 for pulp production schedule and 6002 to 122 for pulp blending schedule.

The Dantzig-Wolfe's algorithm employs iterative approach to solve the so-called master problem which is equivalent to the original one. p subproblems are solved to provide new basis for the master problem at each iteration. The subproblem formulation was made on the basis of time section for pulp blending problem and pulp line for pulp production problem, since we found the interferance between subproblems become less during the course of factory test.

(2) Revised simplex method with bounded variables [2)]

The revised simplex method with bounded variables is used to remove simple upper and lower limits constraints from coefficient matrix. Consequently, the dimension of the matrix and calculation time are greatly reduced. Fig.4 shows the simplified flow chart of the algorithm.

(3) Double buffering

The elements of matrices and vectors are stored in hard disc memory and transfered to the main memory when needed. The double buffering technique is applied to virtually eliminate data transfer time from problem solving. This can be done since the data transfer is done by separate processor. Fig.6 shows the concept of double buffering. Data for (i+1)-th subproblem is transfered when the i-th subproblem is being solved.

(4) Forecasting cushion chest level

The initial cushion chest level must be given in both scheduling problems as in 3.3. Initial cushion chest level is forecasted by production monitoring function by using the latest cushion chest level and production schedule. This is necessary because the scheduling function is turned on two hours before the actual scheduling period

(5) Dynamic modification of upper and lower limit

There are two types of modification. One of them is to change values of upper and lower limits to cope with external condition and process state changes.

The other is to make upper and lower limit equal to avoid problem structure change due to machine stoppage, production line stoppage and constant operation of some of the process variables.

Fig.5 shows the relation between requirements and countermeasures for real-time optimization.

4.3.Effect of the countermeasures

Table 2 shows the effect of memory reduction compared with ordinaly simplex method. The size of basis matrix and coefficient matrix is reduced by more than 90% in pulp blending schedule. The program size

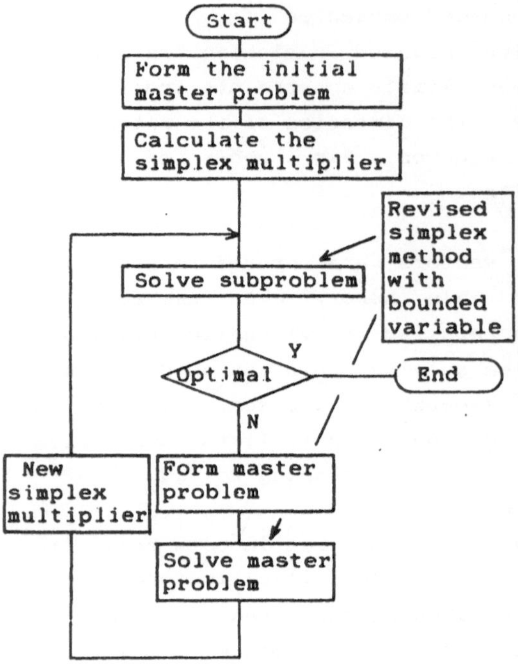


Fig 4 Simplified flow chart of the algorithm

Table 2 Reduction of memory size

| | Main memory (basis matrix) | | | Aux. memory (coef.mat.A) | | |
|---|---|---|---|---|---|---|
| | MB A | KB B | % B/A | MB A | KB B | % B/A |
| Pulp prod. sche. | 1.49 | 105 | 6.9 | 1.2 | 0.4 | 33.3 |
| Pulp blend sche. | 137 | 59 | 0.04 | 55 | 1.3 | 2.36 |

A· ordinary simplex method
B: applied method

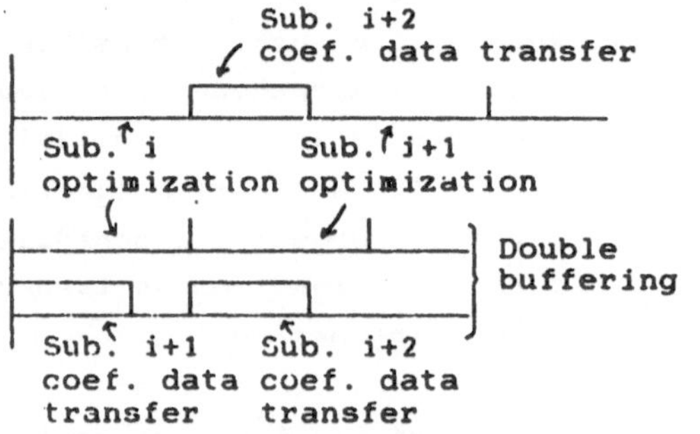


Fig 6 Concept of double buffering

<Requirements> <Countermeasures>

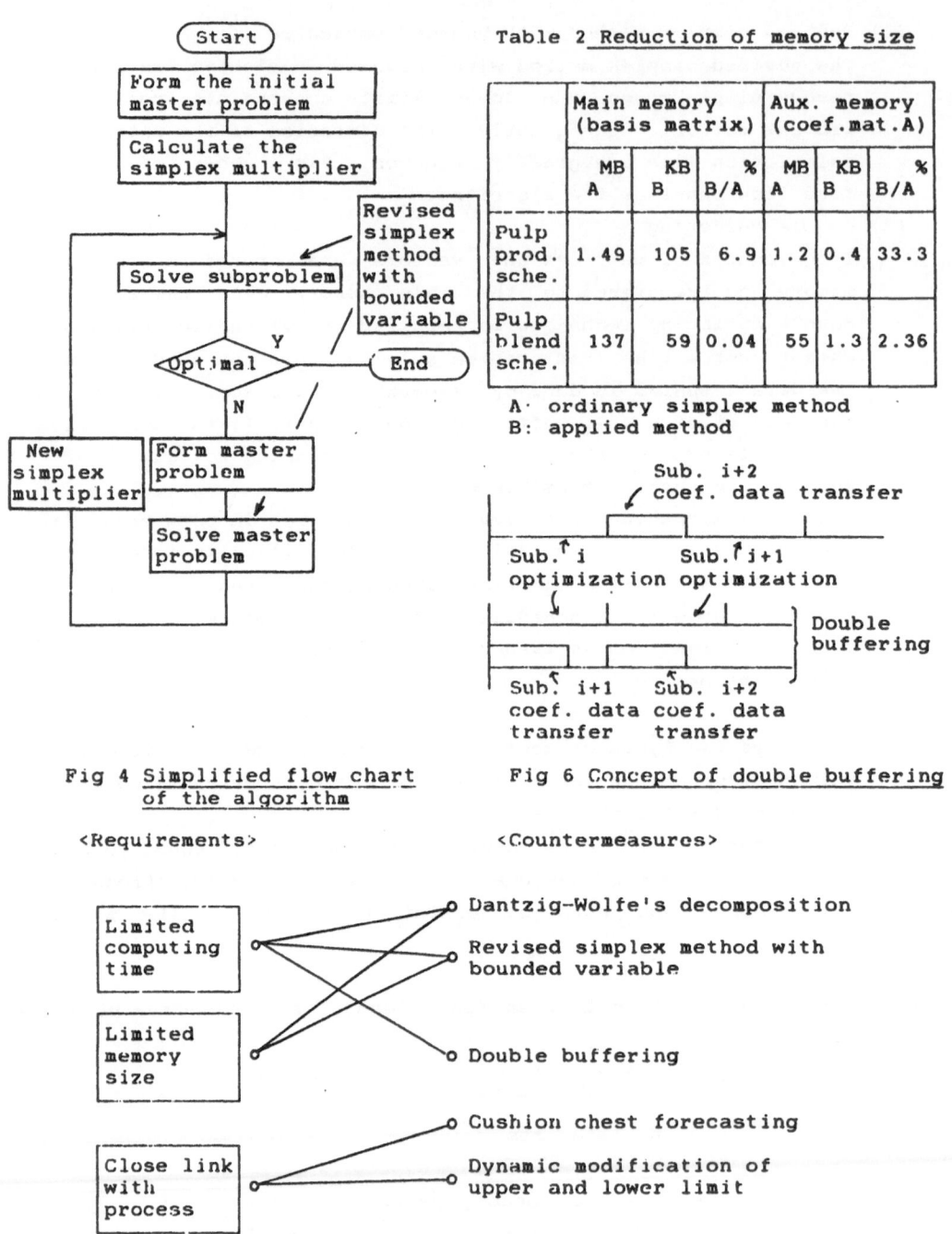

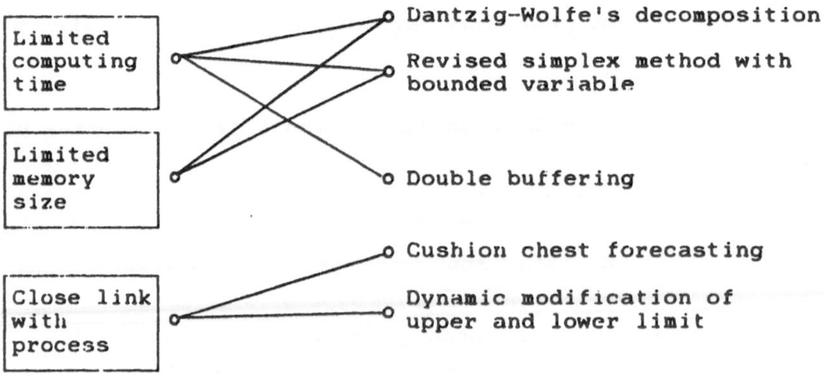


Fig 5 Requirement and measures for real-time optimization

has become under 600 KB on main memory, in both scheduling problems. Average computing time is about 5 minutes for pulp production schedule and about 30 minutes for pulp blending schedule. It is estimated that the computing time is 10 200 times more by ordinary simplex method. This argorithm also improves the computational accuracy, as the size of the problem is greatly reduced.

The scheduling function has been providing the optimal solutions in more than 90% of the executions. The reasons for unsuccessful execution are as follows. The conditions fed in do not allow feasible solution to exist. The other reason is that computing time exceeds the time limit of one hour, which seldom happens.

When the optimal solution is not found, alternative solution is used, althogh it is no longer optimal. However, providing alternative solution is inevitable to maintain continuous operation of the production process. In pulp blending schedule, the solution is given based on basic pulp blending rate and fine adjustment is made by operators. In pulp production schedule, a simple algorithm is used to provide alternative solution at the expence of optimality.

## 5.Conclusion

The paper has presented a practical application result of large scale linear programming to real-time optimal production scheduling in actual pulp and paper mill. Real-time application brings in several restrictions and requirements. The system based on a 32-bit minicomputer has been in operation since March 1986, providing optimal production schedules every 8 hours. The reduction in production cost is estimated to be more than 1% on the average. We believe that this is one of a few practical and successful real-time application cases of large scale linear programming realized on a mini-computer which also runs several hundreds real-time tasks.

## Reference


[1] Leon S. Lasdon, Optimization Theory for Large Systems, London : MacMillan, 1970

[2] Hiramoto, I., and Nagaya, A., Linear Programming, Japan : Baifukan, 1973

INVESTIGATION AND OPTIMIZATION OF RELIABILITY

OF TRANSPORT NETWORKS

Garliauskas A.. Koval V.. Vvgovskava R.*
Institute of Mathematics and Cvbernetics.
Lithuanian Academv of Sciences.
4 Academiios St.. 232012 Vilnius. USSR
*Computing Centre of the USSR Academv,
40 Vavilov St.. 117967 Moscow. USSR

Adoption of the proiect and operation decisions including providing of reliabilitv with regard for various territorial and temporary conditions and varietv of facts having influence on reliabilitv is bound up with necessity to solve series of problems.

Reliabilitv optimization problems include the choice of the rational levels of reliabilitv and of the svstem reserves. Economic effectiveness which allows to distribute the reserves among the subsystems and the elements of the svstems in the best way serves as a criterion for providing with reliability. The optimization problems are closelv associated with estimation ones. that's whv the investigation of functioning and development of various technical and economic svstems representing the net structures (transport.power.pipe.information.communication svstems. etc) is bound up with the determination of reliability of component elements - net groups and branchs. In the capacity of branch it is natural to consider power and telephone lines. main roads. pipe-lines. etc. and as net group - telephone exchanges. railway junctions.current sources and users.compressor stations. etc.

From the point of view of reliabilitv optimization and estimation transport nets are complex, capable to be in various probable defined conditions. The networks of pipes we were occupied with take the important place among such svstems. Further we shall consider range of problems connected with the foundation of gas-pipe net reliabilitv.

Gas-mains are either separated svstems of the long distant

gas transport consisting of part of pipe and compressor stations or their complex nets. The condition of the following main parameters and factors influences on gas-main reliability:

- length and diameter of the gas-pipe,number of the compress stations, number of gas pumping over aggregates and their type-dimensions on the stations;
- schemes of the equipment dispositions on the compress stations and presence of the crosspieces on the line part of gas-pipe;
- irregularity of the technological process of the gas transport with regard for underground repository activities;
- increasment of gas-main productivity at the expense of forced work regime of the compress stations.

Investigation of reliability of gas-mains and their nets is best to do in accordance with following stages:

1. Estimation of reliability parameters of the main equipment (elements): pipes proper, compress aggregates, etc. Choice of models and calculation of distribution of state probabilities for pipe parts and compress stations (links).

2. Calculation of reliability optimum indices for the gas-main part including a few compress stations and a few pipe parts between them.

3. Foundation of reliability optimum levels of the complex gas-pipe nets.

4. Complex investigation of the optimization problems of currents,of irregularity of gas-use, of resources estimation and of reliability foundation problems.

In this work for the gas-main part reliability estimation the initial reliability indices are changing both in the course of the parts length (depending on pressure change, local conditions and other factors) and over the separate gas- pipes which may be projected from various diameters, pipe walls thichness, may differ in constructive peculiarity of pipe location, metal solidity, etc. So, for example, failure rait will be larger on the districts where the gas pressure is higher than mean, on the bending places, when crossing the water and marsh-ridden places, in everfrozen soils, etc.

The compress stations are represented as parallel-successive structures of the elements (aggregates) - working and reserve. Reserve is cold, i.e. the elements are not loaded.

Defining the optimum reliability for the gas transport systems it is expedient to use a maximum of the transport sold gas income as a criterion of the variant estimations. Then reliability optimization problem is formulated in the following way: with given reliability $\pi^*$ and fixed planned productivity $Q_0$ it is necessary to distribute reserve elements in such a way and to define such constructive and conditional parameters that the expectation of the maximum of the transport sold gas income would be provided:

$$M\{\Delta\} = \max_{\substack{R\in\Omega_R,\, r\in\Omega_r,\\ P\in\Omega_P}} [KM\{Q_f\} - CQ_0 - C_j M\{Q_0 - Q_f\} - C_R M\{Q_0 - Q_f\}] \quad (1)$$

where $K$ - realization cost of outcome unit; $M\{Q_f\}$, $M\{Q_0 - Q_f\}$ - mathematical expectations of actual productivity and of not given gas; $Q_0$ - planned productivity; C, $C_R$ - specific bringing expenditures and expenditures in reserve; $C_j$ - specific damage for a user because of deficiency of gas; R, r, P - reserve, conditional and constructive parameters; $\Omega_R$, $\Omega_r$, $\Omega_P$ - permissible sets of the corresponding parameters.

In such statement one can change reliability level $\pi^*$ in the direction of its decrease in range $\pi_0 \le \pi^* \le \pi_K$ . It is necessary for joint reliability optimization problem of the separate branches with the total problem for the net in whole.

Using estimation criterion according to (1) the solution of the total problem of the defining reliability optimum level is expedient to fulfill in two stages. On the first stage it is necessary to solve reliability optimization problem for gas-main elements and links, then on the second stage it is necessary to carry out the search for reliability optimum level of the gas transport system in whole.

For solution of the problem (1) in general, i.e. when reliability parameters change into the space and in time, it is necessary to build suitable calculation procedures in order to find state probabilities. Then for the separate states it is necessary to carry out hydraulic calculation and find quantities $M(Q_f)$, $M(Q_0 - Q_f)$. The first problem as well as the second one are complex enough and difficult to realize without definite device.

For building suitable calculation procedures with the purpose to find state probabilities the part consisting of a few compress stations and a few multi-thread parts between them was considered.

Every multi-thread part is divided into $K \cdot m$ links, links are joined in $m$ section in K's (fig.1). Refusals and repair time of every link are not dependent on refusals and repair time of

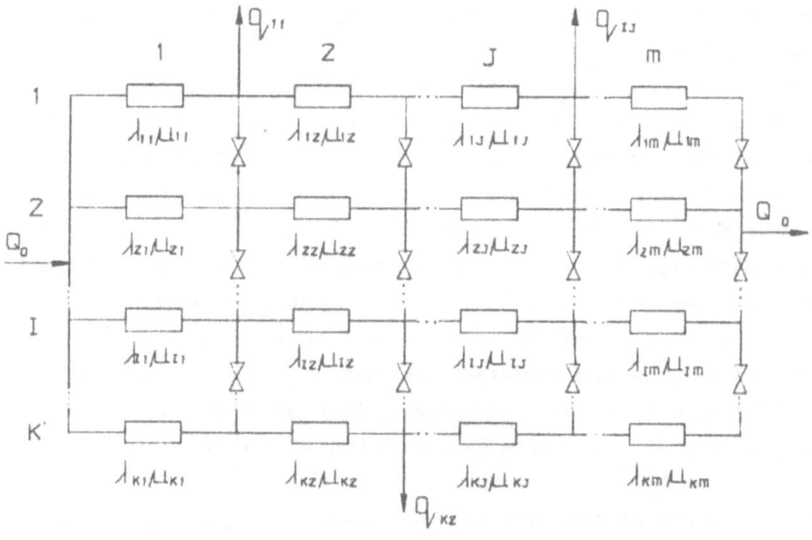

FIG.1. SCHEME OF THE MULTI-THREAD PART OF THE GAS-PIPE WITH THE VARIABLE FAILURE AND REPAIR RATES.

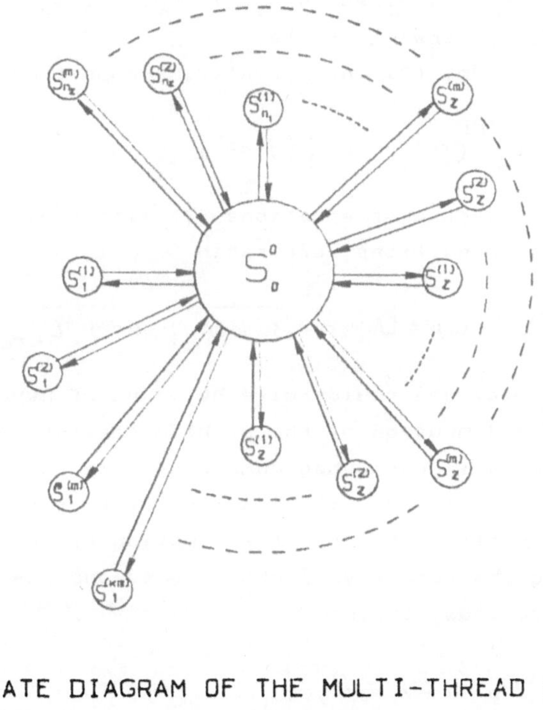

FIG.2. STATE DIAGRAM OF THE MULTI-THREAD PART OF THE GAS-PIPE .

successive and parallel joint links. Refusals and repair time streams are markov.

In order to present all possible states of the part the special diagram was built (fig.2) where the state $S_o^o$, i.e. all the links work, is represented in the centre. The first ring corresponds to states $S_r^{(1)}$ when only one link isn't working, the second - $S_r^{(2)}$ - two links aren't working and so on till $S_r^{(k\cdot m)}$ when all the links aren't working. It was supposed that only one repair team worked all the time. For this for every change from one state to another the generalized refusal and repair intensities were found using following method: the generalized intensities of two and more refusals and repair are considered as mean refusal and repair intensities of the system consisting of those two and more different elements [1].

Then in the case of uniform Markov chain the state probabilities are found according to Kolmogorov-Chepmen equations:

$$\frac{dP_z^{(i)}(t)}{dt} = -\mu_z^{(i)} P_z^{(i)}(t) + \lambda_z^{(i)} P_o(t) \tag{2}$$

where $i = \overline{1, k\cdot m}$, $z = n_1, \ldots, n_{km}$, where $n_1 = \overline{1, C_{km}^1}, \ldots, n_{km} = \overline{1, C_{k\cdot m}^{km}}$, $P_z^{(i)}(t)$ - the probability of the $S_z^{(i)}$ state.

Besides the equations (2) the normalizing condition was used

$$\sum_{r=1}^{k\cdot m} \sum_{z=1}^{n_r} P_z^{(r)}(t) + P_o(t) = 1 \tag{3}$$

For solution the system of equations (2) with regard for (3) the following initial conditions were taken

$$P_o(0) = 1, \; P_z^{(r)}(0) = 0, \; r = \overline{1, k\cdot m}\; ; \; z = \overline{1, C_{k\cdot m}^r} \tag{4}$$

The problem (2)-(4) one could solve by means of numeral methods but the large dimension of the problem creates serious troubles and expecially in case when number of such complex parts reaches ten and more. That's why it is necessary to resort to possible devices. The first device was assumption of the ordinarity condition along the succesive links. The second one was the stational conditions. Thus, taking

$$\lim_{t \to \infty} P_z^{(r)}(t) = P_z^{(r)} \tag{5}$$

the right parts of system (2) are equal to zero, and from system

(2) together with the condition (5) we shall obtain algebraical linear equation system. But even in this case it is impossible to calculate all the possible state probabilities in acceptable interval of time for long gas-pipes. That's why the way of the further simplification is the way of the problem decomposition. For that all the gas-pipe was devided into N sections. A part of section was either parallel links within the one succesive link of the gas-pipe part or compress station.

The procedure way was following. The system was devided into sections. The state probabilities and productivities corresponding to them were found for every section. Then the succesive decomposition of the distribution laws and states grouping to cut their number were made. And so step by step the probabilities of all sections were calculated. At the same time the hydraulic calculation of productivity and selection of the versions according to criterion (1) with regard for given (fixed) level of reliability $\pi^*$ must be done.

Before such procedure realization we proved the correspondence between defineded state probabilities with regard for decomposition and state probabilities without regard for decomposition which is possible with the accepted assumption of the ordinarity condition along the succesive links that from point of view of practice is possible. It was made by the following theoreme [2]:

Theoreme. Distribution of the state probabilities of net part with $\kappa$ parallel and $m$ succesive links with refusal and repair intensities changing in space equals to distribution of probabilities obtained on the basis of decomposition of the net into $m$ section with $\kappa$ parallel links with following sections distribution state probabilities composition under condition of ordinarity along the succesive links to within members of higher order minutes.

Thus, states probabilities were found for every section on the basis of formulas (2)-(4) in the stationary case, and according to hydraulic calculations finding productivities of the appropriate states the distribution laws were formulated.

Assuming the condition of the independent random streams composition of distribution laws was fulfilled

$$P_j'(Q_j') = \int_0^{\infty} P_j(Q_j)\, P_{j+1}(Q_j', Q_j)\, dQ_j \qquad (6)$$

where the stream combination was accomplished according to this logical rule

$$Q_j' = \sigma Q_j + \delta Q_{j+1} = \begin{cases} \sigma = 1, \delta = 0 & \text{when } Q_j \le Q_{j+1} \\ \sigma = 0, \delta = 1 & \text{when } Q_j > Q_{j+1} \end{cases} \quad (7)$$

The smaller stream was the determining one in the succesive links.

Further, in order to dicrease number of the discrete states the states were joined into groups according to closeness degree of the productivity values.

In the analogous way the compress station section was joined in instead of the next one. Contrary to other sections the state probabilities in it were determined according to classic scheme of the reliability theory with regard to reserve aggregates [3] and productivities - according to the special algorithm of the compress station regime calculation. After defining the distribution of probabilities of the whole gas-pipe $P_K(Q_K)$ we found

$$M\{Q_t\} = \sum_{K=1}^{M} P_K(Q_K) \times Q_K \quad (8)$$

where M - number of states,

and also

$$M\{Q_0 - Q_t\} = Q_0 - M\{Q_t\} \quad (9)$$

Then defining reliability by way of productivities ratio we examined the condition

$$\pi = \frac{M\{Q_t\}}{Q_0} \geq \pi^* \quad (10)$$

where $\pi^*$ - fixed level of reliability.

Varing reserves on the compress station we found the set of gas-pipe permissible realizations with appropriate values of the criterion (1) $\Omega(\pi \geq \pi^* : M\{A_\pi\})$ . Further sorting out $M\{D\}$ the expectation of the maximum of the income was found

$$M\{A\} = \max_{var\ \sigma \in \Omega} M\{A_\pi\} \quad (11)$$

where $\sigma$ - gas-pipe variants.

Changing a fixed level of reliability $\pi^*$ ranging from $\pi_0$ to $\pi_K$ and every time defining incomes (11), equivalent reliability characteristics necessary for the optimization of reliability of transport net in whole are constructed.

Pushing away from reliability model of the gas main part consisting of a few compress stations and line parts, one can turn to reliability optimization of the complex transport net. The maximum of the expectation of the transport product income remains as an optimization criterion. The optimization way is proposed to be following.

The project or real gas-pipe system is divided into parts which will be considered as net branches. The optimization problem (1) involving first of all procedure of succesive calculation of the part state probabilities according to productivity $P_K^{(i)}(Q_K^{(i)})$, $i=\overline{1,N}$ - the number of branch, $K=\overline{1,M^{(i)}}$ - the number of state, $M^{(i)}$ - number of states of i-th part, is solved for every part. As the part may be of a different structure so this structure type must be taking into account while choosing the composition of the state probabilities distributions. Then the constructive and conditional parameters of the part are defined, the reserve elements are distributed to provide the expectation of the maximum of the transport product income by the variation method. Planned productivity $Q_0$ and reliability level $\pi^*$ which must be provided are considered to be given.

In order to find the expectation of the maximum productivity it is necessary to present the considered net in every possible state. Because of the great number of every possible state of separate parts as well as of the whole gas-pipe system it is expedient to use the method of statistical trials for imitation of part states. However, in our case, when the probabilities of refusals of compress stations in whole as well as of gas-pipe parts are rather small the use of Monte-Karlo method does't give an expected result because all the states can't be realized in acceptable number of trials. Otherwise the increase of imitation number will lead to increase of the machine time and impossibility to solve the given problem.

That's why it is expedient to carry out the purposeful succesive net branch sorting out with refusal imitation on them. For this for every system realization the maximum flow (productivity) is counted by Ford-Fulkerson method.

As a result we receive

$$MQ_{max} = \frac{1}{n} \sum_{i=1}^{n} \left( \sum_{K=1}^{m_i} P_i^{(K)} Q_{i_{max}}^{(K)} \right) \tag{12}$$

where $MQ_{max}$ - the expectation of the maximum productivity;
$P_i^{(K)}$ - the probability of the K-th state of the i-th branch;
$Q_{i_{max}}^{(K)}$ - the maximum system flow in the K-th state of the i-th branch.

Here n is number of branchs in the net. $m_i$ - number of the i-th branch state.

Formula (12) is rightful with normalizing condition

$$\sum_{k=1}^{m_i} P_i^{(k)} = 1 \qquad \text{for every } i \qquad (13)$$

Then defining reliability by ratio of the productivities we examine the condition of satisfaction of the given reliability

$$\pi = \frac{M Q_{max}}{Q_o} \geq \pi^* \qquad (14)$$

where $\pi^*$- fixed level of reliability.
$Q_o$ - planned productivity.

The above-stated method was approved basing reliability of gas-main net of the Lithuanian SSR which out of all Baltic Soviet republics in respect of length development of gas-mains and of used gas firing volume is the most developed.

## Conclusions

General reliability optimization problem on a maximum of the realization for gas-main with the variable failure and repair rates is formulated. The states of the gas-pipe part are presented by means of the special diagram. The state probabilities are proposed to calculate on the basis of the section decomposition. The algorithmic realization way of the formulated problem is given, on the basis of which the expectation of the maximum of transport net income is found.

## References


1. Billinton R.,Alam M.Effect of restricted repair on system reliability indeces. - IEEE.Trans. Reliability, vol R-27, 1978 Decem. pp. 376-379.

2.Koval V.Reliability optimization of a net part with varing indices of intensivities. - Dep. in Lith.NIINTI Novem. 24.1984, N 1324 Li-84 Dep.

3. Volskij E.,Garliauskas A.,Gerchikov S.Reliability and optimum reservation of gas-mines and gas-mains. - 'Nedra',Moscow, 1980.

## AN OPTIMAL SOLUTION OF THE ASSEMBLY LINE BALANCING PROBLEM WITH THE BALANCING INDEX AS A PRIMARY OBJECTIVE FUNCTION

Shigeji Miyazaki and Hiroshi Ohta
Dep. of Indus. Engr., Univ. of Osaka Prefecture
Sakai, Osaka 591 JAPAN

### 1. INTRODUCTION

Much research has been devoted to the assembly line balancing problems since Salveson [11] and Jackson [7]. The majority of previous research has stressed developing more efficient algorithms or extending manufacturing environments to various situations such as assembly lines with stochastic operation times, with paralleling of work stations, mixed models assembly, and assigning work elements to particular types of stations.

In contrast to this, the objectives used in problem formulations have mainly concentrated on minimizing the number of work stations under the constraint that the upper limit of cycle time is given. A few papers [1, 7, 14] have dealt with minimizing the cycle time for a given number of stations. Recently, however, the minimization of total costs composed of facility layout costs, in-process inventory costs, labour costs, etc. has become an objective for line balancing problems [9, 10, 13]. Okamura, Yamashina and Ohno [8] have formulated a problem to improve the smoothness of station times after minimizing the number of work stations in a mixed models assembly line.

This basic and most frequently used objective, "minimizing the number of work stations," does not necessarily attain the best balance of station times, which is considered to be the real sense of line balancing problems. Moodie and Young [7] have presented a heuristic algorithm composed of two phases: the first phase is to solve the minimization of the number of work stations under an upper limit of cycle time and the second phase is to reallocate work elements to improve the smoothness of station times.

In this paper the smoothness index of Moodie and Young is modified to apply to a main objective function. The solution based on the proposed index (called the balancing index) can approach both the best balance of station times as well as the minimum number of stations in a single phase. The optimal algorithm to minimize the index is presented through Branch and Bound approach based on an investigation into optimal algorithms for the previous problem. The following numerical example shows the difference between solutions by the conventional and proposed formulations for line balancing problems.

## 2. PROBLEM FORMULATION

The basic assembly line balancing problem can be formulated as:

**[Problem 1]**

Min. $K$

sub. to

1) All the precedence relations between work elements should be satisfied.

2) $$\sum_{i=1}^{n_k} t_i = p_k \leq H/R, \quad k = 1,2,\ldots,K.$$

3) $$\sum_{k=1}^{K} \sum_{i=1}^{n_k} t_i = \sum_{k=1}^{K} p_k = T.$$

where $K$ : the number of work stations

$t_i$ : the operation time of i-th element asssigned to each station

$n_k$ : the number of elements assigned to the k-th station

$p_k$ : k-th station time (the total of operation times assigned to k-th station)

$H$ : the length of production period available for the product

$R$ : the quantity of the product to be produced during period $H$

$H/R$ : the upper limit of station times (= the upper limit of cycle time)

$T$ : the total operation time to complete the product

The minimization of total idle time, the minimization of total balance delay and the maximization of efficiency on line can all be entered into Problem 1.

Each station time of the optimal solution through Problem 1 will satisfy the restriction 2), yet the best balance of station time is not necessarily realized. Therefore, the maximum station time (the cycle time) of the optimal solution occasionally has room for reduction, while the number of stations is maintained. The incompleteness stated above can be overcome so that several alternative solutions of Problem 1 are produced under various upper limits of station times (H/R) and the best solution selected from among them. Some difficulty may be encountered in performing such multiple solution processes, since computer time is a major restriction in solving a combinatorial problem like line balancing. In addition to this, a part of the attempts to obtain an alternative solution might fail, since the feasible lower limit of station time can not be anticipated until the solution procedure has been performed.

The method proposed by Moodie and Young [7] is composed of two phases: the first phase is to obtain an approximate optimal solution of Problem 1 and the second phase is to heuristically reallocate the work elements of the solution of Problem 1 so as to improve the smoothness index SI defined by:

$$SI = \sqrt{\sum_{k=1}^{K} (p_{max} - p_k)^2} \quad ; \; p_{max} = \max_{1 \le k \le K} p_k. \tag{1}$$

The smoothness index works to minimize the maximum station time (equivalently for balancing the station times), which is a secondary objective. However, it is impossible for the smoothness index to be a main objective function, since it contains two unknown variables K and $p_{max}$.

This paper introduces a function given by:

$$BI = \sqrt{\sum_{k=1}^{K} (H/R - p_k)^2}, \tag{2}$$

where H/R is substituted with $p_{max}$ in the smoothness index and H/R is known a priori. This function (by eq. (2)) will be called the balancing index. A problem can be formulated using the balancing index as:

**[Problem 2]**

Min. BI

sub. to the restrictions 1), 2) and 3) in Problem 1.

The solution of Problem 2 can approach the minimum number of stations (objective of Problem 1) and the best balance of station times (objective of phase 2 by Moodie and Young), simultaneously.

## 3. BRANCH AND BOUND APPROACH FOR PROBLEM 2

Van Assche and Herroelen [15] proposed a Branch and Bound method for an optimal solution of Problem 1, and reported an average computational time of 37.5 seconds on the IBM 370/158 on seventeen problems composed of 70 elements each. Johnson [3] has made a computational comparison between six algorithms (including Johnson's algorithm) for Problem 1. The results show that Johnson's Branch and Bound approach succeeds in solving a 40-element problem with approximately four elements per station within a computational time of 20 seconds on the IBM 360/91. As Johnson [4] has pointed out, these two approaches demonstrate significant breakthroughs for efficiency of the optimal solution for line balancing problems.

Kao and Queyranne [5] remarkably improved the computational performance of Dynamic Programming (DP) approaches to Problem 1; however, the DP approach can hardly be adapted to Problem 2, since a feasible set of elements does not uniquely correspond to a state in DP formulation for this problem. A feasible set is defined as a subset of elements such that all the predecessors of every element in S are also in S.

On the basis of the discussion above, a Branch and Bound (B & B) approach is employed for an optimal solution for Problem 2. For the B & B approach to Problem 2, let r denote the branching level corresponding to the number of stations to which the work elements have been assigned. A node $N_r$ represents the assignment of work elements to the r-th work station. The assignment of elements to each station preceding the r-th station can be retrieved by tracing the nodes along with a branching path from the initial node $N_0$ to the current $N_r$.

A branching procedure (generation of nodes) can be performed by choosing feasible sets having total element times less than or equal to the upper limit of station times, H/R. Therefore, a systematic procedure for listing all the feasible sets at each stage is desirable in order to realize the effective branching. The method for generating these sets will be discussed in the next section.

### 3.1 Generation of Feasible Sets

Schrage and Baker [12] and Lawler [6] have presented effective procedures for generating feasible sets of jobs with precedence relations for scheduling problems. Kao and Queyranne [5] have adapted these to a DP approach for an assembly line balancing problem [Problem 1]. The adaptation to Schrage and Baker is called lexicographic order set generation and Lawler's method is called cardinality order set generation. Kao and Queyranne [5] compared DP algorithms employing these two procedures and label order set generation in terms of processing time and storage requirement on a computer. It has been concluded that, of the three, Lawler's approach is the best.

As Lawler has pointed out [6, p.7], the data concerning all the feasible sets of size m must be outputed on a hard copy or a peripheral device. The core memory for them must be abandon as soon as all the feasible sets of size m+1 have been listed in order to maintain the effective calculations. Label order set generation also requires the same procedure.

These authors prefer the lexicographic order set generation (LEXO), since the B & B approach should maintain the core memory for all the feasible sets corresponding to nodes until the optimal solution is obtained. LEXO for the B & B approach requires core memory spaces bounded by a function of the total number of feasible sets for the initial node, although LEXO for DP approach is obliged to use some wasteful spaces.

The detail steps of LEXO algorithm can be well documented in the literature [5]. A numerical example illustrated in Fig. 1 (Example 1) has been solved by LEXO and the list of all the feasible sets for the initial node $N_0$ is shown in Table 1. Fig. 1 shows the precedence relations among the elements $O_i$ (i=1,2,...,6), and the element times. The upper limit of cycle time is H/R = 9.

### 3.2 Branching Procedure

A method for branching a node (generation of new nodes) can be proposed on the basis of the feasible sets generated by LEXO. The nodes to be generated must correspond to the feasible sets of work elements which are remaining at the current stage and have a total operation time less than or equal to H/R. For Example 1 the initial node $N_0$ (branching level 0) representing the empty work station can be branched into $\{O_1\}$, $\{O_1,O_2\}$, and $\{O_1,O_3\}$, since the total operation time of these feasible sets are less than or equal to H/R = 9 as shown in Table 1.

In order to branch the node $N_1 = \{O_1,O_3\}$ (branching level 1) for the next step, the elements of $O_1$ and $O_3$ should be eliminated from the original list of feasible sets (Table 1) to create a new list (Table 2). In the process of revising a list, duplications of feasible sets

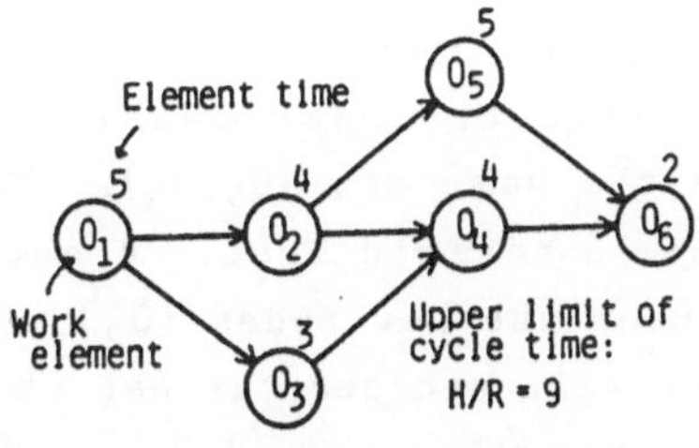


Fig. 1. Data for Example 1.

Table 1: The List of Feasible Sets for Node $N_0$.

| N(S) | $O_i$ / $\sum t_i$ \ $t_i$ | $O_1$ | $O_2$ | $O_3$ | $O_4$ | $O_5$ | $O_6$ |
|---|---|---|---|---|---|---|---|
| | | 5 | 4 | 3 | 4 | 5 | 2 |
| 1 | 5 | 0 | | | | | |
| 2 | 9 | 0 | 0 | | | | |
| 3 | 8 | 0 | | 0 | | | |
| 4 | 12 | 0 | 0 | 0 | | | |
| 5 | 16 | 0 | 0 | 0 | 0 | | |
| 6 | 14 | 0 | 0 | | | 0 | |
| 7 | 17 | 0 | 0 | 0 | | 0 | |
| 8 | 21 | 0 | 0 | 0 | 0 | 0 | |
| 9 | 23 | 0 | 0 | 0 | 0 | 0 | 0 |

N(S) is the number of feasible set S.

Symbols "0" indicate the work elements included in each feasible set.

$\sum t_i$ is the total element time of the feasible set S.

Table 2: The List of Feasible Sets for Node $\{O_1,O_3\}$.

| N(S) | $O_i$ / $\sum t_i$ \ $t_i$ | $O_2$ | $O_4$ | $O_5$ | $O_6$ |
|---|---|---|---|---|---|
| | | 4 | 4 | 5 | 2 |
| 1 | 4 | 0 | | | |
| 2 | 8 | 0 | 0 | | |
| 3 | 9 | 0 | | 0 | |
| 4 | 13 | 0 | 0 | 0 | |
| 5 | 15 | 0 | 0 | 0 | 0 |

may occur. For branching the node $\{O_1,O_3\}$, the sets numbered 2 and 4 in the original list yield the same feasible set $\{O_2\}$. The sets numbered 6 and 7 yield the same set $\{O_2,O_5\}$. The duplicated new sets must be erased to create a revised list. Branching the node $\{O_1,O_3\}$ based on Table 2 generates the new nodes $\{O_2\}$, $\{O_2,O_4\}$, and $\{O_2,O_5\}$. The branching procedures will proceed further while creating new lists of feasible sets; a particular list of feasible sets is attached to each node.

### 3.3 Calculation of a Lower Bound

The sum of the square of the balancing index contributed by the stations already assigned elements at node $N_r$ can be calculated as:

$$F_1 = \sum_{k=1}^{r} (H/R - p_k)^2, \tag{3}$$

where $p_k$ is the station time of the k-th station already assigned elements, and r is the number of such stations and equals the branching level of $N_r$.

The minimum possible number of stations to be required to complete the remaining elements at node $N_r$ is given by:

$$q = \left\lceil \frac{T - \sum_{k=1}^{r} p_k}{H/R} \right\rceil, \tag{4}$$

where $\lceil x \rceil$ is the minimum integer greater than or equal to x. Apart from the precedence relations of work elements and the discreteness of $p_k$, the ideally balanced station time of the remaining work stations to complete the product is:

$$u = ( T - \sum_{k=1}^{r} p_k ) / q , \tag{5}$$

where u is a continuous variable.

As mentioned earlier, the solution having the minimum number of stations can yield the best modified smoothness index among the solutions in which the complete balance of station times is realized (every station time is the same). Therefore, a lower bound on the square of the balancing index contributed by the remaining stations to be generated can be given by:

$$F_2 = (H/R - u)^2 \times q, \tag{6}$$

in which all the station times of q stations (minimum possible number of stations) are equal to u.

Therefore, a lower bound on node $N_r$ is obtained as:

$$LB(N_r) = \sqrt{F_1+F_2} = \sqrt{\sum_{k=1}^{r} (H/R-p_k)^2 + (H/R-u)^2 \times q} \,. \tag{7}$$

### 3.4 The Algorithm

On the basis of the discussion above, an optimal algorithm for Problem 2 is proposed through the B & B approach using a newest node search procedure as follows:

[Step 1] Set the branching level r = 0; the branching node $N_r^o = \phi$; the value of objective function of the trial solution $f'' = \infty$.

[Step 2] Generate all the feasible sets of work elements currently remaining by LEXO.

[Step 3] Perform the branching procedure for $N_r^o$, and set r = r+1. Calculate the lower bound $LB(N_r)$ for each new node, $N_r$.

[Step 4] Choose the minimum lower bound among the nodes newly created. The node having the minimum lower bound is now the objective node, $N_r^o$. If no work element is remaining at $N_r^o$, then go to Step 5. If there are more than one element remaining at $N_r^o$ and $LB(N_r^o) < f''$, then go to Step 2. If there are more than one element remaining at $N_r^o$ and $LB(N_r^o) \geq f''$, then go to Step 6.

[Step 5] If $LB(N_r^o) < f''$, the current $N_r^o$ is to be a new trial solution. Set $f'' = LB(N_r^o)$, and go to Step 6. If $LB(N_r^o) \geq f''$, then delete $N_r^o$ and go to Step 6.

[Step 6] If there exist undeleted (active) nodes in which $LB(N_r) < f''$, then choose the node nearest to (requiring the minimum number of branches to trace back from) the current $N_r^o$ among them. If more than one nearest nodes exists, break the tie by choosing the minimum lower bound. Set the chosen node as the new $N_r^o$. Make r equal the branching level of the current $N_r^o$. Delete all the newest nodes, and return to Step 2. If no objective node exists, then go to Step 7.

[Step 7] The present trial solution is the optimal solution. (END)

## 4. NUMERICAL EXAMPLE

The numerical example shown in Fig. 1 is solved through three algorithms: Jackson's algorithm based on Problem 1, Moodie and Young's algorithm (MY algorithm), and the proposed algorithm based on Problem 2. The solutions through the three algorithms are then compared.

### 4.1 Jackson's Algorithm

Jackson's algorithm yields two alternative optimal solutions, Solution 1 $\{O_1,O_2\}^9 - \{O_3,O_4\}^7 - \{O_5,O_6\}^7$ and Solution 2 $\{O_1,O_2\}^9 - \{O_3,O_5\}^8 - \{O_4,O_6\}^6$, where the superscript indicates the station time. The other optimal solution (Solution 3 $\{O_1,O_3\}^8 - \{O_2,O_4\}^8 - \{O_5,O_6\}^7$)

can not be obtained by Jackson's algorithm, since it has deleted the nodes with the same elements cumulatively assigned on the other node created earlier.

The maximum station time of Solution 3 is 8, which is smaller than the other two, while the number of stations remains the same (3). The maximum station time corresponds to the cycle time of the line, which dominates the efficiency of the line (output per unit time). Solution 3 is the best among the three with regard to both the number of stations and the balance of station times, yet Jackson's could not derive it. This shortcoming is not due to the algorithm itself but inherent in the formulation of Problem 1. In this connection the optimal algorithm by Kao and Queyranne for Problem 1 also yields Solution 1 instead of Solution 3.

### 4.2 Moodie and Young's Algorithm

The first phase of Moodie and Young's algorithm derives the solution, $\{O_1,O_3\}^8 - \{O_2,O_5\}^9 - \{O_4,O_6\}^6$. Then the second phase heuristically trades element $O_5$ in the second station for element $O_4$ in the third and yields the final solution, $\{O_1,O_3\}^8 - \{O_2,O_4\}^8 - \{O_5,O_6\}^7$, of which the smoothness index (by eq. (1)) is 1. This solution coincides the optimal solution for the smoothness index by chance, although the MY algorithm can not guarantee the optimal.

### 4.3 Proposed Algorithm

Now Example 1 is solved through the proposed algorithm. The first trial solution $\{O_1,O_3\}^8 - \{O_2,O_4\}^8 - \{O_5,O_6\}^7$ turns to the optimal and the algorithm is terminated. The total number of nodes created for this problem is eight. The optimal solution by the proposed algorithm coincides with Solution 3 shown in the previous section, and also the final solution by MY algorithm.

The value of objective function BI for this optimal solution is $\sqrt{6}$. The required number of stations is 3 and the maximum station time is 8, which becomes the cycle time. The cycle time obtained is less than the upper limit of cycle time, 9, which can achieve the quantity of the product to be produced during the planned period. The allowance between the actual cycle time and the upper limit of it can effectively be used for shortening the production period for the product or can remain as an allowance of the workforces.

## 5. CONCLUDING REMARKS

In this paper a line balancing problem with the balancing index [Problem 2] was formulated so as to approach both the minimum number of work stations and the complete balance of station times, simultaneously, under restrictions of the cycle time and precedence relations. An optimal algorithm employing Branch and Bound approach for the formulated problem is presented based on a discussion into the characteristics of optimal algorithms for previous problems.

## REFERENCES


[1] Dar-El, E. M., "MALB –A Heuristic Technique for Balancing Large Single-Model Assembly Lines," AIIE Trans., 5, 4, 343-356 (1973).

[2] Jackson, J. R., "A Computing Procedure for a Line Balancing Problem," Management Science, **2**, 3, 261-271 (1956).

[3] Johnson, R. V., "Assembly Line Balancing Algorithms: Computation Comparisons," International Journal of Production Research, **19**, 3, 277-287 (1981).

[4] Johnson R. V., "A Branch and Bound Algorithm for Assembly Line Balancing Problems with Formulation Irregularities," Management Science, **29**, 11, 1309-1324 (1983).

[5] Kao, E. P. C. and Queyranne, M., "On Dynamic Programming Methods for Assembly Line Balancing," Oper. Res. **30**, 2, 375-390 (1982).

[6] Lawler, E. L., "Efficient Implementation of Dynamic Programming Algorithms for Sequencing Problems," Report BW 106/79, Stichting Mathematisch Centrum, Amsterdam (1979).

[7] Moodie, C. L. and Young, H. H., "A Heuristic Method of Assembly Line Balancing for Assumptions of Constant or Variable Work Element Times," J. Industrial Engineering, **16**, 1, 23-29 (1965).

[8] Okamura, K.,Yamashina, H., and Ohno, H., "Balancing Problem of Mixed-model Flow Lines," (in Japanese), Journal of the Japan Society of Precision Engineering, **50**, 11, 75-80 (1984).

[9] Pinto, P. A., Dannenbring, D. G., and Khumawala, B. M., "Assembly Line Balancing with Processing Alternatives: An Application," Management Science, **29**, 7, 817-830 (1983).

[10] Rosenblatt, M. J. and Carlson, R. C., "Designing a Production Line to Maximize Profit," IIE Trans., **17**, 2, 117-122 (1985).

[11] Salveson, M. E., "The Assembly Line Balancing Problem," Journal of Industrial Engineering, **6**, 3, 18-25 (1955).

[12] Schrage, L. and Baker, K. R., "Dynamic Programing Solution of Sequencing Problems with Precedence Constraints," Operations Research, **26**, 3, 444-449 (1978).

[13] Shtub, A., "The Effect of Incompletion Cost on Line Balancing with Multiple Manning of Work Stations," International Journal of Production Research, **22**, 2, 235-245 (1984).

[14] Tonge, F. M., "Summary of Heuristic Line Balancing Procedure," Management Science, **7**, 1, 21-42 (1960).

[15] Van Assche, F. and Herroelen, W. S., "An Optimal Procedure for the Single-Model Deterministic Assembly Line Balancing Problem," European Journal of Operations Research, **3**, 2, 142-149 (1978).

# SCHEDULING FLOW-SHOPS WITH PARALLEL MACHINES AND FINITE IN-PROCESS BUFFERS BY MULTILEVEL PROGRAMMING

Tadeusz J. Sawik
University of Mining and Metallurgy, 30-059 Cracow, Poland

Scheduling of a multi-product flow-shop production system is considered with parallel machines in each stage and finite interstage buffers with nonzero initial inventories. The objective is to determine an assignment of products to machines over a scheduling horizon, which minimizes the maximum lateness of a multiproduct production order with the in-process inventory holding costs as low as possible. The problem is formulated as a multilevel integer programme where each level concerns scheduling of a different production stage. The interdependence of each stage is modelled by intermediate constraints that take into account the semi-finished products availability and the interstage buffers limited capacity. An illustrative example for a three stage production is provided with data from real world machining system.

## 1. Introduction

Many production systems have a flowshop structure with several manufacturing stages in series, separated by finite interstage buffers, where each stage is made up of several identical machines in parallel. A product has to be successively processed through all the stages, at each stage on any machine at a fixed production rate. The in-process inventories maintained in the buffers enable the successive stages to be operated relatively independently. However, their limited storage space may restrict the efficient utilization of production capacities of all stages.

Usually a production order is given that specifies the types and the required amounts of differents products to be simultaneously produced in the system as well as their corresponding due-dates. The scheduling objective for the multistage system considered is to complete the multiproduct production order with the maximum lateness or the completion time minimized as well as the in-process inventories holding costs.

The scheduling problem is a combination of lot-sizing and sequencing within the multistage production-inventory system. A similar case but for a single stage production system has been considered in Dorsey et.al. (1975). Lot-size scheduling for the three machines in series with unlimited intermediate buffers is presented by Cadambo and Venkoba Rao (1984). On the other hand the classical flowshop sequencing with finite intermediate buffers has been considered by Dutta and Cunningham (1975), and Papadimitriou and Kanellakis (1980), for example. The multistage production scheduling however, naturally leads to multilevel programming formulation, e.g., Bard and Falk (1982), Bard (1985), Bialas and Karwan (1982). The multilevel approach to production scheduling has been presented by Sawik (1982, 1984, 1987). The case of two-stage and multistage production system with the objective of minimizing the total processing time at each production stage is considered in Sawik (1982) and (1984), respectively. In Sawik (1987) the optimization of the completion time of a multiproduct production order and the in-process inventories holding costs is presented for the case of the multistage system with no initial inventories in the interstage buffers.

The paper presents the multistage scheduling problem as a multilevel integer programme, where each level problem is a single stage parallel scheduling subject to additional constraints representing the interdependence of each production stage. The constraints take into account the requirements and availability of the semi-finished products as well as the interstage buffers limited capacity. An approximation hierarchical scheduling algorithm is described and an illustrative example for a three-stage production system is provided.

## 2. Problem Formulation

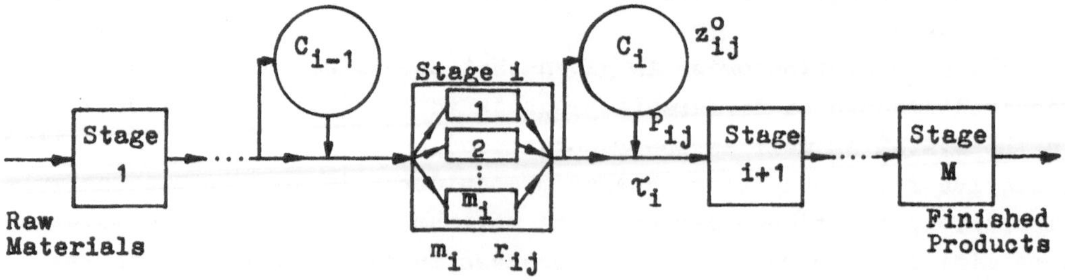


Fig.1. Flow-shop with parallel machines and finite in-process buffers

In this section the problem of scheduling the multistage production with finite in-process buffers and nonzero initial inventories is formulated as a multilevel integer programme.

## 2.1. Definition of Problem Variables

Table 1 gives the notation used in this paper

TABLE 1

Notation Used to Describe the Problem

| Symbol (Listed Alphabetically) | Definition |
|---|---|
| $C_i$ | Capacity of in-process buffer between stages i and i+1 |
| $d_j$ | Due-date for production task $w_j$ |
| $i$ | Stage number ($i=1,\dots,M$) |
| $j$ | Product number ($j=1,\dots,N$) |
| $m_i$ | Number of identical parallel machines in stage i |
| $M$ | Number of production stages in the system |
| $N$ | Number of products |
| $P_{ij}$ | Semi-finished product j made in stage i |
| $r_{ij}$ | Number of units of $P_{ij}$ produced on a machine in stage i during one period |
| $t$ | Period number ($t=1,\dots,T$) |
| $T$ | Scheduling horizon |
| $T_{Mj}$ | Completion time of production task $w_j$ |
| $w_j$ | Number of required machine-periods of production of the finished product j |
| $x_{ijt}$ | Control variable - total number of machines assigned in stage i to produce product j in period t |
| $z^o_{ij}$, $z^f_{ij}$ | Respectively, the beginning and the required ending inventory of semi-finished product $P_{ij}$ |
| $\alpha_{ij}$ | In-process inventory holding cost per unit of $P_{ij}$ carried in buffer for one period |
| $\tau_i$ | A constant delay between stages i and i+1 |

The production system (Fig.1) is assumed to have the following features:

(a) During any period at most one product can be scheduled on each machine, and when any product j is assigned to a machine in stage i for one period, all $r_{ij}$ units of $P_{ij}$ must be made.
(b) During any period the assignment of products to machines is considered fixed, and hence semi-finished product inventory changes during any period occur linearly.
(c) Each product $P_{i+1,j}$ produced in stage i+1 in period t is made of the semi-finished product $P_{ij}$ produced in stage i in period $t-\tau_i$ at the latest.
(d) Semi-finished product $P_{ij}$ (i=1,...,M-1) made in stage i is transferred either to the buffer between the stages i and i+1 or directly to stage i+1.
(e) There are nonzero initial inventories $z^o_{ij}$ (i=1,...,M-1, j=1,...,N) in the interstage buffers.

The production schedule for stage i can be written as

$$x^i = \left\{ x_{ijt} : j=1,\dots,N,\ t=1,\dots,T \right\} \tag{1}$$

The following auxiliary variables will also be used in the sequel:
$v_{ijt}$ (i=1,...,M-1, j=1,...,N, t=1,...,T) - the minimum number of machine-periods of production of $P_{ij}$ that must occur during the first t periods in order to meet the requirements for $P_{ij}$

$$v_{ijt} = \max\left\{0, \left\lceil \left(r_{i+1j} \sum_{k=1}^{t+\tau_i} x_{i+1jk} - z^o_{ij}\right)/r_{ij} \right\rceil \right\} \quad , \ t \leqslant T-1 \tag{2}$$

and $\quad v_{ijT} = w_{ij}$

where $w_{ij}$ - the required number of machine-periods of production of $P_{ij}$

$$w_{ij} = \max\left\{0, \left\lceil \left(r_{i+1j} w_{i+1j} + z^f_{ij} - z^o_{ij}\right)/r_{ij} \right\rceil \right\} \quad , \ i \leqslant M-1 \tag{3}$$

and $\quad w_{Mj} = w_j,\ j=1,\dots,N$

$\lceil a \rceil$ denotes the smallest integer not less than a.

$c_{it}$ (i=1,...,M-1, t=1,...,T) - the maximum production in stage i that can be carried in the buffer and/or used up in stage i+1 during the first t periods

$$c_{it} = C_i - \sum_{j=1}^{N} z^o_{ij} + \sum_{j=1}^{N} \sum_{k=1}^{t} r_{i+1j} x_{i+1jk} \tag{4}$$

$b_{ipt}$ (i=2,...,M, p=1,...,i-1, t=1,...,T) - the maximum number of machine-periods of production in stage p (plus the initial in-process inventories) that can be processed in stage i (i> p) during the first t periods

$$b_{ipt} = \max\left\{0,\ m_p\left(t - \sum_{k=p}^{i-1}\tau_k\right)\right\} + \sum_{j=1}^{N}\sum_{k=p}^{i-1}(z^o_{kj}/r_{pj})1\left(t - \sum_{s=k+1}^{i-1}\tau_s\right) \quad (5)$$

$b_{ijpt}$ ($i=2,\ldots,M$, $j=1,\ldots,N$, $p=1,\ldots,i-1$, $t=1,\ldots,T$) - the maximum number of machine-periods of production of product j in stage p (plus the initial in-process inventories) that could be processed in stage i ($i > p$) into $P_{ij}$ during the first t periods, if the other products were not produced

$$b_{ijpt} = (r_{pj}/r_{ij})\max\left\{0, m_p\left(t - \sum_{k=p}^{i-1}\tau_k\right)\right\} + \sum_{k=p}^{i-1}(z^o_{kj}/r_{ij})1\left(t - \sum_{s=k+1}^{i-1}\tau_s\right) \quad (6)$$

where $1(a) = 1$ if $a > 0$, and $1(a) = 0$ if $a \leqslant 0$.

## 2.2. The Mathematical Model of the Scheduling Problem

In general, each production stage i of the system can be characterized by an individual objective function $f_i(x^i)$ which is to be minimized by the appropriate choice of schedule $x^i$.
For the first M-1 stages the minimum in-process inventory holding costs are chosen to be the optimality criteria with the following $f_i(x^i)$ to be minimized (Sawik, 1987)

$$f_i(x^i) = -\sum_{j=1}^{N}\sum_{t=1}^{T}\alpha_{ij}r_{ij}tx_{ijt}\ ,\quad i=1,\ldots,M-1 \quad (7)$$

For the final stage M however, the maximum lateness of a feasible schedule $x^M$ is to be minimized, that is,

$$f_M(x^M) = L_{max} = \max_j\{T_{Mj} - d_j\} \quad (8)$$

where $T_{Mj} - d_j$ is lateness of $w_j$ (the completion time minus due-date).
A criterion of the secondary importance is the maximum completion time or makespan of $x^M$ (Sawik, 1987) defined by

$$T_{max} = \max_j\{T_{Mj}\} \quad (9)$$

The problem of production scheduling in stage i can be formulated as follows

minimize $\qquad f_i(x^i) \qquad (10)$

subject to

1) the assignment feasibility constraints

$$\sum_{j=1}^{N}x_{ijt} \leqslant m_i,\quad t=1,\ldots,T \quad (11)$$

$$x_{ijt} \geqslant 0 \text{ and integer}, \ j=1,\dots,N, \ t=1,\dots,T \qquad (12)$$

2) the semi-finished products availability constraints

$$\sum_{j=1}^{N} \sum_{k=1}^{t} (r_{ij}/r_{pj}) x_{ijk} \leqslant b_{ipt}, \ p=1,\dots,i-1, \ t=1,\dots,T \qquad (13)$$

$$\sum_{k=1}^{t} x_{ijk} \leqslant b_{ijpt}, \ j=1,\dots,N, \ p=1,\dots,i-1, \ t=1,\dots,T \qquad (14)$$

3) the input buffer capacity constraints

$$\sum_{j=1}^{N} (r_{i-1j} \left\lceil (r_{ij}/r_{i-1j}) \sum_{k=1}^{t} x_{ijk} \right\rceil - r_{ij} \sum_{k=1}^{t-\tau_{i-1}} x_{ijk}) \leqslant C_{i-1} + \sum_{j=1}^{N} z_{ij}^{o} ,$$

$$t=\tau_{i-1},\dots,T \qquad (15)$$

4) the output buffer capacity constraints

$$\sum_{j=1}^{N} \sum_{k=1}^{t} r_{ij} x_{ijk} \leqslant c_{it}, \ t=1,\dots,T \qquad (16)$$

5) the semi-finished products requirements constraints

$$\sum_{k=1}^{t} x_{ijk} \geqslant v_{ijt}, \quad j=1,\dots,N, \ t=1,\dots,T-1 \qquad (17)$$

$$\sum_{t=1}^{T} x_{ijt} = w_{ij}, \ j=1,\dots,N \qquad (18)$$

In the above formulation (11) and (12) are the local constraints in stage i. The other constraints represent the interdependence between i and:

- each preceding stage p, $p \leqslant i-1$, - (13) and (14)
- immediately preceding stage i-1 - (15)
- immediately succeeding stage i+1 - (16), (17) and (18).

If i = 1, i.e., when scheduling stage 1, the constraints (13), (14) and (15) should be discarded. Similarly, if i = M, i.e., when scheduling the final stage, the constraints (16) and (17) should be discarded, and in the sum of (18) the upper limit T should be replaced with $L_{max} + d_j$ for each j=1,...,N.

## 3. Multilevel Scheduling Algorithm

In this section a conceptual algorithm for production scheduling in the multistage system is presented (for more detailed description of the algorithm in the case of no initial inventories, see Sawik,

(1987)). The flowshop structure of the system enables the schedule to be determined sequentially beginning with the schedule $x^M$ for the final stage M, followed by schedule $x^{M-1}$ for stage M-1, down through schedule $x^1$ for stage 1. The schedule for a later stage affects the assignments of machines for an earlier stage through the variables $c_{it}$ and $v_{ijt}$ (Fig.2).

The solution procedure consists of three scheduling algorithms:

A1 - for scheduling stage M,

A2 - for scheduling stage i (i=2,...,M-1),

A3 - for scheduling stage 1.

A1 and A3 are period-by-period heuristics in which each production period is considered once, whereas A2 is a recursive procedure capable of making adjustments in the schedules when a buffer is overfilled or shortage of a semi-finished product occurs.

In the algorithm A1 the first period assigned is t=1, while in A2 and A3 the last period $t=T_i$ is assigned as the first one, where $T_i$ is the number of production periods required for completion the production order $(w_{i1},\dots,w_{iN})$ in stage i

$$T_i = \max_{1 \leqslant t \leqslant T_{i+1}-\tau_i} \left\{ t + \max\left\{ 0, \left\lceil \left( \sum_{j=1}^{N} r_{ij} w_{ij} - c_{it} \right) / a_i \right\rceil \right\} \right\} \tag{19}$$

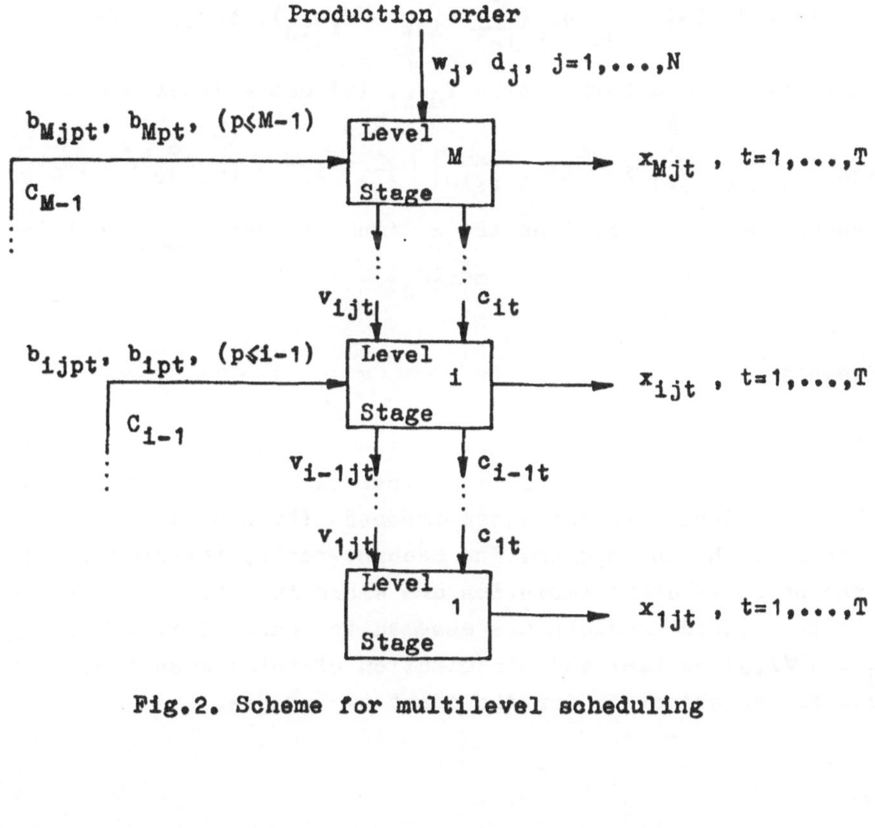


Fig.2. Scheme for multilevel scheduling

where $a_i = \max\{m_i r_{imax}, C_{i-1}/\tau_{i-1}\}$, $r_{imax} = \max_j\{r_{ij}\}$.

In the algorithm A1 the assignment of products to machines in each period is made in the order of nondecreasing due-date $d_j$, i.e., considering the products ordered as follows

$$d_1 \leqslant d_2 \leqslant \ldots \leqslant d_N \tag{20}$$

In the algorithms A2 and A3 however, the semi-finished products $P_{ij}$, $j=1,\ldots,N$ in each stage $i$ $(i=1,\ldots,M-1)$ are considered in the order of nonincreasing holding costs, beginning with $j=N$

$$e_{iN} \geqslant e_{iN-1} \geqslant \ldots \geqslant e_{i1} \tag{21}$$

where $e_{ij} = \alpha_{ij} r_{ij}$.
Hence, the required lots of the semi-finished products with the highest storage costs are scheduled as late as possible.

The computational complexity of the M-level algorithm is $O(MN\log N + MNT^2)$, Sawik (1987).

### 3.1. Lower Bounds on the Objective Functions

It can be easily shown that the optimal values of the objective functions $f_i(x^i)$ for the first M-1 stages are bounded from below as follows

$$f_i(x^i) \geqslant LB_i = \sum_{j=1}^{N} e_{ij}\Big(\sum_{t=1}^{T_i-1} v_{ijt} - T_i w_{ij}\Big), \quad i=1,\ldots,M-1 \tag{22}$$

The maximum completion time $T_{max}$, (9) has a lower bound

$$LBT_{max} = \max_{1\leqslant i\leqslant M}\Big\{\Big\lceil\sum_{j=1}^{N} w_{ij}/m_i\Big\rceil, \max_{1\leqslant p\leqslant i-1}\Big\lceil\sum_{j=1}^{N}\big(r_{ij}w_{ij}-\sum_{k=p}^{i-1} z^o_{kj}\big)/r_{pj}m_p+\sum_{k=p}^{i-1}\tau_k\Big\rceil\Big\} \tag{23}$$

and hence the lower bound on the maximum lateness $L_{max}$, (8) is

$$LB_M = LBT_{max} - \max_j\{d_j\} \tag{24}$$

## 4. Example

In this section an illustrative example is presented for an M=3 - stage production system where N=5 types of products are produced in batches to orders with due-dates imposed. The example comes from the real world machining system. The machine-tools, in-process buffers and the products characteristics are shown in Table 2. In the example the semi-finished products are assumed to cause similar holding costs ($\alpha_{ij} = 1\ \forall i,j$) so that the minimization of total amount of in-process stocks is the objective for the first two stages.

TABLE 2

| | Stage 1 $m_1 = 3$ $C_1=310$ $\tau_1=1$ | | | | Stage 2 $m_2 = 2$ $C_2=330$ $\tau_2=1$ | | | | Stage 3 $m_3 = 4$ | | |
|---|---|---|---|---|---|---|---|---|---|---|---|
| j | $r_{1j}$ | $\alpha_{1j}$ | $z^o_{1j}$ | $z^f_{1j}$ | $r_{2j}$ | $\alpha_{2j}$ | $z^o_{2j}$ | $z^f_{2j}$ | $r_{3j}$ | $w_j$ | $d_j$ |
| 1 | 20 | 1 | 20 | 20 | 40 | 1 | 40 | 40 | 20 | 4 | 9 |
| 2 | 60 | 1 | 60 | 60 | 30 | 1 | 30 | 30 | 60 | 6 | 8 |
| 3 | 50 | 1 | 50 | 50 | 100 | 1 | 100 | 100 | 50 | 8 | 7 |
| 4 | 70 | 1 | 70 | 70 | 50 | 1 | 50 | 50 | 80 | 3 | 10 |
| 5 | 60 | 1 | 60 | 60 | 80 | 1 | 80 | 80 | 70 | 2 | 6 |

For the example data the lower bound on $T_{max}$ is $LBT_{max} = 13$, (23). The production schedule obtained using the multilevel algorithm is

| t | 1 | 2 | 3 | 4 | 5 | 6 | 7 | 8 | 9 | 10 | 11 | 12 | 13 |
|---|---|---|---|---|---|---|---|---|---|---|---|---|---|
| $x_{31t}$ | 1 | 0 | 2 | 1 | 0 | 0 | 0 | 0 | 0 | 0 | 0 | 0 | 0 |
| $x_{32t}$ | 0 | 0 | 1 | 0 | 0 | 0 | 0 | 0 | 1 | 1 | 1 | 1 | 1 |
| $x_{33t}$ | 2 | 0 | 0 | 0 | 0 | 0 | 2 | 2 | 2 | 0 | 0 | 0 | 0 |
| $x_{34t}$ | 0 | 0 | 0 | 1 | 1 | 1 | 0 | 0 | 0 | 0 | 0 | 0 | 0 |
| $x_{35t}$ | 1 | 0 | 0 | 0 | 0 | 0 | 0 | 1 | 0 | 0 | 0 | 0 | 0 |
| $x_{21t}$ | 0 | 1 | 0 | 0 | 0 | 0 | 0 | 0 | 0 | 0 | 0 | 1 | 0 |
| $x_{22t}$ | 1 | 1 | 1 | 0 | 1 | 1 | 0 | 1 | 2 | 2 | 2 | 0 | 0 |
| $x_{23t}$ | 0 | 0 | 0 | 0 | 0 | 1 | 1 | 1 | 0 | 0 | 0 | 0 | 1 |
| $x_{24t}$ | 0 | 0 | 1 | 2 | 1 | 0 | 0 | 0 | 0 | 0 | 0 | 1 | 0 |
| $x_{25t}$ | 0 | 0 | 0 | 0 | 0 | 0 | 1 | 0 | 0 | 0 | 0 | 0 | 1 |
| $x_{11t}$ | 1 | 0 | 0 | 0 | 0 | 0 | 0 | 0 | 0 | 2 | 1 | 0 | 0 |
| $x_{12t}$ | 0 | 1 | 0 | 0 | 1 | 0 | 0 | 1 | 1 | 1 | 0 | 0 | 1 |
| $x_{13t}$ | 0 | 0 | 0 | 0 | 1 | 2 | 2 | 0 | 0 | 0 | 1 | 2 | 0 |
| $x_{14t}$ | 0 | 0 | 2 | 0 | 0 | 0 | 0 | 0 | 0 | 0 | 1 | 0 | 1 |
| $x_{15t}$ | 0 | 0 | 0 | 0 | 0 | 1 | 0 | 0 | 0 | 0 | 0 | 1 | 1 |

The values of the objective functions and their corresponding lower bounds (22), (24) are

$f_1(x^1) = -10720$, $LB_1 = -10900$

$f_2(x^2) = -9510$, $LB_2 = -10500$

$f_3(x^3) = L_{max} = 5$, $LB_3 = 3$, and $T_{max} = LBT_{max} = 13$.

## 5. Conclusion

The multilevel approach enables each production stage of the multistage production system with interstage constraints to be scheduled relatively independently. However, the dynamic interdependence of all production stages should be modelled exactly to avoid infeasibilities in the schedules that may occur when scheduling backwards from the final to the beginning stage. The approach enables also different objective functions for each production stage to be easily incorporated into the model, and the model itself to be extended allowing the inclusion of other aspects of the multistage production.

## REFERENCES


1. J.F.Bard and J.E.Falk (1982) An explicit solution to the multilevel programming problem. Comput.and Ops.Res. 9, 77-100.
2. J.F.Bard (1985) Geometric and algorithmic developments for a hierarchical planning problem. Eur.J.Oper.Res. 19, 372-383.
3. W.F.Bialas and M.H.Karwan (1982) On two-level optimization. IEEE Trans.Automat.Contr. AC-27, 211-214.
4. B.V.Cadambo and T.S. Venkoba Rao (1984) Multiproduct, three-stage production inventory systems. J.Opl.Res.Soc. 35, 105-116.
5. R.C. Dorsey, T.J.Hodgson, and H.D.Ratliff (1975) A network approach to a multi-facility, multi-product production scheduling problem without backordering. Mgmt.Sci. 21, 813-822.
6. S.K.Dutta and A.A.Cunningham (1975) Sequencing two-machine flowshops with finite intermediate storage. Mgmt.Sci. 21, 989-996.
7. C.H.Papadimitriou and P.C.Kanellakis (1980) Flowshop scheduling with limited temporary storage. J.Assoc.Comput.Mach. 27, 533-549.
8. T.J.Sawik (1982) Hierarchical scheduling two-stage, multi-machine production with finite intermediate storage. UMM Scientific Bull. Automatics, no 32, 373-383.
9. T.J.Sawik (1984) Hierarchical scheduling multi-stage production with limited in-process inventory. Proc.3rd Intern.Conf.on Systems Engg. Wright State University, Dayton, Ohio, Sept.5-7.
10. T.J.Sawik (1987) Multilevel scheduling of multistage production with limited in-process inventory. J.Opl.Res.Soc. 38, 651-664.

SOLUTION METHODS OF PLANTS LOCATION DYNAMIC PROBLEMS IN CONTINUOUS-TIME

Prof. Alexander P. Uzdemir
Institute for Systems Studies
9 Prospekt 60 Let Octyabrya
117312 Moscow, USSR

1.Introduction. Statement of the problem. Discreteness and dynamics of economic processes result in emergence of time-dependent integer variables in some optimal planning problems. Below we consider a class of such problems formulated in continuous time and concerned with location of plants in an economic region.

Consider the simplest problem. There are n projects of plants turning out similar products. Let $V_i$ be a production capacity of project i, $K_i$ - capital costs. Besides, demand a(t) for the product at each instant t is assigned for the plan period [0,T].

By denoting the time of project i establishment as $\tau_i$, obtain the following optimization problem:

$$\sum_{i=1}^{n} K_i e^{-\gamma\tau_i}\theta(T-\tau_i) \Longrightarrow \min_{(\tau_i\geq 0)} \quad \text{s.t.} \sum_{i=1}^{n} V_i\theta(t-\tau_i)\geq a(t),\ t\in[0,T]. \quad (1)$$

Here, $\gamma$ is cost discount rate, $\theta(t)$ is Heaviside's function ( $\theta(t)=0$ for $t<0$, $\theta(t)=1$ for $t\geq 0$ ). Problem (1) was first formulated by D.Erlenkotter in 1973 [6] (making no use of $\theta$-function).

By introducing the integer variable $x_i(t)=\theta(t-\tau_i)$ it is possible to reduce (1) to the following dynamic integer optimization problem:

$$\int_0^T \sum_{i=1}^{n} K_i e^{-\gamma t} x_i(t)dt + e^{-\gamma T}\sum_{i=1}^{n} K_i x_i(T) \Longrightarrow \min_{x(.)}$$

$$\text{s.t.} \sum_{i=1}^{n} V_i x_i(t)\geq a(t),\ t\in[0,T], \quad (2)$$

$x_i(t)=0$ or 1, nondecreasing, continuous on the right ($i=\overline{1,n}$).

A much more general problem arises when plants are located in different points of an economic region. This implies transportation of different products; apart from plant projects there are development projects of the transportation network capacities, etc. Consider the following broadly stated dynamic problem of plant location in a region [16] (it generalizes the statement presented in [10]).

Let there be a set of points Q in the considered territory, and a large set of products I manufactured or consumed. In a general case, a set $E_q$ of all plants in point $q\in Q$ breaks into two subsets: $E_q^0$ (running plants) and $E_q^1$ (new plants that can be established). We shall refer to a plant $e \in E_q^1$ as a production development project in point q. The point of the region $q \in Q$ may be a destination point for the external supply $v_{i_1 q}^+$ of product $i_1$, a transit point for product $i_2$, point of product manufacture $i_3$, a point of productive consumption of product $i_4$, point of final consumption of product $i_5$ in the amount $v_{i_5 q}^-(t)$. The product flows between points $q, q' \in Q$ are limited by the capacity of the communication (q,q') which may be increased according to the capacity development projects $e \in E_{qq'}^1$.

The system of constraints at instant t [0,T] is written as follows:

- constraints on production capacities for running plants

$$k_{je}(t)V_{je}(t) \leq \sum_{s \in S_e} d_{jse} v_{se}(t) \leq V_{je}(t), \; j \in J_e \;, \; e \in E_q^0 \;, \; q \in Q^0, \quad (3)$$

where $V_{je}(t)$ is the size of the j production capacity at the running plant e, $J_e$ is a set of capacity types in e, $S_e$ is a set of production modes at plant e, $d_{jse}$ is the level of the employed capacity of type j at plant e for a unit intensity of mode s, $v_{se}(t)$ is intensity of mode s at plant e, $k_{je}(t)$ is a possibly minimum coefficient of capacity j utilization at plant e $(0 \leq k_{je} \leq 1)$, $Q^0 = \{q | q \in Q, \; E_q^0 \neq \emptyset\}$;

- constraints on production capacities for an established production development project e

$$\sum_{s \in S_e} d_{jse} v_{se}(t) \leq V_{je} \, \theta(t - \tau_e), \; j \in J_e, \; e \in E_q^1 \;, \; q \in Q^1, \quad (4)$$

where $V_{je}$ is the size of capacity j for project e, $_e$ is time of capacities introduction for project e, $Q^1 = \{q | q \in Q, \; E_q^1 \neq \emptyset\}$;

- inequalities for products in the region points

$$v_{iq}^+(t) + \sum_{e \in E_q}' \sum_{s \in S_e} (k_{ise} - a_{ise}) v_{se}(t) \geq$$
$$\geq \sum_{q' \in Q_{iq}^+} f_{iqq'}(t) - \sum_{q' \in Q_{iq}^-} f_{iq'q}(t) + v_{iq}^-(t), \; q \in Q \;, \; i \in I_q \;, \quad (5)$$

where $v_{iq}^+(t)$ are external supplies of product i to point q; $k_{ise}$, $a_{ise}$ are output/input of product i respectively for an unit intensity of mode s at plant e; $Q_{iq}^+$, $Q_{iq}^-$ are sets of neighbouring points where product i can be transported from point q, $f_{iqq'}(t)$ is a flow of product i transportation from q to q', $v_{iq}^-(t)$ is a flow of product i final consumption in point q; $I_q$ is a set of products associated with point q;

- constraints on capacities for transportation flows

$$\sum_{i \in I_{qq'}} d_{iqq'} f_{iqq'}(t) \leq F_{qq'}^0(t) + \sum_{e \in E_{qq'}} F_e \theta(t - \tau_e), (q,q') \in R, \quad (6)$$

where $d_{iqq'}$ is employed capacity of communication (q,q') for an unit flow of product i; R is a set of transportation network communications; $I_{qq'}=\{i \mid i\in I_q, q'\in Q^+_{iq}\}$; $F^0_{qq'}(t)$ is the existent communication capacity of (q,q') at time t ; $F_e$ is an increase in capacity according to project e; $\tau_e$ is time of capacity development project e realization;

- constraints on resources required following implementation of each development project

$$\sum_{e\in E^1_q} b_{le}\theta(t-\tau_e)\le v^+_{lq}(t),\ q\in Q^+_l\ ,\ l\in L^+_Q, \tag{7}$$

where $b_{le}$ is an amount of l-type resources required for realization of project e, $v^+_{lq}(t)$ is the size of resource l allocated within [0,t] for implementing the development project in point q; $Q^+_l$ is a set of points resource l is allocated to; $L^+_Q$ is a set of resource types allocated to the region;

- constraints on resources at large allocated to the region

$$\sum_{q\in Q^+_l}\ \sum_{e\in E^1_q} b_{le}\theta(t-\tau_e)\le v^+_l(t),\ l\in L^+, \tag{8}$$

where $v^+_l(t)$ is amount of resource l allocated to the region; $L^+$ is a set of types of resources allocated to the region;

- constraints on the instants of development projects realization

$$\tau_e\in\{[\Delta_e,T],+\infty\},\ e\in E^1, \tag{9}$$

where $E^1=\{e\mid e\in E^1_q, q\in Q^1\}\cup\{e\mid e\in E^1_{qq'}, (q,q')\in R\}$ is a set of development projects, $\Delta_e>0$ is the duration of project e realization.

The minimum discounted production costs for the planning period are selected as an optimality criterion for the dynamic location problem:

$$\Phi=\int_0^T r(t)\sum_{q\in Q}\Big[\sum_{i\in I_q}\ \sum_{q'\in Q^+_{iq}} c_{iqq'}(t)f_{iqq'}(t)+ \sum_{e\in E_q}\ \sum_{s\in S_e} c_{se}(t)v_{se}(t)\Big]dt+\sum_{e\in E^1} K_e(\tau_e)\theta(T-\tau_e) \Longrightarrow \min, \tag{10}$$

where r(t) is a discount function ($r(t)>0$, $\dot r(t)<0$; as a rule $r(t)=e^{-\gamma t}$ and $\gamma>0$), $c_{iqq'}(t)$ is cost of product i unit transportation from q to q'; $c_{se}(t)$ is cost price under an unit intensity of mode s at plant e; $K_e(\tau_e)$ is a discounted capital cost of development project e implementation.

Minimization (10) is carried out with respect to $\{\tau_e\}$, $\{f_{iqq'}(t)\ge 0\}$, $\{v_{se}(t)\ge 0\}$ under constraints (3)-(9).

2. <u>Nonlinear programming problem as a problem of finding an optimal permutation</u>. Represent the formulated problem in the form

$$\Phi=\int_0^T r(t)c^T(t)y(t)dt+K(\tau)\vec\theta(T-\vec\tau) \Longrightarrow \min_{y(.),\tau}\ (=\Phi^*)$$
$$\text{s.t. } Ay(t)\le a(t)+D\vec\theta(t-\vec\tau),\ B\vec\theta(t-\vec\tau)\le b(t), \tag{11}$$
$$\tau_i\in\{[\Delta_i,T],+\infty\}\ (i=\overline{1,n}),\ y(t)\in R^m_+,\ \tau\in R^n,$$

where $K(\tau)=(K_1(\tau_1),\dots,K_n(\tau_n))$, $r(t)\in R^1_+$, $c(t)\in R^m_+$, $a(t)\in R^p$, $b(t)\in R^h$; A,B,D are pxm,hxn,pxn matrices, respectively; $\vec{\theta}(t-\vec{\tau})=(\theta(t-\tau_1),\dots,\theta(t-\tau_n))^T$.

An equivalent writing of (11) as of dynamic integer optimization problem has the form

$$\Phi=\int_0^T r(t)\{c^T(t)y(t)-[\dot{K}(t)/r(t)]x(t)\}dt+K(T)x(T) \implies \min_{y(.),x(.)}$$

$$\text{s.t. } Ay(t)\le a(t)+Dx(t),\ y\in R^m_+,\ Bx(t)\le b(t),\ x(t)\le\vec{\theta}(t-\vec{\Delta}), \tag{12}$$

$x_i(t)=0$ or $1$, nondecreasing, continuous on the right ($i=\overline{1,n}$). The following conditions are assumed to be met: 1) $K_i(\tau_i)>0$, $dK_i/d\tau_i\le 0$ ($i=\overline{1,n}$); 2) $a(t),c(t)$ are continuous; 3) $b(t)\ge 0$ and continuous on the right, nondecreasing; 4) $D\ge 0$, $B\ge 0$, $\tau_i\in[0,T]$ ($i=\overline{1,n}$).

Denote $$\Psi(t,\tau)=\min_y[c^T(t)y \mid Ay\le a(t)+D\vec{\theta}(t-\vec{\tau}),\ y\ge 0], \tag{13}$$

here, it is necessary to introduce an additional condition for $\tau$ (when it is met, problem (13) has a feasible solution):

$\tau\in\Omega^1=\{\tau\mid u[a(t)+D\vec{\theta}(t-\vec{\tau})]\ge 0,\ uA\ge 0,\ u\ge 0\}=\{\tau\mid u^\nu[a(t)+D\vec{\theta}(t-\vec{\tau})]\ge 0,\ \nu\in N\}$, where $u^\nu$ is formative vector of cone $U=\{u\mid uA\ge 0, u\ge 0\}$, $\nu\in N$ (N is a finite set).

Denote a set of conditions for $\tau$, present in (11) as

$$\Omega^0=\{\tau\mid \tau\in R^n,\ B\vec{\theta}(t-\vec{\tau})\le b(t),\ t\in[0,T];\ \tau_i\in([\Delta_i,T],+\infty),\ i=\overline{1,n}\}.$$

Then an optimal solution of (11) is determined by the formulas:

$$\tau^*=\operatorname*{argmin}_\tau\Big[\int_0^T r(t)\Psi(t,\tau)dt+K(\tau)\vec{\theta}(T-\vec{\tau})\mid \tau\in\Omega^0\cap\Omega^1\Big], \tag{14}$$

$$y^*(t)=\operatorname*{argmin}_y[c^T(t)y\mid Ay\le a(t)+D\vec{\theta}(t-\vec{\tau}^*),\ y\ge 0],\ t\in[0,T]. \tag{15}$$

Execute the following narrowness of the feasible set

$$\tau_{i_1}\le\tau_{i_2}\le\dots\le\tau_{i_n}, \tag{16}$$

where $\pi=(i_1,i_2,\dots,i_n)$ is some permutation of component $\tau$ indices. As applied to the equivalent problem (12), condition (16) corresponds

$$x_{i_1}(t)\ge x_{i_2}(t)\ge\dots\ge x_{i_n}(t). \tag{17}$$

In order to find an optimal vector $\tau^*(\pi)$ in such narrowed problem (11),(16) it is necesary to solve $n(\pi)-q(\pi)+1$ problems of the following form (the problem parameter q successively takes values ranging from $q(\pi)$ to $n(\pi)$ ) [17]:

$$\sum_{k=1}^{q}\int_{\tau_{i_{k-1}}}^{\tau_{i_k}} r(t)P_{k-1}(t;\pi)dt+\int_{\tau_{i_q}}^{T} r(t)P_q(t;\pi)dt+\sum_{k=1}^{q}K_{i_k}(\tau_{i_k}) \implies \min_\tau \tag{18}$$

s.t. $\underline{\tau}_k(\pi)\le\tau_{i_k}\le\min[\bar{\tau}_k(\pi),T]$, $k=\overline{1,q}$, $\tau_{i_k}=+\infty$, $k=\overline{q+1,n}$, $\tau_{i_k}\le\tau_{i_{k+1}}$, $k=\overline{1,q-1}$.

The following notations are used here: $\tau_{i_0}=0$, $\underline{\tau}_k(\pi)=\max_{j=\overline{1,h}}\min_t[t: \sum_{l=1}^{k} b_{ji_l} \le b_j(t),\ T\ge t\ge \max_{l=\overline{1,k}} \Delta_{i_l}]$, $\overline{\tau}_k(\pi)=\min_{\nu\in N}\sup_{t\ge 0}(t|\ \sum_{l=1}^{k-1}(u^\nu D)_{i_l}\ge \alpha_\nu(t))$, $\alpha_\nu(t)=\max_{0\le t'\le t}[-u^\nu a(t')]$, $q(\pi)=\max_k[k|\ \overline{\tau}_k(\pi)\le T, k=\overline{1,n}]$, $n(\pi)=\max_k[k|\underline{\tau}_k(\pi)\le T, k=\overline{1,n}]$,

$P_k(t;\pi)=\min_y[c^T(t)y|\ Ay\le a(t)+\sum_{l=1}^{k} D_{i_l}]$.

Note that $\underline{\tau}_{k+1}(\pi)\ge\underline{\tau}_k(\pi)$, $\overline{\tau}_{k+1}(\pi)\ge\overline{\tau}_k(\pi)$ $(k=\overline{1,n-1})$.

In a general case, solution of (18) is confined to finding the shortest path from the starting node to the terminal one in some acyclic graph [17].

Solution is especially simple given for the condition of "large discounted capital costs": $-\dot{K}_{i_k}(\tau)/r(\tau)\ge P_{k-1}(\tau;\pi)-P_k(\tau;\pi)$, $\tau\in[\underline{\tau}_k(\pi), \min(T, \overline{\tau}_k(\pi))]$. In this case, an optimal solution of (18) looks like: $\tau^*_{i_k}=\min[\overline{\tau}_k(\pi), T]$, $k=\overline{1,q}$; $\tau^*_{i_k}=+\infty$, $k=\overline{q+1,n}$.

Solution is found in a similar way given "small discounted capital costs": $-\dot{K}_{i_k}(\tau)/r(\tau)\le P_{k-1}(\tau;\pi)-P_k(\tau;\pi)$, $\tau\in[\underline{\tau}_k(\pi), \min(T,\overline{\tau}_k(\pi))]$. Solution of (18) has form: $\tau^*_{i_k}=\underline{\tau}_k(\pi)$, $k=\overline{1,q}$; $\tau^*_{i_k}=+\infty$, $k=\overline{q+1,n}$.

For these cases, an optimal solution $\tau^*(\pi)$ of narrowed problem (11), (16) corresponds to $q=q(\pi)$.

The feasible set in (11) is an union n! of feasible sets for different narrowed problems. Denote problem (11),(16) in terms of $P_\pi$. Thus, solution of (11) reduces to finding an optimal permutation $\pi^*$ corresponding to problem $P_\pi$ with the least functional.

The following two solution methods are based on an implicit examination of all problems $P_\pi$.

3. Direct method. During the examination one runs into $P_S$ type problems. Here $S=(i_1,i_2,\ldots,i_l)$ is a sequence of different component indices. For $S=\emptyset$ it is the original problem (11). For $l\ge 1$ problem $P_S$ is obtained by adding following inequalities to (11)

$$\tau_{i_1}\le\tau_{i_2}\le\ldots\le\tau_{i_l},\quad \tau_{i_l}\le\tau_i\ (i\ne i_1,\ldots,i_l), \tag{19}$$

which are equivalent to conditions

$$\tau_{i_k}\in[\underline{\tau}_k(\pi_S),\overline{\tau}_k(\pi_S)]\cap([0,T],+\infty),\ k=\overline{1,l}, \tag{20}$$

where $\pi_S$ is permutation the first l elements of which coinside with S.

Problem $P_S$ is equivalent to the dynamic integer problem (12) when introducing additional conditions

$$x_{i_1}(t)\ge x_{i_2}(t)\ge\ldots\ge x_{i_l}(t);\ x_{i_l}(t)\ge x_i(t),\ i\ne i_1,i_2,\ldots,i_l; \tag{21}$$

$$\theta[t-\bar{\tau}_k(\pi_S)]\le x_{i_k}(t)\le\theta[t-\underline{\tau}_k(\pi_S)],\quad k=\overline{1,l}. \tag{22}$$

If in examining $P_S$ one fails to eliminate it from the list of problems, then it divides into n-1 problems $P_{S\circ i}$ where $i\neq i_1,\dots,i_l$ and $S\circ i=(i_1,\dots,i_l,i)$. For $\underline{\tau}_{l+1}(\pi_{S\circ i})>\bar{\tau}_{l+1}(\pi_{S\circ i})$ problem $P_{S\circ i}$ has no solution and is not placed on the list of problems.

If $\underline{\tau}_l(\pi_S)\le T$, $\bar{\tau}_l(\pi_S)=+\infty$ then a feasible solution of $P_S$ corresponds to solution of (18) for $\pi=\pi_S$ , $q=1$ and $\tau_i=+\infty$ $(i\neq i_1,\dots,i_l)$.

The lower estimate of $P_S$ functional optimal value is obtained by solving an evaluation problem $P_S$. The latter differs from $P_S$ in that the conditions of nondecreasing and integerability of variables $x_i(t)$ in (12),(21),(22) are replaced with $0\le x_i(t)\le 1$, $i=\overline{1,n}$. Its solution reduces to solving a parametric linear programming problem with parameter t in which $c^T(t)y-[\dot{K}(t)/r(t)]x$ is minimized with respect to y and x under the conditions at time $t\in[0,T]$, and to solving an ordinary linear programming problem in which $K(T)x$ is minimized under conditions present in $P_S$ at time T (*terminal LP problem*).

By making use of evaluation problems it is possible to arrange elimination of $P_S$ , identified during examination, by comparing the respective estimate with the record.

Thus the direct method is a branch and bound method.

4. Structural method. This method is based on minoranting objective functional in constructing an evaluation problem [15,17]. This is done by utilization of dual solution of the problem of finding "noninteger" vector-function y(t) for a feasible vector $\tau$ (see (15)).

The main idea of the method is the use of structural decomposition. There are two problems: 1) determine of a subset $I^*$ of vector $\tau$ component indices such that in an optimal solution $\tau_i^*\in[\Delta_i,T]$ for $i\in I^*$, $\tau_i^*=+\infty$ for $i\bar{\in} I^*$; 2) finding of $\tau_i^*$, $i\in I^*$.

Such problems are solved following completion of an iterative process of a sequential improvement of the record and elimination of sets of useless versions both in the form of subsets $I=\{1,2,\dots,n\}$ and subsets of feasible set for vector $\tau$.

The current state of the iterative process is characterized by the following additional information: there is a set of feasible vector-functions $u^\mu(t)$, $\mu\in\tilde{M}$, for the problem, being dual problem of finding y (see (15)), and a set of vectors $u^\nu$ , $\nu\in\tilde{N}$, generating certain conditions of problem (15) solvability.

By making use of this information it is possible to formulate the following evaluation problem for (11) with the purpose of finding $\tau$ :

$$\max_{\mu\in\tilde{M}}\{\,\eta^\mu+[K(\tau)-\chi^\mu]\vec{\theta}(T-\vec{\tau})\}\quad\Longrightarrow\quad\min_{\tau}$$

$$\text{s.t. } B\vec{\theta}(t-\vec{\tau})\leq b(t),\ u^{\nu}[a(t)+D\vec{\theta}(t-\vec{\tau})]\geq 0\ (\nu\in\tilde{N}),\ t\in[0,T], \qquad (23)$$
$$\tau_i\in\{[\Delta_i,T],+\infty\},\ i=\overline{1,n},$$

where $\eta^{\mu}=-\int_0^T r(t)u^{\mu}(t)a(t)dt$, $\chi_i^{\mu}=\int_{\Delta_i}^T r(t)[u^{\mu}(t)D]_i dt$, $\chi^{\mu}=(\chi_1^{\mu},\dots,\chi_n^{\mu})$.

Problem (23), by analogy with (11), can be partitioned into a set of problems $\tilde{P}_{\pi}$ each one having an additional condition $\tau_{i_1}\leq\tau_{i_2}\leq\dots\leq\tau_{i_n}$ determined by permutation $\pi=(i_1,i_2,\dots,i_n)$. And the feasible set for $\tilde{P}_{\pi}$ is reduced to $\underline{\tau}_k(\pi)\leq\tau_{i_k}\leq\tilde{\tilde{\tau}}_k(\pi)$, $\tau_{i_k}\in\{[0,T],+\infty\}$ $(k=\overline{1,n})$, $\tau_{i_k}\leq\tau_{i_{k+1}}$ $(k=\overline{1,n-1})$, where $\tilde{\tilde{\tau}}_k(\pi)=\min_{\nu\in\tilde{N}}\sup_{t\geq 0}\{t\mid\sum_{l=1}^{k-1}(u^{\nu}D)_{i_l}\geq\alpha_{\nu}(t)\}$. The following inequalities are correct: $\tilde{\tilde{\tau}}_k(\pi)\geq\underline{\tau}_k(\pi)$ $(k=\overline{1,n})$, $0\leq\tilde{\tilde{\tau}}_k(\pi)\leq\tilde{\tilde{\tau}}_{k+1}(\pi)$ $(k=\overline{1,n-1})$.

Problem $\tilde{P}_{\pi}$ has no solution if $\exists k$: $\underline{\tau}_k(\pi)>\tilde{\tilde{\tau}}_k(\pi)$. Otherwise, due to the fact that $K(\tau)$ is not increasing with respect to all $\tau_i\in[0,T]$ an optimal solution has the form: $\tau^*_{i_k}=\tilde{\tilde{\tau}}_k(\pi)$, $k=1,\dots,\tilde{q}(\pi)$; $\tau^*_{i_k}=T$, $k=\tilde{q}(\pi)+1,\dots,q^*(\pi)$; $\tau^*_{i_k}=+\infty$, $k=q^*(\pi)+1,\dots,n$. The following notations are used here: $\tilde{q}(\pi)=\max_{k=\overline{1,n}}\{k\mid\tilde{\tilde{\tau}}_k(\pi)\leq T\}$, $q^*(\pi)=\arg\min_{q=\tilde{q}(\pi),\dots,n(\pi)}\max_{\mu\in\tilde{M}}\{\eta^{\mu}+\sum_{k=1}^{\tilde{q}(\pi)}[K_{i_k}(\tilde{\tilde{\tau}}_k(\pi)-\chi^{\mu}_{i_k}]+\theta(q-\tilde{q}(\pi)-1)\sum_{k=\tilde{q}(\pi)+1}^{q}[K_{i_k}(T)-\chi^{\mu}_{i_k}]\}$. Obviously, $\tilde{P}_{\pi}$ is an evaluation problem for $P_{\pi}$.

If search for a feasible solution of problem (23) is performed with an objective function smaller than the current value of record $\bar{\Phi}$ in the original problem (11) then the search ends with identification of some problem $\tilde{P}_{\pi}$ with an optimum value smaller than $\bar{\Phi}$. Then, in order to renew the record it is necessary to solve problem $P_{\pi}$. This may entail changes in the value of record $\bar{\Phi}$, record solution $\bar{\tau}$, and sets $\tilde{M},\tilde{N}$.

Elimination of $\tilde{P}_S$ in the process of search leads to elimination of the respective problem $P_S$ too.

Write the conditions in (23) containing $u^{\nu}$ in the form $u^{\nu}D\vec{\theta}(t-\vec{\tau})\geq\alpha_{\nu}(t)$, $\nu\in N$, $t\in[0,T]$. This may be done as $u^{\nu}D\geq 0$. Then relax them providing for the satisfaction of inequalities only for $t=T$. Then, instead of (23) obtain a new evaluation problem for (11). Since the objective function is not increasing with respect to all $\tau_i$ within $[0,T]$ there is such optimal solution for this problem where all $\tau_i=T$ or $+\infty$. By introducing variables $x_i=\theta(T-\tau_i)\in\{0,1\}$ $(i=\overline{1,n})$ write the evaluation problem in the form:

$$\max_{\mu\in\tilde{M}}\{\eta^{\mu}+[K(T)-\chi^{\mu}]x\}\ \Longrightarrow\ \min_x$$
$$\text{s.t. } Bx\leq b(T);\ (u^{\nu}D)x\geq\alpha_{\nu}(T),\ \nu\in\tilde{N};\ x_i\in\{0,1\},\ i=\overline{1,n}. \qquad (24)$$

A sequential utilization of evaluation problems (24),(23) permits a structural decomposition (see Figure 1). The latter consists of three stages: 1)finding a feasible subset of variable $\tilde{I}$ indices by solving a Boolean programming problem (24); 2)finding a feasible permutation $\tilde{\pi}$ of set $\tilde{I}$ components making it possible to differentiate a narrowed problem $\tilde{P}_{\tilde{\pi}}$ with the value of objective function smaller than $\bar{\Phi}$; 3)solution of the narrowed problem $P_{\pi}$ for $\pi=\tilde{\pi}$ and $\tilde{n}=|\tilde{I}|$.

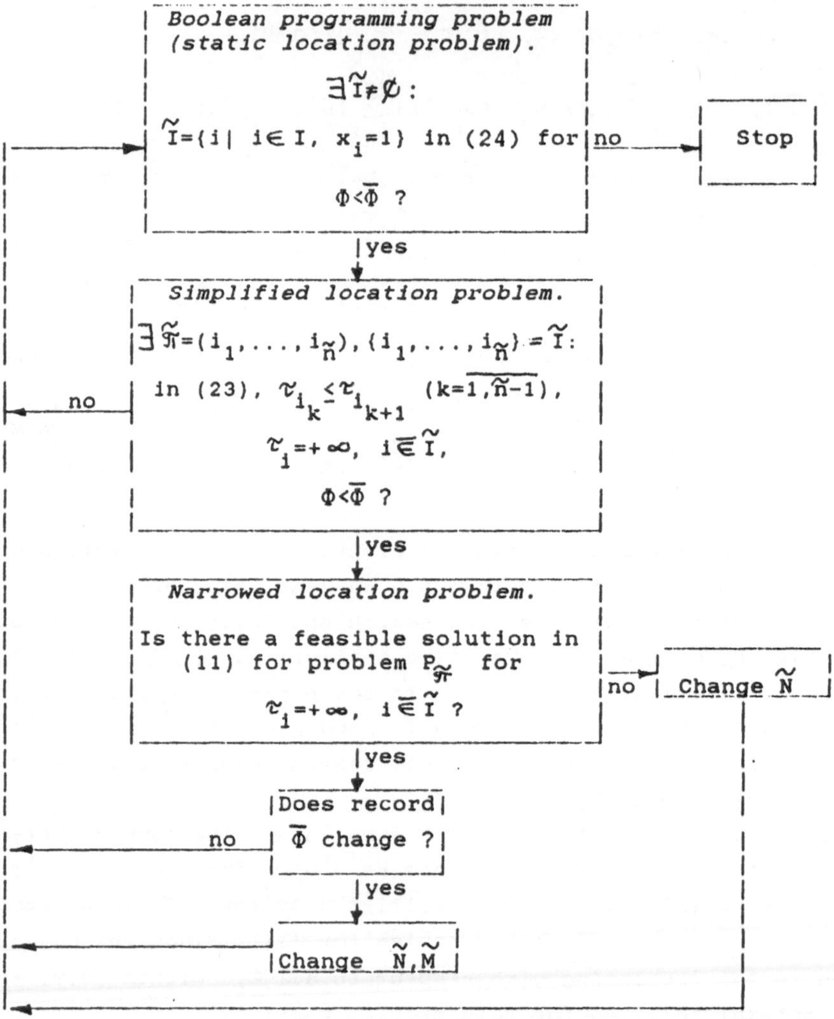


Figure 1. Flow chart of structural decomposition algorithm.

As soon as the third stage is over, return to the first one. Solution of the problem stops if there is no solution at the first stage (then, the current record solution $\bar{\tau}$ is solution of the original problem). If there is no solution at the second stage, the search for solution continues at the first stage (the earlier solution is discarded).

To solve the Boolean programming problem (24) use is made of branch and bound methods [2] or flexible search methods [9]. Elimination takes place under the lower estimate not smaller than $\overline{\Phi}$, solution stops following determination of a feasible solution.

Consider the problem of finding a feasible solution in (23) with an additional condition: $\tau_i \le T$ for $i \in \tilde{I}$; $\tau_i = +\infty$ for $i \in I \setminus \tilde{I}$. Here account is taken of the fact that for $r(t)=e^{-\gamma t}$, $K_i(\tau_i)=\tilde{K}_i e^{-\gamma \tau_i}$ ($\gamma>0$, $\tilde{K}_i>0$). In searching for a solution use is made of a sequential partition of problems. The current problem is that of $P_S(\tilde{I})$, $S=(i_1,\dots,i_l)$, $(i_1,\dots,i_l) \subseteq \tilde{I}$, $l \le \tilde{n} = |\tilde{I}|$. It has the following form:

$$\tilde{\beta}_S(\tilde{I}) + \sum_{i \in \overline{I}} \tilde{K}_i e^{-\gamma \tau_i} \implies \min_{(\tau_i)} \tag{25}$$

$$\text{s.t.} \sum_{i \in \overline{I}} b_{ji} \theta(t-\tau_i) \le b_j^S(t), \; j=\overline{1,h}, \; \sum_{i \in \overline{I}}' \omega_{\nu' i} \theta(t-\tau_i) \ge \alpha_{\nu'}^S(t), \quad \nu' \in N, \; t \in [0,T],$$

$$\max(\Delta_i, \underline{t}^S) \le \tau_i \le T, \; i \in \overline{I}.$$

The following notations are used here: $\overline{I}=\tilde{I} \setminus \{i_1,\dots,i_l\}$, $\tilde{\beta}_S(\tilde{I}) = \max_{\mu \in \tilde{M}} \Big( \eta^{\mu} - \sum_{i \in \tilde{I}} \gamma_i^{\mu} \Big) + \sum_{k=1}^{l} \tilde{K}_{i_k} \exp[-\gamma \tau_{i_k}^{*}(\pi_S)]$; $b_j^S(t)=b_j(t)-\sum_{k=1}^{l} b_{j i_k}$; $\alpha_{\nu'}^S(t)=\alpha_{\nu'}(t)-\sum_{k=1}^{l} \omega_{\nu' i_k}$; $\omega_{\nu' i}=(u^{\nu'} D)_i$; $\underline{t}^S=\tilde{\tilde{\tau}}_{l+1}(\pi_S)$. In writing (25) account was taken of the fact that due to a monotonous decrease in the objective function with respect to all $\tau_i$ for optimal solution $\tilde{\tau}^{*}_{i_k}(\pi_S)=\min(T, \tilde{\tilde{\tau}}_k(\pi_S))$, $k=\overline{1,l}$, and for the remaining $i$ $\tau_i \ge \min(T, \tilde{\tilde{\tau}}_{l+1}(\pi_S))$.

Problem $P_S(\tilde{I})$ for $l<\tilde{n}$ divides into $n-l$ problems $P_{S \circ i}(\tilde{I})$ where $i \in \overline{I}$. Problem $P_{S \circ i}(\tilde{I})$ is eliminated for $\underline{\tau}_{l+1}(\pi_{S \circ i}) > \tilde{\tilde{\tau}}_{l+1}(\pi_{S \circ i})$.

An evaluation problem for a non-ending $P_S(\tilde{I})$ is built as follows. Let $\nu \in \tilde{N}$ such that $\overline{I}_\nu = \{i \mid i \in \overline{I}, \; \omega_{\nu i}>0\} \ne \emptyset$, $\alpha_\nu^S(t) \ne \text{const}$ for $t \in [\underline{t}^S, T]$. Introduce a linear minorant for $\alpha_\nu^S(t)$: $\varkappa_\nu^S + \lambda_\nu^S t \le \alpha_\nu^S(t)$ ($\lambda_\nu^S>0$). Then the the evaluating problem $\tilde{P}_S^{\nu}(\tilde{I})$ takes the following form:

$$\beta_S(\tilde{I}) + \sum_{i \in \overline{I} \setminus \overline{I}_\nu} \tilde{K}_i \exp(-\gamma T) + \sum_{i \in \overline{I}_\nu} K_i \exp(-\gamma \tau_i) \implies \min_{\tau} \; (=\tilde{\Phi}_S^{\nu}(\tilde{I})) \tag{26}$$

$$\text{s.t.} \sum_{i \in \overline{I}_\nu} \omega_{\nu i} \theta(t-\tau_i) \ge \varkappa_\nu^S + \lambda_\nu^S t, \; t \in [\underline{t}^S, \bar{t}_\nu^S], \; \tau_i \in [\underline{t}^S, \bar{t}_\nu^S], \; i \in \overline{I}_\nu.$$

Here $\bar{t}_\nu^S = \Big( \sum_{i \in \overline{I}_\nu} \omega_{\nu i} - \varkappa_\nu^S \Big) / \lambda_\nu^S$. Solution of this problem $\hat{\tau}$ corresponds to permutation $\pi_\nu=(i_1,\dots,i_{\overline{n}(\nu)})$: $\hat{\tau}_{i_1} = -\varkappa_\nu^S / \lambda_\nu^S$, $\hat{\tau}_{i_{k+1}} = \hat{\tau}_{i_k} + \omega_{\nu i_k} / \lambda_\nu^S$, $k=\overline{1,\overline{n}(\nu)-1}$. The permutation $\pi_\nu$ is determined by a chain of inequalities

$$\tilde{K}_{i_k} / [1-\exp(-\gamma \omega_{\nu i_k} / \lambda_\nu^S)] \le \tilde{K}_{i_{k+1}} / [1-\exp(-\gamma \omega_{\nu i_{k+1}} / \lambda_\nu^S)], \; k=\overline{1,\overline{n}(\nu)-1}.$$

The lower estimate for problem (25):

$$\max_{\nu}[\tilde{\Phi}_S^{\nu}(\tilde{I}) \mid \nu\in\tilde{N},\ \bar{I}_\nu\neq\emptyset,\ \alpha_\nu^S(T)>\alpha_\nu^S(\underline{t}^S)].$$

5. Dichotomic method. Consider the problem in the form

$$J = \int_0^T [c^T(t)x(t)+b^T(t)y(t)]dt+g^Tx(T)+f^Ty(T) \Longrightarrow \min_{x(.),y(.)}$$

$$\text{s.t. } Ax(t)+By(t)\leq a(t),\ x(t)\in R^n,\ y(t)\in R_+^m,\ t\in[0,T], \tag{27}$$

$$x_i(t)=0 \text{ or } 1, \text{ nondecreasing, continuous on the right } (i=\overline{1,n}).$$

Here $c(t)\in R^n, b(t)\in R_+^m$, $g\in R^n$, $f\in R_+^m$, $a(t)\in R^l$, A and B are lxn and lxm matrices respectively. Continuity $c(t)$, $b(t)$, continuity on the right $a(t)$, and piecewise linearity $c(t)$, $b(t)$, $a(t)$ and limitation of feasible $y(t)$ components are assumed.

It is suggested to solve problem (27) by a branch and bound method which divides the feasibility set into two subsets (dichotomy). Some "dividing variable" $x_r(t)$ and "dividing instant" $\tau_r$ are selected. Then one subset is determined by an additional condition $x_r(\tau_r)=0$ from where it follows $x_r(t)=0$, $t\in[0,\tau_r]$. Another subset is determined by condition $x_r(\tau_r)=1$ from where it follows $x_r(t)=1$, $t\in[\tau_r,T]$.

As a result of the dividing procedure application, the problem is replaced with a list of problems $\mathcal{P}$, where the current problem $P_S$ is determined by a set $S=(I^0,\tau^0;I^1,\tau^1)$, forming additional conditions in (27) $x_i(t)=0$, $t\in[0,\tau_i^0]$, $i\in I^0$; $x_i(t)=1$, $t\in[\tau_i^1,T]$, $i\in I^1$.

Construction of an evaluation problem $\tilde{P}_S$ is carried out by a substitution of integrity, nondecrease and continuity on the right of all $x_i(t)$ with $x_i(t)\in[0,1]$ whose solution is reduced to solution of a parametric linear programming problem (under the respective constraints at each instant in time) and terminal linear programming problem (for t=T). Denote an optimal value in the evaluation problem as $\tilde{J}_S$.

The dividing rule makes use of penalty values equal to a lower estimate for the objective function increment in the linear programming problem when fixing some variable on 0 or 1. Such penalties are determined with the simplex-table [14].

Solution of a parametric linear programming problem for $x_i(t)$ is followed by computation of the lower penalty $g_i^0(t)$ received when fixing $x_i(t)$ on zero, and the upper penalty $g_i^1(t)$ received when fixing it on unit. The lower terminal penalty $\Gamma_i^0$ and the upper terminal penalty $\Gamma_i^1$ are computed in a similar manner.

Given the dividing variable $x_i$ and the dividing instant $\tau$ division produces two problems: for the first $x_i(t)=0$, $t\in[0,\tau]$, for the second $x_i(t)=1$, $t\in[\tau,T]$. The lower estimate for incrementing $\tilde{J}_S$ when passing to the first problem equals $J_i^0(\tau)=\int_0^{\tau} g_i^0(t)dt$ $(0\leq\tau<T)$, $J_i^0(\tau)=\int_0^T g_i^0(t)dt+\Gamma_i^0$

($\tau$=T) and when passing to the second problem, respectively $J_i^1(\tau)=\int_\tau^T g_i^1(t)dt+\Gamma_i^1$.

Since in solving $P_S$ all $x_i(t)$ must be integer and nondecreasing then, as compared to $\tilde{J}_S$, we get an increment in the estimate in $P_S$: $\Delta\tilde{J}_S=\max_{i=\overline{1,n}}\ \max_{\tau\in T_i}\ \min[J_i^0(\tau),J_i^1(\tau)]$, where $T_i$ is interval in which $x_i(t)$ is not fixed. For $\tilde{J}_S+\Delta\tilde{J}_S\geq\bar{J}$, where $\bar{J}$ is a current record, problem $P_S$ is eliminated from the list. Otherwise, it divides.

Denote $\tau_i=\arg\max_{\tau\in T_i}\min[J_i^0(\tau),J_i^1(\tau)]$, $r=\arg\max_{i=\overline{1,n}}\min[J_i^0(\tau_i),J_i^1(\tau_i)]$.

Treated as a partition variable is $x_r(t)$, and a partition instant - $\tau_r$.

A strong emphasis is placed on determining a feasible solution of $P_S$ which makes it possible to stop partitioning in case the solution is $\varepsilon$-optimal for $P_S$.

Let $\tilde{x}^S(t)$ be an optimal vector-function x(t) in the evaluation problem $\tilde{P}_S$. Introduce $\hat{\tau}_i^S=\inf_{t\in[0,T]}[t|\tilde{x}_i^S(t)>0]$. Determine vector-function $\hat{x}^S(t)$ with components: $\hat{x}_i^S(t)\equiv 0$, if $\tilde{x}_i^S(t)=0$ for $t\in[0,T]$; $\hat{x}_i^S(t)=\theta(t-\hat{\tau}_i^S)$ - otherwise.

As applied to the dynamic location problems, such construction may lead to a feasible solution. Let the respective functional value be $\hat{J}_S$.

If $\hat{J}_S-(\tilde{J}_S+\Delta\tilde{J}_S)\leq\varepsilon(\tilde{J}_S+\Delta\tilde{J}_S)$ where $\varepsilon>0$ is the prescribed accuracy with respect to the functional, then problem $P_S$ is eliminated. For $\hat{J}_S<\bar{J}$ a new record and record solution are formed.

6. Software and computational experience. The first two methods are supported with software. The third method is experimentally tested on an Erlenkotter problem (1).

It was assumed in developing the software that conditions characteristic of the case with large discounted capital costs are met. It was also assumed that the raw material product in the two-stage location problem is unique and a parametric linear programming problem (15) is reduced to a problem of finding a minimal cost circulation in a network (see [3]). The discount function is an exponential curve.

The direct method software makes it possible to solve the problem completely in continuous time [4] and the piecewise-linear approximation is prescribed for a(t), b(t).

An additional decomposition, accompanied by differentiation of a problem of finding a feasible subset of location points (see [11]), was introduced in the structural method software, and a piecewise-constant approximation in a certain time grid (by years) is prescribed for functions a(t), b(t). Solution of the problem is also obtained in this time net (y(t) is piecewise-constant).

Practical problems were solved on a computer with speed of 300,000 operations per sec. Data on some of the solved problems are shown in Table 1.

Table 1

| Srl No. | Number of points: raw material | Number of points: production | Number of points: consumption | Number of products | T (years) | Enterprises: old | Enterprises: new | Computing time (minutes) |
|---|---|---|---|---|---|---|---|---|
| | | | | direct method | | | | |
| 1 | 18 | 22 | 25 | 4 | 6 | 5 | 60 | 31 |
| 2 | 22 | 15 | 24 | 5 | 11 | 4 | 67 | 34 |
| | | | | structural method | | | | |
| 3 | 22 | 17 | 24 | 4 | 11 | 6 | 58 | 21,3 |

The structural method software provided a basis for an interaction optimization system INDUSTRY PLAN [7,8]. It comprises three functional parts: informational support of the problem, software for its solution algorithms, interaction software. The informational support is based on the general systems tools of data base control system INES [1] and contains a data base, forms of information input in the data base, systems of information input and correction in the data base, printing programme, data base access system, programme interface modules for converting information from the data base to the form required for solution of optimization problem. Software consists of two subsystems: subsystem of dynamic location problem solution and service programmes permitting interaction with the terminal user. The interaction software permits the terminal user to control the system informational support and software. Thus, in the course of interaction the user may select, with a view to computation, a desirable version of information characterizing the problem. The interaction is going on in line with the scenario written in the language of interaction scenario description of INES system [5].

Table 2

| Srl. No. | Number of points: raw | Number of points: production | Number of points: consumption | Number products | T (years) | Enterprises: old | Enterprises: new | Time of: computation | Time of: forming arrays |
|---|---|---|---|---|---|---|---|---|---|
| 4 | 3 | 6 | 24 | 4 | 5 | 6 | 17 | 50sec | 50sec |
| 5 | 21 | 29 | 24 | 2 | 15 | 20 | 19 | 15min | 3min |

The characteristics of the two problems solved with the system

INDUSTRY PLAN are shown in Table 2.

7. Methods development prospects. In developing the methods and software was made of the more general scheme of search for solution than the branch and bound technique. The scheme is now referred to as sequential decomposition [18]. Apart from eliminating a problem from the list ,as a result of estimate comparison with the record, the scheme permits elimination with respect to problem dominance (analogous to elimination in dynamic programming). As applied to a dynamic location problem, this method is described in [20] both for the direct method and structural decomposition. The sequential decomposition scheme was used for the development of a solution algorithm for a single-product and one-stage location problem [19].

A dichotomic method is to be preferably employed for solving a narrowed location problem when the condition of "large (or small) discounted capital costs" is fulfilled (see Section 2).

It should be noted, before closing, that in case it is necessary to account for introduction and assimilation process of new production capacities, it is unreasonable to write the dynamic location problem in continuous-time framework as identification of a sufficiently simple narrowed problem is rather difficult. Its formulation in discrete-time framework leads to a Boolean linear programming problem. A good example is provided by a two-criteria dynamic problem of production location and development under the capital investments deficit [13] which is solved with special algorithmic schemes of Boolean programming problem solution based on sequential decomposition [12].

REFERENCES

1. Arlazarov, V.L., Dukalov, A.N., Emel'yanov, N.E., Ivanov, Yu. N., Kochin, Yu.Ya., and Tokarev, V.V. (1979), The INES Economic Information System, *Automation and Remote Control*, 40, 869-879.
2. Breu, R., and Burdet, C.-A. (1974), Branch and Bound Experiments in 0-1 Programming, *Mathematical Programming Study*, 2, 1-50 .
3. Bur'yan, S.B., Serov, S.S., and Uzdemir, A.P. (1976), Dynamic Industrial-Plant-Location Problem and Numerical Method for Solving It. III, *Automation and Remote Control*, 37, 741-750.
4. Bur'yan, S.B. (1978), Direct Method of Solving the Dynamic Problem of Deploying Industrial Plants, *Automation and Remote Control*, 39, 1197-1201.
5. Emel'yanov, N.E., and Chernysheva, I.B. (1983), Language of Dialogue Scenario Description, in *Date Base Management System and Its Environment*, Institute for Systems Studies, Moscow, 75-87 (in Russian).
6. Erlenkotter, D. (1973), Sequencing Expansion Projects, *Opns. Res.*, 21, 542-553.
7. Kovalchuk, G.B., Serov, S.S., and Uzdemir, A.P. (1982), Software for Integer Problems of Plant Development and Location in Industry, in *Software of Optimization Systems* Institute for Systems Studies, Moscow, 106-118 (in Russian).
8. Kovalchuk, G.B. (1982), Software and Dialog System for Planning Problems for a Branch of Industry, Preprint, Institute for Systems

Studies, Moscow (in Russian).
9. Nghiem Ph. Tuan (1971), A Flexible Tree-Search Method for Integer Programming Research, *Opns. Res.*, 19, 115-119 .
10. Serov, S.S., and Uzdemir, A.P. (1976), Dynamic Industrial-Plant-Location Problem and Numerical Method for Solving It. 1, *Automation and Remote Control*, 37, 583-589 .
11. Serov, S.S. (1980), Decomposition Algorithm of Solving Problem of Determining Moments of Introduction of Enterprises, *Automation and Remote Control*, 41, 1351-1357.
12. Shebalkin, V.M. (1985), An Algorithmical Framework for Solving 0-1 Linear Programming Large Scale Problems, in *Methods for Complex Systems Studies*, Institute for Systems Studies, Moscow, 35-40 (in Russian).
13. Shebalkin, V.M. (1985), Dynamic Problem of Development and Location on Conditions the Capital Investments Deficit, in *Dynamics of Heterogeneous Systems*, Institute for Systems Studies, Moscow, 132-140 (in Russian).
14. Tomlin, J.A. (1971), An Improved Branch and Bound Method for Integer Programming, *Opns. Res.*, 19, 1070-1075.
15. Uzdemir, A.P. (1977), Decomposition in Solution of Combinatory Problem of Determining Moments of Introduction of Enterprises, *Automation and Remote Control*, 38, 1519-1528 .
16. Uzdemir, A.P. (1979), Dynamic Location of Industrial Plants. 1., *Automation and Remote Control*, 40, 1671-1678.
17. Uzdemir, A.P. (1979), Dynamic Location of Industrial Plants. 11., *Automation and Remote Control*, 40, 1829-1839.
18. Uzdemir, A.P. (1980), Sequential Decomposition in Optimization Problems, *Automation and Remote Control*, 41, 1630-1639.
19. Uzdemir, A.P., and Shebalkin, V.M. (1984), Application of Sequantial Decomposition Approach for Solving One-Productive Dynamic Location Problem, in *Methods of Optimization*, Institute for Systems Studies, Moscow, 108-116 (in Russian).
20. Uzdemir, A.P. (1985), Sequential and Structural Decomposition in Solving Dynamic Plant Location Problems, *Kibernetika*, 5, 94-101 (in Russian).

# A METHOD OF CONGESTION CONTROL IN THE COMPUTER NETWORK NODE

Dominik Rutkowski
Department of Communication Systems, Technical University of Gdańsk,
Gdańsk, Poland

Krzysztof Chmara
Institute of Telecommunications, Academy of Technology and Agriculture
Bydgoszcz, Poland

## 1. Introduction

The efficiency of a computer communication network operation depends upon the rate and reliability of message transmission among the network nodes. In this connection one of the fundamental quality criteria of network operation is the average time delay. While designing the computer communication network we usually assume a fixed rate of message arrivals and thus the average time delay is fixed for the reliable network. In practice, however, rates of message arrivals vary with time and it may happen that for some users and some time intervals they are larger than the declared rates at the stage of design. Moreover, some channels or nodes may occasionally be down. In these circumstances the congestion in network nodes may arise. As a result of congestion we observe not only an increase of average time delay but also a decrease of network throughput. For these reasons a counteraction of congestion by every possible means is an important matter. Since it is easier to counteract a local congestion separately in each network node due to lesser amount and easier access to auxiliary information, attention has been paid in this paper to that type of congestion.

The rules concerning separate congestion control carried out by each node were examined in [1, 4, 5, 7]. The subject of this paper is a new adaptive method of local congestion control.

## 2. A model of a network node

In our considerations we shall distinguish two classes of messages in each node: local messages generated at the rate $\lambda_l$ msg/sec by sources connected directly to a specific node and transit messages arriving at that node at the rate $\lambda_t$ msg/sec from the neighbouring ones. A model of a node is shown in Fig. 1.

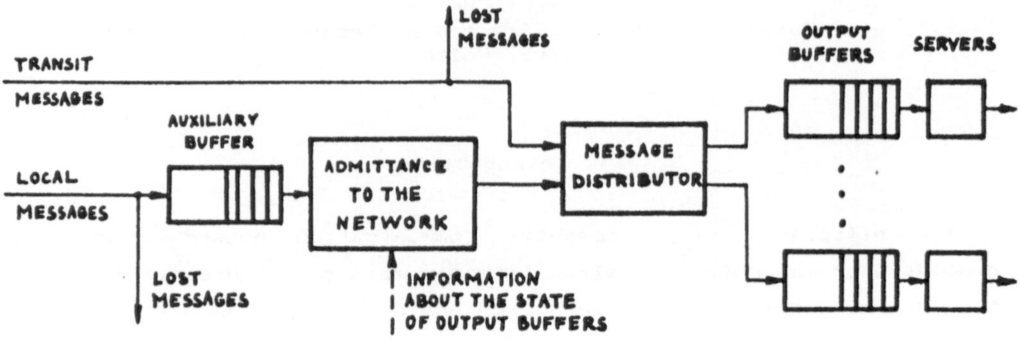


Fig. 1. Computer communication network node from the point of view of transit and local message service

We shall assume that transit messages should have larger chance to be admitted by a node than the local ones as they have traversed some part of its route and should, as quickly as possible, be delivered to their destination nodes. As it results from Fig. 1 transit messages go through a message distributor which directs them to the appropriate output buffer according to the routing rule.

The service of local messages is unlike the service of the transit ones. Local messages are first inputted into an auxiliary buffer from where they can be sent through the message distributor to an appropriate output buffer. However, a decision as to the transfer of a local message into the output buffer is undertaken on the basis of some auxiliary information concerning the state of that buffer.

To continue our considerations we shall make some simplifications in the model of a node given in Fig. 1. Let us assume that there is only one output channel in the node and the service time of a message by the message distributor compared to buffering and service delay can be neglected. As a result the simplified model of a node is shown in Fig. 2.

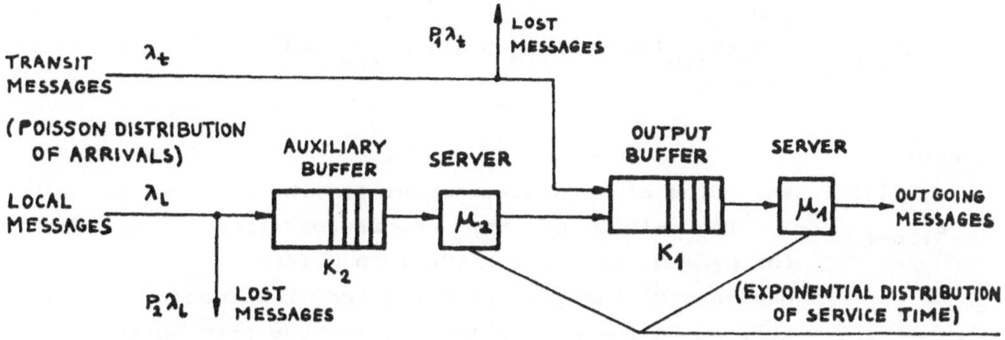


Fig. 2. The simplified model of a node; denotations: $K_i$, i=1,2 - storage capacity of i-th buffer, $P_i$, i=1,2 - probability of i-th buffer overflow, $\mu_i$, i=1,2 - average service rate of i-th server.

## 3. The description of a method of local congestion control

According to the model of the network node shown in Fig. 2 we can distinguish two discrete stochastic chains $\underset{\sim}{K}_1(tj)$ and $\underset{\sim}{K}_2(tj)$ where $t_j = jT$, j=0, ±1, ±2,..., and $\underset{\sim}{K}_i(t_j)$, i = 1,2, describes the states of the i-th buffer with the capacity $K_i$ (the state of each buffer is represented by the number of messages in it). In the proposed method the states of each output buffer are registered on-line and statistically processed. The results of the processing are used to the prediction of buffer state. Statistical processing is carried out with the following assumptions:

1. All measurements of a buffer state are taken periodically every T sec and their processing is carried on every MT sec according to a linear regression function. The period MT is an updating interval of routing tables.
2. The prediction in the period $(MT)_n$ is based on regression function known in the period $(MT)_{n-1}$, and the initial point of a straight line section in the n-th period $(MT)_n$ is identical with the end point of the straight line section in period $(MT)_{n-1}$.

The reason behind the usage of a linear regression function is the simplicity of that approach and already established results both from theoretical and practical point of view (see [6]). Thus, the criterion that will allow us to find the coefficients of the equation of straight line section takes on the form

$$Q(\zeta_{i,n}) = \sum_{j=1}^{M} [K_{i,n}(j) - (\zeta_{i,n-1}\, j + K^{o}_{i,n-1})]^2 \qquad (1)$$

where:

$K_{i,n}(j)$ – the value of j-th measurement for the i-th buffer

$\zeta_{i,n-1}$ – the slope of the regression line section in the period $(MT)_{n-1}$ for the i-th buffer

$K^{o}_{i,n-1}$ – ordinate of the initial point for the regression line section in the period $(MT)_{n-1}$ for the i-th buffer.

We get the minimum of that criterion with respect to $\zeta_{i,n-1}$ for

$$\zeta_{i,n-1} = \frac{\sum_{j=1}^{M} j\,[K_{i,n}(j) - K^{o}_{i,n-1}]}{\sum_{j=1}^{M} j^2} \qquad (2)$$

As a result the equation of the section of linear regression function takes on the form

$$K_{i,n}(j) = \zeta_{i,n-1}\, j + K^{o}_{i,n-1}, \quad i=1,2; \quad j=1,2,\ldots,M \qquad (3)$$

Knowing the parameters of the equation (3) we may find a credibility belt (see Fig. 3) in which with a given probability $(1-\alpha)$ the results of measurements will be located ($\alpha$ is an arbitrary small number greater than zero).

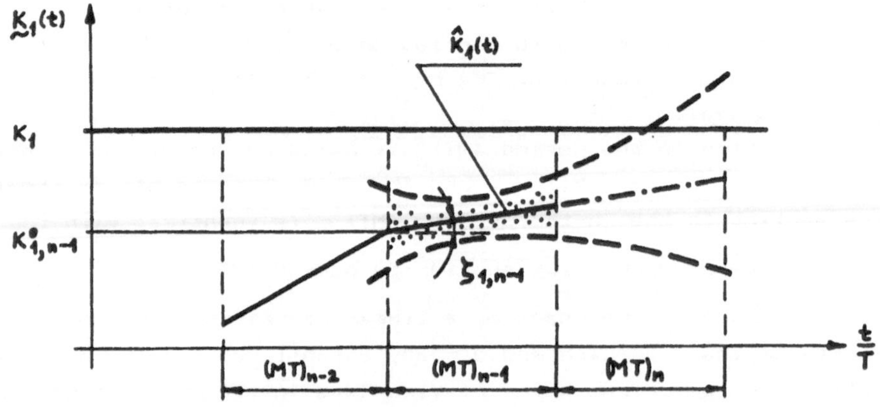


Fig. 3. The illustration of processing and prediction of buffer state measurements; $\hat{K}_1(t)$ is a regression line in the period $(MT)_n$

One can get the curves delimiting that belt by linking accordingly the points, which ordinates we may find from the expression [2]

$$K^{(\alpha)}_{1,n}(j) = \zeta_{1,n-1}\, j + K^{0}_{1,n-1} \pm$$

$$\pm\ s\ t_{\alpha,M-2} \sqrt{\frac{1}{M} + \frac{\left(j - \frac{1+M}{2}\right)^2}{\sum\limits_{j=1}^{M}\left(j - \frac{1+M}{2}\right)^2} + 1}\ , \quad j=1,2,\dots,M \tag{4}$$

where the sign "+" is for the upper curve denoted with the dashed line in Fig. 3 and s is the standard deviation of measurements with respect to conditional average represented by regression line and it is given by

$$s = \sqrt{\frac{1}{M-2}\sum_{j=1}^{M}\left[K_{1,n}(j) - (\zeta_{1,n-1}\, j + K^{0}_{1,n-1})\right]^2} \tag{5}$$

whereas $t_{\alpha,M-2}$ is the value of random variable $\underset{\sim}{T}$ obeying the Student distribution, such that probability $P(|\underset{\sim}{T}| \geq t_{\alpha,M-2}) \leq \alpha$ and (M-2) is the number of freedom degrees. We can use the Student distribution provided the deviations with respect to the average are independent from each other and have the same normal distribution and their sum is zero.

Clearly, the most interesting for us is the curve delimiting the upper bound of credibility belt. Producing this line on the period $(MT)_n$ we can state with probability $(1-\alpha)$ if one may expect the buffer overflow i.e. the upper curve will cut across the straight line $K_1(t) = K_1 = \text{const}$ in Fig. 3.

As it results from Fig. 2 we have

$$\zeta_{1,n-1} = [\, \lambda_{t,n-1}(1 - P_{1,n-1}) + \lambda_{1,n-1}(1 - P_{2,n-1})\,] - \mu_1 \tag{6}$$

where $P_{1,n-1}$ and $P_{2,n-1}$ represent blocking probability for the output and auxiliary buffer respectively.

Let us assume now that the predicted buffer state in period $(MT)_n$

exceeds the buffer capacity. Clearly, by decreasing the rate of introducing local messages into the node we can counteract the congestion.

Similarly we can find the slope of the regression line for auxiliary buffer and we obtain

$$\zeta_{2,n-1} = \lambda_{1,n-1} (1 - P_{2,n-1}) - \mu_2 \qquad (7)$$

Now, we can control the congestion in the node varying the value of that slope suitably with time. For that purpose we first find such a value $\zeta'_{1,n}$ that output buffer capacity with probability $(1 - \alpha)$ will not be exceeded. We obtain this value by solving the equation

$$K^{(\alpha)}_{1,n}(2M, \zeta'_{1,n}) = K_1 \qquad (8)$$

where $K^{(\alpha)}_{1,n}(2M, \zeta'_{1,n-1})$ is given by (4) for the sign ''+'' and $j = 2M$ provided $\zeta_{1,n-1} = \zeta'_{1,n}$ .

Let us define $\Delta\zeta_{1,n}$ as

$$\Delta\zeta_{1,n} = \zeta_{1,n-1} - \zeta'_{1,n} \qquad (9)$$

Let $\gamma_{2,n-1}$ be the rate of introducing local messages into the output buffer

$$\gamma_{2,n-1} = \lambda_{1,n-1} (1 - P_{2,n-1}) \qquad (10)$$

Taking into account (9) and (10) the desired rate of local service becomes

$$\gamma'_{2,n} = \gamma_{2,n-1} - \Delta\zeta_{1,n} \qquad (11)$$

Let $\vartheta_n$ be the parameter characterizing the relative decrease of rate for local messages in the period $(MT)_n$

$$\vartheta_n = \frac{\gamma'_{2,n}}{\gamma_{2,n-1}} \qquad (12)$$

We conclude that by decreasing $\vartheta_n$ times the rate of service for local messages we can avoid with probability $(1-\alpha)$ the overflow of output buffer in the period $(MT)_n$. The decrease of service rate for local messages one can get as a result of a slowing down operation. This operation we can define in the following way provided the process of message arrivals obeys Poisson distribution: each message residing at the head of queue in the auxiliary buffer is served with the rate $\theta_n\mu_2$. The server following the auxiliary buffer in Fig. 2 can execute the slowing down operation. This operation can easily be realized using random number generator obeying the exponential distribution with the parameter $\mu'_{2,n} = \theta_n\mu_2$.

## 4. The influence of the method of congestion control on parameters characterizing the performance of a node

In this section we will consider the operation of a node in a given updating interval, so we will omit the subscripts used to denote the numbers of consecutive periods. We will assume $\mu_1$ and $\theta\mu_2$ as the average service rate of messages waiting in the output and auxiliary buffer respectively.

As the parameters characterizing the performance of the node we shall assume:

- $\overline{T}_t, \overline{T}_l$ — average time delay of transit and local messages respectively
- $P'_1, P'_2$ — probability of output and auxiliary buffer overflow respectively.

One can notice that we have to deal in the node model with two service systems containing finite buffers, so, in the analysis, we can use the results of the investigation of M/M/1/K service system (see [3]). Particularly, we can find the average waiting time in the output buffer and server using the Little's theorem and we obtain

$$\overline{T}_t = \frac{\overline{K}_1}{\lambda_t(1-P'_1) + \theta\lambda_1(1-P'_2)} + \frac{1}{\mu_1} \tag{13}$$

The average waiting time in the auxiliary buffer and server is given by

$$\overline{T}_a = \frac{\overline{K}_2}{\lambda_1(1-P'_2)} + \frac{1}{\theta\mu_2} \tag{14}$$

In turn the average number of messages in the output and auxiliary buffer is as follows:

$$\overline{K}_1 = \frac{\rho_1^{\,2}}{1-\rho_1} \; \frac{1-\rho_1^{K_1}\,(K_1+1-K_1\rho_1)}{1-\rho_1^{K_1+2}} \tag{15}$$

$$\overline{K}_2 = \frac{\rho_2^{\,2}}{1-\rho_2} \; \frac{1-\rho_2^{K_2}\,(K_2+1-K_2\rho_2)}{1-\rho_2^{K_2+2}} \tag{16}$$

where $\rho_i$, $i=1,2$, is the utilization factor of respective buffer defined as the ratio of the intensity of incoming messages to the service intensity. Another convenient measure characterizing the quality of the node operation is the probability of buffer overflow. This parameter may be calculated for the output and auxiliary buffer according to the following expressions:

$$P_1' = \frac{(1-\rho_1)\,\rho_1^{K_1+1}}{1-\rho_1^{K_1+1}} \tag{17}$$

$$P_2' = \frac{(1-\rho_2)\,\rho_2^{K_2+1}}{1-\rho_2^{K_2+2}} \tag{18}$$

Using the above given results we can easily find the average time delay for transit messages that is given by (13). However, the average time delay for local messages $\overline{T}_l$ one can obtain by summing (13) and (14).

## 5. Numerical example

To illustrate the influence of transit and local message rates and of the parameter $\vartheta$ on parameters characterizing the quality of congestion control method presented above, the model of a network node shown in the Fig. 2 has been studied. For that model the calculations of average time delays $\overline{T}_t$ and $\overline{T}_l$, as well as the probabilities of buffer overflow for output and auxiliary buffer $P_1'$ and $P_2'$ have been carried out provided $\vartheta$ is fixed and the total arrival rate $\lambda = \lambda_t + \lambda_l$ of incoming messages is varying. We have assumed $K_1 = 20$, $K_2 = 4$, $\mu_1 = 10^2$, $\mu_2 = 10^3$, $\lambda_t = 0.75\ \lambda$ and $\lambda_l = 0.25\ \lambda$. The results obtained are shown in Fig. 4 and 5.

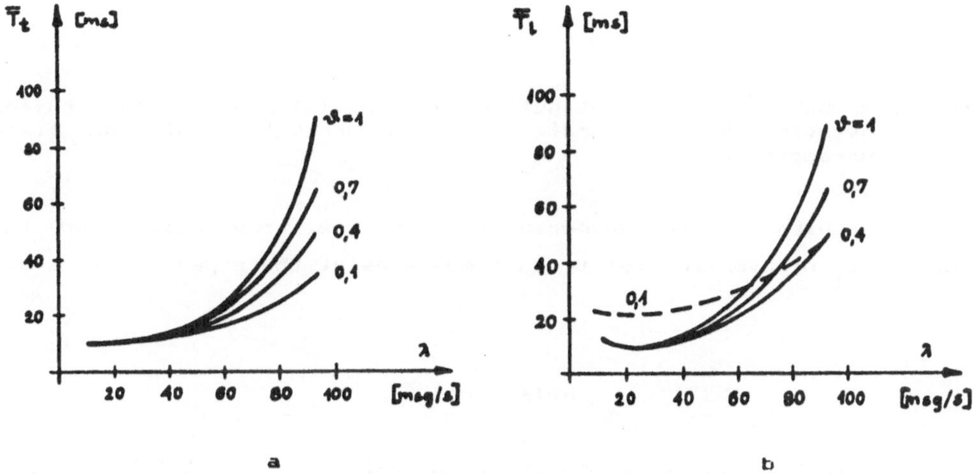


Fig. 4. Average time delay for for transit messages (a) and local message (b) versus rate of message arrivals $\lambda$ at the fixed parameter $\vartheta$

We see from Fig. 4 that average time delays $\overline{T}_t$ and $\overline{T}_l$ decrease while $\vartheta$ decreases. However, for local messages we observe an increase of $\overline{T}_l$ when $\vartheta$ is below some threshold. This behaviour results from the fact that while $\vartheta$ is decreasing local messages are outgoing with smaller and smaller rate from the auxiliary buffer, so in the time delay for local messages the component $\overline{T}_a$ plays an increasing role.

The probabilities of buffer overflow $P_1'$ and $P_2'$ presented in Fig.5 show some opposing trends depending upon $\vartheta$. Clearly, the decrease of probability $P_1'$ is achieved at the expense of the increase of probability $P_2'$. We see that the presented method of congestion

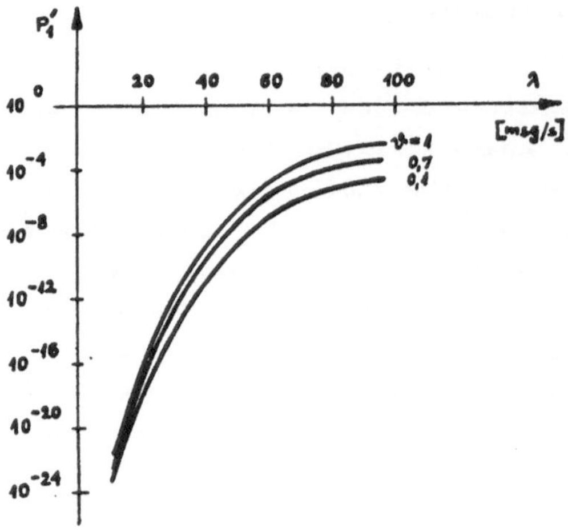


a

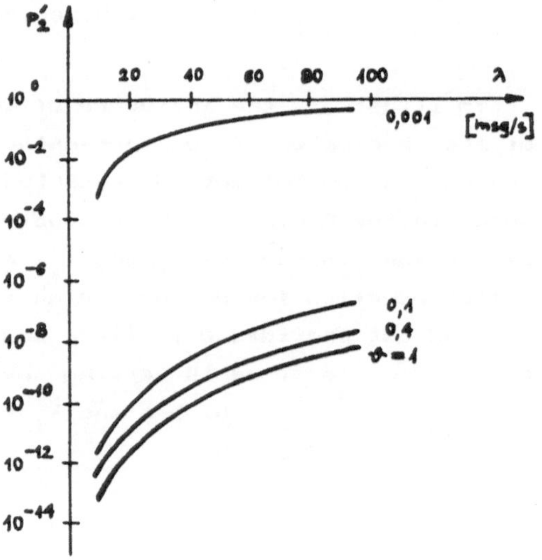


b

Fig. 5. Probability of output buffer overflow (a) and auxiliary buffer overflow (b) versus rate of message arrivals λ at the fixed parameter ϑ

control enables us to decrease the average time delay and the probability of overflow for transit messages at the expense of local ones.

## References


1. van As H.R., Congestion Control in Packet Switching Networks by a Dynamic Foreground-Background Storage Strategy, Proc. of the 2nd International Symposium on the Performance of Computer - Communications Systems, Zurich, 1984, pp. 433-448,
2. Draper N.R., Smith H., Applied Regression Analysis, John Wiley and Sons, New York, 1966,
3. Kleinrock L., Queueing Systems, Vol. I, II, John Wiley and Sons, New York, 1975, 1976,
4. Lam S.S., Reiser M., Congestion Control of Store-and-Forward Networks by Input Buffer Limits, IEEE Trans. on Commun., Vol.COM-27, No. 1, Jan. 1979, pp. 127 - 133,
5. Matsumoto J., Mori H., Analysis of Congestion Control in Packet Switched Networks by Gradual Restrictions of Virtual Calls, The Trans. of the IECE of Japan, Vol. E63, No. 3, March 1980, pp. 184 - 191,
6. Rutkowski D., Algorytm obróbki i weryfikacji pomiarów na bieżąco, Archiwum Automatyki i Telemechaniki, Tom XX, Zeszyt 3, str.275-285, in Polish,
7. Varakulsiripunth R., Shiratori N., Noguchi S., Congestion Control Scheme Based on Traffic Priority in Computer Networks with Finite Nodal Buffer, The Trans. of the IECE of Japan, Vol. E67, No.5 May 1984, pp. 279 -288.

# Program Optimization with Logic Program Transformation

Hiroshi Hoshino, Masahiro Esashi, Kiyoshi Agusa, Yutaka Ohno
Department of Information Science
Kyoto University, Kyoto, 606, JAPAN

**ABSTRACT**

We will introduce a software development method to develop an efficient program for stream data with a logic program transformation technique. The program specification is represented with the graph representation which is transformed to declarative Prolog predicates directly. In the specification, the input/output data are abstracted with list structure and the problem is specified with the relations between input list and output list. Since Prolog program is regarded as not only declarative predicates but also procedural program, we can utilize Prolog description as a declarative specification and a procedural program. In our method, the declarative specification is transformed to an efficient procedural program by the Fold/Unfold program transformation technique. In this paper, we will apply our method to the lexical analysis problem of the compiler program.

## 1.Introduction

We will introduce a software development method to develop an efficient program for stream processing based on logic program transformation. Our method supports design phase and programming phase in the software life cycle. STELLA (Software Transformation Environment for Logic LAnguage) is under development to support our method.

In the early stage of software development process, a software designer specifies a problem in a declarative style to get his idea into shape. This specification is not efficient if the modules are implemented as they are specified. On the other hand, a programmer writes a procedural program which is efficient to execute but complicated and hard to understand. That is, a programmer understands the declarative specification and optimizes it to execute the procedural program efficiently.

Since Prolog program is regarded as not only declarative predicates but also procedural program, we can utilize Prolog description as a declarative specification and a procedural program. We have applied the Fold/Unfold program transformation technique[1][2][3][4] to the Prolog description to optimize the specification.

The Fold/Unfold transformation technique is correctness-preserving transformation and derives an efficient program for stream data processing, that is, the iteration of data processing. Owing to this transformation, we can get an efficient procedural program for the stream data from a declarative specification by a fixed manipulation.

To get a good program, a software designer grasps and specifies a problem without worrying about complicated I/O operations and a programmer introduces a program with I/O operations from the specification after the problem is completely specified. We divide our transformation process into two steps; a transformation without worrying about I/O operations and a transformation considering I/O operations. We can get an efficient procedural program with I/O operations from a declarative specification without I/O operations by the fixed transformation technique.

Our software development process consists of three phases, specification phase, transformation phase which is the transformation step without I/O operations and program derivation phase which is the transformation step with I/O operations.

In the chapter 3, 4 and 5, we will describe each phase in detail with the example of the lexical analysis problem. In the chapter 6, our method is compared with UNIX's 'pipe' mechanism.

## 2.Software Transformation System

The goal of STELLA System is to bridge the gap between design phase and programming phase and to derive efficient programs automatically from the specifications. To achieve this purpose, we adopt logic programming language Prolog for describing program specifications and utilize Fold/Unfold transformation technique to derive procedural programs from the declarative specifications. A declarative description in Prolog is suitable for program transformation. Moreover, not only specifications but also programs can be described in Prolog because Prolog program can be regarded as not only declarative predicates but also procedural programs.

Our software development system STELLA consists of three subsystems; the specification system, the transformation system and the program derivation system. In the follwoing chapters, we explain each system in detail and show the process of transformation from the declarative specifications to the procedural programs with the example of the lexical analysis problem[5].

The lexical analysis program analyzes the input character stream and outputs the word stream. For example, when the following character stream is input,

"the man, who is rich, saw John's watch."

it is analyzed and the next word stream is output:

[the,man,',',who,is,rich,',',saw,'John''s',watch,.]

## 3.Specification System

Program specifications written in predicate logic have several advantages for transformation. The specification system supports the specification phase to specify the problem easily and generates the logical specification which is a declarative Prolog description.

We worked out the specification method with predicates for the stream data processing. The basic idea of our specification method is that the streams of input/output data are represented by list expressions and that the problem is regarded as the relations between the list data. The software designer can define the problem independently of the sequence of input/output operations. The relations between the list data are defined with the following basic operations:

1) *division*

The designer regards the input data as the collection of primitive data and represents the data with the list expression. In other words, to regard the input data as the collection of primitive data is to divide the input list into several parts.

2) *construction*

The designer regards the output data as the construction of the primitive data and represents the data with the list expression.

3) *process*

The relation between primitive parts of the list is regarded as an operation which processes the primitive parts of the input list and generates the primitive parts of the output list.

These operations are represented with predicates. Fig.1 presents a declarative specification for lexical analysis problem. The 'division' and 'construction' operations are represented with 'append' predicate. The former divide a input list into three parts 'Head', 'C' and 'Tail'. The latter construct a output list from two sublists 'HeadWords' and 'TailWords'. The boundary conditions in the figure present termination conditions.

```
getword([],[]).                                          /* boundary condition */
getword(List,Words) :-
        append(Head,[C|Tail],List),                               /* divide */
        single_character(C),
        name(Single,[C]),                                        /* process */
        getword(Head,HeadWords),
        getword(Tail,TailWords),
        append(HeadWords,[Single|TailWords],Words).          /* construct */
getword(List,Words) :-
        append(Head,[C|Tail],List),                               /* divide */
        space_character(C),
        getword(Head,HeadWords),                                 /* process */
        getword(Tail,TailWords),
        append(HeadWords,TailWords,Words).                   /* construct */
getword(List,[Words]) :-                                 /* boundary condition */
        name(Words,List).
```

**Fig.1 Logical Specification of Lexical Analysis Problem**

The information of data structure is used to decide the specification strategy for the basic operations and reserved for following phase. It will be used for the retrieval of the I/O predicates in the program derivation system and also used for the derivation of the procedural program.

## 4.Transformation System

The transformation system supports the transformation phase and transforms the declarative logical specifications into the procedural specifications by Fold/Unfold transformation technique, the rules about the accumulation and some heuristics.

We make good use of the characteristics of 'append' predicate, when we transform the declarative specification to the procedural one. The 'append' predicate connects two list into one. But it is also used to divide one list into two parts. We have used 'append' with the latter meaning to represent the 'division' basic operation.

On the other hand, the definition of 'append' is as follows:

```
append([],L,L).
append([H|T],L,[H|R]):- append(T,L,R).
```

From this definition of 'append', we can find the procedure that the members of list are inspected by turns from the head of list. That is, this definition of 'append' is not declarative but procedural. Therefore, we can get the procedural specification by unfolding of 'append' predicate in the declarative specification.

The logical specification must be transformed so that it is clear when the input/output list are accessed. If the access point to the input/output list is defined exactly in the logical specification, the specification with I/O predicates is transformed into the procedural specification in which it is clear when the input/output data are accessed. Otherwise, the specification with I/O predicate cannot be transformed into the procedural specification in the program derivation phase.

The time when the input data are accessed is the time when the members of the input list are accessed in the logical specification. On the other hand, the time when the output data are fixed is not always the time when the members of the output list are accessed. For example, the access time to the input list is the same in the following predicates; 'sum' and 'sums'.

```
sum([], 0).
sum([H|T], R) :- sum(T, RR), R is RR+H.
sums([], _, []).
sums([H|T],Sum,[HR|R]) :- HR is Sum+H, sums(T,HR,R).
```

The output data of 'sum' is fixed when the input list is exhausted. Each output data of 'sums' is fixed when each member of the input list is accessed.

We have adopted the following predicates with accumulating parameter 'Temp' instead of the 'append' predicate in Fig.1 to make it clear when the output data are fixed.

```
append1(Temp,[C|T],[C|T],Temp) :-
        space_character(C).
append1(Head,T,[H|L],Temp) :-
        append(Temp,[H],Tp), append1(Head,T,L,Tp).
append2(Temp,[C|T],[C|T],Temp) :-
        single_character(C).
append2(Head,T,[H|L],Temp) :-
        append(Temp,[H],Tp), append2(Head,T,L,Tp).
getword([],[],[]).
getword(List,Words,Temp) :-
        append2(Head,[C|Tail],List,Temp),
        getword(Head,HeadWord,[]), name(Single,C),
        getword(Tail,TailWords,[]),
        append(HeadWord,[Single|TailWords],Words).
getword(List,Words,Temp) :-
        append1(Head,[C|Tail],List,Temp),
        getword(Head,HeadWords,[]),
        getword(Tail,TailWords,[]),
        append(HeadWords,TailWords,Words).
getword(List,[Word],[]) :- name(Word,List).
```

**Fig.2 Logical Specification with Accumulating Parameter**

It is important that the accumulating parameter 'Temp' plays a role of an input buffer in the above predicates. Using the predicate which has the accumulating parameter, the predicates which need to look ahead several input data like as the predicate 'getword' can be transformed into the predicates in which the backtrack does not occurred.

Fig.3 shows the transformed logical specification derived from the declarative specification with the transformation. The transformed logical specification is optimized because each element of the input stream is scanned at once and each element of the output stream is fixed as soon as a word boundary 'space_character' or 'single_character' is found. In this transformation process, the accumulation technique is applied.

```
getword([],[],[]).
getword([],[Words],Temp) :-
        name(Words,Temp).
getword([C|Tail],[Single|TailWords],[]) :-
        single_character(C),
        name(Single,[C]),
        getword(Tail,TailWords,[]).
getword([C|Tail],[W,Single|TailWords],Temp) :-
        single_character(C),
        name(Single,[C]),
        name(W,Temp),
        getword(Tail,TailWords,[]).
getword([C|Tail],TailWords,[]) :-
        space_character(C),
        getword(Tail,TailWords,[]).
getword([C|Tail],[W|TailWords],Temp) :-
        space_character(C),
        name(W,Temp),
        getword(Tail,TailWords,[]).
getword([H|L],Words,Temp) :-
        append(Temp,[H],Tmp),
        getword(L,Words,Tmp).
```

**Fig.3 Transformed Logical Specification of Lexical Analysis Problem**

## 5.Program Derivation System

The program derivation system supports the program derivation phase and the program written in the procedural language is derived. This system consists of three subsystems: I/O predicate retrieval system, transformation engine for I/O predicate and procedural program generator.

The I/O predicate retrieval system retrieves the input/output predicates from DB with the key of input/output data structure which is reserved in the specification phase. The input predicate constructs the input list from the real input data and the output predicate converts the output list to the real output data. The input/output predicates are added before/after the transformed logical specification.

Fig.4 shows the input/output predicates and the logical specification with those predicates for the lexical analysis problem.

```
        words(InputFile,OutputFile) :-
              readfile(InputFile,List),
              getword(List,Words,[]),
              writefile(OutputFile,Words).
  /* Predicates for input */
        readfile(InputFile,List) :-
              see(InputFile), getlist(List), seen.
        getlist(Rest) :-
              get0(C),
              conschar(C,Rest).
        conschar(eof,[]).
        conschar(C,[C|Rest]) :-
              getlist(Rest).
  /* Predicates for output */
        writefile(OutputFile,Words) :-
              tell(OutputFile), writelist(Words), told.
        writelist([]).
        writelist([Word|Words]) :-
              write(Word),nl,
              writelist(Words).
```

**Fig.4 Logical Specification with I/O Predicates for Lexical Analysis Problem**

The input predicate 'readfile' reads characters from 'InputFile' and constructs the input list 'List'. The output predicate 'writefile' writes the members of the output list 'Words' into 'OutputFile'. These input/output predicates are added before/after the predicate 'getword' which is the transformed logical specification and the new predicate 'words' is defined.

The transformation engine for I/O predicate transforms the logical specification with I/O predicate into the procedural specification in which the order of processing for the real data is obvious. Fig.5 shows the procedural specification and its control flow of the lexical analysis problem. The loop structure is found in the control flow. This transformation rule is basically Fold/Unfold but differs in the point that the predicates have the side effects of the input/output.

```
words(InputFile,OutputFile) :-
      see(InputFile),tell(OutputFile),
      getwrite([]),
      seen,told.
getwrite(Temp) :-                    getwrite:        Control Flow
      get0(C),
      readwrite(C,Temp).             ┌─────────┐
                                     │ get0(C) │◄──────────────────────┐
                                     └─────────┘                       │
readwrite(eof,[]).                    ──► readwrite(eof,[]).           │
readwrite(eof,Temp) :-                ──► readwrite(eof,Temp):- ...     │
      name(Word,Temp),                                                 │
      write(Word),nl.                                                  │
readwrite(C,[]) :-                    ──► readwrite(C,[]):-            │
      single_character(C),                  single_character(C),... ──►│
      name(Single,[C]),                                                │
      write(Single),nl,                                                │
      getwrite([]).                                                    │
readwrite(C,Temp) :-                  ──► readwrite(C,Temp):-          │
      single_character(C),                  single_character(C),... ──►│
      name(Single,[C]),                                                │
      name(W,Temp),                                                    │
      write(W),nl,                                                     │
      write(Single),nl,                                                │
      getwrite([]).                                                    │
readwrite(C,[]) :-                    ──► readwrite(C,[]):-            │
      space_character(C),                   space_character(C),... ──►│
      getwrite([]).                                                    │
readwrite(C,Temp) :-                  ──► readwrite(C,Temp):-          │
      space_character(C),                   space_character(C),... ──►│
      name(W,Temp),                                                    │
      write(W),nl,                                                     │
      getwrite([]).                                                    │
readwrite(C,Temp) :-                  ──► readwrite(C,Temp):- ... ────►┘
      append(Temp,[C],Tmp),
      getwrite(Tmp).
```

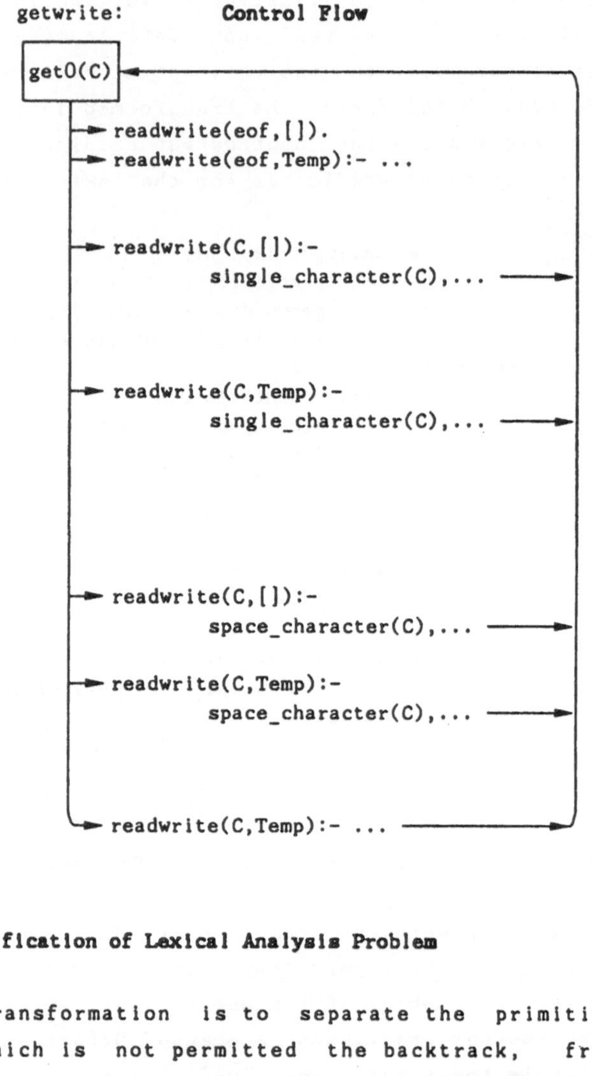

**Fig.5 Procedural Specification of Lexical Analysis Problem**

The point of this transformation is to separate the primitive input predicate 'get0', which is not permitted the backtrack, from other predicates, and to construct the two-step recursion like 'getwrite' and 'readwrite'. After this transformation, we can get the predicates in which the backtrack does not occurred.

The procedural program generator generates the program written in the procedural language C. The tail recursive structure is transformed to the loop structure and the input/output predicates are realized with the procedural input/output functions obtained from DB. Fig.6 shows the procedural program written in language C. The tail recursive structure represented in Fig.5 is transformed to 'while' loop structure. The predicate 'get0' and 'write' is realized with the function 'getc' and 'fprintf'.

```
getwrite(Temp)
char        *Temp;
{int        c;
        while(1) {
                if ((c = getc(ifp)) == EOF && *Temp == '¥0') {
                        return;
                } else if (c == EOF && *Temp) {
                        fprintf(ofp,"%s¥n",Temp); return;
                } else if (*Temp == '¥0' && single_character(c)) {
                        fprintf(ofp,"%c¥n",c); *Temp = '¥0';
                } else if (*Temp && single_character(c)) {
                        fprintf(ofp,"%s¥n%c¥n",Temp,c); *Temp = '¥0';
                } else if (*Temp == '¥0' && space_character(c)) {
                        continue;
                } else if (*Temp && space_character(c)) {
                        fprintf(ofp,"%s¥n",Temp); *Temp = '¥0';
                } else {
                        char        cc[2];
                        cc[0] = c, cc[1] = '¥0', strcat(Temp,cc);
                }
        }
}
```

**Fig.6 Procedural Program of Lexical Analysis Problem**

## 6.Discussion

In our research, the predicates can be recognized as the software parts and the transformation can be considered the customization of the software parts. For example, we can get the word count program by the connection of 'getword' and the next predicate 'count'.

```
count([], 0).
count([H|T], R) :- count(T, RR), R is RR+1.
```

In the UNIX operation system, there are many commands which process the stream data. These commands are connected by the 'pipe' mechanism and are reused for the prototype of a new command. The new command is going to be written in the language C for an efficient execution after the specification of the command is decided. In our method, we can construct a new predicate from the existing predicates which process the stream data and can get the program written in the language C by the transformation technique. Our predicates are more general, because those predicate process the list data and the special predicates for other data structures are separated. That is, our software parts are more general than the UNIX commands.

The logical specification is executable by itself, because it is a Prolog program in our method. The logical specification is regarded as the prototype of the target system and the designer can make sure of the specification with the execution of it. Our prototype grows into the procedural program by the transformation methods after the specification is confirmed. That is, our method is the rapid prototyping method which brings up the prototype to the target system.

## 7.Conclusion

In this paper, we have described the program development method with the logic programming language. The declarative specification for the stream data processing is transformed and optimized to the procedural program by the fixed manipulation. Our transformation technique is basically Fold/Unfold transformation technique, but it is different from the Fold/Unfold at the point that other strategies need to be applied; the strategies for recursion with the accumulating parameters and the strategies for I/O predicates.

Our method is regarded not only as the reusing method of software but also as the rapid prototyping method, because the logical specification is general and reusable and is executable by itself.

**References**

[1] John Darlington,
The Structured Description of Algorithm Derivation
Algorithmic Languages, de Bakker/van Vliet(eds.)
IFIP, North-Holland Publishing Company, 1981, pp221-250

[2] Keith Clark, Sharon Sickel
PREDICATE LOGIC: A CALCULUS FOR DERIVING PROGRAMS
5th IJCAI, 1977

[3] C.J.HOGGER
Derivation of Logic Programs
JACM, Vol.28, No.2, April 1981, pp.372-392

[4] Taisuke SATO, Hisao TAMAKI
TRANSFORMATIONAL LOGIC PROGRAM SYNTHESIS
Proc. of the Int. Conf. on FGCS, 1984

[5] W.F.Clocksin, C.S.Mellish
Programming in Prolog
Springer-Verlag Berlin Heidelberg 1981

# THE SENSITIVITY OF SOME RECOGNITION ALGORITHMS FOR PRINTED ARABIC CHARACTERS

M.F. Tolba, A. Goned, A. Salem and E. Shaddad
Scientific Computing Center, Ain Shams University, Cairo, Egypt

**ABSTRACT**

The sensitivities of four proposed algorithms (A1 - A4), used for the recognition of printed Arabic characters, have been investigated for different square normalization matrices with a dimension of "m" binary dots. The performed analysis shows that, for all algorithms, the sensitivity increase with the value of "m", however, m = 4 to 8 may be adopted as a satisfactory values for recognition. The algorithm A1 has the best recognition sensitivity and A4 is useful for detecting the fine structure of some Arabic characters. It is concluded that a combined feature vector based on (A1 and A4) or (A2 and A4) will produce a new powerful algorithm. This algorithm is expected to be much more sensitive, as an Arabic recognition tool, than any of the four investigated algorithms.

## 1 - Introduction

To the best of our knowledge, the different algorithms proposed for the recognition of printed (or handwritten) Arabic scripts did not solve all facing problems, and many critical points have to be investigated before achieving any practical Arabic Reading Machine. The segmentation of cursive words, the recognition dependence on "dots"- "Hamza" and the possible presence of diacritical signs in the Arabic scripts are the maim problems to be deeply investigated.

Two main approaches have been used for Arabic character recognition: a) A statistical approach (e.g.[ 1 ]-[ 4 ]). b) A structural approach (e.g.[ 5 ]-[ 9 ]). For printed Arabic scripts, we believe that the statistical oriented algorithms can deal properly with all designed fonts beside their expected time and space econamy. The structural approach is convenient for handwritten scripts which will certainly need a syntatical and a semantical analysis for the Arabic sentences to increase the recognition efficiency.

According to our proposed algorithm [ 1 ] , [ 2 ] the physical image is transfered in the form of a binary matrix to the computer memory and then the following procedure is performed :

a) The removal of "dots" and "Hamza" by the vertical projection procedure [10 ] or by two dimensional filter with proper coefficients [ 1 ].
b) The window narrowing process, where the four sides of the matrix are allocated tangentially to the fetched character.
c) The normalization process, where the character is transfered into a standard matrix. This process is necessary for size independent recognition and should be done with minimum distortion to the transfered characters [ 11 ].
d) The extraction of the feature vector from the normalized matrix.
e) The calculation of the recognition probability "P" which is based on the difference between the feature vector of the recognized character and the prestored vectors for all other characters.

In the present work, the sensitivity of four algorithms to the recognition of Arabic characters has been studied. The sensitivity of the algorithm means its ability to recognize the investigated character, efficiently, among all other prestored Arabic characters. This will, comparatively, evaluate these algorithm and will, certainly, throw some light on one of the possible tools for measuring the power and efficiency of any proposed recognition algorithm. On the other hand, the recognition speed and required space for the four algorithms are discussed. The proposed analysis is implemented, on one of the familiar Arabic fonts of letter quality computer printer, using personal computer with 80286 microprocessor.

## 2 - Extraction of Feature Vector for Different Algorithms.

As a first approximation, the normalization process is performed using a square matrix with dimension "m" binary dots. The elements of the normalization matrix are given by e(i,j) for the ith row and jth column, where e(i,j) = 0 or 1. The feature vector is extracted from the normalization matrix for the four investigated algorithms as follows:

a) Algorithm 1 (A1):
The first selected algorithm is the "Aligned Four Side Profile "algorithm [ 1 ],[ 2 ]. The Feature vector $\underline{V}$ is given as:

$$\underline{V} = [\, a(1), a(2), \ldots a(m), b(1), b(2), \ldots b(m), c(1), c(2), \ldots c(m), d(1), d(2), \ldots d(m) \,] \quad \ldots\ (1)$$

where a(j), as an example, is given by:

$$a(j) = \sum_{i=1}^{m} k(i,j) \quad ; \; j = 1,2,3,\ldots m \quad \ldots\ (2)$$

where k(i,j)= 1 if e(i,j)=0 and k(i-1,j) = 1. The first value k(0,j) is always = 1. Identically, b(i), c(j)and d(i) for the four sides of the normalization matrix are evaluated.

b) Algorithm 2 (A2):

The second algorithm is the "Matrix Projection" algorithm [ 4 ]. The Feature vector $\underline{V}$ is given by:

$$\underline{V} = [\, a(1), a(2), \ldots a(m), b(1), b(2), \ldots . b(m) \,] \qquad \ldots . \quad (3)$$

where a(j) and b(i) are given for vertical and horizontal directions, respectively, as follows:

$$a(j) = \sum_{i=1}^{m} e(i,j) \quad ; j = 1,2,3, \ldots . m \qquad \ldots . \quad (4)$$

$$b(i) = \sum_{j=1}^{m} e(i,j) \quad ; i = 1,2,3, \ldots . m \qquad \ldots . \quad (5)$$

c) Algorithm 3 (A3):

The third algorithm the "Linear Prediction" algorithm (e.g. [ 12 ]) where the feature vector may be expressed as:

$$\underline{V} = [\, e(1,1), e(2,1), e(3,1), \ldots e(m,m) \,] \qquad \ldots . \quad (6)$$

d) Algorithm 4 (A4):

The fourth selected algorithm is the "Vertical - Horizontal Intersections" algorithm [ 3 ] where the feature vector is given as:

$$\underline{V} = [\, a(1), a(2), \ldots a(m), b(1), b(2), \ldots b(m) \,] \qquad \ldots . \quad (7)$$

where a(j) is the number of intersections in the jth column and b(i) is the number of intersections in the ith row.

## 3 - The Algorithm Sensitivity

The recognition parameter "P" is defined as:

$$P = 1/(1+D) \qquad \ldots . \quad (8)$$

where D is the length of the difference vector between the feature vector $\underline{V}$, of a given character, and the prestored feature vectors of other characters. The full recognition will be achieved if P=1 (i.e. D=0).

Any recognition algorithm is said to be sensitive if the value of "P" is very high (e.g.=1) for the desired character and very low (e.g. near to zero) for all other characters. For a given font, and a given dimension of normalization matrix "m", the sensitivity matrix (for a particular set of characters) can simply be performed by calculating the value of "P" for each character with all other characters (Tables 1-4). The diagonal of the sensitivity matrix, when presented for a given font with itself, should have values of P=1. If the nondiagonal values of "P" are very small then the adopted algorithm is very powerful as a recognition tool. The sensitivity parameter "S" for a given set of characters may be defined as:

$$S = \sum_{i=1}^{n} \sum_{j=1}^{n} P(i,j)/n^2 - \sum_{i=1}^{n} P(i,i)/n \qquad \ldots. \quad (9)$$

where n is the number of the investigated characters and P(i,j) is the sensitivity parameter as defined by equation (8) for ith row and jth column of the sensitivity matrix. According to equation(9), a small value of the parameter "S" means a high sensitivity of the used algorithm.

## 4 - Classification of Arabic Characters

Based on the classification proposed by [ 1 ], the Arabic characters are classified according to the following groups: a) Single stroke characters. b) Characters with "dots" or "Hamza". Each group is also classified according to: a) Start-Middle characters.
b) End-Isolated characters. The resulting four groups are given as:

Group 1 - Single stroke characters (Start-Middle):
This group contain the following characters:

ـطـ ـسـ ـصـ ـعـ ـهـ

كـ لـ ـمـ ـهـ ـمـ

Group 2 - Single stroke characters (End-Isolated):
This group contain the following characters:

ا ح ح د ر س ص ط ط ع ـع ك ل م ه و لا ى

١ ٢ ٣ ٤ ٥ ٦ ٧ ٨ ٩

Group 3 - Characters with "dots" and "Hamza" (Start-Middle):
This group contain the following characters:

بـ تـ ثـ جـ خـ شـ ضـ

غـ ـغـ فـ قـ نـ يـ ئـ

Group 4 - Characters with "dots" and "Hamza" (End-Isolated):
This group contain the following characters:

أ إ آ ب ت ث خ ذ ز ش ض غ

ـغ ق ن ـة ؤ لأ لا لإ ي ئ ى

It is clear that further classification can be made according to the number of "dots" and zone location of "dots" or "Hamza". However, any minor distortion in these critical identifiers will lead to a recognition failure. Therefore, the sensitivity matrices are developed for the defined four groups with a considerable assurance that no collision may exist between characters in different groups.

## 5 - Analysis and Results

The sensitivity matrices (for Group 1 - Group 4) are constructed for the four defined algorithms using values of m = 3 - 14. Table(1) is an example of the sensitivity matrix for Group 4 and algorithm 1 (A1) using m = 4 . The circles show the values of P(i,j) greater than 0.6 . Idendically, Table(2) shows P(i,j) for the same group and the same algorithm (A1) but for m = 8 . Comparing Table(1) with Table(2), it is clear that the nondiagonal values of P(i, j) (i.e. $i \neq j$) are reduced for m = 8 . This is attributed to the high recognition sensitvity for large values of "m". On the other hand, for both cases the diagonal values of P(i,j) (i.e. i=j) is equal to unity. The same behaviour is also observed for the other three algorithms. Table (3) and Table (4) show the sensitivity matrices for m =4 and m = 8, respectively, according to the algorithm A4 (Circles represent values greater than 0.7).

From equations (1,3,6,7) it is clear that the required space for storing the "Featur Lookup Table" and also the processing time is approximately proportional to 4m, 2m, $m^2$ and 2m for algorithms A1, A2, A3 and A4, respectively. It seems that space and speed economy is in favour of A2 and A4 while it is not bad in A1 as it is in A3.

**Table (1)**
**The sensitivity matrix for Group 4, algorithm A1 and m=4**

| | ا | لا | ب | ه | د | ف | ح | ر | س | ص | ع | ـع | ق | ن | و | ى |
|---|---|---|---|---|---|---|---|---|---|---|---|---|---|---|---|---|
| ا | 1.0 | 0.6 | 0.6 | 0.7 | 0.6 | 0.5 | 0.5 | 0.6 | 0.7 | 0.7 | 0.6 | 0.6 | 0.5 | 0.7 | 0.6 | 0.6 |
| لا | 0.6 | 1.0 | 0.6 | 0.6 | 0.6 | 0.6 | 0.5 | 0.5 | 0.6 | 0.7 | 0.7 | 0.7 | 0.6 | 0.6 | 0.6 | 0.6 |
| ب | 0.6 | 0.6 | 1.0 | 0.6 | 0.6 | 0.6 | 0.5 | 0.5 | 0.6 | 0.5 | 0.5 | 0.5 | 0.6 | 0.7 | 0.5 | 0.6 |
| ه | 0.7 | 0.6 | 0.6 | 1.0 | 0.7 | 0.6 | 0.6 | 0.6 | 0.7 | 0.7 | 0.7 | 0.7 | 0.6 | 0.6 | 0.6 | 0.6 |
| د | 0.6 | 0.6 | 0.6 | 0.7 | 1.0 | 0.6 | 0.5 | 0.6 | 0.6 | 0.6 | 0.7 | 0.8 | 0.6 | 0.6 | 0.6 | 0.6 |
| ف | 0.5 | 0.6 | 0.6 | 0.6 | 0.6 | 1.0 | 0.5 | 0.6 | 0.5 | 0.5 | 0.6 | 0.6 | 1.0 | 0.6 | 0.6 | 0.8 |
| ح | 0.5 | 0.5 | 0.5 | 0.6 | 0.5 | 0.5 | 1.0 | 0.5 | 0.6 | 0.6 | 0.6 | 0.6 | 0.5 | 0.5 | 0.5 | 0.5 |
| ر | 0.6 | 0.5 | 0.5 | 0.6 | 0.6 | 0.6 | 0.5 | 1.0 | 0.6 | 0.6 | 0.6 | 0.6 | 0.6 | 0.5 | 0.6 | 0.6 |
| س | 0.7 | 0.6 | 0.6 | 0.7 | 0.6 | 0.5 | 0.6 | 0.6 | 1.0 | 0.7 | 0.6 | 0.7 | 0.5 | 0.6 | 0.6 | 0.6 |
| ص | 0.7 | 0.7 | 0.5 | 0.7 | 0.6 | 0.5 | 0.6 | 0.6 | 0.7 | 1.0 | 0.6 | 0.6 | 0.5 | 0.6 | 0.6 | 0.6 |
| ع | 0.6 | 0.7 | 0.5 | 0.7 | 0.7 | 0.6 | 0.6 | 0.6 | 0.6 | 0.6 | 1.0 | 0.8 | 0.6 | 0.6 | 0.6 | 0.6 |
| ـع | 0.6 | 0.7 | 0.5 | 0.7 | 0.8 | 0.6 | 0.6 | 0.6 | 0.7 | 0.6 | 0.8 | 1.0 | 0.6 | 0.6 | 0.7 | 0.6 |
| ق | 0.5 | 0.6 | 0.6 | 0.6 | 0.6 | 1.0 | 0.5 | 0.6 | 0.5 | 0.5 | 0.6 | 0.6 | 1.0 | 0.6 | 0.6 | 0.8 |
| ن | 0.7 | 0.6 | 0.7 | 0.6 | 0.6 | 0.6 | 0.5 | 0.5 | 0.6 | 0.6 | 0.6 | 0.6 | 0.6 | 1.0 | 0.5 | 0.5 |
| و | 0.6 | 0.6 | 0.5 | 0.6 | 0.6 | 0.6 | 0.5 | 0.6 | 0.6 | 0.6 | 0.6 | 0.7 | 0.6 | 0.5 | 1.0 | 0.6 |
| ى | 0.6 | 0.6 | 0.6 | 0.6 | 0.6 | 0.8 | 0.5 | 0.6 | 0.6 | 0.6 | 0.6 | 0.6 | 0.8 | 0.5 | 0.6 | 1.0 |

**Table (2)**
**The sensitivity matrix for Group 4, algorithm A1 and m=8**

| | ا | لا | ب | ه | د | ف | ح | ر | س | ص | ع | ـع | ق | ن | و | ى |
|---|---|---|---|---|---|---|---|---|---|---|---|---|---|---|---|---|
| ا | 1.0 | 0.4 | 0.3 | 0.4 | 0.4 | 0.3 | 0.4 | 0.4 | 0.5 | 0.4 | 0.4 | 0.4 | 0.3 | 0.4 | 0.4 | 0.4 |
| لا | 0.4 | 1.0 | 0.4 | 0.4 | 0.4 | 0.5 | 0.4 | 0.4 | 0.4 | 0.5 | 0.4 | 0.4 | 0.5 | 0.5 | 0.4 | 0.4 |
| ب | 0.3 | 0.4 | 1.0 | 0.4 | 0.4 | 0.4 | 0.3 | 0.4 | 0.4 | 0.4 | 0.3 | 0.4 | 0.4 | 0.5 | 0.3 | 0.4 |
| ه | 0.4 | 0.4 | 0.4 | 1.0 | 0.6 | 0.5 | 0.4 | 0.4 | 0.5 | 0.5 | 0.4 | 0.5 | 0.5 | 0.4 | 0.5 | 0.5 |
| د | 0.4 | 0.4 | 0.4 | 0.6 | 1.0 | 0.4 | 0.3 | 0.4 | 0.4 | 0.4 | 0.4 | 0.4 | 0.4 | 0.4 | 0.5 | 0.4 |
| ف | 0.3 | 0.5 | 0.4 | 0.5 | 0.4 | 1.0 | 0.3 | 0.4 | 0.4 | 0.4 | 0.3 | 0.4 | 1.0 | 0.4 | 0.4 | 0.5 |
| ح | 0.4 | 0.4 | 0.3 | 0.4 | 0.3 | 0.3 | 1.0 | 0.3 | 0.4 | 0.3 | 0.4 | 0.4 | 0.3 | 0.3 | 0.3 | 0.3 |
| ر | 0.4 | 0.4 | 0.4 | 0.4 | 0.4 | 0.4 | 0.3 | 1.0 | 0.4 | 0.4 | 0.3 | 0.4 | 0.4 | 0.3 | 0.6 | 0.4 |
| س | 0.5 | 0.4 | 0.4 | 0.5 | 0.4 | 0.4 | 0.4 | 0.4 | 1.0 | 0.6 | 0.4 | 0.4 | 0.4 | 0.4 | 0.4 | 0.4 |
| ص | 0.4 | 0.5 | 0.4 | 0.5 | 0.4 | 0.4 | 0.3 | 0.4 | 0.6 | 1.0 | 0.4 | 0.4 | 0.4 | 0.4 | 0.4 | 0.4 |
| ع | 0.4 | 0.4 | 0.3 | 0.4 | 0.4 | 0.3 | 0.4 | 0.3 | 0.4 | 0.4 | 1.0 | 0.6 | 0.3 | 0.3 | 0.3 | 0.3 |
| ـع | 0.4 | 0.4 | 0.4 | 0.5 | 0.4 | 0.4 | 0.4 | 0.4 | 0.4 | 0.4 | 0.6 | 1.0 | 0.4 | 0.3 | 0.4 | 0.3 |
| ق | 0.3 | 0.5 | 0.4 | 0.5 | 0.4 | 1.0 | 0.3 | 0.4 | 0.4 | 0.4 | 0.3 | 0.4 | 1.0 | 0.4 | 0.4 | 0.5 |
| ن | 0.4 | 0.5 | 0.5 | 0.4 | 0.4 | 0.4 | 0.3 | 0.3 | 0.4 | 0.4 | 0.3 | 0.3 | 0.4 | 1.0 | 0.3 | 0.4 |
| و | 0.4 | 0.4 | 0.3 | 0.5 | 0.5 | 0.4 | 0.3 | 0.6 | 0.4 | 0.4 | 0.3 | 0.4 | 0.4 | 0.3 | 1.0 | 0.4 |
| ى | 0.4 | 0.4 | 0.4 | 0.5 | 0.4 | 0.5 | 0.3 | 0.4 | 0.4 | 0.4 | 0.3 | 0.3 | 0.5 | 0.4 | 0.4 | 1.0 |

**Table (3)**
**The sesitivity matrix for Group 4,algorithm A4 and m=4**

| | ا | لا | ب | ه | د | ف | ح | ر | س | ص | ع | ـع | ق | ن | و | ى |
|---|---|---|---|---|---|---|---|---|---|---|---|---|---|---|---|---|
| ا | 1.0 | 0.7 | 0.7 | 0.7 | 0.7 | 0.7 | 0.7 | 0.8 | 0.7 | 0.7 | 0.7 | 0.7 | 0.7 | 0.7 | 0.7 | 0.7 |
| لا | 0.7 | 1.0 | 0.7 | 0.7 | 0.7 | 0.7 | 0.7 | 0.7 | 0.7 | 0.7 | 0.7 | 0.7 | 0.7 | 0.7 | 0.7 | 0.7 |
| ب | 0.7 | 0.7 | 1.0 | 0.7 | 0.7 | 0.7 | 0.7 | 0.7 | 0.8 | 0.7 | 0.7 | 0.7 | 0.8 | 0.8 | 0.7 | 0.7 |
| ه | 0.7 | 0.7 | 0.7 | 1.0 | 1.0 | 0.7 | 0.7 | 0.7 | 0.7 | 0.7 | 0.8 | 1.0 | 0.7 | 0.7 | 0.8 | 0.7 |
| د | 0.7 | 0.7 | 0.7 | 1.0 | 1.0 | 0.7 | 0.7 | 0.7 | 0.7 | 0.7 | 0.8 | 1.0 | 0.7 | 0.7 | 0.8 | 0.7 |
| ف | 0.7 | 0.7 | 0.7 | 0.7 | 0.7 | 1.0 | 0.7 | 0.7 | 0.7 | 0.7 | 0.7 | 0.7 | 0.8 | 0.7 | 0.7 | 0.8 |
| ح | 0.7 | 0.7 | 0.7 | 0.7 | 0.7 | 0.7 | 1.0 | 0.7 | 0.7 | 0.7 | 0.7 | 0.7 | 0.7 | 0.7 | 0.8 | 0.7 |
| ر | 0.8 | 0.7 | 0.7 | 0.7 | 0.7 | 0.7 | 0.7 | 1.0 | 0.7 | 0.7 | 0.7 | 0.7 | 0.7 | 0.7 | 0.8 | 0.7 |
| س | 0.7 | 0.7 | 0.8 | 0.7 | 0.7 | 0.7 | 0.7 | 0.7 | 1.0 | 0.8 | 0.7 | 0.7 | 0.7 | 1.0 | 0.7 | 0.7 |
| ص | 0.7 | 0.7 | 0.7 | 0.7 | 0.7 | 0.7 | 0.7 | 0.7 | 0.8 | 1.0 | 0.7 | 0.7 | 0.7 | 0.8 | 0.7 | 0.7 |
| ع | 0.7 | 0.7 | 0.7 | 0.8 | 0.8 | 0.7 | 0.7 | 0.7 | 0.7 | 0.7 | 1.0 | 0.8 | 0.7 | 0.7 | 0.7 | 0.7 |
| ـع | 0.7 | 0.7 | 0.7 | 1.0 | 1.0 | 0.7 | 0.7 | 0.7 | 0.7 | 0.7 | 0.8 | 1.0 | 0.7 | 0.7 | 0.8 | 0.7 |
| ق | 0.7 | 0.7 | 0.8 | 0.7 | 0.7 | 0.8 | 0.7 | 0.7 | 0.7 | 0.7 | 0.7 | 0.7 | 1.0 | 0.7 | 0.7 | 0.7 |
| ن | 0.7 | 0.7 | 0.8 | 0.7 | 0.7 | 0.7 | 0.7 | 0.7 | 1.0 | 0.8 | 0.7 | 0.7 | 0.7 | 1.0 | 0.7 | 0.7 |
| و | 0.7 | 0.7 | 0.7 | 0.8 | 0.8 | 0.7 | 0.8 | 0.8 | 0.7 | 0.7 | 0.7 | 0.8 | 0.7 | 0.7 | 1.0 | 0.7 |
| ى | 0.7 | 0.7 | 0.7 | 0.7 | 0.7 | 0.8 | 0.7 | 0.7 | 0.7 | 0.7 | 0.7 | 0.7 | 0.7 | 0.7 | 0.7 | 1.0 |

**Table (4)**
**The sesitivity matrix for Group 4,algorithm A4 and m=8**

| | ا | لا | ب | ه | د | ف | ح | ر | س | ص | ع | ـع | ق | ن | و | ى |
|---|---|---|---|---|---|---|---|---|---|---|---|---|---|---|---|---|
| ا | 1.0 | 0.7 | 0.7 | 0.6 | 0.7 | 0.6 | 0.6 | 0.7 | 0.6 | 0.6 | 0.6 | 0.6 | 0.6 | 0.7 | 0.6 | 0.7 |
| لا | 0.7 | 1.0 | 0.7 | 0.6 | 0.7 | 0.7 | 0.7 | 0.7 | 0.6 | 0.6 | 0.7 | 0.7 | 0.6 | 0.7 | 0.6 | 0.6 |
| ب | 0.7 | 0.7 | 1.0 | 0.7 | 0.7 | 0.6 | 0.7 | 0.7 | 0.6 | 0.6 | 0.7 | 0.6 | 0.6 | 0.7 | 0.6 | 0.6 |
| ه | 0.6 | 0.6 | 0.7 | 1.0 | 0.8 | 0.6 | 0.7 | 0.6 | 0.6 | 0.6 | 0.6 | 0.6 | 0.6 | 0.7 | 0.6 | 0.7 |
| د | 0.7 | 0.7 | 0.7 | 0.8 | 1.0 | 0.6 | 0.6 | 0.7 | 0.6 | 0.6 | 0.6 | 0.6 | 0.6 | 0.7 | 0.6 | 0.7 |
| ف | 0.6 | 0.7 | 0.6 | 0.6 | 0.6 | 1.0 | 0.7 | 0.6 | 0.6 | 0.6 | 0.6 | 0.6 | 0.7 | 0.7 | 0.6 | 0.7 |
| ح | 0.6 | 0.7 | 0.7 | 0.7 | 0.6 | 0.7 | 1.0 | 0.6 | 0.6 | 0.6 | 0.7 | 0.7 | 0.7 | 0.7 | 0.6 | 0.7 |
| ر | 0.7 | 0.7 | 0.7 | 0.6 | 0.7 | 0.6 | 0.6 | 1.0 | 0.6 | 0.6 | 0.6 | 0.6 | 0.6 | 0.7 | 0.7 | 0.7 |
| س | 0.6 | 0.6 | 0.6 | 0.6 | 0.6 | 0.6 | 0.6 | 0.6 | 1.0 | 0.7 | 0.6 | 0.6 | 0.6 | 0.7 | 0.6 | 0.6 |
| ص | 0.6 | 0.6 | 0.6 | 0.6 | 0.6 | 0.6 | 0.6 | 0.6 | 0.7 | 1.0 | 0.6 | 0.5 | 0.7 | 0.6 | 0.6 | 0.6 |
| ع | 0.6 | 0.7 | 0.7 | 0.6 | 0.6 | 0.6 | 0.7 | 0.6 | 0.6 | 0.6 | 1.0 | 0.8 | 0.6 | 0.7 | 0.6 | 0.7 |
| ـع | 0.6 | 0.7 | 0.6 | 0.6 | 0.6 | 0.6 | 0.7 | 0.6 | 0.6 | 0.5 | 0.8 | 1.0 | 0.6 | 0.6 | 0.6 | 0.7 |
| ق | 0.6 | 0.6 | 0.6 | 0.6 | 0.6 | 0.7 | 0.7 | 0.6 | 0.6 | 0.7 | 0.6 | 0.6 | 1.0 | 0.6 | 0.7 | 0.7 |
| ن | 0.7 | 0.7 | 0.7 | 0.7 | 0.7 | 0.7 | 0.7 | 0.7 | 0.7 | 0.6 | 0.7 | 0.6 | 0.6 | 1.0 | 0.7 | 0.6 |
| و | 0.6 | 0.6 | 0.6 | 0.6 | 0.6 | 0.6 | 0.6 | 0.7 | 0.6 | 0.6 | 0.6 | 0.6 | 0.7 | 0.7 | 1.0 | 0.7 |
| ى | 0.7 | 0.6 | 0.6 | 0.7 | 0.7 | 0.7 | 0.7 | 0.7 | 0.6 | 0.6 | 0.7 | 0.7 | 0.7 | 0.6 | 0.7 | 1.0 |

The different values of "S" are calculated according to equation (9) for all groups, all algorithms and m = 3 -14. Table(5) shows these results selected for Group 4. Again the dependence of the sensitivity on the value of "m" is very clear for all algorithms.

It is interesting to compare the sensitivity of the four algorithms from "S" parameters given in Table(5). It seems that A1 is the most sensitive algorithm, however, A2 is also acceptable when considering the required space and speed. The large value of "S" for A4 is attributed to the low sensitivity of this algorithm to differentiate between identical characters.

For example, the feature vectors of Ghein - middle ( ـغـ ) and Feh - middle ( ـفـ ) are exactly identical when using A4. Similarly, Heh - end (ه) identical to Five (٥) and Alif - start (ا) identical to One (١) .... etc. As seen from sensitivity tables, other algorithms can significantly differentiate between them. Preliminary analysis, for other populer Arabic fonts, shows identical results. On the other hand, A4 can be very useful in detecting some detailed structure in Arabic characters (e.g. ٢-٣).

**Table (5)**
The different values of the sensitivity parameter "S" for algorithm A1 - A4 and m = 3 - 14

| m | A1 | A2 | A3 | A4 |
|---|---|---|---|---|
| 3 | .71 | .78 | .78 | .85 |
| 4 | .64 | .70 | .75 | .79 |
| 6 | .53 | .63 | .73 | .67 |
| 8 | .45 | .53 | .72 | .66 |
| 10 | .43 | .46 | .72 | .66 |
| 12 | .37 | .43 | .72 | .65 |
| 14 | .33 | .38 | .71 | .66 |

## 6 - Conclusion

The technique developed in the present work is very useful in testing the sensitivity of any character recognition algorithm based on statistical approach. The analysis shows that the recognition sensitivity, for all groups (Group 1 - Group 4) and all algorithms (A1 - A4), increases by increasing the value of "m". A satisfactory value may be adopted as m = 4 - 8. On the other hand, the investigated Arabic font, using the four proposed algorithms, shows that the algorithm A1 has the best recognition sensitivity. When considering the space and time economy, algorithm A2 is also sensitive and fast algorithm. Although A4 is not sensitive to differentiate between many Arabic characters, it is useful in detecting the fine structure of some characters (preliminary analysis, for other popular Arabic fonts, shows identical results).

From the present results, it can be concluded that a combined feature vector based on (A1 and A4) or (A2 and A4) will certainly produce a new powerful algorithm. This algorithm is expected to be much more sensitive, as an Arabic recognition tool, than any of the four investigated algorithms.

**References**

[ 1 ] M. F. Tolba, S. A. Wahab and A. Salem, "A Recognition Algorithm for Arabic Printed Characters", 11th International Conference for Statistics and Computer Sciences, Cairo, Vol.2, (1986).

[ 2 ] M. F. Tolba, A. Goned, A. Salem and E. Shaddad, "On the Sensivity of the Recognition Algorithm for Printed Arabic Characters", 12th International Conference for Statistics and Computer Sciences, Cairo, Vol.5, P231-241, (1987).

[ 3 ] M. Y. Mahmoud, M. A. El-Hamalaway and A. A. Fahmy, "A Statistical Aproach for Arabic Character Recognition", 12th International Conference for Statistics and Computer Sciences, Cairo, Vol.5, p 243-250, (1987).

[ 4 ] F. W. Zaki, S. H. El-Konyaly, A. I. Abd EL-Fattah and Y. Enab, "A New Technique for Arabic Handwriting Recognition", 11th International Conference for Statistics and Computer Sciences, Cairo, Vol.2, P171-180, (1986).

[ 5 ] K. Badie and M. Shimura, "A Classification Method of Arabic Alphabets", IECE of Japan, P78, (1978).

[ 6 ] K. Badie and M. Shimura, "Machine Recognition of Arabic Handprinted Scripts", Transaction of the IECE of Japan, E65, P107, (1982).

[ 7 ] A. Amin, "Machine Recognition of Handwritten Arabic by the IRAC II System", 6th International Conference on Pattern Recognition IJCPR, Munich, P34, (1982).

[ 8 ] A. Amin, "Arabic Handwriting and Understanding", Kuwait Conference on Computer Processing of the Arabic Language, Vol.1, (1985).

[ 9 ] S. Saadallah and S. G. Yacu, "Design of an Arabic Character Reading Machine" ,Kuwait Conference on Computer Processing of the Arabic Language. Vol.1, (1985).

[ 10] F. Haj-Hassan, "Arabic Character Recognition", 7th Summer Session Arab School of Science and Technology, Syria, P2.A5-1, (1985).

[ 11] H. K. Mahdi, "Small Size Image Transformations and Representations", 12 th International conference for Statistics and Computer Sciences, Cairo, VOL.5, P 1-14, (1987).

[ 12] G. Siromoney, R. Chandrasekaran and M. Chandrasekaran, "Machine Recognition of Brahmi Script", IEEE Trans. on System, Man, and Cybernetics, Vol. SMC-13, NO.4, P 648-654, (1983).

# ANALYSIS OF A TIME-SHARED CENTRAL SERVER SYSTEM WITH HETEROGENEOUS DISTRIBUTED STATIONS

Huanxu Pan, Hidenori Morimura and Masaaki Kijima
Department of Information Sciences, Tokyo Institute of Technology
Oh-okayama, Meguro-ku, Tokyo 152, Japan

**Abstract** We consider a service system consisting of a time-shared central server (CPU) and a number of geographically distributed stations. Customers send service requirements to the CPU through terminals at the stations. Examples of such a system are cash dispenser systems in banks, etc. A notable characteristic of such a system is that two kinds of queues are formed, one at the CPU and the others at the stations. This characteristic was first taken into account for analysis explicitly by Pan and Morimura (1986). In this paper, we generalize their model by removing the restrictive assumptions that all the stations have the same arrival rate and an identical number of terminals. Two approximation formulas for the mean system time of customers are obtained. Extensive numerical experiments reveal that our approximations are quite good. Some possible applications of our results are also discussed.

## 1. Introduction

We consider a service system consisting of a time-shared central server and a number of geographically distributed stations. Customers send service requirements to the central server through terminals located at the stations. Many service systems in the real world are of this fashion. Examples of such systems are cash dispenser systems in banks, information retrieval systems in libraries, airline reservation systems, etc. The central server is, in many cases, a CPU of a computer system. As the number of customers who simultaneously access the time-shared CPU grows, the conflict among them also grows. This conflict then gives rise to a queue at the CPU. On the other hand, a customer has to occupy a terminal in order to access the CPU. Therefore queues are also formed at the stations. See Figure 1.1.

A number of studies have been conducted on time-sharing systems. A general model for such systems is shown in Figure 1.2 (see, e.g., [5,6]). Its counterpart for finite population case that takes the finiteness of the number of terminals into account is depicted in Figure 1.3 [6,8].

Neither of the basic models, however, concern queues at stations. Recently, Pan and Morimura [7] considered a system which contains both the queue of jobs at CPU and the queues of customers at stations (they called the system an <u>on-line network system</u>).

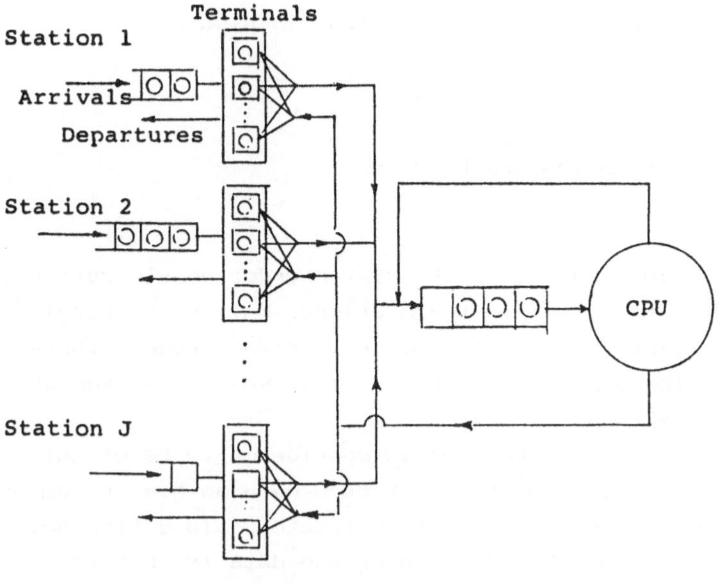


Figure 1.1

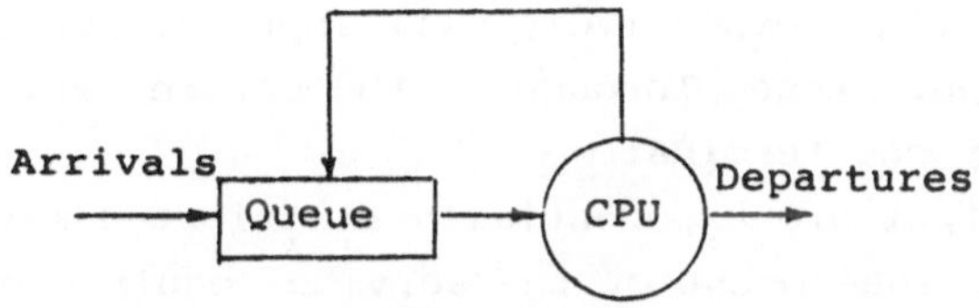


Figure 1.2

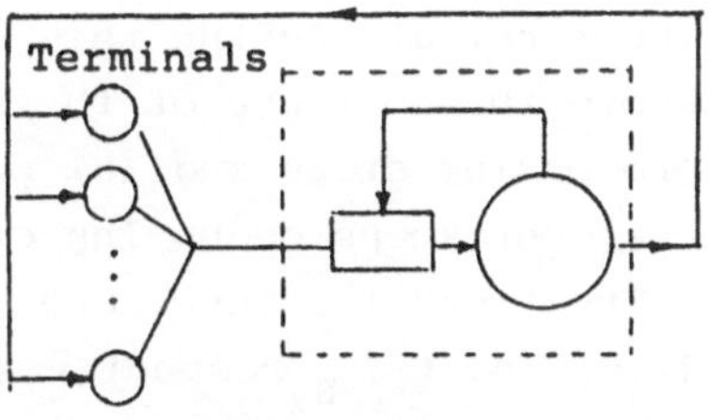


Figure 1.3

In this paper we generalize the model in [7] by removing the restrictive assumptions that all the stations have the same arrival rate and an identical number of terminals. The approach used here is similar to that in [7] but more refined. As in [7], we divide the entire system into a computer subsystem which describes state transitions of terminals, and some queueing subsystems which concern queueing behaviors of customers at stations. The computer subsystem is formulated as a multi-chain closed queueing network and the queueing subsystems are approximated as M/G/m queues. In terms of these, approximate steady-state probabilities in both subsystems are obtained via an iterative procedure. Then system performance measures such as the expected turn-around-time of a job and the mean waiting time of customers are approximately calculated. Numerical results show that the approximations are quite good. Our results can be extended easily to, for example, the case when some of the stations get overloaded and the case that different classes of customers are contained. A discussion on the application of our study is also given.

## 2. Model Description and Notation

As shown in Figure 1.1, the system under consideration consists of a single CPU and J distributed stations. Unlike the model in [7], station j has $m_j$ available terminals and the arrival process there is Poisson with rate $\lambda_j$. The service discipline associated with each station is first-come-first-served.

The service for a customer is considered to be of three stages. First he needs a random time to input such data as his ID number, password, the content of his service requirement, etc., via a terminal. At this stage, the CPU is not used. The bunch of the data is then sent to the CPU. The CPU processes all the service requirements that have been sent to it according to the processor sharing discipline. When the CPU processing is completed, another random time is needed to receive a desired output from the terminal. During this stage, the CPU is not used either. We call the above three stages of the service respectively, (1) input stage, (2) CPU processing stage and (3) output stage. Throughout the entire service, the customer keeps occupying only one terminal.

The times for input and output are exponentially distributed with means $1/\mu_1$ and $1/\mu_3$ respectively. The amount of the service requirement a customer brings is also a random variable and is distributed exponentially with mean $1/\mu_2$. We assume that the CPU processing rate is dependent on the number of jobs in the CPU. Let $s(n)$ be the CPU processing rate when in total n customers are in the CPU processing stage. Then, at that time, each customer receives CPU processing at rate $s(n)/n$.

We divide the entire system into a computer subsystem and queueing subsystems at stations. The computer subsystem consists of the CPU and all the terminals while the queueing subsystem at a station is a standard queue where the terminals at that station are referred to as servers. The state of the queueing subsystem at a station is, as usual, defined as the number of customers in the station. For the computer subsystem, a multichain queueing network can be used to represent the behaviors of terminals (see Figure 2.1). The queueing network has J closed chains where chain j is composed of the terminals at station j. There are four nodes each of which corresponds to one of the four possible states of terminals. We say that a terminal is in node i (i=1, 2, 3) if the terminal is in use and the service through it is in the i-th stage. Node 4 represents the idle state of terminals. A terminal moves from node 1 to node 2 and then to node 3 in an obvious way. When a service through a terminal is completed, the terminal moves from node 3 to node 1 if any customer is waiting for it, or to node 4 if there is no such customer. A movement of a terminal from node 4 to node 1 corresponds to that a newly arriving customer occupies the terminal with no waiting and begins with his input stage. We define the state of the computer subsystem as

$$n = (n_1, n_2, n_3, n_4), \tag{2.1}$$

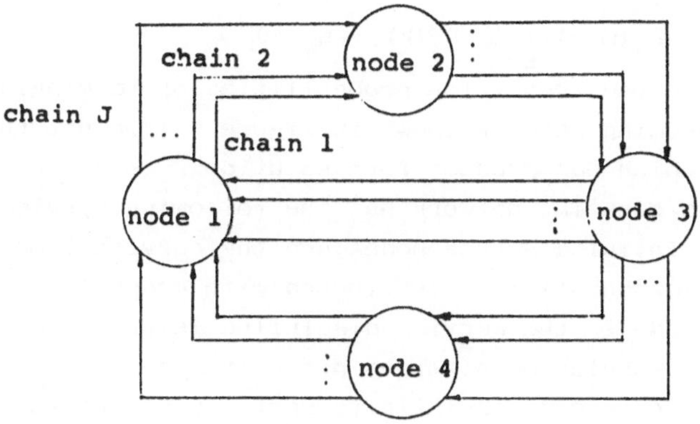


Figure 2.1

where $n_i=(n_{i1}, n_{i2}, \ldots, n_{iJ})$ and $n_{ij}$ $(i=1, 2, 3, 4; j=1, 2, \ldots, J)$ denotes the number of terminals of chain j in node i. It should be kept in mind that

$$n_{1j} + n_{2j} + n_{3j} + n_{4j} = m_j \quad (j=1, 2, \ldots, J). \tag{2.2}$$

We denote throughout the paper the steady-state probabilities of the computer subsystem and the queueing subsystem at station j by $\pi(n)$ and $P_j(k)$ respectively. The following notation will also be used:

$$n_i = \sum_{j=1}^{J} n_{ij} \quad (i=1, 2, 3, 4), \quad M = \sum_{j=1}^{J} m_j \quad \text{and} \quad \Lambda = \sum_{j=1}^{J} \lambda_j.$$

## 3. The Steady-State Probabilities

We consider the model described in the previous section under equilibrium condition.

Suppose that the computer subsystem is in state $n=(n_1, n_2, n_3, n_4)$. As mentioned earlier, a terminal of chain j leaving node 3 will move to node 1 or to node 4 according to whether or not there is any customer waiting for the terminal at station j. It is then clearly dependent on $n_{4j}$. Note that the event that a terminal of chain j leaving node 3 moves to node 4 corresponds to that a customer departing from station j leaves the terminal he used idle. Denote this event by D. Clearly $P_j(D \mid n_{4j})=1$ if $n_{4j}>0$. $P_j(D \mid 0)$ can be represented in terms of the steady-state probabilities $P_j(k)$ $(k \geqq m_j-1)$ in the following way. The event $(n_{4j}=0)$ means that there are at least $m_j$ customers at station j. Then in this state, a departing customer leaves at least $m_j-1$ customers behind. In order for a terminal to become idle given this event, the customer must see exactly $m_j-1$ customers behind when he leaves. Since the arrival process at station j is Poisson, we have

$$P_j(D \mid n_{4j}) = \begin{cases} 1, & n_{4j}>0, \\ P_j(m_j-1) / \sum_{k=m_j-1}^{\infty} P_j(k), & n_{4j}=0. \end{cases} \tag{3.1}$$

We note that, since the transition probabilities of terminals are state dependent, the queueing network shown in Figure 2.1 is not the well-known BCMP network [1] which has product form solution.

Each node of the queueing network has the following service aspects. Nodes 1 and 3 are infinite-server nodes and the service times of a terminal at them are exponentially distributed with means $1/\mu_1$ and $1/\mu_3$ respectively. At node 2, the service discipline is processor-sharing and the mean number of service completions per unit time is $\mu_2 s(n_2)$, conditional on $n_2$. Node 4 is also an infinite-server node but the service rate is state dependent. The mean total number of service completions per unit time for terminals of chain j at node 4 is $\lambda_j P_j(A \mid n_{4j})$, where A denotes the event that a customer arriving at station j finds an idle terminal. It is evident that

$$P_j(A \mid n_{4j}) = \begin{cases} 1, & n_{4j}>0, \\ 0, & n_{4j}=0. \end{cases} \tag{3.2}$$

Then we obtain the local balance equations [3], which equate the rate of flow into a state by a terminal entering a node to the rate of flow out of that state due to a terminal of the same chain leaving the node. Specifically,

$$\begin{aligned} &\pi(n_1-1_j, n_2, n_3, n_4+1_j)\lambda_j P_j(A \mid n_{4j}+1) \\ &\quad + \pi(n_1-1_j, n_2, n_3+1_j, n_4)\mu_3(n_{3j}+1)[1-P_j(D \mid n_{4j})] \\ &\quad = \pi(n)\mu_1 n_{1j}, \qquad (n_{1j} \geqq 1), \ (1 \leqq j \leqq J), \end{aligned} \tag{3.3}$$

$$\begin{aligned} &\pi(n_1+1_j, n_2-1_j, n_3, n_4)\mu_1(n_{1j}+1) \\ &\quad = \pi(n)\mu_2 s(n_2) n_{2j}/n_2, \qquad (n_{2j} \geqq 1), \ (1 \leqq j \leqq J), \end{aligned} \tag{3.4}$$

$$\begin{aligned} &\pi(n_1, n_2+1_j, n_3-1_j, n_4)\mu_2 s(n_2+1)(n_{2j}+1)/(n_2+1) \\ &\quad = \pi(n)\mu_3 n_{3j}, \qquad (n_{3j} \geqq 1), \ (1 \leqq j \leqq J), \end{aligned} \tag{3.5}$$

$$\pi(n_1, n_2, n_3+1_j, n_4-1_j)\,\mu_3(n_{3j}+1)P_j(D \mid n_{4j}-1)$$
$$= \pi(\mathbf{n})\,\lambda_j P_j(A \mid n_{4j}), \qquad (n_{4j} \geqq 1),\ (1 \leqq j \leqq J), \tag{3.6}$$

where $1_j$ is the J-dimensional vector with the j-th component being 1 and others being 0. It is not hard to verify that the following expression for $\pi(\mathbf{n})$ is the solution of the above equations:

$$\pi(\mathbf{n}) = \frac{C\, n_2! \prod_{j=1}^{J} [\Sigma^{\infty}_{k=m_j-1} P_j(k)/P_j(m_j-1)]^{\delta(n_{4j})}}{[\prod_{r=1}^{n_2} s(r)]\, \mu_1^{n_1}\, \mu_2^{n_2}\, \mu_3^{n_3} \prod_{j=1}^{J} (n_{1j}!\, n_{2j}!\, n_{3j}!\, \lambda_j^{n_{4j}})}, \tag{3.7}$$

where C is the normalizing constant and $\delta(n)$ takes the value 1 when n=0 and 0 otherwise. Here we get an example of product form solution in queueing networks with state dependent transition probabilities. Of particular importance is the steady-state probabilities for the number of terminals at CPU, which we denote by $\pi_2(n_2)$. They are used to evaluate the mean sojourn time of customers at this central server. After some algebraic manipulation, we obtain

$$\pi_2(n_2) = \frac{C\, n_2!}{\mu_2^{n_2} (\prod_{r=1}^{n_2} s(r))} \sum_{\substack{\Sigma_{j=1}^{J} n_{2j} = n_2 \\ 0 \leq n_{2j} \leq m_j\ (1 \leq j \leq J)}} \prod_{j=1}^{J} \Bigl\{ \frac{1}{n_{2j}!\, \lambda_j^{m_j - n_{2j}}}$$
$$\cdot \Bigl[ \sum_{\ell_j=0}^{m_j - n_{2j} - 1} \frac{(\lambda_j(\frac{1}{\mu_1} + \frac{1}{\mu_3}))^{\ell_j}}{\ell_j!} + \frac{(\lambda_j(\frac{1}{\mu_1} + \frac{1}{\mu_3}))^{m_j - n_{2j}}}{(m_j - n_{2j})!} \cdot \frac{\sum_{k=m_j-1}^{\infty} P_j(k)}{P_j(m_j-1)} \Bigr] \Bigr\}. \tag{3.8}$$

Note that the quantities $P_j(k)$ $(k \geqq m_j - 1)$ contained in the above expression are yet unknown. If $P_j(k)$ are known by some means, the solutions (3.7) and (3.8) are exact.

Consider next the queueing subsystem at station j. As we have described earlier, customers arrive at station j according to Poisson process with rate $\lambda_j$ and each customer receives service through one of the $m_j$ terminals. The service of a customer is composed of three stages and the time spent in the second stage depends on the total number of customers at all the stations who are in this stage. Consequently the service times of customers at a station are dependent on each other and are dependent on the state of that station. The dependences seem weak, however, when the number of stations, J, is large. Therefore the queueing subsystem at station j is approximately considered to be an ordinary M/G/m queue in isolation. Let B(s) be the Laplace transform of the pdf for the service time of a customer. It is expressible in terms of $\{\pi_2(n_2)\}$ as

$$B(s) = \sum_{n_2=1}^{M} \frac{\pi_2(n_2)}{1-\pi_2(0)} \cdot \frac{\mu_1}{s+\mu_1} \cdot \frac{\mu_2 s(n_2)/n_2}{s+\mu_2 s(n_2)/n_2} \cdot \frac{\mu_3}{s+\mu_3}. \tag{3.9}$$

The mean service time ES is then given by

$$ES = \frac{1}{\mu_1} + \sum_{n_2=1}^{M} \frac{\pi_2(n_2)}{1-\pi_2(0)} \cdot \frac{n_2}{\mu_2 s(n_2)} + \frac{1}{\mu_3}. \tag{3.10}$$

Here $\pi_2(n_2)/(1-\pi_2(0))$ represents the probability that in total $n_2$ customers are in the CPU processing stage given that at least one customer is in this stage.

No exact formula for the steady-state probabilities P(k) in the M/G/m queue has been known. Tijms et al. [10] derived three approximations for P(k). By either of the first two ones, the unknown quantities $\sum_{k=m_j-1}^{\infty} P_j(k)/P_j(m_j-1)$ $(1 \leqq j \leqq J)$ in (3.8) are approximately calculated as

$$\frac{\sum_{k=m_j-1}^{\infty} P_j(k)}{P_j(m_j-1)} \simeq \frac{1}{1-\rho_j}, \tag{3.11}$$

where $\rho_j = \lambda_j ES/m_j < 1$. Note that (3.11) is exact for the M/M/m queue.

Approximate values for both ES and $(\pi_2(n_2))$ are calculated iteratively through (3.8), (3.10) and (3.11) as follows:

Step 1: Set k=0 and $(ES)_0=0$.

Step 2: Increment k by one and calculate $((\pi_2(n_2))_k)$ via (3.8) and (3.11).

Step 3: Obtain $(ES)_k$ via (3.10).

Step 4: If $|(ES)_k-(ES)_{k-1}| < \varepsilon$ for a pre-specified $\varepsilon > 0$, then stop. Otherwise go to step 2.

We note that if $((ES)_k)$ in the above procedure is a nondecreasing sequence with an upper bound then the procedure converges and results in a solution for ES and $(\pi_2(n_2))$. Now we show informally that $((ES)_k)$ satisfies such conditions. First we note that $(ES)_1 \geqq (ES)_0$ in an obvious way. Then we try to prove that $(ES)_{k+1} \geqq (ES)_k$ if $(ES)_k \geqq (ES)_{k-1}$. In order to do so, the following proposition may be helpful.

Proposition 1. Let $(a_n)$ and $(b_n)$ $(1 \leqq n \leqq M)$ be two finite sequences of real numbers. If (1) $a_1 \leqq a_2 \leqq \ldots \leqq a_M$; (2) $\sum_{n=i}^{M} b_n \geqq 0$ $(2 \leqq i \leqq M)$ and $\sum_{n=1}^{M} b_n = 0$, then $\sum_{n=1}^{M} a_n b_n \geqq 0$.

Proof. One has

$$\sum_{n=1}^{M} a_n b_n = \sum_{n=1}^{M-1} a_n \Big( \sum_{i=n}^{M} b_i - \sum_{i=n+1}^{M} b_i \Big) + a_M \sum_{i=M}^{M} b_i = a_1 \sum_{i=1}^{M} b_i + \sum_{n=2}^{M} (a_n - a_{n-1}) \sum_{i=n}^{M} b_i.$$

From (1), $a_n - a_{n-1} \geqq 0$. From (2), $\sum_{i=1}^{M} b_i = 0$ and $\sum_{i=n}^{M} b_i \geqq 0$ for $n \geqq 2$. Thus $\sum_{n=1}^{M} a_n b_n \geqq 0$. □

From (3.10),

$$(ES)_{k+1} - (ES)_k = \frac{1}{\mu_2} \sum_{n_2=1}^{M} \frac{n_2}{s(n_2)} \left[ \frac{(\pi_2(n_2))_{k+1}}{1-(\pi_2(0))_{k+1}} - \frac{(\pi_2(n_2))_k}{1-(\pi_2(0))_k} \right] \tag{3.12}$$

Note that $(\pi_2(n_2))_k/(1-(\pi_2(0))_k)$ is the conditional distribution of $n_2$ given that $n_2>0$, which is calculated by using $(ES)_{k-1}$. Clearly,

$$\sum_{n_2=1}^{M} \left[ \frac{(\pi_2(n_2))_{k+1}}{1-(\pi_2(0))_{k+1}} - \frac{(\pi_2(n_2))_k}{1-(\pi_2(0))_k} \right] = 0 . \tag{3.13}$$

From the hypothesis, $(ES)_k \geqq (ES)_{k-1}$. The difference between them, if there is any, results from the second term of the right-hand side in (3.10), which represents the mean time spent in the second service stage by a customer. It must hold that $n_2$ (given that $n_2>0$) increases stochastically as the mean time in the second service stage increases. Thus for all n,

$$\sum_{n_2=n}^{M} \left[ \frac{(\pi_2(n_2))_{k+1}}{1-(\pi_2(0))_{k+1}} - \frac{(\pi_2(n_2))_k}{1-(\pi_2(0))_k} \right] \geqq 0 . \tag{3.14}$$

On the other hand, recall that $n_2/(\mu_2 s(n_2))$ is the mean CPU processing time for a customer when in total $n_2$ customers are in the CPU processing stage. This quantity does not decrease as $n_2$ increases in any real computer system. From Proposition 1 we then have that $\{(ES)_k\}$ is a nondecreasing sequence. That $\{(ES)_k\}$ has an upper bound is obvious from (3.10). Thus, we conclude that the iterative procedure developed for calculating ES and $\{\pi_2(n_2)\}$ converges so long as $\rho_j<1$ for all j.

From the obtained distribution $\{\pi_2(n_2)\}$, we can specify the service time of a customer quantitatively by using (3.9). The mean waiting time at each station can then be calculated by applying an appropriate approximation formula. Some approximation formulas for the mean waiting time in the M/G/m queue are given in [2,4,9].

Remarks.

1. In calculating $\{\pi_2(n_2)\}$ of (3.8) via a computer, the summation of combinatorial terms may well be divided into several levels. For example, if we denote the inside of the curly braces in (3.8) by $u_j(n_{2,j})$ and let J' be a positive integer smaller than J, then one has a two-level summation:

$$\sum_{\substack{\sum_{j=1}^{J} n_{2j}=n_2 \\ 0\leqslant n_{2j}\leqslant m_j \\ (0\leqslant j\leqslant J)}} \prod_{j=1}^{J} u_j(n_{2j}) = \sum_{\substack{n'+n''=n_2 \\ 0\leqslant n'\leqslant \sum_{j=1}^{J'} m_j \\ 0\leqslant n''\leqslant \sum_{j=J'+1}^{J} m_j}} \Big[ \sum_{\substack{\sum_{j=1}^{J'} n_{2j}=n' \\ 0\leqslant n_{2j}\leqslant m_j \\ (0\leqslant j\leqslant J')}} \prod_{j=1}^{J'} u_j(n_{2j}) \Big] \Big[ \sum_{\substack{\sum_{j=J'+1}^{J} n_{2j}=n'' \\ 0\leqslant n_{2j}\leqslant m_j \\ (J'+1\leqslant j\leqslant J)}} \prod_{j=J'+1}^{J} u_j(n_{2j}) \Big].$$

This sort of manipulation facilitates the complicated calculations and accelerates the computational speed. In actual computation, we employed the above manipulation.

2. The result of the analysis for the computer subsystem can be extended easily to the case when some of the stations get overloaded. In this case, the terminals at an overloaded station, say, station j, move only between nodes 1, 2 and 3 in the corresponding chain (see Figure 2.1). Then it can be shown that the factor of $\left[\sum_{k=m_j-1}^{\infty} P_j(k)/P_j(m_j-1)\right]^{\delta(n_{4,j})} / \lambda_j^{n_{4,j}}$ is removed from (3.7). The steady-state probabilities for the computer subsystem can be used to calculate the throughput of the CPU though the mean waiting times for some stations become meaningless.

3. Further extensions are possible along the following directions. First, classes for customers can be incorporated into our model to concern distinct service time distributions. Second, the distributions of service requirement and time lengths for input and output need not be exponential but are only needed to have rational Laplace transforms. In order to get the local balance equations for these extended models, it needs only to define more detailed states, and it is omitted here.

## 4. Simplified Approximation

In this section, we derive a simplified approximation for $\{\pi_2(n_2)\}$ which is much simpler than the one in Section 3. As we shall see in the next section, this simple formula is quite good except for the heavy traffic case.

Redefine the state of the computer subsystem as

$$n = (n_1, n_2, n_3, n_4), \tag{4.1}$$

where $n_i$ $(i=1,2,3,4)$ is the total number of terminals in node i of the queueing network represented in Figure 2.1. Then we get another set of local balance equations:

$$\begin{aligned} &\pi(n_1-1, n_2, n_3, n_4+1) \sum_{j=1}^{J} \lambda_j P_j(A \mid n_4+1) \\ &\quad + \pi(n_1-1, n_2, n_3+1, n_4)\, \mu_3(n_3+1) \sum_{j=1}^{J} (\lambda_j/\Lambda)[1-P_j(D \mid n_4)] \\ &\quad = \pi(n)\,\mu_1 n_1, \qquad (n_1 \geqq 1) \end{aligned} \tag{4.2}$$

$$\pi(n_1+1, n_2-1, n_3, n_4)\,\mu_1(n_1+1) = \pi(n)\,\mu_2 s(n_2), \qquad (n_2 \geqq 1) \tag{4.3}$$

$$\pi(n_1, n_2+1, n_3-1, n_4)\,\mu_2 s(n_2+1) = \pi(n)\,\mu_3 n_3, \qquad (n_3 \geqq 1) \tag{4.4}$$

$$\begin{aligned} &\pi(n_1, n_2, n_3+1, n_4-1)\,\mu_3(n_3+1) \sum_{j=1}^{J} (\lambda_j/\Lambda) P_j(D \mid n_4-1) \\ &\quad = \pi(n) \sum_{j=1}^{J} \lambda_j P_j(A \mid n_4), \qquad (n_4 \geqq 1) \end{aligned} \tag{4.5}$$

Unlike the situations of $P_j(A \mid n_{4,j})$ and $P_j(D \mid n_{4j})$ in (3.1) and (3.2), $P_j(A \mid n_4)$ and $P_j(D \mid n_4)$ can not be specified easily. It is not hard, however, to imagine that $P_j(A \mid n_4)$ and $P_j(D \mid n_4)$ are very close when $n_4$ is large. Note that $n_4$ is most likely large if the number of stations is large and most of them are not heavy. Suppose now that $P_j(A \mid n_4)=P_j(D \mid n_4)$. We then obtain a very simple approximate solution for the equations of (4.2) through (4.5):

$$\pi(n) \simeq \frac{C}{\mu_1^{n_1}\,\mu_2^{n_2}\,\mu_3^{n_3}\,\Lambda^{n_4}\, n_1!\,[\prod_{r=1}^{n_2} s(r)]\, n_3!}. \tag{4.6}$$

Further, some derivations lead to

$$\pi_2(n_2) \simeq \frac{C\,\Lambda^{n_2}}{\mu_2{}^{n_2}\,[\,\prod_{r=1}^{n_2} s(r)\,]} \sum_{k=0}^{M-n_2} \frac{1}{k!}\,[\,\Lambda\,(\frac{1}{\mu_1} + \frac{1}{\mu_3})\,]^k . \tag{4.7}$$

Here, unknown quantities are no longer included so that $\{\pi_2(n_2)\}$ is calculated directly.

## 5. Numerical Comparisons

In this section, we compare the numerical results calculated by A: the iterative scheme in Section 3 and B: the simplified approximation in Section 4, with S: the simulation results. The values of the mean total system time, T, in A and B are calculated via an approximation formula for the mean waiting time of the M/G/m queue given in [4], namely,

$$W(M/G/m) \simeq \frac{1+c_s^2}{2}\,[\,\frac{1-c_s^2}{1+(1-\rho)(m-1)\frac{\sqrt{4+5m}-2}{16m\rho}} + c_s^2\,]^{-1}\cdot W(M/M/m). \tag{5.1}$$

Here the coefficient of variation of service time $c_s$ and the traffic intensity $\rho$ are calculated straightforwardly via (3.9) provided that $\{\pi_2(n_2)\}$ has been obtained. Note that the numerical results hinge also on the accuracy of this approximation stage.

The parameters are given as: $J=32$; $m_j=4$, 5, 6 and 7 for $1 \leqq j \leqq 8$, $9 \leqq j \leqq 16$, $17 \leqq j \leqq 24$ and $25 \leqq j \leqq 32$ respectively; $\mu_1=0.5$; $\mu_2=1$; $\mu_3=1$; the values of $\lambda_j$ are set identical for each group of $1 \leqq j \leqq 8$, $9 \leqq j \leqq 16$, $17 \leqq j \leqq 24$ and $25 \leqq j \leqq 32$, and they are denoted by $\lambda=(\lambda_{j_1}, \lambda_{j_2}, \lambda_{j_3}, \lambda_{j_4})$. The CPU processing rate function is given as $s(n)=32-(n-96)^2/288$ for $1 \leqq n \leqq 190$ and $s(n)=1$ otherwise. This is the case that the processing rate goes down when the number of jobs in the CPU exceeds 96 and is kept not below the unit rate by some congestion control.

Table 5.1 compares A, B and S in four cases for altered values of $\lambda$. In each of the four cases, $\lambda_j/m_j$ is invariant with j so that each station has the same traffic intensity $\rho$. Here $\bar{T}$ denotes the average value of $(T_j)$. We see that both the results of A and B are close to that of S in the light traffic while B gradually goes far from A and S as $\rho$ gets large. This is the expected result from the underlying assumption in constructing (4.6) and (4.7). Table 5.2 shows the results of two other cases where the total arrival rate $\Lambda$ is the same as the second case of Table 5.1 with the quantities $\lambda_j/m_j$ being different so that the traffics are unbalanced among the stations. In Table 5.2, $\rho_j$ and $T_j$ are shown for each group of $1 \leqq j \leqq 8$, $9 \leqq j \leqq 16$, $17 \leqq j \leqq 24$ and $25 \leqq j \leqq 32$. One can again observe good accuracy for both A and B despite the unbalanced traffic intensities. We note that $\pi_2(n_2)$ of (4.7) depends only on $\Lambda$ and M but not on individual $\lambda_j$ and $m_j$. This means that the mean service times ES calculated by B for the two cases in Table 5.2 are the same as that of the second case in Table 5.1. A little surprise to us is that A also shows a similar phenomenon.

From the above numerical experiences together with others, we conclude that A seems a good approximation for general cases while B is good when many of the stations are not in the heavy traffic, and that the unbalance of the traffics among the stations does not much affect the accuracy for our approximations A and B. All computations were done on a VAX-11/780 computer using FORTRAN as the programming language. By using the manipulation described in Remark 1 of Section 3, the calculation for one case by B usually takes about a half second of CPU time in contrast to about one minute of that by A. The simulation results were obtained by the popular simulation language, SLAM Π, running on a personal computer.

Table 5.1

| $\lambda$ | (0.400, 0.500, 0.600, 0.700) | | | (0.600, 0.750, 0.900, 1.050) | | | (0.680, 0.850, 1.020, 1.190) | | | (0.696, 0.870, 1.044, 1.218) | | |
|---|---|---|---|---|---|---|---|---|---|---|---|---|
| | ES | $\rho$ | $\bar{T}$ | ES | $\rho$ | $\bar{T}$ | ES | $\rho$ | $\bar{T}$ | ES | $\rho$ | $\bar{T}$ |
| A | 4.80 | 0.48 | 4.95 | 5.13 | 0.77 | 6.60 | 5.43 | 0.92 | 12.95 | 5.54 | 0.96 | 23.47 |
| B | 4.80 | 0.48 | 4.95 | 5.15 | 0.77 | 6.64 | 5.38 | 0.91 | 11.89 | 5.42 | 0.94 | 15.96 |
| S | 4.81 | | 4.95 | 5.15 | | 6.50 | 5.44 | | 12.3 | 5.57 | | 25.5 |

Table 5.2

| $\lambda$ | (0.350, 0.625, 1.025, 1.300) | | | (0.750, 0.825, 0.825, 0.9(0) | | |
|---|---|---|---|---|---|---|
| | ES | $\rho$ | T | ES | $\rho$ | T |
| A | 5.132 | (0.45, 0.64, 0.88, 0.95) | (5.36, 5.73, 8.43, 14.39) | 5.134 | (0.96, 0.85, 0.71, 0.66) | (26.66, ..16, 5.85, 5.51) |
| B | 5.15 | (0.45, 0.64, 0.88, 0.96) | (5.37, 5.76, 8.55, 15.08) | 5.15 | (0.96, 0.85, 0.71, 0.66) | (28.34, 8.24, 5.88, 5.53) |
| S | 5.13 | | (5.30, 5.72, 8.00, 14.87) | 5.16 | | (25.98, 9.13, 5.83, 5.44) |

## 6. Application to Terminal Control

In many practical situations, the CPU processing rate s(n) decreases when n increases beyond some point. That is, if we have too many customers in the CPU processing stage, the CPU efficiency will get bad and this may further lead to a longer mean waiting time. Hence in this case, the number of running terminals should be controlled appropriately. Given the arrival rates and the other system characteristics, the problem is to decide on the optimal number of running terminals and the best allocation of them among the stations. In what follows, we demonstrate through a numerical example how to use the results in this paper solving the above problem.

The example considered here has the same parameters and CPU processing rate function s(n) as those given in the previous section, except that the number of terminals, $\{m_j\}$, is changed for various cases while $\{\lambda_j\}$ is fixed to (0.696, 0.870, 1.044, 1.218) as in the last case in Table 5.1. We only consider such cases that $m_j$ are the same for each group of $1 \leqq j \leqq 8$, $9 \leqq j \leqq 16$, $17 \leqq j \leqq 24$ and $25 \leqq j \leqq 32$. Denote them by $m=(m_{j_1}, m_{j_2}, m_{j_3}, m_{j_4})$.

Suppose now that m=(4, 5, 6, 7) as in Section 5. Consider then the problem: Which group of stations classified above ought to be chosen if we intend to increase a terminal for each station of only one group? The results for all of the four possible choices are shown in the top of Table 6.1 where they are calculated by the iterative scheme in Section 3 as well as (5.1). If we take $\bar{T}$ as the criterion, then we see that all these four cases improve the original case and the best one is m=(5, 5, 6, 7). Executing the same work further from the point m=(5, 5, 6, 7), then we get the results of another four cases. Here we observe an interesting phenomenon that m=(6, 5, 6, 7) gets worse than m=(5, 5, 6, 7) due to the increase in ES which results in more defects from others than the improvements of some stations. The other three cases still show improvements for $\bar{T}$. Restarting from the best one of these, m=(5, 6, 6, 7), we finally get the optimal choice, m=(5, 6, 7, 7), for the numbers of terminals at the stations. Further calculations have shown that any more increase of terminals in such a way will lead to the situation that some stations get overloaded.

Table 6.1

| m | (5, 5, 6, 7) | (4, 6, 6, 7) | (4, 5, 7, 7) | (4, 5, 6, 8) |
|---|---|---|---|---|
| ES | 5.542 | 5.543 | 5.544 | 5.544 |
| T | ( 7.26, 25.13, 21.72, 19.29) | (30.37, 7.30, 21.78, 19.35) | (30.46, 25.28, 7.30, 19.39) | (30.53, 25.33, 21.89, 7.30) |
| $\bar{T}$ | 18.35* | 19.70 | 20.61 | 21.26 |

| m | (6, 5, 6, 7) | (5, 6, 6, 7) | (5, 5, 7, 7) | (5, 5, 6, 8) |
|---|---|---|---|---|
| ES | 5.547 | 5.550 | 5.551 | 5.551 |
| T | ( 6.03, 25.72, 22.21, 19.71) | ( 7.29, 7.32, 22.46, 19.93) | ( 7.29, 26.10, 7.33, 19.99) | ( 7.29, 26.27, 22.59, 7.33) |
| $\bar{T}$ | 18.42 | 14.25* | 15.18 | 15.85 |

| m | (6, 6, 6, 7) | (5, 7, 6, 7) | (5, 6, 7, 7) | (5, 6, 6, 8) |
|---|---|---|---|---|
| ES | 5.556 | 5.557 | 5.560 | 5.560 |
| T | ( 6.04, 7.34, 23.07, 20.45) | ( 7.31, 6.09, 23.19, 20.56) | ( 7.32, 7.36, 7.37, 20.80) | ( 7.32, 7.36, 23.55, 7.37) |
| $\bar{T}$ | 14.23 | 14.29 | 10.71* | 11.40 |

We observe from this example that it is the most desirable to install more terminals at the stations which have the worst situation unless any station gets overloaded. It seems however that the influences to the other stations from the installments can not be estimated without the results in this paper.

In actual systems as described in Section 1, the arrival rates at the stations may vary from time to time. In such cases, the number of running terminals at each station should be controlled accordingly.

### References


1. Baskett, F., K. M. Chandy, R. R. Muntz and F. G. Palacios, "Open, Closed and Mixed Networks of Queues with Different Classes of Customers," Journal of the Association for Computing Machinery, Vol. 22 (1975), 248-260.
2. Boxma, O. J., J. W. Cohen and N. Huffels, "Approximations of the Mean Waiting Time in an M/G/s Queueing System," Operations Research, Vol. 27 (1979), 1115-1127.
3. Chandy, K. M., J. H. Howard and D. F. Towsley, "Product Form and Local Balance in Queueing Networks," Journal of the Association for Computing Machinery, Vol. 24 (1977), 250-263.
4. Kimura, T., "A Two-Moment Approximation for the Mean Waiting Time in the GI/G/s Queue," Management Science, Vol. 32 (1986), 751-763.
5. Kleinrock, L., "Analysis of a Time-Shared Processor," Naval Research Logistics Quarterly, Vol. 11 (1964), 59-73.
6. Kleinrock, L., Queueing Systems, Volume II: Computer Applications, John Wiley & Sons, 1976.
7. Pan, H. and H. Morimura, "Approximate Analysis of an On-Line Network System," Journal of the Operations Research Society of Japan, Vol. 29 (1986), 286-304.
8. Scherr, A. A., An Analysis of Time-Shared Computer Systems, MIT Press, 1967.
9. Takahashi, Y., "An Approximation Formula for the Mean Waiting Time of an M/G/c Queue," Journal of the Operations Research Society of Japan, Vol. 20 (1977), 150-163.
10. Tijms, H. C., M. H. V. Hoorn and A. Federgruen, "Approximations for the Steady-State Probabilities in the M/G/c Queue," Advances in Applied Probability, Vol. 13 (1981), 186-206.

# Parallel Ray Tracing on a Cellular Array Processor

*Koichi Murakami, Katsuhiko Hirota, Mitsuo Ishii*

Computer-based Systems Laboratory
Fujitsu Laboratories LTD. Kawasaki
1015, Kamikodanaka, Nakahara-ku,
Kawasaki, 211, Japan
TEL. 044-777-1111 (ext.2-6292)

*ABSTRACT*

We present a method to shorten the computation time of ray tracing, which involves two approaches.

The first approach uses an algorithm which partition the environment with voxels (volume cells) to reduce the number of ray-object intersection calculations, which is the most computationally intensive part of ray tracing. We present a new traversal algorithm, called the *Parametric 3DDDA (3 Dimensional Digital Differential Analyzer)*, which can traverse the voxel data structure in an efficient manner. Performance increases with the number of partitions until the traversal overhead becomes too expensive.

For the second approach, ray tracing algorithm was implemented on the *CAP (Cellular Array Processor)* parallel processor. The parallel ray tracing algorithm capitalizes on the computational independency of the individual rays that pass through the different screen pixels. We will present the static load distribution algorithm which allows the load to be balanced without additional overhead. Each cell can process rays that pass through the pixel elements assigned to the cell. This static load distribution algorithm demonstrates that load can be evenly balanced without communication.

The results demonstrate that performance increases with the number of cells, which implies that our method is suitable for a parallel machine with many processors. Also, we compared the processing time with that of a large scale computer, the *Fujitsu M-380*, which shows performance with 64 cells is comparable to that of a large scale computer. Ray tracing becomes practical for some computer graphics applications with the presented approaches.

## 1. Introduction

Recent CAD systems such as those used in industrial design have required the generation of high-quality images. Optical effects such as reflection, refraction, and shadow have become very important in the modeling. Although ray tracing can generate images of excellent quality, it suffers from lengthy computation times that limit its practical use. The image generation time of conventional ray tracing may take several hours[1].

A number of studies have been proposed to accelerate ray tracing. One group of studies uses coherence [2] to reduce computation time, especially for ray-object intersection calculations, which is the most computationally intensive part of ray tracing. Space partition techniques use object-space coherence that capitalizes on the relative spatial position of the objects. There are two types of space partition schemes, octree and voxel. We[3] and Glassner[4] presented the algorithm that traverses octree data structures.

Although these method improved processing time, the speed is still less than ten times faster than conventional ray tracing. There are two reasons why the octree scheme did not cause much improvement. First, the overhead for traversal of the octree data requires extra computation in addition to the traversal itself. Backtracking is necessary when traversal fails in one child node. Second, since the space is dynamically divided when the octree is built, the size of the octant (cell of octree) can not be known except for the current one. It does not allow the algorithm to incrementally calculate the next voxel element. For these reasons, we chose the voxel data structure as the representation in the space partition scheme[5],[6]. We will present a voxel traversal algorithm that efficient determines the next voxel element for the ray to proceed.

Another group of studies involves the parallel computation of the ray tracing algorithm. The parallelization of ray tracing capitalizes on the computational independency of individual rays that pass through the different screen pixels. Load balancing must be considered so that performance increases with the number of cells. In Links [7] the task of each processor is allocated dynamically during the processing. When a processor finishes its work, the host processor reallocates a task to it. The problem with this method is that it suffers from overhead due to communication. We propose two static load distribution schemes that allow the load to be balanced between the processors without overhead due to communication. In Chapter 3, the implementation of the parallel ray tracing algorithm on the CAP that we developed is discussed. The result shown in Chapter 4 demonstrates performance improvements as compared to conventional ray tracing.

## 2. Traversal algorithm

### 2.1. Concept of the space partition method

In conventional ray tracing, each object in the environment is tested for intersection. Most of these tests useless because only a few objects intersect with a given ray. The space partition approach is more efficient since intersection calculations are done only for the object candidates.

The key idea behind the space partition method can be considered as *divide and conquer*. Intersection calculation time is significantly reduced by subdividing the environment into small orthogonal volumes, (voxel) and performing intersection calculations for only the objects that are likely to intersect with the ray. For each voxel object identifiers are recoded if the objects exist within the voxel. As a ray propagates from one voxel to the next, the objects in each voxel become candidates for ray-intersection calculation. Rejecting the objects that are not likely to intersect with the ray, the algorithm avoids many ray-object intersection calculations. The method can be applied to not only primary rays (the original rays from the view point) but also to shadows and reflected and refracted rays.

### 2.2. Voxel traversal algorithm

The movement operation that finds the next voxel element for the ray to proceed must be fast so that its calculation does not cause additional overhead. We present a new traversal algorithm that efficiently determines the next voxel element for the ray to enter from the current voxel element in an efficient way. The method is called *the Parametric 3DDDA (3 Dimensional Digital Differential Analyzer)* since it uses the parameters of the ray ( $t$ in $X = \alpha t + \beta$). The DDA algorithm [8] was originally used to obtain a rasterized straight line in two dimensions, where the successive pixels along the line are efficiently determined using only incremental addition. This concept can be applied to the traversal algorithm over the voxel data structure. The 3DDDA algorithm determines the next voxel element by using only incremental addition as well as the original DDA algorithm, although the 3DDDA algorithm performs in three dimensions rather than two.

The key idea behind this algorithm is to retain the points on the boundary planes of the voxel element. These points are then used to guide the next voxel element for the ray to proceed. We refer to points on the boundary along the ray with the parameter $t$. The value of $t$ increases as we move away from the origin, where $t$ is 0. Three $t$ parameters, $t_x$,$t_y$, and $t_z$ that indicate the points $X_x$,$X_y$, and $X_z$ on the boundary with which the ray will intersect, are provided as shown in equation 1.

$$X_i = \alpha_i \cdot t_i + \beta_i \qquad (i=x,y,z) \qquad (1)$$

The selection which component of the voxel index changes is based on the following observation. Three $t$ parameters predicate the next voxel element with respect to the current one. The minimum value out of three causes the component to change. Figure 1 illustrates this in two dimensions. Parameters, $t_x$ and $t_y$ indicate the points on the boundary where the ray will intersected. When $t_x$ is smaller than $t_y$, the ray traverses the boundary of x ahead of y. This means that the x component of the voxel index changes by the value +1 or -1 according to the sign of the direction of the ray ($\alpha_x$). In this manner, the determination of component to changes is made by only examining the minimum value of the $t$ parameters.

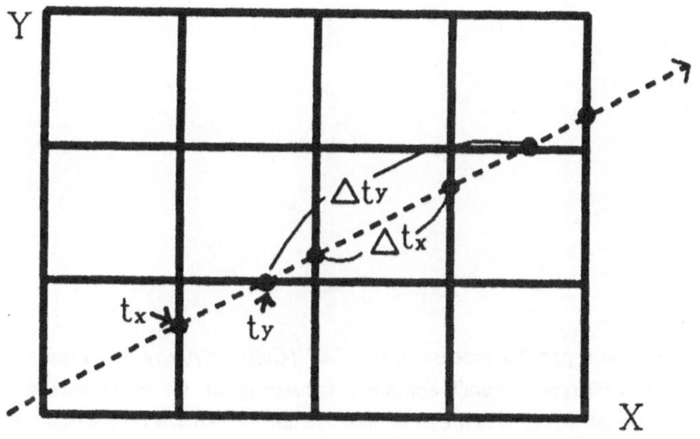


Figure 1. Scheme of Voxel Traverse.

After determining the minimum value, the parameter has to be updated for the next evaluation. The precomputed step, $\Delta_x$,$\Delta_y$, and $\Delta_z$ that represents the distance between the boundary planes along the ray is added to the $t$ parameter which is just selected. These precomputed values are determined as $\Delta x = \frac{\textit{unitsize of voxel}}{\alpha_x}$.

The cost for one voxel movement requires one integer incremental addition for updating the voxel index, one floating incremental addition for updating the $t$ parameter, and two floating comparisons for selecting the minimum value out of three values.

For an exception, the ray may not intersect with one of any boundary planes. For example, the ray with $\alpha_x = 0$ does not intersect YZ-planes of the voxel boundary. Setting the initial value $t_x = \infty$ can avoid the ray from intersecting with YZ-planes automatically because it never can be chosen as the minimum value. The following algorithm is used for traversal of voxel data structures.

Algorithm - 1: 3DDDA

```
int vx,vy,vz;                          /* voxel indices  */
int incre_x,incre_y,incre_z;           /* +1 or -1            */
float tx,ty,tz;                        /* t parameters        */
float delta_x,delta_y,delta_z          /* step                          */
int component;

while(in the voxel space) {
/*     get component that corresponds the minimum value out of t_x,t_y,t_z;  */
      component = GetMinimumComponent(tx,ty,tz);
      switch(component) {
            case X:
                  vx += incre_x;
                  tx += delta_tx;
                  break;
            case Y:
                  vy += incre_y;
                  ty += delta_ty;
                  break;
            case Z:
                  vz += incre_z;
                  tz += delta_tz;
                  break;
      }
}
```

## 3. Parallel ray tracing on CAP

### 3.1. CAP architecture

This algorithm is implemented on a parallel processor, *the CAP (Cellular Array Processor)* [9], which we developed. The CAP is a MIMD-type parallel computer consisting of 64 processors called cells configured in an 8 × 8 two dimensional array. Each cell consists of an i80186 MPU, an i8087 NDP, and 2M-bytes of local memory. Figure 2 shows the CAP system. Every cell is connected by two buses. The Command Bus is used to download data and programs from the host computer. It can also be used for broadcast communication from the host computer to the cells or from a particular cell to the host computer or other cells. Each cell is connected to four neighboring cells to enable local communication. Cells at the boundaries are connected to cells at the opposite sides, so the network topology resembles be seen the surface of a torus.

For the CAP's graphical facility each cell has video memory within it. Images in video memory in cells are displayed on a monitor through the Video Bus. A significant feature of CAP's capability for display is a variety of screen partitioning modes, in which any cell can output its results, usually an intensity value, at any pixel on the screen. Each cell knows its identifier and the pixel location on the screen that the cell handles. Optimal load balancing of ray tracing can be achieved with this facility. The task is distributed to many cells and the corresponding image that the cell produces is displayed.

### 3.2. Load balancing

The ray tracing algorithm is analyzed to consider the parallel ray tracing. Since the ray is traced backward from the view point to the light source, the computation for the rays that pass through the different pixels can be done independently. The parallel ray tracing algorithm can capitalize on the above computational independency of the individual ray.

In parallel ray tracing each processor has a copy of the object database and processes rays that pass through the pixel element assigned to the cell in parallel.

Load balance must be considered so that performance increases with the number of cells. Our method uses static load distribution to allocate part of the screen to each cell instead of dynamic load distribution that uses communication. In the dynamic load distribution method, overhead may become too expensive when a large number of processors are used, since amount of reallocation and communication increases with the number of processors. The static load distribution algorithm allows the load to be balanced without communication.

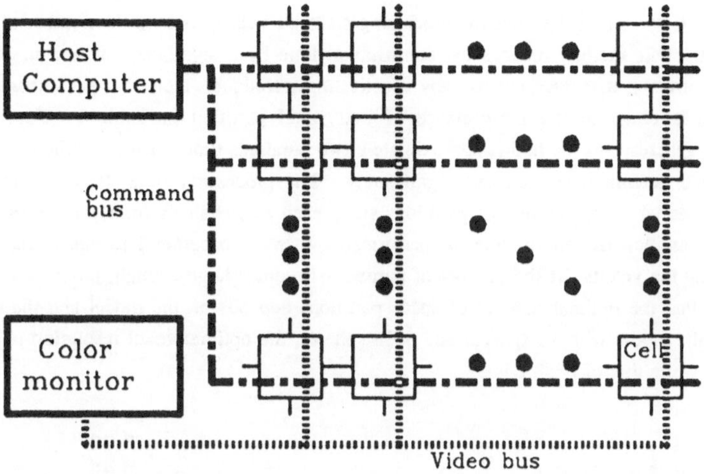


Figure 2. CAP Configuration.

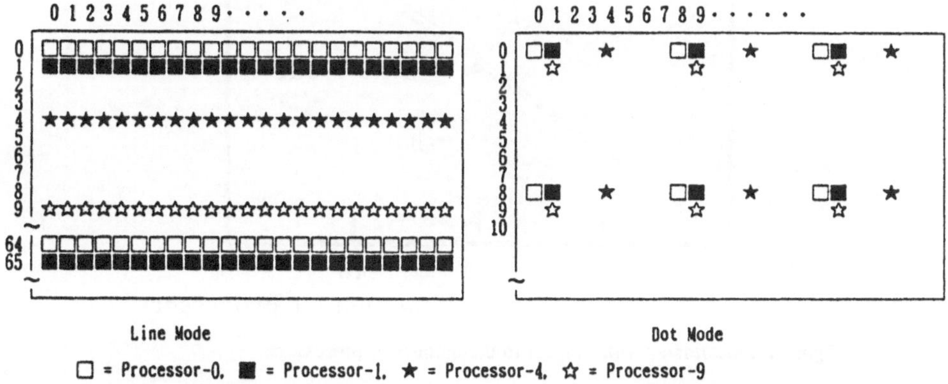


Figure 3. Line and Dot Distribution Modes.

We evaluates two modes of static load distribution method, the line mode and the dot mode. The mark of each figure 3 shows the pixels which are processed by a processor. In the line-mode, cells are allocated each 64th scan line. Cell-0 handles the 0th scanline, the 64th scanline, and so on. Cell-1 handles the first scanline, the 65th scanline, and so on. In the dot mode, each cell performs ray tracing on every eighth pixel horizontally and vertically. Cell-0 handles the top left pixel of the screen. Next it handles the eighth pixel to the right. After completing the scan, the process is repeated eight pixels lower. Cell-1 process the right neighboring pixels that cell-0 handles. In this way, the whole screen is processed by the 64 cells.

## 4. Results and discussion

### 4.1. The results of the voxel partition scheme

Figure 5-a shows the model SPHERES with 125 spheres used to demonstrate the performance of the space partition method. The test image was calculated at a resolution of 512 × 384 pixels. No antialiasing was performed for the tests. Table 1 shows the processing statistics with respect to the number of the partitions along each axis. This method reduces the computation time by a significant amount in comparison with conventional ray tracing that does not use any accelerating technique. Note that performance of conventional ray tracing is equal to the performance of one voxel partition in Table 1. For the model SPHERES with 6858 spheres, this method is approximately two hundreds times faster. While the number of partitions is small, the performance increases significantly. The processing time decreases from 13094 seconds to 2856 seconds when the number of partitions along each axis changes from 1 to 2. However, as the number of partitions increases the amount of performance increase becomes less due to the additional overhead for traversing the voxels. If the number of partitions becomes large enough, processing time actually increases. Note that the optimal number of space partitions depends on the model and the number of objects. Comparing the model with 16 spheres and 6858 spheres, the optimal voxel resolution of the model with 16 spheres is 8, while that of 6858's is 32.

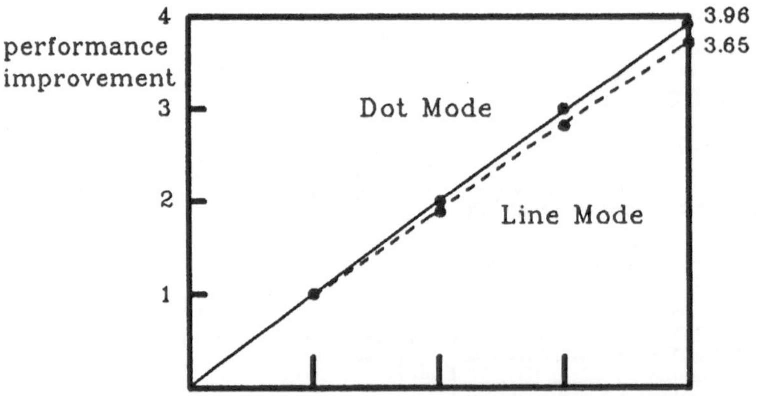


Figure 4. Performance with respect to the number of processors.

### 4.2. Load balancing

Table 2 lists the statistics obtained from the line and the dot distribution modes. The difference ratio is defined as the ratio of difference between the maximum and minimum processing time compared with the maximum processing time. The difference ratio in the dot mode is within a few percent while that of the line distribution mode ranges more than ten percent. This implies that the dot mode distribution scheme works effectively. Figure 4 shows how performance increases with the number of cells for both dot and line load distribution modes. The graph is normalized by the processing time of 16 cells. In the dot mode, the performance factor of 64 cells compared with that of 16 cells is 3.96, which is almost proportional to the number of cells. On the other hand, that of the line mode is 3.65. This implies that performance in the dot mode is expected to increases for a large number of processors.

### 4.3. Comparison with large scale computers

To demonstrate the performance and effectiveness of the ray tracing algorithm on CAP, we compare the performance of the CAP to that of a large scale computer, *Fujitsu M-380 (15 MISP)*. The processing time is measured on the CAP and M380 for the same algorithm. Table 3 lists the processing time for the models shown in Figure 5, which demonstrates that the performance of the CAP is comparable with the M380. Furthermore, ray tracing on the CAP has several advantages over a large scale computer. First, cpu time is equal to the ellapsed time in dedicated machines such as the CAP, while the ellapsed time is usually much longer than the cpu time in a large scale computer. Since interactive image generation requires high-speed processing this characteristic becomes very important. Second, the performance increases with the number of processors in the CAP. Much higher performance will be achieved when a large number of processors are used.

## 5. Conclusion

We have developed the fast ray tracing algorithm using two techniques. First, the ray-object intersection calculation is reduced using the space partition technique and efficient traversal algorithm. It demonstrates that significant improvement is obtained, from several dozen to 200 times faster than that of conventional ray tracing. Second, this algorithm is implemented on parallel processor, the CAP. We proposed efficient static load balancing method. The dot mode static distribution algorithm demonstrates that load can be evenly balanced without communication. We have tried up to 64 processors. It is expected that the performance should increase linearally with the number of processors within a certain ranges. Also, we compared the processing time of CAP with that of a large scale computer, the Fujitsu M-380, which shows the performance with 64 cells is comparable to that of a large scale computer. Ray tracing becomes practical for some computer graphics applications with the presented approaches.

As mentioned in the implementation section, each cell has the same object database. This limits the number of objects which the system can handle. Data can be distributed between every cell and accessed through local communication.

Table 1. Processing time vs. number of voxel partitions.

| Number of objects | 1 | 2 | 4 | 8 | 16 | 32 | 48 |
|---|---|---|---|---|---|---|---|
| 16 | 102 | 39 | 23 | 20 | 21 | 25 | 30 |
| 512 | 825 | 212 | 68 | 34 | 28 | 31 | 36 |
| 1000 | 1537 | 450 | 126 | 59 | 38 | 38 | 43 |
| 4096 | 13094 | 2856 | 678 | 211 | 104 | 96 | 101 |
| 6858 | 32621 | 7306 | 1557 | 482 | 219 | 162 | 163 |

The model is SPHERES. Time is in seconds.

Table 2. Load balance statistics

| Model | Line mode<br>min - max (sec)<br>difference ratio(%) | Dot mode<br>min - max (sec)<br>difference ratio(%) |
|---|---|---|
| SPHERES with 125 sphere | 84 - 93<br>9.7% | 88 - 86<br>2.3% |
| Chess | 259 - 324<br>20.0% | 281 - 294<br>4.4% |
| CAD | 637 - 792<br>19.6% | 667 - 691<br>3.5% |

(*) difference ratio = (max - min) / max × 100

Table 3. Performance comparision of CAP and M380

| Model (number of objects) | CAP (sec) | M380 (sec) |
|---|---|---|
| SPHERES (125) | 72 | 60 |
| CHESS (144) | 155 | 128 |
| CAD (178) | 20 | 20 |
| PISTON (179) | 40 | 33 |

(a) SPHERES

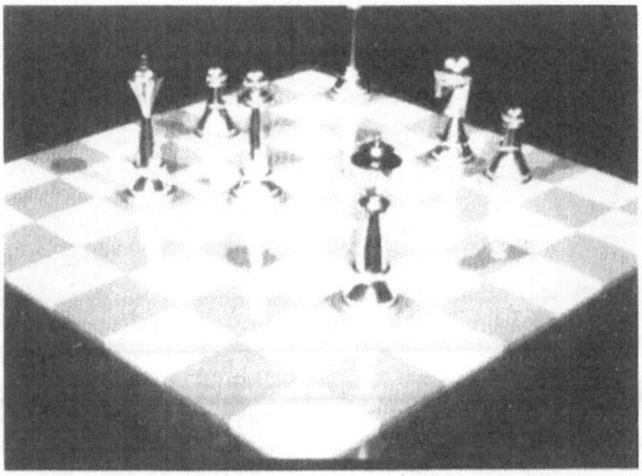

(b) CHESS

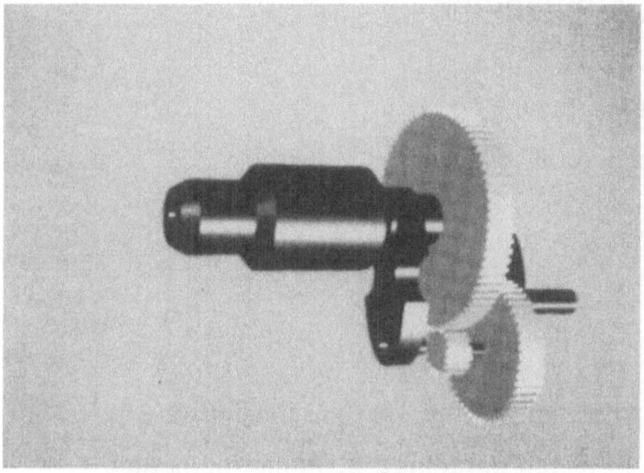

(c) CAD

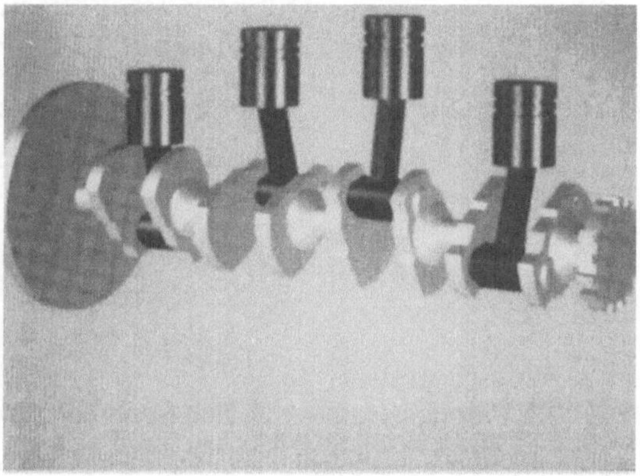

(d) PISTON

Figure 5. Models

**Reference**

[1] T.Whitted, "An Improved Illumination Model for Shaded Display," *Comm. ACM* Vol.23, No.6, 1980, pp.343-349.

[2] P. Atheron, "A Scan-line Hidden Surface Removal Procedure for Constructive Solid Geometry," *Computer Graphics*, 17(3),July 1983,73-82

[3] Matsumoto, H., and Murakami, K., "Fast Ray Tracing using the octree partitioning," *27th Information Processing Conference Proceedings* October 1983,1537-1538,(In Japanese)

[4] A. Glassner, "Space Subdivision for Fast Ray Tracing," *IEEE Computer Graphics and Applications*, 4(10),October 1984,15-22

[5] K. Murakami, K. Hirota, "Ray Tracing System using the voxel partitioning on a Cellular Array Processor," *Graphics and CAD* 22(2),July 1986,1537-1538,(In Japanese)

[6] A. Fujimoto, T. Tanaka and K. Iwata, "ARTS:Accelerated Ray Tracing System," *IEEE Computer Graphics and Applications*, 6(4),April 1986,16-26

[7] Nishimura, H., Ohno, H., Kawata, T., Shirakawa, I., and Omura, K., "LINKS-1: A Parallel Pipelined Multimicrocomputer System for Image Creation," Proceedings of the 10th Symposium on Computer Architecture, SIGARCH(1983), pp.387-394.

[8] J.E. Bresenham, "Algorithm for Computer Control of a Digital Plotter," *IBM System Journal, Vol.4,1965,25-30*

[9] H.Ishihata, M.Kakimoto, "VLSI for the Cellular Array Processor," Proc. ICCD,October 1987,320-323

# The FPS T Series Supercomputer

S. Hawkinson
Floating Point Systems, Inc.
Beaverton, Oregon, 97005

## Abstract

The FPS T Series Parallel Vector Supercomputer consists of multiple processing nodes interconnected as a binary *n*-cube. Each node is a self-contained scientific computer consisting of a control/communication processor, fast interprocessor links, and a powerful vector processing engine. All of these features of a single node can operate in parallel, enhancing the performance of the system as a whole by allowing vector computations and internode communications to proceed concurrently. Each node is implemented on a single printed circuit board, contains one megabyte of data storage memory, and has a peak vector computational speed of 12 MFLOPS. Eight vector nodes are grouped together with a system support node and a system disk to form a module. The module is the homogeneous unit on which larger T Series configurations are based. This paper presents an overview of the architecture and performance capabilities of the FPS T Series.

## I. Introduction

As the performance limits of sequential processors are approached, research and development in the field of parallel processing is increasing. The utilization of data level parallelism in the form of vector processing has proven to be a viable method of obtaining performance gains [4]. The advent of parallel functional units to provide for the concurrent execution of instructions has also resulted in significant performance gains in both scalar and vector processing [1]. Further considerable performance improvements can be achieved through the effective implementation of an ensemble of self-contained processing nodes interconnected by high-speed links. The FPS T Series employs parallelism at all three levels, with the finer grains of parallelism providing the basis for the efficient utilization of the larger grain, or processor level, parallelism.

Each T Series node is a parallel computer comprised of a control/communication processor with multiple bi-directional serial links, one megabyte of dual-ported data storage memory, and a 12 MFLOP Vector Processing Unit (VPU). The VPU is a complete arithmetic unit consisting of a microcode control store, a microsequencer, a floating-point adder, and a floating-point multiplier. The VPU operates in parallel with, and is serviced by, the control/communication processor. The control/communication processor can drive up to four inter-processor links bidirectionally and perform address calculations all in parallel. Each T Series node is thus capable of performing addressing, communication, and vector computation concurrently. This intranode parallelism provides the potential for full utilization of the VPU on each compute node.

Eight T Series compute nodes, one system support node, and a system disk are combined to form a *module*. The compute nodes in a module are connected as a binary 3-cube. Each compute node is also linked to the system node and thereby to the system disk. T Series modules can be interconnected to form binary *n*-cubes of higher dimension. Each module has an aggregate vector performance of 96 MFLOPS, and configurations of tens or even hundreds of modules can deliver compute speeds in the GIGAFLOP range [5].

## II. Processor Node Architecture

An individual processor is called a *node*. It contains a control processor, floating-point arithmetic, dual-ported memory, and communication links to other nodes (see Figure 1).

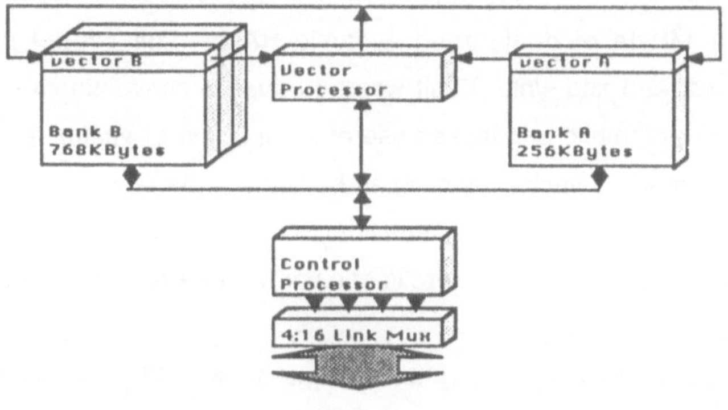


Figure 1: Vector Node Architecture

The FPS T Series design provides all of these functions on a single printed-circuit board. Each of the major elements of the node has been implemented with advanced, cost-effective VLSI technology, in contrast with more traditional bit-slice designs.

## Control

The ability to interpret and execute programs resides in the central control unit. The T Series control unit is a 32-bit CMOS microprocessor (the Inmos Transputer) with the following functional features:

- 1 - 7 MIPS instruction rate (highest when on-chip RAM is used)
- Byte addressability (4 GByte address space)
- 2048 bytes of on-chip RAM (single processor cycle)
- 3-cycle minimum access time for off-chip memory
- Four bidirectional serial communications links
- Stack-oriented instruction set with variable operand sizes
- Two-level process priority and interrupt services

The control processor executes system and user applications code and it also serves to arrange vector operands to be sent to the vector arithmetic hardware. The control processor can execute integer arithmetic and gather/scatter operations in parallel with the vector unit, and it provides inter-node communications via the serial links.

## Memory

An essential feature of a computer's architecture is its central memory, which supplies both instructions and operands to the processing units. The main memory of each FPS T Series node consists of 1 MByte of dual-ported dynamic RAM. The control processor and communications links read and write 32-bit words through a conventional random-access port, while the vector arithmetic unit makes use of a collection of *vector registers* closely coupled with main memory. A vector register can be loaded with an entire 1024-byte row of memory, in parallel (see Figure 1), in the same time that it would have taken to read or write a single 32-bit word: 0.467 $\mu$seconds. There is one parity bit for each byte in memory.

The control processor views the memory as a single bank of 256K words (32-bit). The vector arithmetic unit views memory as two banks of vectors, with 256 vectors in one bank and 768 vectors in the other, aligned on 1024-byte boundaries. Thus, for 32-bit operations, the vectors are 256 elements long, while for 64-bit operations, the vectors are 128 elements in length. The division of memory into two banks permits two inputs in parallel to the arithmetic unit on each cycle (125 ns). The output of the arithmetic unit shifts results into either or both banks. Hence, operations such as SAXPY, Vector Add, and Vector Multiply proceed at the full speed of the arithmetic components, without being limited by available memory bandwidth. This dual-bank memory organization allows the node to function without the need for auxiliary data registers or data cache.

The control processor can access a 4-byte word in 467 ns. Its effective bandwidth to RAM is therefore (4 bytes) / (0.467 μs) = 8.6 MB/s

A primary use for the control processor is to gather operands into a contiguous vector, and scatter results back to random locations in memory. To move a 64-bit operand from one memory location to another with simple addressing takes a minimum of 8.6 μs. This is the minimum gather-scatter time within a node; it does not include the time to *read* the actual addresses of the desired destination or source from a table. For 32-bit operands, the minimum gather-scatter time is 4.3 μs per element.

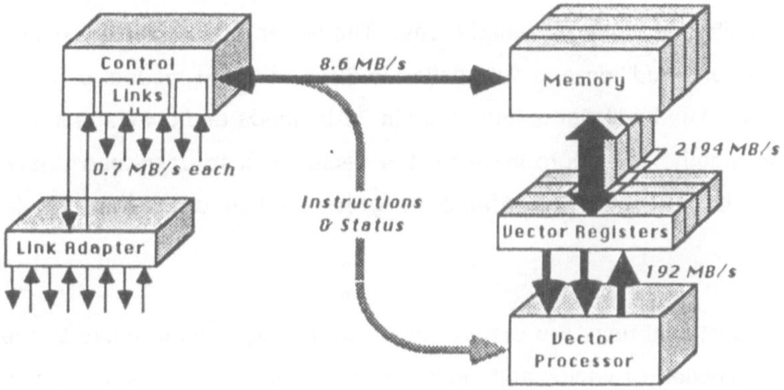


Figure 2: Node bandwidths

A row of data can be moved to or from a vector register in only 467 ns; thus the effective bandwidth between memory and a vector register is (1024 bytes) / (0.467 μs) = 2194 MB/s. An application might make use of this extraordinary speed by moving data physically, rather than keeping linked lists of pointers to vectors, as for example, in pivoting rows of a matrix or sorting records.

The vector registers each supply data to the arithmetic unit at a maximum rate of one 32-bit word every 62.5 ns, or one 64-bit word every 125 ns. The vector register bandwidth supports two vector inputs and one vector output every 125 ns in 64-bit mode. Thus, its bandwidth is (3 words)x(8 bytes/word)/(0.125 μs) = 192 MB/s.

## Arithmetic

The ability to perform high-speed arithmetic is essential in scientific computing. The arithmetic hardware in the FPS T Series consists of a floating-point adder, floating-point multiplier, interconnection hardware, and a micro-sequencer. The adder and multiplier each can produce a 32- or 64-bit result every 125 ns, yielding *nominal* performance of 16 MFLOPS per node. Overhead inherent in the architecture reduces the effective peak to 12 MFLOPS. Floating-point operations are performed using the proposed IEEE floating-point standard format; however, gradual underflow is not supported. In 64-bit mode, the mantissa has approximately 15 decimal digits of precision (53 bits) and a dynamic range of roughly $10^{-308}$ to $10^{+308}$ (11-bit binary exponent).

The arithmetic units operate in pipelined mode. The adder has a six-stage pipeline. It can perform floating-point addition and subtraction in 32- and 64-bit modes, comparisons and data conversions. The multiplier is four-stage in 32-bit mode and six-stage in 64-bit mode. These pipeline lengths are appropriate for the vector access described above. Scalar operations can be efficiently performed by grouping like operations for level-order evaluation.

The arithmetic functional units are supervised by a preprogrammed micro-sequencer that implements a collection of vector arithmetic operations referred to as *vector forms*. The programmer only needs to describe the input and output vectors and the vector form desired. This frees the control processor for other tasks while vector operations are being executed. Scalars can be held in the input registers on each floating-point functional unit, and outputs from the functional units can be fed directly back as inputs to perform operations such as dot products and sums. This provides a wide range of useful vector forms without memory reference limitations. The complete arithmetic unit operates in parallel with the node control processor. The arithmetic unit only interrupts the controller when a vector operation has completed.

## Communications

In a distributed computer system, communications channels are required for passing data between processors participating in a common computational process. The control processors of the T Series contain drivers for four serial, bidirectional links each with an effective rate of 0.666 MByte/s. Every 8-bit byte is sent with two synchronization bits and one stop bit, and requires two acknowledge bits from the receiver. When the receiver is not also sending data, the unidirectional rate increases to 0.691 MByte/s. The total bandwidth of the four links is thus about 5 MB/s. With all links operating, the control processor performance is degraded only slightly. The links operate via DMA transfers with a total startup time of about 9 μs on each end of the transmission.

Each link is switched four ways to provide a total of 16 bidirectional sublinks per node. Two sublinks are used for system communication, and two will often be utilized for mass storage and/or external I/O. This will typically leave 12 sublinks available for connection to other compute nodes, of which four can be simultaneously active. Switching link multiplexers takes from 0.5 to 0.3 msec per link, depending on how many links (1 to 4) are switched simultaneously.

A convenient way to interpret the relative bandwidths is with respect to the processing time for 64-bit vector operations (128 elements):

| (Arithmetic Time) : | (Gather Time) : | (Link Transfer Time) |
|---|---|---|
| 21.3 μs | 1100 μs | 1500 μs |

that are in the approximate ratios

1 : 52 : 70

Thus, a vector should enter into about 50 operations while gathering the next vector into an aligned, contiguous order. With this restriction, the gather time matches the vector arithmetic time, and the node can approach peak speed. Of course, if vectors are always aligned and elements contiguous, no such restriction applies. Similarly, roughly 70 operations should result from every 64-bit word that must be moved between nodes over a link. Moving a collection of data over four links simultaneously would reduce this requirement to about 20 operations per 64-bit word.

## III. System Description

A T Series configuration consists of a number of node processors connected as a binary $n$-cube. There are $2^n$ processors, with $n$ connections per node. Numbering the processors from 0 to $2^n-1$, each processor is directly connected to all others whose numbers differ in only one binary digit. The binary $n$-cube can be mapped onto many important applications topologies, including meshes (up to dimension $n$), rings, cylinders, toroids, and even FFT butterfly connections of radix 2 [2,3]. Since the maximum number of connections between any two processors is $n$, long-range communication costs grow only as $O(\log_2 n)$.

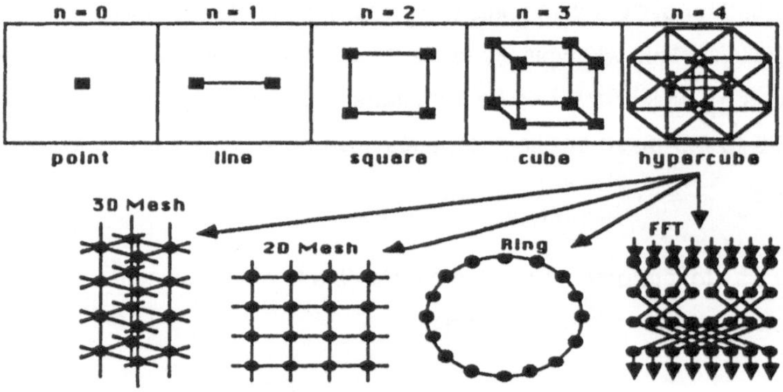


Figure 3: Hypercube Mappings

### Modules and System Ring

A processor node is constructed on a single etched circuit board. Eight nodes are combined with disk storage and a system board to form a *module*. Such a module has 96 MFLOPS peak vector floating-point performance, and 8 MB of user RAM. The internode communications bandwidth internal to the module is about 16 MB/s, while the system board can support 0.7 MB/s to an external connection.

The system board provides input/output and management functions. It is connected directly to the nodes by a tri-state bus made out of one of the sublinks. The system boards are directly connected by communications links to form a ***system ring*** that is independent of the binary *n*-cube network (connecting the processor nodes). The primary function of the system disk is to record memory "snapshots" which checkpoint computations for error recovery, and to back up snapshots from other modules. The user controls the interval between snapshots. About 10 minutes provides a good compromise between time spent to record memory and interval between restart points. It takes about 15 seconds to take a snapshot, regardless of configuration.

The module requires three links for intramodule hypercube network communications, while the system board connections require one link from each processor node. One link of the 16 is not driven. This reduces remaining links to 11 for hypercube and external communications.

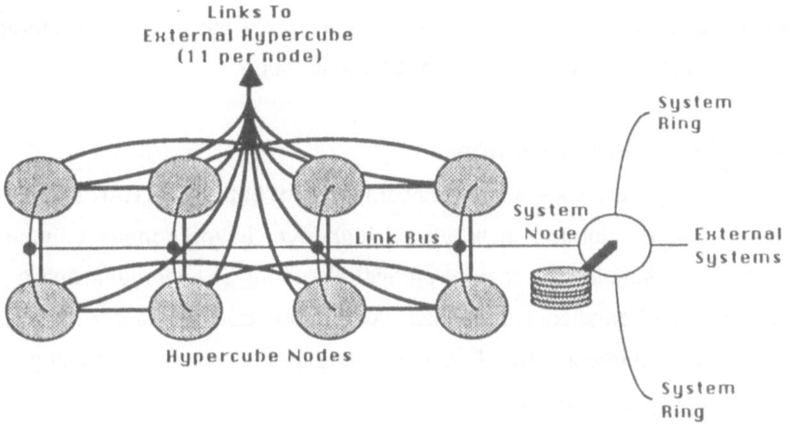


Figure 4: Eight Processor Module

Two modules (16 nodes) form a *cabinet*, or 4-cube (a tesseract). A cabinet is modular and self-contained in a standard 19-inch rack-mounted assembly. With power supplies, system disks, and air cooling fans, it does not require any special "computer room" facilities. Larger systems are simply assembled from these units by interconnecting cables. Connections up to 40 feet can be made without special consideration.

## Larger Configurations

A four-cabinet (64-node) system has an aggregate peak vector speed of 0.768 GFLOPS and total user memory of 64 MBytes. Eight system-disk units provide backup and restart capability. This configuration of the T Series can be located in many laboratory and computing facilities. The air-cooled unit requires no special facilities beyond normal air conditioning, and the power requirements are supplied by typical 220 VAC services.

There are enough links per node to permit a 14-cube to be constructed as the largest T Series configuration. Using two links per node for external I/O and mass storage systems, a maximum-sized 12-cube consists of 4096 nodes arranged as 256 cabinets (4-cubes). Such a system has over 49 GFLOPS vector processing capability and 4 GBytes of primary RAM storage. Special facilities would be required to house the largest configurations.

Because the system is homogeneous, i.e., each module is identical and contains identical connections to other modules, programming is greatly simplified. This homogeneity also ensures that the balance between computing speed, main storage, mass storage, and external I/O can be preserved as configurations become large.

## Programming and Performance

The new T Series software package includes compilers for both FORTRAN and C, tools for data partitioning and distribution, a library of high-level asynchronous communication routines which greatly simplify the task of parallel algorithm development, and a powerful library of vectorized mathematical subroutines. All program development is done within the familiar ULTRIX® environment. The T Series can also be programmed using a parallel programming language called *occam.*

The FPS T Series has been shown to achieve a high percentage of peak performance on important kernel algorithms coded in occam [5]. The advent of the communications library has made the power of the T Series accessible from the more traditional scientific programming languages of FORTRAN and C. The following table summarizes the performance of a T100 (64 nodes) on five of these algorithms coded in both C and occam.

| | Operations (x10 6) | MFLOPS | |
|---|---|---|---|
| | | C | OCCAM |
| MATRIX MULTIPLY | 2146 | 511 | 596 |
| 2D CONVOLUTION | 1309 | 607 | 579 |
| 3D N-BODY | 1153 | 269 | 268 |
| 2D WAVE | 1307 | 285 | 232 |
| LINEAR SOLVER | 716 | 63 | 135 |

Figure 5: T Series Performance

## IV. Conclusion

The FPS T Series is configurable over three orders of magnitude. The intranode parallelism of the T Series complements and enhances the utility of large processor configurations. The high-speed links, dual-ported memory, and Vector Processing Unit strike a careful balance between communication, memory, and arithmetic speed through cost-effective VLSI. The homogeneity of the T Series module allows all system facilities to expand naturally with the number of compute nodes. This integration, inherent in the parallel/vector architecture of the T Series, provides the capability to achieve a high percentage of peak performance on important applications.

## References

[1] Charlesworth, A.E., "An Approach to Scientific Array Processing: the Architectural Design of the AP-120B/FPS-164 Family." *IEEE Computer*, **14**, 9, (1981), 18-27.

[2] Pease, M.C., "The Indirect Binary *n*-Cube Microprocessor Array." *IEEE Transactions on Computers*, **C-26**, 5 (1977), 458-473.

[3] Saad, Y., and Schultz, M.H., "Topological Properties of Hypercubes." Technical Report YALEU/DCS/RR-389, Computer Science Dept., Yale Univ., 1985.

[4] Hwang, K., and Briggs, F.,*Computer Architecture and Parallel Processing* (1984), 212-229.

[5] Miles D., Kinney P., Groshong J., and Fazzari R., "Specification and Performance Analysis of Six Benchmark Programs for the FPS T Series." *Proceedings of the 1987 ARRAY Conference*, (1987), 113-128.

# CONFERENCE PROGRAMME

## PLENARY SEESIONS

**MONDAY, AUG. 31**

J. J. Moré, USA

Trust Regions and Projected Gradients

H. Karatsu, Japan

How to Cope with Grey Part of Management : Sysyem Modelling and Optimization in Japanese Industry

**TUESDAY, SEPT. 1**

B. Korte, FRG

Large Scale Applications of Combinatorial Optimization

T. Asano, Japan

Computational Geometry in Japan

**WEDNESDAY, SEPT. 2**

J. Zabczyk, Poland (Presented by A. Ichikawa, Japan)

Stochastic Distributed Systems : Regularity and Asymptotics

Y. Murotsu, Japan

Probability-based Optimum Design of Structural Systems

**THURSDAY, SEPT. 3**

J. Doležal, Czechoslovakia

Optimal Control of Discrete-Time Systems

## PARALLEL SESSIONS

| Subjects | Sessions |
| --- | --- |
| 1. CONTROL AND ESTIMATION | A1, A2, A3, A4, A5, A6 |
| 2. DISTRIBUTED PARAMETER SYSTEMS | C7, C8, C9 |
| Session Organizers : I. Lasiecka, USA, M. Koda, Japan | |
| 3. CONTINUOUS OPTIMIZATION | D1, D2, D3, D4, D5 |
| 4. COMBINATORIAL OPTIMIZATION | B3, B4, B5, B10, B11 |
| Session Organizers : M. Lucertini, Italy, B. Korte, FRG | |
| 5. COMPUTATIONAL GEOMETRY | B6, B7, B8 |
| Session Organizer: T. Asano, Japan | |
| 6. SIMULATION | B9 |
| 7. OPTIMIZATION AND RELIABILITY OF STRUCTURAL SYSTEMS | C10, C11 |
| Session Organizers : P. Thoft-Christensen, Denmark, Y. Murotsu, Japan | |
| 8. TRANSPORTATION SYSTEMS | C1, C2 |
| 9. ENERGY SYSTEMS | D9 |
| 10. BIOLOGICAL SYSTEMS | A8, A9 |
| 11. OPERATIONAL RESEARCH IN FORESTRY | B1, B2 |
| Session Organizer: L. R. Foulds, New Zealand | |

## PROGRAMME

MONDAY, AUG. 31

Session A1. Control and Estimation

- Chairperson: S. Hitotumatu -

- **A New Necessary Condition for Optimality of Singular Controls** (A1-1)
  Hansheng Wu, China, K. Mizukami, Japan
- **Necessary Conditions for Impulsive Control Problems with State Constraints** (A1-2)
  F. M. F. L. Pereira, Portugal, R. B. Vinter, United Kingdom
- **Application of Optimization-Based Design Methods to Control System Design** (A1-3)
  W. Y. Ng, United Kingdom

Session B1. Operational Research in Forestry

- Chairperson: L. R. Foulds -

- **Increasing O. R. Applications in Forest Operations : The DSS Approach** (B1-1)
  E. Robak, Canada
- **A Resource Planning and Management System for Pine Plantations in Fiji** (B1-2)
  A. G. D. Whyte, New Zealand
- **A Stochastic Growth Model for Even-Aged, Single-Species Forest Stands** (B1-3)
  T. Sweda, Japan

Session C1. Transportation Systems

- Chairperson: K. Tone -

- **Transportation Supply and Economic Growth in a Multiregional System** (C1-1)
  D. Campisi, A. La Bella, Italy (Presented by A. Germani, Italy)

Session D1. Continuous Optimization

- Chairperson: Y. Evtushenko -

- **New Homotopy Methods for Solving Convex Analytic Programming Problems** (D1-1)
  G. Sonnevend, Hungary, J. Stoer, FRG
- **Duality and Regularization for Certain Infsup Problems** (D1-2)
  W. Oettli, FRG
- **Higher Order Sensitivity of Solutions to Convex Programming Problems without Strict Complementarity** (D1-3)
  K. Malanowski, Poland

Session A2. Control and Estimation

- Chairperson: K. Malanowski -

- **Modelling Support System for System Dynamics** (A2-1)

M. Kishi, K. Noda, Y. Yamashita, K. Taguchi, Japan

Session B2. Operational Research in Forestry
- Chairperson: J. Tsunekawa -

- **Modelling Forest Mosaics** (B2-1)
  D. G. Green, Australia
- **Forestry Applications of Operations Research** (B2-2)
  L. R. Foulds, New Zealand

Session C2. Transportation Systems
- Chairperson: K. Tone -

- **Modelling and Optimization of Public Traffic Large Scale Systems** (C2-1)
  Zhang Qi Ren, Xiong Gui Lin, China
- **Quadratic Programming and Its Impoved Computation : An Application in Urban Traffic System** (C2-2)
  Wang Yuji, Huang Dazhan, China

Session D2. Continuous Optimization
- Chairperson: J. Stoer -

- **Distributed Nonlinear Optimization Algorithms for General Network Problems** (D2-1)
  D. K. Subramanian, India
- **A Multiplex Algorithm for Linear Programming Problems** (D2-2)
  J. A. Snyman, M. van Rooyen
- **Newton-Type Algorithms with Nonmonotone Line Search for Large-Scale Unconstrained Optimization** (D2-3)
  L. Grippo, F. Lampariello, S. Lucidi, Italy
- **Centered Newton Method for Mathematical Programming** (D2-4)
  K. Tanabe, Japan

**TUESDAY, SEPT. 1**

Session A3. Control and Estimation
- Chairperson: K. Malanowski -

- **A New Structural Identification Method for Discrete Linear Multivariable Systems** (A3-2)
  Zhang Qi Ren, Long Mao Lin, China
- **Stability and Control of Stochastic Periodic Delay Systems** (A3-3)
  A. Ichikawa, Japan

Session B3. Combinatorial Optimization
- Chairperson: B. Korte -

- **Recent Results on Perfect Graphs** (B3-1)
  V. Chvatal, USA
- **On Cutting-Plane Proofs in Mixed Integer Programming** (B3-2)
  W. Cook, USA
- **Combinatorial Optimization and VLSI-Design** (B3-3)
  B. Korte, FRG

Session C3. Modelling and Methodology in Social and Economic Systems

- Chairperson: P. Kall -

- **A Primary Application of the Managerial Cybernetics—Viable System Model** (C3-1)
  K. Hoshino, S. Sekiya, K. Araya, Japan
- **Mathematical Modelling in the Educational System of Kuwait University** (C3-3)
  A. A. Elimam, Kuwait

Session D3. Continuous Optimization
- Chairperson: J. Stoer -

- **Predicting Method for Solving Quadratic Programming Problem** (D3-1)
  Yang Zihou, Li Baozhe, Gao Zhiyuan, China by Wu Han-Shen'g
- **Flow Management in Flexible Manufacturing System** (D3-2s)
  M. Lucertini, F. Nicole, Italy
- **Application of Generalized Raiffa Solution to Multicriteria Bargaining Support** (D3-3)
  P. Bronisz, L. Krus, Poland (Presented by T. Romanowicz, Poland)

Session A4. Control and Estimation
- Chairperson: S. Hitotumatu -

- **On Stochastic Control Problems : An Algorithm for the Value Function and the Optimal Policy. Some Applications** (A4-1)
  R. Gonzalez, E. Rofman, Argentina
- **Solving SLP Recourse Problems. The Case of Stochastic Tenders, RHS and Objective** (A4-2)
  K. Frauendorfer, Switzerland
- **Optimal Input for Autoregressive Model Discrimination Based on Kullback's Divergence** (A4-3)
  K. Uosaki, T. Hatanaka, Japan

Session B4. Combinatorial Optimization
- Chairperson: M. Lucertini -

- **Modular Dimensioning Algorithms for Circuit Switched Networks** (B4-1)
  M. Shinohara, Japan
- **A Variable Priority Queue with Applications to Multicommodity Flows** (B4-2)
  H. Suzuki, T. Nishizeki, N. Saito, Japan
- **Some Opitimization Problems on Tree-Structures** (B4-3)
  J. Oberklammer, Austria

Session C4. Modelling and Methodology in Social and Economic Systems
- Chairperson: J. Doležal -

- **A Causal Model of Computers, Education and Development** (C4-1)
  S. Herath, Sri Lanka, M. Kojima, J. Hoken, Japan
- **Incentive Stackelberg Strategies in Linear Quadratic Differential Games with Two Noncooperative Followers** (C4-3)
  K. Mizukami, Japan, Wu Hansheng, China

Session D4. Continuous Optimization
- Chairperson: J. Szelezsan -

- **New Algorithm for Structural Decomposition of Sparse Linear Models** (D4-2)
  T. Romanowicz, Poland
- **Recent Results on Nondifferentiable Exact Penalty Functions** (D4-2s)
  G. Di Pillo, L. Grippo, Italy
- **A Penalty Function Formulation for Interactive Multiobjective Programming Problems** (D4-3s)
  T. Sunaga, Japan M. A. Mazeed, India, E. Kondo, Japan

Session A5. Control and Estimation
- Chairperson: S. Hitotumatu -

- **Identification of Linear Discrete-time Stochastic System Using Covariance Information** (A5-1)
  S. Nakamori, H. Tokumaru, Japan
- **Optimum State Estimation Using a New Class of Robust Doubly Optimized Recursive Estimators** (A5-2)
  P. J. Radonja, Yugoslavia
- **Chandrasekhar Filtering and Smoothing Methods Given Covariance Model and Properties of X and Y Functions** (A5-3)
  M. Sugisaka, Japan
- **Optimal Control system under Integral Constraints** (A5-4s)
  S. J. Lee, A. N. V. Rao, USA

Session B5. Combinatorial Optimization
- Chairperson: M. Lucertini -

- **An LP-based Approach to the Quadratic Assignment Problem** (B5-1)
  S. Walukiewicz, Poland
- **A Parallel Algorithm for the Machine Scheduling Problem** (B5-2)
  M. Nakamori, Japan
- **Optimal Tradeoff among Delay, Throughput and Fairness in Flow Controlled Networks** (B5-4)
  Suk-gwon Chang, Korea

Session C5. Modelling and Methodology in Social and Economic Systems
- Chairperson: K. Wakayama -

- **Stackelberg Game Model Incorporating Public Participation for Electric Generating System Development** (C5-2)
  K. Ogino, Japan
- **Cooperative Fuzzy Games with Side-Payments** (C5-3)
  A. Yamada, Japan

Session D5. Continuous Optimization
- Chairperson: T. Romano Szelezsan -

**WEDNESDAY, SEPT. 2**
Session A6. Control and Estimation
- Chairperson: M. Koda -

Y. Murotsu, H. Okubo, F. Terui, Japan

- **Modelling of Flexible Manipulator Arms** (C9-3)
  S. Tsujio, Japan

Session D9. Energy Systems
- Chairperson: K. Yamashita -

- **Energy Optimization in Industrial Models** (D9-1)
  R. Kummel, H.-M. Groscurth, FRG, W. van Gool, Netherlands
- **Modelling Continuous State Variables into Discrete State Variables in Dynamic Programming : Impacts on the Optimal Operations of the French PWR** (D9-2)
  M. Moatti, France
- **Power Generator Control by Variable Structure Control Theory** (D9-3)
  S. Omatu, K. Matsushita, K. Isaka, Japan
- **Automatic Generator Control in Hydro-Thermal Electric Power System** (D9-4)
  D. Dervisevic, E. Soljanin, Yugoslavia

FRIDAY, SEPT. 4

Session A10. Computers
- Chairperson: T. Kawai -

- **Knowledge Based Discrete Central Problems** (A10-3s)
  H. -J. Sebastian, GDR
- **A Method of Congestion Control in the Computer Network Node** (A10-1)
  D. Rutkowski, K. Chmara, Poland
- **Program Optimization with Logic Program Transformation** (A10-2)
  H. Hoshino, M. Esashi, K. Agusa, Y. Ohno, Japan

Session B10. Combinatotial Optimization
- Chairperson: J. Szelezsan -

- **Distributed Computing of Stochastic Algorithm for Combinatorial Optimization Problems** (B10-2)
  Zhao Yue, China, T. Fukao, Japan

Session C10. Optimization and Reliability of Structural Systems
- Chairperson: P. Thoft-Chrstensen -

- **Application of Multicriteria Optimization to Structural Systems** (C10-1)
  M. Weck, F. Fortsch, FRG
- **Reliability Based Optimization of Strutural Systems** (C10-2)
  J. D. Sorensen, Denmark
- **Application of a Multiplication Factor Method to the Identification of Dominant Structural Failure Modes** (C10-3s)
  Y. Murotsu, S. Matsuzaki, H. Arima, Japan

Session D10.
- Chairperson: J. Doležal -

- **Interactive System for Multicriteria Optimization** (D10-1)
  Y. Evtushenko, V. Ratkin, USSR

# Author Index

# Lecture Notes in Control and Information Sciences

Edited by M. Thoma and A. Wyner

Vol. 62: Analysis and Optimization
of Systems
Proceedings of the Sixth International
Conference on Analysis and Optimization
of Systems
Nice, June 19–22, 1984
Edited by A. Bensoussan, J. L. Lions
XIX, 591 pages. 1984.

Vol. 63: Analysis and Optimization
of Systems
Proceedings of the Sixth International
Conference on Analysis and Optimization
of Systems
Nice, June 19–22, 1984
Edited by A. Bensoussan, J. L. Lions
XIX, 700 pages. 1984.

Vol. 64: Arunabha Bagchi
Stackelberg Differential Games
in Economic Models
VIII, 203 pages, 1984

Vol. 65: Yaakov Yavin
Numerical Studies
in Nonlinear Filtering
VIII, 273 pages, 1985.

Vol. 66: Systems and Optimization
Proceedings of the Twente Workshop
Enschede, The Netherlands, April 16–18, 1984
Edited by A. Bagchi, H. Th. Jongen
X, 206 pages, 1985.

Vol. 67: Real Time Control of Large Scale Systems
Proceedings of the First European Workshop
University of Patras, Greece, Juli 9–12, 1984
Edited by G. Schmidt, M. Singh, A. Titli,
S. Tzafestas
XI, 650 pages, 1985.

Vol. 68: T. Kaczorek
Two-Dimensional Linear Systems
IX, 397 pages, 1985.

Vol. 69: Stochastic Differential Systems –
Filtering and Control
Proceedings of the IFIP-WG 7/1 Working Conference
Marseille-Luminy, France, March 12-17, 1984
Edited by M. Metivier, E. Pardoux
X, 310 pages, 1985.

Vol. 70: Uncertainty and Control
Proceedings of a DFVLR International Colloquium
Bonn, Germany, March, 1985
Edited by J. Ackermann
IV, 236 pages, 1985.

Vol. 71: N. Baba
New Topics in Learning Automata
Theory and Applications
VII, 231 pages, 1985.

Vol. 72: A. Isidori
Nonlinear Control Systems:
An Introduction
VI, 297 pages, 1985.

Vol. 73: J. Zarzycki
Nonlinear Prediction
Ladder-Filters for Higher-Order
Stochastic Sequences
V, 132 pages, 1985.

Vol. 74: K. Ichikawa
Control System Design based on
Exact Model Matching Techniques
VII, 129 pages, 1985.

Vol. 75: Distributed Parameter
Systems
Proceedings of the 2nd International
Conference, Vorau, Austria 1984
Edited by F. Kappel, K. Kunisch,
W. Schappacher
VIII, 460 pages, 1985.

Vol. 76: Stochastic Programming
Edited by F. Archetti, G. Di Pillo,
M. Lucertini
V, 285 pages, 1986.

Vol. 77: Detection of
Abrupt Changes in Signals
and Dynamical Systems
Edited by M. Basseville,
A. Benveniste
X, 373 pages, 1986.

Vol. 78: Stochastic
Differential Systems
Proceedings of the 3rd Bad Honnef
Conference, June 3–7, 1985
Edited by N. Christopeit, K. Helmes,
M. Kohlmann
V, 372 pages, 1986.

Vol. 79: Signal
Processing for Control
Edited by K. Godfrey, P. Jones
XVIII, 413 pages, 1986.

Vol. 80: Artificial Intelligence
and Man-Machine Systems
Edited by H. Winter
IV, 211 pages, 1986.

# Lecture Notes in Control and Information Sciences

Edited by M. Thoma and A. Wyner

Vol. 81: Stochastic Optimization
Proceedings of the International Conference, Kiew, 1984
Edited by I. Arkin, A. Shiraev, R. Wets
X, 754 pages, 1986.

Vol. 82: Analysis and Algorithms of Optimization Problems
Edited by K. Malanowski, K. Mizukami
VIII, 240 pages, 1986.

Vol. 83: Analysis and Optimization of Systems
Proceedings of the Seventh International Conference of Analysis and Optimization of Systems
Antiba, June 26-27, 1986
Edited by A. Bensoussan, J. L. Lions
XVI, 901 pages, 1986.

Vol. 84: System Modelling and Optimization
Proceedings of the 12th IFIP Conference
Budapest, Hungary, September 2–6, 1985
Edited by A. Prékopa, J. Szelezsán, B. Strazicky
XII, 1046 pages, 1986.

Vol. 85: Stochastic Processes in Underwater Acoustics
Edited by Charles R. Baker
V, 205 pages, 1986.

Vol. 86: Time Series and Linear Systems
Edited by Sergio Bittanti
XVII, 243 pages, 1986.

Vol. 87: Recent Advances in System Modelling and Optimization
Proceedings of the IFIP-WG 7/1 Working Conference
Santiago, Chile, August 27-31, 1984
Edited by L. Contesse, R. Correa, A. Weintraub
IV, 199 pages, 1987.

Vol. 88: Bruce A. Francis
A Course in $H_\infty$ Control Theory
XI, 156 pages, 1987.

Vol. 88: Bruce A. Francis
A Course in $H_\infty$ Control Theory
X, 150 pages, 1987.
Corrected - 1st printing 1987

Vol. 89: G. K. H. Pang/A. G. J. McFarlane
An Expert System Approach to Computer-Aided Design of Multivariable Systems
XII, 223 pages, 1987.

Vol. 90: Singular Perturbations and Asymptotic Analysis in Control Systems
Edited by P. Kokotovic, A. Bensoussan, G. Blankenship
VI, 419 pages, 1987.

Vol. 91 Stochastic Modelling and Filtering
Proceedings of the IFIP-WG 7/1 Working Conference
Rome, Italy, Decembre 10-14, 1984
Edited by A. Germani
IV, 209 pages, 1987.

Vol. 92: L. T. Grujić, A. A. Martynyuk, M. Ribbens-Pavella
Large-Scale Systems Stability Under Structural and Singular Perturbations
XV, 366 pages, 1987.

Vol. 93: K. Malanowski
Stability of Solutions to Convex Problems of Optimization
IX, 137 pages, 1987.

Vol. 94: H. Krishna
Computational Complexity of Bilinear Forms
Algebraic Coding Theory and Applications to Digital Communication Systems
XVIII, 166 pages, 1987.

Vol. 95: Optimal Control
Proceedings of the Conference on Optimal Control and Variational Calculus
Oberwolfach, West-Germany, June 15-21, 1986
Edited by R. Bulirsch, A. Miele, J. Stoer and K. H. Well
XII, 321 pages, 1987.

Vol. 96: H. J. Engelbert/W. Schmidt
Stochastic Differential Systems
Proceedings of the IFIP-WG 7/1 Working Conference
Eisenach, GDR, April 6-13, 1986
XII, 381 pages, 1987.

# Lecture Notes in Control and Information Sciences

Edited by M. Thoma and A. Wyner

Vol. 97: I. Lasiecka/R. Triggiani (Eds.)
Control Problems for Systems
Described by Partial Differential Equations
and Applications
Proceedings of the IFIP-WG 7.2
Working Conference
Gainesville, Florida, February 3-6, 1986
VIII, 400 pages, 1987.

Vol. 98: A. Aloneftis
Stochastic Adaptive Control
Results and Simulation
XII, 120 pages, 1987.

Vol. 99: S. P. Bhattacharyya
Robust Stabilization Against
Structured Perturbations
IX, 172 pages, 1987.

Vol. 100: J. P. Zolésio (Editor)
Boundary Control and Boundary Variations
Proceedings of the IFIP WG 7.2 Conference
Nice, France, June 10-13, 1987
IV, 398 pages, 1988.

Vol. 101: P. E. Crouch,
A. J. van der Schaft
Variational and Hamiltonian
Control Systems
IV, 121 pages, 1987.

Vol. 102: F. Kappel, K. Kunisch,
W. Schappacher (Eds.)
Distributed Parameter Systems
Proceedings of the 3rd International Conference
Vorau, Styria, July 6–12, 1986
VII, 343 pages, 1987.

Vol. 103: P. Varaiya, A. B. Kurzhanski (Eds.)
Discrete Event Systems:
Models and Applications
IIASA Conference
Sopron, Hungary, August 3-7, 1987
IX, 282 pages, 1988.

Vol. 104: J. S. Freudenberg/D. P. Looze
Frequency Domain Properties of Scalar
and Multivariable Feedback Systems
VIII, 281 pages, 1988.

Vol. 105: Ch. I. Byrnes/A. Kurzhanski (Eds.)
Modelling and Adaptive Control
Proceedings of the IIASA Conference
Sopron, Hungary, July 1986
V, 379 pages, 1988.

Vol. 106: R. R. Mohler (Editor)
Nonlinear Time Series and
Signal Processing
V, 143 pages. 1988.

Vol. 107: Y. T. Tsay, L.-S. Shieh, St. Bar
Structural Analysis and Design
of Multivariable Systems
An Algebraic Approach
VIII, 208 pages, 1988.

Vol. 108: K. J. Reinschke
Multivariable Control
A Graph-theoretic Approach
274 pages, 1988.

Vol. 109: M. Vukobratović/R. Stojić
Modern Aircraft Flight Control
VI, 288 pages, 1988.

Vol. 110: In preparation

Vol. 111: A. Bensoussan, J. L. Lions (Ed
Analysis and Optimization
of Systems
XIV, 1175 pages, 1988.

Vol. 112: In preparation

Vol. 113: M. Iri, K. Yajima (Eds.)
System Modelling and Optimization
Proceedings of the 13th IFIP Confere
Tokyo, Japan, Aug. 31 – Sept. 4, 1987
IX, 787 pages, 1988.